中国生物多样性红色名录：脊椎动物

China's Red List of Biodiversity: Vertebrates

主编 蒋志刚
Chief Editor: Zhigang Jiang

第一卷 哺乳动物（上册）
Volume I, Mammals (I)

主编 蒋志刚
Chief Editor: Zhigang Jiang

副主编 吴　毅　刘少英　蒋学龙
　　　 周开亚　胡慧建

Vice-Chief Editors: Yi Wu　Shaoying Liu　Xuelong Jiang
　　　　　　　　　 Kaiya Zhou　Huijian Hu

科学出版社
北　京

内 容 简 介

本书是"中国生物多样性红色名录：脊椎动物"的"第一卷 哺乳动物"，全书分为总论和各论两部分。总论介绍了哺乳动物的演化与现状、中国哺乳动物多样性与保护现状、本红色名录评估对象的分类系统、中国哺乳动物编目（2021）、中国哺乳动物分布格局和受保护状况；介绍了红色名录评估过程、依据的评估等级和标准，还介绍了咨询专家、评估队伍，以及建立数据库和开展初评、通讯评审、形成评估报告的过程；总结分析了评估结果，介绍了中国哺乳动物濒危状况，分析了野外灭绝的和区域灭绝的物种，分析了受威胁物种比例、哺乳动物的分布和濒危原因等，并将评估结果与《IUCN 受威胁物种红色名录》（2020-2）进行了比较分析；最后，分析了中国哺乳动物保护成效与远景。各论是图书的主体，对评估的 700 种中国哺乳动物物种的相关信息，即分类地位、评估信息、地理分布、种群状况、生境与生态系统、威胁因子、保护级别与保护行动及相关文献进行了详细叙述。

本书适合从事野生动物研究与保护的科研人员参考，适合自然保护区、环境保护、进出口对外贸易、检验检疫等相关各级行政管理部门作为行使管理职能的参照资料，适合作为高年级研究生教学参考书，适合国内大中型图书馆馆藏。

审图号：GS (2020) 2858号

图书在版编目（CIP）数据

中国生物多样性红色名录. 脊椎动物. 第一卷，哺乳动物. 上册 = China's Red List of Biodiversity: Vertebrates, Volume I, Mammals (I) : 汉英对照 / 蒋志刚主编. —北京：科学出版社，2021.3
国家出版基金项目
ISBN 978-7-03-065664-3

Ⅰ. ①中… Ⅱ. ①蒋… Ⅲ. ①珍稀动物–中国–名录–汉、英 ②珍稀植物–中国–名录–汉、英 ③哺乳动物纲–中国–名录–汉、英 Ⅳ. ①Q958.52-62 ②Q948.52-62 ③Q959.808

中国版本图书馆CIP数据核字（2020）第131620号

责任编辑：马 俊 孙 青 郝晨扬／责任校对：严 娜
责任印制：肖 兴／排版设计：北京鑫诚文化传播有限公司

科学出版社 出版
北京东黄城根北街16号
邮政编码：100717
http://www.sciencep.com

中国科学院印刷厂 印刷
科学出版社发行 各地新华书店经销

*

2021年3月第 一 版 开本：889×1194 1/16
2021年3月第一次印刷 印张：108 1/4
字数：3 117 000

定价：1428.00元（全三册）
（如有印装质量问题，我社负责调换）

China's Red List of Biodiversity: Vertebrates, Volume I, Mammals (I)

Chief Editor: Zhigang Jiang
Vice-Chief Editors: Yi Wu   Shaoying Liu   Xuelong Jiang   Kaiya Zhou   Huijian Hu

Abstract

This book is the first volume of the *"China's Red List of Biodiversity: Vertebrates"*, *i.e.* *"Volume I, Mammals"*, and is divided into two parts of "General Introduction" and "Species Monograph". General Introduction contains evolution and status of mammals, diversity and conservation status of mammals in China, taxonomic system of this red list, inventory of China's mammals (2021), the distribution patterns and conservation status of China's mammals; the evaluation process, categories and criteria that evaluation refers to, building the database, evaluation teams and how the evaluation teams to establish database, to carry out preliminary evaluation, to review the preliminary results by correspondence and to formulate the evaluation report; analyzes and summarizes the evaluation results, introduces the status of endangered mammal, analyzes the species extinct in the wild and regionally extinct, analyzes the proportion of threatened mammal species in different groups of mammals, in different habitats and its provincial distribution and the threats to endangered mammals, and compares the evaluation results with the *IUCN Red List of Endangered Species* (2020-2). Finally, the implications and prospects of mammal conservation in China are discussed. Species Monograph is the main part of this book, this part elaborates the detailed evaluating information of 700 Chinese mammal species, including taxonomic status, assessment information, geographical distribution, population situation, habitat and ecosystem, threats factor, protection category and conservation action, citations and references.

This book can be used as a reference for wild animal protect staffs and researchers, and can be used as an important data for decisions of government management department, *e.g.* nature reserve, environment protect, overseas trade, inspection and quarantine. This book is appropriate for senior grade graduated students in school, and appropriate to be collected by large- or medium-sized library.

ISBN 978-7-03-065664-3

Copyright© 2021, Science Press (Beijing)
All rights reserved. No part of this publication may be reproduced, stored in a retrieval system, or transmitted in any form or by any means, mechanical, photocopying, recording or otherwise, without the prior written permission of the copyright owner.

Acknowledgement
Illustrations (with mark of "Lynx Edicions") by Toni Liobet from: Wilson, D. E., Lacher, T. E. Jr & Mittermeier, R. A. eds. (2009-2018). *Handbook of the Mammals of the World*. Volumes 1 to 8. Lynx Edicions, Barcelona.

《中国生物多样性红色名录：脊椎动物》
编委会

顾 问

陈宜瑜　郑光美　张亚平　金鉴明　马建章　曹文宣

主 编

蒋志刚

副主编

江建平　王跃招　张　鹗　张雁云

编委会成员

蔡　波　曹　亮　车　静　陈小勇　陈晓虹　丁　平
董　路　胡慧建　胡军华　计　翔　江建平　蒋学龙
蒋志刚　李　成　李春旺　李家堂　李丕鹏　梁　伟
刘　阳　刘少英　卢　欣　马　鸣　马　勇　马志军
饶定齐　史海涛　王　斌　王剑伟　王英永　王跃招
吴　华　吴　毅　吴孝兵　谢　锋　杨道德　杨晓君
曾晓茂　张　鹗　张　洁　张保卫　张雁云　张正旺
赵亚辉　周　放　周开亚

责任编辑

马　俊　李　迪　郝晨扬　孙　青

Editorial Committee of China's Red List of Biodiversity: Vertebrates

Consultants of the Editorial Committee

Yiyu Chen   Guangmei Zheng   Yaping Zhang   Jianming Jin   Jianzhang Ma   Wenxuan Cao

Chief Editor of the Editorial Committee

Zhigang Jiang

Vice-Chief Editors of the Editorial Committee

Jianping Jiang   Yuezhao Wang   E Zhang   Yanyun Zhang

Members of the Editorial Committee

Bo Cai   Liang Cao   Jing Che   Xiaoyong Chen   Xiaohong Chen   Ping Ding
Lu Dong   Huijian Hu   Junhua Hu   Xiang Ji   Jianping Jiang   Xuelong Jiang
Zhigang Jiang   Cheng Li   Chunwang Li   Jiatang Li   Pipeng Li   Wei Liang
Yang Liu   Shaoying Liu   Xin Lu   Ming Ma   Yong Ma   Zhijun Ma
Dingqi Rao   Haitao Shi   Bin Wang   Jianwei Wang   Yingyong Wang
Yuezhao Wang   Hua Wu   Yi Wu   Xiaobing Wu   Feng Xie   Daode Yang
Xiaojun Yang   Xiaomao Zeng   E Zhang   Jie Zhang   Baowei Zhang
Yanyun Zhang   Zhengwang Zhang   Yahui Zhao   Fang Zhou   Kaiya Zhou

Responsible Editors

Jun Ma   Di Li   Chenyang Hao   Qing Sun

序 一

地球进入了一个崭新的地质纪元——人类世（Anthropocene），而地球上的人口仍呈指数增长。人类社会进入了全球化、信息化时代。人类的生态足迹日益扩大，人类对自然资源的消耗、对环境的污染达到了一个前所未有的水平，人类的影响已经遍及地球各个角落，导致了全球变化，影响了地球生物圈的结构与功能，危及了许多野生动植物的生存，造成全球范围的生物多样性危机，影响了人类社会的可持续发展。

中国是一个生物多样性大国，是地球上生物多样性最丰富的国家之一。根据《中国生物物种名录》(2019)，中国已经记载生物达 106,509 个物种及种下单元，其中物种 94,260 个，种下单元 12,249 个。中国南北纬度跨度大，海拔跨度也大。中国还有多种气候类型、多样生境类型，栖息着丰富的高等生物。这些生物物种是国家重要的战略资源，是社会经济可持续发展中不可替代的物质基础。

濒危物种红色名录已经成为重要的生物多样性保护研究工具。目前，《世界自然保护联盟受威胁物种红色名录》（《IUCN 受威胁物种红色名录》）评估了 98,500 多个物种，发现其中 32,000 多个面临灭绝威胁，包括 41% 的两栖动物、34% 的针叶树、33% 的造礁珊瑚、26% 的哺乳动物和 14% 的鸟类。然而，《IUCN 受威胁物种红色名录》对物种的生存状况的评估是基于全球资料所做的，并不代表物种在各个分布国的生存状况。国家是生物多样性保护的主体，各国须开展自己的濒危物种红色名录研究，对其生物物种的生存状况进行评估。在某种程度上，可以说濒危物种红色名录研究反映了一个国家生物多样性综合研究的能力。

早在 20 世纪 90 年代，中国即引入 IUCN 受威胁物种红色名录等级标准开展了物种濒危状况评估工作，如 1991 年，我国学者发表了《中国植物红皮书》。1998 年，国家环境保护总局联合国家濒危物种科学委员会发表了《中国濒危动物红皮书·鱼类》《中国濒危动物红皮书·两栖类和爬行类》《中国濒危动物红皮书·鸟类》《中国濒危动物红皮书·兽类》等著作。另外，2004 年和 2009 年，相关领域专家开展了不同类群物种的濒危状况评估工作，先后发表了《中国物种红色名录

（第一卷）红色名录》和《中国物种红色名录（第二卷）脊椎动物》。

物种生存状况是变化的，于是，IUCN每年定期更新IUCN红色名录。IUCN红色名录并不反映一个跨越国家分布的物种在一个分布国家的生存状况。国家是濒危物种的管理主体，各国需要应用国际标准进行红色名录评估。鉴于此，为全面评估中国野生脊椎动物濒危状况，中国研究人员于2013年启动了"中国生物多样性红色名录——脊椎动物卷"的物种评估和报告编制工作。这次评估组织全国鱼类、两栖类、爬行类、鸟类与哺乳类专家收集数据，采用综合分析和专家评估相结合的方法，依据中国鱼类、两栖类、爬行类、鸟类和哺乳类野生种群与生境现状，利用IUCN红色名录标准第3.1版，编制了"中国生物多样性红色名录——脊椎动物卷"。该卷红色名录于2015年5月6日通过环境保护部和中国科学院的联合验收，并于5月22日以环境保护部、中国科学院2015年第32号公告形式正式发布。

2016年以来，中国脊椎动物红色名录工作组组织中国研究人员再次厘定了中国脊椎动物多样性，重新评估了中国脊椎动物的生存状况。完成了《中国生物多样性红色名录：脊椎动物》(2021)。本次脊椎动物红色名录评估发现中国脊椎动物生存状况严峻，中国脊椎动物的灭绝风险高于世界平均水平。中国脊椎动物哺乳类、鸟类、两栖类、爬行类和鱼类等各个类群都发现了野外灭绝或区域灭绝的物种，有许多物种处于极危、濒危和易危的受威胁状态。

中国正处于发展期，人口众多，地貌复杂，区域发展程度差异大。如何拯救这些濒危物种是中国生物多样性保护面临的一项艰巨任务。中国政府十分重视生物多样性保护，缔结了《生物多样性公约》《濒危野生动植物种国际贸易公约》《关于特别是作为水禽栖息地的国际重要湿地公约》（简称《湿地公约》）等国际公约，并积极主动履约。中国大力开展了以自然保护区为主体、以国家公园为龙头的保护地建设。目前，我国建立的各类保护地已达1.18万处，面积占国土面积的18%以上。其中有474个国家级自然保护区。自然保护区保护了90.5%的陆地生态系统类型、85%的野生动植物种类、65%的高等植物群落。中国的森林覆盖率逐年增加，为濒危物种的种群与栖息地恢复奠定了基础。

生物多样性研究既是一项综合性研究，也是一项组合型研究。生物物种编目与受威胁状态评估需要不同学科的联合研究。为了生物多样性科学研究与保护，来自不同学科的学者走到一起，完成中国生物多样性红色名录研究。这项研究是中国整体生物多样性研究的重要组成部分。《中国生物多样性红色名录：脊椎动物》各卷在《生物多样性公约》第15次缔约方大会即将在中国召开之前出版发行，为中国生物多样性保护提供了基础数据，为监测中国生物多样性现状、开展阶段性IUCN红色名录指数研究积累了参数，也是中国保护生物学研究成果的展示。

中国科学院院士
国家自然科学基金委员会前主任
国家濒危物种科学委员会主任
2020年6月26日

Foreword I

The earth has entered a new geological epoch, the Anthropocene, while the human population on the earth is still increasing exponentially. Human society has entered the era of globalization and information. The human ecological footprint is enlarging while the human consumption of natural resources increases; environmental pollution created by human being has reached an unprecedented level. Impact of human reaches throughout all corners of the earth, causes the global change and affects the structure and function of the earth's biosphere, threatens the survival of many wild animals and plants, causing a global biodiversity crisis, consequently, influences the sustainable development of human society.

China is a country with great biodiversity and one of the countries with the richest biodiversity on the earth. According to the *Species Catalogue of China* (2019), China has already recorded 106,509 species and subspecies taxa, including 94,260 species and 12,249 subspecies. The territory of China spans a large latitude and has huge elevation differences. China also has a variety of climate types, diverse habitat types, rich niches of higher organisms. These biological species are important strategic resources of the country and irreplaceable material basis for sustainable social and economic development.

The red list of endangered species has become an important tool for biodiversity conservation research. Currently, the *IUCN Red List of Threatened Species* has assessed more than 98,500 species and found that more than 32,000 of them are threatened with extinction, including 41% of amphibians, 34% of conifers, 33% of reef-building corals, 26% of mammals and 14% of birds. However, the *IUCN Red List of Threatened Species* is based on global data and does necessarily not represent the status of species in each range country. A country is the main sovereign body of biodiversity conservation, and each country needs to carry out the red list study of its endangered species to assess the survival status of its biological species. To some extent, red list study reflects the comprehensive research capacity of biodiversity in a country.

The red list of endangered species has become an important tool for biodiversity conservation research. As early as in the 1990s, China introduced the IUCN red list criteria for threatened species to carry out the assessment of status of endangered species. For example, in 1991, Chinese scholars published the *Red Data Book of Chinese Plants*. In 1998, Environmental Protection Administration, together with the National Scientific Committee on Endangered Species, published such books as *China Red Data Book of Endangered Animals*: *Pisces*; *China Red Data Book of Endangered Animals*: *Amphibia & Reptilia*; *China Red Data Book of Endangered Animals*: *Aves*; *China Red Data Book of Endangered Animals*: *Mammalia*. In addition, in 2004 and 2009, experts in related fields carried out the assessment of the endangered status of different taxa of species, and published the *China Species Red List Vol I Red List* and the *China Species Red List Vol II Vertebrates*.

Status of species is changing, therefore the IUCN updates its red list every year. However, the IUCN red list does not reflect the status of particular species in a range country if a species distributes in multi-countries. Countries are the main management bodies of endangered species; thus, each country needs to apply international standards for red list assessment. Therefore, in order to comprehensively assess the endangered status of wild vertebrates in China, Chinese researchers launched the compilation of the "China's Red List of Biodiversity: Volume of Vertebrates" in 2013. During that evaluation, by adopting the combination of comprehensive analysis and expert evaluation method, the experts were coordinated to collect data for assessing wild population and habitat status of China's fishes, amphibians, reptiles, birds and mammals and compiled the *China's Red List of Biodiversity*: *Volume of Vertebrates*. The red list was approved by Ministry of Environmental Protection and the Chinese Academy of Sciences on May 6, 2015, and was officially released on May 22 in the form of Announcement No. 32 of 2015 by the Ministry of Environmental Protection and Chinese Academy of Sciences.

Since 2016, the China's vertebrate red list working group has organized Chinese researchers to reassess the diversity and the status of Chinese vertebrates. The working group completed the *China's Red List of Biodiversity: Vertebrates* (2021). The assessment found that China's vertebrate survival situation is still grim, and the risk of extinction of Chinese vertebrates is higher than the world average. Various groups of vertebrates, including mammals, birds, amphibians, reptiles and fishes in China have found species of Extinct in the Wild or Regionally Extinct, and many species are in a state of Critically Endangered, Endangered and Vulnerable.

China is in the process of rapid development. China has the largest human population, complex landforms and huge differences in regional development levels. How to save these endangered species is a difficult task for China's biodiversity conservation. The Chinese government attaches great importance to the protection of biological diversity, and has signed and actively implemented international conventions such as the *Convention on Biological Diversity*, the *Convention on International Trade in Endangered Species of Wild Fauna and Flora*, and the *Convention on Wetlands of International Importance, Especially as Waterfowl Habitats* (*Convention on Wetlands* for short). China has vigorously established protected areas with the nature reserves as the main body and national parks as the leading part. At present, China has set up 11,800 protected areas of various types, accounting for more than 18% of the country's land area. There are 474 national nature reserves. The nature reserves protect 90.5% of terrestrial ecosystem types, 85% of wildlife species, and 65% of higher plant communities. On the other hand, China's forest coverage is increasing year by year, laying a sound foundation for the population and habitat restoration of endangered species.

Biodiversity research is not only a comprehensive research, but also a combined study. The inventory of biological species and the assessment of threatened status require joint efforts of different disciplines. For the scientific research and conservation of biodiversity, scholars from different disciplines came together to complete the red list of China's biodiversity. The study is an important part of China's overall biodiversity research. All volumes of *China's Red List of Biodiversity*: *Vertebrates* are published before the fifteenth meeting of the Conference of the Parties to the *Convention on Biological Diversity* which will be held in Kunming, China. The set of books will provide basic data for the biodiversity conservation in China, for monitoring the current situation of biological diversity, for accumulating parameters to conduct periodic IUCN red list index research, and is also an outcome of the Chinese conservation biology research.

Yiyu Chen

Member of the Chinese Academy of Sciences

Former Director of National Natural Science Foundation of China

Director of Endangered Species Scientific Commission, P. R. China

June 26, 2020

序 二

1948年，在法国枫丹白露举行的一次由23个政府、126个国家组织和8个国际组织参与的国际会议上，世界自然保护联盟（International Union for the Protection of Nature，IUPN）成立了。当时，这个组织没有财政来源、没有长期预算，甚至没有永久雇员，但是IUPN成为世界政府与非政府组织（Governmental and Nongovernmental Organization，GONGO）的发端。世界自然保护联盟成立后的第一个重大举措是在1950年建立了"生存服务（Survival Service）机构"。"生存服务机构"利用当时筹集到的2,500美元，召集全球的科学家、志愿者为全球濒危物种撰写评估报告，要求各国政府保护其境内的濒危物种。1964年，世界自然保护联盟正式发布《濒危物种红皮书》（*Endangered Species Red Book*）。今天，世界自然保护联盟已经完全改变了它自己，包括其名称也改变为International Union for Conservation of Nature（IUCN）。IUCN已经成为联合国的观察员、世界范围内主要保护组织。IUCN"生存服务机构"已经演化为物种存续委员会（Species Survival Commission，SSC），IUCN《濒危物种红皮书》也演化为《IUCN受威胁物种红色名录》。现在，IUCN物种存续委员会每年发布《IUCN受威胁物种红色名录》。《IUCN受威胁物种红色名录》已发展成为世界上关于动物、植物和真菌物种全球保护状况最全面的信息源。

世界各国是生物多样性保护的主体。一个物种的IUCN红色名录等级并不一定等同于其在一个国家红色名录中的等级，除非这一物种是该国特有的物种。于是，世界各国也在制定各自的濒危物种红色名录。通过濒危物种红色名录的研究，各国对其境内分布的植物与动物物种的分布、生存状况和保护状况进行调查，然后，对物种的生存状况进行全面评估。因此，濒危物种红色名录是一份物种及其分布的清单，是对物种生存状况、保护状况的客观评估，是生物多样性健康状况的一个重要指标。

濒危物种红色名录被各国政府、自然保护地与野生动植物主管部门、与保护有关的非政府组织、研究人员、自然资源规划人员、教育机构使用。红色名录为生物多样性保护和政策变化提供了信息和促进行动的有力工具，对保护我们赖以生存的自然资源至关重要。通过网络应用，濒危物种红色名录也成为濒危物种信息库，成为保护工作者与研究人员的工具。

中国是 IUCN 的成员，中国也是联合国《生物多样性公约》的最早缔约方之一。中国一直走在生物多样性保护的前沿。中国还是世界上生物多样性最丰富的国家之一，有 7,300 余种脊椎动物，约占全球脊椎动物总数的 11%。中国动物区系组成复杂，空间分布格局差异显著，起源古老，拥有生物演化系统中的各种类群，如有"活化石"之称的大熊猫（*Ailuropoda melanoleuca*）、白鱀豚（*Lipotes vexillifer*）和扬子鳄（*Alligator sinensis*）等。此外，中国还是许多家养动物的起源中心。中国也是生物多样性受威胁最严重的国家之一。人类活动造成的资源过度利用、生境丧失与退化、环境污染及气候变化等因素导致脊椎动物多样性受到严重的威胁。

近年来，党中央和国务院高度重视生物多样性保护工作，将生物多样性保护上升为国家战略，发布了《中国生物多样性保护战略与行动计划（2011—2030 年）》，建立了生物物种资源保护部际联席会议制度，成立了中国生物多样性保护国家委员会，制定和实施了一系列生物多样性保护规划和计划，取得了积极进展。然而，中国生物多样性下降的总体趋势尚未得到有效遏制，保护形势依然严峻，特别是由于目前对中国物种受威胁状况缺乏全面的了解，影响了生物多样性的有效保护。因此，评估物种的受威胁状况，制定红色名录，从而提出针对性的保护策略，对于推动实施《中国生物多样性保护战略与行动计划（2011—2030 年）》和生态文明建设具有重要意义。

到目前，在中国科学家的努力下，中国哺乳动物、鸟类、两栖动物、爬行动物、淡水鱼类都得到了全面的评估。除了评估新发现的物种，中国濒危脊椎动物红色名录还重新评估了一些现存物种的状况，如大熊猫、藏羚（*Pantholops hodgsonii*）等物种，由于中国的保护努力，这些物种的中国濒危脊椎动物红色名录濒危等级下降。然而，中国生物多样性濒危局面仍然严峻。

尽管中国受威胁物种的比例很高，但中国政府正在加强生态环境保护，加强自然保护区、国家公园、世界遗产地及其他类型的保护地建设，加强荒漠化治理、湿地恢复、植树造林，努力扭转或至少制止生物多样性的下降。《生物多样性公约》第 15 次缔约方大会即将在中国昆明召开之际，中国科学家发表最新版《中国生物多样性红色名录：脊椎动物 第一卷 哺乳动物》、《中国生物多样性红色名录：脊椎动物 第二卷 鸟类》、《中国生物多样性红色名录：脊椎动物 第三卷 爬行动物》、《中国生物多样性红色名录：脊椎动物 第四卷 两栖动物》和《中国生物多样性红色名录：脊椎动物 第五卷 淡水鱼类》，全面更新了中国脊椎动物生存状况与种群和栖息地保护状况，从而为确定哪些物种须有针对性地努力恢复，为确定须保护的关键种群和栖息地提供了依据，有助于鉴别未来的濒危脊椎动物保护重点。这套图书的出版是中国自然保护史上的一件大事。

IUCN 主席

2020 年 6 月 26 日

Foreword II

The International Union for the Protection of Nature (IUPN) was established in Fontainebleau of France in 1948 at an international conference that 23 governments, 126 national organizations and 8 international organizations participated in. At that time, the organization had no financial resources, no long-term budget or not even a permanent employee, but it marked the born of the first world Governmental and Nongovernmental Organization (GONGO). Its first move was to establish the "Survival Service" in 1950. Using the $2,500 it raised at the time, the Survival Services called on scientists and volunteers from all around the world to write assessments of the world's endangered species, asking governments to protect those species within their borders. In 1964, the IUCN officially released the *Endangered Species Red Book*. Today, the International Union for Conservation of Nature (IUCN) has completely changed itself, including its name. The IUCN has become an observer of the United Nations and a leading conservation organization worldwide. IUCN Survival Service has evolved into the Species Survival Commission (SSC), the IUCN *Endangered Species Red Book* has been expanded into the website of *IUCN Red List of Threatened Species*, which is now renewed annually by the IUCN Species Survival Committee. The *IUCN Red List of Threatened Species* is the world's most comprehensive source of information on the status of global conservation of animal, plant and fungal species.

Sovereignty countries in the world are the main body of biodiversity protection. The status of a species in IUCN red list is not the affirmatively same in a country's red list except that the species is an endemic species in that country; countries around the world are also developing their own red lists of endangered species. Through the study of the red list of endangered species, countries conduct surveys on the distribution, survival and conservation status of plant and animal species in their territory and then conduct a comprehensive assessment of the survival status of species. Therefore, the red list of endangered species is a list of species and their distribution, an objective assessment of the survival and conservation status of species, and an important indicator of the health status of biodiversity. The red list is used by governments, natural protected areas and wildlife authorities, conservation NGOs, researchers, natural resource planners and educational institutions. It provides information and powerful tools for promoting action on biodiversity conservation and policy formation and is critical to safely guarding the natural resources on which we depend. Through the internet, the red list of endangered species has also become an information base of endangered species and a tool for conservation workers and researchers.

China is a member of the IUCN and one of the earliest parties to the UN *Convention on Biological Diversity*. China has been at the forefront of biodiversity conservation. China is also one of the most biodiverse countries in the world, with more than 7,300 vertebrate species, accounting for about 11% of the total number of vertebrates in the world. China has complex fauna composition, significant differences in spatial distribution pattern, ancient origins and various groups in the biological evolution system, such as Giant Panda (*Ailuropoda melanoleuca*), Baiji (*Lipotes vexillifer*) and Yangtze Alligator (*Alligator sinensis*). In addition, China is the origin center of many domestic animals and plants. China is also one of the countries where biodiversity is most threatened. Due to the overuse of resources, habitat loss and degradation caused by human activities, environmental pollution, climate change and other factors, vertebrate diversity is seriously threatened.

During recent years, the CPC Central Committee and the State Council attach great importance to the protection of biodiversity. Biodiversity conservation is announced as the national strategy, *China's Biodiversity Conservation Strategy and Action Plan (2011-2030)* is issued, the Joint Inter-Ministerial Meeting for Biological Species Resources Protection is regularly held, the China National Committee for Biodiversity Conservation has been set up, a series of biological diversity protection programs and plans have been formulated and implemented, and positive progress in the field has been made. However, the overall trend of biodiversity decline in China has not been effectively stopped, and the conservation situation is still pressing, especially, lack of comprehensive understanding of threatened species in China has hindered the effective conservation of biodiversity. Therefore, it is of great significance to assess the threatened status of species and to formulate the red list of endangered species, and to propose targeted conservation strategies for promoting the implementation of *China's Biodiversity Conservation Strategy and Action Plan (2011-2030)* and the construction of ecological civilization.

Now, the status of Chinese mammals, birds, reptiles, amphibians and freshwater fishes have all been comprehensively assessed by Chinese scientists. In addition to assessing newly discovered species, China's red list of endangered vertebrates has also reassessed the status of some species, including the Giant Panda and Tibetan Antelope (*Pantholops hodgsonii*), whose status on the red list of endangered vertebrates in China has been downgraded due to conservation efforts. However, the overall situation of endangered biodiversity in China is still serious.

Despite the high proportion of threatened species in China, the Chinese government is strengthening

Foreword II

ecological protection, stepping up the construction of nature reserves, national parks, World Heritage Sites and other protected areas, working on desertification control, wetland restoration and afforestation, and trying to reverse or at least stop the trend of biodiversity decline. On the occasion that the fifteenth meeting of the Conference of the Parties to the *Convention on Biological Diversity*, which will be held in Kunming, China, Chinese scientists published the latest edition of the "*China's Red List of Biodiversity: Vertebrates, Volume I, Mammals*", "*China's Red List of Biodiversity: Vertebrates, Volume II, Birds*", "*China's Red List of Biodiversity: Vertebrates, Volume III, Reptiles*", "*China's Red List of Biodiversity: Vertebrates, Volume IV, Amphibians*" and "*China's Red List of Biodiversity: Vertebrates, Volume V, Freshwater Fishes*". This set of books comprehensively update the survival status, population and habitat protection of vertebrate, determine the recovery efforts needed for the targeted species, identify key populations and habitats that need to be protected, thus provide a basis for identifying future priorities for endangered vertebrates conservation. Publication of these books is an important event in the history of Chinese nature conservation.

Xinsheng Zhang

President of IUCN,

the International Union for Conservation of Nature

June 26, 2020

总前言

物种的濒危现状和濒危机制是保护生物学的核心研究内容，其研究目标是评估人类对生物多样性的影响，提出防止物种灭绝及保护的策略，通过保护生物物种的种群和栖息地，避免物种受到灭绝的威胁。保护生物学研究既关注全球性问题，又具有鲜明的地域特色。中国具有世界上最多的人口，国土面积为世界第三，监测和评估其生物多样性、保护濒危物种，将为中国实现可持续发展提供科学支撑。中国研究人员通过濒危物种红色名录研究，量化了物种灭绝风险，预警了潜在的生态危机，为中国履行《生物多样性公约》等提供科技支撑。

世界自然保护联盟（International Union for Conservation of Nature，IUCN）成立后的第一个重大举措是在1950年建立了"生存服务（Survival Service）机构"。"生存服务机构"利用当时募集的2,500美元，召集科学家、志愿者评估全球濒危物种灭绝风险，发表有灭绝风险的物种研究报告，呼吁各国政府保护其境内的濒危物种，这是《IUCN受威胁物种红色名录》的发端。直到1964年，世界自然保护联盟才正式发布《濒危物种红皮书》。今天，世界自然保护联盟完全改变了它自己，"生存服务机构"已经演化成为物种存续委员会（Species Survival Commission，SSC）。IUCN《濒危物种红皮书》已经演变为网络版的《IUCN受威胁物种红色名录》。物种存续委员会从不定期发布IUCN濒危物种红皮书发展到现在每年发布更新的《IUCN受威胁物种红色名录》。

《IUCN受威胁物种红色名录》是世界生物多样性健康状况的重要指标。它是世界上最全面的一份动物、植物和真菌物种濒危状况清单。濒危物种红色名录基于物种种群数量、种群数量下降速率、生境破碎程度、生境面积及下降速率、预测灭绝概率等指标估测物种灭绝概率。《IUCN受威胁物种红色名录》（2020-2）发现地球上的32,000多个物种面临着灭绝的风险，占所有被评估物种的27%。其中，41%的两栖类、26%的哺乳类、34%的针叶树、14%的鸟类、30%的鲨鱼、33%的造礁珊瑚，以及28%的特定甲壳类物种面临灭绝风险。《IUCN受威胁物种红色名录》为保护我们赖以生存的自然资源提供了至关重要的信息。

项目背景

自1980年以来，中国经济步入高速发展期。目前，中国已经成为世界第二大经济体。在人口增长、经济发展、全球变化的背景下，中国的生物多样性正面临着前所未有的城镇化、乡村和社会基础设施建设及全球变化的压力，野生生物的生存受到威胁。许多证据显示地球上的生物正面临生物进化中的第六次大灭绝。保护濒危物种是生物多样性保护的核心问题。评估物种濒危等级是生物多样性监测与保护的迫切需要。

虽然《IUCN受威胁物种红色名录》没有国际法和国家法律的效力，但它是专家对全部物种生存状况的评估，它不仅限于评估濒危物种和明星物种，而是最大限度地涵盖了已知的物种，它不仅仅指导世界范围的濒危物种保护，也是指导生物多样性研究的有用工具。《IUCN受威胁物种红色名录》对于政府间组织和非政府组织的保护决策及各国自然与自然保护法律法规的制定都产生了重要影响。

综上所述，濒危物种红色名录是物种灭绝风险的测度，IUCN定期更新其濒危物种红色名录，预警全球物种的生存危机。同时，各国也开展了本国濒危物种红色名录研究。那么，既然已经有《IUCN受威胁物种红色名录》，为什么还要开展国家濒危物种红色名录研究？

《IUCN受威胁物种红色名录》与国家濒危物种红色名录都是物种灭绝风险的测度，前者是全球性评估，后者则是依国别的研究，两者的研究空间尺度不同。《IUCN受威胁物种红色名录》预警了全球物种的濒危状况，为全球生物多样性研究提供了大数据；各国红色名录则确定了各国物种受威胁状况，填补了前者的知识空缺，两份红色名录互为补充。

基于如下原因，应当重视依国别的濒危物种红色名录。①国家是濒危物种保护的行为主体，物种在一个国家的生存状况是确定其保护级别、开展濒危物种保育的依据。②对于仅分布于一个国家的特有物种来说，其按国别的濒危物种红色名录等级即是其全球濒危等级。③《IUCN受威胁物种红色名录》只提供了全球范围的物种濒危信息，并没有评估每一个国家所有物种的生存状况，特别是一些特有物种和跨越国境分布的物种。一些物种跨越国界分布，全球的生存状况并不反映其在个别国家的生存状况，一些全球无危的物种在其边缘分布区的国家里却是极度濒危的或受威胁的物种。世界各国的物种濒危状况有待各国科学家的研究。对于跨国境分布的物种来说，依国别的濒危物种红色名录等级则确定了该物种在本国的生存状况。④结合《IUCN受威胁物种红色名录》，依国别的濒危物种红色名录为建立跨国保护地、保护迁徙物种的栖息地与跨国迁徙洄游通道提供依据。⑤依国别的濒危物种红色名录所特有的"区域灭绝"等级，反映了一个物种边缘种群在该国的区域灭绝，对于一个国家来说，事关重大；恢复"区域灭绝"物种是该物种原分布国家重新引入的相关保育工作的重点。⑥物种濒危状况是不断变化的。近年来，新种、新记录不断被发现。随着人们对生命世界认识的深入，脊椎动物分类系统也发生了变化。依国别的濒危物种红色名录提供了该国物种编目、分类、分布和生存状况的最新信息（蒋志刚等，2020）。

国家红色名录的重要性在许多情况下被忽视了。在研究报告和科普作品中，对国家濒危物种红色名录重视不够。论及物种濒危属性时，作者通常言必《IUCN受威胁物种红色名录》濒危等级而不提其国家级的红色名录濒危等级。目前正值全球新型冠状病毒肺炎大流行，人们正在重新审视人与野生动物的关系。我国将修订有关野生动物保护与防疫法规和法律、重点保护野生物种名录，防控新的人与野生动物共患疾病再次暴发。对于确定《国家重点保护野生动物名录》而言，物种受威胁程度是物种列为国家重点保护野生物种的特征之一。重视依国别的红色名录有特别的意义。于是，生态环境部（原环境保护部）与中国科学院联合开展了中国生物多样性红色名录研究。

中国动物学家掌握了中国动物分布和生存状况的第一手资料，有必要组织全国淡水鱼类、两栖类、爬行类、鸟类与哺乳类专家及时更新中国脊椎动物分类系统，提供中国脊椎动物多样性的全面、完整的信息；有必要应用统一的国际物种濒危等级标准评估物种生存状况。在国家层面，定期组织全国淡水鱼类、两栖类、爬行类、鸟类与哺乳类专家应用IUCN受威胁物种红色名录等级标准和IUCN区域受威胁物种红色名录标准，全面评估更新的中国脊椎动物生物多样性红色名录，提供与国际红色名录研究可对比的结果，为红色名录指数的研究积累数据。

经过系统评审制定的中国生物多样性红色名录，由国家权威机构发布。中国生物多样性红色名录淡水鱼类、两栖类、爬行类、鸟类与哺乳类各卷将为监测中国生物多样性现状、为开展阶段性IUCN红色名录指数研究和履行《生物多样性公约》提供数据。

中国在1998年首次出版了《中国濒危动物红皮书》，2004年，又出版了《中国物种红色名录》，2009年，环境保护部组织开展了"中国陆栖脊椎动物物种濒危等级评估"。时隔多年，有必要重新全面评估中国生物多样性的濒危状况。于是，环境保护部委托中国科学院组织有关专家开展了"中国生物多样性红色名录——脊椎动物卷"的研究。在环境保护部和中国科学院的领导下，我们依据IUCN受威胁物种红色名录等级标准和IUCN区域受威胁物种红色名录标准，全面评估了中国哺乳动物生存状况。

2015年，环境保护部与中国科学院联合发布了"中国生物多样性红色名录——脊椎动物卷"。现在，历时6年，我们全面编研、更新、丰富了此名录，形成了此2021版的《中国生物多样性红色名录：脊椎动物》。

项目目标

通过脊椎动物各类群的研究，收集整理中国脊椎动物现有物种种群、生境研究数据、资源监测数据，充实数据库；组织专家，采用综合分析和专家评估相结合的方法，依据中国脊椎动物野生种群与生境现状，利用IUCN受威胁物种红色名录等级标准第3.1版和IUCN区域受威胁物种红色名录标准第4.0版综合评价中国脊椎动物濒危状况，编制2021版《中国生物多样性红色名录：脊椎动物》。全面评价中国脊椎动物的灭绝风险，对中国濒危物种保护及时提供基础信息。

编研过程

2013年5月16日，在中国科学院动物研究所召开了研究启动会。项目聘请陈宜瑜院士、郑光美院士、张亚平院士、金鉴明院士、马建章院士、曹文宣院士为咨询专家，并成立了哺乳类、鸟类、

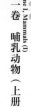

爬行类、两栖类、淡水鱼类课题组。各课题组就评估程序和规范展开了研讨，对典型物种进行了评估并听取了专家委员会的意见。会后总结了专家意见，完善了中国脊椎动物红色名录评估程序和规范。

针对哺乳类、鸟类、爬行类、两栖类和淡水鱼类分别建立了工作组、核心专家组和咨询专家组。工作组负责按照预定的红色名录判定规程开展工作，工作包括资料收集与整理、红色名录初步评定、与通讯评审专家联络及通讯评估结果汇总。核心专家组对红色名录评估的方法、标准使用、数据来源等重要科学问题进行界定，讨论审核有关物种的受威胁等级。工作组在全国范围遴选咨询专家，建立咨询专家库。咨询专家参加了红色名录的通讯评审和会议评审。评审结束后，工作组按照统一格式，整理每个物种包含的信息，形成最终的物种评估说明书。物种评估说明书的内容包括物种的学名、中文名、评估受威胁等级及 IUCN 红色名录等级。"中国生物多样性红色名录——脊椎动物卷"于 2015 年 5 月 6 日通过环境保护部和中国科学院的联合验收，并于 5 月 22 日以环境保护部、中国科学院 2015 年第 32 号公告形式发布。

红色名录评估的信息来源主要有研究积累、标本数据、文献数据和专家咨询。项目各课题组相关研究团队是工作在中国淡水鱼类、两栖类、爬行类、鸟类和哺乳类研究一线的研究团队，在数十年的研究中积累了大量的科学数据，各分卷主持人还是国家濒危物种科学机构，以及淡水鱼类、两栖类、爬行类、鸟类和哺乳类学术团体的骨干，所在单位是有关动物物种分类、标本收藏、研究的信息交换所，并各自建立了数据库。各分卷主持人还主持或参与了国家有关物种资源本底调查、科学评估、自然保护区生物多样性考察及相关的保护政策制定。

实践意义

《中国生物多样性红色名录：脊椎动物》的出版是一项重大的系统工程。这次生物多样性红色名录评估是迄今评估对象最广、涉及信息最全、参与专家人数最多的一次评估。通过 2015 版红色名录研究，我们更新了中国脊椎动物编目。中国有 2,854 种陆生脊椎动物，其中，有 407 种两栖类，402 种爬行类，1,372 种鸟类，673 种哺乳类。在 2021 版《中国生物多样性红色名录：脊椎动物》的编研中，我们再次更新了中国脊椎动物分类系统和编目。中国有 3,147 种陆生脊椎动物，其中，有 475 种两栖类，527 种爬行类，1,445 种鸟类，700 种哺乳类，比 2015 年的统计数据增加了 293 种。我们发现，中国是全球哺乳动物物种数最多的国家。中国陆生脊椎动物中，特有种超过 20%。我们还分析了中国脊椎动物的分布格局和特有类群，探讨了其濒危种类的空间分布规律。

中国濒危脊椎物种濒危模式与分布格局在我国生物多样性和生态系统保护中具有指导意义，这将为我国重点保护物种确定、国土空间开发和生态功能区的划分及各类保护地规划设计提供重要参考依据。也是确定中国物种多样性保护热点的依据之一。我们发现，中国脊椎动物生存危机依然严重，中国濒危脊椎物种的分布格局不均衡。物种的空间分布是一种立体格局，除了水平纬度上的物种分布格局，我们也需要物种多样性和濒危种类的垂直分布格局，这些格局对物种多样性保护具有重要参考价值。我们发现，高海拔地区受威胁哺乳动物的比例比低海拔地区高，高海拔地区的濒危物种应受到更多的关注。

展望与致谢

《中国生物多样性红色名录》的编制和发布为生物多样性保护政策和规划的制定提供了科学依据，发挥了中国科学家作为中国《生物多样性公约》履约"智库"的功能，同时，为开展生物多样性科学基础研究积累了基础数据，更新了脊椎动物分类系统与编目，为公众参与生物多样性保护创造了必要条件。《中国生物多样性红色名录》的编制是贯彻实施《中国生物多样性保护战略与行动计划（2011—2030 年）》和积极履行《生物多样性公约》的具体行动。通过《中国生物多样性红色名录》的编制，中国在生物多样性评价方面已经在全球先行一步，使我国在履行《生物多样性公约》方面走在世界的前列。

本项目得到了生态环境部（原环境保护部）、中国科学院、国家林业与草原局（原国家林业局）、中国科学院大学、科学出版社的关怀、指导和大力支持；得到了国家出版基金的大力支持。课题组还得到了如下项目的资助：中国科学院战略性先导科技专项（A类）"地球大数据科学工程"（项目编号：XDA19050204）、国家重点研发计划项目（项目编号：2016YFC0503303）、国家科技基础性工作专项（项目编号：2013FY110300）的资助。在此谨致感谢！

蒋志刚
中国科学院动物研究所研究员
中国科学院大学岗位教授
国家濒危物种科学委员会前常务副主任
2020 年 6 月 6 日

Series' Preface

The current status and threats to species are the key issues in conservation biology. The primary goal of putting forward an endangered species red list is to evaluate human impact on biodiversity, identifying key threats and preventing species extinction by protecting the populations and habitats of threatened species. Conservation research not only pays attention to global issues, but also must focus attention on regional and national problems. China has the largest human population and the third largest terrestrial area in the world. Monitoring and evaluating the country's biodiversity and protecting endangered species will provide scientific support for China's sustainable development. Therefore, Chinese researchers are working to quantify the risk of extinctions through studies related to the red list of threatened species, providing early warning about potential ecological hazards, thus offering scientific support for implementation of the *Convention on Biological Diversity* in China.

The first major move for the International Union for Conservation of Nature (IUCN) after its establishment was to launch the Survival Service in 1950. Using the $2,500 raised at that time, the Survival Service called on scientists and volunteers to dedicate their expertise and time to assess the extinction risk of globally threatened species. The Survival Service then publicized their research reports on species at risk of extinction and called on governments to protect endangered species within their borders. Such an act marked the beginning of the *IUCN Red List of Threatened Species*. However, it was not until 1964 that IUCN officially published its first *Endangered Species Red Book*. Today, IUCN has completely changed itself, the Survival Service has been renamed as the Species Survival Commission (SSC). The IUCN

Endangered Species Red Book has evolved into the online version of *IUCN Red List of Threatened Species*. The Species Survival Commission refreshes and revises the *IUCN Red List of Threatened Species* periodically, and updates the *IUCN Red List of Threatened Species* annually.

The *IUCN Red List of Threatened Species* is an important indicator of the health of the world's biodiversity. It is the world's most comprehensive list of rare and threatened animal, plant and fungal species. The *IUCN Red List of Threatened Species* estimates the extinction probability of species based on population size, population decline rate, degree of habitat fragmentation, rate of decline of habitat area and other indicators. The *IUCN Red List of Threatened Species* (2020-2) estimated that more than 32,000 species on the earth are at risk of extinction, accounting for 27% of all assessed species globally. 41% of amphibians, 26% of mammals, 34% of conifers, 14% of birds, 30% of sharks, 33% of corals, and 28% of certain crustaceans are presently at risk of extinction. The *IUCN Red List of Threatened Species* provides vital information for protecting the biodiversity and natural resources on which we all collectively depend.

The Background

Since the 1980s, China has embarked on a fast track of socioeconomic development. China has become the world's second largest economy. Against a backdrop of population growth, rapid economic development and many global changes, China's biodiversity is under unprecedented pressure from urbanization, infrastructure development and a wide range of other factors, and the survival of wildlife is under threat. Ample evidence shows that life on the earth is facing its Sixth Mass Extinction in its long evolutionary history. Protecting endangered species is the core issue for biodiversity conservation. Thus, assessing the endangerment level of species is the primary and most urgent need that biodiversity monitoring and protection measures seek to address.

Though the *IUCN Red List of Threatened Species* does not possess the power of international or national laws, it is an expert assessment of the survival status of all species, not only endangered or charismatic species, but all known species to the greatest extent possible. It thus serves as a most useful tool not only to guide worldwide protection of endangered species but also for the study of biodiversity. The *IUCN Red List of Threatened Species* has significant impact on the conservation decisions of intergovernmental and non-governmental organizations as well as for the formulation of national laws and regulations regarding wildlife and nature conservation.

As stated above, the red list of threatened species provides a measure of the risk of extinction of species. IUCN regularly updates its global red list of endangered species in order to raise public awareness of the global status of wildlife and the species survival crisis. At the same time, countries also conduct national-level studies on the status of endangered species. However, since there is already an *IUCN Red List of Threatened Species*, the question may arise, why bother to conduct research at country level to produce national red lists of endangered species?

Both the *IUCN Red List of Threatened Species* and country red lists of threatened species assess species' risk of extinction, with the former being global in scope while the latter are regional assessments. The

Series' Preface

IUCN Red List of Threatened Species alerts the world to the status of endangered species, and also serves as a database of global biodiversity. Country red lists, on the other hand, ascertain the status of species in particular countries, filling knowledge gaps in the former. The two lists are thus complementary to each other.

Country-level red lists should be given greater attention for at least the following reasons: (i) A sovereign country is the main authority for taking conservation action in regard to wildlife species within its boundaries, based on the level of endangerment (conservation status) of the species; (ii) For endemic species in a country, the country red list status constitutes its global status; (iii) The *IUCN Red List of Threatened Species* provides only the information on species at risk worldwide and does not assess the status of all species in each country, especially endemic species and those species with transboundary distribution. Some species are distributed across national boundaries and the global conservation status does not entirely reflect the survival status of the species in any particular country. Some global non-threatened species are critically endangered or threatened in the countries where they have peripheral ranges. The endangered status of species in different countries of the world thus remains to be studied by scientists in relation to specific countries. For species whose ranges cross national borders, the country's red list status reflects the survival status of the species in the country; (iv) Combined with the global *IUCN Red List of Threatened Species*, country red lists provide a basis from which to consider the establishment of transnational protected areas, the protection of important habitats for migratory species, and the protection of international migration corridors; (v) The category "Regionally Extinct" is unique to country (regional) red lists of endangered species as it refers only to a subset of the broader geographic distribution of the species, yet the national status is still indicative of the species' overall risk of extinction, this matters a lot for a country, and the restoration of "regionally extinct" species is the focus of conservation efforts for reintroduction in countries where the species originated; (vi) Country red lists provide updated information about endangered species with national inventories as well as with national reviews of classification, geographic distribution, and status of species at national level, which are also relevant for global species descriptions and assessments (Jiang *et al*., 2020).

Despite these benefits, the significance of country-level red lists is often overlooked. Following onset of the global COVID-19 pandemic, however, people's outlook has been changing in regard to the relationship between people and wildlife. Consequently, China is amending its national laws on wildlife protection, epidemic prevention, and the list of state key protected wild species, in order to better prevent and control emerging zoonoses. The status of wildlife species included in China's red list of threatened species should be one of the defining elements for identifying and updating species on the *List of State Key Protected Wild Animal Species* in China. It is therefore critical to duly recognize the significance of the country red list at this special moment in time. For this purpose, the Ministry of Ecology and Environment (former Ministry of Environmental Protection) and the Chinese Academy of Sciences have jointly launched China's biodiversity red list.

Chinese zoologists have obtained first-hand information on the distribution and living status of animals in China. It is necessary to organize national experts on freshwater fishes, amphibians, reptiles, birds and

mammals to update the taxonomy of vertebrates in China in a timely manner and to provide systematic and comprehensive information on the diversity of vertebrates in China. It is necessary to apply standard international criteria for threatened species to assess the status of species. At the national level, it is necessary to coordinate national experts on freshwater fishes, amphibians, reptiles, birds and mammals to apply the IUCN red list criteria for threatened species and the IUCN regional red list criteria for threatened species, and through this process also to comprehensively update the *China's Red List of Biodiversity: Volume of Vertebrates* and thus to provide a country red list that is comparable to international red lists, and to enable index studies of red lists.

The red list of China's biodiversity, which has been systematically reviewed and formulated, shall be issued by the state authorities. The volumes of freshwater fishes, amphibians, reptiles, birds and mammals of the red list of china's biodiversity will provide data for the implementation of the *Convention on Biological Diversity*, for monitoring the state of biodiversity in China, as well as for conducting periodic IUCN red list index studies in the country.

China firstly published its *China Red Data Book of Endangered Animals* in 1998, followed by the *China Species Red List* in 2004 and in 2009, the Ministry of Environmental Protection coordinated the assessment and publishing of the "Assessment of the Red List of Endangered Species of Terrestrial Vertebrates in China", which is a multi-year project for the comprehensive re-assessment of the threatened status of China's biodiversity. Therefore, the Ministry of Environmental Protection entrusts the Chinese Academy of Sciences to organize experts to carry out research on the *China's Red List of Biodiversity: Volume of Vertebrates*. Under the leadership of the Ministry of Environmental Protection and the Chinese Academy of Sciences, we have conducted a comprehensive assessment of the living status of the vertebrates in China based on the IUCN red list criteria for threatened species and the IUCN regional red list criteria for endangered species.

In 2015, the Ministry of Environmental Protection and the Chinese Academy of Sciences jointly released the *China's Red List of Biodiversity: Volume of Vertebrates*. Now, six years on, we have thoroughly updated and compiled the series of books of *China's Red List of Biodiversity: Vertebrates* (2021).

The Goal

Through the study and preparation for each volume of vertebrates in China's biodiversity red list, we collected and sorted existing information on the population and habitat status of vertebrates in China and completed the database. We systematically and comprehensively evaluated the status of vertebrates in China, using the *IUCN Red List Categories and Criteria* (version 3.1) and the *Guidelines for Application of IUCN Red List Criteria at Regional and National Levels* (version 4.0) based on the status of the species' wild population and habitat. Combined with the empirical analysis and expert evaluation, we compiled the *China's Red List of Biodiversity: Vertebrates* (2021), which is a comprehensive assessment of extinction risk of China's vertebrates, providing the basic information pertinent for the protection of endangered species over the coming years.

The Assessment

The project launch meeting was held at the Institute of Zoology, Chinese Academy of Sciences on May 16, 2013. Academicians Yiyu Chen, Guangmei Zheng, Yaping Zhang, Jianming Jin, Jianzhang Ma and Wenxuan Cao were invited to participate in the meeting as consulting experts. Mammals, birds, reptiles, amphibians and freshwater fishes research groups were formed at the meeting. Each research group held a discussion on evaluation procedures and norms, assessed the typical species of their own taxonomic group, and consulted the opinion of the expert committee. After the meeting, all experts' opinions were summarized and the evaluation procedures and norms of the red list of vertebrates in China were finalized.

Working groups, core expert groups and communication expert groups were formed for each research group, focused respectively on mammals, birds, reptiles, amphibians and fresh water fishes. The working groups were responsible for carrying out work in accordance with the red list category assessment procedures, including data collection and classification, preliminary red list category evaluation, liaison with experts by correspondence, and providing summaries and communicating evaluation results. The core expert group defined the methods, standards, data sources and other important scientific issues of the red list assessment, discussed and reviewed the status of species. The working group selected consulting experts nationwide and established a database of consulting experts. Each consulting expert participated in the red list evaluation and conference review. After the review, the information about every species was sorted and summarized in a unified format to provide the final species evaluation specifications. The species description and assessment includes scientific name, Chinese name, threat level assessment and IUCN red list category criteria. The "*China's Red List of Biodiversity: Volume of Vertebrates*" was jointly approved by the Ministry of Environmental Protection and the Chinese Academy of Sciences on May 6, 2015, and officially released on May 22, 2015, through the Announcement No. 32 of the Ministry of Environmental Protection and the Chinese Academy of Sciences, on International Biodiversity Day of 2015.

The information sources for the evaluation of vertebrates in the red list of China's biodiversity included published and unpublished literature, specimen data, and expert consultation. Experts from the red list working groups for freshwater fishes, amphibians, reptiles, birds and mammals are experts who have accumulated a large amount of scientific data over decades. The coordinators of working groups are people from state endangered species scientific authorities and established academics from scientific communities focused on freshwater fishes, amphibians, reptiles, birds and mammals from across the country. The research institutions are the centers of taxonomy, specimen collections, and databases for animal species. The principal scientists of freshwater fishes, amphibians, reptiles, birds and mammals also often coordinated or participated in background investigations on national species resources, scientific assessments, biodiversity investigations of nature reserves, and the formulation of relevant government conservation policies.

The significance

The publishing of the *China's Red List of Biodiversity: Vertebrates* (2021) is a major systematic project. The

biodiversity red list assessment covered the widest range of subjects, providing the most complete information and involving the largest number of experts so far in the country. During the process of developing the 2015 edition of the red list, we updated the Chinese vertebrate inventory. On this basis, it was found that China has 2,854 terrestrial vertebrates, of which 407 are amphibians, 402 are reptiles, 1,372 are birds and 673 are mammals. In the preparation and research of the 2021 edition, we have once again updated the classification system and produced an updated inventory of vertebrates in China. Altogether there are 3,147 terrestrial vertebrates in China, among which there are 475 species of amphibians, 527 reptiles, 1,445 birds and 700 mammals. A further 293 vertebrate species were assessed for preparing the 2021 edition. China is now found to have the largest number of mammal species in the world. Among land vertebrates in China, more than 20% are endemic. We also analyzed the distribution pattern and endemic groups of vertebrates in China and discussed the spatial distribution pattern of endangered species.

The conservation status and distribution patterns of threatened vertebrates in China that are shared in this book provide an important reference for identification of key protected vertebrate species, planning and development of national strategic land use blueprints, the design of ecological functional zones and various protected sites, which are of great significance for biodiversity and ecosystem conservation in China. This information is also one of the criteria for determining hotspot locations for species diversity in China. We have found that the survival crisis of vertebrates is still present in China and the distribution pattern of endangered vertebrates remains uneven. The spatial distribution of species presents a three-dimensional pattern, including their geographic distribution (two dimensions) as well as vertical or elevational dimension where species including threatened species are generally situated. In particular, we found that the proportions of threatened mammals in different families and orders were greater at higher altitudes than those found at lower altitudes, and additionally we found that endangered species that live at higher altitudes often should receive more attention from the public as well as from the government.

The Outlook and Appreciation

The compiling and publishing of *China's Red List of Biodiversity* provides a scientific basis for biodiversity conservation planning and policy in China, based on the long-standing work of Chinese scientists, who constitute a *de facto* "think-tank" for research and implementation of the *Convention on Biological Diversity* in China. At the same time, the red list study has updated the vertebrate taxonomy and inventory in China, accumulated data for basic zoological research in biodiversity science both nationally and globally, and created the necessary conditions for public participation in biodiversity conservation. Producing the *China's Red List of Biodiversity* has been a concrete action in the implementation of *China's Biodiversity Conservation Strategy and Action Plan (2011-2030)* and has also been a key step in implementing the *Convention on Biological Diversity* in its territory. Through the compilation of the *China's Red List of Biodiversity*, China has demonstrated its leading role in the global assessment of the current status of biodiversity.

The project that enabled development and publication of this red list book received guidance and support from the Ministry of Ecology and Environment (former Ministry of Environmental Protection), the Chinese

Academy of Sciences, the National Forestry and Grassland Administration (former National Forestry Administration), the University of Chinese Academy of Sciences, and the Science Press (Beijing). The project received funding from the National Publication Foundation, and each individual research group also received support through the following projects: "Earth Big-Data Scientific Project" (XDA19050204) of the Strategic Leading Science and Technology Project, Chinese Academy of Sciences (Category A); National Key Research and Development Project (2016YFC0503303); Basic Science Special Project of the Ministry of Science and Technology of China (2013FY110300). We express our most sincere gratitude to all of these governmental bodies, institutions and funding agencies for their many different forms of support.

Zhigang Jiang, Ph.D.

Professor of Institute of Zoology, Chinese Academy of Sciences

Professor of University of Chinese Academy of Sciences

Former Executive Director of the Endangered Species Scientific Commission, P. R. China

前 言

中国哺乳动物区系有鲜明的特色：中国是世界上哺乳动物种类最多的国家之一，也是特有哺乳动物丰富的国家。进入 21 世纪后，中国哺乳动物研究得到了长足的发展。由于种种原因，中国的哺乳动物编目工作一直到 21 世纪初才由王应祥先生完成。王应祥先生在《中国哺乳动物种和亚种分类名录与分布大全》一书报道中国有哺乳动物 603 种。从 2008 年起，我们开始评估中国濒危哺乳动物生存状况。我们首先补充了新种与新记录种，删去了无效种，采用最新的哺乳动物分类系统，开展了中国哺乳动物编目研究。我们 2015 年报道了中国有哺乳动物 12 目 55 科 245 属 673 种（蒋志刚等，2015a）。2017 年，我们再次更新了分类系统，补充了新种与新记录种，删去了无效种，更新了这一数据，记录哺乳动物 13 目 56 科 248 属 693 种（蒋志刚等，2017）。在本次《中国生物多样性红色名录：脊椎动物 第一卷 哺乳动物》(2021) 的评估中，经过查遗补缺，再次更新中国哺乳动物为 13 目 56 科 248 属 700 种。我们共评估了除智人外的 700 种哺乳动物。

本书为中英文双语，面向全球读者。为了方便读者提纲挈领，掌握本红色名录研究方法和中国哺乳动物的生存与保护状况。本书在前面部分有介绍中国哺乳动物生存状况评估的"总论"。在总论中，介绍了哺乳动物的演化与现状，中国哺乳动物多样性与保护现状，本次红色名录评估对象的分类系统、分布格局、受保护状况；介绍了中国哺乳动物红色名录的评估过程、依据的评估等级标准，还介绍了咨询专家顾问、评估队伍，以及评估团队建立数据库和开展初步评定、通讯评审、形成评估报告的过程。最后，总结分析了本次红色名录评估结果，比较了《中国生物多样性红色名录：脊椎动物 第一卷 哺乳动物》2021 版与 2015 版的异同，介绍了中国哺乳动物濒危状况，分析了野外灭绝的和区域灭绝的物种，分析了不同哺乳动物类群、不同生境的受威胁哺乳动物物种比例及其省区哺乳动物的分布和濒危原因，并将本次评估结果与《IUCN 受威胁物种红色名录》(2020-2) 的评估结果进行了比较分析。最后，分析展望了中国哺乳动物保护成效与远景。

本书"各论"中的物种编排顺序基本按照 IUCN 受威胁物种红色名录濒危等级："极危""濒危""易危""近危""无危""野外灭绝""区域灭绝"（本书特有）"数据缺乏"排列。"各论"编排列出了每一物种中文名 Chinese Name 和其他物种的分类信息如目 Order、科 Family、学名 Scientific Name、命名人 Species Authority、英文名 English Name(s)、同物异名 Synonym(s)、种下单元评估 Infra-specific Taxa Assessed，并配有彩绘插图或照片。各论还列出了评估信息 Assessment Information、红色名录等级 Red List Category（评估标准版本，已列在物种标题上）、评估年份 Year Assessed、评定人 Assessor(s)、审定人 Reviewer(s) 和其他贡献人 Other Contributor(s)。并对评估对象进行了评估理由 Justification、地理分布 Geographical Distribution（国内分布 Domestic Distribution 和世界分布 World Distribution），以及是否特有种的分布标注 Distribution Note 描述；还给出了国内分布图 Map of Domestic Distribution 和种群数量 Population Size、种群趋势 Population Trend、所在生境与生态系统 Habitat(s) and Ecosystem(s)，以及威胁 Threat(s)、保护级别与保护行动 Protection Category and Conservation Action(s) [国家重点保护野生动物等级（2021）Category of National Key Protected Wild Animals（2021）、IUCN 红色名录（2020-2）IUCN Red List（2020-2）、CITES 附录（2019）CITES Appendix（2019），是否开展了"保护行动 Conservation Action(s)"]。最后，列出了相关文献 Relevant References。书末附有检索表和参考文献。

在本书的编研过程中，有关工作得到了中国科学院、生态环境部（原环境保护部）、国家林业和草原局（原国家林业局）、中国科学院大学、国家濒危物种科学委员会、中国科学院动物研究所的精心指导、大力支持与帮助。在本书的编辑出版过程中，得到了科学出版社的细心帮助，得到国家出版基金的资助。在本书的编研过程中，我们执行了原环境保护部生物多样性专项：中国脊椎动物红色名录研究（2012—2018 年），国家科技基础性工作专项（项目编号：2013FY110300）、国家重点研发计划项目（项目编号：2016YFC0503303）、中国科学院战略性先导科技专项（A 类；项目编号：XDA19050204）、"美丽中国"生态文明建设科技工程项目（项目编号：XDA23100203）、全国第二次陆生野生动物资源调查项目、国家自然科学基金项目等项目。

本书的编研历时数年，由于研究团队知识与时间有限，我们在编研过程中深深领会到"吾生也有涯，而知也无涯。以有涯随无涯，殆已"。保护生物学是一门处理危机的学科，濒危物种红色名录是生物多样性预警报告。为了及时拯救濒危物种，尽管缺点、疏漏在所难免，我们仍将这本图书呈示于世，以期有关方面及时采取行动，保护人类赖以生存的基础。希望有关专家、读者对本书存在的问题不吝指正。

蒋志刚

2020 年 6 月 6 日

Preface

China has distinct characteristics in regard to its mammalian fauna: China has amongst the richest mammal diversity in the world, and it also has one of the greatest levels of endemism of mammal species globally. Since the beginning of the new millennium, mammal research in China has made significant progress. For various reasons, the inventory of mammals in China was not completed until the early 21st century. In his book *Taxonomy and Distribution of Mammal Species and Subspecies in China*, Professor Yingxiang Wang reported that there were 603 mammal species in China. Following this, since 2008, we have been systematically assessing the survival status of threatened mammals in China. First, we have added new species and new records of species, deleted invalid species, and adopted the latest mammal taxonomic system to carry out the inventory research of mammals in China. By 2015, we reported 673 species from 245 genera, 55 families and 12 orders of mammals in China (Jiang *et al.*, 2015a). In 2017, we renewed the classification system, added several more species and new records of species, deleted additional invalid species, and through this process updated the data and inventory, recording in total 693 species from 248 genera, 56 families, 13 orders of mammals (Jiang *et al.*, 2017). In this assessment, in the Mammals Volume of *China's Red List of Biodiversity*: *Vertebrates* (2021), we now recognize a total of 700 species of mammals (in 248 genera, 56 families, 13 orders) in the national inventory and evaluated 700 species except the *Homo sapiens*.

This book is bilingual in both Chinese and English, and is intended for a global audience, aiming to introduce readers to the basic concepts and grasp research methods concerning the red list of threatened species, and more specifically, to main findings from long-term research about the living conditions and conservation status of mammals in China. This book is preceded by a General Introduction regarding the assessment of the survival or conservation status of mammals in China. In the General Introduction, the evolution and current state of mammals are introduced including their diversity, their taxonomy, and their distribution patterns, along with their conservation or red list status. The General Introduction also describes the evaluation process that was adopted for the mammals of China's red list of biodiversity, including the evaluation criteria and consulting experts, and how the evaluation team established the mammal database, carried out preliminary evaluation, reviewed preliminary results by correspondence, and formalized the evaluation report. Finally, the evaluation results are analyzed and summarized for the readers. The authors compared the Mammal Volume of *China's Red List of Biodiversity: Vertebrates* 2021 edition with the 2015 edition, analyzed the status of threatened mammals, including special note of species that are Extinct in the Wild and Regionally Extinct, as well as the proportion of threatened species in different mammal groups, in different habitats and range provinces, and the threats that are faced by endangered mammals. A comparison of the results of mammals in *China's Red List of Biodiversity: Vertebrates* (2021) with the *IUCN Red List of Threatened Species* (2020-2) also is provided. Finally, the implications and prospects for mammal conservation in China are discussed.

In the book, the species are arranged in "Species Monograph" according to their status in the *IUCN Red List of Threatened Species*: "Critically Endangered", "Endangered", "Vulnerable", "Near Threatened", "Not Threatened", "Extinct in the Wild", "Regionally Extinct" (unique only in this book), and "Data Deficient". In each "Species Monograph", with colored illustrations or photographs, the Chinese Name and taxonomy of the species such as: Order, Family, Scientific Name, Species Authority, English Name(s), Synonym(s), Infra-specific Taxa Assessed of the species are given. In each "Species Monograph", Assessment Information, Red List Category (evaluation criteria version, listed in the title), Year Assessed, Assessor(s), Reviewer(s) and Other Contributor(s) are listed, plus the assessment Justification, Domestic Distribution in China and World Distribution, as well as the Distribution Note of whether the Species is Endemic or not. Map of Domestic Distribution, Population Size and Trend, Habitat(s) and Ecosystem(s) of species are also given. And more, "Species Monograph" presents the Threats, Protection Category and Conservation Action taken such as Category of National Key Protected Wild Animals (2021), IUCN Red List (2020-2), CITES Appendix (2019), and Whether there is a "Conservation Action(s)" for species. Finally, Relevant References are listed. At the end of the book, Index and Reference are listed.

During the compilation and research of the book, the work has been carefully guided, supported and helped by the Chinese Academy of Sciences, the Ministry of Ecology and Environment (formerly the Ministry of Environmental Protection), the State Forestry and Grassland Administration (formerly the State Forestry Administration), the University of Chinese Academy of Sciences, the National Endangered Species Scientific Commission, and the Institute of Zoology of the Chinese Academy of Sciences. During the process of editing and publishing this book, we have received help from Science Press in Beijing, and financial support from the National Publication Foundation. During the process of research for the book, we performed the biodiversity special project: China's Vertebrate Red List of 2012–2018 of the former Ministry of Environmental Protection,

Basic Science Special Project of the Ministry of Science and Technology of China (2013FY110300), National Key Research and Development Project (2016YFC0503303), Strategic Leading Science and Technology Project, Chinese Academy of Sciences (Category A, XDA19050204), Science and Technology Project of Ecological Civilization Construction of "The Beautiful China" (XDA23100203), as well as the special projects of the Second National Survey on Terrestrial Wild Animals and projects of the Natural Science Foundation, *etc*.

    Although the compilation of this book took several years, due to the limited knowledge and time of the research team, we deeply understood an ancient proverb "My life is limited, and the knowledge is limitless. To use the limited time to explore the boundlessness, it will be exhausted", during the compilation and research process. Conservation biology is a crisis management discipline, and the red list of endangered species is an early-warning report. In order to save endangered species in a timely manner, despite the inevitable shortcomings, we present this book to the world so that action can be taken to protect the very foundation on which humanity depends. I hope that experts and readers will not hesitate to point out the problems in this book.

Zhigang Jiang

June 6, 2020

目 录 Contents

序 一 ············ i	**Foreword I** ············ iii
序 二 ············ vii	**Foreword II** ············ ix
总前言 ············ xiii	**Series' Preface** ············ xix
前 言 ············ xxvii	**Preface** ············ xxix

总 论 General Introduction

1 哺乳动物的演化与现状 ············ 2	1 Evolution and Status of Mammals ············ 3
2 中国哺乳动物多样性与保护现状 ············ 2	2 Diversity and Conservation Status of Mammals in China ············ 3
3 分类系统 ············ 6	3 Taxonomic System ············ 7
4 中国哺乳动物编目（2021） ············ 10	4 Inventory of China's Mammals (2021) ············ 11
5 分布格局 ············ 22	5 Distribution Patterns ············ 23
6 受保护状况 ············ 24	6 Conservation Status ············ 25
7 评估过程 ············ 26	7 Evaluation Process ············ 27
8 评估等级和标准 ············ 28	8 Categories and Criteria ············ 29
9 建立数据库 ············ 36	9 Building the Database ············ 37
10 初步评定 ············ 36	10 A Preliminary Assessment ············ 37
11 通讯评审 ············ 36	11 Review by Correspondence ············ 37
12 形成评估报告 ············ 40	12 Forming the Evaluation Report ············ 41
13 评估结果分析 ············ 40	13 Analysis of the Assessment Results ············ 41
14 受威胁状况 ············ 42	14 Threatened Status ············ 43
15 野外灭绝和区域灭绝物种分析 ············ 58	15 Analysis of Extinct in the Wild or Regionally Extinct Species ············ 59
16 受威胁物种分析 ············ 72	16 Analysis of Threatened Species ············ 73
17 受威胁物种省级区域分布 ············ 76	17 Distribution of Threatened Species in Provincial Region ············ 77
18 受威胁原因分析 ············ 78	18 Analysis of Threats ············ 79
19 与中国历年红皮书、红色名录评估结果的比较 ············ 82	19 Comparison with the Results of Historical Red List Assessments in the Country ············ 83

20	与《IUCN 受威胁物种红色名录》(2020-2) 的比较 ·········· 88	20	Comparison with the *IUCN Red List of Threatened Species* (2020-2) ·········· 89
21	保护成效 ·········· 90	21	Conservation Achievements ·········· 91
22	结束语 ·········· 94	22	Conclusions ·········· 95

各论（上册） Species Monograph (I)

极危		CR	
高氏缺齿鼩	100	*Chodsigoa caovansunga*	100
倭蜂猴	102	*Nycticebus pygmaeus*	102
北豚尾猴	104	*Macaca leonina*	104
白颊猕猴	106	*Macaca leucogenys*	106
长尾叶猴	108	*Semnopithecus schistaceus*	108
白头叶猴	110	*Trachypithecus leucocephalus*	110
黔金丝猴	112	*Rhinopithecus brelichi*	112
缅甸金丝猴	114	*Rhinopithecus strykeri*	114
西白眉长臂猿	116	*Hoolock hoolock*	116
东白眉长臂猿	118	*Hoolock leuconedys*	118
高黎贡白眉长臂猿	120	*Hoolock tianxing*	120
白掌长臂猿	122	*Hylobates lar*	122
西黑冠长臂猿	124	*Nomascus concolor*	124
海南长臂猿	126	*Nomascus hainanus*	126
北白颊长臂猿	128	*Nomascus leucogenys*	128
东黑冠长臂猿	130	*Nomascus nasutus*	130
印度穿山甲	132	*Manis crassicaudata*	132
马来穿山甲	134	*Manis javanica*	134
穿山甲	136	*Manis pentadactyla*	136
马来熊	138	*Helarctos malayanus*	138
江獭	140	*Lutrogale perspicillata*	140
小爪水獭	142	*Aonyx cinerea*	142
大斑灵猫	144	*Viverra megaspila*	144
大灵猫	146	*Viverra zibetha*	146
熊狸	148	*Arctictis binturong*	148
小齿狸	150	*Arctogalidia trivirgata*	150
缟灵猫	152	*Chrotogale owstoni*	152
荒漠猫	154	*Felis bieti*	154
丛林猫	156	*Felis chaus*	156

云豹	158	Neofelis nebulosa	158
虎	160	Panthera tigris	160
儒艮	164	Dugong dugon	164
亚洲象	166	Elephas maximus	166
野骆驼	168	Camelus ferus	168
林麝	170	Moschus berezovskii	170
马麝	172	Moschus chrysogaster	172
黑麝	174	Moschus fuscus	174
原麝	176	Moschus moschiferus	176
塔里木马鹿	178	Cervus hanglu	178
坡鹿	180	Panolia siamensis	180
麋鹿	182	Elaphurus davidianus	182
驼鹿	184	Alces alces	184
印度野牛	186	Bos gaurus	186
蒙原羚	188	Procapra gutturosa	188
贡山羚牛	190	Budorcas taxicolor	190
长尾斑羚	192	Naemorhedus caudatus	192
塔尔羊	194	Hemitragus jemlahicus	194
阿尔泰盘羊	196	Ovis ammon	196
戈壁盘羊	198	Ovis darwini	198
白鱀豚	200	Lipotes vexillifer	200
长江江豚	204	Neophocaena asiaeorientalis	204
比氏鼯鼠	206	Biswamoyopterus biswasi	206
河狸	208	Castor fiber	208
海南兔	210	Lepus hainanus	210

濒危 / EN

海南新毛猬	212	Neohylomys hainanensis	212
琉球狐蝠	214	Pteropus dasymallus	214
安氏长舌果蝠	216	Macroglossus sobrinus	216
琉球长翼蝠	218	Miniopterus fuscus	218
彩蝠	220	Kerivoula picta	220
蜂猴	222	Nycticebus bengalensis	222
台湾猴	224	Macaca cyclopis	224
达旺猴	226	Macaca munzala	226
印支灰叶猴	228	Trachypithecus crepusculus	228
黑叶猴	230	Trachypithecus francoisi	230

戴帽叶猴	232	Trachypithecus pileatus ... 232
萧氏叶猴	234	Trachypithecus shortridgei ... 234
滇金丝猴	236	Rhinopithecus bieti ... 236
豺	238	Cuon alpinus ... 238
懒熊	240	Melursus ursinus ... 240
石貂	242	Martes foina ... 242
貂熊	244	Gulo gulo ... 244
白鼬	246	Mustela erminea ... 246
纹鼬	248	Mustela strigidorsa ... 248
虎鼬	250	Vormela peregusna ... 250
缅甸鼬獾	252	Melogale personata ... 252
水獭	254	Lutra lutra ... 254
椰子狸	256	Paradoxurus hermaphroditus ... 256
野猫	258	Felis silvestris ... 258
兔狲	260	Otocolobus manul ... 260
猞猁	262	Lynx lynx ... 262
云猫	264	Pardofelis marmorata ... 264
金猫	266	Pardofelis temminckii ... 266
金钱豹	268	Panthera pardus ... 268
雪豹	272	Panthera uncia ... 272
威氏鼷鹿	274	Tragulus williamsoni ... 274
安徽麝	276	Moschus anhuiensis ... 276
喜马拉雅麝	278	Moschus leucogaster ... 278
黑麂	280	Muntiacus crinifrons ... 280
贡山麂	282	Muntiacus gongshanensis ... 282
马鹿	284	Cervus canadensis ... 284
梅花鹿	286	Cervus nippon ... 286
西藏马鹿	288	Cervus wallichii ... 288
白唇鹿	290	Przewalskium albirostris ... 290
普氏原羚	292	Procapra przewalskii ... 292
赤斑羚	294	Naemorhedus baileyi ... 294
喜马拉雅斑羚	296	Naemorhedus goral ... 296
天山盘羊	298	Ovis karelini ... 298
喜马拉雅鬣羚	300	Capricornis thar ... 300
北太平洋露脊鲸	302	Eubalaena japonica ... 302
塞鲸	304	Balaenoptera borealis ... 304
蓝鲸	306	Balaenoptera musculus ... 306

中文名	页码	学名	页码
长须鲸	308	*Balaenoptera physalus*	308
东亚江豚	310	*Neophocaena sunameri*	310
恒河豚	312	*Platanista gangetica*	312
中华白海豚	314	*Sousa chinensis*	314
滇攀鼠	316	*Vernaya fulva*	316
耐氏大鼠	318	*Leopoldamys neilli*	318
四川毛尾睡鼠	320	*Chaetocauda sichuanensis*	320
扁颅鼠兔	322	*Ochotona flatcalvariam*	322
伊犁鼠兔	324	*Ochotona iliensis*	324
柯氏鼠兔	326	*Ochotona koslowi*	326
原仓鼠	328	*Cricetus cricetus*	328
粗毛兔	330	*Caprolagus hispidus*	330

易危 / VU

中文名	页码	学名	页码
峨眉鼩鼹	332	*Uropsilus andersoni*	332
宽齿鼹	334	*Euroscaptor grandis*	334
短尾鼹	336	*Euroscaptor micrura*	336
小齿鼹	338	*Euroscaptor parvidens*	338
大鼩鼱	340	*Sorex mirabilis*	340
扁颅鼩鼱	342	*Sorex roboratus*	342
水鼩鼱	344	*Neomys fodiens*	344
灰腹水鼩	346	*Chimarrogale styani*	346
小臭鼩	348	*Suncus etruscus*	348
抱尾果蝠	350	*Rousettus amplexicaudatus*	350
短耳犬蝠	352	*Cynopterus brachyotis*	352
球果蝠	354	*Sphaerias blanfordi*	354
长舌果蝠	356	*Eonycteris spelaea*	356
印度假吸血蝠	358	*Megaderma lyra*	358
单角菊头蝠	360	*Rhinolophus monoceros*	360
云南菊头蝠	362	*Rhinolophus yunnanensis*	362
莱氏蹄蝠	364	*Hipposideros lylei*	364
无尾蹄蝠	366	*Coelops frithii*	366
金黄鼠耳蝠	368	*Myotis formosus*	368
小巨足鼠耳蝠	370	*Myotis hasseltii*	370
渡濑氏鼠耳蝠	372	*Myotis rufoniger*	372
大黑伏翼	374	*Arielulus circumdatus*	374
黄喉黑伏翼	376	*Arielulus torquatus*	376

亚洲宽耳蝠	378	*Barbastella leucomelas*	378
短尾猴	380	*Macaca arctoides*	380
熊猴	382	*Macaca assamensis*	382
藏酋猴	384	*Macaca thibetana*	384
菲氏叶猴	386	*Trachypithecus phayrei*	386
棕熊	388	*Ursus arctos*	388
黑熊	390	*Ursus thibetanus*	390
大熊猫	392	*Ailuropoda melanoleuca*	392
小熊猫	394	*Ailurus fulgens*	394
北海狗	396	*Callorhinus ursinus*	396
黄喉貂	398	*Martes flavigula*	398
紫貂	400	*Martes zibellina*	400
艾鼬	402	*Mustela eversmanii*	402
伶鼬	404	*Mustela nivalis*	404
斑海豹	406	*Phoca largha*	406
环斑小头海豹	408	*Pusa hispida*	408
斑林狸	410	*Prionodon pardicolor*	410
爪哇獴	412	*Herpestes javanicus*	412
食蟹獴	414	*Herpestes urva*	414
豹猫	416	*Prionailurus bengalensis*	416
蒙古野驴	418	*Equus hemionus*	418
獐	420	*Hydropotes inermis*	420
海南麂	422	*Muntiacus nigripes*	422
野牦牛	424	*Bos mutus*	424
鹅喉羚	426	*Gazella subgutturosa*	426
秦岭羚牛	428	*Budorcas bedfordi*	428
四川羚牛	430	*Budorcas tibetanus*	430
不丹羚牛	432	*Budorcas whitei*	432
中华斑羚	434	*Naemorhedus griseus*	434
帕米尔盘羊	436	*Ovis polii*	436
中华鬣羚	438	*Capricornis milneedwardsii*	438
抹香鲸	440	*Physeter macrocephalus*	440
印太江豚	442	*Neophocaena phocaenoides*	442
巨松鼠	444	*Ratufa bicolor*	444
复齿鼯鼠	446	*Trogopterus xanthipes*	446
红背鼯鼠	448	*Petaurista petaurista*	448
小飞鼠	450	*Pteromys volans*	450

沟牙田鼠	452	*Proedromys bedfordi*	452
小狨鼠	454	*Hapalomys delacouri*	454
褐尾鼠	456	*Niviventer cremoriventer*	456
阿尔泰鼢鼠	458	*Myospalax myospalax*	458
长白山鼠兔	460	*Ochotona coreana*	460
大巴山鼠兔	462	*Ochotona dabashanensis*	462
黑鼠兔	464	*Ochotona nigritia*	464
峨眉鼠兔	466	*Ochotona sacraria*	466

总 论
General Introduction

1 哺乳动物的演化与现状

哺乳动物是地球上出现较晚的高等生物类群，属于温血羊膜类 (endothermic amniotes) 动物。最早的哺乳类从下孔类 (synapsids) 演化而来，在 3 亿年前的石炭纪 (Carboniferous Period) 末才出现 (Kemp, 2005)，然而，早期的哺乳类是地球上一个占据边缘生态位的生物类群。在恐龙灭绝之后，哺乳动物迅速辐射进化，占据了恐龙消失后腾出的中型与大型生态位，其中一些种类演化成地球陆地上和海洋中体型最大的动物 (Smith *et al*., 2010)，成为世界上最成功的动物类群之一。Wilson 和 Reeder (2005) 的 *Mammal Species of the World: A Taxonomic and Geographic Reference* 第 3 版记录了世界上 5,416 种哺乳动物种类，分属于 1,229 属 153 科 29 目。2008 年，世界自然保护联盟 (IUCN) 濒危物种红色名录完成了一项为期 5 年的全球哺乳动物评估，该名录评估了 5,488 个哺乳动物物种，分属于 1,267 属 159 科 27 目。根据 Burgin *et al*. (2018) 在 *Journal of Mammalogy* (《哺乳动物学杂志》，英文版) 上发表的一项研究，目前地球上已知的哺乳动物种类为 6,495 种，分属于 1,314 属 167 科 27 目。其中有新热带界 (Neotropic Realm)1,617 种，非洲热带界 (Afrotropic Realm)1,572 种，古北界 (Palearctic Realm) 1,162 种，埃塞俄比亚新北界 (Nearctic Realm) 697 种，澳洲界 (Australian Realm) 527 种，海洋 (Marine) 种类 124 种，家养种类 17 种，还有近期灭绝的 96 种。哺乳动物能够利用从深海到高山、从雨林到沙漠、从地下到树冠、从极地到火山口的各种各样的栖息地，产生了相应的遗传、形态、生理、生化、行为的适应。其食性有植食性、肉食性、杂食性，其行走方式可分为陆地行走型、攀援型、飞行型、掘洞与地下生活型和游水型。哺乳动物的陆地行走型可以细分为蹄行性、趾行性、跖行性。哺乳动物拥有大脑，产生了复杂的通讯行为、个性和社会性。尽管啮齿目 Rodentia、翼手目 Chiroptera、鼩形目 Soricomorpha 种类占地球哺乳动物种数目的 70% 以上。作为灵长类一员的智人 (*Homo sapiens*) 的出现，改变了地球地表覆盖和生物圈，影响了地球上众多其他物种的生存。

2 中国哺乳动物多样性与保护现状

哺乳动物是高等脊椎动物，是地球上适应能力很强的动物类群。从戈壁到深海，从"世界屋脊"的雪山到海滨湿地，到处都有哺乳动物的分布。中国疆域辽阔，南北跨越近 50 个纬度，东西跨越 60 余个经度，是世界国土面积第三大的国家。中国地质环境经历了海陆变迁、高原隆升等重大地质事件，境内地貌多样、气候迥异。气温、日照与降水的悬殊差异和地质历史决定了中国脊椎动物的分布格局 (Luo *et al*., 2012)。境内有冲积平原、海岸盐沼、黄土高原、喀斯特地貌、丘陵山地、极旱荒漠、永久冻土、内蒙古草原和青藏高原等地理单元。存在多种多样的哺乳动物生境，如喀斯特洞穴、悬崖峭壁、热带雨林、亚热带常绿阔叶林、暖温带针阔混交林、寒温带针叶林、沼泽湿地、红树林湿地、干旱草原、

1 Evolution and Status of Mammals

Mammals, belonging to the endothermic amniotes, are a relatively late-coming higher life form on Earth. The earliest mammals evolved from synapsids at the end of the Carboniferous Period around 300 million years ago (Kemp, 2005). The early mammals on Earth were organisms living in marginal niches. After the extinction of dinosaurs, mammals radiated rapidly and diversified into new species which now occupy the medium and large ecological niches left over by the dinosaurs. Some of these mammal species evolved into the largest animals on land and in the sea (Smith *et al.*, 2010). Therefore, mammals are one of the most successful animal groups on Earth. According to the third edition of the *Mammal Species of the World: A Taxonomic and Geographic Reference* (Wilson and Reeder, 2005), there are 5,416 mammalian species divided into 1,229 genera, 153 families and 29 orders. In 2008, IUCN's red list of threatened species working group completed a 5-year global mammal assessment and assessed 5,488 species in 1,267 genera, 159 families and 27 orders. According to research of Burgin *et al.* (2018) published in *Journal of Mammalogy*, there are 6,495 mammal species in 1,314 genera, 167 families and 27 orders currently known on Earth, including 1,617 species in the Neotropic Realm, 1,572 in the Afrotropic Realm, 1,162 in the Palearctic Realm, 697 in the Nearctic Realm, 527 in the Australian Realm, plus 124 marine species, 17 domesticated species and 96 recently extinct species. Mammals are able to live in and use a wide variety of habitats ranging from the deep sea to high mountains, from the rainforest to deserts, from underground environments to tree canopies, from polar regions to the craters of volcanoes, bringing with this level of diversity corresponding genetic, morphological, physiological, biochemical and behavioral adaptation mechanisms. Their feeding habits include vegetarians, carnivorous and omnivorous. Locomotive modes include walking, arboreal climbing, flying, fossorial and subterranean life, and swimming. Terrestrial walking may also be subdivided into different running styles: unguligrade, digitigrade, and plantigrade. Mammals have brains and complex communication behaviors, personalities, and in some instances also high levels of sociality. The orders Rodentia, Chiroptera and Soricomorpha include most of the mammal species, accounting for over 70% of the total species on Earth. The arrival of humans, *Homo sapiens*, a member of the primates, has brought the most substantial changes to Earth's land cover and biosphere, in turn affecting the conditions and survival of many other species on Earth.

2 Diversity and Conservation Status of Mammals in China

Mammals are higher vertebrates and they are a successful adaptable group of animals on Earth. Mammals are found everywhere from the Gobi Desert to the deep sea, from mountains with snow-covered peaks on the "Roof of the World" to coastal wetlands. China has a vast territory, spanning nearly 50 degrees of latitudes from north to south and over 60 degrees of longitudes from east to west; it is the third largest country by land area in the world. China's environment has undergone major geological events such as sea-land transformations and the uplifting of high plateaus. Huge differences in temperature, sunlight and precipitation and the dramatic geological histories in the country have shaped the distribution pattern of vertebrates in China (Luo *et al.*, 2012). There are alluvial plains, coastal salt marshes, loess plateaus,

高寒草甸、苔原、流石滩、冰川、沙漠、寒漠、戈壁等生境，栖息着高度多样化的哺乳动物区系（蒋志刚等，2015a）。

中国哺乳动物区系的形成可以追溯到第三纪末，由古北界 (Palearctic Realm) 和东洋界 (Oriental Realm) 组成 (Wallace, 1876)，或古北界和中国-日本界 (Sino-Japanese Realm) (Holt et al., 2013) 组成。中国有从第四纪冰期避难所存活下来的第三纪孑遗动物——大熊猫科 Ailuropodidae 和白暨豚科 Lipotidae (Zhou and Qian, 1985)。在中国的多山地理环境中还隔离分化产生了许多特有灵长类动物，如仰鼻猴属 (Rhinopithecus) 是东亚特有猴科森林树栖动物，栖息于海拔 1,500~3,300m 亚热带山地针阔混交林、落叶阔叶林和常绿针叶林，集群生活，金丝猴属 5 种金丝猴中有 4 种在中国有分布，其中川金丝猴 (R. roxellana，图1)、滇金丝猴 (R. bieti) 和黔金丝猴 (R. brelichi) 是中国的特有灵长类，川金丝猴群体有典型群层社会结构 (Grueter et al., 2020)。2011 年在中缅边境地区发现了缅甸金丝猴 (R. strykeri)，2012 年，中国动物学家证实缅甸金丝猴在中国高黎贡山也有分布 (Geissmann et al., 2020)。

在第三纪新构造运动中，青藏高原快速隆升 (Corlett, 2014)。因此，中国的哺乳动物区系发生分化并产生了许多特有种，如藏羚 (Pantholops hodgsonii)（图2）、藏原羚 (Procapra picticaudata)、藏野驴 (Equus kiang)、

图 1　湖北神农架的川金丝猴（蒋志刚摄）。湖北神农架的川金丝猴湖北亚种 (Rhinopithecus roxellana hubei) 属于川金丝猴的 3 个亚种 R. r. roxellana, R. r. qinglin 和 R. r. hubei 之一，栖息于神农架海拔 1,000m 以上的亚热带山地针阔混交林

Figure 1　Golden Snub-nosed Monkey of Shennongjia, Hubei (Photographed by Zhigang Jiang). The Hubei subspecies of Golden Snub-nosed Monkey (Rhinopithecus roxellana hubei) is one of the three subspecies of Golden Snub-nosed Monkey: R. r. roxellana, R. r. qinglin 和 R. r. hubei. It lives in the Mixed Coniferous and Broad-leaved Forest in the subtropical mountain area at 1,000m a.s.l. (above sea level)

图 2　藏羚 (*Pantholops hodgsonii*, 蒋志刚摄)。栖息在海拔 4,000m 的青藏高原腹地的藏羚是青藏高原动物区系的代表性哺乳动物，是哺乳动物适应青藏高原高寒低氧环境的特有属。长期以来，人们一直将藏羚归类于羚羊，然而，根据分子生物学研究，事实上藏羚是一类特化的山羊 (Lei *et al.*, 2003)

Figure 2　Tibetan Antelope (*Pantholops hodgsonii*, Photographed by Zhigang Jiang). The Chiru, which lives in the hinterland of Qinghai-Tibet (Xizang) Plateau nearly 4,000m a.s.l., is the representative mammal of the plateau fauna. For a long time, people would take the Chiru as an antelope. However, according to the research with molecular biology means, the Chiru is actually a kind of specialized wild goat (Lei *et al.*, 2003)

karst landscapes, low hills and high mountains, extremely dry deserts, permafrost lands, steppes and grasslands, the Qinghai-Tibet (Xizang) Plateau and other geographical units in the country. There are thus a great variety of habitats for mammals, such as karst caves, rocky cliffs, tropical rain forest, subtropical evergreen broad-leaved forest, warm and temperate coniferous and broad-leaved mixed forest, cool coniferous forest, marshes, mangrove wetland, arid grassland, alpine meadow, tundra, glacier, desert, cold desert, gobi, *etc.*, which are inhabited by a highly diverse mammalian fauna (Jiang *et al.*, 2015a).

图3 白唇鹿（*Przewalskium albirostris*，蒋志刚摄）。白唇鹿分布于青海、西藏和四川，是青藏高原的特有鹿种

Figure 3 White-lipped Deer (*Przewalskium albirostris*, Photographed by Zhigang Jiang). White-lipped Deer distributes in Qinghai, Tibet (Xizang) and Sichuan, is endemic to the Qinghai-Tibet (Xizang) Plateau

野牦牛 (*Bos mutus*)、白唇鹿 (*Przewalskium albirostris*) 等 (图3)（冯祚建等，1986）。中国兔形目特有种比例高达33%。鼠兔科 (Ochotonidae) 以青藏高原为分布中心，中国分布的31种鼠兔中48%以上为中国特有种 (Liu et al., 2017)。

3 分类系统

中国哺乳动物学研究起步较晚，《中国动物志》的编研工作迄今尚未完成，对中国哺乳动物的分类与编目工作仍在继续。关于中国哺乳动物物种数的最早记录是1938年G. Allan的研究，他记录了中国（不包括东北和台湾）哺乳动物8目30科97属314种。随后，新的哺乳动物不断被发现，中国哺乳动物名录不断被更新。寿振黄 (1962) 记录了中国哺乳动物12目52科180属405种。张荣祖 (1979) 记录了12目44科183属414种。王应祥 (2003) 记录了中国哺乳动物13目55科235属607种。Wilson 和 Reeder (2005) 记录了中国哺乳动物13目54科245属572种。此后，潘清华等 (2007) 在《中国哺乳动物彩色图鉴》中记录了中国哺乳动物13目58科242属645种。史密斯等 (2009) 在《中国兽类野外手册》一书中记录了中国哺乳动物14目53科240属585种。

The formation of China's mammalian fauna can be traced back to the end of the Tertiary Period. Fauna in China is composed of species of the Palearctic Realm and the Oriental Realm (Wallace, 1876), or the Palearctic Realm and the Sino-Japanese Realm (Holt *et al.*, 2013). There are Tertiary relict mammals surviving from Quaternary glacial shelters in China, such as the giant panda family—Ailuropodidae and the fresh water dolphins, the Baiji family—Lipotidae (Zhou and Qian, 1985). In the mountainous topographical environment of China, many endemic primates are differentiated during geographic isolation, such as the snub-nosed monkey genus (*Rhinopithecus*), which is a unique genus of arboreal monkeys in East Asia. Snub-nosed monkeys live in subtropical montane coniferous and broad-leaved mixed forests, deciduous broad-leaved forest and evergreen coniferous forest between 1,500 ~ 3,300m. Four of the five snub-nosed monkeys are distributed in China, in which Golden Snub-nosed Monkey (*R. roxellana*, **Figure 1**), Black Snub-nosed Monkey (*R. bieti*) and Gray Snub-nosed Monkey (*R. brelichi*) are endemic to China, and Golden Snub-nosed Monkey has a multilevel organization of sociality (Grueter *et al.*, 2020). The Stryker's Snub-nosed Monkey (*R. strykeri*) was discovered in the China-Myanmar border region in 2011, and Chinese zoologists confirmed that the Stryker's Snub-nosed Monkey is also distributed in the Gaoligong Shan, Yunnan, China in 2012 (Geissmann *et al.*, 2020).

During the Tertiary neotectonic movement, the Qinghai-Tibet (Xizang) Plateau was rapidly uplifted (Corlett, 2014). China's mammalian fauna thus has differentiated, and many endemic species such as Chiru (*Pantholops hodgsonii*, **Figure 2**), Tibetan Gazelle (*Procapra picticaudata*), Kiang (*Equus kiang*), Wild Yak (*Bos mutus*) and White-lipped Deer (*Przewalskium albirostris*) (**Figure 3**) evolved in the highlands (Feng *et al.*, 1986). The proportion of lagomorphs that are endemic to China is as high as 33%. The distribution of the pika family, Ochotonidae, is centered on the Qinghai-Tibet (Xizang) Plateau, and over 48% of the 31 pika species found in China are endemic to the country (Liu *et al.*, 2017).

3 Taxonomic System

Mammalian research in China started rather late. The compilation of *Fanna Sinica* has not yet been completed. Classification and inventory of mammals in China are still in progress. The earliest record of the number of mammal species in China was a study in 1938 by Garry Allan, who recorded 314 mammals species in 8 orders, 30 families, 97 genera, in China (excluding northeast China and Taiwan). Since then, new species have been discovered and new records of species have been recorded in the country; the checklist of Chinese mammals had been updated. Shou (1962) recorded 405 species of mammals in China, belonging to 12 orders, 52 families, and 180 genera. Zhang (1979) recorded 414 species, 183 genera, 44 families and 12 orders of China's mammals. Wang (2003) recorded 607 species of 235 genera, 55 families, 13 orders of mammals in China. Wilson and Reeder (2005) recorded 572 species of mammals of 13 orders, 54 families, and 245 genera in China.

动物学是一门地域特色鲜明的科学。近年来，由于新一代哺乳动物学家的成长，野外考察范围的扩大，网络数据库的建设与运行，新技术手段在野外考察中的运用，分子生物学技术在分类鉴定与谱系地理学中的广泛应用，人们发现了中国哺乳动物的新种、新记录。同时，随着动物学知识的积累，人们还对哺乳动物分类系统进行了修订，提出了新的哺乳动物分类系统（蒋志刚，2016a）。例如，《IUCN 受威胁物种红色名录》将鲸目 (Cetacea) 与偶蹄目 (Artiodactyla) 合并为鲸偶蹄目 (Cetartiodactyla)，因为这两个目的相同进化起源。

在完成原环境保护部生物多样性专项"中国生物多样性红色名录——哺乳动物卷"的编研过程中，我们参照最新的哺乳动物分类系统，分第一阶段 (2012~2015 年)、第二阶段 (2016~2019 年) 两个阶段，对中国哺乳动物多样性进行了整理。在第一阶段，我们首先收集了正式出版文献中的中国哺乳动物资料。以 Wilson 和 Reeder (2012) 的 *Handbook of the Mammals of the World*，Wilson 和 Mittermeier (2005) 的 *Mammal Species of the World: A Taxonomic and Geographic Reference*（第 3 版）和王应祥 (2003)《中国哺乳动物种和亚种分类名录与分布大全》的分类系统为基础，前两者是目前世界公认的哺乳动物分类权威著作，由 IUCN 和《濒危野生动植物种国际贸易公约》(*Convention on International Trade in Endangered Species of Wild Fauna and Flora*, CITES) 推荐的世界哺乳动物分类体系，后者是一部集 20 世纪中国哺乳动物分类和分布之大全的分类著作。此外，还参考了 *IUCN Red List of Threatened Species* (2014)、史密斯等 (2009)《中国兽类野外手册》、刘瑞玉 (2008)《中国海洋生物名录》、潘清华等 (2007)《中国哺乳动物彩色图鉴》、盛和林等 (1998)《中国野生哺乳动物》及张荣祖 (1997)《中国哺乳动物分布》，灵长类分类还参考了夏武平和张荣祖 1995 年的分类系统，海兽类参考了周开亚 (2004) 的分类系统，获得了一份全新的中国哺乳动物编目。收集整理了中国所有哺乳动物种类，包括发表的中国新哺乳动物种和新记录种，对中国哺乳动物物种多样性编目进行了全面增补与修订，在《生物多样性》杂志发表了《中国哺乳动物多样性》（蒋志刚等，2015a），当时，依据《IUCN 受威胁物种红色名录》的分类系统，将鲸目 (Cetacea) 与偶蹄目 (Artiodactyla) 合并为鲸偶蹄目 (Cetartiodactyla)，记录了中国哺乳动物 12 目 55 科 245 属 673 种（图 4）。

Since then, Pan *et al.* (2007) recorded 645 species, 242 genera, 58 families, 13 orders of Chinese mammals in their book—*A Field Guide to Mammals of China*. Smith *et al.* (2009) recorded 585 species of 240 genera, 53 families, 14 orders of Chinese mammals in their book, *A Guide to the Mammals of China*.

Zoology is a science with distinctive regional geographic features. During recent years, as the growth of a new generation of mammologists, the expansion of scope of field investigation, the construction and operation of database, the application of new technical means in field investigation, the wide application of molecular biology technology in species identification and phylogeny, new species and new species records of mammals in China have been discovered or reported. At the same time, with the accumulation of zoological knowledge, people also revised the mammalian taxonomy and proposed a new mammalian taxonomy (Jiang, 2016a). For example, the *IUCN Red List of Threatened Species* combined Cetacea and Artiodactyla into a new order Cetartiodactyla because of their single evolutionary origin.

In the process of completing the special biodiversity project of former Ministry of Environmental Protection—"*China's Red List of Biodiversity: Volume of Mammals*", we sorted out the inventory of mammals in China during a two phase study: the 1st phase, from 2012 to 2015 and the 2nd phase, from 2016 to 2019, using the latest mammal classification system. During the 1st phase of the study, we first collected the Chinese mammal data in the published literature and built a preliminary checklist of China's mammals, based on the taxonomic system of *Handbook of the Mammals of the World* by Wilson and Mittermeier (2012), *Mammal Species of the World: A Taxonomic and Geographic Reference* (3rd edition) by Wilson and Reeder (2005) and *A Complete Checklist of Mammal Species and Subspecies in China-A Taxonomic and Geographic Reference* by Wang (2003). Wilson's taxonomy is currently recognized as the authoritative work of mammal classification in the world, which is recommended by the *Convention on International Trade in Endangered Species of Wild Fauna and Flora* (CITES), Yinxiang Wang's work is a comprehensive taxonomic work on the classification and distribution of mammals in China in the 20th century. In addition, we referred to *IUCN Red List of Threatened Species* (2014), *A Guide to the Mammals of China* (Smith *et al.*, 2009), *The Checklist of China's Marine Life* (Liu, 2008), *A Field Guide to Mammals of China* (Pan *et al.*, 2007), *Wild Mammals of China* (Sheng *et al.*, 1998) and *Distribution of Mammals in China* (Zhang, 1997) for additional information. For the classification of primates, we also consulted classification system of Wuping Xia and Rongzu Zhang in 1995, for the sea mammals we made reference to the classification system of Zhou (2004). We collected all mammal species, including published new mammal species and new species records in China. We finalized a new inventory of mammals in China and published the *China's Mammal Diversity* in the journal of *Biodiversity Science* (Jiang *et al.*, 2015a). According to the taxonomy in *IUCN Red List of Threatened Species* (2014), Cetacea and Artiodactyla are combined into Cetartiodactyla. We recorded 12 orders, 55 families, 245 genera and 673 species of mammals in China in 2015 (**Figure 4**).

4 中国哺乳动物编目 (2021)

"中国生物多样性红色名录——脊椎动物卷"(2015)发布后，2016~2019年，我们在生态环境部(原环境保护部)与中国科学院的资助下，开展了第二阶段的工作，继续更新了红色名录。

中国哺乳动物红色名录工作组继续上一版红色名录研究的程序，基于哺乳动物分类系统的修订、中国哺乳动物的新发现，继续收集整理中国哺乳动物新种、新记录种信息。我们在第二阶段的研究中首先更新

总论 **General Introduction**

图 4 除鲸目以外的中国哺乳动物 12 个目的代表种类。从左上角开始，由左向右，由上到下，依次为劳亚食虫目（Eulipotyphla）的东北刺猬（*Erinaceus amurensis*）（蒋志刚摄），兔形目（Lagomorpha）的高原鼠兔（*Ochotona curzoniae*）（徐爱春摄），长鼻目（Proboscidea）的亚洲象（*Elephas maximus*）（S. K. Yathin 摄），攀鼩目（Scandentia）的北树鼩（*Tupaia belangeri*）（马晓锋摄），奇蹄目（Perissodactyla）的野马（*Equus ferus*）（蒋志刚摄），灵长目（Primates）的川金丝猴（*Rhinopithecus roxellana*）（蒋志刚摄），偶蹄目（Artiodactyla）的西藏盘羊（*Ovis hodgsoni*）（蒋志刚摄），食肉目（Carnivora）的金钱豹（*Panthera pardus*）（蒋志刚摄），翼手目（Chiroptera）的尖耳鼠耳蝠（*Myotis blythii*）（黄元骏摄），啮齿目（Rodentia）的灰旱獭（*Marmota baibacina*）（蒋志刚摄），鳞甲目（Pholidota）的穿山甲（*Manis pentadactyla*）（蒋志刚摄），海牛目（Sirenia）的儒艮（*Dugong dugon*）（蒋志刚摄）

Figure 4 Representative species of the 12 mammalian orders of China except cetaceans. Starting at the top left, from left to right and from top to bottom, *Erinaceus amurensis* of Eulipotyphla (Photographed by Zhigang Jiang) and *Ochotona curzoniae* of Lagomorpha (Photographed by Aichun Xu), *Elephas maximus* of Proboscidea (Photographed by S. K. Yathin), *Tupaia belangeri* of Scandentia (Photographed by Xiaofeng Ma), *Equus ferus* of Perissodactyla (Photographed by Zhigang Jiang), *Rhinopithecus roxellana* of Primates (Photographed by Zhigang Jiang), *Ovis hodgsoni* of Artiodactyla (Photographed by Zhigang Jiang), and *Panthera pardus* of Carnivora (Photographed by Zhigang Jiang), *Myotis blythii* of Chiroptera (Photographed by Yuanjun Huang), *Marmota baibacina* of Rodentia (Photographed by Zhigang Jiang), *Manis pentadactyla* of Pholidota (Photographed by Zhigang Jiang), *Dugong dugon* of Sirenia (Photographed by Zhigang Jiang)

4 Inventory of China's Mammals (2021)

We continued the 2nd phase of the red list study, and continued to update the red list checklist after *China's Red List of Biodiversity: Volume of Vertebrates* (2015) published, with the support of the Ministry of Ecology and Environment (former Ministry of Environmental Protection) and the Chinese Academy of Sciences.

The working group of China's red list of mammals continued the previous red list research procedure, and continued to collect and sort information of China's new mammal species and new records, basing on the taxonomic

了中国哺乳动物多样性编目。事实上，我们在此期间两次更新了中国哺乳动物编目：第一次是 2017 年出版的《中国哺乳动物多样性》(第 2 版)，第二次是本书所采用的在《中国哺乳动物多样性》(第 2 版)基础之上更新的中国哺乳动物编目系统。

《中国哺乳动物多样性》(第 2 版)的新编目系统　进入 21 世纪后，动物学家们仍不断发现新种。如 Sinha et al. (2005) 报道了分布在我国藏南地区的猕猴实为一个新猕猴种，命名为达旺猴 (Macaca munzala)。白眉长臂猿分成两种——以缅甸亲敦江为界，亲敦江以东的为东白眉长臂猿 (Hoolock leuconedys)，而亲敦江以西的为西白眉长臂猿 (H. hoolock)。范朋飞带领的研究团队发现，分布在高黎贡山的白眉长臂猿与东白眉长臂猿有明显差别，于是，将高黎贡山的东白眉长臂猿命名为一个新种——高黎贡白眉长臂猿 (天行长臂猿 H. tianxing) (Fan et al., 2017，图 5)。高黎贡白眉长臂猿分布区狭小，仅分布于中国云南省，种群数量不足 200

图 5　高黎贡白眉长臂猿 (天行长臂猿，Hoolock tianxing, 范朋飞摄)

Figure 5　Tianxing Hoolock Gibbon (Skywalker Hoolock Gibbon, Hoolock tianxing, Photographed by Pengfei Fan) in the Gaoligong Shan

system revision and new discovery of China's mammals. First of all, we carried out the 2nd phase of the red list study, and updated the inventory of mammalian diversity in China. In fact, we updated the inventory twice: *China's mammal diversity* (2nd edition) published in 2017 is the first time, and the second time is the present renewed China's mammalian inventory system in this book.

The new inventory of mammals in *China's Mammal Diversity* (2nd edition) Zoologists are still discovering new species in the 21st century. For examples, Sinha *et al.* (2005) reported that *Macaca munzala* is a new macaque species distributed in southern Tibet (Xizang) in Zangnan area, China. There are two types of *Hoolock* gibbons—their boundary defined by the Chindwin River in Myanmar, with *Hoolock leuconedys* living east of the river and *H. hoolock* living west of the river. Furthermore, the research team led by Pengfei Fan found that the morphology and DNA of *Hoolock leuconedys* distributed in the Gaoligong Shan in Yunnan, China were significantly different from that of other populations, therefore a new species was named, *i.e.* Tianxing Hoolock Gibbon (also known as Skywalker Hoolock Gibbon; *H. tianxing*) (Fan *et al.*, 2017, **Figure 5**). This new gibbon species is only narrowly distributed in Yunnan Province, China with a total population fewer than 200 individuals patchily distributed across their range, including around 60~70 individuals in fewer than 20 groups in the Gaoligong Shan National Nature Reserve. Other populations of the gibbon are distributed in Sudian and Zhina Townships of Yingjiang County and Houqiao Township of Tengchong County on the border of China and Myanmar.

New records of mammals have also been reported. For example, during their field expedition, Hu *et al.* (2017) recorded the new presence of *Trachypithecus pileatus* for the first time, which was originally noted only in the southern foothills of the Himalayas, in the Shannan region of Tibet (Xizang) in China, and they have suggested to rename Shortridge's Langur (*T. shortridgei*) as "灰叶猴" in Chinese and to translate the Chinese name of *T. pileatus* as "戴帽叶猴".

We have collected new mammal records from the literature since March 2015, including both known and new mammal species, and we have conducted our own independent research. On this basis, we have adopted the new taxonomy presented herein with the rules for the Chinese names of new species set by Jiang *et al.* (2017). We also have added mammal species from the Zangnan area, Tibet (Xizang), China. In these and other ways, we have both renewed and expanded the inventory of mammalian diversity in the country, and published the *China's Mammal Diversity* (2nd edition). The major changes in this updated inventory of China's mammal species include the following details: (i) The order Cetartiodactyla has been elevated to the super order Cetartiodactyla, allowing for due recognition of the split of the (former) super order Cetartiodactyla into two separate orders, Cetacea and Artiodactyla. (ii) The species *Chodsigoa hoffmanni*, *C. furva* were added to the order Eulipotyphla. (iii) In the order Chiroptera, *Murina fanjingshanensis*, *Myotis rufoniger* and *Rhinolophus subbadius* were added to the mammal inventory, while *Myotis hirsutus* and *Miniopterus fuscus* were removed.

只，种群呈斑块化分布。高黎贡山国家级自然保护区内现有高黎贡白眉长臂猿不到 20 群，数量 60~70 只，其他种群分布在盈江县苏典乡、支那乡和中缅边境的腾冲猴桥镇。

人们还不断报道哺乳动物的新分布种记录。如胡一鸣等 (Hu et al., 2017) 在西藏山南地区考察时记录到原分布在喜马拉雅山脉南麓的叶猴 (*Trachypithecus pileatus*) 在中国西藏山南地区的新分布，并建议将 Shortridge's Langur (*T. shortirdgei*) 更名为"灰叶猴"，将 *T. pileatus* 的中文名翻译为"戴帽叶猴"。

在收集整理 2015 年 3 月以来发表的中国哺乳动物新种和新分布记录种的基础上，我们采用新的分类系统 (蒋志刚等，2017)，综合作者的最新研究，补充了以前知之甚少的藏南地区的哺乳动物信息，确定了新增物种中文名的规则，更新了中国哺乳动物多样性编目，发表了《中国哺乳动物多样性》（第 2 版）。这一版编目的主要修改有如下几个。①将鲸偶蹄类 (Cetartiodactyla) 列为总目，将鲸类与偶蹄类恢复为鲸目 (Cetacea) 和偶蹄目 (Artiodactyla)。②劳亚食虫目增加了新种霍氏缺齿鼩 (*Chodsigoa hoffmanni*) 及由亚种提升为种的烟黑缺齿鼩 (*C. furva*)。③翼手目增补了梵净山管鼻蝠 (*Murina fanjingshanensis*)、渡濑氏鼠耳蝠 (*Myotis rufoniger*) 和葛氏菊头蝠 (*Rhinolophus subbadius*)，删除了毛须鼠耳蝠 (*Myotis hirsutus*) 和琉球长翼蝠 (*Miniopterus fuscus*)。④灵长目增补了高黎贡白眉长臂猿 (*Hoolock tianxing*)、戴帽叶猴 (*Trachypithecus pileatus*)、懒猴 (*Nycticebus coucang*) 和西白眉长臂猿 (*H. hoolock*)。⑤食肉目增补了分布在中国藏南的懒熊 (*Melursus ursinus*)、亚洲胡狼 (*Canis aureus*)（图 6）、孟加拉狐 (*Vulpes bengalensis*)、灰獴 (*Herpestes edwardsii*) 和渔猫 (*Prionailurus viverrinus*)。⑥依据 Wilson 和 Mittermeier 的 *Handbook of the Mammals of the World, Vol. 2, Ungulates* (2012) 的偶蹄类分类系统重新厘定了中国偶蹄目动物分类。偶蹄目增加了阿尔泰盘羊 (*Ovis ammon*)、哈萨克盘羊 (*O. collium*)、贡山羚牛 (*Budorcas taxicolor*) 和印度麂 (*Muntiacus muntjak*)。将中国境内的梅花鹿合并为 *Cervus nippon*、驼鹿合并为 *Alces alces*。删去了阿拉善马鹿 (*Cervus alashanicus*)、四川马鹿 (*C. macneilli*) 和矮岩羊 (*Pseudois sharferi*)。将分布在西双版纳的小鼷鹿定为鼷鹿未定种 (*Tragulus sp.*)。⑦鲸目增加了恒河豚 (*Platanista gangetica*)。⑧啮齿目增加了小板齿鼠 (*Bandicota bengalensis*)、小猪尾鼠 (*Typhlomys nanus*)、墨脱松田鼠 (*Neodon medogensis*)、聂拉木松田鼠 (*N. nyalamensis*) 以及由亚种提升为种的大猪尾鼠 (*T. daloushanensis*)；还增加了甘肃鼢鼠 (*Myospalax cansus*)、比氏鼯鼠 (*Biswamoyopterus biswasi*)、白腹鼠 (*Niviventer niviventer*)、印度小鼠 (*Mus booduga*)。删去了休氏壮鼠 (*Hadromys humei*)。同时厘清了我国田鼠亚科 Arvicolini 族的分类。⑨兔形目增加了粗毛兔 (*Caprolagus hispidus*) 和尼泊尔黑兔 (*Lepus nigricollis*)。整理了鼠兔属 (*Ochotona*) 的分类，降级了 5 个鼠兔种，提升了 4 个鼠兔亚种为种，增加了 5 个新种。中国有 29 种鼠兔分布，北美鼠兔 (*O. princeps*)、斑颈

(iv) In the order Primates, *Hoolock tianxing*, *Trachypithecus pileatus*, *Nycticebus coucang* and *H. hoolock* were added to the inventory. (v) The Golden Jackal (*Canis aureus*), along with Sloth Bear (*Melursus ursinus*) in Zangnan area, Tibet (Xizang), China, are both added to the order Carnivora (**Figure 6**). *Vulpes bengalensis*, *Herpestes edwardsii* and *Prionailurus viverrinus* were also added. (vi) Furthermore, we reclassified the species in the order Artiodactyla according to the taxonomy used in the *Handbook of the Mammals of the World, Vol. 2, Ungulates* by Wilson and Mittermeier. On this basis, *Ovis ammon*, *O. collium*, *Budorcas taxicolor* and *Muntiacus muntjak* were added to the order Artiodactyla. All sika deers in China were grouped as a single species, *Cervus nippon*, and all moose in China were also grouped as a single species, *Alces alces*. On the other hand, *Cervus alashanicus*, *C. macneilli* and *Pseudois sharferi*. The taxonomic status of *Tragulus* in Xishuangbanna, Yunnan, was unclear based on the field survey; it was listed as *Tragulus* sp. (vii) In the order Cetacea, *Platanista gangetica* was added to the inventory. (viii) In

图 6 亚洲胡狼 (*Canis aureus*)。亚洲胡狼分布在中南半岛、西亚和南亚次大陆。蒋志刚等在 2017 年的《中国哺乳动物多样性》(第 2 版) 新增了亚洲胡狼。董磊等 (2019) 在西藏吉隆沟考察中发现了亚洲胡狼的存在 (Charles J. Sharp 摄影,CC BY-SA 4.0,https://commons.wikimedia.org)

Figure 6 Golden Jackal (*Canis aureus*) is found in Indochina Peninsula, west Asia and the subcontinent of South Asia. Jiang *et al.* (2017) added Golden Jackal to *China's Mammal Diversity* (2nd ed.). Dong Lei *et al.* (2019) discovered the existence of jackals in Tibet during their expedition in Jilong valley (Charles J. Sharp photography, CC BY-SA 4.0, https://commons.wikimedia.org)

鼠兔 (*O. collaris*)、荷氏鼠兔 (*O. hoffinanni*)、阿富汗鼠兔 (*O. rufescens*) 和草原鼠兔 (*O. pusilla*) 在中国没有分布。与 2015 版的《中国哺乳动物多样性》比较，本编目删去了 20 个种，新增了 40 个种，其中，新增了藏南地区分布的哺乳动物 15 种。截至 2017 年 8 月底，中国记录哺乳动物 13 目 56 科 248 属 693 种 (包括人科人属智人种)。比《中国哺乳动物多样性》(2015 版) 多 1 目 1 科 3 属 20 种。人们对 18 种中国哺乳动物的分类地位尚存在争议。中国有 146 种特有哺乳动物，占中国哺乳动物的 21%。特有种分别占兔形目、劳亚食虫目和偶蹄目的 37%、35% 和 25%。

  本书的中国哺乳动物编目系统   2017 年后，在写作《中国生物多样性红色名录：脊椎动物 第一卷 哺乳动物》期间，我们再次更新了中国哺乳动物编目系统。在本书的写作过程中，工作组进一步更新了《中国哺乳动物多样性》(第 2 版) 的哺乳动物编目体系，根据 *Handbook of the Mammals of the World* 和 IUCN 红色名录采用的分类系统，整理了中国马鹿的分类系统。我们同时继续收集整理了中国哺乳动物新种、新记录种。增补了新种：高黎贡林猬 (*Mesechinus wangi*) (Ai et al., 2018)，榕江管鼻蝠 (*Murina rongjiangensis*) (Cheng et al., 2017)、荔波管鼻蝠 (*Murina liboensis*) (Zeng et al., 2018)、锦矗管鼻蝠 (*Murina jinchui*) (Yu et al., 2020)、剑纹小社鼠 (*Niviventer gladiusmaculus*) (Ge et al., 2018) 和黑姬鼠 (*Apodemus nigrus*) (Ge et al., 2019)。周旭明等于 2018 年将窄脊江豚 (*Neophocaena asiaeorientalis*) 分为东亚江豚 (*Neophocaena sunameri*) 和长江江豚 (*Neophocaena asiaeorientalis*)，2021 版收录了东亚江豚和长江江豚。2021 版还增加了新记录种红鬣羚 (*Capricornis rubidus*)、印支鼠耳蝠 (*Myotis indochinensis*)、华南扁颅蝠 (*Tylonycteris fulvidus*)、道氏东京鼠 (*Tonkinomys davovantien*)、草原鼠兔 (*Ochotona pusilla*) 和越南鼬獾 (*Melogale cucphuongensis*) (Li et al., 2019)，并查遗补漏，根据专家意见增加了贺兰山鼠兔 (*Ochotona argentata*)。从《中国哺乳动物多样性》(第 2 版) 删去了帕米尔鼩鼱 (*Sorex buchariensis*)、扁颅蝠 (*Tylonycteris pachypus*)、纳氏鼠耳蝠 (*Myotis nattereri*)、长喙真海豚 (*Delphinus capensis*)、东亚屋顶鼠 (*Rattus brunneusculus*)、普通田鼠 (*Microtus arvalis*) 和窄脊江豚。将华南菊头蝠 (*Rhinolophus huananus*) 更名为清迈菊头蝠 (*Rhinolophus siamensis*)，将华南扁颅蝠 (*Tylonycteris fulvidus*) 的学名更正为 *Tylonycteris fulvida*。仍将鲸偶蹄类 (Cetartiodactyla) 列为总目，下列鲸目 (Cetacea) 和偶蹄目 (Artiodactyla)。根据新的分类系统，将斑林狸 (*Prionodon pardicolor*) 从灵猫科 Viverridae 分离出来，置于新的林狸科 Prionodontidae 之下。

  在 2009~2019 年，由 Don E. Wilson 和 Russell Mittermeier 主编的九卷本 *Handbook of the Mammals of the World* 各分册陆续出版。该书由世界著名动物学家集体编著，由 Conservation International, Texas A & M University 和 IUCN 赞助出版。在此期间，还有 *Wild Cats of the World* (Hunter, 2015)，*Bovids of the World* (Castelló, 2016)，*Pangolins: Science,*

the order Rodentia, several new species were added to the national inventory including *Bandicota bengalensis*, *Typhlomys nanus*, *Neodon medogensis*, *Neodon nyalamensis*, *Myospalax cansus*, *Biswamoyopterus biswasi*, *Niviventer niviventer* and *Mus booduga*, whereas *T. daloushanensis* was elevated from its previous status as subspecies to full species and *Hadromys humei* was deleted from the mammal inventory. The taxonomies of Arvicolini were renewed. (iv) In the order Lagomorpha, *Caprolagus hispidus* and *Lepus nigricollis* were added to the inventory and the taxonomy of *Ochotona* was renewed with 5 species being downgraded to subspecies and 4 subspecies elevated to the level of species. Altogether, there are now 29 pika species, *O. princeps*, *O. collaris*, *O. hoffinanni*, *O. rufescens* and *O. pusilla* didn't exist in China. Compared with the 2015 edition *China's Mammal Diversity*, this inventory removed 20 species and added 40, including 15 mammal species distributed in Zangnan area of Tibet (Xizang). Until the end of August in 2017, 693 mammal species of 248 genera, 56 families, 13 orders (including *Homo sapiens*, *Homo*, Hominidae) were recorded in China. In 2017 edition, there are 1 order, 1 family, 3 genera and 20 species more than 2015 edition of *China's Mammal Diversity*. The taxonomic status of 18 Chinese mammal species remains controversial. There are 146 endemic species in China, 21% of the total species number. Endemic ratios in Lagomorpha, Eulipotyphla and Artiodactyla are 37%, 35% and 25%.

The inventory of mammals in China in this book During the period of composing *China's Red List of Biodiversity: Vertebrates*, *Volume I*, *Mammals*, we updated the inventory of mammals in China once again, and the working group carried a further updating of the inventory of *China's Mammal Diversity* (2nd edition). Following the *Handbook of the Mammals of the World* (HMW), the taxonomic system used by the IUCN red list, the classification of *Cervus* in China was updated. At the same time, we continued to identify new species and new records of known species across China. Wang's Hedgehog (*Mesechinus wangi*) (Ai *et al.*, 2018), Rongjiang Tube-nosed Bat (*Murina rongjiangensis*) (Cheng *et al.*, 2017), Libo Tube-nosed Bat (*Murina liboensis*) (Zeng *et al.*, 2018), Jinchu's Tube-nosed Bat (*Murina jinchui*) (Yu *et al.*, 2020), Least White-bellied Rat (*Niviventer gladiusmaculus*) (Ge *et al.*, 2018) and Black Field Mouse (*Apodemus nigrus*) (Ge *et al.*, 2019) has been added in the updated mammal inventory. In light of research of Xuming Zhou *et al.* in 2018, Narrow-ridged Finless Porpoise (*Neophocaena asiaeorientalis*) was split into the East Asian Finless Porpoise (*Neophocaena sunameri*) and Yangtze Finless Porpoise (*Neophocaena asiaeorientalis*), and the East Asian Finless Porpoise and Yangtze Finless Porpoise were added to the new checklist of China's mammals. The 2021 edition also includes newly recorded species such as *Capricornis rubidus*, *Myotis indochinensis*, *Tylonycteris fulvidus*, *Tonkinomys davovantien* and *Ochotona pusilla* and *Melogale cucphuongensis* (Li *et al.*, 2019). We also added the pika species *Ochotona argentata* according to expert's opinion. Conversely, based on expert opinion, the *Sorex buchariensis*, *Tylonycteris pachypus*, *Myotis nattereri*, *Delphinus capensis*, *Rattus brunneusculus*, *Microtus arvalis* and Narrow-ridged Finless Porpoise were removed from the checklist. Additionally, *Rhinolophus huananus* was renamed as *Rhinolophus siamensis* and *Tylonycteris fulvidus*

Society and Conservation (Challender et al., 2019) 等专著出版，Burgin et al. (2018) 提出了全球哺乳动物的编目系统。这些重要文献为进一步增补修订中国哺乳动物编目提供了基础。

在此期间，研究团队对各自的研究类群开展了研究。蒋志刚研究团队在新疆考察了河狸 (*Castor fiber*)、天山盘羊、野马 (*Equus ferus*) 和阿勒泰地区其他特有哺乳动物，同时还承担了全国第二次陆生野生动物资源调查的鼷鹿 (*Tragulus* sp.)、印度野牛 (*Bos gaurus*)、爪哇野牛 (*Bos javanicus*) 以及豚鹿 (*Axis porcinus*) 等专项考察；在甘肃、内蒙古、新疆考察了高鼻羚羊 (*Saiga tatarica*) 的原有分布区与潜在分布区，在喜马拉雅山南坡开展了哺乳动物调查；并继续进行了普氏原羚 (*Procapra przewalskii*)、麋鹿 (*Elaphurus davidianus*) 的研究。吴毅团队继续进行了翼手目研究，刘少英团队继续开展了鼠兔、啮齿类的研究，蒋学龙团队继续开展了食虫类的研究，周开亚团队继续开展了水兽的研究。这些研究为《中国生物多样性红色名录：脊椎动物》(2021) 评估相关物种的生存状况提供了新的信息。例如，研究结果发现分布在中国西双版纳的鼷鹿是威氏鼷鹿 (*Tragulus williamsoni*)，爪哇野牛和豚鹿在中国已经区域灭绝，野马尚未形成真正的野生种群，而普氏原羚的生存状况有了改善。工作组按照《中国生物多样性红色名录》的编研程序，逐一对每一物种的生存状况进行了再评估。《中国生物多样性红色名录：脊椎动物 第一卷 哺乳动物》(2021) 共评估了除智人之外的中国哺乳动物 13 目 56 科 248 属 700 种。加上智人，中国哺乳动物种数为 701 种 (表 1)。

由于高山峡谷和江河海洋的地理隔离，中国有 156 种特有哺乳

图 7 大熊猫 (*Ailuropoda melanoleuca*, 蒋志刚摄)。号称活化石，为中国特有食肉目动物，也是两种中国特有食肉目动物之一

Figure 7 The Giant Panda (*Ailuropoda melanoleuca*, Photographed by Zhigang Jiang). Known as a living fossil, the Giant Panda is one of two endemic carnivores in the country

renamed as *Tylonycteris fulvida*, and the orders Cetacea and Artiodactyla are now both included in the Super Order, Cetartiodactyla. According to the new taxonomy, Spotted Linsang (*Prionodon pardicolor*) is split from Viverridae and is now placed in a new family, Prionodontidae.

During the period from 2009 to 2019, the nine-volume series *Handbook of the Mammals of the World* (HMW), edited by Don E. Wilson and Russell Mittermeier was published. Edited by Don E. Wilson and Russell Mittermeier and sponsored by Conservation International, Texas A & M University and IUCN, this significant publication represents the collaborative effort of numerous world-renowned mammalogists. During the same period, several important monographs such as *Wild Cats of the World* (Hunter, 2015), *Bovids of the World* (Castelló, 2016) and *Pangolins: Science, Society and Conservation* (Challender *et al.*, 2019) were published, and Burgin *et al.* (2018) proposed a global inventory of mammals. These critical references provided the basis for our further study, and ultimately revision of the inventory of mammals in China.

During this period, the research teams of the editors of the book carried out research on different species. For instance, Zhigang Jiang's team investigated *Castor fiber*, Tianshan Argali, *Equus ferus*, and other mammal species endemic to the Altay region of Xinjiang, China. They also undertook the special projects of the Second National Survey on Terrestrial Wild Animals in China, including the special projects on *Tragulus* sp., *Bos gaurus*, *Bos javanicus*, and *Axis porcinus*. They also investigated the original and potential distribution ranges of *Saiga tatarica* in Gansu, Inner Mongolia (Nei Mongol) and Xinjiang, and surveyed the mammal on southern slope of the Himalayas. They continued studies of *Procapra przewalskii* and *Elaphurus davidianus*. Yi Wu's team continued to study bats, Shaoying Liu's team continued to study pikas and rodents, Xuelong Jiang's team continued their study on insectivorous and Kaiya Zhou's team continued to study aquatic animals. These studies provide new information for the 2021 edition of *China's Red List of Biodiversity: Vertebrates* to assess the status of relevant species. For example, field study results showed that the mouse deer in Xishuangbanna is *Tragulus williamsoni*. On the far side, Banteng and Hog Deer are Regionally Extinct in China. No true wild population of Przewalski's Horse has been formed, and the status of Przewalski's Gazelle has been improved. The working group reassessed each species on a case-by-case basis in accordance with the compilation process of the *China's Red List of Biodiversity*. A total of 700 mammal species in 13 orders, 56 families, 248 genera, excluding *Homo sapiens*, have been assessed in the 2021 edition of the *China's Red List of Biodiversity: Vertebrates, Volume I, Mammals* in China. There are 701 species of mammals in China including *Homo sapiens* (**Table 1**).

Due to the ecological niches created by the geographical isolation of mountains, valleys, rivers and oceans, 156 species of mammals are endemic to China (**Table 2**). Among them, there are world-renowned endemic animals such as Giant Panda *Ailuropoda melanoleuca* (**Figure 7**), Golden Snub-nosed Monkey *Rhinopithecus roxellanae*, Wild Yak *Bos mutus*, White-lipped Deer *Przewalskium albirostris*, and Baiji *Lipotes vexillifer*. The proportion

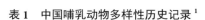

表1 中国哺乳动物多样性历史记录[1]

目	科	属	种	文献
8	30	97	314	Allen (1938~1940)
			382	郑作新 (1947)
12	40	148	327	Elleman and Morrison (1951)
13			c. 400	寿振黄 (1962)
12	52	180	405	郑作新 (1963)
12	44	183	414	张荣祖 (1979)
			418	郑作新 (1982)
13	43	154	390	Honacki et al. (1982)
14	45		430	盛和林 (1983)
13	46	169	467	中国科学院动物研究所 (1986)[2]
13	54	210	509	郑昌琳 (1986)[3]
			394	McNeely et al. (1990)
14	44	155	405	Corbet and Hill (1991)
13	43	154	405	Wilson and Reeder (1993)
14	52	220	510	张荣祖 (1997)
13	55	235	607	王应祥 (2003)
13	54	245	572	Wilson and Reeder (2005)
13	58	242	645	潘清华等 (2007)
14	53	240	585	史密斯等 (2009)
12	55	246	673	蒋志刚等 (2015a)
13	56	248	693	蒋志刚等 (2017)
13	56	248	701	蒋志刚等 (2020)[4]

注：1. 表格空白处无数据；2. 中国科学院动物研究所油印稿；3. 包括外来哺乳动物物种；4. 包括智人种

动物（**表2**）。其中，有举世闻名的特有动物，如大熊猫（*Ailuropoda melanoleuca*，**图7**）、川金丝猴（*Rhinopithecus roxellanae*）、野牦牛、白唇鹿、白鱀豚（*Lipotes vexillifer*）等。中国哺乳动物特有种比例为22%。中国哺乳动物中，特有种比例最高的类群是兔形目（40%）和劳亚食虫目（36%）。中国灵长目多样性高，其中，24%为中国特有种，而扩散能力较弱的、需要特殊生境的如鼹科Talpidae（18种中58%为中国特有种）、猬科Erinaceidae（18种中50%为中国特有种）的特有种比例也较高。中国拥有世界上最丰富的偶蹄类，共有65种，栖息于森林、草原、荒漠和高山悬崖生境，其中中国特有种比例为20%。此外，啮齿目Rodentia和翼手目Chiroptera特有种比例分别为25%和20%（**表2**）。

Table 1 Historical records of the mammal species in China[1]

Order	Family	Genus	Species	Source
8	30	97	314	Allen (1938~1940)
			382	Zheng (1947)
12	40	148	327	Elleman and Morrison (1951)
13			c. 400	Shou (1962)
12	52	180	405	Zheng (1963)
12	44	183	414	Zhang (1979)
			418	Zheng (1982)
13	43	154	390	Honacki et al. (1982)
14	45		430	Sheng (1983)
13	46	169	467	IoZ / CAS (1986)[2]
13	54	210	509	Zheng (1986)[3]
			394	McNeely et al. (1990)
14	44	155	405	Corbet and Hill (1991)
13	43	154	405	Wilson and Reeder (1993)
14	52	220	510	Zhang (1997)
13	55	235	607	Wang (2003)
13	54	245	572	Wilson and Reeder (2005)
13	58	242	645	Pan et al. (2007)
14	53	240	585	Smith et al. (2009)
12	55	246	673	Jiang et al. (2015a)
13	56	248	693	Jiang et al. (2017)
13	56	248	701	Jiang et al. (2020)[4]

Notes: 1. Cells left blank if no data is available; 2. Manuscript of Institute of Zoology, Chinese Academy of Sciences; 3. Included alien mammal species; 4. Included *Homo sapiens*

of mammals that are endemic to China account for 22% of the national total of mammals. Among mammals in China, the groups with the highest proportion of endemic species are Lagomorpha (40%) and Eulipotyphla (36%), along with Talpidae (58% of 18 species endemic to China) and Erinaceidae (50% of 18 species are endemic to China) which are even more endemic due to species in these families having particularly weak dispersal abilities as well as highly specialized habitat requirements. 24% of China's primates are endemic to China. China also has the world's greatest diversity of ungulates with a total of 65 species, inhabiting forests, grasslands, deserts, alpine and cliff habitats; 20% of ungulates are endemic to China. In addition, endemic ratios are 25% in Rodentia and 20% in Chiroptera (**Table 2**).

表2　中国哺乳动物科属种数与特有哺乳动物种数

目	科数	属数	种数	特有种数	特有种比例 /%
劳亚食虫目	3	24	88	32	36
攀鼩目	1	1	1	0	0
翼手目	7	33	139	28	20
灵长目	4	9	29	7	24
鳞甲目	1	1	3	0	0
食肉目	10	40	64	2	3
鲸目	9	26	39	2	5
偶蹄目	6	29	65	13	20
海牛目	1	1	1	0	0
长鼻目	1	1	1	0	0
奇蹄目	2	3	6	0	0
啮齿目	9	78	221	55	25
兔形目	2	2	43	17	40
总计	56	248	700	156	22

5　分布格局

与高等植物、两栖动物、爬行动物、鸟类的物种分布格局相似，中国哺乳动物分布格局与温度、降水有关。中国哺乳动物分布极不均匀，青藏高原隆升这一地质事件打破了中国哺乳动物分布的纬度格局。以 50km×50km 的栅格统计，云南西北部和四川西部的横断山区是中国哺乳动物物种密度最高的区域，特别是云南西北地区，其哺乳动物密度最高地区物种密度可达每 $2,500km^2$ 164 种。云贵高原、秦巴山区、东南沿海、海南、台湾的物种密度也较高。中国北部，特别是青藏高原、蒙古高原西部和新疆南部地区哺乳动物物种密度较低（**图 8**）。

中国灵长目物种主要集中分布于中国南部地区，以横断山区、喜马拉雅山南坡和中国喀斯特地区灵长目物种多样性较高。云贵高原、四川盆地周缘、中部地区啮齿目物种多样性高，华南和新疆北部啮齿目物种多样性较高，东北地区和西北干旱区啮齿目物种多样性次之，青藏高原腹地啮齿目物种密度低。青藏高原东南边缘和横断山区的偶蹄类物种密度最高，云贵高原、秦巴山区、青藏高原东部、南岭山脉次之。青藏高原中西部再次之，新疆南部、华北地区和东北地区偶蹄目物种数量较低。食肉目的分布密度与偶蹄目相仿。青藏高原东南沿、云南西南部和横断山区的食肉目物种密度最高，云贵高原、秦巴山系、南岭山脉次之，天山、东北地区食肉目也较丰富，华北、内蒙古和青藏高原西部的食肉目物种数量较少。青藏高原东部及其横断山脉、帕米尔高原兔形目物种多样性较高，中国东部季风带兔形目物种数量少。中国翼手目动物集中

Table 2 Numbers of Families, Genera, Species, Endemic (Species) and Endemic (Species) ratio in China

Order	No. Families	No. Genera	No. Species	No. Endemic	Endemic Ratio/%
Eulipotyphla	3	24	88	32	36
Scandentia	1	1	1	0	0
Chiroptera	7	33	139	28	20
Primates	4	9	29	7	24
Pholidota	1	1	3	0	0
Carnivora	10	40	64	2	2
Cetacea	9	26	39	2	5
Artiodactyla	6	29	65	13	20
Sirenia	1	1	1	0	0
Proboscidea	1	1	1	0	0
Perissodactyla	2	3	6	0	0
Rodentia	9	78	221	55	25
Lagomorpha	2	2	43	17	40
Total	56	248	700	156	22

5 Distribution Patterns

Similar to the distribution patterns of higher plants, amphibians, reptiles and birds, the distribution pattern of mammals in China is primarily related to temperature and precipitation. The distribution of mammals in China is very uneven, the geological event of uplifting of the Qinghai-Tibet (Xizang) Plateau intercepted the latitudinal pattern of mammal species distribution pattern in China. With a grid system of 50km×50km, the Hengduan Shan region in northwest Yunnan Province and western Sichuan Province harbor the highest density of mammal species in the country, especially in northwest Yunnan, where species richness reaches 164 species per 2,500km^2. The densities of mammal species are also high in the Yunnan-Guizhou Plateau, Qinling-Bashan Mountains, the southeast coastal region, Hainan and Taiwan. The density of mammal species is relatively low in northern China, especially on the Qinghai-Tibet (Xizang) Plateau and in western parts of the Mongolian Plateau and southern Xinjiang (**Figure 8**).

Primates are mainly distributed in southern China, with high density in the Hengduan Shan Mountains, on the south slope of the Himalayas and in the karst area of China. Rodent species diversities in Yunnan-Guizhou Plateau, the fridge of Sichuan Basin, and central China are high, followed by the rodent densities in southern China and northern Xinjiang whereas rodent species density is low in the hinterlands of the Qinghai-Tibet (Xizang) Plateau. Densities of even-toed ungulate species are high in the eastern part of Qinghai-Tibet (Xizang) Plateau and in the Hengduan Shan Mountains, followed by the densities of even-toed ungulate species in Yunnan-Guizhou Plateau, the Qinling and Bashan Mountains, eastern Qinghai-Tibet (Xizang)

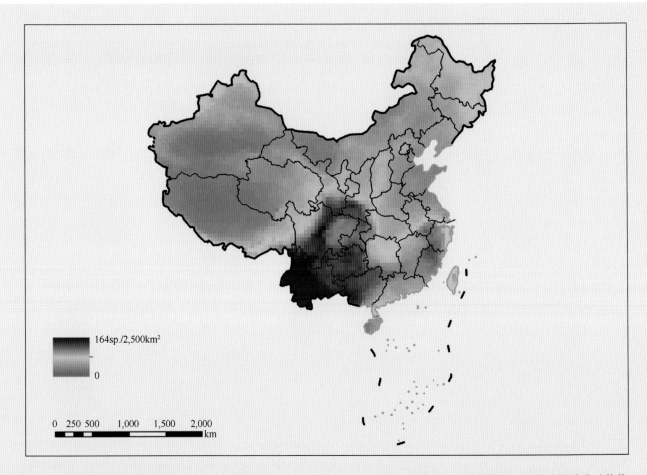

图 8 中国哺乳动物物种密度分布。中国哺乳动物分布极不均匀。云南西北部和四川西部的横断山区是中国哺乳动物物种密度最高的地区，云贵高原、秦巴山区、东南沿海、海南、台湾的哺乳动物物种密度也很高。中国北部，特别是青藏高原、蒙古高原西部与南疆地区哺乳动物物种密度低

Figure 8 Mammalian species density distribution of China. Distribution of mammals in China is clumpy. The Hengduan Shan region in northwest Yunnan and western Sichuan harbor the highest density of mammal species in the country. The density of mammal species in Yunnan-Guizhou Plateau, Qinling-Bashan Mountains regions, the southeast coastal regions, Hainan and Taiwan are also high. The density of mammal species is relatively low in northern China, especially specially on the Qinghai-Tibet (Xizang) Plateau and in western parts of the Mongolian Plateau and southern Xinjiang

分布在南方喀斯特地区，云南、贵州、广西、重庆、海南、台湾和东南沿海地区翼手目物种多样性较高，长江以北及青藏高原地区翼手目物种少（图9）。

6 受保护状况

由于栖息地丧失、过度利用等，许多中国哺乳动物面临生存危机，已经被列为国际和国家保护物种。中国有 59 种哺乳动物（含亚种）被列入 CITES 附录 I，严格禁止其国际商业贸易；有 52 种哺乳动物（含亚种）被列入 CITES 附录 II，仅在可持续贸易的前提下，允许国际商业贸易；还有 20 种哺乳动物被列入 CITES 附录 III，其国际商业贸易需要有关缔约方监测（中华人民共和国濒危野生动植物进出口办公室和中华人

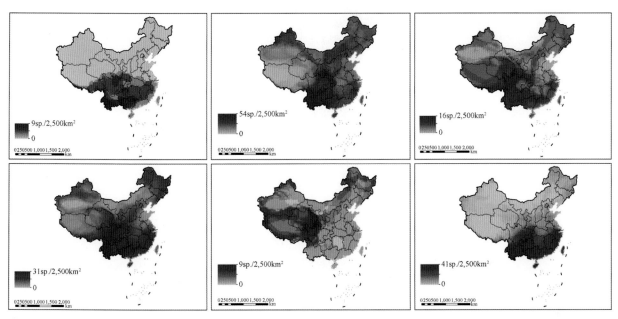

图9 哺乳纲灵长目（上左）、啮齿目（上中）、偶蹄目（上右）、食肉目（下左）、兔形目（下中）和翼手目（下右）在中国的物种数量密度分布

Figure 9 Species density distribution of Primates (upper left), Rodentia (upper middle), Artiodactyla (middle right), Carnivora (lower left), Lagomorpha (lower middle) and Chiroptera (lower right) in China

Plateau, the Nanling Mountains, and the central and western parts of Qinghai-Tibet (Xizang) Plateau. The numbers of Artiodactyla species are much lower in southern Xinjiang and northern and northeastern China. The distribution pattern of Carnivora diversity is similar to that of Artiodactyla. Diversity of Carnivore species in the southeastern part of Qinghai-Tibet (Xizang) Plateau, southwestern Yunnan and Hengduan Shan Mountains are high, followed by the Carnivore species densities in Yunnan-Guizhou Plateau, Qinling-Bashan Mountains and Nanling Mountains, then those in Tian Shan and northeastern China, while densities of Carnivore species in northern China, Inner Mongolia (Nei Mongol) and western Qinghai-Tibet (Xizang) Plateau are low. The eastern part of Qinghai-Tibet (Xizang) Plateau and the Hengduan Shan Mountains, Pamir Plateau have high diversities of Lagomorpha species. There are few Lagomorpha species in monsoon zone of eastern China. Bats in China are mainly found in the south karst area. Yunnan, Guizhou, Guangxi, Chongqing, Hainan, Taiwan and southeast coastal areas have high diversities of bat species. There are few bat species in areas north of the Changjiang River and on the Qinghai-Tibet (Xizang) Plateau (**Figure 9**).

6 Conservation Status

Due to habitat loss and overuse, many mammals in China are threatened and thus have been listed as species protected by international laws and national laws. China has 59 species of mammals (including subspecies) listed in CITES Appendix I, for which international commercial trade is strictly forbidden; 52 species of mammals (including subspecies) listed in CITES Appendix II, for which only sustainable international commercial trade is allowed; and 20 species of mammals listed in CITES Appendix III, for which international

民共和国濒危物种科学委员会，2017)。57 种哺乳动物被列为《国家重点保护野生动物名录》(1989) 的 I 级重点保护野生动物，74 种被列为 II 级重点保护野生动物。根据《中华人民共和国野生动物保护法》，国家林业局 2000 年颁布了《国家保护的有益的或者有重要经济、科学研究价值的陆生野生动物名录》(2000)，88 种哺乳动物被列入该名录。《国家保护的有益的或者有重要经济、科学研究价值的陆生野生动物名录》(2000) 列入了未列入国家重点保护野生动物名录但仍被视为具有明显保护价值的物种。

2018 年，《IUCN 受威胁物种红色名录》(2018) 评估了全球哺乳动物濒危状况。但是，该红色名录低估了中国濒危物种，仅 87 种哺乳动物被列为受威胁物种，其中，17 种为 IUCN 红色名录收录的中国特有哺乳动物（共 84 种）。54 种《中国哺乳动物多样性》(第 2 版) 中记载的哺乳动物，IUCN 红色名录工作组从未对其进行过濒危等级评估；另有 130 种中国有分布的哺乳动物，在 IUCN 红色名录中却错误地记载为中国无分布。

7 评估过程

红色名录评估过程包括数据收集、初评、复审、形成评估报告等步骤。《中国生物多样性红色名录：脊椎动物 第一卷 哺乳动物》(2021) 项目组组成了以中国科学院动物研究所蒋志刚研究员、南京师范大学周开亚教授、广州大学吴毅教授、中国科学院昆明动物研究所蒋学龙研究员和四川林业科学研究院刘少英研究员为成员的核心专家组，中国科学院动物研究所野生动物与行为生态研究组的研究助理承担了工作组的任务。项目组组织了全国有关专家，按照 IUCN 受威胁物种红色名录评估标准（第 3.1 版，2012)，经过会议研讨、通讯评审，在认真修订与增补哺乳动物名录之后，完成了哺乳类红色名录的评定。按照 IUCN 物种红色名录的惯例，以下分析不包括智人这一物种。

commercial trade should be monitored by the parties of CITES (Office of the Import and Export of Endangered Species of Wild Fauna and Flora of the People's Republic of China, Endangered Species Scientific Commission of the People's Republic of China, 2017). 57 mammal species are listed as species protected in China under the First Category of *State Key Protected Wild Animals* (1989), and 74 mammal species are listed as Second Category of State Key Protected Wild Animals. In accordance with the Wild Animal Protection Law of the People's Republic of China, the former State Forestry Administration promulgated the *List of Terrestrial Wildlife That Is Useful or Has Important Economic and Scientific Research Value* (2000) in 2000. Eighty-eight species of mammals are included in the list. The *List of State Protected Wild Animals of Ecological, Important Economic and Scientific Research Values* includes those species not listed in the State Key Protected Wild Animal Species but that are nonetheless deemed to have significant merit for protection.

In 2018, the *IUCN Red List of Threatened Species* (2018) renewed the status of threatened mammals worldwide. However, the IUCN red list underestimates the number of endangered species in China. In the IUCN red list, only 87 mammal species in China are listed as threatened, and amongst these only 17 of the 84 species are recognized in the red list as being endemic to China. Furthermore, 54 species noted in *China's Mammal Diversity* (2nd edition) were not assessed by the IUCN red list team, and 130 species of China's mammals listed in the global IUCN red list are erroneously listed as not occurring in China.

7 Evaluation Process

The red list evaluation process includes data collection, preliminary evaluation, peer review, and formulation of the evaluation report. The working group of *China's Red List of Biodiversity: Vertebrates, Volume I, Mammals* (2021) was comprised of myself (Zhigang Jiang) and my team of research assistants in the Wildlife and Behavioral Ecology Research Group of the Institute of Zoology, Chinese Academy of Sciences. The leading expert group was formed by Professor Kaiya Zhou (Nanjing Normal University), Professor Yi Wu (Guangzhou University), Professor Xuelong Jiang (Kunming Institute of Zoology, Chinese Academy of Sciences), Professor Shaoying Liu (Sichuan Forestry Science Academy) and myself. We coordinated experts from across the country during the assessment, according to the IUCN red list categories and criteria, version 3.1 (2012). After several panel meetings and reviews by correspondence, we carefully mandated the inventory of China's mammals and assessed the status of each species. According to the tradition of the *IUCN Red List of Threatened Species*, *Homo sapiens* was not included in the assessment and is not included in the following analysis.

8 评估等级和标准

本次评估依据 IUCN 受威胁物种红色名录等级标准和 IUCN 区域受威胁物种红色名录标准，对中国哺乳动物的濒危等级进行评估。

本次生物多样性红色名录评估依据的是如下 IUCN 红色名录标准，包括：

(1) *IUCN Red List Categories and Criteria*, version 3.1 (2001).

(2) *Guidelines for Using the IUCN Red List Categories and Criteria*, version 8.1 (2010).

(3) *Guidelines for Application of IUCN Red List Criteria at Regional and National Levels*, version 4.0 (2012).

20 世纪 60 年代，IUCN 通常依据一位作者所收集到的物种分布信息，咨询 IUCN 物种生存委员会专家组，根据物种受威胁程度和估计灭绝风险将物种列为不同的濒危等级，开始编制全球濒危物种红皮书。IUCN 发布濒危物种红皮书有 3 个目的：

(1) 不定期地推出濒危物种红皮书，以唤起世界对野生物种生存现状的关注；

(2) 提供有关濒危物种信息，供各国政府和立法机构参考；

(3) 为全球科学家提供有关物种濒危现状和生物多样性基础数据。

最初，IUCN 受威胁物种红皮书仅包括陆栖脊椎动物。早期 IUCN 红皮书的编制仅根据一位专家所收集到的物种分布信息制定，后来发展到通过网络咨询 IUCN 物种生存委员会专家组专家意见形成红色名录。之后 IUCN 红皮书发生了很大的变化，开始收录无脊椎动物和植物，内容逐年增多，参与工作的专家也逐年增多。逐步发展到每年定期发布《IUCN 受威胁物种红色名录》(*IUCN Red List of Threatened Species*)。此外，一些国家也开始编制本国的濒危物种红皮书。中国在 1991 年出版了《中国植物红皮书》，1998 年出版了《中国濒危动物红皮书·鱼类》《中国濒危动物红皮书·两栖类和爬行类》《中国濒危动物红皮书·鸟类》《中国濒危动物红皮书·兽类》。然而，每一类群仅仅评估了一部分物种的生存状况。

人们从 20 世纪 60 年代开始研究制定物种濒危等级标准。其中比较成熟的，在国内外濒危物种的濒危等级划分上应用较为广泛的是《IUCN 受威胁物种红色名录》标准。早期，IUCN 于 1988 年使用的濒危物种等级系统包括灭绝 (Extinct)、濒危 (Endangered)、易危 (Vulnerable)、稀有 (Rare)、未定 (Indeterminate) 和欠了解 (Insufficiently Known)。上述标准存在很大的主观性。在 60 年代和 70 年代，编写濒危物种红皮书的工作由一位作者完成时，当评估的物种数目有限时，作者尚能掌握濒危标准。80 年代以来，开始由多位作者来制定《IUCN 受威胁物种红色名录》。于是，迫切需要一份可量化的濒危物种等级标准。

8 Categories and Criteria

According to the IUCN red list criteria for threatened species and the IUCN regional red list criteria for endangered species, the endangerment status of mammals in China were evaluated.

The assessment of this red list is based on the following IUCN red list documents:

(1) *IUCN Red List Categories and Criteria*, version 3.1 (2001).

(2) *Guidelines for Using the IUCN Red List Categories and Criteria*, version 8.1 (2010).

(3) *Guidelines for Application of IUCN Red List Criteria at Regional and National Levels*, version 4.0 (2012).

In the 1960s, the IUCN began to compile the global red data book on endangered species by consulting with the volunteer expert group of IUCN Species Survival Commission based on the species distribution information collected by one author and classifying species into categories of endangerment according to the degree of threat and the estimated risk of extinction. At that time, IUCN occasionally published the red books of threatened species for three purposes: (i) to raise worldwide attention about the status and likelihood of extinction of wildlife; (ii) to provide information on endangered species for reference by governments and legislative bodies; (iii) to provide global scientists with reliable basic data on the current status of species in danger and the status of biodiversity.

Initially, the IUCN red book for endangered species included only terrestrial vertebrates. In the early stage, the compilation of IUCN red books was based on species population and distribution information collected by single experts. Later, it evolved into a global red list of endangered species based on consultations with a broader suite of experts from the IUCN Species Survival Commission, made possible as communications improved with internet. Since then, great changes have taken place in the IUCN red data book, which came to include invertebrates and plants, more recently also fungi, and with the document's overall content and scope increasing year by year and the number of experts involved in assessments also increasing. Now IUCN regularly renews the *IUCN Red List of Threatened Species*. In addition, some countries have begun to compile their own red books on endangered species. For instance, China published the *Red Data Book of Chinese Plants* in 1996. Later, in 1998, it published the *China Red Data Book of Endangered Animals: Pisces*; *China Red Data Book of Endangered Animals: Amphibia & Reptilia*; *China Red Data Book of Endangered Animals: Aves*; and *China Red Data Book of Endangered Animals: Mammalia*. However, in each case, the status of species' endangerment was assessed for only a fraction of the flora or fauna targeted in each volume.

People began to study and formulate criteria of endangered species in the 1960s. Among them, the criteria for the *IUCN Red List of Threatened Species* is relatively mature and widely used in the classification of endangered species at home and abroad. The hierarchy used by IUCN in 1988 in its first phase of endangered species categorization included Extinct, Endangered, Vulnerable, Rare, Indeterminate, and Insufficiently Known. While useful, these criteria nevertheless remain highly subjective. In the 1960s and 1970s, when the red book on endangered species was written by a single author, the author was

1991年，Mace和Lande第一次提出了根据在一定时间内物种的灭绝概率来确定物种濒危等级的思想，并据此制定了一套物种濒危标准(Mace and Lande, 1991)。随后，人们在一些生物类群中尝试应用Mace-Lande物种濒危等级标准划分濒危等级。1994年11月，IUCN第40次理事会会议正式通过了经过修订的Mace-Lande物种濒危等级标准作为新的IUCN濒危物种等级标准系统。1996年、2001年，IUCN应用了此新的濒危物种等级标准来评估全球物种濒危状态。我们应用了IUCN受威胁物种红色名录评估等级标准来确定中国生物多样性红色名录（图10）。

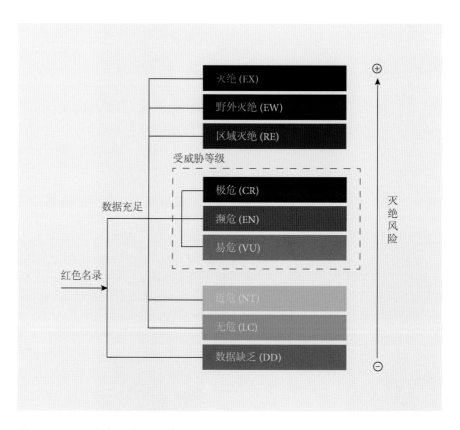

图10　IUCN受威胁物种红色名录评估等级标准 (IUCN Standards and Petitions Subcommittee, 2017)

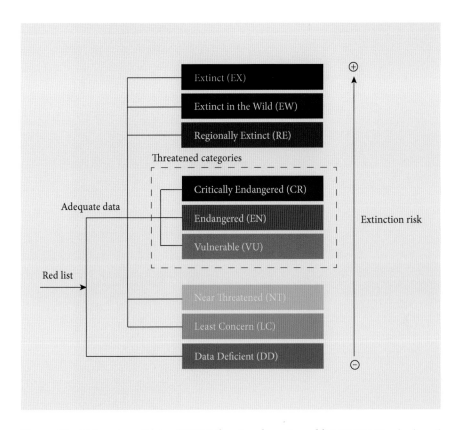

Figure 10 Categories criteria of IUCN threatened species red list (IUCN Standards and Petitions Subcommittee, 2017)

able to master the criteria for endangered species as only a limited number of species were assessed. Since the 1980s, however, *IUCN Red Lists of Threatened Species* have been drawn up by multiple authors. Thus, more quantitative criteria for classification of endangered species were urgently needed.

Mace and Lande (1991) first proposed the idea of determining the level of endangered species based on the extinction probability of a species over a specified period of time and they formulated a set of discrete criteria to assess wildlife species. Subsequently, researchers have applied Mace-Lande criteria to classify endangered species in certain taxonomic groups. At the 40th Council Meeting of the IUCN in November 1994, the revised Mace-Lande criteria for threatened species was formally adopted as the new standard and criteria for producing IUCN red lists. IUCN applied the new criteria in 1996 and 2001 to assess the global status of endangered species. We applied the IUCN categories criteria for red list of threatened species for classifying China's wildlife species and producing national biodiversity red lists (**Figure 10**).

Mace-Lande criteria for threatened species use population size, population decline rate, distribution range size, the expected population decline rate in future and the probability of extinction of endangered species as quantitative thresholds. During the assessment, the taxa (normally species) are, firstly, divided into two categories: Evaluated and Not Evaluated (NE). When there is insufficient data to apply the IUCN criteria for threatened species to evaluate

Mace-Lande 物种濒危等级利用濒危物种的种群下降速率、分布范围大小、预计种群下降速率及灭绝概率等作为数量标准。评估时，评估对象分为评估与未评估 (Not Evaluated, NE) 两大类，未应用有关 IUCN 濒危物种标准评估的分类单元列为未评估，否则即对该分类单元进行评估。

Mace-Lande 物种濒危等级使用了如下等级：①灭绝 (Extinct, EX)，如果一个生物分类单元的最后一个个体已经死亡，列为灭绝；②野外灭绝 (Extinct in the Wild, EW)，如果一个生物分类单元的个体仅生活在人工栽培和人工圈养状态下，列为野外灭绝；③极危 (Critically Endangered, CR)，野外状态下一个生物分类单元灭绝概率很高时，列为极危；④濒危 (Endangered, EN)，一个生物分类单元，虽未达到极危，但在可预见的不久的将来，其野生状态下灭绝的概率高，列为濒危；⑤易危 (Vulnerable, VU)，一个生物分类单元虽未达到极危或濒危的标准，但在未来一段时间中其在野生状态下灭绝的概率较高，列为易危；⑥当一物种未达到极危、濒危或易危标准，但在未来一段时间内，接近符合或可能符合受威胁等级，则该种为近危 (Near Threatened, NT)；⑦无危 (Least Concern, LC)，一个生物分类单元，经评估不符合列为极危、濒危或易危任一等级，则列为无危；⑧数据缺乏 (Data Deficient, DD)，对于一个生物分类单元，若无足够的资料对其灭绝风险进行直接或间接的评估时，可列为数据缺乏。

各等级的含义和评估标准如下：

灭绝 (EX)。如果一个物种的最后一个个体已经死亡，则该种"灭绝"。

野外灭绝 (EW)。如果一个物种的所有个体只生活在人工养殖状态下，则该种"野外灭绝"。

区域灭绝 (RE)。如果一个物种在某个区域内的最后一个个体已经死亡，则该物种已经"区域灭绝"。

极危 (CR)、濒危 (EN) 和易危 (VU)(表3)。这三个等级统称为受威胁等级 (Threatened Categories)，灭绝的风险由高到低。当某一物种符合表3中 A~E 的任一标准时，该种被列为相应的濒危等级。如果根据不同标准评定的濒危等级不同，则该种应被归于风险最高的濒危等级。

近危 (NT)。当一物种未达到极危、濒危或易危标准，但在未来一段时间内，接近符合或可能符合受威胁等级，则该种为"近危"。

无危 (LC)。当某一物种未达到任何灭绝等级，且评估也未达到"极危"、"濒危"、"易危"或"近危"标准，则该种为"无危"。广泛分布和个体数量多的物种都属于该等级。

数据缺乏 (DD)。当缺乏足够的信息对某一物种的灭绝风险进行评估时，则该种属于"数据缺乏"。

a taxon, then that taxon is classified as Not Evaluated (NE); otherwise the taxon is evaluated.

Applying the Mace-Lande criteria for threatened species proceeds as follows, sequentially: (i) If the last individual in a taxon (normally a species) has died, the taxon is classified as Extinct (EX); (ii) If all remaining individuals of a taxon live only in a state of artificial cultivation or in captivity, it is classified as Extinct in the Wild (EW); (iii) If the probability is very high that a taxon could soon be extinct in the wild, it is listed as Critically Endangered (CR); (iv) If a taxon is deemed to have a high probability of becoming extinct in the wild in the foreseeable future, even if not currently considered Critically Endangered, it is listed as Endangered (EN); (v) Although a taxon may not meet all the criteria of Critically Endangered or Endangered, if it has a high probability of extinction in the wild at some time in the future, it is classified as Vulnerable (VU); (vi) When a taxonomic unit fails to meet the criteria for Critically Endangered, Endangered, or Vulnerable, but is close to meeting or deemed likely to meet such criteria in the near future, the taxon is classified as Near Threatened (NT); (vii) If a taxonomic unit is assessed and found not to meet the above criteria, it is classified as Least Concern (LC); (viii) Finally, a taxon may be classified as Data Deficient (DD) if there is insufficient direct or indirect information available to assess its risk of extinction in the wild.

The definitions and evaluation criteria for each category are as follows:

Extinct (EX). If the last individual of a species has died, the species is Extinct.

Extinct in the Wild (EW). A species is extinct in the wild if all its individuals live only in an artificial environment or in captivity.

Regionally Extinct (RE). If the last individual of a species in an area dies, then the species is Regionally Extinct.

Critically Endangered (CR), Endangered (EN) and Vulnerable (VU) (Table 3). These three categories are collectively recognized as threatened categories. When a species meets any of the criterion A~E in **Table 3**, it is classified as an endangered species. If an endangered species matches the criteria from several different categories, then the species should be assigned to the category of highest risk.

Near Threatened (NT). When a species does not meet the criteria for Critically Endangered, Endangered or Vulnerable, but is near or likely to meet these three threatened categories in the future, it is considered as Near Threatened.

Least Concern (LC). When a species does not meet the criteria for any extinction category, and also is assessed as being neither Threatened (meeting criteria of Critically Endangered, Endangered, or Vulnerable) nor Near Threatened, it is considered as Least Concern. Species that are widely distributed and with large populations fall into this category.

Data Deficient (DD). When there is insufficient information available to assess the risk of extinction of a species, it is considered to be Data Deficient.

表 3　IUCN 红色名录濒危等级标准中 A~E 的生物学指标与数量阈值 (IUCN, 2012)

A. 种群数量减少. 基于任意 A1~A4 的种群下降（测算时间超过 10 年或 3 个世代）			
	极危 CR	濒危 EN	易危 VU
A1	≥ 90%	≥ 70%	≥ 50%
A2, A3 & A4	≥ 80%	≥ 50%	≥ 30%

A1. 过去 10 年或 3 个世代内种群减少的比例，其减少的原因是可逆转的且被理解和已经停止的

A2. 观察、估计、推断或猜测到已经发生种群下降，这些种群下降的原因可能不会停止，或不被理解，或不可逆

A3. 预期、推断或猜测到未来将会发生的种群下降（时间上限为 100 年）[易危一列的"(a)"不适用于此条]

A4. 观察、估计、推断、预测或怀疑的种群减少，其时间周期必须包括过去和未来（未来时间上限 100 年），并且这些种群下降的原因可能不会停止，或不被理解，或不可逆

基于以下任意方面：
(a) 直接观察（A3 除外）
(b) 适合该分类单元的丰富度指数
(c) 占有面积 (AOO) 减少，分布范围 (EOO) 减少和（或）栖息地质量下降
(d) 实际的或潜在的开发水平
(e) 外来物种、杂交、病原体、污染物、竞争者或寄生物的影响

B. 以分布范围 (B1) 和（或）占有面积 (B2) 体现的地理范围			
	极危 CR	濒危 EN	易危 VU
B1. 分布范围 (EOO)	< 100km²	< 5,000km²	< 20,000km²
B2. 占有面积 (AOO)	< 10km²	< 500km²	< 2,000km²

以及以下 3 个条件中的至少 2 个：

(a) 严重片断化或分布地点数	= 1	≤ 5	≤ 10
(b) 在以下方面观察、估计、推断或预期持续下降：(i) 分布范围；(ii) 占有面积；(iii) 占有面积、分布范围和（或）栖息地质量；(iv) 分布地点或亚种群数；(v) 成熟个体数			
(c) 以下任何方面的极度波动：(i) 分布范围；(ii) 占有面积；(iii) 分布地点或亚种群数；(iv) 成熟个体数			

C. 小种群的规模和下降情况			
	极危 CR	濒危 EN	易危 VU
成熟个体数	< 250	< 2,500	< 10,000

和至少 C1 或 C2 其一

	极危 CR	濒危 EN	易危 VU
C1. 观察、估计或预期的持续下降的最小比例（未来时间上限 100 年）	3 年或 1 个世代内 25%（以较长时间为准）	5 年或 2 个世代内 20%（以较长时间为准）	10 年或 3 个世代内 10%（以较长时间为准）
C2. 观察、估计或预期的持续下降和至少以下 3 个条件之一			
(a) (i) 每个亚种群中的成熟个体数	≤ 50	≤ 250	≤ 1,000
(a) (ii) 亚种群中的成熟个体数的比例	90%~100%	95%~100%	100%
(b) 成熟个体数量极度波动			

D. 种群数量极小或分布范围局限			
	极危 CR	濒危 EN	易危 VU
D. 成熟个体数	< 50	< 250	D1. < 1,000
D2 仅适用于易危等级：占有有限区域的面积或分布点数目，并在未来很短时间内有一个可信的、可能驱动该分类单元走向极危或灭绝的威胁	—	—	D2. 一般情况下：AOO < 20km² 或分布点数目 ≤ 5

E. 定量分析			
	极危 CR	濒危 EN	易危 VU
使用定量模型评估的野外灭绝率：	未来 10 年或 3 代内 ≥ 50%（以较长时间为准，上限为 100 年）	未来 20 年或 5 代内 ≥ 20%（以较长时间为准，上限为 100 年）	未来 100 年内 ≥ 10%

Table 3 Biological indicators and quantitative thresholds of A~E in the IUCN red list for threatened species (IUCN, 2012)

A. Population size reduction. Population reduction (measured over the longer of 10 years or 3 generations) based on any of A1 to A4			
	Critically Endangered	Endangered	Vulnerable
A1	≥ 90%	≥ 70%	≥ 50%
A2, A3 & A4	≥ 80%	≥ 50%	≥ 30%

A1. Population reduction observed, estimated, inferred, or suspected in the past where the causes of the reduction are clearly reversible AND understood AND have ceased

A2. Population reduction observed, estimated, inferred, or suspected in the past where the causes of reduction may not have ceased OR may not be understood OR may not be reversible

A3. Population reduction projected, inferred or suspected to be met in the future (up to a maximum of 100 years) [*(a) cannot be used for A3*]

A4. An observed, estimated, inferred, projected or suspected population reduction where the time period must include both the past and the future (up to a max. of 100 years in future), and where the causes of reduction may not have ceased OR may not be understood OR may not be reversible

Based on any of the following:

(a) direct observation (*except A3*)
(b) an index of abundance appropriate to the taxon
(c) a decline in area of occupancy (AOO), extent of occurrence (EOO) and/or habitat quality
(d) actual or potential levels of exploitation
(e) effects of introduced taxa, hybridization, pathogens, pollutants, competitors or parasites

B. Geographic range in the form of either B1 (extent of occurrence) AND/OR B2 (area of occupancy)			
	Critically Endangered	Endangered	Vulnerable
B1. Extent of occurrence (EOO)	< 100km²	< 5,000km²	< 20,000km²
B2. Area of occupancy (AOO)	< 10km²	< 500km²	< 2,000km²

AND at least 2 of the following 3 conditions:

(a) Severely fragmented OR Number of locations	=1	≤ 5	≤ 10

(b) Continuing decline observed, estimated, inferred, or projected in any of: (i) extent of occurrence; (ii) area of occupancy; (iii) area, extent and/or quality of habitat; (iv) number of locations or subpopulations; (v) number of mature individuals

(c) Extreme fluctuations in any of: (i) extent of occurrence; (ii) area of occupancy; (iii) number of locations or subpopulations; (iv) numbers of mature individuals

C. Small population size and decline			
	Critically Endangered	Endangered	Vulnerable
Number of mature individuals	< 250	< 2,500	< 10,000

AND at least one of C1 or C2

C1. An observed, estimated or projected continuing decline of at least (up to a max of 100 years in future)	25% in 3 years or 1 generation (whichever is longer)	20% in 5 years or 2 generations (whichever is longer)	10% in 10 years or 3 generations (whichever is longer)
C2. An observed, estimated, projected or inferred continuing decline AND at least 1 of the following 3 conditions:			
(a) (i) Number of mature individuals in each subpopulation	≤ 50	≤ 250	≤ 1,000
(a) (ii) %of mature individuals in one subpopulation	90%~100%	95%~100%	100%
(b) Extreme fluctuations in the number of mature individuals			

D. Very small or restricted population			
	Critically Endangered	Endangered	Vulnerable
D. Number of mature individuals	< 50	< 250	D1. < 1,000
D2 *Only applies to the VU category* Restricted area of occupancy or number of locations with a plausible future threat that could drive the taxon to CR or EX in a very short time	—	—	D2. Typically: AOO < 20km² or number of locations ≤ 5

E. Quantitative Analysis			
	Critically Endangered	Endangered	Vulnerable
Indicating the probability of extinction in the wild to be:	≥ 50% in 10 years or 3 generations, whichever is longer (100 years max)	≥ 20% in 20 years or 5 generations, whichever is longer (100 years max)	≥ 10% in 100 years

9 建立数据库

在本研究的第一阶段（2012~2015 年），哺乳动物红色名录工作组首先更新了中国陆栖脊椎动物数据库。在红色名录编研期间，工作组收集了国内外博物馆数据库，包括 Smithsonian Institution (USA)、Natural History Museum (British)、Field Museum of Natural History (USA)，以及国内标本库中有关中国哺乳类的标本信息。在本研究的第二阶段（2016~2019 年），根据本书所采用的在《中国哺乳动物多样性》（第 2 版）基础之上更新的中国哺乳动物编目系统（2021），重点收集了国内外有关中国哺乳类的最新研究文献，并利用了 IUCN red list、ITIS 数据库中的中国哺乳类的信息，再次更新了数据库，作为本次红色名录评估的评估对象。

10 初步评定

在评估中，依据 IUCN 红色名录评估标准、数据库和专家经验，工作组初步评估了中国哺乳类红色名录。收集了评估对象的种群数量、分布区、分布点、种群数量变化趋势、生存状况、保护状况。工作组在物种信息、标本、文献等本底资料的基础上开展初评。必要时参考利用评估软件 (RAMAS Red List) 生成的结果及前人的评估结果。必要时，对特殊物种还利用种群生存力分析软件 VORTEX 9.98 (Lacy et al., 2003) 进行了种群生存力分析，外推了物种适宜分布区。

对每一物种信息收集包括下列内容（中英文对照）：种名、同物异名（拉丁名、英文名、中文名）、照片（如果有）、分类地位、分布区、生境描述、生存状况和保护状况、IUCN 红色名录等级、红色名录等级评定结果，以及评定时依据的标准、评定人、审定人等信息。专家组对初步判定结果进行审定并汇总结果，整理了中国哺乳动物红色名录初评稿，以向全国有关哺乳动物专家征求意见。

11 通讯评审

按照 IUCN 红色名录评估工作规程，每个物种的评估结果，须由评估人以外的人员来担任复审。复审安排包括通讯评审和会议复审两种方式。工作组建立了中国哺乳类专家库，在全国范围内邀请了 60 余位哺乳动物的专家，对中国哺乳动物红色名录初评等级结果在指定时间内进行了通讯评审。通讯评审时，邀请专家对中国哺乳动物红色名录初评等级结果进行复审，完善评估理由，补充了物种致危因素，以保证评估结果的准确性。

此外，采取了"会议复审"的方式进行评估。红色名录工作组成员召集会议，集中专家统一讲解 IUCN 评估方法、标准和其他相关指南及复审要求，然后针对专家熟悉的类群开展复审，及时根据专家的意见重新调整和修改初评结果并补充相关信息。共有近 10 家单位的 30 名专家参与了会议复审。会议复审与函评初审、复审覆盖了初评中全部的哺

9 Building the Database

During the first phase of the study (2012~2015), the mammals red list working group updated the database of China's terrestrial vertebrates. During the compilation of the red list, the working group collected information about Chinese mammal specimens from domestic and foreign museums, including the Smithsonian Institution (USA), Natural History Museum (British), and Field Museum of Natural History (USA) as well as domestic specimen databases. During the second phase of the red list study (2016~2019), we focused to collect the latest literatures for Chinese mammals in and abroad in light of the inventory of China's mammals (2021) that renewed from *China's Mammal Diversity* (2nd edition). And then, based on Chinese mammals' information in the IUCN red list、ITIS database, we renewed our database once again, and thus, the evaluation subjects occur.

10 A Preliminary Assessment

During the assessment, the working group made a preliminarily evaluation of the red list of China's mammals based on criteria used in IUCN red list, on the database that we built, and on specialist expertise. Population size, distribution area, points of occurrence, population trend, survival status and protection status were collected for all the evaluated species. Initial evaluations were carried out on basis of the background information available for each species, including both specimens and literature. At the same time, results generated from the evaluation software RAMAS Red List and results of previous evaluations are referenced, and VORTEX 9.98 (Lacy *et al.*, 2003) was used to analyze population viability for selected species if deemed necessary.

Information collected for each species included the following (in Chinese and English): species name and synonyms in Latin, English and Chinese; color plates or photo (as available); taxonomic status; distribution area; primary habitat; survival status; conservation status; IUCN red list category; assessment result of this red list; and assessment justification, assessor (s) and reviewer (s), *etc*. Preliminary assessments based on the above criteria and information were then reviewed and gathered by the expert group, and information and initial results were then shared with relevant mammal experts nationwide to solicit further opinions.

11 Review by Correspondence

Following the IUCN red list assessment protocol, assessment results for each species shall be reviewed by experts other than the primary evaluator (s). The process includes reviews by correspondence and expert workshops. The working group developed an expert database for evaluation of China's mammals. Over 60 experts were invited to review assessments within a specified timeframe and provide initial ratings for China's red list of mammals. During the correspondence-based reviews, selected mammal experts were invited to review preliminary red list assessments of mammals, to judge the rationales provided and to supplement the risk factors attributed to each species in order to ensure complete and accurate assessment results.

In addition, a panel review was organized during the assessment. Members of the red list working group invited national experts to convene and focus on the red list by firstly explaining the IUCN assessment methodology, criteria and standards, and other guidelines and review requirements. Review was

乳动物。复审过程重点关注的对象是初评过程中被评为受威胁等级的物种以及各科、属可能遗漏的受威胁物种。

工作组对通讯评审专家返回的中国哺乳动物红色名录初评稿的评审意见逐项进行了研究，对中国哺乳动物红色名录等级进行汇总评定。补充完成了各类群重要物种的发生区、占有区以及必要的种群生存力分析研究，以提升《中国哺乳动物红色名录评审稿》的评估结果（图11）。

图11 《中国哺乳动物红色名录》研究与发布流程

then carried out by groups of experts familiar with the initial assessment of specific taxonomic groups. Experts discussed and modified preliminary evaluation results as necessary. The opinions of experts and other relevant information about the assessments were summarized and incorporated into documents as appropriate. In total, 30 experts from 10 universities and research institutes across the country participated in the workshop review process. Both the expert panel meeting and the review by correspondence covered all mammal species in China that were assessed in the preliminary assessment, with focus on species that were rated as threatened in the initial assessment as well as proactively looking for other threatened species that may have been missed from each family and genus.

The working group carefully studied the evaluation results presented in the first draft of the red list of China's mammals returned by the experts, one by one, and it calculated Extent of Occurrence (EOO) and Area of Occupancy (AOO). Population viability analyses were also conducted for important species in each taxonomic group as necessary, to further enhance the final evaluation of their status to be reported in the *Reviewing Draft of China's Red List of Mammals* (**Figure 11**).

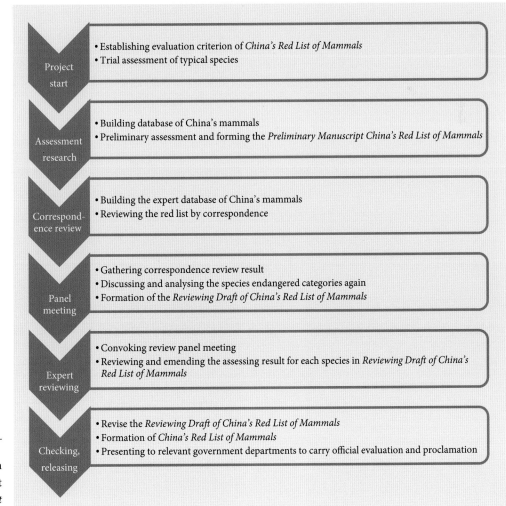

Figure 11 Research and publish flow chart of the *China's Red List of Mammals*

12 形成评估报告

复审结束后，在每个物种（主要是受威胁等级）包含信息的基础上，形成了最终的评估报告。评估报告的内容包括本次评估的背景、必要性和意义、评估对象、依据的 IUCN 受威胁物种等级及标准、评估过程、受威胁种类分析、重要物种介绍、威胁因子分析与《IUCN 受威胁物种红色名录》的受威胁等级比较及未来展望等内容。

13 评估结果分析

2015 年，最终完成的"中国生物多样性红色名录——脊椎动物卷"（哺乳动物红色名录是其中的一个重要组成部分）的内容包括中国哺乳动物的中文名、拉丁名及它们所属的科和特有种信息等，以及它们相应的其他评估信息。"中国生物多样性红色名录——脊椎动物卷"及"评估报告"通过专家评审后，2015 年 5 月 22 日国际生物多样性日由环境保护部与中国科学院联合发布。在后文中，将 2015 年发布的"中国生物多样性红色名录——脊椎动物卷"称之为《中国生物多样性红色名录：脊椎动物卷》(2015)。

2016 年 3 月 25 日至 26 日，IUCN 在北京举办了"东北亚红色名录"培训班，我们应邀交流了中国生物多样性红色名录研究。我被邀请介绍了《中国生物多样性红色名录：脊椎动物卷》(2015) 的研究。参加会议的除了 IUCN 红色名录工作组的专家和中国的代表之外，还有蒙古、韩国、朝鲜、日本等国的代表。当时，我特别介绍了在 2015 年版中国生物多样性红色名录编研中，基于种群下降速率、栖息地面积减小速率、小种群数目、生境斑块数目的评估，我们将大熊猫濒危等级从"濒危"下调为"易危"。我还同时展示了我们推动的"中国生物多样性红色名录——脊椎动物卷"，此后该成果由环境保护部与中国科学院以环境保护部 2015 年第 32 号公告正式颁布。半年以后，IUCN 红色名录工作组也采纳了我们的观点，2016 年 9 月在美国夏威夷，IUCN 在世界保护大会上宣布将大熊猫的濒危等级从"濒危"降到"易危"。

工作组自 2017 年来又进一步更新了《中国哺乳动物多样性》（第 2 版）的哺乳动物编目体系，根据 Handbook of the Mammals of the World (HMW)、IUCN 红色名录采用的分类系统，整理了分类系统。我们同时继续收集整理了中国哺乳动物新种、新记录种。工作组按照《中国生物多样性红色名录：脊椎动物卷》(2015) 的编研程序，逐一对每一物种的生存状况进行了再评估，重复了《中国生物多样性红色名录：脊椎动物卷》(2015) 中哺乳类群的编研过程。完成了《中国生物多样性红色名录：脊椎动物 第一卷 哺乳动物》(2021)，现对评估结果分析如下。

《中国生物多样性红色名录：脊椎动物 第一卷 哺乳动物》(2021) 再次更新了中国哺乳动物记录。在 2018 版的《IUCN 受威胁物种红色名录》(*IUCN Red List of Threatened Species*) 中，中国大陆的哺乳动物物种

12 Forming the Evaluation Report

At the end of the review process, a final assessment report was prepared based on the information available concerning each species (mainly based on their status). The final report included the background and significance of the assessment, the categories and criteria of the assessment, the process undertaken for the assessment, and most importantly, the analyses of important taxa and associated threats, a comparison of this study's results with those of the *IUCN Red List of Threatened Species*, and an outlook on the future of mammal species in China.

13 Analysis of the Assessment Results

"China's Red List of Biodiversity: Volume of Vertebrates" (in which the red list of mammals is an integral part) includes Chinese and Latin names of Chinese mammal species and their families, their status as designated through the assessment, and whether they are endemic in the country. "China's Red List of Biodiversity: Volume of Vertebrates" was jointly released by the former Ministry of Environmental Protection and the Chinese Academy of Sciences on International Biodiversity Day on May 22, 2015. This important document is formally referred to as the *China's Red List of Biodiversity: Volume of Vertebrates* (2015).

At an IUCN workshop on the "Red List in Northeast Asia" that was held in Beijing, China, on March 25~26, 2016, we were invited to present our research on China's biodiversity red list. More specifically, I (Zhigang Jiang) was requested to introduce the research that we undertook to write the *China's Red List of Biodiversity: Volume of Vertebrates* (2015). In addition to international experts from the IUCN red list working group and a range of representatives from across China, representatives from Mongolia, Korea (the Republic of), Korea (the Democratic People's Republic of) and Japan also attended the meeting. I highlighted how we down-listed the conservation status of Giant Panda from Endangered to Vulnerable in the 2015 edition *China's Red List of Biodiversity: Volume of Vertebrates* based on our assessments of reducing rates of population increase and increase of total habitat area, along with the growing number of small populations and habitat patches. I also announced the launch of "China's Red List of Biodiversity: Volume of Vertebrates", which was officially proclaimed by the former Ministry of Environmental Protection and the Chinese Academy of Sciences in the government's Official Announcement No. 32 in 2015. Six months later, the IUCN red list working group followed suit and similarly announced at the World Conservation Congress in Hawaii, USA, in September 2016 that Giant Panda was now newly recognized as Vulnerable, rather than Endangered in the *IUCN Red List of Threatened Species*.

In the process of writing this book, the Working Group has since 2017 further updated the mammal inventory system earlier put forward in *China's Mammal Diversity* (2nd edition), broadly following the taxonomic system used in the *Handbook of the Mammals of the World* (HMW), which is now also used in IUCN red lists. Over this period, we continued to collect new mammal records information and occasionally even identified new mammal species in China. Finally, following the development of a comprehensive and updated inventory, the status of each species was individually reassessed

数记录为 551 种，名列世界第 3 名，加上中国台湾的哺乳动物，中国哺乳动物也只有 566 种，仍名列世界第 3。通过本次研究订正和增删，中国哺乳动物种数超过了原来名列世界第一的印度尼西亚 (670 种)。中国已经成为世界哺乳动物多样性第一丰富的国家 (表 4)。

表 4 中国哺乳动物多样性与名列世界哺乳动物多样性前 10 位国家的比较 *

IUCN (2018) 数据			本次评估数据		
序号	国家	种数	序号	国家	种数
1	印度尼西亚	670	1	**中国**	**700**
2	巴西	648	2	印度尼西亚	670
3	**中国**	**566**	3	巴西	648
4	墨西哥	523	4	墨西哥	523
5	秘鲁	467	5	秘鲁	467
6	哥伦比亚	442	6	哥伦比亚	442
7	美国	440	7	美国	440
8	刚果（金）	430	8	刚果（金）	430
9	印度	412	9	印度	412
10	肯尼亚	376	10	肯尼亚	376

注：* 智人除外

14 受威胁状况

IUCN 红色名录中极危 (CR)、濒危 (EN)、易危 (VU) 三个等级的物种被称为受威胁物种。本次评估结果显示，受威胁中国哺乳动物共计 181 种，约占评估物种总数的 25.9%。此外，属于近危等级 (NT) 的哺乳动物有 138 种，占评估物种总数的 19.7%；属于"数据缺乏"(DD) 的有 105 种，占所有哺乳动物的 15.0%（图 12）。中国有 10 种哺乳动物野外灭绝或区域灭绝，其中，4 种哺乳动物属于"野外灭绝"，6 种属于"区域灭绝"。

according to evaluation procedures agreed for the development of *China's Red List of Biodiversity: Volume of Vertebrates* (2015). Here we offer the key results and analyses from *China's Red List of Biodiversity: Vertebrates, Volume I, Mammals* (2021).

China's Red List of Biodiversity: Vertebrates, Volume I, Mammals (2021) has once again updated the mammal records of China. In the *IUCN Red List of Threatened Species* (2018), the recorded number of mammals in the Chinese mainland was 551 species, ranking third in the world for overall mammalian diversity. Adding the mammal species from Taiwan, China, the total number increased to 566 species, yet China still ranked third in the world. With the revised inventory used in this study, the number of mammals species in China has surpassed Indonesia (670 species) which previously ranked in first positions in terms of mammal diversity. **China has thus become the country with the richest mammalian diversity in the world** (Table 4).

Table 4　Diversity of China's mammal species and a comparison with world's top 10 countries with the richest mammal species diversities*

IUCN (2018)			Data of this Study		
Rank	Country	No. species	Rank	Country	No. species
1	Indonesia	670	1	China	700
2	Brazil	648	2	Indonesia	670
3	China	566	3	Brazil	648
4	Mexico	523	4	Mexico	523
5	Peru	467	5	Peru	467
6	Colombia	442	6	Colombia	442
7	USA	440	7	USA	440
8	Congo, D. R.	430	8	Congo, D. R.	430
9	India	412	9	India	412
10	Kenya	376	10	Kenya	376

Note: * Excluding *Homo sapiens*

14 Threatened Status

Species listed as Critically Endangered (CR), Endangered (EN) and Vulnerable (VU) in the IUCN red list are collectively known as Threatened Species. The results of this assessment also show that 181 mammals in China are considered as threatened, accounting for 25.9% of the total number of species that were assessed in this study. In addition, 138 species have been classified as Near Threatened (NT), accounting for a further 19.7% of all mammal species in the country, and 105 species were considered as Data Deficient (DD), accounting for 15.0% of mammals (**Figure 12**). 10 mammal species are considered as Extinct in the Wild or Regionally Extinct, in which 4 mammal species are Extinct in the Wild across their entire range and 6 species have been assessed as Regionally Extinct.

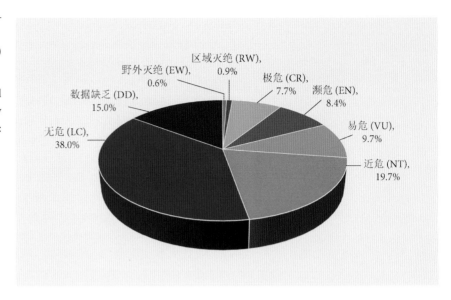

图 12 《中国生物多样性红色名录：脊椎动物 第一卷 哺乳动物》(2021) 中各濒危等级物种的比例

Figure 12 Proportion of mammal species in each red list status category in *China's Red List of Biodiversity: Vertebrates, Volume I, Mammals* (2021)

中国鳞甲目动物群全部列为极危 (CR)。中国有穿山甲 (*Manis pentadactyla*)、印度穿山甲 (*M. crassicaudata*，图 13)、马来穿山甲 (*M. javanica*)。其中，穿山甲 (也称为中华穿山甲) 曾是中国南方红壤丘陵地带的广布种，多在山麓地带草丛中或丘陵杂灌丛中筑洞，以白蚁、蚂蚁、蜜蜂、胡蜂和其他昆虫幼虫等为食物。虽然自 1989 年以来，穿山甲受到法律的保护。然而，在巨大市场需求和经济利益的刺激下，野外穿山甲的种群数量急剧下降。《濒危野生动植物种国际贸易公约》第 17 次缔约方大会通过决议，于 2017 年将所有穿山甲列入该公约附录 I，禁止穿山甲的任何国际商业贸易活动。尽管目前在广东象头山国家级自然保护区发现了穿山甲繁殖个体，但其分布区与种群极为有限。基于其种群与分布地现状，中国分布的 3 种穿山甲的濒危等级全部为极危 (CR)。

中国特有哺乳动物计 156 种，占中国哺乳动物总数的 22%，这些特有种具有巨大的种质资源价值和遗传多样性价值。一个国家的特有种在一个国家的生存状况即是其在全球的生存状况。本次评估结果显示，33 种中国哺乳动物特有种属于受威胁物种。其中，"极危"特有种 9 种，"濒危"特有种 11 种，"易危"特有种 13 种 (表 5)。

中国哺乳动物各目中濒危物种比例是非均匀分布的 (表 6)。翼手目 Chiroptera 没有极危的种类。劳亚食虫目 Eulipotyphla 仅一种为极危种，绝大多数种没有灭绝风险 (表 6)。啮齿目 Rodentia 是中国种数最多的哺乳动物目，有 221 种。然而，啮齿目受威胁物种却最少，只有河狸 (*Castor fiber*) 和比氏鼯鼠 (*Biswamoyopterus biswasi*) 两种被评为极危 (CR)。目前，河狸仅分布在新疆阿尔泰山南坡的布尔根河、大小清河流域，种群数量约为 700 只，它在中国的分布区是其边缘分布区 (图 14)。比氏鼯鼠分布在西藏，目前根据采集的标本，估计其分布区面积为 10~100km^2。

All of the species of Pholidota in China are listed as Critically Endangered (CR), including the Chinese Pangolin (*Manis pentadactyla*), Indian Pangolin (*M. crassicaudata*, **Figure 13**) and Sunda Pangolin (*M. javanica*). The Chinese Pangolin was once a common species in hilly terrains in southern China, especially in the krasnozem (red loamy or red clay) hills. Pangolins dig burrows in the piedmont zone covered by grasses or shrubs and feed on termites, ants, bees, wasps, and others insect larvae. Although pangolin species have been legally protected since 1989, they have long been hunted due to huge market demands and their populations have sharply declined in the wild. A resolution was therefore adopted at the 17th Conference of the Parties of CITES in 2017 to include all pangolins in the Appendix I of CITES, banning any international commercial trade in pangolins. Although breeding pangolin individuals in Xiangtou Shan National Nature Reserve, Guangdong Province have been recorded recently, their distribution and population are

图 13 印度穿山甲 (*Manis crassicaudata*，蒋志刚摄)。在巨大市场需求和经济利益的刺激下，野外穿山甲的种群数量急剧下降。《濒危野生动植物种国际贸易公约》于2017年禁止任何穿山甲的国际商业贸易活动。中国分布的三种穿山甲的生存状况全部极危

Figure 13 The population of Indian Pangolin (*Manis crassicaudata*, Photographed by Zhigang Jiang) has declined sharply in the wild under pressures of market demand. CITES banned all international commercial trade in pangolins in 2017. All three species of pangolins found in China are Critically Endangered

表5 中国哺乳动物红色名录等级中的种数与特有种数

红色名录等级	种数	比例/%	特有种
野外灭绝	4	0.6	0
区域灭绝	6	0.9	0
极危	54	7.7	9
濒危	59	8.4	11
易危	68	9.7	13
近危	138	19.7	40
无危	266	38.0	42
数据缺乏	105	15.0	41
小计	700	100	156

表6 中国哺乳动物各目物种濒危等级数

目	种数	野外灭绝	区域灭绝	极危	濒危	易危	近危	无危	数据缺乏	受胁种数
劳亚食虫目	88	0	0	1	1	9	31	33	13	11
攀鼩目	1	0	0	0	0	0	0	1	0	0
翼手目	139	0	0	0	4	15	43	42	35	19
灵长目	29	0	0	15	8	4	1	1	0	27
鳞甲目	3	0	0	3	0	0	0	0	0	3
食肉目	64	0	0	12	17	15	13	2	5	44
海牛目	1	0	0	1	0	0	0	0	0	1
长鼻目	1	0	0	0	1	0	0	0	0	1
奇蹄目	6	1	3	0	0	1	1	0	0	1
偶蹄目	65	3	3	16	14	10	10	3	6	40
鲸目	39	0	0	2	7	2	0	14	14	11
啮齿目	221	0	0	2	4	8	33	145	29	14
兔形目	43	0	0	1	4	4	6	25	3	9
总计	700	4	6	54	59	68	138	266	105	181

still rather limited. Based on the status of their populations and distributions, all three pangolin species distributed in China are Critically Endangered (CR).

There are 156 species of mammals in China that are endemic to the country, accounting for 22% of the total number of species. These endemic species have great value as germplasm resources as well as the intrinsic value of their genetic diversity. The status of an endemic species in the country represents its global survival status. This assessment shows that 33 species of China's endemic mammal are Threatened. Among these, 9 of the species are Critically Endangered, 11 are Endangered, and 13 are Vulnerable (**Table 5**).

Table 5 Numbers of mammal species and endemic mammal species in each category of the China's red list of mammals

Red List Category	No. Species	Ratio/%	Endemic Species
EW	4	0.6	0
RW	6	0.9	0
CR	54	7.7	9
EN	59	8.4	11
VU	68	9.7	13
NT	138	19.7	40
LC	266	38.0	42
DD	105	15.0	41
Sum	700	100	156

The proportion of endangered species in each order of mammals in China is not randomly distributed (**Table 6**). No species of Chiroptera has been assessed as Critically Endangered. Most insectivores are safe, with only one species of Eulipotyphla assessed as Critically Endangered (**Table 6**). Amongst the largest order of mammals in China, Rodentia, only two of its 221species is considered at risk of extinction (Critically Endangered, CR), Eurasian Beaver, *Castor fiber* and Namdapha Flying Squirrel, *Biswamoyopterus biswasi*. Currently, the beaver lives in marginal range along the Bulgen River and Major and Minor Qinghe Rivers in the foothills of the Altay Shan of Xinjiang, with a total population around 700 individuals (**Figure 14**). Namdapha Flying Squirrel is only found in Xizang in the country, its range is estiamated $10\sim100km^2$ according to the specimens collected.

The Black-bellied Hamster (*Cricetus cricetus*, **Figure 15**) is distributed in Eurasia continent, however, in recent years, the range and number of the Black-bellied Hamster in the whole Europe have decreased. There have been local extinctions in Belgium, Netherland, France, Austria, Germany, Poland, Czechia, Hungary, Slovakia, Ukraine, Belarus and Russia, with only a few hundred of the Black-bellied Hamster now living in France. The Black-

图 14 河狸 (*Castor fiber*, 蒋志刚摄)。河狸在中国的分布区为其边缘分布，目前，河狸仅分布在新疆阿尔泰山地区布尔根河、大小青河流域，种群数量约为 700 只。其生存面临着人类活动的严重压力，为"极危"物种

Figure 14 Beaver (*Castor fiber*, Photographed by Zhigang Jiang). The beaver is marginally distributed in China. At present, the dens of beaver are scattered along the Bulgen River and the Major and Minor Qinghe Rivers flowing in the southern slope of the Altay Shan in Xinjiang, with a population of around 700 individuals. The beaver's survival is seriously threatened by human activities and it is therefore assessed as being Critically Endangered

    原仓鼠 (*Cricetus cricetus*，图 15) 分布在欧亚大陆，在整个欧洲，原仓鼠的分布区范围和数量都在下降。比利时、荷兰、法国、奥地利、德国、波兰、捷克、匈牙利、斯洛伐克、乌克兰、白俄罗斯和俄罗斯都发生了局部的原仓鼠种群灭绝，目前生活在法国的原仓鼠只剩下几百只。原仓鼠在瑞士已经灭绝，可能在格鲁吉亚也灭绝了。曾经有分布的卢森堡也未再发现过原仓鼠。自 2016 年上一次 IUCN 红色名录评估以来，原仓鼠大部分分布区中种群数量减少 50%，估计原仓鼠的发生区面积减少了 468,634km^2，占有区面积减少了 226,835km^2。基于繁殖率和分布区的快速下降，Banaszek *et al.* (2020) 将原仓鼠列为极危 (CR, A3a) 等级。中国新疆是原仓鼠的边缘分布区。刘少英团队近年曾 3 次前往新疆采样，未能获得原仓鼠标本。本次评估中将原仓鼠列为濒危。

    中国兔形目代表性濒危物种有中国特有种伊犁鼠兔 (*Ochotona iliensis*，图 16) 与海南兔 (*Lepus hainanus*，图 17)。1983 年，伊犁鼠兔由马勇与李维东发现并命名，其仅在中国天山山脉有发现。自 2000 年以来，伊犁鼠兔的栖息地减少至原本分布区面积的 17.05%。种群成熟个体数少于 2,500 只，已经处于濒危状态 (Li and Smith, 2005)。

Table 6 Numbers of species amongst the 13 orders of mammals in China in each red list status category

Order	No. Species	EW	RE	CR	EN	VU	NT	LC	DD	No. Threatened
Eulipotyphla	88	0	0	1	1	9	31	33	13	11
Scandentia	1	0	0	0	0	0	0	1	0	0
Chiroptera	139	0	0	0	4	15	43	42	35	19
Primates	29	0	0	15	8	4	1	1	0	27
Pholidota	3	0	0	3	0	0	0	0	0	3
Carnivora	64	0	0	12	17	15	13	2	5	44
Sirenia	1	0	0	1	0	0	0	0	0	1
Proboscidea	1	0	0	1	0	0	0	0	0	1
Perissodactyla	6	1	3	0	0	1	1	0	0	1
Artiodactyla	65	3	3	16	14	10	10	3	6	40
Cetacea	39	0	0	2	7	2	0	14	14	11
Rodentia	221	0	0	2	4	8	33	145	29	14
Lagomorpha	43	0	0	1	4	4	6	25	3	9
Sum	700	4	6	54	59	68	138	266	105	181

图 15 原仓鼠 (*Cricetus cricetus*, Katanski 摄)。自 2016 年，欧亚大陆的原仓鼠种群数量下降、分布区缩小，在其分布国家发生了区域灭绝。原仓鼠在中国新疆的分布是其边缘分布。刘少英团队最近 3 次新疆野外考察，均未能采集到原仓鼠标本

Figure 15 Black-bellied Hamster (*Cricetus cricetus*, Photographed by Katanski). Since 2016, the populations of Black-bellied Hamster in Eurasian Continent has declined, its distribution area has shrunk, regional extinction has occurred in its range countries. The distribution of Black-bellied Hamster in Xinjiang of China is its marginal range. Shaoying Liu's team failed to collect specimens of Black-bellied Hamster during three recent field trips in Xinjiang

图 16 伊犁鼠兔 (*Ochotona iliensis*, 李维东摄) 仅在中国新疆发现，栖息在海拔 2,800~4,100m 的天山裸岩区。自 2000 年以来，伊犁鼠兔的栖息地减少至 1983 年原本分布区面积的 17.05%。种群数量的成熟个体数少于 2,500 只。伊犁鼠兔已经处于濒危状态

Figure 16 Ili Pika (*Ochotona iliensis*, Photographed by Weidong Li). Ili Pika is found only in the Tian Shan Mountains in Xinjiang, China, inhabiting bare rocky habitat between 2,800 and 4,100 meters above sea level. Since 2000, the habitat of Ili Pika has reduced to 17.05% of its original area. The number of mature Ili Pika is less than 2,500. The Ili Pika has already been Endangered

bellied Hamster is extinct in Switzerland and probably in Georgia, and it has not been found in Luxembourg, where it was once distributed. Since the last IUCN red list assessment on the status of Black-bellied Hamster in 2016, populations in most ranges of the Black-bellied Hamster have declined by 50 percent, it is estimated that its Area of Occurrence has decreased by 468,634km^2 and the Area of Occupancy has decreased by 226,835km^2. Based on the rate of population and range decline, Banaszek *et al.* (2020) listed the Black-bellied Hamster as Critically Endangered (CR, A3a). Xinjiang of China is a marginal range of the Black-bellied Hamster. Shaoying Liu's team went to Xinjiang three times in recent years to sample the Black-bellied Hamster but failed to obtain any specimen in their field trips. Therefore, the Black-bellied Hamster is listed as Endangered in this assessment.

Representative endangered species amongst lagomorphs in China include the highly endemic Ili Pika *Ochotona iliensis* (**Figure 16**) and Hainan Hare *Lepus hainanus* (**Figure 17**). Discovered by Yong Ma and Weidong Li in 1983, Ili pika is found only in the Tian Shan Mountains of China. Since 2000, the habitat of Ili pika has reduced to 17.05% of its original area. The population of adult Ili Pika is now fewer than 2,500 individuals and the species is considered as Endangered (Li and Smith, 2005).

Hainan Hare (*Lepus hainanus*) is an endemic species mainly distributed in the semi-arid shrubs of hilly terraces in the eastern part of Hainan, China. This hare is a typical species of the tropical lowland ecosystem in Hainan and may serve as indicator of ecosystem health in this type of habitat (**Figure 17**). Hainan Hare have high fecundity with litter sizes of 3~5 offspring. However, Haisheng Jiang's research team has found that due to the clear-cutting of shrub grasslands for developing farmland and construction land, Hainan Hare's habitat is continuously being fragmented and shrinking in area. Additionally, its population has decreased sharply due to excessive hunting and its survival status is now Critically Endangered. Siliang Lin and Haisheng Jiang compared wildlife survey data of Hainan Hare from 1997~1998 and 2017~2018 (during the Second National Survey on Terrestrial Wild Animals in China) and they found that the average population density of Hainan Hare reduced over the intervening two decades from 2.08/km^2 to only 0.19/km^2. In view of Hainan Hare's population trend over 20 years, its status is now rated as Critically Endangered.

Primates, Carnivora and Artiodactyla have the most mammal species at risk of extinction in China. 27 of the 29 non-human Primates are now assessed as threatened. 15 of these species are at risk of extinction and thus classified as Critically Endangered (CR), including *Macaca leonina*, *Macaca leucogenys*, *Rhinopithecus bieti*, *R. brelichi*, *R. strykeri*, *Semnopithecus schistaceus* (**Figure 18**), *Trachypithecus leucocephalus*, as well as *Hoolock leuconedys*, *Hylobates lar*, *Nomascus concolor*, *N. hainanus*, *N. leucogenys*, *N. nasutus*, *Nycticebus pygmaeus* and the 2017 identified Tianxing Hoolock Gibbon (Skywalker Hoolock Gibbon, *H. tianxing*) (**Figure 5**). The status of primates in China has also been similarly reported by Pan *et al.* (2016).

图 17 海南兔 (*Lepus hainanus*，袁喜才摄)。海南兔是兔形目动物中唯一的"极危"种。尽管海南兔的繁殖能力强，但是，其在海南东部丘陵台地灌草丛的栖息地被开垦，种群密度较 20 年前大幅下降

Figure 17　Hainan Hare (*Lepus hainanus*, Photographed by Xicai Yuan). Hainan Hare is the only Critically Endangered species in the order of Lagomorpha. Although Hainan Hare have high fecundity, they inhabit semi-arid shrub and grasslands on hilly terraces in the eastern part of Hainan Island, much of which has recently been reclaimed as new farmland or for construction. Its population density is now much lower than it was only 20 years ago

海南兔 (*Lepus hainanus*) 是中国海南的特有种，主要分布于海南东部丘陵台地的半干旱的灌草丛中，是海南热带低地陆地生态系统的典型物种，是该类生境的生态系统健康状况的指示物种 (图 17)。海南兔繁殖力强，每窝平均产仔 3~5 只。然而，根据江海生团队的考察结果，由于大量的灌丛草地被开垦为耕地、建设用地，海南兔栖息地面积持续缩减，同时海南兔栖息地出现严重斑块化、破碎化的趋势。加之过度捕猎，海南兔种群数量锐减，其生存现状为极危。林思亮和江海生在全国第二次陆生野生动物资源调查专项调查中，比较了 1997~1998 年与 2017~2018 年海南兔的调查数据，1997~1998 年海南兔全岛平均种群密度为 2.08 只 /km^2，而 2017~2018 年仅为 0.19 只 / km^2。鉴于海南兔 20 年来的种群数量变化，其濒危等级被评定为极危。

在中国有灭绝风险的哺乳动物种主要分布在灵长目 Primates、食肉目 Carnivora 与偶蹄目 Artiodactyla。29 种非人类灵长目动物中有 27 种的生存受到威胁。其中 15 种面临灭绝风险，列为"极危 (CR)"等级：北豚尾猴 (*Macaca leonina*)、白颊猕猴 (*Macaca leucogenys*) 滇金丝

Artiodactyla is another group of mammals with many threatened species. Sixteen (16) of the 65 even-toed hooved species in China are in imminent danger of extinction and listed as Critically Endangered (CR), while a further 14 of these species are listed as Endangered (EN). Among these are the Bactrian Camel (*Camelus ferus*), Forest Musk Deer (*Moschus berezovskii*), Williamson's Mouse Deer (*Tragulus williamsoni*, **Figure 19**), Himalayan

图 18 分布在喜马拉雅山南坡的长尾叶猴 (*Semnopithecus schistaceus*, 蒋志刚摄)。长尾叶猴在中国的分布区是其边缘分布区，仅分布在喜马拉雅山南坡部分地区

Figure 18 Nepal Gray Langur (*Semnopithecus schistaceus*, Photographed by Zhigang Jiang), distributed on the southern slope of the Himalayas. The distribution range of Nepal Gray Langur in China is in its marginal distribution area

猴 (*Rhinopithecus bieti*)、黔金丝猴 (*Rhinopithecus brelichi*)、缅甸金丝猴 (*Rhinopithecus strykeri*)、长尾叶猴 (*Semnopithecus schistaceus*)(图 18)、白头叶猴 (*Trachypithecus leucocephalus*)、东白眉长臂猿 (*Hoolock leuconedys*)、白掌长臂猿 (*Hylobates lar*)、西黑冠长臂猿 (*Nomascus concolor*)、海南长臂猿 (*Nomascus hainanus*)、北白颊长臂猿 (*Nomascus leucogenys*)、东黑冠长臂猿 (*Nomascus nasutus*)、倭蜂猴 (*Nycticebus pygmaeus*) 以及 2017 年新报道的高黎贡白眉长臂猿 (图 5)。中国灵长类濒危状况与潘汝亮等报道的中国绝大部分灵长类濒危研究结果相似 (Pan *et al.*, 2016)。

偶蹄目动物也是生存受到威胁比例较高的动物类群，65 种偶蹄类中有 16 种面临灭绝，列为"极危 (CR)"，14 种列为"濒危 (EN)"。其中有野骆驼 (*Camelus ferus*)、林麝 (*Moschus berezovskii*)、威氏鼷鹿 (*Tragulus williamsoni*, 图 19)、喜马拉雅麝 (*Moschus leucogaster*)、黑麝 (*Moschus fuscus*)、原麝 (*Moschus moschiferus*)、贡山麂 (*Muntiacus gongshanensis*)、梅花鹿 (*Cervus nippon*)、麋鹿 (*Elaphurus davidianus*)、驼鹿 (*Alces alces*)、东方坡鹿 (*Panolia siamensis*)、印度野牛 (*Bos gaurus*)、蒙原羚 (*Procapra gutturosa*)、长尾斑羚 (*Naemorhedus caudatus*)、塔尔羊 (*Hemitragus jemlahicus*) 和戈壁盘羊 (*Ovis darwini*)。

64 种食肉目物种中有 12 种面临灭绝并被列为"极危 (CR)"，包括：荒漠猫 (*Felis bieti*)、云豹 (*Neofelis nebulosa*)、虎 (*Panthera tigris*)、云猫 (*Pardofelis marmorata*)、金猫 (*Pardofelis temminckii*)、马来熊 (*Helarctos malayanus*) 和缟灵猫 (*Chrotogale owstoni*) 等。

图 19　2015 年，在中国云南西双版纳的热带雨林中拍摄到的威氏鼷鹿 (*Tragulus williamsoni*, 胡一鸣等摄)。威氏鼷鹿在中国的分布区是其边缘分布区，目前其仅在云南南部被发现

Figure 19　Williamson's Mouse Deer (*Tragulus williamsoni*, Photographed by Yiming Hu *et al.* in 2015), in the tropical forest of Xishuangbanna, Yunnan, China. Mouse deer' distribution area in China is its marginal range area and it is only found in southern Yunnan

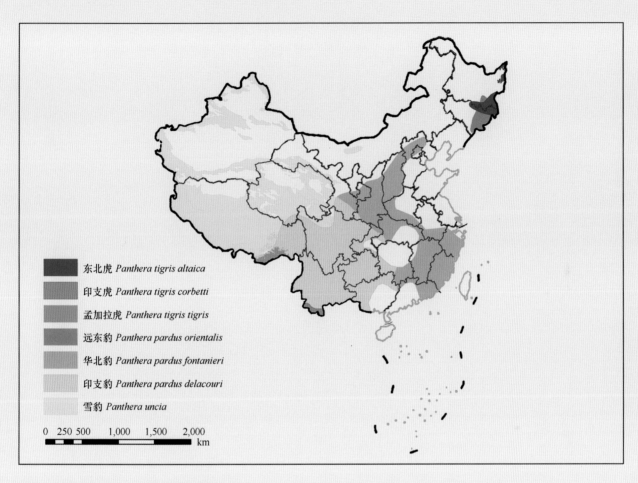

图 20　豹属 (*Panthera*) 动物在中国的分布

Figure 20　Distribution of *Panthera* in China

Musk Deer (*Moschus leucogaster*), Black Musk Deer (*Moschus fuscus*), Siberian Musk Deer (*Moschus moschiferus*), Gongshan Muntjac (*Muntiacus gongshanensis*), Sika Deer (*Cervus nippon*), Père David's Deer (*Elaphurus davidianus*), Moose (*Alces alces*), Eastern Eld's Deer (*Panolia siamensis*), Gaur (*Bos gaurus*), Mongolian Gazelle (*Procapra gutturosa*), Long-tailed Goral (*Naemorhedus caudatus*), Himalayan Tahr (*Hemitragus jemlahicus*) and Gobi Argali (*Ovis darwini*).

12 of the 64 species in Carnivora in China are facing the danger of extinction and thus are classified as Critically Endangered (CR). These species include the Chinese Mountain Cat (*Felis bieti*), Clouded Leopard (*Neofelis nebulosa*), Tiger (*Panthera tigris*), Marbled Cat (*Pardofelis marmorata*), Asiatic Golden Cat (*P. temminckii*), Sun Bear (*Helarctos malayanus*) and Owston's Palm Civet (*Chrotogale owstoni*), etc.

图 21 长江江豚 (*Neophocaena asiaeorientalis*, 张先锋摄)。长江江豚为中国特有水生哺乳类。近年来，由于长江航运船只螺旋桨误伤、误捕、食物稀少、水体污染等，长江江豚种群数量下降，濒临灭绝

Figure 21 Yangtze Finless Porpoise (*Neophocaena asiaeorientalis*, Photographed by Xianfeng Zhang). The Yangtze Finless Porpoise is an endemic aquatic mammal in China. In recent years, due to the accidental injury by the propellers of shipping ships along the Changjiang River, the by-catching, the scarcity of food resource and the water pollution, the population of Yangtze Finless Porpoise declines and it is on the verge of extinction

首次评估了金钱豹 (*Panther pardus*) 亚种在中国的濒危情况。由于豹分布地区广泛，地理亚种分化多，种下分类单元有争论。中国有远东豹 (*P. p. orientalis*)、华北豹 (*P. p. fontanieri*)、印度豹 (*P. p. fusea*) 和印支豹 (*P. p. delacouri*) 等金钱豹亚种分布 (图 20)。近年来远东豹数量有所增加，但是在中国东北地区仅发现 40 余只，故其濒危等级定为极危 (CR)。近年来，华北豹在山西、陕西、河北、河南、甘肃等地被红外相机拍摄到，在野外仍有华北豹的可生存种群，故其濒危等级定为濒危 (EN)。印度豹仅分布于喜马拉雅山脉局部低海拔区域，在西藏吉隆曾被发现记录，故其濒危等级定为极危 (CR)。印支豹现在中国南方野外稀少，极少见到，故其濒危等级定为极危 (CR)。此外，目前分布在青海及青藏高原东缘西藏昌都、四川甘孜的豹，其分类地位未定，可能是介于华北豹和印支豹之间的中间类型。

长江江豚 (*Neophocaena asiaeorientalis*，图 21) 主要生活在长江及通江湖泊，其性成熟年龄为 4~9 年，一个世代长度为 5~10 年。由于遭受滥捕，加之沿岸开发、水源污染，以及长江生态环境变化，长江鱼类资源下降，包括受到船只误伤的损害，中国淡水豚类的生存现状严峻。根据 1984~1991 年的考察结果，长江江豚的种群数量约 2,700 头；2006 年 11~12 月，由农业部和中国科学院水生生物研究所组织七国专家共同进行了"长江豚类考察"活动，发现长江江豚种群数量为 1,200~1,400 头；2012 年，长江淡水豚考察结果表明长江江豚种群数量约为 1,040 头 (中华人民共和国农业部，2014)。而 2006~2012 年长江江豚种群的下降速率为 25%，年种群下降速率高于 4%，高于《IUCN 受威胁物种红色名录》的极危标准 C1。长江江豚已处于极危状态。

The status of the different subspecies of Leopard (*Panther pardus*) was evaluated for the first time in China in this study. Due to the species wide distribution range and geographic differentiation, the taxon of subspecies in the Leopard has remained controversial. There are several subspecies recognized in China, including the Amur Leopard (*P. p. orientalis*), North China Leopard (*P. p. fontanieri*), Indian Leopard (*P. p. fusea*) and Indochinese Leopard (*P. p. delacouri*) (**Figure 20**). The population of Amur Leopard in northeastern China has increased in recent years, however there are still only more than 40 individuals that have been found in the northeastern China region. The Amur Leopard is therefore classified as Critically Endangered (CR). In recent years, North China Leopard were frequently photographed by infrared cameras in Shanxi, Shaanxi, Hebei, Henan, Gansu and other provinces, indicating that this subspecies still has a fairly large range, and consequently a viable population, the North China Leopard has therefore been classified as Endangered (EN). Indian Leopard is only found in some low-altitude regions of the Himalayas, particularly in Jilong, Tibet (Xizang), China. Therefore, it is classified as Critically Endangered (CR). The Indochinese Leopard is now very rarely seen in the wild in southern China, thus, it is classified as Critically Endangered (CR). In addition, the taxonomy of the Leopard currently distributed in Qinghai and the eastern part of the Qinghai-Tibet (Xizang) Plateau, including Qamdo in Tibet (Xizang) and Ganzi in Sichuan, is still largely unknown; it may be a median type between the North China Leopard and the Indochinese Leopard.

Yangtze Finless Porpoise (*Neophocaena asiaeorientalis*, **Figure 21**) mainly inhabits the Changjiang River and lakes connected with the River. The age of sexual maturity of Yangtze Finless Porpoise is 4~9 years, and the generation span is 5~10 years. Declining fish resources due to excessive fishing in the Changjiang River, the exploitation of riverine banks and water pollution, and deterioration of the ecological environment of the Changjiang River have all greatly harmed the Yangtze Finless Porpoise. The porpoises also may be accidentally injured by ship propellers. The Yangtze Finless Porpoises in China are clearly in peril. From 1984 to 1991, their population was estimated to be around 2,700 individuals. From November to December 2006, the Ministry of Agriculture and Institute of Hydrobiology of Chinese Academy of Sciences together with experts from seven countries jointly conducted the "Changjiang River Freshwater Dolphin Survey" and estimated the population size at around 1,200~1,400 individuals. A follow-up survey in 2012 indicated that the population of Yangtze Finless Porpoise was about 1,040 individuals (Ministry of Agriculture of the People's Republic of China, 2014). From 2006 to 2012, the population thus declined by 25%, representing an average annual decline rate over 4%, which is higher than Criterion C1 threshold in the Criteria of *IUCN Red List of Threatened Species*. On this basis, Yangtze Finless Porpoise is assessed to be Critically Endangered.

15 野外灭绝和区域灭绝物种分析

中国有蹄类动物面临着"第6次大灭绝"(Barnosky et al., 2011)。评估结果中有10种哺乳动物属于"野外灭绝"或"区域灭绝"等级，全部是大中型有蹄类动物(表7)。

表7 中国"野外灭绝"或"区域灭绝"的哺乳动物

种名	学名	红色名录等级
野马	*Equus ferus*	野外灭绝 ver. 3.1
驯鹿	*Rangifer tarandus*	野外灭绝 ver. 3.1
大额牛	*Bos frontalis*	野外灭绝 ver. 3.1
高鼻羚羊	*Saiga tatarica*	野外灭绝 ver. 3.1
双角犀	*Dicerorhinus sumatrensis*	区域灭绝 ver. 3.1
爪哇犀	*Rhinoceros sondaicus*	区域灭绝 ver. 3.1
大独角犀	*Rhinoceros unicornis*	区域灭绝 ver. 3.1
爪哇野牛	*Bos javanicus*	区域灭绝 ver. 3.1
野水牛	*Bubalus arnee*	区域灭绝 ver. 3.1
豚鹿	*Axis porcinus*	区域灭绝 ver. 3.1

15.1 野外灭绝种分析

在中国，野马 (*Equus ferus*)、驯鹿 (*Rangifer tarandus*)、高鼻羚羊 (*Saiga tartarica*) 和大额牛 (*Bos frontalis*) 4个物种属于野外灭绝等级 (EW)。

15.1.1 野马

20世纪初，野马在中国新疆灭绝。中国20世纪80年代从国外重新引入了野马，分别在新疆、甘肃建立了野马圈养种群。2001年，27匹野马被放归到准噶尔盆地。但是目前野放的野马在冬季仍需要人工补饲，在野外尚未形成可生存种群 (Jiang et al., 2015; 王渊等, 2016; Jiang and Zong, 2019; 图22)。

15.1.2 驯鹿

中国现有约700头驯鹿，但这些驯鹿基本处于人工驯养状态。在中国大兴安岭林区，鄂温克族人使用驯鹿作为生产工具与交通工具，目前没有在野外发现驯鹿。驯鹿已被列入国家林业局2000年发布的《国家保护的有益的或者有重要经济、科学研究价值的陆生野生动物名录》。2017~2019年，内蒙古根河从芬兰引入了179头驯鹿 (图23)。野生驯鹿在中国灭绝。本次评估将驯鹿的濒危等级定为"野外灭绝"。

15.1.3 大额牛

1807年，大额牛 (*Bos frontalis*) 由 Aylmer Bourke Lambert 根据一只

15 Analysis of Extinct in the Wild or Regionally Extinct Species

The ungulate fauna is facing the "Sixth Mass Extinction" (Barnosky *et al.*, 2011). Ten of China's mammal species were classified as either Extinct in the Wild (EW) or Regionally Extinct (RE) in this assessment, all of them are large or median sized ungulates (**Table 7**).

Table 7 Mammal species "Extinct in the Wild" or "Regionally Extinct" in China

English Name	Scientific Name	Red List Category
Przewalski's Horse	*Equus ferus*	EW ver. 3.1
Reindeer	*Rangifer tarandus*	EW ver. 3.1
Gayal	*Bos frontalis*	EW ver. 3.1
Saiga	*Saiga tatarica*	EW ver. 3.1
Sumatran Rhinoceros	*Dicerorhinus sumatrensis*	RE ver. 3.1
Javan Rhinoceros	*Rhinoceros sondaicus*	RE ver. 3.1
Indian Rhinoceros	*Rhinoceros unicornis*	RE ver. 3.1
Banteng	*Bos javanicus*	RE ver. 3.1
Asian Buffalo	*Bubalus arnee*	RE ver. 3.1
Hog Deer	*Axis porcinus*	RE ver. 3.1

15.1 Analysis of Extinct in the Wild Species

Przewalski's Horse (*Equus ferus*), Reindeer (*Rangifer tarandus*), Saiga (*Saiga tartarica*) and Gayal (*Bos frontalis*) are the four species that are now Extinct in the Wild (EW).

15.1.1 Przewalski's Horse

Although Przewalski's Horses were initially reintroduced from abroad in the 1980s and captive populations were established in Xinjiang and Gansu, they have not yet successfully survived following reintroduction into the wild, 27 individuals were released into the Junggar Pendi, northern Xinjiang in 2001, however, the partially re-wilded Przewalski's Horse still needs on-going supplementary feeding in winter and it has not formed a viable population in the wild (Jiang *et al.*, 2015; Wang *et al.*, 2016; Jiang and Zong, 2019; **Figure 22**).

15.1.2 Reindeer

China has about 700 Reindeers in total, but they are largely domesticated. Reindeers are used as farming animals and means of conveyance by the Ewenkis in China's Da Hinggan Ling forest area, and no reindeer have been found in the wild now. Reindeer has been included in the *List of Terrestrial Wildlife That Is Useful or Has Important Economic and Scientific Research Value* (2000). From 2017~2019, 179 Reindeers were introduced into Genhe, Inner Mongolia (Nei Mongol) from Finland (**Figure 23**). However, wild

图 22 野放到中国新疆准噶尔盆地的野马（蒋志刚摄）

Figure 22 The Przewalski's Horse that are released into the Junggar Pendi, Xinjiang, China (Photographed by Zhigang Jiang)

家养个体标本命名。2003 年，国际动物命名法规委员会认定 *Bos frontalis* 为该种 (包括野生个体) 的正式名称。该种名被学术界采纳 (Wilson and Reeder, 2005; Castelló, 2016; Burgin *et al*., 2018)。有研究者通过线粒体 DNA 比较研究，发现大额牛是与印度野牛、瘤牛与家牛平行演化的一种牛科物种。大额牛在原分布区已经被人类驯化饲养，而其野生种群已经灭绝。再现了"动物驯化的悲剧"(蒋志刚等, 2017)。在中国，大额牛目前仅分布在云南高黎贡山独龙河流域，故又称之为"独龙牛"（图 24）。独龙牛已经被当地牧民驯化，在夏季被放养在高海拔山区，任其自由采食。冬季牧民能召唤独龙牛集群返回低海拔地区的营地。在高黎贡山地区，被驯养的独龙牛数量仍不到 1,000 只。大额牛在野外灭绝，故评定为"野外灭绝"。

15.1.4 高鼻羚羊

20 世纪 60 年代初，高鼻羚羊在中国新疆北部灭绝。80 年代中国从国外重新引进高鼻羚羊，在甘肃建立了圈养种群。中国重新引入的高鼻羚羊种群经历了一个缓慢增长的阶段，在 30 多年的圈养过程中出现过几次种群崩溃 (Cui *et al*., 2017; Jiang *et al*., 2020)。目前，高鼻羚羊仍生活在甘肃武威濒危动物繁育中心的圈养环境中，种群增长缓慢 (图 25)。

Reindeer is extinct in China, and thus Reindeer (*Rangifer tarandus*) is classified as Extinct in the Wild.

15.1.3 Gayal

Gayal (*Bos frontalis*) was named by Aylmer Bourke Lambert in 1807 on the basis of an individual domestic specimen. *Bos frontalis* was formally adopted in 2003 by the International Commission on Zoological Nomenclature as the official name for the species, including both domestic and wild individuals, and this name is now most commonly used by academic researchers (Wilson and Reeder, 2005; Castelló, 2016; Burgin *et al.*, 2018). Through a comparative study of mitochondrial DNA, researchers found that *Bos frontalis* is a bovine species that evolved independently, in parallel with the gaur, zebu and domestic cattle. *Bos frontalis* has now been domesticated by humans across its original range

图 23 驯鹿在中国野外灭绝。2017~2019 年，内蒙古根河林业局从芬兰引入了 179 头驯鹿，图为饲养在内蒙古根河林业局的驯鹿（蒋志刚摄）

Figure 23 Reindeer has been Extinct in the Wild (EW). From 2017 to 2019, The Genhe Forestry Bureau of Inner Mongolia (Nei Mongol) introduced 179 Reindeers from Finland. In the figure, the Reindeers of the Genhe Forestry Bureau are kept in the paddock (Photographed by Zhigang Jiang)

图 24 大额牛 (*Bos frontalis*，蒋志刚摄)。大额牛已经被人类驯化。在中国，大额牛目前仅分布在云南高黎贡山独龙河流域，故也称之为"独龙牛"，数量不到 1,000 只。在野外已经没有野生大额牛。由于大额牛在野外灭绝，故评定为"野外灭绝"

Figure 24 Gayal (*Bos frontalis*, Photographed by Zhigang Jiang). Gayal is domesticated across its entire range. In China, *Bos frontalis* are only found in the Dulong (Trung) River basin in the Gaoligong Shan area of Yunnan, where the domesticated *Bos frontalis* is known as Dulong Cattle. The number of Dulong Cattle in the Gaoligong Shan area is less than 1,000 individuals. As no wild *Bos frontalis* have been found, this species is now assessed as Extinct in the Wild

15.2 区域灭绝种分析

在中国区域灭绝的哺乳动物有双角犀 (*Dicerorhinus sumatrensis*)、爪哇犀 (*Rhinoceros sondaicus*)、大独角犀 (*Rhinoceros unicornis*)、爪哇野牛 (*Bos javanicus*)、野水牛 (*Bubalus arnee*) 和豚鹿 (*Axis porcinus*)。

15.2.1 双角犀、爪哇犀、大独角犀

双角犀 (*Dicerorhinus sumatrensis*) 和爪哇犀 (*Rhinoceros sondaicus*) 在 20 世纪 50 年代前在中国云南消失。此后，中国境内再没有发现它们。20 世纪初，大独角犀 (*Rhinoceros unicornis*，图 26) 也已在中国灭绝 (蒋志刚等，2017)。2018 年，尼泊尔政府赠送给中国两对大独角犀，分别安置在上海和广州的两家动物园 (图 26)。在本次评估中，双角犀、爪哇犀、大独角犀均列为"区域灭绝"等级。

图 25 高鼻羚羊 (*Saiga tatarica*, 蒋志刚 摄)。高鼻羚羊在 19 世纪曾分布于新疆伊犁、哈密和准噶尔盆地 (Cui *et al*., 2017)。20 世纪中叶在新疆灭绝。图为 20 世纪 80 年代重引入甘肃武威濒危动物繁育中心的高鼻羚羊

Figure 25 Saiga (*Saiga tatarica*, Photographed by Zhigang Jiang). Saigas were distributed in Ili, Hami and the Junggar Pendi in Xinjiang in the 19th century (Cui *et al*., 2017). However, the Saiga was extirpated in Xinjiang in the mid-20th century. Saigas were reintroduced to China in the 1980s and now live in captivity in the Endangered Animal Breeding Center in Wuwei, Gansu as this figure shows

and its wild populations are extinct, repeating *The Tragedy of Domestication* (Jiang *et al*., 2017). In China, *Bos frontalis* are only distributed in the Dulong (Trung) River basin in the Gaoligong Shan area of Yunnan Province, where domesticated *Bos frontalis* is called Dulong Cattle (**Figure 24**). Having been domesticated by local herdsmen, Dulong Cattle are left to graze in high-altitude mountain summer pastures. For winter, herdsmen bring Dulong Cattle back to their lower-altitude camps. The number of Dulong Cattle in the Gaoligong Shan is still less than 1,000 individuals. As no wild *Bos frontalis* have been found, it is assessed as Extinct in the Wild.

15.1.4 Saiga

Saiga was extinct in northern Xinjiang, China in the early 1960s. Saiga was initially reintroduced from abroad in the 1980s and captive populations were established in Gansu. For its part, China's reintroduced Saiga population

图 26 大独角犀 (*Rhinoceros unicornis*, 蒋志刚摄) 在中国已经区域灭绝。2018 年，尼泊尔政府赠送给中国两对大独角犀，分别安置在上海和广州的两家动物园。A 图：生活在尼泊尔奇旺国家公园的大独角犀；B 图：上海野生动物园的从尼泊尔引入的大独角犀

Figure 26 Indian Rhinoceros (*Rhinoceros unicornis*, Photographed by Zhigang Jiang) has been extirpated in China. In 2018, the Nepalese government presented two pairs of this rhinoceros to China. These pairs are now hosted in two zoos, in Shanghai and Guangzhou. Photo A: *R. unicornis* in Chitwan National Park, Nepal; Photo B: *R. unicornis* in the Shanghai Wild Animal Park

15.2.2 爪哇野牛

中国科学院动物研究所蒋志刚研究团队在 2015~2016 年承担了全国第二次陆生野生动物资源调查的"爪哇野牛 (*Bos javanicus*) 调查专项"，经过系统的野外考察，在野外没有发现爪哇野牛，爪哇野牛已经在中国区域灭绝 (丁晨晨等，2018；图 27)。

15.2.3 野水牛

野水牛在西藏东南部的米什米山曾有分布记录 (王应祥，2003)。近年来，野水牛由于近交、猎杀、栖息地丧失而种群数量急剧下降，并已经与家水牛或野外的家水牛杂交。目前，未获得米什米山发现野水牛踪迹的资料。按照 Wilson 和 Mittermeier (2012) 及 Castelló (2016) 的意见，

has experienced a slow growth phase with a few crashes over more than 30 years of captivity (Cui *et al*., 2017; Jiang *et al*., 2020). Currently, Saiga is still living only in a captive setting in the Endangered Animal Breeding Center in Wuwei, Gansu Province (**Figure 25**).

15.2 Analysis of Regionally Extinct Species

Regionally Extinct Species in China include Sumatran Rhinoceros (*Dicerorhinus sumatraensis*), Javan Rhinoceros (*Rhinoceros sondaicus*), Indian Rhinoceros (*Rhinoceros unicornis*), Banteng (*Bos javanicus*) , Asian Buffalo (*Bubalus arnee*) and Hog Deer (*Axis porcinus*).

15.2.1 Sumatran Rhinoceros, Javan Rhinoceros and Indian Rhinoceros

Sumatran Rhinoceros (*Dicerorhinus sumatraensis*) and Javan Rhinoceros (*Rhinoceros sondaicus*) vanished in Yunnan, China prior to the 1950s. Since then, neither of these species have been found in the country. Indian Rhinoceros (*Rhinoceros unicornis*, **Figure 26**) was also extirpated in China toward the beginning of the last century (Jiang *et al*., 2017). In 2018, the Nepalese government presented two pairs of Indian Rhinoceros to China and these are now hosted in the Shanghai and Guangzhou zoos (**Figure 26**). In this assessment, all the three species above are listed as Regionally Extinct (RE).

15.2.2 Banteng

During the period from 2015 to 2016, the research team led by Zhigang Jiang of the Institute of Zoology, Chinese Academy of Sciences conducted "A special survey of the Banteng" for the Second National Survey on Terrestrial Wild Animals in China. Despite systematic investigation, no Banteng was found in the wild and it is now classified as Regionally Extinct (Ding *et al*., 2018; **Figure 27**).

15.2.3 Asian Buffalo

Asian Buffaloes were once recorded in Mishmi Shan, southeast area of Tibet (Xizang) of China (Wang, 2003), but they have interbred with domestic and feral domestic buffaloes and thus have suffered a sharp decline in population in recent years due to inbreeding as well as hunting and habitat loss. No current information was found on the status of Asian Buffalo in Mishmi Shan area. According to Wilson and Mittermeier (2012) and Castelló (2016), the domestic buffaloes in China are the subspecies *Bubalus arnee bubalis*, but domesticated. Since 1970s, domestic water buffalo have been set free by the farmers on the Lantau Island, Hong Kong due to industry transformation in local community. In 2006 and 2009, two feral water buffaloes were introduced from Lantau Island to the Mipu Nature Reserve in Hong Kong. The reserve managers hoped to create habitat for water birds through the grazing of buffaloes and to increase the diversity of water birds in the wetland of the reserve. In 2012, two feral buffaloes were introduced to the reserve from Kam Tin in Hong Kong from Mipu Nature Reserve. These feral buffaloes are

图 27 爪哇野牛 (*Bos javanicus*, 蒋志刚摄)。2015~2016 年，中国科学院动物研究所蒋志刚研究团队承担了全国第二次陆生野生动物资源调查的"爪哇野牛调查专项"，经过系统考察，该团队在野外没有发现爪哇野牛

Figure 27 Banteng (*Bos javanicus*, Photographed by Zhigang Jiang). Zhigang Jiang's team researchers from the Institute of Zoology, Chinese Academy of Sciences, undertook a special investigation on Banteng during the Second National Survey on Terrestrial Wild Animals in China from 2015~2016. After a systematic survey, the team found no Banteng in the wild

中国分布的家养水牛是野水牛沼泽亚种 (*Bulalus arnee bubalis*)。20 世纪 70 年代以来，由于产业转型，香港大屿山岛 (Lantau Island) 农民的家养水牛被释放到野外，成为野化家养水牛。2006 年和 2009 年，两头野化家养水牛分别从大屿山岛被引入香港米埔自然保护区，公园管理者希望通过水牛啃食改造湿地环境，引入水鸟。2012 年，该自然保护区又从香港锦田 (Kam Tin) 引入了两头野化家养水牛。但是，考虑野化家养水牛尚未完全野化为野水牛，故仍将野水牛列为区域灭绝。

not wild yet. Thus, the Asian Buffalo (*Bubalus arnee*) is listed as Regionally Extinct.

15.2.4 Hog Deer

Hog Deer (**Figure 28**) are distributed in South and Southeast Asia, with two subspecies are reported. Hog Deer is common in the grasslands of the Gangetic Plain and it was once widely distributed in northwest and northeast India. Peng *et al.* (1962) found antlers and skins of Hog Deer in local markets in Gengma and Ximeng in Yunnan, China, and estimated there were about more than 10 individuals in Gengma County. However, Dehua Yang *et al.* only found four in the same area several years later, in 1965. In the mid and late 1970s, a state-owned farm was set up in the Nanding River drainage of Mengding county and much Hog Deer habitat was reclaimed for farming land. In addition, hunting and trapping led to the extinction of the Hog Deer in Gengma County in the late 1980s. In 1989, Hog Deer was listed as Category Ⅰ State Key Protected Wild Animal. According to our recent field investigation in the species' former range in Lincang, Yunnan, in 2018, along with the monitoring results of 80 infrared cameras between October 2018 and July 2020 in Nangunhe Nature Reserve (with a total of 15,120 camera survey days, based on photographs from 82 camera stations along 39.4km of transects) and in Daxueshan Nature Reserve (13,554 camera survey days, with photographs taken at 68 camera stations along 41.1km of transects), no Hog Deer is found in the area. Currently, only the Chengdu Zoo and Shanghai Zoo have Hog Deer in captivity. Thus, this deer species is now recognized as being Regionally Extinct in China.

图 28 豚鹿（蒋志刚摄）在中国已经区域灭绝。图为上海动物园饲养的豚鹿

Figure 28 Hog Deer (*Axis porcinus*, Photographed by Zhigang Jiang) is Regionally Extinct in China. Photo shows Hog Deer kept in the Shanghai Zoo

15.2.4 豚鹿

豚鹿（图 28）分布在南亚和东南亚地区，有两个亚种。豚鹿是恒河平原草地常见的物种，曾广泛分布于印度西北和东北部。中国是豚鹿的边缘分布区，仅发现其印支亚种。彭鸿授等(1962)在云南耿马和西盟的集市上发现了豚鹿角和皮，估计当时在耿马地区有10余只豚鹿。几年后，杨德华等于1965年调查时仅发现4只豚鹿。20世纪70年代中后期，云南孟定南丁河地区开办农场，豚鹿的栖息生境被开垦，加之猎捕，80年代末期，豚鹿在云南耿马地区绝迹。1989年，豚鹿被列为国家Ⅰ级重点保护野生动物。根据我们2018年在云南临沧原豚鹿分布区的野外调查和从2018年10月至2020年7月在豚鹿原分布区云南南滚河自然保护区39.4km长样线上架设的82台红外相机的15,120个摄影日，以及在云南大雪山自然保护区41.1km长样线上架设的68台红外相机的13,554个摄影日的监测，均未发现豚鹿。目前，中国仅成都动物园与上海动物园有饲养的豚鹿种群。豚鹿在中国已经区域灭绝。

15.3 其他灭绝边缘的物种

中国还有几种动物处于灭绝的边缘，如虎 (*Panthera tigris*) 和白鱀豚 (*Lipotes vexillifer*)。

15.3.1 虎

虎 (*Panthera tigris*) 曾是中国森林生态系统的顶级捕食者。中国曾有5个虎亚种，本次评估对每一个中国分布的虎亚种的生存状况进行了评估。其中，新疆虎，即里海虎 (*P. t. virgata*) 已于1916年在中国灭绝。20世纪末，西藏墨脱曾有孟加拉虎 (*P. t. tigris*) 的报道，但在21世纪初该地孟加拉虎踪迹难觅。然而，王渊等(2019)最近又在雅鲁藏布江南岸发现了孟加拉虎的踪迹，墨脱仅存2~3只孟加拉虎游荡个体，旱季在背崩乡、墨脱镇雅鲁藏布江南岸区、格当乡金珠藏布南岸区活动。冯利民曾在西双版纳用红外相机拍摄到印支虎 (*P. t. corbetti*)。近年来该地区未见印支虎报道，如果当地仍有印支虎的话，其数量也已经十分稀少。冯利民等(2013)报道西双版纳自然保护区的尚勇子保护区可能有3只印支虎。在黑龙江省和吉林省中俄边境地区，尚有20只东北虎 (*P. t. altaica*) 迁徙游荡于中国、俄罗斯边境地区。东北虎面临的威胁有栖息地破碎化、种群分布区隔离，老爷岭南部中俄边境区域东北虎种群密度较高，但缺乏向内陆扩散的通道。中俄两国建立了东北虎保护区，加强了东北虎通道的建设和保护。中国最近还建立了东北虎豹国家公园，以期进一步加强对东北虎的保护。华南虎 (*P. t. amoyensis*) 是中国特有的虎亚种，模式标本产于福建厦门（图 29）。华南虎曾经广泛分布于湖南、江西、贵州、福建、广东、广西、安徽、浙江、湖北、四川、河南、陕西、山西等地。20世纪50年代，估计华南虎分布区有4,000余只华南虎。20世纪50~70年代，华南虎

15.3 Other Animals on the Brink of Extinction

Several other animals are also on the brink of extinction, including Tiger (*Panthera tigris*) and Baiji (*Lipotes vexillifer*).

15.3.1 Tiger

Tigers were once the top predators in China's forest ecosystems. The status of each of the tiger's five subspecies in China were assessed in this red list assessment. The *P. t. virgate*, also known as the Caspian Tiger, became extinct in China in 1916. *P. t. tigris* were reported in Mêdog, Tibet (Xizang) at the end of the 20th century, but they seemed to vanish in the area near the beginning of this century. However, Wang *et al.* (2019) recently found traces of *P. t. tigris* on the south bank of the Yarlung Zangbo River with their foot prints indicating 2~3 individuals wandering in Mêdog. Other tigers also were active in Beibeng, Gedang and Mêdog Townships on the south bank of the Yarlung Zangbo River during the dry season. Three *P. t. corbetti* were photo-trapped by Liming Feng in Shangyong Sub-Nature Reserve of Xishuangbanna Nature Reserve, Yunnan several years ago (Feng *et al.*, 2013). Around 20 individuals of *P. t. altaica* still roam along the China-Russia border in Heilongjiang and Jilin Provinces. Threats to the *P. t. altaica* include habitat fragmentation and the isolation of its sub-populations. The population density of *P. t. altaica* in the southern part of the Laoyeling Mountains is relatively high, but there are no dispersal corridors connecting them with areas further inland. China and Russia have established nature reserves focused on *P. t. altaica* and have strengthened the construction of ecological corridors to link their fragmented habitats to aid in their protection. China has also recently established the Northeast Tiger and Leopard National Park to further strengthen protection for both of these species. The *P. t. amoyensis* is an endemic subspecies in the country, with its type specimen coming from Xiamen, Fujian Province (**Figure 29**). The *P. t. amoyensis* was once widely distributed in Hunan, Jiangxi, Guizhou, Fujian, Guangdong, Guangxi, Anhui, Zhejiang, Hubei, Sichuan, Henan, Shaanxi, Shanxi and other places in China. In the 1950s, it was estimated that there were more than 4,000 individuals in its range. From 1950s to 1970s, however, the *P. t. amoyensis* was hunted as a pest animal and more than 3,000 of them were culled in 30 years. Due to habitat loss, human-tiger conflict and disrupted food chains as a consequence of over-hunting of tiger prey, no wild *P. t. amoyensis* individual has been found in its original range since 1980. From 1990 to 2001, the State Forestry Administration carried out several special expeditions focused on *P. t. amoyensis* in its original range in southern parts of the country, but they found no solid evidence of its continued existence in the wild. Traces of suspected activities of *P. t. amoyensis* were scattered and there were gaps in the spatio-temporal distributions of the evidence chain. Tilson *et al.* (2004) thus concluded that the *P. t. amoyensis* is at least functionally extinct in the wild. Although a wild population of the *P. t. amoyensis* cannot be entirely excluded, the investigations of an expert group from the State Forestry

图29 华南虎 (*Panthera tigris amoyensis*, 蒋志刚摄)。虎的华南虎亚种已经野外灭绝。为拯救虎, 实现2010年 (虎年) 圣彼得堡虎峰会设定的到下一个虎年 (2022年), 野生虎种群数量与栖息地面积翻番的战略目标, 中国投入了大量的人力与物力保护野生虎

Figure 29 South China Tiger (*Panthera tigris amoyensis*, Photographed by Zhigang Jiang). The South China Tiger subspecies is extinct in the wild. In order to achieve the strategic goal of doubling both the population size and habitat area of wild tigers by 2022 (Year of the Tiger), a target set in 2010 (previous Year of the Tiger) at the Tiger Summit held in St. Petersburg of Russia, China has invested a great amount of human and financial resources for the protection of wild tigers

被当成"害兽"捕杀。在30年中约有3,000只华南虎被猎杀。由于过度捕杀和栖息地丧失等, 1980年后, 在华南虎分布区再没有发现野生华南虎。1990~2001年, 国家林业局曾在原华南虎分布区开展过多次华南虎专项调查, 均未发现华南虎存在的确切证据。所收集到的疑似华南虎的活动痕迹也十分零散, 在空间和时间上出现了极大的断裂。Tilson *et al.* (2004) 据此提出了华南虎已经在野外功能性灭绝的观点。虽然目前还不能认定华南虎野外种群已功能性灭绝, 但2011年笔者带领国家林业局华南虎重引入专家组对华南虎分布区的江西、湖南、湖北、福建、重庆和广东六省 (直辖市) 的考察, 也没有发现野生华南虎的踪迹。目前, 我国人工饲养的华南虎已经有100余只, 分散在全国十几家动物园等单位, 受近亲繁殖、管理水平和饲养条件等因素的影响, 其种群发展缓慢。故华南虎亚种为"野外灭绝"。综上所述, 在本次评估中, 虎的生存状况仍定为"极危"。

Figure 30 Baiji (*Lipotes vexillifer*, photo used with permission from the Institute of Hydrobiology, Chinese Academy of Sciences). The Baiji mainly lived in the Changjiang River. The Baiji is facing a survival crisis due to water pollution and noise from shipping vessels, along with injuries from propellers and fishing gear. No Baiji has been found in the Changjiang River for several years, and it is therefore generally considered to be extinct

Administration in 2011 in its former range in Jiangxi, Hunan, Hubei, Fujian, Chongqing and Guangdong found no traces of the tiger in the wild. There are, however, over 100 individuals in captivity, scattered amongst dozens of zoos and other breeding centers across the country. Yet, due to inbreeding and other factors, the captive population of this subspecies is growing but slowly and thus, it is classified as Extinct in Wild. Collectively, the status of the Tiger (*Panthera tigris*) is classified as Critically Endangered in this assessment.

15.3.2 Baiji

The IUCN listed the Baiji (*Lipotes vexillifer*, **Figure 30**) as a Data Deficient (DD) species in 1970s. In 1979, China conducted the first survey on the population of Baiji in the Changjiang River, and it was estimated that there were only about 400 Baijis in the river at that time (Zhou, 1982). In 1986,

15.3.2 白鱀豚

20世纪70年代，IUCN将白鱀豚(图30)列为"数据缺乏"物种。1979年，中国首次开展了长江白鱀豚种群考察，估算当时长江白鱀豚只有400头左右(周开亚，1982)。1986年，野外考察估计长江白鱀豚种群数量少于300头。IUCN将白鱀豚列为"濒危"等级物种。90年代，估计长江白鱀豚种群数量少于200头。1996年，IUCN将白鱀豚列为"极危"等级物种。1997年，野外考察中仅发现23头白鱀豚，估计长江白鱀豚种群数量少于50头。1998年，考察中所发现的白鱀豚数量只剩7头，估计种群数量少于15头。2006年12月4日，为时六周的长江淡水豚类考察在长江中没有发现白鱀豚。2006年12月13日，专家们宣布白鱀豚物种可能已经灭绝，即使还有少数白鱀豚个体存在，也不能保证种群成功地繁衍，即白鱀豚已经"功能性灭绝"(Turvey et al., 2007)。但是，2007年8月19日，安徽铜陵一位市民在铜陵淡水豚类国家级自然保护区江段发现了一头疑似白鱀豚的动物。2011年7月6日，渔民在长江中发现了3头白鱀豚。这些事实说明长江中仍有极少数白鱀豚生存。但是，很长时间过去了，再也没有在长江中发现白鱀豚实体。尽管本次评估将白鱀豚列为"极危"，但是白鱀豚有可能已经灭绝。

16 受威胁物种分析

除野外灭绝和区域灭绝的物种之外，本次评估中受威胁(极危、濒危、易危)哺乳动物共计181种。中国哺乳动物受威胁物种在科级、属级分类单元之间的分布为非随机分布，各个哺乳动物科的受威胁比例相差很大。其中，儒艮科Dugongidae(1种)、懒猴科Lorisidae(2种)、长臂猿科Hylobatidae(8种)、熊科Ursidae(4种)、大熊猫科Ailuropodidae(1种)、小熊猫科Ailuridae(1种)、林狸科Prionodontidae(1种)、露脊鲸科Balaenidae(1种)、白鱀豚科Lipotidae(1种)、鼠海豚科Phocoenidae(3种)、骆驼科Camelidae(1种)、鼷鹿科Tragulidae(1种)、河狸科Castoridae(1种)、象科Elephantidae(1种)、麝科Moschidae(6种)等的所有种类全部为受威胁种，受威胁比例为100%。獴科Herpestidae(3种，2种受威胁)、猫科Felidae(13种，12种受威胁)、猴科Cercopithecidae(19种，17种受威胁)、灵猫科Viverridae(8种，6种受威胁)的受威胁比例达60%或以上。牛科Bovidae(34种，20种受威胁)、鹿科Cervidae(22种，12种受威胁)、鼬科Mustelidae(20种，13种受威胁)、假吸血蝠科Megadermatidae(2种，1种受威胁)、狐蝠科Pteropodidae(11种，6种受威胁)、海狮科Otariidae(2种，1种受威胁)、睡鼠科Gliridae(2种，1种受威胁)的受威胁种比例均在50%~60%。而豪猪科Hystricidae(3种)、刺山鼠科Platacanthomyidae(4种)、树鼩科Tupaiidae(1种)、鞘尾蝠科Emballonuridae(2种)、犬吻蝠科Molossidae(3种)、猪科Suidae(1种)、灰鲸科Eschrichtiidae(1种)、喙鲸科Ziphiidae(6种)和跳鼠科Dipodidae(18种)没有受威胁种(**表8**)。

the number of Baiji in the Changjiang River was estimated to be fewer than 300 individuals. Based on this low population estimate, IUCN listed the dolphin as Endangered (EN). In the 1990s, the population of Baiji was estimated to be fewer than 200 individuals, and in 1996 IUCN listed it as Critically Endangered (CR). Only 23 Baijis were discovered during a field survey in 1997, and the total population in the entire Changjiang River was estimated to be fewer than 50 individuals. Only 7 individuals were discovered during the 1998 survey, with the total population estimated at fewer than 15 individuals. A 6 weeks survey that started on December 4, 2006 found no Baiji in the Changjiang River, and on December 13, 2006 experts announced that the Baiji may has become extinct. Even if a few individuals still live in the Changjiang River, it is unsure that the population could successfully reproduce, therefore the Baiji was considered as Functionally Extinct (Turvey *et al.*, 2007). However, a local resident in Tongling, Anhui Province, saw a Baiji-like dolphin in the Tongling Freshwater Dolphin National Nature Reserve on August 19, 2007, and three Baijis were discovered by fishermen in the Changjiang River on July 6, 2011. This indicates that there may be still a few Baiji dolphins living in the wild. However, none has been found in the Changjiang River since 2011. Although classified as Critically Endangered, the Baiji may in reality be extinct.

16 Analysis of Threatened Species

In addition to species that are Extinct in the Wild (EW) and Regionally Extinct (RE), a total of 181 mammal species are classified as threatened (either Critically Endangered, Endangered or Vulnerable) in this assessment. The distribution of threatened species of China's mammals is non-random among taxa, the proportion of threatened species in each family varying greatly. Among them, all of the species in Dugongidae (1 species), Lorisidae (2 species), Hylobatidae (8 species), Ursidae (4 species), Ailuropodidae (1 species), Ailuridae (1 species), Prionodontidae (1 species), Balaenidae (1 species), Lipotidae (1 species), Phocoenidae (3 species), Camelidae (1 species), Tragulidae (1 species), Castoridae (1 species), Elephantidae (1 species) and Moschidae (6 species), *etc.*, are threatened (the threatened ratio is 100%). The proportions of threatened species in Herpestidae (3 species, 2 threatened), Felidae (13 species, 12 threatened), Cercopithecidae (19 species, 17 threatened) and Viverridae (8 species, 6 threatened) all reach or are higher than 60%. 50%~60% of the species are threatened in Bovidae (34 species, 20 threatened), Cervidae (22 species, 12 threatened), Mustelidae (20 species, 13 threatened), Megadermatidae (2 species, 1 threatened), Pteropodidae (11 species, 6 threatened), Otariidae (2 species, 1 threatened) and Gliridae (2 species, 1 threatened). At the other end of the spectrum, no species are considered threatened in Hystricidae (3 species), Platacanthomyidae (4 species), Tupaiidae (1 species), Emballonuridae (2 species), Molossidae (3 species), Suidae (1 species), Eschrichtiidae (1 species), Ziphiidae (6 species) and Dipodidae (18 species) (**Table 8**).

表 8　中国哺乳动物各科的濒危比例

科名	拉丁科名	物种数目	特有种数	特有种比例 /%	受威胁数	受威胁比例 /%
猬科	Erinaceidae	10	5	50	1	10
鼹科	Talpidae	18	10	56	4	22
鼩鼱科	Soricidae	60	17	28	6	10
树鼩科	Tupaiidae	1	0	0	0	0
狐蝠科	Pteropodidae	11	0	0	6	55
鞘尾蝠科	Emballonuridae	2	0	0	0	0
假吸血蝠科	Megadermatidae	2	0	0	1	50
菊头蝠科	Rhinolophidae	21	6	29	2	10
蹄蝠科	Hipposideridae	9	0	0	2	22
犬吻蝠科	Molossidae	3	0	0	0	0
蝙蝠科	Vespertilionidae	91	22	24	8	9
懒猴科	Lorisidae	2	0	0	2	100
猴科	Cercopithecidae	19	6	32	17	89
长臂猿科	Hylobatidae	8	1	13	8	100
鲮鲤科	Manidae	3	0	0	3	100
犬科	Canidae	8	0	0	1	13
熊科	Ursidae	4	0	0	4	100
大熊猫科	Ailuropodidae	1	1	100	1	100
小熊猫科	Ailuridae	1	0	0	1	100
海狮科	Otariidae	2	0	0	1	50
鼬科	Mustelidae	20	0	0	13	65
海豹科	Phocidae	3	0	0	2	67
灵猫科	Viverridae	8	0	0	6	75
林狸科	Prionodontidae	1	0	0	1	100
獴科	Herpestidae	3	0	0	2	67
猫科	Felidae	13	1	7.7	12	92
儒艮科	Dugongidae	1	0	0	1	100
象科	Elephantidae	1	0	0	1	100
犀科	Rhinocerotidae	3	0	0	0	0
马科	Equidae	3	0	0	1	33
猪科	Suidae	1	0	0	0	0
骆驼科	Camelidae	1	0	0	1	100
鼷鹿科	Tragulidae	1	0	0	1	100
麝科	Moschidae	6	2	33	6	100
鹿科	Cervidae	22	4	18	12	55
牛科	Bovidae	34	7	21	20	59
露脊鲸科	Balaenidae	1	0	0	1	100
灰鲸科	Eschrichtiidae	1	0	0	0	0
须鲸科	Balaenopteridae	7	0	0	3	43
白鱀豚科	Lipotidae	1	1	100	1	100
抹香鲸科	Physeteridae	3	0	0	1	33
喙鲸科	Ziphiidae	6	0	0	0	0
鼠海豚科	Phocoenidae	3	1	33	3	100
恒河豚科	Platanistidae	1	0	0	1	100
海豚科	Delphinidae	16	0	0	1	6
松鼠科	Sciuridae	50	9	18	5	10
河狸科	Castoridae	1	0	0	1	100
仓鼠科	Cricetidae	66	22	33	2	3
鼠科	Muridae	64	13	20	4	6
刺山鼠科	Platacanthomyidae	4	3	75	0	0
鼹型鼠科	Spalacidae	13	6	46	1	8
睡鼠科	Gliridae	2	1	50	1	50
跳鼠科	Dipodidae	18	1	6	0	0
豪猪科	Hystricidae	3	0	0	0	0
鼠兔科	Ochotonidae	31	15	48	7	23
兔科	Leporidae	12	2	17	2	17
总计		700	156	22	181	26

Table 8 The ratio of threatened species in each family of Mammalia

Family Name in Latin	No. of Species	No. of Endemic	Ratio of Endemic/%	No. of Threatened	Ratio of Threatened/%
Erinaceidae	10	5	50	1	10
Talpidae	18	10	56	4	22
Soricidae	60	17	28	6	10
Tupaiidae	1	0	0	0	0
Pteropodidae	11	0	0	6	55
Emballonuridae	2	0	0	0	0
Megadermatidae	2	0	0	1	50
Rhinolophidae	21	6	29	2	10
Hipposideridae	9	0	0	2	22
Molossidae	3	0	0	0	0
Vespertilionidae	91	22	24	8	9
Lorisidae	2	0	0	2	100
Cercopithecidae	19	6	32	17	89
Hylobatidae	8	1	13	8	100
Manidae	3	0	0	3	100
Canidae	8	0	0	1	13
Ursidae	4	0	0	4	100
Ailuropodidae	1	1	100	1	100
Ailuridae	1	0	0	1	100
Otariidae	2	0	0	1	50
Mustelidae	20	0	0	13	65
Phocidae	3	0	0	2	67
Viverridae	8	0	0	6	75
Prionodontidae	1	0	0	1	100
Herpestidae	3	0	0	2	67
Felidae	13	1	7.7	12	92
Dugongidae	1	0	0	1	100
Elephantidae	1	0	0	1	100
Rhinocerotidae	3	0	0	0	0
Equidae	3	0	0	1	33
Suidae	1	0	0	0	0
Camelidae	1	0	0	1	100
Tragulidae	1	0	0	1	100
Moschidae	6	2	33	6	100
Cervidae	22	4	18	12	55
Bovidae	34	7	21	20	59
Balaenidae	1	0	0	1	100
Eschrichtiidae	1	0	0	0	0
Balaenopteridae	7	0	0	3	43
Lipotidae	1	1	100	1	100
Physeteridae	3	0	0	1	33
Ziphiidae	6	0	0	0	0
Phocoenidae	3	1	33	3	100
Platanistidae	1	0	0	1	100
Delphinidae	16	0	0	1	6
Sciuridae	50	9	18	5	10
Castoridae	1	0	0	1	100
Cricetidae	66	22	33	2	3
Muridae	64	13	20	4	6
Platacanthomyidae	4	3	75	0	0
Spalacidae	13	6	46	1	8
Gliridae	2	1	50	1	50
Dipodidae	18	1	6	0	0
Hystricidae	3	0	0	0	0
Ochotonidae	31	15	48	7	23
Leporidae	12	2	17	2	17
Total	700	156	22	181	26

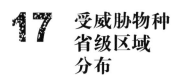

17 受威胁物种省级区域分布

中国各省级区域受威胁哺乳动物在空间上的分布也是不均匀的（表9）。西南地区哺乳动物受威胁比例高于其他地区。西藏（受威胁种数59种）、重庆（31种）和云南（91种）的受威胁哺乳动物比例高于全国平均值。多数省级区域的受威胁哺乳动物物种占本省级区域哺乳动物数目的20%~30%。湖北（17种）、福建（22种）、河北（12种）、宁夏（12种）、香港（7种）、北京（7种）、天津（2种）、山东（1种）和澳门（0种）等受威胁哺乳动物比例低于20%。

表9 中国哺乳动物受威胁物种按省级区域统计

省级区域	总种数	受威胁种数	受威胁种数占省级行政区种数的比例/%	受威胁种数占全国受威胁种数比例/%
西藏	178	59	33	33
重庆	106	31	29	17
云南	318	91	29	50
青海	97	25	26	14
吉林	83	20	24	11
广西	152	37	24	20
黑龙江	91	22	24	12
上海	29	7	24	4
新疆	147	36	24	20
江西	97	23	24	13
内蒙古	128	30	23	17
江苏	61	14	23	8
海南	92	21	23	12
四川	228	51	22	28
甘肃	177	39	22	22
湖南	97	21	22	12
广东	134	29	22	16
辽宁	67	14	21	8
安徽	92	19	21	10
浙江	92	19	21	10
贵州	158	32	20	18
河南	64	13	20	7
山西	74	15	20	8
台湾	104	21	20	12
陕西	147	29	20	16
湖北	88	17	19	9
福建	115	22	19	12
河北	65	12	18	7
宁夏	69	12	17	7
香港	51	7	14	4
北京	52	7	13	4
天津	21	2	10	1
山东	34	1	3	1
澳门	12	0	0	0

17 Distribution of Threatened Species in Provincial Region

Spatially, the distribution of threatened mammals in provincial level district are also uneven (**Table 9**). The proportion of threatened mammals is higher in southwest China than in other regions of the country. The proportions of threatened mammals in Tibet (Xizang) (59 threatened species), Chongqing (31) and Yunnan (91) are higher than the national average. Threatened mammal species in most provincial level district ranges between 20% and 30%. For their part, the proportions of threatened mammal species in Hubei (17), Fujian (22), Hebei (12), Ningxia (7), Hong Kong (7), Beijing (7), Tianjin (2), Shandong (1), Macao (0), *etc.* are all less than 20%.

Table 9 Statistics of Chinese threatened mammal species by provincial region

Provincial Region	Total No. Species	No. Threatened Species	Ratio of Threatened Species of the Provincial Total/%	Ratio of Threatened Species of the National Total/%
Xizang (Tibet)	178	59	33	33
Chongqing	106	31	29	17
Yunnan	318	91	29	50
Qinghai	97	25	26	14
Jilin	83	20	24	11
Guangxi	152	37	24	20
Heilongjiang	91	22	24	12
Shanghai	29	7	24	4
Xinjiang	147	36	24	20
Jiangxi	97	23	24	13
Inner Mongolia (Nei Mongol)	128	30	23	17
Jiangsu	61	14	23	8
Hainan	92	21	23	12
Sichuan	228	51	22	28
Gansu	177	39	22	22
Hunan	97	21	22	12
Guangdong	134	29	22	16
Liaoning	67	14	21	8
Anhui	92	19	21	10
Zhejiang	92	19	21	10
Guizhou	158	32	20	18
Henan	64	13	20	7
Shanxi	74	15	20	8
Taiwan	104	21	20	12
Shaanxi	147	29	20	16
Hubei	88	17	19	9
Fujian	115	22	19	12
Hebei	65	12	18	7
Ningxia	69	12	17	7
Hong Kong	51	7	14	4
Beijing	52	7	13	4
Tianjin	21	2	10	1
Shandong	34	1	3	1
Macao	12	0	0	0

18 受威胁原因分析

物种灭绝的主要原因是生境丧失和退化（图31）。全新世以来，人类活动改变了土地性质，使野生哺乳动物生境转变为农田林地、人工种植园和建筑用地。人类居住地和基础设施建设、路网和管网建设造成野生哺乳动物生境破碎化，甚至生境丧失，导致野生哺乳动物种群数量减少甚至消失。"生境丧失"名列受威胁哺乳动物的致危因子之首，是276种受威胁哺乳动物的威胁因子，占所有受威胁哺乳动物的所有致危因子的25%。"人工利用"和"人类干扰"名列受威胁哺乳动物致危因子的第2位，分别占所有受威胁哺乳动物的所有致危因子的19%。"未知因素"占所有受威胁哺乳动物的所有致危因子的14%。"自然灾害"和"种群波动"分别占所有受威胁哺乳动物的所有致危因子的6%与7%。在所有受威胁哺乳动物的所有致危因子中，"环境污染"、"气候变化"、"人为毒杀"、"疾病"和"意外死亡"占所有致危因子的比例均在4%以下（图31）。图31说明目前中国哺乳动物的濒危主要是人类活动造成的，而物种本身的进化原因对物种的影响基本可以忽略不计。

针对不同主要生境中的哺乳动物种及生境中受威胁的物种数目统计分析，笔者发现："森林区域（包括灌丛）"是哺乳动物的主要生境，有576种哺乳动物利用森林区域作为生境，其中受威胁物种166种，受威胁比例29%；188种哺乳动物利用"草原（包括草甸）"作为生境，其中受威胁物种45种，受威胁比例24%；159种哺乳动物（其中多为啮齿类）生活在"人类景观（包括农田、人工建筑等）"生境，生活在人类景观之中的哺乳类多数已经适应人类社会，成为"伴人动物"，人类景观中有受威胁物种25种，受威胁比例16%；74种哺乳动物（多为翼手目种类）利用"洞穴"作为生境，其中受威胁物种11种，受威胁比例15%；

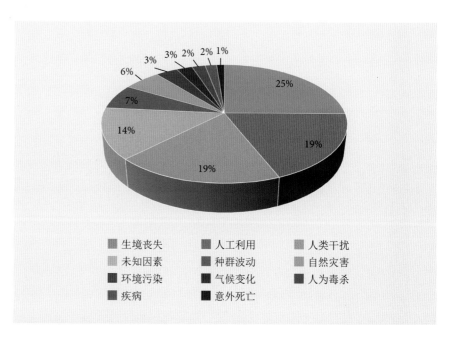

图31 中国哺乳动物濒危原因分析

18 Analysis of Threats

The main causes of species extinction are habitat loss and habitat degradation (**Figure 31**). Since the Holocene, human activities have changed the landscape, transforming habitats of wild mammals into farmlands, artificial plantations and construction lands. Furthermore, the habitat of wild mammals is fragmented by human settlement, infrastructure networks and pipelines, resulting in the parcelization of land and a loss of mammal habitats, consequently leading to reductions or even extinctions of the populations of wild mammals. "Habitat loss" is ranked as the first among all of the threats to wildlife. It is a threat to 276 of the threatened mammal species, accounting for 25% of all of the noted threats. "Human exploitation" and "Human interference" both ranked second place, accounting for 19% of all the noted threats. "Unknown factor" accounted for a further 14% of threats, while "Natural catastrophe" and "Population fluctuation" accounted for 6% and 7%, respectively. For their parts, "Environment pollution", "Climate change", "Poisoning", "Disease", "Accidental death" each accounted for less than 4% of the threats (**Figure 31**). This figure shows that mammals in China are mainly threatened by human activities, and that natural or evolutionary causes of species loss may be ignored as primary threats.

Based on the analyses of the number of species and threatened species inhabit different habitat types, the author find: "Forested area (including shrubland)" is the main habitat of mammals. Of the 576 mammal species that use "Forest", 166 species are threatened, the threatened ratio is 29%; of the 188 mammal species that use "Grassland (including meadow)" as habitat, 45 species are threatened, the threatened ratio is 24%; of the 159 mammal species that use "Anthropogenic landscape (including farmland, man-made buildings, *etc.*)" as habitat, 25 species are threatened, the threatened ratio is 16%; of the 74 mammal species that use "Cave" as habitat, 11 species are threatened, the

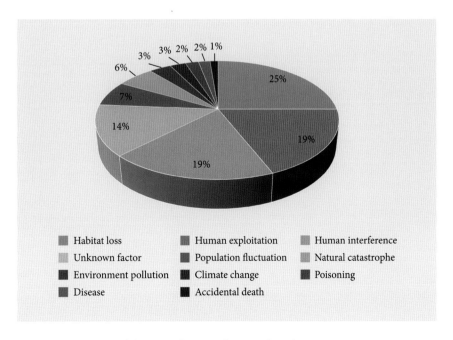

Figure 31 Analysis of threats to threatened mammal in China

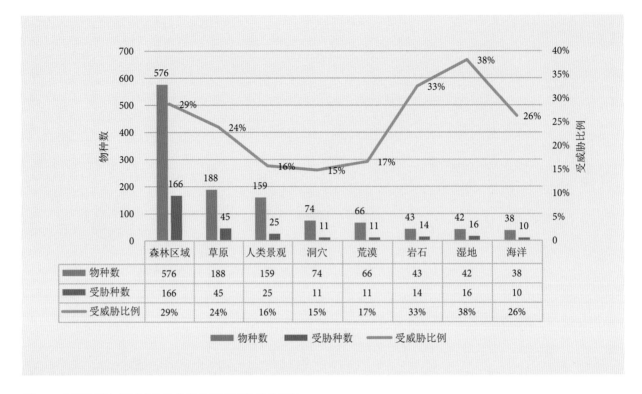

图 32　主要生境中的哺乳动物物种数和受威胁种数统计

66 种哺乳动物生活在干旱半干旱荒漠地区，利用"荒漠（包括荒漠化草原）"作为生境，其中受威胁物种 11 种，受威胁比例 17%；43 种哺乳动物利用"岩石（包括高海拔流石滩、冰缘、裸岩悬崖等）"作为生境，其中受威胁物种 14 种，受威胁比例 33%；42 种哺乳动物利用"湿地（包括内陆水体、沼泽、河谷、滩涂等）"作为生境，其中受威胁物种 16 种，受威胁比例 38%；38 种哺乳动物利用"海洋"作为生境，其中受威胁物种 10 种，受威胁比例 26%（图 32）。

以上的分析反映了一个现象：尽管在一些生境中物种多样性较低，但其受威胁的物种比例却较高，如"岩石""湿地"等的哺乳类物种数相对较少，然而受威胁物种比例却分别达到了 33% 和 38%。

中国自西向东有三级地理台阶。第一级地理台阶是青藏高原，平均海拔 4,000 多米；第二级地理台阶平均海拔 1,000~2,000m，包括蒙古高原、云贵高原和黄土高原；第三级地理台阶主要由中国东部海拔 200m 以下的平原组成，点缀着一些丘陵和低山。中国哺乳动物种类多分布在中国第二级地理台阶，以分布在海拔 500~1,000m、1,000~1,500m 和 1,500~2,000m 生境中的哺乳动物种类为多，其中，受威胁物种比例随着海拔升高而增高，分布在海拔 500~1,000m、1,000~1,500m、1,500~2,000m 生境中哺乳动物种类受威胁比例依次为栖息在该海拔区间的哺乳动物种数的 24%、24% 和 27%（图 33）。然而，出乎意料的是，高海拔地区的哺乳动物种数虽少，但是受威胁哺乳动物种类比例却很高。生活在海拔

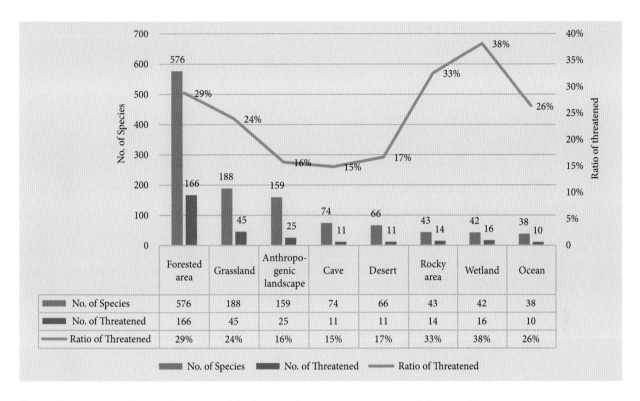

Figure 32 Statistics of mammal species and the threatened species in various major habitats in China

threatened ratio is 15%; of the 66 mammal species that use "Desert (including the arid and semi-arid areas with sparse grasses)" as habitat, 11 species are threatened, the threatened ratio is 17%; of the 43 mammal species that use "Rocky area (including the high elevation stream rocky beach, ice edge and bare rock cliff)" as habitat, 14 species are threatened, the threatened ratio is 33%; of the 42 mammal species that use "Wetland (including inland waterbodies, swamps, river valleys and coasts)" as habitat, 16 species are threatened, threatened ratio is 38%; of the 38 mammal species that use "Ocean" as habitat, 10 species are threatened, the threatened ratio is 26% (**Figure 32**).

The above analysis also reveals a fact: although the number of species is relatively small in some habitats, the proportion of threatened species is higher, such as "Rocky area" and "Wetland" area, where the number of mammalian species is relatively small, but the proportion of threatened species reaches 33% and 38%, respectively.

China has a three-stepped terraced terrain topography. The first geographic terrace is the Qinghai-Tibet (Xizang) Plateau, with an average elevation of over 4,000m. The second geographic terrace is averaging 1,000~2,000m above sea level, including the Mongolian Plateau, the Yunnan-Guizhou Plateau and the Loess Plateau. The third geographic terrace is composed mainly of plains below 200m above sea level, dotted with some hills and low mountains in eastern China. China's mammal species are mostly distributed in the second terrace (of three terraces). Most mammal species live in habitats distributed in altitudes between 500~1,000m, 1,000~1,500m, 1,500~2,000m above sea level, in which the proportion of threatened species increases with altitude

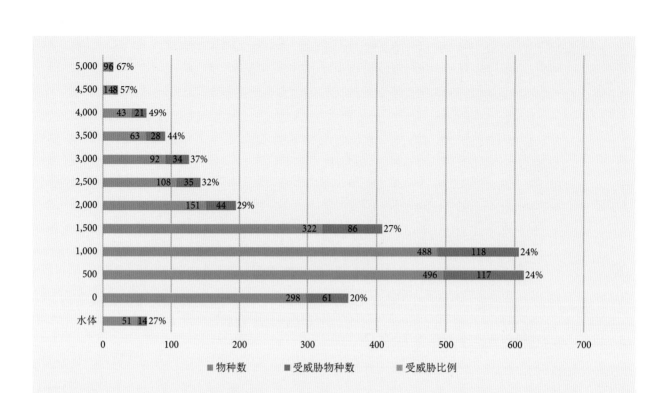

图 33　不同海拔生境中的哺乳动物与受威胁哺乳动物物种数

3,000~3,500m、3,500~4,000m、4,000~4,500m、4,500~5,000m 及海拔 5,000m 以上生境中的受威胁哺乳动物种数分别占该海拔区间哺乳动物种数的 37%、44%、49%、57% 和 67%（图 33）。

19　与中国历年红皮书、红色名录评估结果的比较

中国的濒危物种红色名录研究起步较早，在 1998 年即出版了《中国濒危动物红皮书》。2004 年和 2009 年，分别出版了《中国物种红色名录（第一卷）红色名录》和《中国物种红色名录（第二卷）脊椎动物卷》（汪松和解焱，2004，2009）。2015 年，环境保护部与中国科学院联合发布了"中国生物多样性红色名录——脊椎动物卷"。历次评估中，被评定的中国哺乳动物数目从 1998 年的 137 种，上升到 2004 年的 579 种、2015 的 673 种，到这次评估的 700 种，评估的受威胁物种等级从 1998 年的"野生绝迹 (Ex)"、"国内绝迹 (Et)"、"濒危 (E)"、"易危 (V)"、"稀有 (R)"、"未定 (I)" 等级到应用《IUCN 受威胁物种红色名录标准》来评估全部受威胁物种红色名录等级（图 34）。

本次评估的 2021 版红色名录哺乳动物部分与 2015 版比较，有很多变化（表 10）。

(1) 2021 版反映了中国哺乳动物分类系统的新变化。两版共评定 732 种，其中 60 种为新增的新种、新记录种，34 种为被降为亚种、分

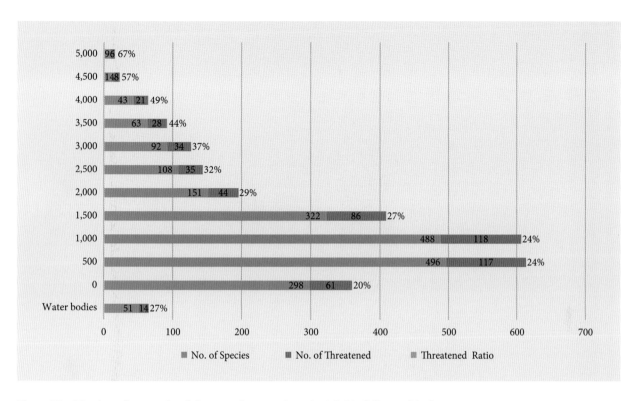

Figure 33 Number of mammal and threatened mammal species inhabit different altitudes

as 24%, 24% and 27% respectively (**Figure 33**). Surprisingly, although fewer mammal species live at high altitudes, the proportion of these species that are classified as threatened is significantly higher, with the number of threatened species living at altitudes of 3,000~3,500m, 3,500~4,000m, 4,000~4,500m, 4,500~5,000m and over 5,000m above sea level accounting for 37%, 44%, 49%, 57% and 67% of the total mammal species living at those altitude ranges, respectively (**Figure 33**).

19 Comparison with the Results of Historical Red List Assessments in the Country

The study on red list of threatened species started early in China. In 1998, the *China Red Data Book of Endangered Animals* was published. Song Wang *et al.* edited the *China Species Red List Vol I Red List* and the *China Species Red List Vol II Vertebrates* in 2004 and 2009, respectively (Wang and Xie, 2004, 2009). "China's Red List of Biodiversity: Volume of Vertebrates" was officially released by the Ministry of Environment Protection and China Academy of Sciences in 2015. The number of mammal species assessed in those China's red list studies increased from 137 in 1998, to 579 in 2004, 673 in 2015, and 700 in this assessment, the threatened species categories changed from using categories of "Extinct (Ex)", "Extirpated (Et)", "Endangered (E)", "Vulnerable (V)", "Rare (R)", "Indeterminate (I)" in 1998 to using the *IUCN Red List Categories and Criteria* (**Figure 34**).

There are many changes in the 2021 edition red list of China's mammals, compared to the 2015 edition (**Table 10**). Six main changes are presented here.

(1) The 2021 edition reflects new changes in the mammal taxonomy in China. A total of 732 species were assessed in the two editions, among which 60 species had new records and 34 species were removed for reasons such as

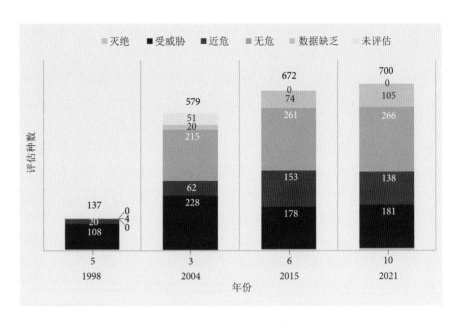

图 34 中国历年红皮书与红色名录哺乳类部分评估结果比较

布区待核实、被合并等原因而被剔除的物种。

(2) 2021 版评估了 700 种哺乳动物，而 2015 版评估了 672 种，2021 版比 2015 版多评估了 28 物种。

(3) 首次开展了一些重要物种，如虎、金钱豹、普氏原羚等的种下单元的濒危等级评定。

(4) 2021 版与 2015 版相比较，其中 537 种（占 2021 版评估物种的 77%）在两版红色名录中的濒危等级相同；在过去 5 年中，中国哺乳动物生存发生了一些变化：2021 版中 35 种 (5%) 的濒危状况比在 2015 版的濒危状况上升，如从无危上升到近危、易危、濒危或极危，从近危上升到易危、濒危或极危，从易危上升到濒危或极危等；2021 版中 35 种 (5%) 的濒危状况比在 2015 版的濒危状况下降，如从极危下降到濒危，从濒危下降到易危、近危等。

(5) 2015 版中 24 种 (3%) 的濒危状况，经过再评定，在 2021 版的濒危状况评估为数据缺乏。

(6) 总体来看，2021 版的受威胁物种（极危、濒危和易危）为 181 种，2015 版为 178 种；2021 版的受威胁物种比例最低估计值约为 26%，最佳估计值约为 38%，最高估计值约为 41%；2015 版最低估计值约为 27%，最佳估计值约为 30%，最高估计值约为 41%（计算方法见下方框中的描述）。然而，根据最新的野外考察结果，与 2015 版比较，2021 版中的区域灭绝、野外灭绝种从 6 种增加到 10 种。尽管区域灭绝、野外灭绝的种并没有灭绝，然而，必须将区域灭绝种从分布国重新引入中国、恢复野外灭绝的种的野生种群提到生物多样性保护的议事日程。

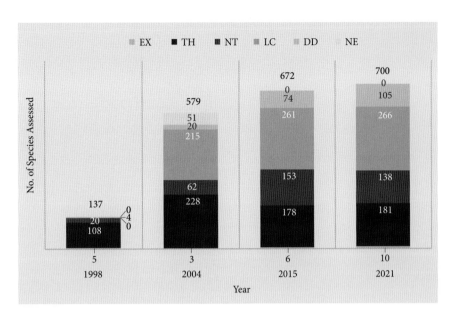

Figuer 34 Comparison of the evaluation results of mammal parts in the red list and red data book over the years in China

being degraded to subspecies, geographic distributions to be further verified, merging of species, *etc.*

(2) Seven hundred (700) mammal species were assessed in the 2021 edition, while 672 were assessed the 2015 edition; an increase of 28 species.

(3) For the first time, the status of several infra-specific taxa were assessed in important species such as Tiger, Leopard and Przewalski's Gazelle.

(4) Compared with the 2015 edition, 537 species (accounting for 77% of the species assessed in the 2021 edition) have retained the same endangerment status in both editions of the red lists. However, there have been some significant changes in the status of mammals in China during the past five years. 35 species (5%) in the 2021 edition have been elevated from their 2015 status, such as from LC to NT, VU, EN or CR; from NT to VU, EN or CR; and from VU to EN or CR, *etc.* Additionally, in the 2021 edition, 35 species (5%) are deemed to be safer now as compared to five years ago, therefore their status has been down-listed from CR to EN, from EN to VU, or from VU to NT.

(5) Compared to the 2015 edition, 24 species (3%) were reassessed as Data Deficient (DD) in the 2021 edition due to lack of adequate population or habitat data for these species.

(6) In total, 181 species of mammals are listed as threatened (Critically Endangered, Endangered, or Vulnerable) in 2021, compared with 178 assessed as threatened in 2015. The lower limit for the proportion of threatened species is 26%, the best estimate is about 38%, and the upper estimate is about 41% in 2021, whereas in the 2015 edition, the lower limit was about 27%, the best estimate was about 30%, and the upper estimate was about 41% (The calculation method is described in the box below). At the same time, according to the latest findings, the number of species assessed as Regionally Extinct (RE) and Extinct in the Wild (EW) in the 2021 edition increased to 10 mammal species, compared with only 6 species in the 2015 edition. Although

濒危物种的最低、最佳和最高估计 (IUCN, 2020)

受威胁物种的比例的核算报告只针对评价较完全的群体进行（即 > 80% 的物种得到评价）。此外，报告的每一组受威胁物种的百分比被作为一个最佳估计，在一系列可能的值范围内，以较低和较高的估计为界。

(1) 如果所有"数据缺乏 (DD)"物种都没有受到威胁，则受威胁现存物种最低估计值计算公式为

受威胁现存物种最低估计值 = (CR + EN + VU) / (总评估物种数 – EX)

(2) 如果所有"数据缺乏 (DD)"物种受威胁程度与评估数据充足的物种相同，则受威胁现存物种最佳估计值计算公式为

受威胁现存物种最佳估计值 = (CR + EN + VU) / (总评估物种数 – EX– DD)

(3) 如果所有"数据缺乏 (DD)"物种都受到威胁，则受威胁现存物种最高估计值计算公式为

受威胁现存物种最高估计值 = (CR + EN + VU + DD) / (总评估物种数 – EX)

表 10　中国生物多样性红色名录哺乳类部分 2015 版与 2021 版评定物种数、评定物种濒危等级变化

	等级和总计	2021 版								
		EW	RE	CR	EN	VU	NT	LC	DD	NI	总计
2015 版	EW	2	1	0	0	0	0	0	0	0	3
	RE	0	3	0	0	0	0	0	0	0	3
	CR	2	2	35	0	0	1	0	1	8	58
	EN	0	0	7	39	2	1	1	2	1	53
	VU	0	0	2	3	52	7	1	0	2	67
	NT	0	0	0	0	5	119	12	14	3	153
	LC	0	0	1	2	3	7	227	7	14	261
	DD	0	0	2	1	1	0	6	60	4	74
	NI	0	0	7	5	5	3	19	21	0	60
	总计	4	6	54	59	68	138	266	105	32	732

注：NI 含义为未列入

The Lowest, Best and Upper Estimates of Threatened Species (IUCN, 2020)

Proportion of threatened species is only reported for the more completely evaluated groups (*i.e.*, >80% of species have been evaluated). Also, the reported percentage of threatened species for each group is presented as a best estimate within a range of possible values bounded by lower and upper estimates:

(1) If all DD (Data Deficient) species are not threatened,

Lower estimate = (CR + EN + VU) / (Total assessed − EX)

(2) If all DD species are equally threatened as data sufficient species,

Best estimate = (CR + EN + VU) / (Total assessed − EX − DD)

(3) If all DD species are threatened,

Upper estimate = (CR + EN + VU + DD) / (Total assessed − EX)

species that are Regionally Extinct and Extinct in the Wild are not yet globally extinct, the reintroduction of Regionally Extinct mammal species from other range countries to China and the restoration of Extinct in the Wild species into the wild with viable populations must be placed high on the biodiversity conservation agenda in China.

Table 10 Number of species and changes of endangered status in the 2015 and 2021 editions of the red list of China's biodiversity for mammals' part

	Levels and Sum	EW	RE	CR	EN	VU	NT	LC	DD	NI	Sum
2015 edition	EW	2	1	0	0	0	0	0	0	0	3
	RE	0	3	0	0	0	0	0	0	0	3
	CR	2	2	35	9	0	1	0	1	8	58
	EN	0	0	7	39	2	1	1	2	1	53
	VU	0	0	2	3	52	7	1	0	2	67
	NT	0	0	0	0	5	119	12	14	3	153
	LC	0	0	1	2	3	7	227	7	14	261
	DD	0	0	2	1	1	0	6	60	4	74
	NI	0	0	7	5	5	3	19	21	0	60
	Sum	4	6	54	59	68	138	266	105	32	732

Note: NI means "Not included"

20 与《IUCN受威胁物种红色名录》(2020-2)的比较

与《IUCN 受威胁物种红色名录》(2020-2) 比较，《中国生物多样性红色名录：脊椎动物 第一卷 哺乳动物》(2021) 覆盖了已知所有中国哺乳动物，评估了 61 种《IUCN 受威胁物种红色名录》(2020-2) 没有评估的哺乳动物，其中有 25 种被 Burgin *et al.* (2018) 收录。《IUCN 受威胁物种红色名录》(2020-2) 中 67 种中国哺乳动物被列为"数据缺乏 (DD)"种，在本次评估中有 34 个物种根据物种信息和专家知识确定了它们的濒危等级。总的来说，这次《中国生物多样性红色名录：脊椎动物 第一卷 哺乳动物》(2021) 与《IUCN 受威胁物种红色名录》(2020-2) 比较，濒危等级在相当程度上是吻合的，328 种哺乳动物 (包括数据缺乏物种，约占中国哺乳动物总数的 47%) 的濒危等级相同 (**表 11**)。由于中国动物学家掌握了中国哺乳动物的完整信息，通过对中国哺乳动物多样性的查遗补缺，订正完善了大部分中国哺乳动物的濒危状况。与 2020-2 版 IUCN 红色名录比较，《中国生物多样性红色名录：脊椎动物 第一卷 哺乳动物》(2021) 无"未评估 (NE)"物种。

表 11 《中国生物多样性红色名录：脊椎动物 第一卷 哺乳动物》(2021) 与《IUCN 受威胁物种红色名录》(2020-2) 的比较

		IUCN 红色名录 (2020-2)									
	等级和总计	EX	EW	RE	CR	EN	VU	NT	LC	DD	NE	总计
中国的红色名录 (2021)	EX	0	0	0	0	0	0	0	0	0	0	0
	EW	0	0	0	1	3	0	0	0	0	0	4
	RE	0	0	0	4	1	1	0	0	0	0	6
	CR	0	1	0	8	13	15	4	6	1	6	54
	EN	0	0	0	1	16	10	7	13	4	8	59
	VU	0	0	0	0	3	10	5	37	4	9	68
	NT	0	0	0	0	2	0	11	106	15	4	138
	LC	0	0	0	0	0	0	5	249	9	3	266
	DD	0	0	0	0	0	2	1	36	34	32	105
	总计	0	1	0	14	38	38	33	447	67	62	700

20 Comparison with the *IUCN Red List of Threatened Species* (2020-2)

Compared with the *IUCN Red List of Threatened Species* (2020-2), the *China's Red List of Biodiversity: Vertebrates, Volume I, Mammals* (2021) covers all known mammals in China, including the evaluation of 61 species not assessed by the IUCN red list even though 25 of these species are listed in Burgin *et al.* (2018). 34 of the 67 species were listed as DD in the *IUCN Red List of Threatened Species* (2020-2) have been assessed during this national assessment. In this assessment, species evaluations were based on published information about species and expert knowledge and *China's Red List of Biodiversity: Vertebrates, Volume I, Mammals* (2021) is in good agreement overall with the *IUCN Red List of Threatened Species* (2020-2). The status of 328 species (about 47% of the total number of mammal species in China) are the same in both red lists (**Table 11**). Working with zoologists in China that hold extensive knowledge about the country's mammals, the endangered status of most species have now been revised and improved through this study. Compared with the IUCN red list (2020-2), there are now no Not Evaluated (NE) species in *China's Red List of Biodiversity: Vertebrates, Volume I, Mammals* (2021).

Table 11 Comparison between *China's Red List of Biodiversity: Vertebrates, Volume I, Mammals* (2021) and *IUCN Red List of Threatened Species* (2020-2)

	Levels and Sum	EX	EW	RE	CR	EN	VU	NT	LC	DD	NE	Sum
China's red list (2021)	EX	0	0	0	0	0	0	0	0	0	0	0
	EW	0	0	0	1	3	0	0	0	0	0	4
	RE	0	0	0	4	1	1	0	0	0	0	6
	CR	0	1	0	8	13	15	4	6	1	6	54
	EN	0	0	0	1	16	10	7	13	4	8	59
	VU	0	0	0	0	3	10	5	37	4	9	68
	NT	0	0	0	0	2	0	11	106	15	4	138
	LC	0	0	0	0	0	0	5	249	9	3	266
	DD	0	0	0	0	0	2	1	36	34	32	105
	Sum	0	1	0	14	38	38	33	447	67	62	700

21 保护成效

中国对部分濒危物种的保护措施已见成效。例如，中国采取就地保护与迁地保护相结合的方式加强对大熊猫的保护，现已建设了 67 处大熊猫自然保护区，覆盖了 66.8% 的野生大熊猫种群和 53.8% 的大熊猫栖息地。"全国第四次大熊猫调查"发现，野生大熊猫种群数量比第三次大熊猫调查增长 16.8%，达到 1,864 只。国家组织启动了大熊猫野化放归工作，先后将 7 只人工繁育的大熊猫放归自然。此外，2019 年全球人工圈养繁殖的大熊猫已经超过 600 只。2018 年，大熊猫国家公园正式挂牌。鉴于大熊猫种群的恢复，2015 年的中国生物多样性红色名录评估中，大熊猫的濒危等级降为"易危 (VU)"，本次评估中仍将大熊猫的濒危等级定为"易危 (VU)"。2018 年初，IUCN 宣布雪豹的濒危等级从"濒危"降为"易危"。2018 年 9 月，在深圳雪豹保护国际会议上，中国代表指出基于有效种群大小，雪豹在中国仍处于濒危状态 (图 35)。

20 世纪末，由于严重的盗猎，藏羚数量仅余 6 万 ~7 万只。20 世纪末以来，中国政府在藏羚分布区先后建立了西藏羌塘，青海可可西里、三江源、新疆阿尔金山、中昆仑及西昆仑等国家级和省级自然保护区，形成了中国面积最大的自然保护区群。藏羚分布区各县也组建了森林公安派出所，加强了藏羚的保护和执法力度，有效地遏制了盗猎藏羚活动，

图 35　2018 年 9 月 4 日，蒋志刚在深圳国际雪豹保护研讨会 (International Conference on Snow Leopard Conservation) 上报告中国雪豹红色名录评估等级，根据中国雪豹的有效种群数量，雪豹在中国仍是濒危物种 (李欣海摄)

Figure 35　Zhigang Jiang reported on the China's red list status of Snow Leopard to the International Conference on Snow Leopard Conservation (Shenzhen, China) on September 4, 2018. According to the effective population number, the Snow Leopard in China is still an endangered species (Photographed by Xinhai Li)

21 Conservation Achievements

Some of China's measures to protect endangered species have been proven effective, such as the country's great effort to strengthen the protection of Giant Pandas through a combination of on-site (*in situ*) and off-site (*ex situ*) conservation approaches. Sixty-seven (67) Giant Panda nature reserves have been built, which collectively contain almost two-thirds (66.8%) of the panda populations and more than half (53.8%) of the giant panda's habitat. The "Fourth National Giant Panda Survey" found the number of wild Giant Panda population to have increased by 16.8% compared with that reported from the third survey, with the number of wild panda reaching 1,864 individuals. The state has started the work of releasing and habituating Giant Panda in the wild, and seven artificially bred pandas have been released into the wild. In parallel, the number of Giant Panda bred in captivity worldwide reached over 600 in 2019. The Giant Panda National Park was officially launched in 2018. Given the recovery of its population, the status of Giant Panda was down-listed to "Vulnerable" (VU) in the assessment of China's red list of biodiversity in 2015 and Giant Panda has maintained the same status in this the assessment. Early in 2018, IUCN also announced that the Snow Leopard was now down-listed from Endangered to Vulnerable. However, at the "International Conference on Snow Leopard Conservation" held in Shenzhen in September 2018, the Chinese representative pointed out that Snow Leopard is still endangered in China, based on recent assessments of effective population size (**Figure 35**).

At the end of the 20th century, the number of Chiru (Tibetan Antelope) living on the Qinghai-Tibet (Xizang) Plateau had reduced dramatically, down to only 60,000~70,000 individuals, due to severe poaching. Since the end of the 20th century, however, the Chinese government has established a cluster of national and provincial nature reserves such as the AlTun Shan, Central Kunlun Shan and Western Kunlun Shan nature reserves in Xinjiang, the Qiangtang (Changtang) Nature Reserve in Tibet (Xizang), and the Kekexili and Sanjiangyuan nature reserves in Qinghai, each encompassing substantial portions of the range of Chiru, together forming the single largest network of nature reserves in China. Every county in the range of Chiru also established a forest police station, which helped strengthen law enforcement, effectively curbed the poaching of Chiru and protecting their habitat. Now, the number of Chiru on the Qinghai-Tibet (Xizang) Plateau has recovered to more than 200,000 individuals. In view of the status of the Chiru population, this assessment down-listed the endangered status of Chiru to Near Threatened (NT), thus removing it from the list of threatened species.

Przewalski's Gazelle (*Procapra przewalskii*, **Figure 36**) is an endangered ungulate endemic to the Qinghai-Tibet (Xizang) Plateau. It is an example of successful conservation of a formerly endangered species in China. Type specimens of Przewalski's Gazelle were collected by Nikolai Przewalski from the Ordos of Inner Mongolia (Nei Mongol) in 1875. The *State Key Protected Wild Animal list* promulgated in 1989 listed this gazelle species as a Category I State Key Protected Wild Animal. In 2001, Przewalski's Gazelle was then included in the *National Master Plan for Wildlife Protection and Nature Reserve Construction* as one of the 15 species requiring urgent action in China. In December 2002,

图 36 普氏原羚 (*Procapra przewalskii*, 蒋志刚摄) 是青藏高原地区特有的濒危有蹄类动物，曾称之为最濒危的有蹄类动物。经过四分之一世纪的保护，青藏高原的普氏原羚种群数量增长。其濒危等级下调为"濒危"

Figure 36 Przewalski's Gazelle (*Procapra przewalskii*, Photographed by Zhigang Jiang) is an endangered ungulate species (hoofed animal) endemic to the Qinghai-Tibet (Xizang) Plateau. It was once called the most endangered ungulate in the world. However, the population of Przewalski's Gazelle on the Qinghai-Tibet (Xizang) Plateau increased during the last quarter of the 20th century, due largely to official protection measures, and the status of Przewalski's Gazelle has now been down-listed to Endangered

加强了藏羚生境的保护。目前，青藏高原的藏羚数量已经回升到 20 万余只。鉴于藏羚的种群恢复状况，本次评估将藏羚降为近危 (NT) 等级，将其从受威胁物种名单中剔除。

普氏原羚 (*Procapra przewalskii*, 图 36) 是青藏高原地区特有的濒危有蹄类动物，是中国濒危物种保育比较成功的例子。1875 年，尼可拉·普热瓦尔斯基 (Nikolai Przewalski) 在内蒙古鄂尔多斯采集到普氏原羚模式标本。1989 年颁布的《国家重点保护野生动物名录》将其列为国家 I 级重点保护野生动物。2001 年制定的《全国野生动植物保护和自然保护区建设工程总体规划》中将其列为全国 15 个亟须拯救的物种之一。2002 年 12 月，国家林业局编制了《全国普氏原羚保护工程总体规划》。IUCN 红色名录 1996 年将其列为极危 (CR)；由于种群恢复及新种群的发现，2009 年将其濒危等级由极危 (CR) 调整为濒危 (EN)。中国生物多样性红色名录评估结果中仍将其列为极危 (CR) (蒋志刚等，2015a)。本次评估中，鉴于普氏原羚的种群恢复到 1,000 余只 (平晓鸽等，2018)，其濒危等级下调为"濒危"。

麋鹿 (图 37) 曾经一度在中国野外灭绝。20 世纪初，中国最后一群麋鹿在中国北京南苑皇家猎苑毁于洪灾与战火。20 世纪 80 年代从英国重新引入麋鹿，现在引种到 24 个省 81 个地点，已经分别建立了江苏大丰、北京南苑、湖北天鹅洲等迁地保育种群 (Jiang *et al.*, 2013)。2020 年，中国麋鹿种群数量达 8,000 多头。目前，在江苏大丰黄海海滨和洞庭湖地区分别通过人工释放和自然野化建立了野化种群。由于麋鹿的生存状

the State Forestry Administration formulated the *National Master Plan for the Protection of Przewalski's Gazelle*. The IUCN red list listed the species as Critically Endangered (CR) in 1996. With recovery of known populations as well as the discovery of new populations, the IUCN red list down-listed the status of Przewalski's Gazelle from Critically Endangered (CR) to Endangered (EN) in 2009, though China's red list of biodiversity continued to consider it as Critically Endangered (CR) (Jiang *et al.*, 2015a). In this assessment, however, in

图 37 麋鹿 (*Elaphurus davidianus*, 蒋志刚摄)。麋鹿是中国通过重引入而恢复的灭绝动物。20 世纪初，麋鹿在中国毁于战火与洪灾。20 世纪 80 年代中期从英国重新引入。目前，已经被引入 24 个省的 81 处自然保护区、湿地公园、繁育中心、野生动物园和动物园，形成了大丰、天鹅洲、南海子和洞庭湖等种群。2020 年，种群数量达 8,000 多头。在苏北黄海海滨和洞庭湖已经形成野生种群。图为苏北黄海海滨的麋鹿

Figure 37 Père David's Deer (Milu, *Elaphurus davidianus*, Photographed by Zhigang Jiang). Père David's Deer is an animal that was extinct in the wild, but that has recovered through several re-introduction measures undertaken in China. The last herd of Père David's Deer in the Nanyuan Imperial Hunting Garden was destroyed by floods and wars in the early 20th century. Père David's Deer was later reintroduced from Britain in the mid-1980s. Père David's Deer has been relocated to 81 nature reserves, wetland parks, breeding centers, safaris and zoos in 24 provinces. At the present time, the wild populations of Père David's Deer have been established in Dafeng, Tian'e Zhou, Nanhaizi and Dongting Lake. In 2018, the total number in the country reached more than 8,000 individuals. Wild populations of Père David's Deer have also established in the Yellow Sea coastal areas. Père David's Deer near the Yellow Sea coastal area are shown in above photograph

图 38 亚洲象 (*Elephas maximus*, 蒋志刚摄)。亚洲象在中国的分布区为其分布区的北缘，亚洲象分布在云南西双版纳和沧源。图为云南西双版纳"野象谷"的野生亚洲象

Figure 38 Asian Elephant (*Elephas maximus*, Photographed by Zhigang Jiang). The distribution range of Asian Elephant in China is the north fringe of its range. Asian Elephants are found in Xishuangbanna and Cangyuan in Yunnan, China. This picture shows the wild Asian Elephant in the "Wild Elephant Valley", Xishuangbanna, Yunnan

况改善，在本次评估中，麋鹿的濒危等级由"野外灭绝"降为"极危"。

然而，中国人口众多，在一些地区人兽冲突问题凸显，如亚洲象 (*Elephas maximus*，图 38) 与人类的冲突。历史上，亚洲象在中国分布区由于气候变化和人类的影响而向南退缩。目前，亚洲象的分布区从云南西双版纳和临沧扩大到了普洱。经过 20 多年的保护，尽管存在尖锐的人象冲突和热带雨林砍伐等问题，目前亚洲象在中国的种群数量基本稳定，数量为 200 余头。如何平衡经济发展与濒危物种保育问题，仍是一项重大议题 (Zhang et al., 2015)。

22 结束语

《中国生物多样性红色名录：脊椎动物 第一卷 哺乳动物》(2021) 研究具有重要的意义。第一，在红色名录的编研过程中，再次厘定了中国哺乳动物多样性，确定了中国是世界上哺乳动物最丰富的国家，中国哺乳动物中 22% 的种类是中国特有种类。第二，《中国生物多样性红色名录：脊椎动物 第一卷 哺乳动物》评估了除智人以外的全部 13 目 56 科 248 属 700 种哺乳动物，是中国历年红色名录研究评估哺乳动物种数最多的一次，还评估了 61 种《IUCN 受威胁物种红色名录》(2020-2) 没有评估的哺乳动物。首次开展了一些重要物种如虎、金钱豹、普氏原羚等的种下单元的濒危等级评定。第三，本次评估发现大中型有蹄类面临生存危机，有 4 种哺乳动物属于"野外灭绝"，6 种属于"区域灭绝"。尽管一个物种的"野外灭绝"或"区域灭绝"并不是该物种的灭绝，但是，应当重视这些"野外灭绝"或"区域灭绝"种的重引入、恢复该物种的野外种群。第四，白鱀豚、华南虎与白掌长臂猿已经多年未在其原生境中发现踪迹，处于灭绝边缘。第五，本次评估结果显示，受威胁中国哺乳动物共计 181 种，约占评估物种总数的 26%，然而，这仅是中国受威胁物种比例最低估计值。中国受威胁物种比例最佳估计值为 38%，

view of the recovery of the Przewalski's Gazelle populations (Ping *et al.*, 2018), the status of this species is now down-listed to Endangered.

Père David's Deer (**Figure 37**) was once extinct in the wild in China. In the early 20th century, the last group of Père David's Deer was destroyed by a combination of floods and wars. In the 1980s, Père David's Deer was reintroduced from Britain. The reintroduced Père David's Deer has been relocated to 81 localities in 24 provinces in the country. Now, several populations including the Dafeng population in Jiangsu, the Nanyuan population in Beijing and the Tian'e zhou population in Hubei have been established, bringing the total number to more than 8,000 individuals in 2020. At present, wild populations of Père David's Deer have been established in the Yellow Sea coastal area in Dafeng and the Dongting Lake region through artificial release and natural rewilding (Jiang *et al.*, 2003). With the increase of its population size, the conservation status of Père David's Deer has been down-listed in this assessment from "Extinct in the Wild" to "Critically Endangered".

However, China has a large human population and there remain many conflicts between people and wildlife, quite serious in some cases, such as the on-going human-wildlife conflicts with Asian Elephants (*Elephas maximus*, **Figure 38**). Historically, Asian Elephant populations in China retreated south due to climate change and human impact. At present, the distribution of Asian Elephant has expanded from Xishuangbanna and Lincang to Pu'er. With over 20 years of protection, the population of Asian Elephants in China is now generally stable, increased to about 200 individuals, despite some acute conflicts as well as the impact of deforestation. Balancing economic development and endangered species conservation remains a major challenge (Zhang *et al.*, 2015).

22 Conclusions

The assessment presented in *China's Red List of Biodiversity: Vertebrates, Volume I, Mammals* (2021) holds great national and global significance. First, during the compilation of the red list, the diversity of mammals in China was again updated. China is the country with the greatest diversity of mammal species in the world, and 22% of the mammal species in China are endemic. Second, all known mammal species in China were assessed in this red list study, including 700 species (excluding *Homo sapiens*) in 13 orders, 56 families and 248 genera. The study is the largest red list study ever conducted in China in term of the number of mammal species assessed, it also evaluated 61 species not previously assessed in the IUCN red list (2020-2). This study has for the first time assessed the endangerment status of the infra-specific taxa of Tiger, Leopard and Przewalski's Gazelle. Third, ungulates in the country are facing imminent danger of mass extinction. 4 mammal species in China are now classified as Extinct in the Wild and other six species as Regionally Extinct. Although the categories Extinct in the Wild and Regionally Extinct do not mean these species are globally extinct, nevertheless the reintroduction and restoration of wild populations of these species are clearly conservation priorities for the country. Fourth, the Baiji, the South China Tiger and the Lar Gibbon have not been found in their habitats in the

最高估计值为41%，中国哺乳动物的灭绝风险高于世界平均水平。第六，537个中国哺乳动物种（占2021版评估物种的77%）在两版红色名录中的濒危等级相同；由于近年来发现的新种，其种群和分布区多不清楚，2021版中，"数据缺乏"的种类增多。第七，中国哺乳动物各目中的受威胁比例是非均匀分布的，受威胁比例最高的目是灵长目、食肉目与偶蹄目。在空间上，中国哺乳动物生存状况也不相同。多数省区的受威胁哺乳动物物种为本省区的哺乳动物数目的20%~30%。西藏、重庆和云南的受威胁比例高于全国平均值。第八，发现森林区域（包括灌丛）、草原（包括草甸）、人类景观（包括农田、人工建筑等）是哺乳动物的主要生境，生活在这些生境中的受威胁哺乳动物比例分别为29%、24%和16%，但是，生活在岩石（包括高海拔流石滩、冰缘、裸岩悬崖等）、湿地（包括内陆水体、沼泽、河谷、滩涂等）生境的哺乳动物受威胁比例更高，分别为33%和38%。第九，中国哺乳动物种类主要分布在中国第二级地理台阶。以分布在海拔500~2,000m生境中的哺乳动物种类为多；生活在高海拔地区的哺乳动物虽然种类少，但是受威胁哺乳动物种类比例却高，随着海拔的升高，生存受威胁的哺乳动物比例增加。生活在海拔4,500~5,000m以及海拔5,000m以上的受威胁哺乳动物分别占该海拔区间哺乳动物种数的57%和67%。第十，发现"生境丧失"和"人工利用"、"人类干扰"三种致危因子名列受威胁哺乳动物所有致危因子的前三位。

中国正处于发展期。中国人口分布不均匀，90%以上的人口分布在"黑河－腾冲线"以东地区，区域发展程度差异大。东部地区人类文明历史悠久，开发规模大，开发强度高，西部荒漠、高原开发程度低，新疆仅部分区域开发，西藏地区几乎没有开发区。2012年，中国消费了世界一半的钢材、水泥和煤炭。中国的二氧化碳和二氧化硫的排放量世界第一。中国生态环境脆弱，中度以上生态脆弱区域占全国陆地国土空间的55%。目前，中国生态环境的基本状况是：局部改善，总体恶化，治理能力赶不上破坏速度，生态赤字逐渐扩大（石玉林等，2015）。

鉴于中国哺乳动物区系的独特性、多样性和重要性，许多哺乳动物濒临灭绝，拯救这些濒危物种是中国生物多样性保护的一项艰巨任务。自从《中华人民共和国野生动物保护法》实施以来，一些中国濒危哺乳动物的生存状况得以改善，特别是大熊猫与藏羚的生存状况有所好转。然而，由于中国地形地貌的复杂性和区域经济发展不平衡性，基于现状，有些物种如钓鱼岛鼹、野水牛、达旺猴的濒危等级，只能依据国外已经发表的文献资料评估。红色名录评定是一项物种灭绝风险评估，通过这一评定，预警了物种的生存危机。这种对未来生态危机的预警是"天际线扫描"的核心内容之一（蒋志刚，2014）。预测中国未来的环境将督促我们尽快采取行动(Jiang and Ma, 2014)。如此多的哺乳动物濒危，如何拯救它们仍然是中国每个人乃至全世界所有人面临的紧迫挑战。

country for many years and are on the verge of extinction. Fifth, according to this assessment, 181 species of mammals are threatened, accounting for around 26 percent of the species assessed, which is only the minimum estimate for the threatened species ratio (the best estimate of this ratio is 38%, and the maximum estimate of the ratio is 41%). The risk of extinction of mammals in China (based on the number of mammals listed as "threatened species") is found to be higher than the world average. Sixth, compared with the 2015 edition, 537 species (accounting for 77% of the species assessed in the 2021 edition) have been found to have the same endangerment status. Due to many new species discovered in recent years, the population and distribution area of these species are largely unclear; the number of "Data Deficient" species increased in the 2021 edition. Seventh, the proportions of threatened mammals in each order are uneven. The orders with the highest proportion of threatened mammals are Primates, Aarnivores and Artiodactyla. The status of different orders of mammals also differs across their geographic distributions. Threatened mammal species in most provinces of China account for 20%~30% of all mammal species, but in Tibet (Xizang), Chongqing and Yunnan, threatened species ratios are higher than national average. Eighth, Forested area (including shrubland), Grassland (including meadow) and Anthropogenic landscape (including farmland, man-made buildings, *etc.*) habitats in China are found to be the main habitats of mammals, and the ratio of threatened species (compared to all species) in these habitats are 29%, 24% and 16%, respectively. And the mammals living in the Rocky area (including the high elevation stream rocky beach, ice edge and bare rock cliff) and Wetland (including inland waterbodies, swamps, river valleys and coasts) are threatened, accounting for 33% and 38%, respectively. Ninth, China's mammal species are mostly distributed in the second geographical terrace of the country, with the majority of mammal species living in habitats between 500~2,000m above sea level. Although fewer mammal species live at high altitudes, the proportion of threatened mammals is higher at higher altitudes. Threatened mammals living at high altitudes between 4,500~5,000m and over 5,000m account for 57% and 67%, respectively, of the species in those altitudinal ranges. Finally, this assessment found that "Habitat loss", "Human exploitation" and "Human interference" were the top three threats for all threatened mammals across the country.

China is in a period of rapid transformation and development. China's human population is unevenly distributed, with over 90% of people living in areas east of the "Heihe-Tengchong Line". Regional socioeconomic development also varies greatly in China. The eastern part of the country has a long history of human civilization together with large scale and high intensity of modern development. In 2012, China consumed half of the world's steel, cement and coal, and the country is the world's largest emitter of carbon dioxide and sulfur dioxide. Yet, China's ecological environment is fragile and therefore at risk of degradation. Moderately to highly fragile ecological regions account for 55% of China's land area. At present, the state of China's ecological environment can be summarized as follows: while local environments are improving, the overall environment is deteriorating; the governance capacity for protection cannot match the pace of destruction; and the ecological deficit is steadily increasing (Shi *et al.*, 2015).

Given the unique nature, diversity and importance of China's mammalian fauna and the reality that so many mammal species are threatened, determining how best to protect and save these threatened species proves to be a difficult task for those engaged with conserving China's biodiversity. Since implementation of the *Wildlife Protection Law of the People's Republic of China* began, the living conditions of some threatened mammals in China have improved, especially for the Giant Panda and Chiru. However, given the diversity, remoteness and topographic complexity of many of China's threatened species' habitats, though some of them could only be assessed based on published literature. The red list assessment presented here is a critical assessment of the risk of extinction of China's mammal species. Through this assessment, we provide early warning of the crisis of species' survival in China. At a broader scale, this warning of ecological crisis also constitutes a core element of "Horizon Scanning" about conservation and future sustainability more generally Jiang, 2014. Working to predict the future state of China's environment urges us to take action as soon as possible (Jiang and Ma, 2014). With so many mammal species clearly recognized as endangered, determining how to save them remains a pressing challenge for everyone in the country, and indeed for all people around the world.

各 论（上册）
Species Monograph (I)

高氏缺齿鼩
Chodsigoa caovansunga

极危 CR B1+2a, b(i, ii, iii); C1

| 数据缺乏 DD | 无危 LC | 近危 NT | 易危 VU | 濒危 EN | **极危 CR** | 区域灭绝 RE | 野外灭绝 EW | 灭绝 EX |

分类地位 Taxonomic Status

动物界 Animalia	脊索动物门 Chordata	哺乳纲 Mammalia	劳亚食虫目 Eulipotyphla	鼩鼱科 Soricidae
学　　名 Scientific Name		*Chodsigoa caovansunga*		
命 名 人 Species Authority		Lunde, Musser and Son, 2003		
英 文 名 English Name(s)		Van Sung's Long-tailed Shrew		
同物异名 Synonym(s)		Van Sung's Shrew, Cao Van Sung Mountain Shrew		
种下单元评估 Infra-specific Taxa Assessed		无 / None		

评估信息 Assessment Information

评 估 年 份 Year Assessed	2020
评 定 人 Assessor(s)	蒋志刚 / Zhigang Jiang
审 定 人 Reviewer(s)	蒋学龙、冯祚建 / Xuelong Jiang, Zuojian Feng
其他贡献人 Other Contributor(s)	李立立、丁晨晨 / Lili Li, Chenchen Ding

理由 Justification: 高氏缺齿鼩首次于 2003 年记录于越南北部河江省 (Lunde et al., 2003)，并以越南学者高文充 (Cao Van Sung) 名字命名。由于发现不久，迄今已知该物种的分布仅限于模式产地 Tay Con Linh II 山。何锴等 (2012) 在云南南部采到的标本不仅是中国新记录，也是其在模式产地以外的首个分布记录。根据其占有区和发生区面积，该种定为极危 / Van Sung's Long-tailed Shrew was firstly recorded in northern Hà Giang Province of Viet Nam in 2003, and was named after Vietnamese scholar Cao Van Sung (Lunde et al., 2003). Due to the short discovery, the distribution of the species is known to be limited to the type specimen collection site (Tay Con Linh II Shan). He et al. (2012) collected the samples in southern Yunnan, which was not only a new record in China, but also the first record of the species outside Viet Nam. According to its extent of occurrence and area of occupancy, it is listed at Critically Endangered

地理分布 Geographical Distribution

国内分布 Domestic Distribution
云南 / Yunnan
世界分布 World Distribution
中国、越南 / China, Viet Nam
分布标注 Distribution Note
非特有种 / Non-endemic

国内分布图
Map of Domestic Distribution

种群 Population

种群数量 Population Size	未知 / Unknown
种群趋势 Population Trend	未知 / Unknown

生境与生态系统 Habitat(s) and Ecosystem(s)

生境 Habitat(s)	亚热带湿润山地森林 / Subtropical Moist Montane Forest
生态系统 Ecosystem(s)	森林生态系统 / Forest Ecosystem

威胁 Threat(s)

主要威胁 Major Threat(s)	未知 / Unknown

保护级别与保护行动 Protection Category and Conservation Action(s)

国家重点保护野生动物等级 (2021) Category of National Key Protected Wild Animals (2021)	无 / NA
IUCN 红色名录 (2020-2) IUCN Red List (2020-2)	数据缺乏 / DD
CITES 附录 (2019) CITES Appendix (2019)	无 / NA
保护行动 Conservation Action(s)	无 / None

相关文献 Relevant References

Burgin *et al*., 2018; Jiang *et al*. (蒋志刚等), 2017; He *et al*. (何锴等), 2012; Lunde *et al*., 2003

高氏缺齿鼩 *Chodsigoa caovansunga*

倭蜂猴
Nycticebus pygmaeus

中国生物多样性 红色名录

极危 CR B2bc; C2a(i)

| 数据缺乏 DD | 无危 LC | 近危 NT | 易危 VU | 濒危 EN | 极危 CR | 区域灭绝 RE | 野外灭绝 EW | 灭绝 EX |

分类地位 Taxonomic Status

动物界 Animalia	脊索动物门 Chordata	哺乳纲 Mammalia	灵长目 Primates	懒猴科 Lorisidae

学名 Scientific Name	*Nycticebus pygmaeus*
命名人 Species Authority	Bonhote, 1907
英文名 English Name(s)	Pygmy Slow Loris
同物异名 Synonym(s)	*Nycticebus intermedius* (Dao Van Tien, 1960)
种下单元评估 Infra-specific Taxa Assessed	无 / None

评估信息 Assessment Information

评估年份 Year Assessed	2020
评定人 Assessor(s)	蒋志刚、蒋学龙 / Zhigang Jiang, Xuelong Jiang
审定人 Reviewer(s)	黄乘明、蒋学龙、范朋飞 / Chengming Huang, Xuelong Jiang, Pengfei Fan
其他贡献人 Other Contributor(s)	李立立、丁晨晨 / Lili Li, Chenchen Ding

理由 Justification: 倭蜂猴在中国仅分布在云南，分布区非常狭窄，目前种群数量仅90只左右，野外调查很难发现。生存状况堪忧。根据其占有区、发生区面积及种群数量，该种定为极危 / In China, Pygmy Slow Loris only distributed narrowly in Yunnan Province. Its population size is extremely small, about 90 individuals live in the wild. Now it is extremely difficult to be found in the wild. According to its extent of occurrence and area of occupancy, as well its population size, it is listed as Critically Endangered

地理分布 Geographical Distribution

国内分布 Domestic Distribution
云南 / Yunnan
世界分布 World Distribution
中国；东南亚 / China; Southeast Asia
分布标注 Distribution Note
非特有种 / Non-endemic

国内分布图
Map of Domestic Distribution

种群 Population

种群数量 Population Size	90 只 / 90 individuals
种群趋势 Population Trend	下降 / Decreasing

生境与生态系统 Habitat (s) and Ecosystem (s)

生境 Habitat(s)	热带湿润低地森林、次生林、灌丛 / Tropical Moist Lowland Forest, Secondary Forest, Shrubland
生态系统 Ecosystem(s)	森林生态系统、灌丛生态系统 / Forest Ecosystem, Shrubland Ecosystem

威胁 Threat (s)

主要威胁 Major Threat(s)	狩猎、耕种 / Hunting, Plantation

保护级别与保护行动 Protection Category and Conservation Action (s)

国家重点保护野生动物等级 (2021) Category of National Key Protected Wild Animals (2021)	一级 / Category I
IUCN 红色名录 (2020-2) IUCN Red List (2020-2)	濒危 / EN
CITES 附录 (2019) CITES Appendix (2019)	I
保护行动 Conservation Action(s)	在自然保护区内的种群及栖息地得到保护 / Populations and habitats in nature reserves are protected

相关文献 Relevant References

Burgin *et al.*, 2018; Jiang *et al.* (蒋志刚等), 2017; Chen *et al.* (陈敏杰等), 2014; Yu *et al.* (余梁哥等), 2013; Duan *et al.* (段艳芳等), 2012; Wilson and Mittermeier, 2012; Pan *et al.*, 2007

倭蜂猴 *Nycticebus pygmaeus*　　　　By David Haring

北豚尾猴
Macaca leonina
极危 CR A2bd; B2

| 数据缺乏 DD | 无危 LC | 近危 NT | 易危 VU | 濒危 EN | **极危 CR** | 区域灭绝 RE | 野外灭绝 EW | 灭绝 EX |

分类地位 Taxonomic Status

动物界 Animalia	脊索动物门 Chordata		哺乳纲 Mammalia	灵长目 Primates	猴科 Cercopithecidae
学 名 Scientific Name		*Macaca leonina*			
命 名 人 Species Authority		Blyth, 1863			
英 文 名 English Name(s)		Northern Pig-tailed Macaque			
同物异名 Synonym(s)		豚尾猴; *Macaca adusta* (Miller, 1906); *andamanensis* (Bartlett, 1869); *blythii* (Pocock, 1931); *coininus* (Kloss, 1903); *indochinensis* (Kloss, 1919); *insulana* (Miller, 1906)			
种下单元评估 Infra-specific Taxa Assessed		无 / None			

评估信息 Assessment Information

评估年份 Year Assessed	2020
评 定 人 Assessor(s)	蒋志刚、蒋学龙 / Zhigang Jiang, Xuelong Jiang
审 定 人 Reviewer(s)	黄乘明、蒋学龙、范朋飞 / Chengming Huang, Xuelong Jiang, Pengfei Fan
其他贡献人 Other Contributor(s)	李立立、丁晨晨 / Lili Li, Chenchen Ding

理由 Justification: 北豚尾猴在中国分布极为狭窄，仅分布在云南。冯利民（北京师范大学）和蒋志刚研究组分别曾在西双版纳和沧源拍摄到北豚尾猴红外照片，此外，鲜有其他报道。基于其种群数量，北豚尾猴的濒危等级列为极危等级 / Northern Pig-tailed Macaque is narrowly distributed in Yunnan Province in China. Limin Feng (Beijing Normal University) and Zhigang Jiang's research group camera trapped the species in Xishuangbanna and Cangyuan, respectively. Based on its population status, Northern Pig-tailed Macaque is listed as Critically Endangered

地理分布 Geographical Distribution

国内分布 Domestic Distribution
云南 / Yunnan
世界分布 World Distribution
中国；南亚、东南亚 / China; South Asia, Southeast Asia
分布标注 Distribution Note
非特有种 / Non-endemic

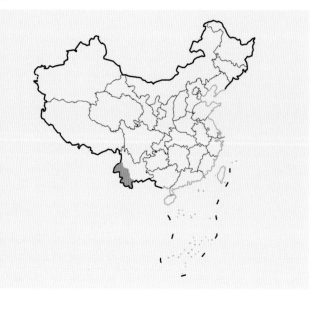

国内分布图
Map of Domestic Distribution

种群 Population

种群数量 Population Size	1,700 只 / 1,700 individuals
种群趋势 Population Trend	下降 / Decreasing

生境与生态系统 Habitat(s) and Ecosystem(s)

生　　境 Habitat(s)	热带和亚热带湿润低地森林，耕地 / Tropical and Subtropical Moist Lowland Forest, Arable Land
生态系统 Ecosystem(s)	森林生态系统、农田生态系统 / Forest Ecosystem, Cropland Ecosystem

威胁 Threat(s)

主要威胁 Major Threat(s)	自然生态系统破坏，伐木、火灾、公路、铁路、堤坝及水道修建、狩猎 / Nature Ecosystem Modification, Logging, Fire, Road and Railroad, Dam and Cannal Construction, Hunting

保护级别与保护行动 Protection Category and Conservation Action(s)

国家重点保护野生动物等级 (2021) Category of National Key Protected Wild Animals (2021)	一级 / Category I
IUCN 红色名录 (2020-2) IUCN Red List (2020-2)	易危 / VU
CITES 附录 (2019) CITES Appendix (2019)	II
保护行动 Conservation Action(s)	在自然保护区内的种群及栖息地得到保护 / Populations and habitats in nature reserves are protected

相关文献 Relevant References

Burgin *et al*., 2018; Jiang *et al*. (蒋志刚等), 2017; Wilson and Mittermeier, 2012; Zhang (张荣祖), 2002; Gippoliti, 2001; Groves, 2001

北豚尾猴 *Macaca leonina*　　　　　蒋志刚 提供　By Zhigang Jiang

白颊猕猴
Macaca leucogenys

中国生物多样性 红色名录 / China's Red List of Biodiversity

极危 CR A2bd; B2

DD	LC	NT	VU	EN	**CR**	RE	EW	EX
数据缺乏	无危	近危	易危	濒危	**极危**	区域灭绝	野外灭绝	灭绝

分类地位 Taxonomic Status

动物界 Animalia	脊索动物门 Chordata	哺乳纲 Mammalia	灵长目 Primates	猴科 Cercopithecidae

学 名 Scientific Name	*Macaca leucogenys*
命 名 人 Species Authority	Li, 2015
英 文 名 English Name(s)	White-cheeked Macaque
同物异名 Synonym(s)	无 / None
种下单元评估 Infra-specific Taxa Assessed	无 / None

评估信息 Assessment Information

评 估 年 份 Year Assessed	2020
评 定 人 Assessor(s)	蒋志刚、蒋学龙 / Zhigang Jiang, Xuelong Jiang
审 定 人 Reviewer(s)	黄乘明、蒋学龙、龙勇诚、范朋飞 / Chengming Huang, Xuelong Jiang, Yongcheng Long, Pengfei Fan
其他贡献人 Other Contributor(s)	李立立、丁晨晨 / Lili Li, Chenchen Ding

理由 Justification: 白颊猕猴为近期发现的一个新种，仅边缘分布于西藏地区，种群数量极少，目前尚未采到标本。基于其种群和分布区现状，白颊猕猴列为极危等级 / White-cheeked Macaque is a species discovered lately in China, only distributed in the southern border region of Tibet (Xizang). Its population size is extremely small. So far, no specimen has been collected yet. Based on its population and habitat status, White-cheeked Macaque is listed as Critically Endangered

地理分布 Geographical Distribution

国内分布 Domestic Distribution	西藏 / Tibet (Xizang)
世界分布 World Distribution	中国 / China
分布标注 Distribution Note	特有种 / Endemic

国内分布图 Map of Domestic Distribution

种群 Population

种群数量 Population Size	<1,000 只 / <1,000 individuals
种群趋势 Population Trend	未知 / Unknown

生境与生态系统 Habitat (s) and Ecosystem (s)

生　　境 Habitat(s)	阔叶林、针阔混交林 / Broad-leaved Forest, Coniferous and Broad-leaved Mixed Forest
生态系统 Ecosystem(s)	森林生态系统 / Forest Ecosystem

威胁 Threat (s)

主要威胁 Major Threat(s)	人类活动干扰 / Human Disturbance

保护级别与保护行动 Protection Category and Conservation Action (s)

国家重点保护野生动物等级 (2021) Category of National Key Protected Wild Animals (2021)	二级 / Category II
IUCN红色名录(2020-2) IUCN Red List (2020-2)	未列入 / NA
CITES 附录 (2019) CITES Appendix (2019)	II
保护行动 Conservation Action(s)	无 / None

相关文献 Relevant References

Burgin *et al*., 2018; Jiang *et al*. (蒋志刚等), 2017; Li *et al*., 2015a

白颊猕猴 *Macaca leucogenys*　　　　　　　　　　　　　　　　　刘务林 摄　By Wulin Liu

中国生物多样性红色名录 China's Red List of Biodiversity

长尾叶猴
Semnopithecus schistaceus

极危　CR A2bd; B2

| 数据缺乏 DD | 无危 LC | 近危 NT | 易危 VU | 濒危 EN | **极危 CR** | 区域灭绝 RE | 野外灭绝 EW | 灭绝 EX |

分类地位 Taxonomic Status

动物界 Animalia	脊索动物门 Chordata	哺乳纲 Mammalia	灵长目 Primates	猴科 Cercopithecidae
学　名 Scientific Name		*Semnopithecus schistaceus*		
命名人 Species Authority		Hodgson, 1840		
英文名 English Name(s)		Nepal Gray Langur		
同物异名 Synonym(s)		喜山长尾叶猴；*Semnopithecus achilles* (Pocock, 1928); *lania* (Elliot, 1909); *nipalensis* (Hodgson, 1840)		
种下单元评估 Infra-specific Taxa Assessed		无 / None		

评估信息 Assessment Information

评估年份 Year Assessed	2020
评定人 Assessor(s)	蒋志刚、蒋学龙 / Zhigang Jiang, Xuelong Jiang
审定人 Reviewer(s)	黄乘明、龙勇诚、范朋飞 / Chengming Huang, Yongcheng Long, Pengfei Fan
其他贡献人 Other Contributor(s)	李立立、丁晨晨 / Lili Li, Chenchen Ding

理由 Justification: 长尾叶猴虽然在国外分布较广，种群大，但在中国仅分布在西藏南部部分地区，数量极少。基于其种群与分布区，长尾叶猴列为极危等级 / Although Nepal Gray Langur is widely distributed in the world with fairly large populations, however, it is only distributed in limited area in the south Tibet (Xizang), China with extremely small population size. Based on its population and habitat status, Nepal Gray Langur is listed as Critically Endangered

地理分布 Geographical Distribution

国内分布 Domestic Distribution
西藏 / Tibet (Xizang)
世界分布 World Distribution
中国；南亚 / China; South Asia
分布标注 Distribution Note
非特有种 / Non-endemic

国内分布图 Map of Domestic Distribution

种群 Population

种群数量 Population Size	760 只 / 760 individuals
种群趋势 Population Trend	下降 / Decreasing

生境与生态系统 Habitat(s) and Ecosystem(s)

生　　境 Habitat(s)	亚热带湿润山地森林、亚热带高海拔灌丛 / Subtropical Moist Montane Forest, Subtropical High Altitude Shrubland
生态系统 Ecosystem(s)	森林生态系统、灌丛生态系统 / Forest Ecosystem, Shrubland Ecosystem

威胁 Threat(s)

主要威胁 Major Threat(s)	伐木、火灾、狩猎、人类活动干扰 / Logging, Fire, Hunting, Human Disturbance

保护级别与保护行动 Protection Category and Conservation Action(s)

国家重点保护野生动物等级 (2021) Category of National Key Protected Wild Animals (2021)	一级 / Category I
IUCN 红色名录 (2020-2) IUCN Red List (2020-2)	无危 / LC
CITES 附录 (2019) CITES Appendix (2019)	I
保护行动 Conservation Action(s)	在自然保护区内的种群及栖息地得到保护 / Populations and habitats in nature reserves are protected

相关文献 Relevant References

Burgin *et al.*, 2018; Jiang *et al.* (蒋志刚等), 2017; Hu *et al.* (胡一鸣等), 2014; Wilson and Mittermeier, 2012; Wang *et al.* (王应祥等), 1999

长尾叶猴 *Semnopithecus schistaceus* 　　　　　　　　　　　　　　　　　　　　蒋志刚 摄　By Zhigang Jiang

白头叶猴
Trachypithecus leucocephalus

中国生物多样性红色名录 China's Red List of Biodiversity

极危 CR A2bd; B1ab(i, ii, iii); C2a(i)

| 数据缺乏 DD | 无危 LC | 近危 NT | 易危 VU | 濒危 EN | **极危 CR** | 区域灭绝 RE | 野外灭绝 EW | 灭绝 EX |

分类地位 Taxonomic Status

动物界 Animalia	脊索动物门 Chordata	哺乳纲 Mammalia	灵长目 Primates	猴科 Cercopithecidae

学 名 Scientific Name	*Trachypithecus leucocephalus*
命 名 人 Species Authority	Tan, 1955
英 文 名 English Name(s)	White-headed Langur
同物异名 Synonym(s)	*Trachypithecus poliocephalus leucocephalus* (IUCN Redlist, 2018)
种下单元评估 Infra-specific Taxa Assessed	无 / None

评估信息 Assessment Information

评估年份 Year Assessed	2020
评 定 人 Assessor(s)	蒋志刚、蒋学龙 / Zhigang Jiang, Xuelong Jiang
审 定 人 Reviewer(s)	黄乘明 / Chengming Huang
其他贡献人 Other Contributor(s)	李立立、丁晨晨 / Lili Li, Chenchen Ding

理由 Justification: 白头叶猴的分布区狭窄，仅生活在广西崇左喀斯特石山区，种群数量约 1,000 只，尽管白头叶猴分布区降水充沛，但是石山无法保留水分。白头叶猴生活在缺水的喀斯特山地，其生境被稻田包围且其受到人类活动的影响。基于其种群与分布区现状，白头叶猴列为极危等级 / Distribution range of White-headed Langur is narrow. It is only found in the karst rocky hills in Chongzuo County, Guangxi. There are about 1,000 individuals live in the wild. The karst hills cannot hold water though the area is rich in precipitation. The habitats of White-headed Langur lack of water source besides its habitats are encroached by rice paddies and disturbed by firewood collection. Based on its population and habitat status, White-headed Langur is listed as Critically Endangered

地理分布 Geographical Distribution

国内分布 Domestic Distribution
广西 / Guangxi
世界分布 World Distribution
中国 / China
分布标注 Distribution Note
特有种 / Endemic

国内分布图 Map of Domestic Distribution

种群 Population

种群数量 Population Size	1,000 只 / 1,000 individuals
种群趋势 Population Trend	稳定 / Stable

生境与生态系统 Habitat(s) and Ecosystem(s)

生境 Habitat(s)	喀斯特地貌中的灌丛、森林 / Shrubland, Forest in Karst Landscape
生态系统 Ecosystem(s)	森林生态系统、喀斯特生态系统 / Forest Ecosystem, Karst Ecosystem

威胁 Threat(s)

主要威胁 Major Threat(s)	栖息地隔离、人为干扰和缺水 / Habitat Fragmentation, Human Disturbance and Limited Water Source

保护级别与保护行动 Protection Category and Conservation Action(s)

国家重点保护野生动物等级 (2021) Category of National Key Protected Wild Animals (2021)	一级 / Category I
IUCN 红色名录 (2020-2) IUCN Red List (2020-2)	极危 / CR
CITES 附录 (2019) CITES Appendix (2019)	II
保护行动 Conservation Action(s)	在自然保护区内的种群及栖息地得到保护 / Populations and habitats in nature reserves are protected

相关文献 Relevant References

Burgin *et al.*, 2018; Jiang *et al.*（蒋志刚等），2017; Wilson and Mittermeier, 2012; Sun *et al.*（孙涛），2010; Chen *et al.*（陈智等），2008; Hu（胡刚），1998; Tan（谭邦杰），1955

白头叶猴 *Trachypithecus leucocephalus*　　　　蒋志刚 摄　By Zhigang Jiang

黔金丝猴
Rhinopithecus brelichi

极危　CR B2; C2a(ii)

| 数据缺乏 DD | 无危 LC | 近危 NT | 易危 VU | 濒危 EN | **极危 CR** | 区域灭绝 RE | 野外灭绝 EW | 灭绝 EX |

分类地位 Taxonomic Status

| 动物界 Animalia | 脊索动物门 Chordata | 哺乳纲 Mammalia | 灵长目 Primates | 猴科 Cercopithecidae |

学　　名 Scientific Name	*Rhinopithecus brelichi*
命　名　人 Species Authority	Thomas, 1903
英　文　名 English Name(s)	Gray Snub-nosed Monkey
同物异名 Synonym(s)	Brelich's Snub-nosed Monkey, Guizhou Snub-nosed Monkey, Guizhou Golden Monkey
种下单元评估 Infra-specific Taxa Assessed	无 / None

评估信息 Assessment Information

评估年份 Year Assessed	2020
评定人 Assessor(s)	蒋志刚、蒋学龙 / Zhigang Jiang, Xuelong Jiang
审定人 Reviewer(s)	黄乘明、蒋学龙、龙勇诚、范朋飞 / Chengming Huang, Xuelong Jiang, Yongcheng Long, Pengfei Fan
其他贡献人 Other Contributor(s)	李立立、丁晨晨 / Lili Li, Chenchen Ding

理由 Justification: 黔金丝猴的分布区极其狭窄，仅分布在贵州梵净山地区。建立保护区之前，当地森林采伐和采矿等人类活动对其栖息地的破坏还未完全恢复。建立保护区后，黔金丝猴种群增长缓慢，数量不到1,000只。基于其种群、分布点与分布区现状，黔金丝猴列为极危等级 / The distribution range of Gray Snub-nosed Monkey is extremely narrow, limited to Fanjing Shan in Guizhou Province. Its habitats were destroyed by human activities such as over-logging and quarrying before the establishment of nature reserve and its habitats are not yet fully recovered. The populations of this species currently include no more than 1,000 individuals. Thus, based on its population, habitat and area of occupancy, it is listed as Critically Endangered

地理分布 Geographical Distribution

国内分布 Domestic Distribution	贵州 / Guizhou
世界分布 World Distribution	中国 / China
分布标注 Distribution Note	特有种 / Endemic

国内分布图
Map of Domestic Distribution

种群 Population

种群数量 Population Size	700 只 / 700 individuals
种群趋势 Population Trend	下降 / Decreasing

生境与生态系统 Habitat(s) and Ecosystem(s)

生境 Habitat(s)	亚热带湿润山地森林 / Subtropical Moist Montane Forest
生态系统 Ecosystem(s)	森林生态系统 / Forest Ecosystem

威胁 Threat(s)

主要威胁 Major Threat(s)	采集陆生植物、旅游、耕种 / Collection of Terrestrial Plant, Tourism, Plantation

保护级别与保护行动 Protection Category and Conservation Action(s)

国家重点保护野生动物等级 (2021) Category of National Key Protected Wild Animals (2021)	一级 / Category I
IUCN 红色名录 (2020-2) IUCN Red List (2020-2)	濒危 / EN
CITES 附录 (2019) CITES Appendix (2019)	I
保护行动 Conservation Action(s)	已经建立自然保护区 / Nature reserves established

相关文献 Relevant References

Burgin *et al*., 2018; Jiang *et al*. (蒋志刚等), 2017; Kolleck *et al*., 2013; Wilson and Mittermeier, 2012; Niu *et al*., 2010; Yang *et al*. (杨海龙等), 2010; Yang *et al*. (杨业勤等), 2002

黔金丝猴 *Rhinopithecus brelichi*　　梵净山国家级自然保护区 提供　By Fanjing Shan National Nature Reserve

中国生物多样性红色名录

缅甸金丝猴
Rhinopithecus strykeri

极危　CR B2; C2a

| 数据缺乏 DD | 无危 LC | 近危 NT | 易危 VU | 濒危 EN | **极危 CR** | 区域灭绝 RE | 野外灭绝 EW | 灭绝 EX |

分类地位 Taxonomic Status

动物界 Animalia	脊索动物门 Chordata	哺乳纲 Mammalia	灵长目 Primates	猴科 Cercopithecidae

学　名 Scientific Name	*Rhinopithecus strykeri*
命名人 Species Authority	Geissmann, Ngwe Lwin, Saw Soe Aung, Thet Naing Aung, Zin Myo Aung, Tony Htin Hla, Grindley and Momberg, 2011
英文名 English Name(s)	Stryker's Snub-nosed Monkey
同物异名 Synonym(s)	怒江金丝猴；Myanmar Snub-nosed Monkey, Black Snub-nosed Monkey
种下单元评估 Infra-specific Taxa Assessed	无 / None

评估信息 Assessment Information

评估年份 Year Assessed	2020
评定人 Assessor(s)	蒋志刚、蒋学龙 / Zhigang Jiang, Xuelong Jiang
审定人 Reviewer(s)	李明、龙勇诚 / Ming Li, Yongcheng Long
其他贡献人 Other Contributor(s)	李立立、丁晨晨 / Lili Li, Chenchen Ding

理由 Justification: 缅甸金丝猴是在缅甸和中国新发现的灵长类物种，其在中国的分布极其狭窄且数量稀少，为在中国边缘分布的物种。基于其种群大小与占有区面积，缅甸金丝猴被列为极危等级 / Stryker's Snub-nosed Monkey is a newly discovered primate species, distributed in both Myanmar and China. Its distribution range is narrow and its population is small. It is a marginal distribution species in China. Thus, based on its population size and area of occupancy, Stryker's Snub-nosed Monkey is listed as Critically Endangered

地理分布 Geographical Distribution

国内分布 Domestic Distribution
云南 / Yunnan
世界分布 World Distribution
中国、缅甸 / China, Myanmar
分布标注 Distribution Note
非特有种 / Non-endemic

国内分布图
Map of Domestic Distribution

种群 Population

种群数量 Population Size	490~620 只 / 490~620 individuals
种群趋势 Population Trend	未知 / Unknown

生境与生态系统 Habitat(s) and Ecosystem(s)

生境 Habitat(s)	湿润常绿阔叶林 / Moist Evergreen Broad-leaved Forest
生态系统 Ecosystem(s)	森林生态系统 / Forest Ecosystem

威胁 Threat(s)

主要威胁 Major Threat(s)	伐木、堤坝及水道修建 / Logging, Dam and Cannal Construction

保护级别与保护行动 Protection Category and Conservation Action(s)

国家重点保护野生动物等级 (2021) Category of National Key Protected Wild Animals (2021)	一级 / Category I
IUCN 红色名录 (2020-2) IUCN Red List (2020-2)	极危 / CR
CITES 附录 (2019) CITES Appendix (2019)	I
保护行动 Conservation Action(s)	大部分栖息地已经建立自然保护区 / Most of its habitats are protected by nature reserves

相关文献 Relevant References

Burgin *et al.*, 2018; Jiang *et al.* (蒋志刚等), 2017; Long *et al.*, 2012; Wilson and Mittermeier, 2012; Peng *et al.* (彭燕章等), 1988

缅甸金丝猴 *Rhinopithecus strykeri* 　　董磊 摄（西南山地供图）　By Lei Dong (Swild.cn)

西白眉长臂猿
Hoolock hoolock

极危 CR A2acd+3cd+4acd

| 数据缺乏 DD | 无危 LC | 近危 NT | 易危 VU | 濒危 EN | **极危 CR** | 区域灭绝 RE | 野外灭绝 EW | 灭绝 EX |

分类地位 Taxonomic Status

动物界 Animalia	脊索动物门 Chordata	哺乳纲 Mammalia	灵长目 Primates	长臂猿科 Hylobatidae

学 名 Scientific Name	*Hoolock hoolock*
命名人 Species Authority	Harlan, 1834
英文名 English Name(s)	Western Hoolock Gibbon
同物异名 Synonym(s)	Hoolock Gibbon, Western Hoolock, *Bunopithecus hoolock* ssp. *hoolock* (Harlan, 1834), *Hylobates hoolock* (Harlan, 1834)
种下单元评估 Infra-specific Taxa Assessed	无 / None

评估信息 Assessment Information

评估年份 Year Assessed	2020
评定人 Assessor(s)	蒋志刚、蒋学龙 / Zhigang Jiang, Xuelong Jiang
审定人 Reviewer(s)	李明、龙勇诚、蒋学龙 / Ming Li, Yongcheng Long, Xuelong Jiang
其他贡献人 Other Contributor(s)	李立立、丁晨晨 / Lili Li, Chenchen Ding

理由 Justification: 西白眉长臂猿在中国已经很难发现其踪迹，很长时间未见报道。因此，基于其种群现状，将西白眉长臂猿列为极危等级 / In China, it is very difficult to find Western Hoolock Gibbon in the wild, there is no report about the species for a long period. Thus, based on its population status, Western Hoolock Gibbon is listed as Critically Endangered

地理分布 Geographical Distribution

国内分布 Domestic Distribution
西藏 / Tibet (Xizang)
世界分布 World Distribution
中国；东南亚 / China; Southeast Asia
分布标注 Distribution Note
非特有种 / Non-endemic

国内分布图
Map of Domestic Distribution

种群 Population

种群数量 Population Size	未知 / Unknown
种群趋势 Population Trend	下降 / Decreasing

生境与生态系统 Habitat (s) and Ecosystem (s)

生　　境 Habitat(s)	海拔 2,000 ～ 2,500m 的中山湿润常绿阔叶林 / Mid-mountain Moist Evergreen Broad-leaved Forest at 2,000~2,500m a.s.l.
生态系统 Ecosystem(s)	森林生态系统 / Forest Ecosystem

威胁 Threat (s)

主要威胁 Major Threat(s)	采集陆生植物、公路铁路建设、伐木 / Collection of Terrestrial Plant, Road and Railroad Construction, Logging

保护级别与保护行动 Protection Category and Conservation Action (s)

国家重点保护野生动物等级 (2021) Category of National Key Protected Wild Animals (2021)	一级 / Category I
IUCN 红色名录 (2020-2) IUCN Red List (2020-2)	濒危 / EN
CITES 附录 (2019) CITES Appendix (2019)	I
保护行动 Conservation Action(s)	大部分栖息地已经建立自然保护区 / Most of its habitats are protected by nature reserves

相关文献 Relevant References

Burgin *et al*., 2018; Jiang *et al*. (蒋志刚等), 2017; Fan (范朋飞), 2012; Wilson and Mittermeier, 2012; Grueter *et al*., 2009; Li and Lin (李致祥和林正玉), 1983

西白眉长臂猿 *Hoolock hoolock* © Programme HURO

中国生物多样性 红色名录

东白眉长臂猿
Hoolock leuconedys

极危　CR B2; C2a(ii)

| 数据缺乏 DD | 无危 LC | 近危 NT | 易危 VU | 濒危 EN | **极危 CR** | 区域灭绝 RE | 野外灭绝 EW | 灭绝 EX |

分类地位 Taxonomic Status

动物界 Animalia	脊索动物门 Chordata	哺乳纲 Mammalia	灵长目 Primates	长臂猿科 Hylobatidae
学名 Scientific Name		*Hoolock leuconedys*		
命名人 Species Authority		Groves, 1967		
英文名 English Name(s)		Eastern Hoolock Gibbon		
同物异名 Synonym(s)		*Bunopithecus hoolock leuconedys* (Groves, 1967)		
种下单元评估 Infra-specific Taxa Assessed		无 / None		

评估信息 Assessment Information

评估年份 Year Assessed	2020
评定人 Assessor(s)	蒋志刚、蒋学龙 / Zhigang Jiang, Xuelong Jiang
审定人 Reviewer(s)	龙勇诚、范朋飞、江海声 / Yongcheng Long, Pengfei Fan, Haisheng Jiang
其他贡献人 Other Contributor(s)	李立立、丁晨晨 / Lili Li, Chenchen Ding

理由 Justification: 新的长臂猿分类系统以缅甸的亲敦江为界，将白眉长臂猿分成两个种，中国的白眉长臂猿为东白眉长臂猿，其在中国分布区极其狭窄且数量极其稀少，故东白眉长臂猿列为极危 / The latest taxonomic system divides *Bunopithecus hoolock* into two species: Western Hoolock Gibbon *Hoolock hoolock* and Eastern Hoolock Gibbon *H. leuconedys* with the Chindwin River in Myanmar as their boundary. Eastern Hoolock Gibbon is marginally distributed in China. It has narrow distribution range and small population size. Thus, it is listed as Critically Endangered

地理分布 Geographical Distribution

国内分布 Domestic Distribution
云南、西藏 / Yunnan, Tibet (Xizang)
世界分布 World Distribution
中国、缅甸 / China, Myanmar
分布标注 Distribution Note
非特有种 / Non-endemic

国内分布图
Map of Domestic Distribution

种群 Population

种群数量 Population Size	50～300 只 / 50~300 individuals
种群趋势 Population Trend	下降 / Decreasing

生境与生态系统 Habitat(s) and Ecosystem(s)

生　　境 Habitat(s)	热带和亚热带湿润山地森林 / Tropical and Subtropical Moist Montane Forest
生态系统 Ecosystem(s)	森林生态系统 / Forest Ecosystem

威胁 Threat(s)

主要威胁 Major Threat(s)	人类活动干扰 / Human Disturbance

保护级别与保护行动 Protection Category and Conservation Action(s)

国家重点保护野生动物等级 (2021) Category of National Key Protected Wild Animals (2021)	一级 / Category I
IUCN 红色名录 (2020-2) IUCN Red List (2020-2)	易危 / VU
CITES 附录 (2019) CITES Appendix (2019)	I
保护行动 Conservation Action(s)	在大部分栖息地已经建立自然保护区 / Most of its habitats are protected by nature reserves

相关文献 Relevant References

Burgin *et al*., 2018; Fan *et al*.（范朋飞等），2017; Jiang *et al*.（蒋志刚等），2017; Fan（范朋飞），2012; Mootnick *et al*., 2012; Wilson and Mittermeier, 2012; Chan *et al*., 2008

东白眉长臂猿 *Hoolock leuconedys* 　　　　董磊 摄（西南山地供图）　　By Lei Dong (Swild.cn)

高黎贡白眉长臂猿
Hoolock tianxing

极危 CR C1+2; D1

| 数据缺乏 DD | 无危 LC | 近危 NT | 易危 VU | 濒危 EN | **极危 CR** | 区域灭绝 RE | 野外灭绝 EW | 灭绝 EX |

分类地位 Taxonomic Status

动物界 Animalia	脊索动物门 Chordata	哺乳纲 Mammalia	灵长目 Primates	长臂猿科 Hylobatidae
学 名 Scientific Name		*Hoolock tianxing*		
命 名 人 Species Authority		Fan *et al.*, 2017		
英 文 名 English Name(s)		Tianxing Hoolock Gibbon		
同物异名 Synonym(s)		天行长臂猿，Skywalker Hoolock Gibbon, Gaoligong Hoolock Gibbon		
种下单元评估 Infra-specific Taxa Assessed		无 / None		

评估信息 Assessment Information

评 估 年 份 Year Assessed	2020
评 定 人 Assessor(s)	蒋志刚、蒋学龙 / Zhigang Jiang, Xuelong Jiang
审 定 人 Reviewer(s)	黄乘明、蒋学龙、龙勇诚、范朋飞 / Chengming Huang, Xuelong Jiang, Yongcheng Long, Pengfei Fan
其他贡献人 Other Contributor(s)	李立立、丁晨晨 / Lili Li, Chenchen Ding

理由 Justification: 新的长臂猿分类系统将白眉长臂猿分成了两个种（以缅甸亲敦江为界），我国的白眉长臂猿为东白眉长臂猿，Fan *et al.* (2017) 将分布在高黎贡山的东白眉长臂猿命名为新种——高黎贡白眉长臂猿（天行长臂猿）。高黎贡白眉长臂猿在我国分布区狭窄，且数量极其稀少。基于其占有区和种群数量，将高黎贡白眉长臂猿列为极危等级 / The latest taxonomic system divides *Bunopithecus hoolock* into two species: *Hoolock hoolock* and *H. leuconedys* with the Chindwin River in Myanmar as boundary. *H. leuconedys* is distributed in China. Fan *et al.* (2017) discovered *H. leuconedys* in the Gaoligong Shan is morphologically and genetically different from *H. leuconedys*, thus named a new species *H. tianxing*. *H. tianxing* has narrow distribution, and its populations are small. Based on its area of occupancy and population status, Tianxing Hoolock Gibbon is listed as Critically Endangered

地理分布 Geographical Distribution

国内分布 Domestic Distribution
云南 / Yunnan
世界分布 World Distribution
中国、缅甸 / China, Myanmar
分布标注 Distribution Note
非特有种 / Non-endemic

国内分布图
Map of Domestic Distribution

种群 Population

种群数量 Population Size	50~300 只 / 50~300 individuals
种群趋势 Population Trend	下降 / Decreasing

生境与生态系统 Habitat (s) and Ecosystem (s)

生境 Habitat(s)	湿润山地森林 / Moist Montane Forest
生态系统 Ecosystem(s)	森林生态系统 / Forest Ecosystem

威胁 Threat (s)

主要威胁 Major Threat(s)	人类活动干扰、生境丧失 / Human Disturbance, Habitat Loss

保护级别与保护行动 Protection Category and Conservation Action (s)

国家重点保护野生动物等级 (2021) Category of National Key Protected Wild Animals (2021)	一级 / Category I
IUCN 红色名录 (2020-2) IUCN Red List (2020-2)	未列入 / NA
CITES 附录 (2019) CITES Appendix (2019)	无 / NA
保护行动 Conservation Action(s)	大部分栖息地已经建立自然保护区 / Most of its habitats are protected by nature reserves

相关文献 Relevant References

Burgin *et al*., 2018; Fan *et al*., 2017, 2011; Jiang *et al*. (蒋志刚等), 2017; Fan (范朋飞), 2012

高黎贡白眉长臂猿 *Hoolock tianxing* 范朋飞 摄 By Pengfei Fan

白掌长臂猿
Hylobates lar

极危 CR B1ab(i, ii, iii); C2a(i)

| DD | LC | NT | VU | EN | **CR** | RE | EW | EX |

🦌 分类地位 Taxonomic Status

| 动物界 Animalia | 脊索动物门 Chordata | 哺乳纲 Mammalia | 灵长目 Primates | 长臂猿科 Hylobatidae |

学 名 Scientific Name	*Hylobates lar*
命 名 人 Species Authority	Linnaeus, 1771
英 文 名 English Name(s)	Lar Gibbon
同物异名 Synonym(s)	*Hylobates albimana* (Vigors and Horsfield, 1828); *longimana* (Schreber, 1774); *variegates* (É. Geoffroy, 1812); *varius* (Latreille, 1801); White-handed Gibbon
种下单元评估 Infra-specific Taxa Assessed	无 / None

🦌 评估信息 Assessment Information

评 估 年 份 Year Assessed	2020
评 定 人 Assessor(s)	蒋志刚、蒋学龙 / Zhigang Jiang, Xuelong Jiang
审 定 人 Reviewer(s)	黄乘明、龙勇诚、范朋飞 / Chengming Huang, Yongcheng Long, Pengfei Fan
其他贡献人 Other Contributor(s)	李立立、丁晨晨 / Lili Li, Chenchen Ding

理由 Justification: 20 世纪 80 年代中期，白掌长臂猿在野外仅余 5 群，19～27 只，此后，多次考察没有发现该种踪迹 (Grueter *et al*., 2009)。蒋志刚等 (2018) 在南滚河考察时，护林员反映班老老石头寨的深山中仍偶尔听到长臂猿吼叫。白掌长臂猿灭绝风险极高，故列为极危等级 / In the mid-1980s, there were only 19~27 individuals living in five groups in the wild. Since then, several field surveys have not revealed any Lar Gibbon in the wild (Grueter *et al*., 2009). Duirng their field survey in Nangunhe, Jiang *et al*. (2018) heard from the forest rangers that the howling of gibbon were still occasionaly heard in mountains near Banlao Laoshitouzhai. The risk of extinction of this species is very high and it is listed as Critically Endangered

🦌 地理分布 Geographical Distribution

国内分布 Domestic Distribution	
云南 / Yunnan	
世界分布 World Distribution	
中国；东南亚 / China; Southeast Aisa	
分布标注 Distribution Note	
非特有种 / Non-endemic	

国内分布图
Map of Domestic Distribution

种群 Population

种群数量 Population Size	25 只 / 25 individuals
种群趋势 Population Trend	下降 / Decreasing

生境与生态系统 Habitat (s) and Ecosystem (s)

生境 Habitat(s)	湿润常绿阔叶林 / Moist Evergreen Broad-leaved Forest
生态系统 Ecosystem(s)	森林生态系统 / Forest Ecosystem

威胁 Threat (s)

主要威胁 Major Threat(s)	人类活动干扰 / Human Disturbance

保护级别与保护行动 Protection Category and Conservation Action (s)

国家重点保护野生动物等级 (2021) Category of National Key Protected Wild Animals (2021)	一级 / Category I
IUCN 红色名录(2020-2) IUCN Red List (2020-2)	濒危 / EN
CITES 附录 (2019) CITES Appendix (2019)	I
保护行动 Conservation Action(s)	大部分栖息地已经建立自然保护区 / Most of its habitats are protected by nature reserves

相关文献 Relevant References

Burgin *et al*., 2018; Fan *et al*., 2017; Jiang *et al*.（蒋志刚等）, 2017; Fan（范朋飞）, 2012; Wilson and Mittermeier, 2012; Grueter *et al*., 2009

白掌长臂猿 *Hylobates lar*　　　　蒋志刚 摄　By Zhigang Jiang

中国生物多样性 红色名录
China's Red List of Biodiversity

西黑冠长臂猿
Nomascus concolor

极危　CR A2bd; B1ab(i, ii, iii); C2a(i)

| 数据缺乏 DD | 无危 LC | 近危 NT | 易危 VU | 濒危 EN | **极危 CR** | 区域灭绝 RE | 野外灭绝 EW | 灭绝 EX |

分类地位 Taxonomic Status

动物界 Animalia	脊索动物门 Chordata	哺乳纲 Mammalia	灵长目 Primates	长臂猿科 Hylobatidae

学　名 Scientific Name	*Nomascus concolor*
命名人 Species Authority	Harlan, 1826
英文名 English Name(s)	Black Crested Gibbon
同物异名 Synonym(s)	*Hylobates concolor* (Harlan, 1826); *Nomascus harlani* (Lesson, 1827); *N. henrici* (de Pousargues, 1897); *N. niger* (Ogilby, 1840)
种下单元评估 Infra-specific Taxa Assessed	无 / None

评估信息 Assessment Information

评估年份 Year Assessed	2020
评定人 Assessor(s)	蒋志刚、蒋学龙 / Zhigang Jiang, Xuelong Jiang
审定人 Reviewer(s)	黄乘明、龙勇诚、范朋飞 / Chengming Huang, Yongcheng Long, Pengfei Fan
其他贡献人 Other Contributor(s)	李立立、丁晨晨 / Lili Li, Chenchen Ding

理由 Justification: 由于人口增长和人类活动，西黑冠长臂猿原有的栖息地丧失了75%，剩余栖息地严重破碎化，且种群分布不连续。在云南无量山分布有98群，约1,300只西黑冠长臂猿。鉴于其栖息地与种群现状，故列为极危等级 / Three quarters of the original habitats of Black Crested Gibbon had been lost and its remaining habitat and populations are fragmented due to increasing of human population and human activities. There are 98 groups about 1,300 Black Crested Gibbons in Wuliang Shan, Yunnan. Based on the population and habitat status, Black Crested Gibbons is listed as Critically Endangered

地理分布 Geographical Distribution

国内分布 Domestic Distribution
西藏 / Tibet (Xizang)
世界分布 World Distribution
中国；东南亚 / China; Southeast Asia
分布标注 Distribution Note
非特有种 / Non-endemic

国内分布图
Map of Domestic Distribution

种群 Population

种群数量 Population Size	1,000~1,300 只 / 1,000~1,300 individuals
种群趋势 Population Trend	下降 / Decreasing

生境与生态系统 Habitat(s) and Ecosystem(s)

生　　境 Habitat(s)	湿润山地森林 / Moist Montane Forest
生态系统 Ecosystem(s)	森林生态系统 / Forest Ecosystem

威胁 Threat(s)

主要威胁 Major Threat(s)	未知 / Unknown

保护级别与保护行动 Protection Category and Conservation Action(s)

国家重点保护野生动物等级 (2021) Category of National Key Protected Wild Animals (2021)	一级 / Category I
IUCN 红色名录 (2020-2) IUCN Red List (2020-2)	极危 / CR
CITES 附录 (2019) CITES Appendix (2019)	I
保护行动 Conservation Action(s)	大部分栖息地已经建立自然保护区 / Most of its habitats are protected by nature reserves

相关文献 Relevant References

Burgin *et al.*, 2018; Fan *et al.*（范朋飞等），2017; Zhao *et al.*（赵启龙等），2016; Hua *et al.*（华朝朗等），2013; Fan（范朋飞），2012; Sun *et al.*（孙国政等），2012; Wilson and Mittermeier, 2012; Li *et al.*（李国松等），2011; Luo（罗忠华），2011

西黑冠长臂猿 *Nomascus concolor* 　　董磊 摄（西南山地供图）　By Lei Dong (Swild.cn)

海南长臂猿
Nomascus hainanus

极危 CR B1ab(i, ii, iii); C2a(i); D

| 数据缺乏 DD | 无危 LC | 近危 NT | 易危 VU | 濒危 EN | **极危 CR** | 区域灭绝 RE | 野外灭绝 EW | 灭绝 EX |

分类地位 Taxonomic Status

动物界 Animalia	脊索动物门 Chordata	哺乳纲 Mammalia	灵长目 Primates	长臂猿科 Hylobatidae

学名 Scientific Name	*Nomascus hainanus*
命名人 Species Authority	Thomas, 1892
英文名 English Name(s)	Hainan Gibbon
同物异名 Synonym(s)	Hainan Black Crested Gibbon
种下单元评估 Infra-specific Taxa Assessed	无 / None

评估信息 Assessment Information

评估年份 Year Assessed	2020
评定人 Assessor(s)	蒋志刚、蒋学龙 / Zhigang Jiang, Xuelong Jiang
审定人 Reviewer(s)	黄乘明、蒋学龙、龙勇诚、范朋飞 / Chengming Huang, Xuelong Jiang, Yongcheng Long, Pengfei Fan
其他贡献人 Other Contributor(s)	李立立、丁晨晨 / Lili Li, Chenchen Ding

理由 Justification: 海南长臂猿为中国特有种，仅分布在海南霸王岭。由于天然雨林消失和其生境中的人类活动，海南长臂猿的生存受到严重威胁，尽管经过 20 多年的保护，目前海南长臂猿种群数量仍只有 4 群 30 只。基于其占有区和发生区面积以及种群数量，海南长臂猿列为极危等级 / Hainan Gibbon is an endemic species in China, only distributed in Bawangling, Hainan Province. The survival of the Hainan Gibbon is threatened due to loss of natural rainforests and human activities in its range. Although after decades of protection, there are still only 30 individuals living in 4 groups. Thus, according to its area of occupancy and extent of occurrence, as well its population size, Hainan Gibbon is listed as Critically Endangered

地理分布 Geographical Distribution

国内分布 Domestic Distribution	海南 / Hainan
世界分布 World Distribution	中国 / China
分布标注 Distribution Note	特有种 / Endemic

国内分布图
Map of Domestic Distribution

种群 Population

种群数量 Population Size	23 只 / 23 individuals
种群趋势 Population Trend	稳定 / Stable

生境与生态系统 Habitat(s) and Ecosystem(s)

生境 Habitat(s)	热带湿润森林 / Tropical Moist Forest
生态系统 Ecosystem(s)	森林生态系统 / Forest Ecosystem

威胁 Threat(s)

主要威胁 Major Threat(s)	栖息地丧失、近交衰退 / Habitat Loss, Inbreeding Depression

保护级别与保护行动 Protection Category and Conservation Action(s)

国家重点保护野生动物等级 (2021) Category of National Key Protected Wild Animals (2021)	一级 / Category I
IUCN 红色名录 (2020-2) IUCN Red List (2020-2)	极危 / CR
CITES 附录 (2019) CITES Appendix (2019)	I
保护行动 Conservation Action(s)	已建立自然保护区 / Nature reserves established

相关文献 Relevant References

Chan *et al*., 2020; Liu *et al*., 2020; Burgin *et al*., 2018; Fan *et al*.（范朋飞等），2017; Jiang *et al*.（蒋志刚等），2017; Bryant *et al*., 2016; Fan（范朋飞），2012; Mootnick *et al*., 2012; Li *et al*.（李志刚等），2010; Chan *et al*., 2008; Fellowes *et al*., 2008

海南长臂猿 *Nomascus hainanus* 　　　　唐万玲 摄　By Wanling Tang

北白颊长臂猿
Nomascus leucogenys

极危　CR A2bd; B1ab(i, ii, iii); C2a(i)

| 数据缺乏 DD | 无危 LC | 近危 NT | 易危 VU | 濒危 EN | **极危 CR** | 区域灭绝 RE | 野外灭绝 EW | 灭绝 EX |

分类地位 Taxonomic Status

动物界 Animalia	脊索动物门 Chordata	哺乳纲 Mammalia	灵长目 Primates	长臂猿科 Hylobatidae

学　名 Scientific Name	*Nomascus leucogenys*
命名人 Species Authority	Ogilby, 1840
英文名 English Name(s)	Northern White-cheeked Gibbon
同物异名 Synonym(s)	无 / None
种下单元评估 Infra-specific Taxa Assessed	无 / None

评估信息 Assessment Information

评估年份 Year Assessed	2020
评定人 Assessor(s)	蒋志刚、蒋学龙 / Zhigang Jiang, Xuelong Jiang
审定人 Reviewer(s)	黄乘明、龙勇诚、范朋飞 / Chengming Huang, Yongcheng Long, Pengfei Fan
其他贡献人 Other Contributor(s)	李立立、丁晨晨 / Lili Li, Chenchen Ding

理由 Justification: 20 世纪五六十年代，北白颊长臂猿在云南南部常见。经过多次考察，该种目前在我国仅分布在云南西双版纳，以 2~5 只小群活动，濒临灭绝。基于北白颊长臂猿在中国的分布是其北部边缘分布区这一事实，其栖息地丧失和种群破碎现状，将北白颊长臂猿列为极危等级 / Northern White-cheeked Gibbon was common in south of Yunnan Province, China during 1950s~1960s. Now the gibbon is only found in Xishuangbanna, Yunnan Province in small group of 2~5 gibbons after several field surveys. Base on the fact that distribution range of Northern White-cheeked Gibbon in the country is its northern peripheral range, now its habitats are lost and its population fragmented in small groups; based on the danger of extinction it is facing, it is listed as Critically Endangered

地理分布 Geographical Distribution

国内分布 Domestic Distribution
云南 / Yunnan
世界分布 World Distribution
中国；东南亚 / China; Southeast Asia
分布标注 Distribution Note
非特有种 / Non-endemic

国内分布图
Map of Domestic Distribution

种群 Population

种群数量 Population Size	10 只 / 10 individuals
种群趋势 Population Trend	下降 / Decreasing

生境与生态系统 Habitat (s) and Ecosystem (s)

生　　境 Habitat(s)	热带湿润低地森林 / Tropical Moist Lowland Forest
生态系统 Ecosystem(s)	森林生态系统 / Forest Ecosystem

威胁 Threat (s)

主要威胁 Major Threat(s)	耕种、家畜养殖、放牧、伐木、狩猎、近交衰退、疾病 / Plantation, Livestock Farming, Logging, Hunting, Inbreeding Depression, Disease

保护级别与保护行动 Protection Category and Conservation Action (s)

国家重点保护野生动物等级 (2021) Category of National Key Protected Wild Animals (2021)	一级 / Category I
IUCN 红色名录 (2020-2) IUCN Red List (2020-2)	极危 / CR
CITES 附录 (2019) CITES Appendix (2019)	I
保护行动 Conservation Action(s)	已建立自然保护区 / Nature reserve established

相关文献 Relevant References

Burgin *et al.*, 2018; Fan *et al.*, 2017; Jiang *et al.*（蒋志刚等）, 2017; Fan（范朋飞）, 2012; Mootnick *et al.*, 2012; Wilson and Mittermeier, 2012; Chan *et al.*, 2008

北白颊长臂猿 *Nomascus leucogenys*　　　　　王昌大 摄（西南山地供图）　　By Changda Wang (Swild.cn)

东黑冠长臂猿
Nomascus nasutus

极危 CR A2bd; B1ab(i, ii, iii); C2a(i)

| 数据缺乏 DD | 无危 LC | 近危 NT | 易危 VU | 濒危 EN | **极危 CR** | 区域灭绝 RE | 野外灭绝 EW | 灭绝 EX |

分类地位 Taxonomic Status

| 动物界 Animalia | 脊索动物门 Chordata | 哺乳纲 Mammalia | 灵长目 Primates | 长臂猿科 Hylobatidae |

学 名 Scientific Name	*Nomascus nasutus*
命 名 人 Species Authority	Kunkel d' Herculais, 1884
英 文 名 English Name(s)	Cao-vit Crested Gibbon
同物异名 Synonym(s)	Cao-vit Black Crested Gibbon, Eastern Black Crested Gibbon
种下单元评估 Infra-specific Taxa Assessed	无 / None

评估信息 Assessment Information

评估年份 Year Assessed	2020
评 定 人 Assessor(s)	蒋志刚、蒋学龙 / Zhigang Jiang, Xuelong Jiang
审 定 人 Reviewer(s)	黄乘明、蒋学龙、龙勇诚、范朋飞 / Chengming Huang, Xuelong Jiang, Yongcheng Long, Pengfei Fan
其他贡献人 Other Contributor(s)	李立立、丁晨晨 / Lili Li, Chenchen Ding

理由 Justification： 东黑冠长臂猿生活在中越边界地区喀斯特森林中。其栖息地已经破碎成为人类活动之中的孤岛。且该种种群极小，还受到捕猎的威胁。基于其种群与栖息地现状，故列为极危等级 / Cao-vit Crested Gibbon lives in karst forested areas along the China-Viet Nam board region. The habitats of Cao-vit Crested Gibbon is an island surrounded by human activities. Furthermore, its populations are very small. The poaching is also threatening the survival of this species. Thus, based on its habitat and population status, Cao-vit Crested Gibbon is listed as Critically Endangered

地理分布 Geographical Distribution

国内分布 Domestic Distribution
广西 / Guangxi
世界分布 World Distribution
中国；东南亚 / China; Southeast Asia
分布标注 Distribution Note
非特有种 / Non-endemic

国内分布图
Map of Domestic Distribution

种群 Population

种群数量 Population Size	20~30 只 / 20~30 individuals
种群趋势 Population Trend	下降 / Decreasing

生境与生态系统 Habitat (s) and Ecosystem (s)

生境 Habitat(s)	热带湿润低地森林 / Tropical Moist Lowland Forest
生态系统 Ecosystem(s)	森林生态系统 / Forest Ecosystem

威胁 Threat (s)

主要威胁 Major Threat(s)	耕种、家畜养殖、放牧、伐木、狩猎、近交衰退 / Plantation, Livestock Farming, Logging, Hunting, Inbreeding Depression

保护级别与保护行动 Protection Category and Conservation Action (s)

国家重点保护野生动物等级 (2021) Category of National Key Protected Wild Animals (2021)	一级 / Category I
IUCN 红色名录 (2020-2) IUCN Red List (2020-2)	极危 / CR
CITES 附录 (2019) CITES Appendix (2019)	I
保护行动 Conservation Action(s)	大部分栖息地已经建立自然保护区 / Most of its habitats are protected by nature reserves

相关文献 Relevant References

Burgin *et al*., 2018; Fan *et al*. (范朋飞等), 2017; Jiang *et al*. (蒋志刚等), 2017; Fan (范朋飞), 2012; Mootnick *et al*., 2012; Wilson and Mittermeier, 2012; Chan *et al*., 2008

东黑冠长臂猿 *Nomascus nasutus*　　　　　　　　　　　　　　　　　　　　　　　　　赵超 摄　By Chao Zhao

印度穿山甲
Manis crassicaudata

极危 CR B1ab(i, ii, iii); C2a(i)

| 数据缺乏 DD | 无危 LC | 近危 NT | 易危 VU | 濒危 EN | **极危 CR** | 区域灭绝 RE | 野外灭绝 EW | 灭绝 EX |

分类地位 Taxonomic Status

动物界 Animalia	脊索动物门 Chordata	哺乳纲 Mammalia	鳞甲目 Pholidota	鲮鲤科 Manidae
学名 Scientific Name		*Manis crassicaudata*		
命名人 Species Authority		É. Geoffroy, 1803		
英文名 English Name(s)		Indian Pangolin		
同物异名 Synonym(s)		无 / None		
种下单元评估 Infra-specific Taxa Assessed		无 / None		

评估信息 Assessment Information

评估年份 Year Assessed	2020
评定人 Assessor(s)	蒋志刚 / Zhigang Jiang
审定人 Reviewer(s)	吴诗宝、孟智斌 / Shibao Wu, Zhibin Meng
其他贡献人 Other Contributor(s)	李立立、丁晨晨 / Lili Li, Chenchen Ding

理由 Justification: 印度穿山甲分布范围比较狭窄，仅分布在滇西中缅边境一线的腾冲、云龙等地，最早记载见于王应祥于1987年编著的《云南的兽类资源及其合理利用与保护》，后来再没有关于印度穿山甲的分布、种群及栖息地状况等野外生态学资料的报道，也没有组织过印度穿山甲专项考察。印度穿山甲被多数人误当作中华穿山甲，一直没得到关注。蒋志刚等 (2018) 在南滚河考察时，有边民告知当地曾有黑色与红色两种穿山甲(即中华穿山甲和印度穿山甲)，但是都已经消失。印度穿山甲在我国是一类知之甚少的物种，种群信息极其缺乏，但是其生存处境极度濒危。因此，它被列为极危 / The distribution range of this species is relatively narrow, restricted to Tengchong, Yunlong and other places in the Myanmar border region in west Yunnan Province, China. The earliest records are in the book *Yunnan Mammalian Resources and its Rational Utilization and Protection*, edited by Yingxiang Wang in 1987. Since then, there has been neither any report about the Indian Pangolin's distribution, population, habitat status and field ecology, nor has any formal special investigation been organized to learn about the species. Little attention has been given to this species. Indian Pangolin are often mistaken as Chinese Pangolin. Jiang *et al.* (2018) were told by indigenous people during their field expedition in Nangunhe, Yunnan that two kinds of pangolin were present in the region, one with black scales (Chinese Pangolin) and the other with red scales (Indian Pangolin), but both disappeared. Information about the species in China is lacking, particularly population status, but it's in a Critically Endangered status. Thus, it is listed as Critically Endangered

地理分布 Geographical Distribution

国内分布 Domestic Distribution
云南 / Yunnan
世界分布 World Distribution
中国；南亚 / China; South Asia
分布标注 Distribution Note
非特有种 / Non-endemic

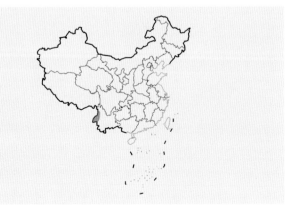

国内分布图
Map of Domestic Distribution

种群 Population

种群数量 Population Size	未知 / Unknown
种群趋势 Population Trend	下降 / Decreasing

生境与生态系统 Habitat (s) and Ecosystem (s)

生　　境 Habitat(s)	森林、灌丛、草地 / Forest, Shrubland, Grassland
生态系统 Ecosystem(s)	森林生态系统、灌丛生态系统、草地生态系统 / Forest Ecosystem, Shrubland Ecosystem, Grassland Ecosystem

威胁 Threat (s)

主要威胁 Major Threat(s)	狩猎 / Hunting

保护级别与保护行动 Protection Category and Conservation Action (s)

国家重点保护野生动物等级 (2021) Category of National Key Protected Wild Animals (2021)	一级 / Category I
IUCN 红色名录 (2020-2) IUCN Red List (2020-2)	濒危 / EN
CITES 附录 (2019) CITES Appendix (2019)	I
保护行动 Conservation Action(s)	无 / None

相关文献 Relevant References

Burgin *et al.*, 2018; Jiang *et al.* (蒋志刚等), 2017; Liu *et al.* (刘曦庆等), 2011; Smith *et al.* (史密斯等), 2009; Wang (王应祥), 2003; Heath, 1995

印度穿山甲 *Manis crassicaudata*

马来穿山甲 Manis javanica

极危 CR B1ab(i, ii, iii); C2a(i)

| 数据缺乏 DD | 无危 LC | 近危 NT | 易危 VU | 濒危 EN | 极危 CR | 区域灭绝 RE | 野外灭绝 EW | 灭绝 EX |

分类地位 Taxonomic Status

动物界 Animalia	脊索动物门 Chordata	哺乳纲 Mammalia	鳞甲目 Pholidota	鲮鲤科 Manidae
学　名 Scientific Name		*Manis javanica*		
命 名 人 Species Authority		Desmarest, 1822		
英文名 English Name(s)		Sunda Pangolin		
同物异名 Synonym(s)		Malayan Pangolin, Javan Pangolin		
种下单元评估 Infra-specific Taxa Assessed		无 / None		

评估信息 Assessment Information

评估年份 Year Assessed	2020
评 定 人 Assessor(s)	蒋志刚 / Zhigang Jiang
审 定 人 Reviewer(s)	吴诗宝、孟智斌 / Shibao Wu, Zhibin Meng
其他贡献人 Other Contributor(s)	李立立、丁晨晨 / Lili Li, Chenchen Ding

理由 Justification: 中国的马来穿山甲由吴诗宝等 (2005) 基于对中国科学院昆明动物研究所的馆藏标本的研究而发现，其在中国的分布区可能是其边缘分布区。目前缺乏其分布、种群及栖息地状况等资料，也没有组织过专项考察，仅有几具标本。马来穿山甲分布狭窄，数量稀少。基于其种群与生存现状，马来穿山甲列为极危 / Sunda Pangolin was reported by Wu *et al.* (2005) based on the specimens in Kunming Institute of Zoology, Chinese Academy of Sciences. Its distributed region in China may be its peripheral range. At present, we lack the information about distribution, population, and habitat status of the species. No special investigation on the species was conducted, only a few specimens have been collected. Sunda Pangolin distributed narrowly and is very rare. Based on its population and living status, it is listed as Critically Endangered

地理分布 Geographical Distribution

国内分布 Domestic Distribution
云南 / Yunnan
世界分布 World Distribution
中国；南亚 / China; South Asia
分布标注 Distribution Note
非特有种 / Non-endemic

国内分布图
Map of Domestic Distribution

种群 Population

种群数量 Population Size	未知 / Unknown
种群趋势 Population Trend	下降 / Decreasing

生境与生态系统 Habitat (s) and Ecosystem (s)

生　　境 Habitat(s)	森林、种植园、乡村庭院 / Forest, Plantation, Rural Garden
生态系统 Ecosystem(s)	森林生态系统、灌丛生态系统、草地生态系统 / Forest Ecosystem, Shrubland Ecosystem, Grassland Ecosystem

威胁 Threat (s)

主要威胁 Major Threat(s)	猎捕 / Hunting

保护级别与保护行动 Protection Category and Conservation Action (s)

国家重点保护野生动物等级 (2021) Category of National Key Protected Wild Animals (2021)	一级 / Category I
IUCN 红色名录 (2020-2) IUCN Red List (2020-2)	极危 / CR
CITES 附录 (2019) CITES Appendix (2019)	I
保护行动 Conservation Action(s)	无 / None

相关文献 Relevant References

Burgin *et al.*, 2018; Jiang *et al.* (蒋志刚等), 2017; Zhou *et al.* (周昭敏等), 2012; Hu *et al.* (胡诗佳等), 2010; Zhang *et al.* (张立等), 2010; Wu *et al.* (吴诗宝等), 2005

马来穿山甲 *Manis javanica*

穿山甲
Manis pentadactyla

极危　CR A2bd; B1ab(i, ii, iii); C2a(i)

| 数据缺乏 DD | 无危 LC | 近危 NT | 易危 VU | 濒危 EN | **极危 CR** | 区域灭绝 RE | 野外灭绝 EW | 灭绝 EX |

分类地位 Taxonomic Status

动物界 Animalia	脊索动物门 Chordata	哺乳纲 Mammalia	鳞甲目 Pholidota	鲮鲤科 Manidae
学　　名 Scientific Name		*Manis pentadactyla*		
命 名 人 Species Authority		Linnaeus, 1758		
英 文 名 English Name(s)		Chinese Pangolin		
同物异名 Synonym(s)		中华穿山甲		
种下单元评估 Infra-specific Taxa Assessed		无 / None		

评估信息 Assessment Information

评 估 年 份 Year Assessed	2020
评　定　人 Assessor(s)	蒋志刚 / Zhigang Jiang
审　定　人 Reviewer(s)	吴诗宝、孟智斌 / Shibao Wu, Zhibin Meng
其他贡献人 Other Contributor(s)	李立立、丁晨晨 / Lili Li, Chenchen Ding

理由 Justification: 20世纪70年代末至今，市场对穿山甲的巨大需求导致穿山甲几乎被猎杀殆尽，估计穿山甲种群数量至少减少了90%。穿山甲易捕捉、易携带运输，也给偷猎活动提供了方便。穿山甲繁殖能力低下，通常每年1胎1仔。预测在未来相当长的一段时间内种群还会继续下降。因此，基于其种群现状，将穿山甲列为极危等级 / Since the later 1970s till now, the mounting demand for Chinese Pangolin as food and traditional medicine, resulting in over-hunting of Chinese Pangolin to the brink of extinction. It is estimated that the population of Chinese Pangolin has declined at least by 90%. It is easy to capture, carry and transport pangolins, thus facilitating the poaching activities. The reproductive rate of Chinese Pangolin is low, usually one fetus per litter. Chinese Pangolin populations are expected to continue declining in the future. Thus, based on its status of population, Chinese Pangolin is listed as Critically Endangered

地理分布 Geographical Distribution

国内分布 Domestic Distribution
湖南、海南、浙江、上海、江苏、安徽、福建、江西、广东、广西、四川、贵州、云南、西藏、台湾、香港、重庆 / Hunan, Hainan, Zhejiang, Shanghai, Jiangsu, Anhui, Fujian, Jiangxi, Guangdong, Guangxi, Sichuan, Guizhou, Yunnan, Tibet (Xizang), Taiwan, Hong Kong, Chongqing
世界分布 World Distribution
中国；南亚、东南亚 / China; South Asia, Southeast Asia
分布标注 Distribution Note
非特有种 / Non-endemic

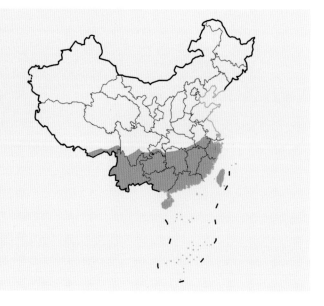

国内分布图
Map of Domestic Distribution

种群 Population

种群数量 Population Size	64,000 只 / 64,000 individuals
种群趋势 Population Trend	下降 / Decreasing

生境与生态系统 Habitat (s) and Ecosystem (s)

生境 Habitat(s)	森林、灌丛、草地 / Forest, Shrubland, Grassland
生态系统 Ecosystem(s)	森林生态系统、灌丛生态系统、草地生态系统 / Forest Ecosystem, Shrubland Ecosystem, Grassland Ecosystem

威胁 Threat (s)

主要威胁 Major Threat(s)	狩猎 / Hunting

保护级别与保护行动 Protection Category and Conservation Action (s)

国家重点保护野生动物等级 (2021) Category of National Key Protected Wild Animals (2021)	一级 / Category I
IUCN 红色名录 (2020-2) IUCN Red List (2020-2)	极危 / CR
CITES 附录 (2019) CITES Appendix (2019)	I
保护行动 Conservation Action(s)	无 / None

相关文献 Relevant References

Burgin *et al*., 2018; Jiang *et al*. (蒋志刚等), 2017; Hu *et al*. (胡一鸣等), 2014; Deng *et al*. (邓可等), 2013; Hu *et al*. (胡诗佳等), 2010; Zhang *et al*. (张立等), 2010

穿山甲 *Manis pentadactyla*

中国生物多样性 红色名录
马来熊
Helarctos malayanus

极危　CR A3d; B1ab(i, ii, iii); C1

| DD | LC | NT | VU | EN | **CR** | RE | EW | EX |

分类地位 Taxonomic Status

动物界 Animalia	脊索动物门 Chordata	哺乳纲 Mammalia	食肉目 Carnivora	熊科 Ursidae

学名 Scientific Name	*Helarctos malayanus*
命名人 Species Authority	Raffles, 1821
英文名 English Name(s)	Sun Bear
同物异名 Synonym(s)	无 / None
种下单元评估 Infra-specific Taxa Assessed	无 / None

评估信息 Assessment Information

评估年份 Year Assessed	2020
评定人 Assessor(s)	蒋志刚 / Zhigang Jiang
审定人 Reviewer(s)	马建章、徐爱春、张明海 / Jianzhang Ma, Aichun Xu, Minghai Zhang
其他贡献人 Other Contributor(s)	李立立、丁晨晨 / Lili Li, Chenchen Ding

理由 Justification: 马来熊在中国的栖息地为其边缘分布区，分布区面积小，数量少，在野外极少发现。且偷猎行为及非法贸易使该物种种群数量不断下降。基于其种群数量、种群数量下降速率和栖息地现状，马来熊的濒危等级已达到极危标准 / Sun Bear is marginally distributed in China. Its habitats are small; it is difficult to find in wild. Its range and population are decreasing due to human encroachment. Poaching and illegal trades caused declining in the population of the species. Now it is rather difficult to find in the field. Thus, based on the population size, the rate of population decline and habitat status, Sun Bear is classified as Critically Endangered

地理分布 Geographical Distribution

国内分布 Domestic Distribution	四川、云南、西藏 / Sichuan, Yunnan, Tibet (Xizang)
世界分布 World Distribution	中国；南亚、东南亚 / China; South Asia, Southeast Asia
分布标注 Distribution Note	非特有种 / Non-endemic

国内分布图
Map of Domestic Distribution

种群 Population

种群数量 Population Size	未知 / Unknown
种群趋势 Population Trend	下降 / Decreasing

生境与生态系统 Habitat(s) and Ecosystem(s)

生　　境 Habitat(s)	热带湿润低地森林、沼泽森林、开垦种植区 / Tropical Moist Lowland Forest, Swamp Forest, Farming Area
生态系统 Ecosystem(s)	森林生态系统、农田生态系统 / Forest Ecosystem, Cropland Ecosystem

威胁 Threat(s)

主要威胁 Major Threat(s)	生境丧失与生境破碎、耕种 / Habitat Loss and Fragmentation, Plantation

保护级别与保护行动 Protection Category and Conservation Action(s)

国家重点保护野生动物等级 (2021) Category of National Key Protected Wild Animals (2021)	一级 / Category I
IUCN 红色名录 (2020-2) IUCN Red List (2020-2)	易危 / VU
CITES 附录 (2019) CITES Appendix (2019)	I
保护行动 Conservation Action(s)	无 / None

相关文献 Relevant References

Burgin *et al.*, 2018; Jiang *et al.*（蒋志刚等），2017; Smith *et al.*（史密斯等），2009; Yang *et al.*, 2004; Wang（王应祥），2003; Ma, 1994; Gao *et al.*（高耀亭等），1987

马来熊 *Helarctos malayanus*

江獭
Lutrogale perspicillata

极危 CR C2a(i)

| 数据缺乏 DD | 无危 LC | 近危 NT | 易危 VU | 濒危 EN | **极危 CR** | 区域灭绝 RE | 野外灭绝 EW | 灭绝 EX |

分类地位 Taxonomic Status

动物界 Animalia	脊索动物门 Chordata	哺乳纲 Mammalia	食肉目 Carnivora	鼬科 Mustelidae

学 名 Scientific Name	*Lutrogale perspicillata*
命 名 人 Species Authority	I. Geoffroy Saint-Hilaire, 1826
英 文 名 English Name(s)	Smooth-coated Otter
同物异名 Synonym(s)	*Lutra perspicillata* (I. Geoffroy Saint-Hilaire, 1826)
种下单元评估 Infra-specific Taxa Assessed	无 / None

评估信息 Assessment Information

评估年份 Year Assessed	2020
评 定 人 Assessor(s)	蒋志刚 / Zhigang Jiang
审 定 人 Reviewer(s)	陈辈乐、李飞 / Bosco P. L. Chan, Fei Li
其他贡献人 Other Contributor(s)	李立立、丁晨晨 / Lili Li, Chenchen Ding

理由 Justification: 江獭曾分布于广东沿海岛屿及云南南部与西南部边境地带。然而，除 19 世纪 70 年代野外采集的标本记录外，中国再无江獭记录。近年嘉道理农场暨植物园在其原分布区进行了访问调查，没有获得该物种任何确切的消息。基于这些事实，故将江獭列为极危 / The Smooth-coated Otter was previously distributed in the coastal islands of Guangdong Province and the border area in southern Yunnan Province in Southwest China. However, there has been no further record of this river otter in China apart from the specimens obtained in the 1870s, and the surveys conducted by Kadoorie Farm and Botanic Garden in its original range provided no additional information. Thus, based on these facts, it is listed as Critically Endangered

地理分布 Geographical Distribution

国内分布 Domestic Distribution
广东、云南 / Guangdong, Yunnan
世界分布 World Distribution
中国；南亚、东南亚 / China; South Asia, Southeast Asia
分布标注 Distribution Note
非特有种 / Non-endemic

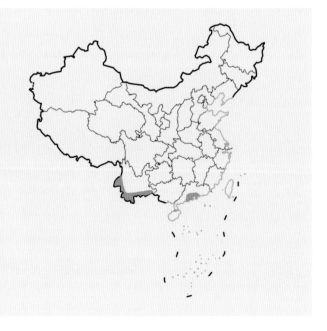

国内分布图
Map of Domestic Distribution

种群 Population

种群数量 Population Size	未知 / Unknown
种群趋势 Population Trend	下降 / Decreasing

生境与生态系统 Habitat (s) and Ecosystem (s)

生　　境 Habitat(s)	红树林、江河、沼泽、农田 / Mangrove, River, Swamp, Cropland
生态系统 Ecosystem(s)	湿地生态系统、湖泊河流生态系统、农田生态系统 / Wetland Ecosystem, Lake and River Ecosystem, Cropland Ecosystem

威胁 Threat (s)

主要威胁 Major Threat(s)	堤坝及水道修建、耕种、食物缺乏、污染、狩猎 / Dam and Cannal Construction, Plantation, Food Shortage, Pollution, Hunting

保护级别与保护行动 Protection Category and Conservation Action (s)

国家重点保护野生动物等级 (2021) Category of National Key Protected Wild Animals (2021)	二级 / Category II
IUCN 红色名录(2020-2) IUCN Red List (2020-2)	易危 / VU
CITES 附录 (2019) CITES Appendix (2019)	II
保护行动 Conservation Action(s)	该种的部分关键栖息地部分位于自然保护区内 / Part of the species' critical habitats are covered by nature reserves

相关文献 Relevant References

Burgin *et al*., 2018; Li and Chan, 2018; Jiang *et al*.（蒋志刚等）, 2017; Shi *et al*.（师蕾等）, 2013; Lau *et al*., 2010; Kruuk *et al*., 1993; Gao *et al*.（高耀亭等）, 1987

江獭 *Lutrogale perspicillata*

中国生物多样性红色名录
China's Red List of Biodiversity

小爪水獭
Aonyx cinerea

极危　CR A3d; C2a

| 数据缺乏 DD | 无危 LC | 近危 NT | 易危 VU | 濒危 EN | **极危 CR** | 区域灭绝 RE | 野外灭绝 EW | 灭绝 EX |

分类地位 Taxonomic Status

| 动物界 Animalia | 脊索动物门 Chordata | 哺乳纲 Mammalia | 食肉目 Carnivora | 鼬科 Mustelidae |

学名 Scientific Name	*Aonyx cinerea*
命名人 Species Authority	Illiger, 1815
英文名 English Name(s)	Asian Small-clawed Otter
同物异名 Synonym(s)	Oriental Small-clawed Otter, Small-clawed Otter
种下单元评估 Infra-specific Taxa Assessed	无 / None

评估信息 Assessment Information

评估年份 Year Assessed	2020
评定人 Assessor(s)	蒋志刚 / Zhigang Jiang
审定人 Reviewer(s)	陈辈乐、李飞 / Bosco P. L. Chan, Fei Li
其他贡献人 Other Contributor(s)	李立立、丁晨晨 / Lili Li, Chenchen Ding

理由 Justification: 由于人类活动，小爪水獭的栖息地不断受到破坏。工业废水和农药的使用可直接毒死小爪水獭。为了入药或者皮毛，人类长期过度捕猎，严重威胁着该物种的生存。这些原因使小爪水獭的种群数量不断下降。基于其种群现状，小爪水獭被列为极危 / The habitat of Asian Small-clawed Otter is constantly being destroyed by human encroachment. Industrial wastewater and agricultural chemicals can fatally poison this otter. People have long hunted the Asian Small-clawed Otter for its fur or for traditional Chinese medicine, threatening the survival of this species. All these reasons have resulted in the dramatic decrease of this species' population. Based on its population status, it is listed as Critically Endangered

地理分布 Geographical Distribution

国内分布 Domestic Distribution	福建、广东、广西、贵州、湖南、江西、云南、西藏、香港、海南 / Fujian, Guangdong, Guangxi, Guizhou, Hunan, Jiangxi, Yunnan, Tibet (Xizang), Hong Kong, Hainan
世界分布 World Distribution	中国；南亚、东南亚 / China; South Asia, Southeast Asia
分布标注 Distribution Note	非特有种 / Non-endemic

国内分布图
Map of Domestic Distribution

种群 Population

种群数量 Population Size	未知 / Unknown
种群趋势 Population Trend	下降 / Decreasing

生境与生态系统 Habitat (s) and Ecosystem (s)

生　　境 Habitat(s)	红树林、沼泽、池塘、淡水湖 / Mangrove, Swamp, Pond, Freshwater Lake
生态系统 Ecosystem(s)	湿地生态系统、湖泊河流生态系统 / Wetland Ecosystem, Lake and River Ecosystem

威胁 Threat (s)

主要威胁 Major Threat(s)	堤坝及水道修建、耕种、食物缺乏、农业或林业污染、狩猎 / Dam and Cannal Construction, Plantation, Food Shortage, Agricultural or Forestry Effluent, Hunting

保护级别与保护行动 Protection Category and Conservation Action (s)

国家重点保护野生动物等级 (2021) Category of National Key Protected Wild Animals (2021)	二级 / Category II
IUCN 红色名录(2020-2) IUCN Red List (2020-2)	易危 / VU
CITES 附录 (2019) CITES Appendix (2019)	II
保护行动 Conservation Action(s)	部分种群位于自然保护区内 / Parts of its populations are covered by nature reserves

相关文献 Relevant References

Burgin *et al*., 2018; Li and Chan (李飞和陈辈乐), 2017; Jiang *et al*. (蒋志刚等), 2017; Wang *et al*. (王丕烈等), 2008; Ma *et al*. (马志强等), 2007; Zhao (赵正阶), 1999; Gao *et al*. (高耀亭等), 1987

小爪水獭 *Aonyx cinerea* © Fir0002 / Flagstaffotos

大斑灵猫
Viverra megaspila

极危 CR B1ab(i, ii, iii); C2a(i)

| 数据缺乏 DD | 无危 LC | 近危 NT | 易危 VU | 濒危 EN | 极危 CR | 区域灭绝 RE | 野外灭绝 EW | 灭绝 EX |

分类地位 Taxonomic Status

动物界 Animalia	脊索动物门 Chordata	哺乳纲 Mammalia	食肉目 Carnivora	灵猫科 Viverridae

学 名 Scientific Name	*Viverra megaspila*
命 名 人 Species Authority	Blyth, 1862
英 文 名 English Name(s)	Large-spotted Civet
同物异名 Synonym(s)	无 / None
种下单元评估 Infra-specific Taxa Assessed	无 / None

评估信息 Assessment Information

评 估 年 份 Year Assessed	2020
评 定 人 Assessor(s)	蒋志刚 / Zhigang Jiang
审 定 人 Reviewer(s)	陈辈乐、李飞 / Bosco P. L. Chan, Fei Li
其他贡献人 Other Contributor(s)	李立立、丁晨晨 / Lili Li, Chenchen Ding

理由 Justification: 大斑灵猫生活在中国云南与广西低海拔热带森林中。由于人类活动，其栖息地面积减少、破碎化严重，野外种群数量稀少。近年来仅在西双版纳的红外相机调查中记录到一次大斑灵猫(Wei *et al*., 2017)。基于其种群数量与种群数量下降速率，将其列为极危 / The Large-spotted Civet lives in the low altitude tropical forest in Yunnan and Guangxi, its habitats are either lost or fragmented due to human activities. Its population density is quite low, only one Large-spotted Civet was recorded in Xishuangbanna by the infrared camera in recent years (Wei *et al*., 2017). Thus, based on its population density and the rate of population decline, Large-spotted Civet Cat is listed as Critically Endangered

地理分布 Geographical Distribution

国内分布 Domestic Distribution	广西、云南、贵州 / Guangxi, Yunnan, Guizhou
世界分布 World Distribution	中国；东南亚 / China; Southeast Asia
分布标注 Distribution Note	非特有种 / Non-endemic

国内分布图
Map of Domestic Distribution

种群 Population

种群数量 Population Size	未知 / Unknown
种群趋势 Population Trend	下降 / Decreasing

生境与生态系统 Habitat (s) and Ecosystem (s)

生　　境 Habitat(s)	亚热带湿润低地山地森林 / Subtropical Moist Lowland Montane Forest
生态系统 Ecosystem(s)	森林生态系统 / Forest Ecosystem

威胁 Threat (s)

主要威胁 Major Threat(s)	狩猎、伐木、耕种 / Hunting, Logging, Plantation

保护级别与保护行动 Protection Category and Conservation Action (s)

国家重点保护野生动物等级 (2021) Category of National Key Protected Wild Animals (2021)	一级 / Category I
IUCN 红色名录 (2020-2) IUCN Red List (2020-2)	易危 / VU
CITES 附录 (2019) CITES Appendix (2019)	III
保护行动 Conservation Action(s)	部分种群位于自然保护区内 / Part of populations are encompassed within nature reserves

相关文献 Relevant References

Burgin *et al*., 2018; Jiang *et al*. (蒋志刚等), 2017; Wei *et al*., 2017; Lau *et al*., 2010; Zhang (张荣祖), 1997; Li and Li (李义明和李典谟), 1994; Gao *et al*. (高耀亭等), 1987

大斑灵猫 *Viverra megaspila*

大灵猫
Viverra zibetha

极危　CR B1ab(i, ii, iii); C2a(i)

| 数据缺乏 DD | 无危 LC | 近危 NT | 易危 VU | 濒危 EN | **极危 CR** | 区域灭绝 RE | 野外灭绝 EW | 灭绝 EX |

分类地位 Taxonomic Status

动物界 Animalia	脊索动物门 Chordata	哺乳纲 Mammalia	食肉目 Carnivora	灵猫科 Viverridae

学　　名 Scientific Name	*Viverra zibetha*
命　名　人 Species Authority	Linnaeus, 1758
英　文　名 English Name(s)	Large Indian Civet
同物异名 Synonym(s)	*Viverra tainguensis* (Sokolov, Rozhnov and Pham Chong, 1997)
种下单元评估 Infra-specific Taxa Assessed	无 / None

评估信息 Assessment Information

评估年份 Year Assessed	2020
评　定　人 Assessor(s)	蒋志刚 / Zhigang Jiang
审　定　人 Reviewer(s)	陈辈乐、李飞 / Bosco P. L. Chan, Fei Li
其他贡献人 Other Contributor(s)	李立立、丁晨晨 / Lili Li, Chenchen Ding

理由 Justification: 大灵猫历史上广泛分布于中国黄河以南地区。中国记录有 4 个亚种。西南亚种 *V. z. ashtoni* 仅在四川马关地区发现一只受伤被救个体，重庆有一次猎杀记录；印度支那亚种 *V. z. surdaster* 目前只在西双版纳 (白德凤等，2018) 以及德宏 (猫科动物保护联盟，私人通信) 记录到；缅北亚种 *V. z. picta* 目前只在西藏墨脱有记录 (温立嘉等，2014)；海南亚种 *V. z. hainana* 自 19 世纪 70 年代后再无确切记录。鉴于该物种历史分布广泛，但近 10 年来记录稀少，种群及分布区明显减少，故列为极危 / Historically, the Large Indian Civet was distributed widely in areas south of the Yellow River in China. Four subspecies are recorded in China. *V. z. ashtoni* is a subspecies in Southwest China, with records of an individual rescued in the Maguan area of Sichuan and one killing noted in Chongqing. *V. z. surdaster* is recorded only in Xishuangbanna (Bai *et al.*, 2018) and Dehong (Cat Protection Alliance, Personal communication) in Yunnan Province. *V. z. picta*, a subspecies found mainly in northern Myanmar, has been recorded in Mêdog County in Tibet (Xizang) (Wen *et al.*, 2014). *V. z. hainan* has not been officially recorded since the 1970s. In view of its widely historical distribution, but sparse records in the past decade, it is apparent that the population and distribution of the Large Indian Civet have significantly reduced. Thus, it is listed as Critically Endangered

地理分布 Geographical Distribution

国内分布 Domestic Distribution
湖南、海南、浙江、上海、江苏、安徽、福建、江西、湖北、广东、广西、四川、贵州、云南、西藏、陕西、甘肃、重庆 / Hunan, Hainan, Zhejiang, Shanghai, Jiangsu, Anhui, Fujian, Jiangxi, Hubei, Guangdong, Guangxi, Sichuan, Guizhou, Yunnan, Tibet (Xizang), Shaanxi, Gansu, Chongqing
世界分布 World Distribution
中国；南亚、东南亚 / China; South Asia, Southeast Asia
分布标注 Distribution Note
非特有种 / Non-endemic

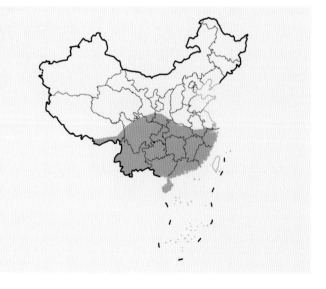

国内分布图
Map of Domestic Distribution

种群 Population

种群数量 Population Size	未知 / Unknown
种群趋势 Population Trend	下降 / Decreasing

生境与生态系统 Habitat (s) and Ecosystem (s)

生　　境 Habitat(s)	森林、灌丛、农田 / Forest, Shrubland, Cropland
生态系统 Ecosystem(s)	森林生态系统、灌丛生态系统、农田生态系统 / Forest Ecosystem, Shrubland Ecosystem, Cropland Ecosystem

威胁 Threat (s)

主要威胁 Major Threat(s)	狩猎、伐木、耕种 / Hunting, Logging, Plantation

保护级别与保护行动 Protection Category and Conservation Action (s)

国家重点保护野生动物等级 (2021) Category of National Key Protected Wild Animals (2021)	一级 / Category I
IUCN 红色名录 (2020-2) IUCN Red List (2020-2)	近危 / NT
CITES 附录 (2019) CITES Appendix (2019)	III
保护行动 Conservation Action(s)	部分种群位于自然保护区内 / Part of populations are covered by nature reserves

相关文献 Relevant References

Bai *et al.* (白德凤等), 2018; Burgin *et al.*, 2018; Jiang *et al.* (蒋志刚等), 2017; Wen *et al.* (温立嘉等), 2014; Lau *et al.*, 2010; Zhong (钟福生), 2001; Zhang (张荣祖), 1997; Gao *et al.* (高耀亭等), 1987

大灵猫 *Viverra zibetha*

熊狸
Arctictis binturong

极危　CR B1ab(i, ii, iii); C2a(i)

| 数据缺乏 DD | 无危 LC | 近危 NT | 易危 VU | 濒危 EN | 极危 CR | 区域灭绝 RE | 野外灭绝 EW | 灭绝 EX |

分类地位 Taxonomic Status

动物界 Animalia	脊索动物门 Chordata	哺乳纲 Mammalia	食肉目 Carnivora	灵猫科 Viverridae
学名 Scientific Name		*Arctictis binturong*		
命名人 Species Authority		Raffles, 1821		
英文名 English Name(s)		Binturong		
同物异名 Synonym(s)		Bearcat		
种下单元评估 Infra-specific Taxa Assessed		无 / None		

评估信息 Assessment Information

评估年份 Year Assessed	2020
评定人 Assessor(s)	蒋志刚 / Zhigang Jiang
审定人 Reviewer(s)	李晟、胡慧建、王大军、王昊、周友兵、陈辈乐、李飞 / Sheng Li, Huijian Hu, Dajun Wang, Hao Wang, Youbing Zhou, Bosco P. L. Chan, Fei Li
其他贡献人 Other Contributor(s)	李立立、丁晨晨 / Lili Li, Chenchen Ding

理由 Justification: 中国为熊狸边缘分布区，种群数量少，仅在云南盈江、西双版纳、河口及广西瑶山有熊狸记录。2014年8月在云南瑞丽猎人手里发现一具熊狸死体的报道是中国20余年来唯一记录 (Huang *et al.*, 2017)。熊狸记录稀少，且适合该种生存的低海拔热带森林破坏严重，故熊狸列为极危 / China is marginal in the global range of the Binturong. Only small populations of Binturong have been recorded in Yingjiang, Xishuangbanna and Hekou in Yunnan and Yao Shan in Guangxi. For over 20 years, the only published record of Binturong in China has been one dead specimen found in the hands of a hunter in Ruili, Yunnan in August 2014 (Huang *et al.*, 2017). Because of its rare field records and the severe destruction of tropical forests that were suitable habitat for the species, Binturong is listed as Critically Endangered

地理分布 Geographical Distribution

国内分布 Domestic Distribution
广西、云南 / Guangxi, Yunnan
世界分布 World Distribution
中国；南亚、东南亚 / China; South Asia, Southeast Asia
分布标注 Distribution Note
非特有种 / Non-endemic

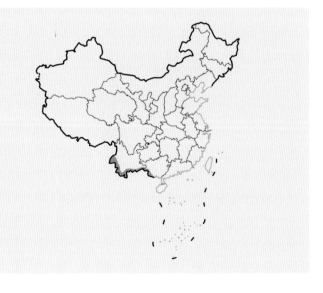

国内分布图
Map of Domestic Distribution

🦌 种群 Population

种群数量 Population Size	未知 / Unknown
种群趋势 Population Trend	下降 / Decreasing

🦌 生境与生态系统 Habitat (s) and Ecosystem (s)

生　　境 Habitat(s)	热带湿润低地森林 / Tropical Moist Lowland Forest
生态系统 Ecosystem(s)	森林生态系统 / Forest Ecosystem

🦌 威胁 Threat (s)

主要威胁 Major Threat(s)	耕种、伐木、狩猎 / Plantation, Logging, Hunting

🦌 保护级别与保护行动 Protection Category and Conservation Action (s)

国家重点保护野生动物等级 (2021) Category of National Key Protected Wild Animals (2021)	一级 / Category I
IUCN 红色名录 (2020-2) IUCN Red List (2020-2)	易危 / VU
CITES 附录 (2019) CITES Appendix (2019)	III (India)
保护行动 Conservation Action(s)	无 / None

🦌 相关文献 Relevant References

Burgin *et al.*, 2018; Huang *et al.*, 2017; Jiang *et al.* (蒋志刚等), 2017; Smith *et al.* (史密斯等), 2009; Wang (王应祥), 2003; Li (李思华), 1989; Gao *et al.* (高耀亭等), 1987

熊狸 *Arctictis binturong*

© Lynx Edicions

小齿狸
Arctogalidia trivirgata

极危 CR B1ab(i, ii, iii); C2a(i)

| 数据缺乏 DD | 无危 LC | 近危 NT | 易危 VU | 濒危 EN | **极危 CR** | 区域灭绝 RE | 野外灭绝 EW | 灭绝 EX |

分类地位 Taxonomic Status

动物界 Animalia	脊索动物门 Chordata	哺乳纲 Mammalia	食肉目 Carnivora	灵猫科 Viverridae
学 名 Scientific Name		*Arctogalidia trivirgata*		
命 名 人 Species Authority		Gray, 1832		
英 文 名 English Name(s)		Small-toothed Palm Civet		
同物异名 Synonym(s)		Three-striped Palm Civet		
种下单元评估 Infra-specific Taxa Assessed		无 / None		

评估信息 Assessment Information

评 估 年 份 Year Assessed	2020
评 定 人 Assessor(s)	蒋志刚 / Zhigang Jiang
审 定 人 Reviewer(s)	李晟、胡慧建、王大军、王昊、周友兵、陈辈乐、李飞 / Sheng Li, Huijian Hu, Dajun Wang, Hao Wang, Youbing Zhou, Bosco P. L. Chan, Fei Li
其他贡献人 Other Contributor(s)	李立立、丁晨晨 / Lili Li, Chenchen Ding

理由 Justification: 小齿狸在中国的记录极少。采于云南西双版纳的三件小齿狸标本是中国仅有的小齿狸记录。2018年3月4日，西双版纳森林公安曾没收一只小齿狸死体，但不确定此个体采自国内还是边境贸易所得。小齿狸适应能力较强，能生活在受干扰的森林中，树栖性强，不易被红外相机调查发现，所以中国很可能尚有残存种群未被调查到。但由于小齿狸在中国分布狭窄且记录稀少，故将其列为极危等级 / Small-toothed Palm Civet are very rarely recorded in China. Three specimens of Small-toothed Palm Civet collected in Xishuangbanna, Yunnan Province, is the only record of this species in China. On March 4, 2018, the Forestry Police Department of Xishuangbanna confiscated the carcass of a Small-toothed Palm Civet, however it was not adequeately ascertained whether the specimen was collected in China or came through international trade. The species is arboreal and well adapted to disturbed forest. It is difficult to photograph through infrared camera surveys, so it is likely that Small-toothed Palm Civet populations remain in China that have not yet been investigated. On the basis of its restricted distribution in China and limited records, it is listed as Critically Endangered

地理分布 Geographical Distribution

国内分布 Domestic Distribution
云南 / Yunnan
世界分布 World Distribution
中国；南亚、东南亚 / China; South Asia, Southeast Asia
分布标注 Distribution Note
非特有种 / Non-endemic

国内分布图
Map of Domestic Distribution

种群 Population

种群数量 Population Size	未知 / Unknown
种群趋势 Population Trend	下降 / Decreasing

生境与生态系统 Habitat (s) and Ecosystem (s)

生境 Habitat(s)	热带湿润低地森林、次生林、种植园。杂食性，通常以昆虫、小型哺乳动物、筑巢鸟类、水果、青蛙和蜥蜴为食。独居、树栖和夜间活动 / Tropical Moist Lowland Forest, Secondary Forest, Plantation. Omnivorous, usually prey on insect, small mammal, nesting bird, fruit, frog and lizard. Solitary, tree-dwelling and nocturnal
生态系统 Ecosystem(s)	森林生态系统、农田生态系统 / Forest Ecosystem, Cropland Ecosystem

威胁 Threat (s)

主要威胁 Major Threat(s)	狩猎、伐木、耕种 / Hunting, Logging, Plantation

保护级别与保护行动 Protection Category and Conservation Action (s)

国家重点保护野生动物等级 (2021) Category of National Key Protected Wild Animals (2021)	一级 / Category I
IUCN 红色名录 (2020-2) IUCN Red List (2020-2)	无危 / LC
CITES 附录 (2019) CITES Appendix (2019)	II
保护行动 Conservation Action(s)	无 / None

相关文献 Relevant References

Burgin *et al.*, 2018; Jiang *et al.*（蒋志刚等），2017; Hunter and Barrett, 2011; Tanomtong *et al.*, 2005; Zhang（张荣祖），1997; Chu（褚新洛），1989; Gao *et al.*（高耀亭等），1987

小齿狸 *Arctogalidia trivirgata*

缟灵猫
Chrotogale owstoni

极危　CR C2a(i)

| 数据缺乏 DD | 无危 LC | 近危 NT | 易危 VU | 濒危 EN | **极危 CR** | 区域灭绝 RE | 野外灭绝 EW | 灭绝 EX |

分类地位 Taxonomic Status

动物界 Animalia	脊索动物门 Chordata	哺乳纲 Mammalia	食肉目 Carnivora	灵猫科 Viverridae
学　　名 Scientific Name			Chrotogale owstoni	
命　名　人 Species Authority			Thomas, 1912	
英　文　名 English Name(s)			Owston's Palm Civet	
同物异名 Synonym(s)			无 / None	
种下单元评估 Infra-specific Taxa Assessed			无 / None	

评估信息 Assessment Information

评估年份 Year Assessed	2020
评　定　人 Assessor(s)	蒋志刚 / Zhigang Jiang
审　定　人 Reviewer(s)	李晟、胡慧建、王大军、王昊、周友兵、陈辈乐、李飞 / Sheng Li, Huijian Hu, Dajun Wang, Hao Wang, Youbing Zhou, Bosco P. L. Chan, Fei Li
其他贡献人 Other Contributor(s)	李立立、丁晨晨 / Lili Li, Chenchen Ding

理由 Justification: 缟灵猫历史上记录于云南东南部以及广西西南部与越南接壤的地区。然而，中国至少已30年没有该种动物的报道。故列为极危 / Owston's Palm Civet was historically recorded in southeast Yunnan and the southwest Guangxi bordering Viet Nam. Owston's Palm Civet has not been found in China for at least 30 years. Based on its population status, Owston's Palm Civet is listed as Critically Endangered

地理分布 Geographical Distribution

国内分布 Domestic Distribution
云南 / Yunnan
世界分布 World Distribution
中国；东南亚 / China; Southeast Asia
分布标注 Distribution Note
非特有种 / Non-endemic

国内分布图
Map of Domestic Distribution

种群 Population

种群数量 Population Size	300 只 / 300 individuals
种群趋势 Population Trend	下降 / Decreasing

生境与生态系统 Habitat(s) and Ecosystem(s)

生 境 Habitat(s)	森林、次生林、竹林 / Forests, Secondary Forest and Bamboo Forest
生态系统 Ecosystem(s)	森林生态系统 / Forest Ecosystem

威胁 Threat(s)

主要威胁 Major Threat(s)	伐木、耕种、狩猎 / Logging, Plantation, Hunting

保护级别与保护行动 Protection Category and Conservation Action(s)

国家重点保护野生动物等级 (2021) Category of National Key Protected Wild Animals (2021)	一级 / Category I
IUCN 红色名录 (2020-2) IUCN Red List (2020-2)	濒危 / EN
CITES 附录 (2019) CITES Appendix (2019)	无 / NA
保护行动 Conservation Action(s)	无 / None

相关文献 Relevant References

Burgin *et al.*, 2018; Jiang *et al.*（蒋志刚等）, 2017; Lau *et al.*, 2010; Pei *et al.*, 2010; Huo（霍晟）*et al.*, 2003; Wang（王应祥）, 2003; Gao *et al.*（高耀亭等）, 1987

缟灵猫 *Chrotogale owstoni*

荒漠猫
Felis bieti

极危 CR A2ab

| 数据缺乏 DD | 无危 LC | 近危 NT | 易危 VU | 濒危 EN | **极危 CR** | 区域灭绝 RE | 野外灭绝 EW | 灭绝 EX |

分类地位 Taxonomic Status

| 动物界 Animalia | 脊索动物门 Chordata | 哺乳纲 Mammalia | 食肉目 Carnivora | 猫科 Felidae |

学 名 Scientific Name	*Felis bieti*
命 名 人 Species Authority	Milne-Edwards, 1892
英 文 名 English Name(s)	Chinese Mountain Cat
同物异名 Synonym(s)	漠猫，Chinese Desert Cat, Chinese Steppe Cat
种下单元评估 Infra-specific Taxa Assessed	无 / None

评估信息 Assessment Information

评估年份 Year Assessed	2020
评 定 人 Assessor(s)	蒋志刚 / Zhigang Jiang
审 定 人 Reviewer(s)	李晟、胡慧建、王大军、王昊、周友兵 / Sheng Li, Huijian Hu, Dajun Wang, Hao Wang, Youbing Zhou
其他贡献人 Other Contributor(s)	李立立、丁晨晨 / Lili Li, Chenchen Ding

理由 Justification： 荒漠猫为中国的特有种，由于人类活动，其栖息地严重退化或丧失。20世纪50年代至1978年，草原地区大规模毒杀草原害鼠，间接毒杀了以鼠类为食的食肉动物。估计荒漠猫种群数量少于1,000只。故将荒漠猫的濒危等级列为极危 / Chinese Mountain Cat is a species endemic to China. However, the distribution range of this cat has been greatly affected by human encroachment, causing severe habitat loss or degradation. Furthermore, from the 1950s to 1978, chemical poisons were used extensively to control small burrowing mammals and inadvertently also killed many of these wild cats, which prey on the targeted (poisoned) species. In parts of the Chinese Mountain Cat distribution range, chemical poisons are still used and leading to the decline of its population, which is estimated to be less than 1,000 individuals in the wild. Thus, Chinese Mountain Cat is listed as Critically Endangered

地理分布 Geographical Distribution

国内分布 Domestic Distribution	青海、四川 / Qinghai, Sichuan
世界分布 World Distribution	中国 / China
分布标注 Distribution Note	特有种 / Endemic

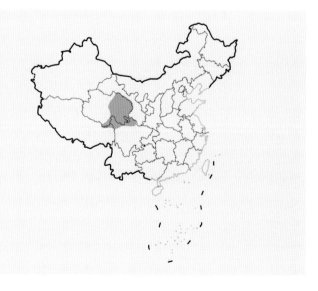

国内分布图
Map of Domestic Distribution

种群 Population

种群数量 Population Size	少于 1,000 只 / Less than 1,000 individuals
种群趋势 Population Trend	下降 / Decreasing

生境与生态系统 Habitat (s) and Ecosystem (s)

生　境 Habitat(s)	草甸、灌丛、泰加林、草地 / Meadow, Shrubland, Taiga Forest, Grassland
生态系统 Ecosystem(s)	森林生态系统、灌丛生态系统、草地生态系统 / Forest Ecosystem, Shrubland Ecosystem, Grassland Ecosystem

威胁 Threat (s)

主要威胁 Major Threat(s)	耕种、狩猎、人类活动干扰 / Plantation, Hunting, Human Disturbance

保护级别与保护行动 Protection Category and Conservation Action (s)

国家重点保护野生动物等级 (2021) Category of National Key Protected Wild Animals (2021)	一级 / Category I
IUCN 红色名录 (2020-2) IUCN Red List (2020-2)	易危 / VU
CITES 附录 (2019) CITES Appendix (2019)	II
保护行动 Conservation Action(s)	部分种群位于自然保护区内 / Part of populations are covered by nature reserve

相关文献 Relevant References

Burgin *et al*., 2018; Jiang *et al*. (蒋志刚等), 2017; Smith *et al*. (史密斯等), 2009; He *et al*., 2004; Liao (廖炎发), 1988; Gao *et al*. (高耀亭等), 1987

荒漠猫 *Felis bieti*

丛林猫 *Felis chaus*

极危 CR A2ab

| 数据缺乏 DD | 无危 LC | 近危 NT | 易危 VU | 濒危 EN | **极危 CR** | 区域灭绝 RE | 野外灭绝 EW | 灭绝 EX |

分类地位 Taxonomic Status

| 动物界 Animalia | 脊索动物门 Chordata | 哺乳纲 Mammalia | 食肉目 Carnivora | 猫科 Felidae |

学名 Scientific Name	*Felis chaus*
命名人 Species Authority	Schreber, 1777
英文名 English Name(s)	Jungle Cat
同物异名 Synonym(s)	Reed Cat, Swamp Cat
种下单元评估 Infra-specific Taxa Assessed	无 / None

评估信息 Assessment Information

评估年份 Year Assessed	2020
评定人 Assessor(s)	蒋志刚 / Zhigang Jiang
审定人 Reviewer(s)	陈辈乐、李飞 / Bosco P. L. Chan, Fei Li
其他贡献人 Other Contributor(s)	李立立、丁晨晨 / Lili Li, Chenchen Ding

理由 Justification: 丛林猫曾经零星分布于中国西部。近年来在中国无任何发现丛林猫的报道，其数量已经十分稀少。基于其种群现状，故将丛林猫列为极危 / Jungle Cat was once scattered distributed throughout western China, but there have been no reports of Jungle Cat in China in recent years, its population density is very low. Thus, Jungle Cat is listed as Critically Endangered

地理分布 Geographical Distribution

国内分布 Domestic Distribution	四川、云南、西藏、新疆 / Sichuan, Yunnan, Tibet (Xizang), Xinjiang
世界分布 World Distribution	中国；东南亚、南亚、中亚及西亚 / China; Southeast Asia, South Asia, Central Asia and West Asia
分布标注 Distribution Note	非特有种 / Non-endemic

国内分布图
Map of Domestic Distribution

种群 Population

种群数量 Population Size	未知 / Unknown
种群趋势 Population Trend	下降 / Decreasing

生境与生态系统 Habitat (s) and Ecosystem (s)

生　　境 Habitat(s)	草地、沼泽、森林、荒漠、乡村花园 / Grassland, Swamp, Forest, Desert, Rural Garden
生态系统 Ecosystem(s)	森林生态系统、草地生态系统、农田生态系统、荒漠生态系统、湿地生态系统 / Forest Ecosystem, Grassland Ecosystem, Cropland Ecosystem, Desert Ecosystem, Wetland Ecosystem

威胁 Threat (s)

主要威胁 Major Threat(s)	耕种、狩猎及采集 / Plantation, Hunting and Collection

保护级别与保护行动 Protection Category and Conservation Action (s)

国家重点保护野生动物等级 (2021) Category of National Key Protected Wild Animals (2021)	一级 / Category I
IUCN 红色名录 (2020-2) IUCN Red List (2020-2)	无危 / LC
CITES 附录 (2019) CITES Appendix (2019)	II
保护行动 Conservation Action(s)	部分种群位于自然保护区内 / Part of populations are covered by nature reserves

相关文献 Relevant References

Burgin *et al.*, 2018; Jiang *et al.*（蒋志刚等），2017; Hu *et al.*（胡一鸣等），2014; Li *et al.*（李云秀等），2012; Smith *et al.*（史密斯等），2009; Liu and Sheng（刘志霄和盛和林），1998; Gao *et al.*（高耀亭等），1987

丛林猫 *Felis chaus*

中国生物多样性 红色名录

云豹
Neofelis nebulosa

极危 CR A2ab; B1ab(i, ii, iii)

数据缺乏 DD	无危 LC	近危 NT	易危 VU	濒危 EN	**极危 CR**	区域灭绝 RE	野外灭绝 EW	灭绝 EX

分类地位 Taxonomic Status

动物界 Animalia	脊索动物门 Chordata	哺乳纲 Mammalia	食肉目 Carnivora	猫科 Felidae

学名 Scientific Name	*Neofelis nebulosa*
命名人 Species Authority	Griffith, 1821
英文名 English Name(s)	Clouded Leopard
同物异名 Synonym(s)	无 / None
种下单元评估 Infra-specific Taxa Assessed	无 / None

评估信息 Assessment Information

评估年份 Year Assessed	2020
评定人 Assessor(s)	蒋志刚 / Zhigang Jiang
审定人 Reviewer(s)	李晟、胡慧建、王大军、王昊、周友兵、陈辈乐、李飞 / Sheng Li, Huijian Hu, Dajun Wang, Hao Wang, Youbing Zhou, Bosco P. L. Chan, Fei Li
其他贡献人 Other Contributor(s)	李立立、丁晨晨 / Lili Li, Chenchen Ding

理由 Justification: 云豹的毛皮柔软，花纹美观，在国际市场上价格比普通豹更高。云豹的骨可入传统中药，因此对该种的过度捕猎和非法贸易，加上人类活动造成的栖息地退化和丧失严重威胁着该物种的生存。云豹曾广泛分布于中国南部地区，但近年的调查显示云豹的种群数量及分布区已经锐减，目前仅在西藏东南部、云南边境地带以及我国中部极少数的保护区中尚有云豹记录。而在台湾、海南两个岛屿的云豹种群似乎都已绝灭。因此，将云豹列为极危 / The fur of the Clouded Leopard is soft and has beautiful pattern, the price of the pelt of Clouded Leopard is therefore higher on the international market than that of the common leopard. The bone of this species also can be used in traditional Chinese medicine. Therefore, poaching and illegal trade along with human-caused degradation and loss of the species, main habitats are severely threatening the survival of the Clouded Leopard. The species used to be distributed widely across southern China. However, surveys in recent years show that the population and distribution area have been reduced sharply. At present, the Clouded Leopard is only recorded in southeast Tibet (Xizang), border areas of Yunnan, and a few protected areas in central China. Local populations of Clouded Leopard appear to be extinct in Taiwan and Hainan. Thus, it is listed as Critically Endangered

地理分布 Geographical Distribution

国内分布 Domestic Distribution	江西、海南、浙江、安徽、福建、湖北、湖南、广东、广西、四川、贵州、云南、西藏、陕西、甘肃、台湾、重庆 / Jiangxi, Hainan, Zhejiang, Anhui, Fujian, Hubei, Hunan, Guangdong, Guangxi, Sichuan, Guizhou, Yunnan, Tibet (Xizang), Shaanxi, Gansu, Taiwan, Chongqing
世界分布 World Distribution	中国；南亚、东南亚 / China; South Asia, Southeast Asia
分布标注 Distribution Note	非特有种 / Non-endemic

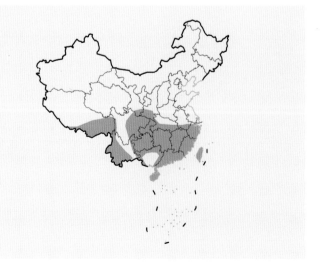

国内分布图
Map of Domestic Distribution

种群 Population

种群数量 Population Size	2,600 只 / 2,600 individuals
种群趋势 Population Trend	下降 / Decreasing

生境与生态系统 Habitat(s) and Ecosystem(s)

生境 Habitat(s)	常绿阔叶林、次生林、泰加林 / Evergreen Broad-leaved Forest, Secondary Forest, Taiga Forest
生态系统 Ecosystem(s)	森林生态系统 / Forest Ecosystem

威胁 Threat(s)

主要威胁 Major Threat(s)	捕猎、非法贸易，人类活动造成栖息地破坏 / Hunting, Illegal Trade; Habitat Destruction by Human Activity

保护级别与保护行动 Protection Category and Conservation Action(s)

国家重点保护野生动物等级 (2021) Category of National Key Protected Wild Animals (2021)	一级 / Category I
IUCN 红色名录 (2020-2) IUCN Red List (2020-2)	易危 / VU
CITES 附录 (2019) CITES Appendix (2019)	I
保护行动 Conservation Action(s)	在自然保护区内的种群得到保护 / Populations in nature reserves are under protection

相关文献 Relevant References

Burgin *et al.*, 2018; Jiang *et al.* (蒋志刚等), 2017; Christiansen and Kitchener, 2011; Piao *et al.* (朴正吉等), 2011; Feng and Jutzeler, 2010; Lau *et al.*, 2010; Gao *et al.* (高耀亭等), 1987

云豹 *Neofelis nebulosa*

中国生物多样性红色名录
China's Red List of Biodiversity

虎
Panthera tigris

极危 CR A2d; B1ab(i, ii, iii); C2(i)

| 数据缺乏 DD | 无危 LC | 近危 NT | 易危 VU | 濒危 EN | **极危 CR** | 区域灭绝 RE | 野外灭绝 EW | 灭绝 EX |

分类地位 Taxonomic Status

动物界 Animalia	脊索动物门 Chordata	哺乳纲 Mammalia	食肉目 Carnivora	猫科 Felidae

学 名 Scientific Name	*Panthera tigris*
命 名 人 Species Authority	Linnaeus, 1758
英 文 名 English Name(s)	Tiger
同物异名 Synonym(s)	*Felis tigris* (Linnaeus, 1758)
种下单元评估 Infra-specific Taxa Assessed	华南虎 (*P. t. amoyensis*)：20 世纪 80 年代以来，在原华南虎生境中没有发现野生华南虎个体存在的证据，故华南虎定为野生灭绝 (EW)。东北虎 (*P. t. altaica*)（转后第 2 页）

评估信息 Assessment Information

评 估 年 份 Year Assessed	2020
评 定 人 Assessor(s)	蒋志刚 / Zhigang Jiang
审 定 人 Reviewer(s)	冯利民、姜广顺、李晟、胡慧建 / Limin Feng, Guangshun Jiang, Sheng Li, Huijian Hu
其他贡献人 Other Contributor(s)	李立立、丁晨晨 / Lili Li, Chenchen Ding

理由 Justification: 虎 (*Panthera tigris*) 是森林生态系统的顶级捕食者。中国曾有 5 个虎亚种，其中，新疆虎，即里海虎 (*P. t. virgata*) 已于 1916 年在中国灭绝。20 世纪末，西藏墨脱曾有孟加拉虎 (*P. t. tigris*) 的报道，然而，尔后很长时间没有再次发现。直到最近，王渊等 (2019) 在墨脱发现了 2～3 只孟加拉虎的足迹，同期，蒋学龙等在墨脱拍摄到孟加拉虎的红外照片。在西双版纳有印支虎 (*P. t. corbetti*) 分布，但是数量十分稀少。在黑龙江和吉林中俄边境地区，尚有 20 只东北虎 (*P. t. altaica*) 迁徙游荡于中国、俄罗斯边境地区。中国、俄罗斯建立了东北虎保护区，加强了东北虎通道的建设和保护。华南虎 (*P. t. amoyensis*) 是中国特有虎亚种，模式标本产于福建。华南虎曾经广泛分布于湖南、江西、贵州、福建、广东、广西、安徽、浙江、湖北、四川、河南、陕西、山西等地。50 年代估计华南虎分布区有 4,000 余只华南虎。50～70 年代，华南虎被当成"害兽"，30 年中被猎杀了约 3,000 只。由于过度捕杀和栖息地丧失等，1980 年后，华南虎分布区再没有发现野生华南虎。1990～2010 年，国家林业局曾在原华南虎分布区开展过多次华南虎专项调查，均未发现华南虎存在的确切证据，所收集到的疑似华南虎活动痕迹十分零散，并在空间和时间上出现了极大的断裂。2004 年有人据此推断华南虎已经在野外功能性灭绝。虽然目前还不能认定华南虎已经灭绝，但有理由确信野生华南虎已经不能形成稳定种群。目前我国人工饲养的 100 余只华南虎，分散在全国十几家动物园等单位，受饲养条件、管理水平和近亲繁殖等因素的影响，种群发展缓慢。因此，虎的濒危等级仍定为极危 / Tiger (*Panthera tigris*) is the top predator in forest ecosystems. Five (5) subspecies of tiger existed in China, but *P. t. virgata* became extinct in China in 1916. At the end of twentieth century, *P. t. tigris* was reported in Mêdog, Tibet (Xizang), however it is now rarely found. And there were no reports about the Bengal Tiger until Wang *et al*. (2019) reported footprints of Bengal Tigers in Tibet (Xizang) and Xuelong Jiang *et al*. camera trapped Bengal Tiger in Mêdog in the same period. Moreover, a few individuals of *P. t. corbetti* are distributed in Xishuangbanna, but this subspecies is rare as well. In the border area between Russia and China (Heilongjiang and Jilin), 20 *P. t. altaica* individuals are migrating and wandering in the region. China and Russia have cooperated together and established a nature reserve to protect them. *P. t. amoyensis* is an endemic subspecies in China, with type specimen from Fujian Province. This subspecies used to be common in Hunan, Jiangxi, Guizhou, Fujian, Guangdong, Guangxi, Anhui, Zhejiang, Hubei, Sichuan, Henan, Shaanxi, Shanxi, *etc*. In the 1950s, around 4,000 *P. t. amoyensis* individuals were known to be extant. However, during the 1950s and into the 1970s, many were killed as vermin – about 3,000 individuals were killed at that time. Since the 1980s, no observations have been reported from its original distribution range due to over hunting and habitat loss. Between 1990 and 2010, the State Forest Administration conducted several surveys on *P. t. amoyensis*, but no firm evidence could verify its continued existence. On this basis, some experts have suggested that *P. t. amoyensis* is functionally extinct in the wild. However, this species cannot yet be identified as Extinct. Nonetheless it is reasonable to recognize that *P. t. amoyensis* is no longer able to maintain a stable population in the wild, and furthermore, its population cannot recover unaided in the wild. At present, China keeps and manages 100 breeding individuals of *P. t. amoyensis*, spread across a dozen breeding centres including zoos. Because of various challenges related to breeding conditions, conservation management, and inbreeding problems, the *in situ* population is growing very slowly. Thus, it is listed as Critically Endangered

🦌 地理分布 Geographical Distribution

国内分布 Domestic Distribution	
云南、吉林、黑龙江、西藏 / Yunnan, Jilin, Heilongjiang, Tibet (Xizang)	
世界分布 World Distribution	
孟加拉国、不丹、柬埔寨、中国、印度、印度尼西亚、老挝、马来西亚、缅甸、尼泊尔、俄罗斯、泰国、越南 / Bangladesh, Bhutan, Cambodia, China, India, Indonesia, Laos, Malaysia, Myanmar, Nepal, Russia, Thailand, Viet Nam	
分布标注 Distribution Note	
非特有种 / Non-endemic	

国内分布图
Map of Domestic Distribution

🦌 种群 Population

种群数量 Population Size	20 只 / 20 individuals
种群趋势 Population Trend	下降 / Decreasing

🦌 生境与生态系统 Habitat (s) and Ecosystem (s)

生境 Habitat(s)	灌丛、森林、红树林、沼泽 / Shrubland, Forest, Mangrove, Swamp
生态系统 Ecosystem(s)	森林生态系统、灌丛生态系统、湿地生态系统 / Forest Ecosystem, Shrubland Ecosystem, Wetland Ecosystem

🦌 威胁 Threat (s)

主要威胁 Major Threat(s)	狩猎、住宅及商业发展、耕种、伐木、人兽冲突 / Hunting, Residential and Commercial Development, Plantation, Logging, Human-Animal Conflict

🦌 保护级别与保护行动 Protection Category and Conservation Action (s)

国家重点保护野生动物等级 (2021) Category of National Key Protected Wild Animals (2021)	一级 / Category I
IUCN红色名录(2020-2) IUCN Red List (2020-2)	濒危 / EN
CITES 附录 (2019) CITES Appendix (2019)	I
保护行动 Conservation Action(s)	在自然保护区内的种群得到保护 / Populations in nature reserves are under protection

🦌 相关文献 Relevant References

Wang *et al.* (王渊等), 2019; Burgin *et al.*, 2018; Jiang *et al.* (蒋志刚等), 2017; Wen (文榕生), 2016; Tian *et al.*, 2014; Feng *et al.* (冯利民等), 2013; Zhang *et al.*, 2013a; Tian *et al.* (田瑜等), 2009; Gao *et al.* (高耀亭等), 1987

虎 *Panthera tigris* © Lynx Edicions

(续前)

在中国东北开展的红外相机调查发现了20余只东北虎；东北虎定为极危(CR)。印支虎(*P. t. corbetti*)：云南南部地区没有发现定居的印支虎，由于周边国家的印支虎数量不断下降，在可预见的未来可能不会有国外的印支虎个体扩散到中国境内，印支虎在中国极危(CR)。孟加拉虎(*P. t. tigris*)：孟加拉虎仍在雅鲁藏布江河谷游荡，西藏是否有定居的孟加拉虎，尚存疑问；孟加拉虎的濒危等级为极危(CR)。里海虎(*P. t. virgata*)：19世纪末20世纪初已在新疆灭绝(EX) / South China Tiger (*P. t. amoyensis*): No evidence of existence has been found in wild in its original habitat in South China since 1980s; it is Extinct in the Wild (EW) in the country; Amur Tiger (*P. t. altaica*): Only dozens of individuals have been infrared camera trapped in northeast China; Amur Tiger is Critically Endangered (CR) in the country; Indochinese Tiger (*P. t. corbetti*): There is no resident Indochinese Tiger in southern Yunnan Province, as the status of Indochinese Tiger population in neighboring country is deteriorating; there may be no more individuals to disperse into China in the foreseeable future; Indochinese Tiger is CR in the country; Bengal Tiger (*P. t. tigris*): Whether there is resident Bengal Tiger in Tibet (Xizang) was still in doubt, however, single Bengal Tiger still reported wandered into the Yarlung Zangbo River (Brahmaputra) Valley, Tibet (Xizang); the Bengal Tiger is Critically Endangered (CR) in the country. Caspian Tiger (*P. t. virgate*): Caspian Tiger was extinct at the end of 19th century and the beginning of 20th century in Xinjiang, it is Extinct (EX) in the country

儒艮
Dugong dugon
极危　CR A1

| 数据缺乏 DD | 无危 LC | 近危 NT | 易危 VU | 濒危 EN | 极危 CR | 区域灭绝 RE | 野外灭绝 EW | 灭绝 EX |

分类地位 Taxonomic Status

动物界 Animalia	脊索动物门 Chordata	哺乳纲 Mammalia	海牛目 Sirenia	儒艮科 Dugongidae

学　名 Scientific Name	*Dugong dugon*
命名人 Species Authority	Müller, 1776
英文名 English Name(s)	Dugong
同物异名 Synonym(s)	无 / None
种下单元评估 Infra-specific Taxa Assessed	无 / None

评估信息 Assessment Information

评估年份 Year Assessed	2020
评定人 Assessor(s)	周开亚 / Kaiya Zhou
审定人 Reviewer(s)	蒋志刚 / Zhigang Jiang
其他贡献人 Other Contributor(s)	李立立、丁晨晨 / Lili Li, Chenchen Ding

理由 Justification: 儒艮专食热带印度洋、太平洋的海草。20世纪50～70年代，在广东、广西和海南沿岸还有数量不多的儒艮。进入21世纪后，只在2008年6月在海南省文昌市椰林湾发现了1头雌性儒艮尸体。因此，儒艮列为极危等级 / Dugong feeds on seagrass in tropical Indo-Pacific Ocean. In the 1950s to the 1970s, a small number of Dugongs lived along the coasts of Guangdong, Guangxi, and Hainan. In the 21st century, only one female Dugong carcass was found in Yelinwan, Wenchang City, Hainan Province in June 2008. Thus, Dugong is listed as Critically Endangered

地理分布 Geographical Distribution

国内分布 Domestic Distribution
南海 / South China Sea
世界分布 World Distribution
印度洋、太平洋 / Indian Ocean, Pacific Ocean
分布标注 Distribution Note
非特有种 / Non-endemic

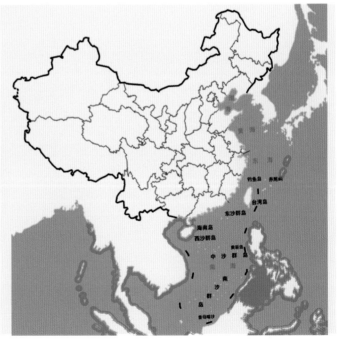

国内分布图
Map of Domestic Distribution

种群 Population

种群数量 Population Size	未知 / Unknown
种群趋势 Population Trend	未知 / Unknown

生境与生态系统 Habitat (s) and Ecosystem (s)

生　　境 Habitat(s)	海洋 - 浅海海域 / Marine-Coastal Water
生态系统 Ecosystem(s)	海洋生态系统 / Ocean Ecosystem

威胁 Threat (s)

主要威胁 Major Threat(s)	人类活动干扰 / Human Disturbance

保护级别与保护行动 Protection Category and Conservation Action (s)

国家重点保护野生动物等级 (2021) Category of National Key Protected Wild Animals (2021)	一级 / Category I
IUCN 红色名录 (2020-2) IUCN Red List (2020-2)	易危 / VU
CITES 附录 (2019) CITES Appendix (2019)	I
保护行动 Conservation Action(s)	在北部湾建立了儒艮自然保护区，但保护区内未发现儒艮 / Dugong Nature Reserve was established in the Beibu Gulf, but dugong are not found in the reserve

相关文献 Relevant References

Burgin *et al*., 2018; Jiang *et al*.（蒋志刚等），2015a; Qiu *et al*.（邱广龙等），2013; Wang *et al*.（王力军等），2010; Zhou（周开亚），2004

儒艮 *Dugong dugon*

中国生物多样性红色名录
China's Red List of Biodiversity

亚洲象
Elephas maximus

极危 CR D1

DD	LC	NT	VU	EN	**CR**	RE	EW	EX
数据缺乏	无危	近危	易危	濒危	**极危**	区域灭绝	野外灭绝	灭绝

分类地位 Taxonomic Status

动物界 Animalia	脊索动物门 Chordata	哺乳纲 Mammalia	长鼻目 Proboscidea	象 科 Elephantidae

学 名 Scientific Name	*Elephas maximus*
命 名 人 Species Authority	Linnaeus, 1758
英 文 名 English Name(s)	Asian Elephant
同物异名 Synonym(s)	Asiatic Elephant
种下单元评估 Infra-specific Taxa Assessed	无 / None

评估信息 Assessment Information

评估年份 Year Assessed	2020
评 定 人 Assessor(s)	蒋志刚 / Zhigang Jiang
审 定 人 Reviewer(s)	张立 / Li Zhang
其他贡献人 Other Contributor(s)	李立立、丁晨晨 / Lili Li, Chenchen Ding

理由 Justification: 亚洲象在中国的分布区是其分布区的北缘，亚洲象分布在西双版纳、普洱和临沧，仅200余头，受到栖息地丧失和猎杀的威胁。其栖息地面积每年正以 33km² 的速度减少。因为亚洲象致人伤亡、损害作物而遭致报复性伤害，也曾有偷猎者为了获取象牙而偷猎。人象冲突加剧。种群间没有交流。因此，亚洲象列为极危等级 / China is the northern part of Asian Elephant's range, more than 200 Asian Elephants roam in Xishuangbanna, Puer and Lincang, where the habitat area of Asian Elephant is decreasing at a speed of 33km² per year. Asian Elephant also is poached for its ivories or for retaliating during escalating human-elephant conflict. There is no gene exchange between populations. Thus, it is listed as Critically Endangered

地理分布 Geographical Distribution

国内分布 Domestic Distribution
云南 / Yunnan
世界分布 World Distribution
中国；南亚、东南亚 / China; South Asia, Southeast Asia
分布标注 Distribution Note
非特有种 / Non-endemic

国内分布图
Map of Domestic Distribution

种群 Population

种群数量 Population Size	4 个种群 293 只 / 293 individuals in 4 populations
种群趋势 Population Trend	稳定 / Stable

生境与生态系统 Habitat(s) and Ecosystem(s)

生　　境 Habitat(s)	热带湿润低地森林 / Tropical Moist Lowland Forest
生态系统 Ecosystem(s)	森林生态系统 / Forest Ecosystem

威胁 Threat(s)

主要威胁 Major Threat(s)	栖息地丧失、人象冲突 / Habitat Loss, Human-Elephant Conflict

保护级别与保护行动 Protection Category and Conservation Action(s)

国家重点保护野生动物等级 (2021) Category of National Key Protected Wild Animals (2021)	一级 / Category I
IUCN 红色名录 (2020-2) IUCN Red List (2020-2)	濒危 / EN
CITES 附录 (2019) CITES Appendix (2019)	I
保护行动 Conservation Action(s)	已经建立自然保护区 / Nature reserves established

相关文献 Relevant References

Burgin *et al*., 2018; Jiang *et al*.（蒋志刚等）, 2017; Wilson and Mittermeier, 2012; Groves and Grubb, 2011; Lin *et al*.（林柳等）, 2011; Zhang, 2011; Feng *et al*.（冯利民等）, 2010; Zhang（张立）, 2006

亚洲象 *Elephas maximus*

野骆驼
Camelus ferus

极危 CR A1acd; B1ab(i, ii, iii)+2ab(i, ii, iii)

| 数据缺乏 DD | 无危 LC | 近危 NT | 易危 VU | 濒危 EN | **极危 CR** | 区域灭绝 RE | 野外灭绝 EW | 灭绝 EX |

分类地位 Taxonomic Status

动物界 Animalia	脊索动物门 Chordata	哺乳纲 Mammalia	偶蹄目 Artiodactyla	骆驼科 Camelidae
学名 Scientific Name		*Camelus ferus*		
命名人 Species Authority		Przewalski, 1878		
英文名 English Name(s)		Bactrian Camel		
同物异名 Synonym(s)		*Camelus bactrianus* (Linnaeus, 1758)		
种下单元评估 Infra-specific Taxa Assessed		无 / None		

评估信息 Assessment Information

评估年份 Year Assessed	2020
评定人 Assessor(s)	蒋志刚 / Zhigang Jiang
审定人 Reviewer(s)	李迪强 / Diqiang Li
其他贡献人 Other Contributor(s)	李立立、丁晨晨 / Lili Li, Chenchen Ding

理由 Justification: 野骆驼仅分布在罗布泊和塔克拉玛干沙漠地区，最近调查估计约有600头野骆驼。由于人类活动挤占了其栖息地，野骆驼被迫退居荒漠。其生存环境严酷，跨国迁徙通道被阻断。开矿、越野旅游也对野骆驼生存造成了一定的影响，因此，基于其栖息地与种群现状，野骆驼列为极危等级 / Bactrian Camel is only distributed in Lop Nor area and in the Taklimakan Desert, there are about 600 individuals according to the latest population survey. Bactrian Camel retreated to deep desert as its original habitat is overtaken by human being and livestock, now it lives in a harsh desert environment, border wire fences obstruct it cross boundary migration. Mining and vehicle travelling transversing in its habitat also interfere the wild camel. Thus, based on its habitat and population status, it is listed as Critically Endangered

地理分布 Geographical Distribution

国内分布 Domestic Distribution	新疆、内蒙古、甘肃 / Xinjiang, Inner Mongolia (Nei Mongol), Gansu
世界分布 World Distribution	中国、蒙古 / China, Mongolia
分布标注 Distribution Note	非特有种 / Non-endemic

国内分布图
Map of Domestic Distribution

种群 Population

种群数量 Population Size	600 只 / 600 individuals
种群趋势 Population Trend	下降 / Decreasing

生境与生态系统 Habitat(s) and Ecosystem(s)

生境 Habitat(s)	荒漠、半荒漠、灌丛 / Desert, Semi-desert, Shrubland
生态系统 Ecosystem(s)	灌丛生态系统、荒漠生态系统 / Shrubland Ecosystem, Desert Ecosystem

威胁 Threat(s)

主要威胁 Major Threat(s)	家畜放牧、与家骆驼杂交、采矿、狩猎 / Livestock Ranching, Hybridization with Domestic Camel, Mining, Hunting

保护级别与保护行动 Protection Category and Conservation Action(s)

国家重点保护野生动物等级 (2021) Category of National Key Protected Wild Animals (2021)	一级 / Category I
IUCN 红色名录 (2020-2) IUCN Red List (2020-2)	极危 / CR
CITES 附录 (2019) CITES Appendix (2019)	无 / NA
保护行动 Conservation Action(s)	自然保护区内种群得到保护；列入《迁徙物种保护公约》的"迁徙物种行动计划" / Populations in nature reserves are protected; included in the *Convention on Migratory Species* (CMS)－Species Action Plan

相关文献 Relevant References

Burgin *et al*., 2018; Jiang *et al*. (蒋志刚等), 2017; Yuan *et al*., 2014; Wilson and Mittermeier, 2012; Groves and Grubb, 2011; Silbermayr *et al*., 2010; Yao (姚积生), 2009

野骆驼 *Camelus ferus* 蒋志刚 摄 By Zhigang Jiang

林麝
Moschus berezovskii

极危　CR A1acd; B1ab(i, ii, iii)

| 数据缺乏 DD | 无危 LC | 近危 NT | 易危 VU | 濒危 EN | **极危 CR** | 区域灭绝 RE | 野外灭绝 EW | 灭绝 EX |

分类地位 Taxonomic Status

动物界 Animalia	脊索动物门 Chordata	哺乳纲 Mammalia	偶蹄目 Artiodactyla	麝科 Moschidae

学名 Scientific Name	*Moschus berezovskii*
命名人 Species Authority	Flerov, 1929
英文名 English Name(s)	Forest Musk Deer
同物异名 Synonym(s)	Chinese Forest Musk Deer, Dwarf Musk Deer
种下单元评估 Infra-specific Taxa Assessed	无 / None

评估信息 Assessment Information

评估年份 Year Assessed	2020
评定人 Assessor(s)	蒋志刚 / Zhigang Jiang
审定人 Reviewer(s)	孟秀祥、胡德夫、鲍毅新 / Xiuxiang Meng, Defu Hu, Yixin Bao
其他贡献人 Other Contributor(s)	李立立、丁晨晨 / Lili Li, Chenchen Ding

理由 Justification: 林麝在我国曾分布广泛，但由于生境丧失，目前林麝分布区萎缩。20 世纪末 21 世纪初，曾经一度发生大规模偷猎，包括林麝在内的麝科动物被钢丝套套杀，导致一些地区的林麝种群数量急剧下降，甚至在一些地区局部灭绝。因此，基于其种群数量下降速率和栖息地现状，将林麝列为极危等级 / Forest Musk Deer was once wildly distributed in the country, but its range is greatly reduced now. A large scale poaching of musk deer, including Forest Musk Deer, occurred during the end of 20th century and early 21st century, which caused rapid population decline of Forest Musk Deer; its population even extirpated in some areas. Therefore, based on its rate of population decline and habitat status, Forest Musk Deer is listed as Critically Endangered

地理分布 Geographical Distribution

国内分布 Domestic Distribution	青海、河南、湖南、西藏、宁夏、湖北、广东、广西、四川、贵州、云南、陕西、甘肃、重庆 / Qinghai, Henan, Hunan, Tibet (Xizang), Ningxia, Hubei, Guangdong, Guangxi, Sichuan, Guizhou, Yunnan, Shaanxi, Gansu, Chongqing
世界分布 World Distribution	中国、越南 / China, Viet Nam
分布标注 Distribution Note	非特有种 / Non-endemic

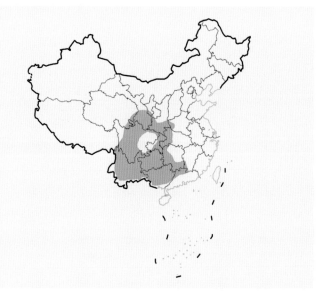

国内分布图
Map of Domestic Distribution

种群 Population

种群数量 Population Size	31,800 只 / 31,800 individuals
种群趋势 Population Trend	下降 / Decreasing

生境与生态系统 Habitat (s) and Ecosystem (s)

生　　境 Habitat(s)	泰加林、针阔混交林 / Taiga Forest, Coniferous and Broad-leaved Mixed Forest
生态系统 Ecosystem(s)	森林生态系统 / Forest Ecosystem

威胁 Threat (s)

主要威胁 Major Threat(s)	偷猎、生境破碎、伐木 / Poaching, Habitat Fragmentation, Logging

保护级别与保护行动 Protection Category and Conservation Action (s)

国家重点保护野生动物等级 (2006) Category of National Key Protected Wild Animals (2006)	一级 / Category I
IUCN 红色名录 (2020-2) IUCN Red List (2020-2)	濒危 / EN
CITES 附录 (2019) CITES Appendix (2019)	II
保护行动 Conservation Action(s)	自然保护区内种群得到保护 / Populations in nature reserves are protected

相关文献 Relevant References

Burgin *et al*., 2018; Jiang *et al*.（蒋志刚等）, 2017; Zhang and Chen（张英和陈鹏）, 2013; Wilson and Mittermeier, 2012; Groves and Grubb, 2011; Wang *et al*.（王淯等）, 2006; Wu and Wang（吴家炎和王伟）, 2006; Zhang *et al*.（章敬旗等）, 2004

林麝 *Moschus berezovskii*

马麝
Moschus chrysogaster

极危 CR A1acd; B1ab(i, ii, iii)

| 数据缺乏 DD | 无危 LC | 近危 NT | 易危 VU | 濒危 EN | 极危 CR | 区域灭绝 RE | 野外灭绝 EW | 灭绝 EX |

🦌 分类地位 Taxonomic Status

| 动物界 Animalia | 脊索动物门 Chordata | 哺乳纲 Mammalia | 偶蹄目 Artiodactyla | 麝科 Moschidae |

学 名 Scientific Name	*Moschus chrysogaster*
命 名 人 Species Authority	Hodgson, 1839
英 文 名 English Name(s)	Alpine Musk Deer
同物异名 Synonym(s)	*Moschus sifanicus* (Büchner, 1891)
种下单元评估 Infra-specific Taxa Assessed	无 / None

🦌 评估信息 Assessment Information

评估年份 Year Assessed	2020
评 定 人 Assessor(s)	蒋志刚 / Zhigang Jiang
审 定 人 Reviewer(s)	孟秀祥、胡德夫、鲍毅新 / Xiuxiang Meng, Defu Hu, Yixin Bao
其他贡献人 Other Contributor(s)	李立立、丁晨晨 / Lili Li, Chenchen Ding

理由 Justification: 马麝在其分布区曾为常见种。由于麝香价值高，市场需求旺盛。偷猎者为获得珍贵的麝香而下钢丝套偷猎马麝，不论雌雄成幼，一律捕杀。普遍的偷猎导致马麝种群数量急剧下降。由于偷猎和栖息地丧失，马麝分布区萎缩，种群数量急剧下降。因此，基于其种群数量下降速率和栖息地现状，将马麝列为极危等级 / Alpine Musk Deer was common in its habitat. Due to market demand, musk is highly profitable. Poachers set up steel wire snares to trap musk deer in wild, killed musk deer, female or male, adult or young indiscriminately, in order to get valuable musk pad; the population density of Alpine Musk Deer decreasing rapidly; the population densities of Alpine Musk Deer are extremely low now in wild. Thus, based on its rate of population decline and habitat status, Alpine Musk Deer is listed as Critically Endangered

🦌 地理分布 Geographical Distribution

国内分布 Domestic Distribution	宁夏、青海、甘肃、四川、云南、陕西、西藏 / Ningxia, Qinghai, Gansu, Sichuan, Yunnan, Shaanxi, Tibet (Xizang)
世界分布 World Distribution	中国；南亚 / China; South Asia
分布标注 Distribution Note	非特有种 / Non-endemic

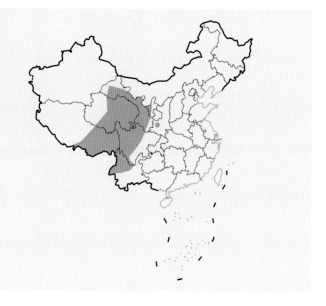

国内分布图
Map of Domestic Distribution

种群 Population

种群数量 Population Size	未知 / Unknown
种群趋势 Population Trend	下降 / Decreasing

生境与生态系统 Habitat (s) and Ecosystem (s)

生境 Habitat(s)	草甸、灌丛、荒漠、森林 / Meadow, Shrubland, Desert, Forest
生态系统 Ecosystem(s)	森林生态系统、灌丛生态系统、草地生态系统 / Forest Ecosystem, Shrubland Ecosystem, Grassland Ecosystem

威胁 Threat (s)

主要威胁 Major Threat(s)	狩猎、生境破碎、伐木 / Hunting, Habitat Fragmentation, Logging

保护级别与保护行动 Protection Category and Conservation Action (s)

国家重点保护野生动物等级 (2004) Category of National Key Protected Wild Animals (2004)	一级 / Category I
IUCN 红色名录 (2020-2) IUCN Red List (2020-2)	濒危 / EN
CITES 附录 (2019) CITES Appendix (2019)	II
保护行动 Conservation Action(s)	自然保护区内种群得到保护 / Populations in nature reserves are protected

相关文献 Relevant References

Burgin *et al.*, 2018; Jiang *et al.*（蒋志刚等）, 2017; Yang *et al.*, 2013; Wilson and Mittermeier, 2012; Groves and Grubb, 2011; Wu and Wang（吴家炎和王伟）, 2006; Xia *et al.*（夏霖等）, 2004; Liu and Sheng（刘志霄和盛和林）, 2000

马麝 *Moschus chrysogaster* 　　　　董磊 摄（西南山地供图）　By Lei Dong (Swild.cn)

黑麝
Moschus fuscus

极危　CR A1acd; B1ab(i, ii, iii)

| 数据缺乏 DD | 无危 LC | 近危 NT | 易危 VU | 濒危 EN | **极危 CR** | 区域灭绝 RE | 野外灭绝 EW | 灭绝 EX |

分类地位 Taxonomic Status

动物界 Animalia	脊索动物门 Chordata	哺乳纲 Mammalia	偶蹄目 Artiodactyla	麝科 Moschidae
学　　名 Scientific Name		*Moschus fuscus*		
命　名　人 Species Authority		Li, 1981		
英　文　名 English Name(s)		Black Musk Deer		
同物异名 Synonym(s)		Dusky Musk Deer		
种下单元评估 Infra-specific Taxa Assessed		无 / None		

评估信息 Assessment Information

评估年份 Year Assessed	2020
评定人 Assessor(s)	蒋志刚 / Zhigang Jiang
审定人 Reviewer(s)	孟秀祥、胡德夫、鲍毅新 / Xiuxiang Meng, Defu Hu, Yixin Bao
其他贡献人 Other Contributor(s)	李立立、丁晨晨 / Lili Li, Chenchen Ding

理由 Justification: 黑麝在我国分布狭窄。由于麝香价值高，市场需求旺盛。偷猎者为获得珍贵的麝香而下钢丝套偷猎包括黑麝在内的麝科动物。不论雌雄成幼，一律捕杀。普遍的偷猎导致黑麝种群数量急剧下降，野外种群数量稀少。因此，基于其种群数量下降速率和栖息地现状，将黑麝列为极危 / Black Musk Deer has a narrow range in China. Due to market demand, musk is highly profitable. Poachers set up steel wire snares to trap musk deer in wild, killed musk deer, female or male, adult or young indiscriminately, in order to get valuable musk pad; the population density of Black Musk Deer decreasing rapidly. Thus, based on its rate of population decline and habitat status, Black Musk Deer is listed as Critically Endangered

地理分布 Geographical Distribution

国内分布 Domestic Distribution
西藏、云南 / Tibet (Xizang), Yunnan
世界分布 World Distribution
中国；南亚 / China; South Asia
分布标注 Distribution Note
非特有种 / Non-endemic

国内分布图
Map of Domestic Distribution

种群 Population

种群数量 Population Size	5,950 只 / 5,950 individuals
种群趋势 Population Trend	下降 / Decreasing

生境与生态系统 Habitat(s) and Ecosystem(s)

生境 Habitat(s)	泰加林、内陆岩石区域 / Taiga Forest, Inland Rocky Area
生态系统 Ecosystem(s)	森林生态系统 / Forest Ecosystem

威胁 Threat(s)

主要威胁 Major Threat(s)	偷猎、生境破碎、伐木 / Poaching, Habitat Fragmentation, Logging

保护级别与保护行动 Protection Category and Conservation Action(s)

国家重点保护野生动物等级 (2021) Category of National Key Protected Wild Animals (2021)	一级 / Category I
IUCN 红色名录 (2020-2) IUCN Red List (2020-2)	濒危 / EN
CITES 附录 (2019) CITES Appendix (2019)	II
保护行动 Conservation Action(s)	自然保护区内种群得到保护 / Populations in nature reserves are protected

相关文献 Relevant References

Burgin *et al.*, 2018; Jiang *et al.*（蒋志刚等）, 2017; Wilson and Mittermeier, 2012; Groves and Grubb, 2011; Wu and Wang（吴家炎和王伟）, 2006; Yang *et al.*, 2003; Li（李致祥）, 1981

黑麝 *Moschus fuscus*

原麝
Moschus moschiferus

极危　CR A1acd; B1ab(i, ii, iii)

| 数据缺乏 DD | 无危 LC | 近危 NT | 易危 VU | 濒危 EN | **极危 CR** | 区域灭绝 RE | 野外灭绝 EW | 灭绝 EX |

分类地位 Taxonomic Status

动物界 Animalia	脊索动物门 Chordata	哺乳纲 Mammalia	偶蹄目 Artiodactyla	麝科 Moschidae
学　名 Scientific Name		*Moschus moschiferus*		
命 名 人 Species Authority		Linnaeus, 1758		
英 文 名 English Name(s)		Siberian Musk Deer		
同物异名 Synonym(s)		无 / None		
种下单元评估 Infra-specific Taxa Assessed		无 / None		

评估信息 Assessment Information

评估年份 Year Assessed	2020
评　定　人 Assessor(s)	蒋志刚 / Zhigang Jiang
审　定　人 Reviewer(s)	孟秀祥、胡德夫、鲍毅新 / Xiuxiang Meng, Defu Hu, Yixin Bao
其他贡献人 Other Contributor(s)	李立立、丁晨晨 / Lili Li, Chenchen Ding

理由 Justification: 原麝分布在我国东北地区。由于麝香价值高，市场需求旺盛。偷猎者为获得珍贵的麝香而下钢丝套偷猎麝科动物，不论雌雄成幼，一律捕杀。普遍的偷猎导致原种群数量急剧下降，野外种群数量稀少。此外，原始林的砍伐也导致了原麝栖息地萎缩。因此，基于其种群与栖息地现状，将原麝列为极危 / Siberian Musk Deer inhabits the forested area in northeast China. Musk is highly profitable due to market demand. Poachers set up steel wire snare to trap musk deer in forests, killed the Siberian Musk Deer indiscriminately, regardless female or male, adult or young, in order to get musk pad; the population density of Siberian Musk Deer decreasing rapidly. Furthermore, the logging of natural forests also reduced the range size of Siberian Musk Deer. Thus, based on its population and habitat status, it is listed as Critically Endangered

地理分布 Geographical Distribution

国内分布 Domestic Distribution	山西、内蒙古、新疆、辽宁、吉林、黑龙江 / Shanxi, Inner Mongolia (Nei Mongol), Xinjiang, Liaoning, Jilin, Heilongjiang
世界分布 World Distribution	亚洲 / Aisa
分布标注 Distribution Note	非特有种 / Non-endemic

国内分布图
Map of Domestic Distribution

种群 Population

种群数量 Population Size	3,500 只 / 3,500 individuals
种群趋势 Population Trend	下降 / Decreasing

生境与生态系统 Habitat (s) and Ecosystem (s)

生　　境 Habitat(s)	泰加林、针阔混交林 / Taiga Forest, Coniferous and Broad-leaved Mixed Forest
生态系统 Ecosystem(s)	森林生态系统 / Forest Ecosystem

威胁 Threat (s)

主要威胁 Major Threat(s)	狩猎、生境破碎、伐木 / Hunting, Habitat Fragmentation, Logging

保护级别与保护行动 Protection Category and Conservation Action (s)

国家重点保护野生动物等级 (2021) Category of National Key Protected Wild Animals (2021)	一级 / Category I
IUCN红色名录 (2020-2) IUCN Red List (2020-2)	易危 / VU
CITES 附录 (2019) CITES Appendix (2019)	II
保护行动 Conservation Action(s)	自然保护区内种群得到保护 / Populations in nature reserves are protected

相关文献 Relevant References

Burgin *et al.*, 2018; Jiang *et al.*（蒋志刚等）, 2017; Zhang *et al.*（张冬冬等）, 2014; Wilson and Mittermeier, 2012; Groves and Grubb, 2011; Zhang *et al.*（张海龙等）, 2008; Wu and Wang（吴家炎和王伟）, 2006

原麝 *Moschus moschiferus*

塔里木马鹿
Cervus hanglu

极危 CR B1ab(i, ii, iii)+2ab(i, ii, iii)

| 数据缺乏 DD | 无危 LC | 近危 NT | 易危 VU | 濒危 EN | **极危 CR** | 区域灭绝 RE | 野外灭绝 EW | 灭绝 EX |

分类地位 Taxonomic Status

动物界 Animalia	脊索动物门 Chordata	哺乳纲 Mammalia	偶蹄目 Artiodactyla	鹿科 Cervidae

学 名 Scientific Name	*Cervus hanglu*
命 名 人 Species Authority	Wagner, 1844
英 文 名 English Name(s)	Tarim Red Deer
同物异名 Synonym(s)	Red Deer, *Cervus elaphus yarkandensis*, *C. yarkandensis*
种下单元评估 Infra-specific Taxa Assessed	Lorenzini 和 Garofolo 在 2015 年基于分子生物学研究结果指出，分布在塔里木盆地至中亚地区的原有 3 个马鹿亚种，即分布在叶尔羌—塔里木区域的 (转下页)

评估信息 Assessment Information

评 估 年 份 Year Assessed	2020
评 定 人 Assessor(s)	蒋志刚 / Zhigang Jiang
审 定 人 Reviewer(s)	张明海、宋延龄 / Minghai Zhang, Yanling Song
其他贡献人 Other Contributor(s)	李晟、李立立、丁晨晨 / Sheng Li, Lili Li, Chenchen Ding

理由 Justification：野生塔里木马鹿种群数量在多数区域锐减，在一些区域塔里木马鹿已经绝迹，原因主要来自山区围栏建设工程、当地鹿场的违法捕捉幼崽、流域治理和新建大型芦苇场所导致的马鹿栖息地急剧丧失和破碎化。因此，塔里木马鹿被列为极危等级 / Tarim Red Deer have been decimated in most areas of its habitats and in some areas they have been fully extirpated, due to habitat rapidly declined and fragmented caused by fence construction in mountain pastures, illegal capture of calves by deer farmers, river basin management with planting of reeds in riverine wetlands. Thus, Tarim Red Deer is listed as Critically Endangered

地理分布 Geographical Distribution

国内分布 Domestic Distribution
新疆 / Xinjiang

世界分布 World Distribution
阿富汗、中国、印度、哈萨克斯坦、塔吉克斯坦、土库曼斯坦、乌兹别克斯坦 / Afghanistan, China, India, Kazakhstan, Tajikistan, Turkmenistan, Uzbekistan

分布标注 Distribution Note
非特有种 / Non-endemic

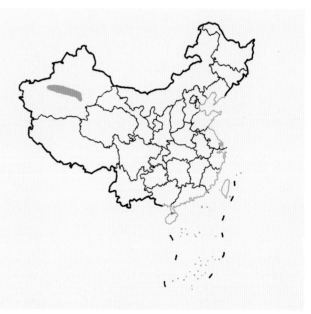

国内分布图
Map of Domestic Distribution

种群 Population

种群数量 Population Size	未知 / Unknown
种群趋势 Population Trend	下降 / Decreasing

生境与生态系统 Habitat (s) and Ecosystem (s)

生　　境 Habitat(s)	从泰加林到高寒草原的一系列生境 / Its habitats range from Taiga Forest to Alpine Steppe
生态系统 Ecosystem(s)	从针叶林生态系统到高寒草原生态系统 / From Coniferous Forest Ecosystem to Alpine Steppe Ecosystem

威胁 Threat (s)

主要威胁 Major Threat(s)	生境改变、生境丧失、猎捕 / Habitat Modification, Habitat Loss and Poaching

保护级别与保护行动 Protection Category and Conservation Action (s)

国家重点保护野生动物等级 (2021) Category of National Key Protected Wild Animals (2021)	一级 (仅限野外种群) / Category I (Only Wild Population)
IUCN红色名录(2020-2) IUCN Red List (2020-2)	未列入 / NA
CITES 附录 (2019) CITES Appendix (2019)	无 / NA
保护行动 Conservation Action(s)	自然保护区内种群得到保护 / Populations in nature reserves are protected

相关文献 Relevant References

Burgin *et al*., 2018; Jiang *et al*. (蒋志刚等), 2017; Groves and Grubb, 2011; Wilson and Mittermeier, 2011; Qin and Zhang (秦瑜和张明海), 2009

(接上页)
Cervus elaphus yarkandensis，乌兹别克斯坦布哈拉区域的 *C. e. bactrianus*，与克什米尔区域的 *C. e. hanglu*，应合并为一个种 Tarim Red Deer (*C. hanglu* Wagner, 1844)。分布在中国的是 *C. h. yarkandensis*。IUCN 红色名录 (2017) 初步将塔里木马鹿视为独立于 *C. elaphus* 和 *C. canadensis* 的一个物种 / Based on the results of molecular biology research of Lorenzini and Garofolo in 2015, the three Red Deer subspecies distribute from Tarim Basin to central Asia region, namely *Cervus elaphus yarkandensis* in Yel Qiang-Tarim area, *C. e. bactrianus* in Bukhara area of Uzbekistan, and *C. e. hanglu* in the Kashmir region, should be classified as a single species: Tarim Red Deer, *Cervus hanglu* Wagner, 1844. The subspecies of Tarim Red Deer is *C. h. yarkandensis*. IUCN red list (2017) regarded provisionally the Tarim Red Deer as a separate species from *C. elaphus* and *C. canadensis*

塔里木马鹿 *Cervus hanglu*

坡鹿
Panolia siamensis

极危　CR B1ab(i, ii, iii)+2ab(i, ii, iii)

| 数据缺乏 DD | 无危 LC | 近危 NT | 易危 VU | 濒危 EN | **极危 CR** | 区域灭绝 RE | 野外灭绝 EW | 灭绝 EX |

分类地位 Taxonomic Status

动物界 Animalia	脊索动物门 Chordata	哺乳纲 Mammalia	偶蹄目 Artiodactyla	鹿科 Cervidae

学　　名 Scientific Name	*Panolia siamensis*
命　名　人 Species Authority	Lydekker, 1915
英　文　名 English Name(s)	Eastern Eld's Deer
同物异名 Synonym(s)	*Panolia eldii siamensis*, Thai Brow-antlered Deer
种下单元评估 Infra-specific Taxa Assessed	无 / None

评估信息 Assessment Information

评估年份 Year Assessed	2020
评　定　人 Assessor(s)	蒋志刚 / Zhigang Jiang
审　定　人 Reviewer(s)	宋延龄 / Yanling Song
其他贡献人 Other Contributor(s)	李立立、丁晨晨 / Lili Li, Chenchen Ding

理由 Justification: 坡鹿在中国仅分布在海南省，种群数量少。由于捕杀和栖息地破坏而濒危。1986年大田自然保护区围栏建成后，围栏内的东方坡鹿种群增长到700～800只，围栏面积限制了坡鹿种群增长。由于缺乏适宜栖息地和偷猎，大田保护区围栏外仅有约300只坡鹿。因此，将该种列为极危等级 / Eastern Eld's Deer only distributed in Hainan Province in China. The species is endangered due to poaching and habitat destruction. Since the paddock fence was constructed in the Datian Nature Reserve in 1986, the population of Eastern Eld's Deer in the enclosure increased to between 700 to 800, however, the carrying capacity of paddock area limited the further growth of the deer population. Deer population outside the enclosure was around 300 due to no suitable habitat and poaching. Thus, it is listed as Critically Endangered

地理分布 Geographical Distribution

国内分布 Domestic Distribution
海南 / Hainan
世界分布 World Distribution
中国；东南亚 / China; Southeast Asia
分布标注 Distribution Note
非特有种 / Non-endemic

国内分布图
Map of Domestic Distribution

种群 Population

种群数量 Population Size	1,000 只 / 1,000 individuals
种群趋势 Population Trend	上升 / Increasing

生境与生态系统 Habitat (s) and Ecosystem (s)

生境 Habitat(s)	半干旱热带稀树草原 / Semi-arid Tropical Savanna
生态系统 Ecosystem(s)	森林生态系统、湿地生态系统 / Forest Ecosystem, Wetland Ecosystem

威胁 Threat (s)

主要威胁 Major Threat(s)	生境丧失、蟒蛇捕食 / Habitat Loss, Predation by Python

保护级别与保护行动 Protection Category and Conservation Action (s)

国家重点保护野生动物等级 (2021) Category of National Key Protected Wild Animals (2021)	一级 / Category I
IUCN 红色名录 (2020-2) IUCN Red List (2020-2)	濒危 / EN
CITES 附录 (2019) CITES Appendix (2019)	无 / NA
保护行动 Conservation Action(s)	自然保护区内种群得到保护 / Populations in nature reserves are protected

相关文献 Relevant References

Burgin *et al*., 2018; Jiang *et al*. (蒋志刚等), 2017; Groves and Grubb, 2011; Zhang *et al*. (张琼等), 2009; Lu *et al*. (卢学理等), 2008

坡鹿 *Panolia siamensis* 蒋志刚 摄 By Zhigang Jiang

麋鹿
Elaphurus davidianus

极危 CR B1ab(i, ii, iii)+2ab(i, ii, iii)

| 数据缺乏 DD | 无危 LC | 近危 NT | 易危 VU | 濒危 EN | **极危 CR** | 区域灭绝 RE | 野外灭绝 EW | 灭绝 EX |

分类地位 Taxonomic Status

动物界 Animalia	脊索动物门 Chordata	哺乳纲 Mammalia	偶蹄目 Artiodactyla	鹿科 Cervidae
学　　名 Scientific Name		*Elaphurus davidianus*		
命　名　人 Species Authority		Milne-Edwards, 1866		
英　文　名 English Name(s)		Père David's Deer		
同物异名 Synonym(s)		Milu, Milu Deer, Elaphure		
种下单元评估 Infra-specific Taxa Assessed		无 / None		

评估信息 Assessment Information

评估年份 Year Assessed	2020
评　定　人 Assessor(s)	蒋志刚 / Zhigang Jiang
审　定　人 Reviewer(s)	杨道德、李春旺 / Daode Yang, Chunwang Li
其他贡献人 Other Contributor(s)	李立立、丁晨晨 / Lili Li, Chenchen Ding

理由 Justification: 20 世纪初，麋鹿在中国灭绝。80 年代中期，麋鹿被重新引入中国，于 1985 年、1986 年、2001 年和 2002 年分别建立了北京、大丰、石首天鹅州和原阳等繁殖种群。1998 年，在大丰麋鹿国家级自然保护区，一群麋鹿被放归到黄海滩涂。经过多次补充放归个体，这群野放的鹿群数量在 2017 年增加到 300 多只。然而，大丰野放鹿群的栖息地被农田、人工林区、居民区、堤坝和高速公路所包围，限制了该种群的进一步发展。1998 年洪水期间，36 只麋鹿意外地逃出了位于长江中游的石首国家级自然保护区的围栏。这些逃逸的麋鹿在洞庭湖地区湿地景观中建立了 300 只鹿的繁殖种群。现在，麋鹿已被迁移到洞庭湖、鄱阳湖等 50 多个地点。考虑到它的野生种群规模和占有区面积，仍将麋鹿列为极危 / Père David's Deer went extinct in China at the beginning of the 20th century. Since then, the endemic deer was reintroduced in the mid-1980s, with the Beijing, Dafeng, Tian'ezhou (Shishou) and Yuanyang populations being established in 1985, 1986, 2001 and 2002, respectively. An initial group of Père David's Deer was released near the coast of the Yellow Sea in the Dafeng Milu National Nature Reserve in 1998. Together with subsequent releases, the group size increased to more than 300 individuals by 2017. However, the habitat available for the reintroduced herd is being engulfed by cropland, artificial forested areas, residential areas, dykes, and highways, which altogether are restricting the deer population's further dispersal. Elsewhere, 36 Père David's Deer accidentally escaped from their enclosure in the Shishou National Nature Reserve in the middle Changjiang River drainage during a major flood in 1998, which allowed them to roam freely in the Dongting Lake area, where they established breeding populations with over 300 deers in the anthropogenic landscape. Now, the Père David's Deer has been relocated to over 50 sites in China. However, considering its wild population size and the area of occupancy, it is still listed as Critically Endangered

地理分布 Geographical Distribution

国内分布 Domestic Distribution
北京、江苏、湖北、河南、湖南、江西 / Beijing, Jiangsu, Hubei, Henan, Hunan, Jiangxi
世界分布 World Distribution
中国 / China
分布标注 Distribution Note
特有种 / Endemic

国内分布图
Map of Domestic Distribution

种群 Population

种群数量 Population Size	6,000 只 (野化种群) / 6,000 individuals (re-wild population)
种群趋势 Population Trend	上升 / Increasing

生境与生态系统 Habitat (s) and Ecosystem (s)

生　　境 Habitat(s)	湿地、草地、灌丛 / Wetland, Grassland, Shrubland
生态系统 Ecosystem(s)	湿地生态系统、灌丛生态系统、草地生态系统 / Wetland Ecosystem, Shrubland Ecosystem, Grassland Ecosystem

威胁 Threat (s)

主要威胁 Major Threat(s)	生境丧失与生境改变 / Habitat Loss and Modification

保护级别与保护行动 Protection Category and Conservation Action (s)

国家重点保护野生动物等级 (2021) Category of National Key Protected Wild Animals (2021)	一级 / Category I
IUCN 红色名录 (2020-2) IUCN Red List (2020-2)	野外灭绝 / EW
CITES 附录 (2019) CITES Appendix (2019)	无 / NA
保护行动 Conservation Action(s)	法定保护对象、建立了自然保护区 / Protected by law, nature reserves established

相关文献 Relevant References

Burgin *et al.*, 2018; Jiang *et al.* (蒋志刚等), 2017, 2015a, 2001; Jiang, 2013; Hou *et al.* (侯立冰等), 2012; Wilson and Mittermeier, 2012; Groves and Grubb, 2011; Smith *et al.* (史密斯等), 2009; Jiang and Harris, 2008; Yang *et al.* (杨道德等), 2007

麋鹿 *Elaphurus davidianus*

驼鹿
Alces alces

极危　CR B1ab(i, ii, iii); D

| 数据缺乏 DD | 无危 LC | 近危 NT | 易危 VU | 濒危 EN | **极危 CR** | 区域灭绝 RE | 野外灭绝 EW | 灭绝 EX |

分类地位 Taxonomic Status

动物界 Animalia	脊索动物门 Chordata	哺乳纲 Mammalia	偶蹄目 Artiodactyla	鹿科 Cervidae
学　名 Scientific Name		*Alces alces*		
命 名 人 Species Authority		Linnaeus, 1758		
英 文 名 English Name(s)		Moose		
同物异名 Synonym(s)		汗达罕		
种下单元评估 Infra-specific Taxa Assessed		无 / None		

评估信息 Assessment Information

评估年份 Year Assessed	2020
评　定　人 Assessor(s)	蒋志刚 / Zhigang Jiang
审　定　人 Reviewer(s)	张明海 / Minghai Zhang
其他贡献人 Other Contributor(s)	李立立、丁晨晨 / Lili Li, Chenchen Ding

理由 Justification: 驼鹿为环北极圈物种，是对气候变化敏感的大型哺乳动物，中国种群是其亚洲分布区的最南缘。随着全球气候变暖，中国境内的驼鹿分布区正在向北退缩。此外，在大小兴安岭盗猎现象严重。在从前驼鹿分布区中已经很难发现其踪迹（姜广顺，个人通信）。由于地理隔离，大兴安岭与阿尔泰山的驼鹿种群无基因交流。驼鹿在我国分布范围小，种群数量低。因此，将驼鹿列为极危 / Moose is a circum-Arctic species and a large mammal species that sensitive to climate change, the range of Moose in China is its south-mast range in Asia. Due to global warming, the distribution of Moose in China is retreating northward. In the Da and Xiao Hinggan Ling, poaching of Moose was common. Now Moose is rather difficult to find in its former range (Guangshun Jiang, Personal communications). Now, its ranges in the country are small and its population densities are low. The populations of the Da and Xiao Hinggan Ling and Altay Shan are isolated without gene exchange. Thus, it is listed as Critically Endangered

地理分布 Geographical Distribution

国内分布 Domestic Distribution
新疆、内蒙古 / Xinjiang, Inner Mongolia (Nei Mongol)
世界分布 World Distribution
中国；环北极圈 / China; Circum-Arctic
分布标注 Distribution Note
非特有种 / Non-endemic

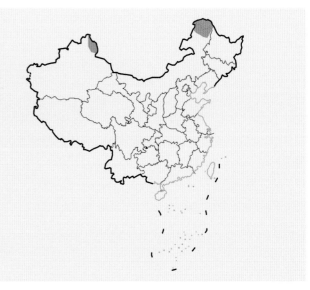

国内分布图
Map of Domestic Distribution

种群 Population

种群数量 Population Size	未知 / Unknown
种群趋势 Population Trend	上升 / Increasing

生境与生态系统 Habitat(s) and Ecosystem(s)

生境 Habitat(s)	针阔混交林 / Coniferous and Broad-leaved Mixed Forest
生态系统 Ecosystem(s)	森林生态系统 / Forest Ecosystem

威胁 Threat(s)

主要威胁 Major Threat(s)	狩猎、生境丧失、全球变暖 / Hunting, Habitat Loss, Global Warming

保护级别与保护行动 Protection Category and Conservation Action(s)

国家重点保护野生动物等级 (2021) Category of National Key Protected Wild Animals (2021)	一级 / Category I
IUCN 红色名录 (2020-2) IUCN Red List (2020-2)	无危 / LC
CITES 附录 (2019) CITES Appendix (2019)	无 / NA
保护行动 Conservation Action(s)	自然保护区内种群得到保护 / Populations in nature reserves are protected

相关文献 Relevant References

Burgin *et al.*, 2018; Jiang *et al.*（蒋志刚等），2017; Wilson and Mittermeier, 2012; Groves and Grubb, 2011; Smith *et al.*（史密斯等），2009; Wang（王应祥），2003

驼鹿 *Alces alces*

印度野牛
Bos gaurus

极危 CR A1acd; B1ab(i, ii, iii); D1

| DD | LC | NT | VU | EN | CR | RE | EW | EX |
| 数据缺乏 | 无危 | 近危 | 易危 | 濒危 | 极危 | 区域灭绝 | 野外灭绝 | 灭绝 |

分类地位 Taxonomic Status

动物界 Animalia	脊索动物门 Chordata	哺乳纲 Mammalia	偶蹄目 Artiodactyla	牛科 Bovidae

学名 Scientific Name	*Bos gaurus*
命名人 Species Authority	C. H. Smith, 1827
英文名 English Name(s)	Gaur
同物异名 Synonym(s)	白肢野牛，野牛；*Bos asseel* (Horsfield, 1851); *B. cavifrons* (Hodgson, 1837); *B. gaur* (Sundevall, 1846); *B. gaurus* (Lydekker, 1907); subsp. *hubbacki*; *B. gour*（转下页）
种下单元评估 Infra-specific Taxa Assessed	无 / None

评估信息 Assessment Information

评估年份 Year Assessed	2020
评定人 Assessor(s)	蒋志刚 / Zhigang Jiang
审定人 Reviewer(s)	蒋学龙 / Xuelong Jiang
其他贡献人 Other Contributor(s)	李立立、丁晨晨 / Lili Li, Chenchen Ding

理由 Justification：印度野牛分布在中国云南南部和藏南地区。中国为其边缘分布区。目前云南地区印度野牛种群由于栖息地丧失和猎杀，仅余180～210头，在高黎贡山最近的调查没有发现印度野牛。在藏南地区，印度野牛面临生境丧失的威胁（丁晨晨等，2018）。因此，根据该种的种群下降速率和发生区面积，印度野牛的濒危等级定为极危 / The range of Gaur in the country is its peripheral range. Gaur is found in southern Yunnan and Zangnan area of Tibet (Xizang). No Gaur was found in a recent field survey in Gaoligong Shan. Due to habitat loss and poaching, there are 180~210 Gaurs hanging on in southern Yunnan and the Gaur in Zangnan are also threatened by habitat loss (Ding *et al*., 2018). Based on its population decline rate and extent of occurrence, it is listed as Critically Endangered

地理分布 Geographical Distribution

国内分布 Domestic Distribution	云南、西藏 / Yunnan, Tibet (Xizang)
世界分布 World Distribution	南亚、东南亚 / South Asia, Southeast Asia
分布标注 Distribution Note	非特有种 / Non-endemic

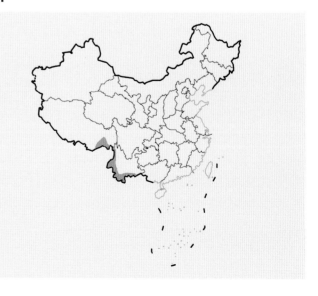

国内分布图
Map of Domestic Distribution

种群 Population

种群数量 Population Size	云南种群180～210头；藏南种群数量未知 / 180~210 individuals in Yunnan; population size in Zangnan area is unkown
种群趋势 Population Trend	下降 / Decreasing

生境与生态系统 Habitat (s) and Ecosystem (s)

生　　境 Habitat(s)	热带湿润森林 / Tropical Moist Forest
生态系统 Ecosystem(s)	森林生态系统 / Forest Ecosystem

威胁 Threat (s)

主要威胁 Major Threat(s)	狩猎、伐木、耕种、公路和铁路 / Hunting, Logging, Plantation, Road and Railroad

保护级别与保护行动 Protection Category and Conservation Action (s)

国家重点保护野生动物等级 (2021) Category of National Key Protected Wild Animals (2021)	一级 / Category I
IUCN 红色名录 (2020-2) IUCN Red List (2020-2)	易危 / VU
CITES 附录 (2019) CITES Appendix (2019)	I
保护行动 Conservation Action(s)	自然保护区内种群得到保护 / Populations in nature reserves are protected

相关文献 Relevant References

Burgin *et al*., 2018; Ding *et al*. (丁晨晨等), 2018; Jiang *et al*. (蒋志刚等), 2017; Groves and Grubb, 2011; Wilson and Mittermeier, 2011; Gan and Hu (甘宏协和胡华斌), 2008; Pan *et al*. (潘清华等), 2007; Zhang *et al*. (张洪亮等), 2000; Yang *et al*. (杨德华等), 1988

(接上页)
(Hardwicke, 1827); *B. subhemachalus* (Hodgson, 1837); *Bubalibos annamiticus* (Heude, 1901); *Gauribos brachyrhinus* (Heude, 1901); *G. laosiensis* (Heude, 1901); *G. mekongensis* (Heude, 1901); *G. sylvanus* (Heude, 1901); *Uribos platyceros* (Heude, 1901)

印度野牛 *Bos gaurus* 　　　　　　　　蒋志刚 摄　By Zhigang Jiang

中国生物多样性 红色名录
蒙原羚
Procapra gutturosa

极危 CR A1acd; B1ab(i, ii, iii)

| 数据缺乏 DD | 无危 LC | 近危 NT | 易危 VU | 濒危 EN | **极危 CR** | 区域灭绝 RE | 野外灭绝 EW | 灭绝 EX |

分类地位 Taxonomic Status

动物界 Animalia	脊索动物门 Chordata	哺乳纲 Mammalia	偶蹄目 Artiodactyla	牛科 Bovidae
学名 Scientific Name		*Procapra gutturosa*		
命名人 Species Authority		Pallas, 1777		
英文名 English Name(s)		Mongolian Gazelle		
同物异名 Synonym(s)		黄羊，Dzeren (Russian), Zeer (Mongolian)		
种下单元评估 Infra-specific Taxa Assessed		无 / None		

评估信息 Assessment Information

评估年份 Year Assessed	2020
评定人 Assessor(s)	蒋志刚 / Zhigang Jiang
审定人 Reviewer(s)	刘丙万、毕俊怀 / Bingwan Liu, Junhuai Bi
其他贡献人 Other Contributor(s)	李立立、丁晨晨 / Lili Li, Chenchen Ding

理由 Justification: 蒙原羚是季节迁徙物种，冬季向南从蒙古迁入我国越冬，春季返回蒙古产羔。由于内蒙古境内的草原围栏、过度放牧和盗猎以及国境铁丝围栏的影响，中国境内几乎已经不存在蒙原羚繁殖种群；仅在中蒙边境附近有少数的蒙原羚游荡群体。因此，根据该种的种群下降速率和发生区面积，将蒙原羚列为极危等级 / Mongolian Gazelle is a migratory species. It moves southward from Mongolia to China in winter and returns to Mongolia in spring for calving. Due to the pasture enclosure, livestock over-grazing and poaching in Inner Mongolia (Nei Mongol) and the impact of border wire fence, there is barely any breeding population of Mongolian Gazelle in China, only some wandering individuals live along the China-Mongolia border. Thus, based on its population decline rate and extent of occurrence, it is listed as Critically Endangered

地理分布 Geographical Distribution

国内分布 Domestic Distribution	内蒙古、甘肃 / Inner Mongolia (Nei Mongol), Gansu
世界分布 World Distribution	中国、蒙古、俄罗斯 / China, Mongolia, Russia
分布标注 Distribution Note	非特有种 / Non-endemic

国内分布图
Map of Domestic Distribution

种群 Population

种群数量 Population Size	1,000 只 / 1,000 individuals
种群趋势 Population Trend	未知 / Unknown

生境与生态系统 Habitat(s) and Ecosystem(s)

生境 Habitat(s)	草地、半荒漠 / Steppe, Semi-desert
生态系统 Ecosystem(s)	草地生态系统、荒漠生态系统 / Steppe Ecosystem, Desert Ecosystem

威胁 Threat(s)

主要威胁 Major Threat(s)	狩猎、疾病、极端天气 / Hunting, Disease, Extreme Weather

保护级别与保护行动 Protection Category and Conservation Action(s)

国家重点保护野生动物等级 (2021) Category of National Key Protected Wild Animals (2021)	一级 / Category I
IUCN 红色名录 (2020-2) IUCN Red List (2020-2)	无危 / LC
CITES 附录 (2019) CITES Appendix (2019)	无 / NA
保护行动 Conservation Action(s)	自然保护区内种群得到保护 / Populations in nature reserves are protected

相关文献 Relevant References

Burgin *et al*., 2018; Jiang *et al*.（蒋志刚等）, 2017; Castelló, 2016; Okada *et al*., 2012; Wilson and Mittermeier, 2012; Groves and Grubb, 2011; Sorokin *et al*., 2005; Jin and Ma（金崑和马建章）, 2004; Li *et al*.（李俊生等）, 2001

蒙原羚 *Procapra gutturosa*　　　　　　　　　　　蒋志刚 摄　By Zhigang Jiang

中国生物多样性 红色名录

贡山羚牛
Budorcas taxicolor

极危　CR A1abc; C1+2; D1

| 数据缺乏 DD | 无危 LC | 近危 NT | 易危 VU | 濒危 EN | 极危 CR | 区域灭绝 RE | 野外灭绝 EW | 灭绝 EX |

分类地位 Taxonomic Status

动物界 Animalia	脊索动物门 Chordata	哺乳纲 Mammalia	偶蹄目 Artiodactyla	牛科 Bovidae

学名 Scientific Name	*Budorcas taxicolor*
命名人 Species Authority	Hodgson, 1850
英文名 English Name(s)	Gongshan Takin
同物异名 Synonym(s)	Takin, Cattle Chamois, Gnu Goat
种下单元评估 Infra-specific Taxa Assessed	无 / None

评估信息 Assessment Information

评估年份 Year Assessed	2020
评定人 Assessor(s)	蒋志刚 / Zhigang Jiang
审定人 Reviewer(s)	蒋学龙 / Xuelong Jiang
其他贡献人 Other Contributor(s)	李立立、丁晨晨 / Lili Li, Chenchen Ding

理由 Justification: 历史上，贡山羚牛在高黎贡山地区广泛分布，且群体之间隔离程度不高。但近期调查显示，其分布区片断化，种群主要集中分布于泸水县片马镇以西及贡山县茨开镇独龙江地区；种群间距离远。调查发现，大约 270 只贡山羚牛分布于高黎贡山区域。因此，贡山羚牛列为极危等级 / Historically, the Gongshan Takin was widely distributed in the Gaoligong Shan and the degree of isolation between populations was limited. More recent sightings, however, indicate that their distribution pattern is fragmented, with populations mainly concentrated west of Pianma Town, Lushui County, and in the Dulongjiang area of Cikai Town, Gongshan County. The distance between these populations is far. According to field survey, there are around 270 takins in the Gaoligong Shan. Thus, Gongshan Takin is listed as Critically Endangered

地理分布 Geographical Distribution

国内分布 Domestic Distribution
云南 / Yunnan
世界分布 World Distribution
中国、不丹、印度、缅甸 / China, Bhutan, India, Myanmar
分布标注 Distribution Note
非特有种 / Non-endemic

国内分布图
Map of Domestic Distribution

种群 Population

种群数量 Population Size	270 只 / 270 individuals
种群趋势 Population Trend	下降 / Decreasing

生境与生态系统 Habitat(s) and Ecosystem(s)

生境 Habitat(s)	草甸、森林 / Meadow, Forest
生态系统 Ecosystem(s)	森林生态系统、草地生态系统 / Forest Ecosystem, Grassland Ecosystem

威胁 Threat(s)

主要威胁 Major Threat(s)	生境丧失 / Habitat Loss

保护级别与保护行动 Protection Category and Conservation Action(s)

国家重点保护野生动物等级 (2021) Category of National Key Protected Wild Animals (2021)	一级 / Category I
IUCN 红色名录 (2020-2) IUCN Red List (2020-2)	易危 / VU
CITES 附录 (2019) CITES Appendix (2019)	无 / NA
保护行动 Conservation Action(s)	自然保护区内种群得到保护 / Populations in nature reserves are protected

相关文献 Relevant References

Burgin *et al*., 2018; Jiang *et al*. (蒋志刚等), 2018; Castelló, 2016; Groves and Grubb, 2011; Wilson and Mittermeier, 2011

贡山羚牛 *Budorcas taxicolor*

长尾斑羚
Naemorhedus caudatus
极危 CR A2cd+3cd

| 数据缺乏 DD | 无危 LC | 近危 NT | 易危 VU | 濒危 EN | **极危 CR** | 区域灭绝 RE | 野外灭绝 EW | 灭绝 EX |

分类地位 Taxonomic Status

动物界 Animalia	脊索动物门 Chordata	哺乳纲 Mammalia	偶蹄目 Artiodactyla	牛科 Bovidae
学 名 Scientific Name		*Naemorhedus caudatus*		
命 名 人 Species Authority		Milne-Edwards, 1867		
英 文 名 English Name(s)		Long-tailed Goral		
同物异名 Synonym(s)		Amur Goral, Chinese Gray Goral, Korean Serow		
种下单元评估 Infra-specific Taxa Assessed		无 / None		

评估信息 Assessment Information

评 估 年 份 Year Assessed	2020
评 定 人 Assessor(s)	蒋志刚 / Zhigang Jiang
审 定 人 Reviewer(s)	姜广顺 / Guangshun Jiang
其他贡献人 Other Contributor(s)	李立立、丁晨晨 / Lili Li, Chenchen Ding

理由 Justification: 长尾斑羚在东北地区几乎灭绝，很少有发现长尾斑羚的存在证据。根据姜广顺等在东北的调查，仅在吉林汪清大荒沟林场雪地上发现一次足迹。在中国其他地区尚无数据。根据该种的种群下降速率，长尾斑羚的濒危等级列为极危 / Long-tailed Goral is nearly extinct in the northeast China. Evidence of occurrence of Long-tailed Goral in field is scarce. According to the survey conducted by Guangshun Jiang *et al.* in the northeast China, only one trail of footprints in snow was discovered in the Wangqing Dahuanggou Forest Farm of Jilin Province. There is not record of this species in other places in China. Thus, based on its population decline rate, it is listed as Critically Endangered

地理分布 Geographical Distribution

国内分布 Domestic Distribution	黑龙江、吉林、辽宁 / Heilongjiang, Jilin, Liaoning
世界分布 World Distribution	中国、朝鲜、韩国、俄罗斯 / China, Korea (the Democratic People's Republic of), Korea (the Republic of), Russia
分布标注 Distribution Note	非特有种 / Non-endemic

国内分布图
Map of Domestic Distribution

种群 Population

种群数量 Population Size	未知 / Unknown
种群趋势 Population Trend	下降 / Decreasing

生境与生态系统 Habitat(s) and Ecosystem(s)

生　　境 Habitat(s)	峭壁、森林 / Cliff, Forest
生态系统 Ecosystem(s)	森林生态系统 / Forest Ecosystem

威胁 Threat(s)

主要威胁 Major Threat(s)	狩猎、家畜养殖或放牧、耕种 / Hunting, Livestock Farming or Ranching, Plantation

保护级别与保护行动 Protection Category and Conservation Action(s)

国家重点保护野生动物等级 (2021) Category of National Key Protected Wild Animals (2021)	二级 / Category II
IUCN 红色名录 (2020-2) IUCN Red List (2020-2)	易危 / VU
CITES 附录 (2019) CITES Appendix (2019)	无 / NA
保护行动 Conservation Action(s)	无 / None

相关文献 Relevant References

Burgin *et al*., 2018; Jiang *et al*.（蒋志刚等）, 2017; Castelló, 2016; Groves and Grubb, 2011; Wilson and Mittermeier, 2011; Smith *et al*.（史密斯等）, 2009; Wang（王应祥）, 2003

长尾斑羚 *Naemorhedus caudatus* 　　　　　　　　　　　　　　　　　　　胡慧建 摄　By Huijian Hu

中国生物多样性 红色名录

塔尔羊
Hemitragus jemlahicus

极危　CR B1ab(i, ii, iii); C1

| 数据缺乏 DD | 无危 LC | 近危 NT | 易危 VU | 濒危 EN | **极危 CR** | 区域灭绝 RE | 野外灭绝 EW | 灭绝 EX |

分类地位 Taxonomic Status

动物界 Animalia	脊索动物门 Chordata	哺乳纲 Mammalia	偶蹄目 Artiodactyla	牛科 Bovidae

学　名 Scientific Name	*Hemitragus jemlahicus*
命 名 人 Species Authority	C. H. Smith, 1826
英 文 名 English Name(s)	Himalayan Tahr
同物异名 Synonym(s)	无 / None
种下单元评估 Infra-specific Taxa Assessed	无 / None

评估信息 Assessment Information

评 估 年 份 Year Assessed	2020
评 定 人 Assessor(s)	蒋志刚 / Zhigang Jiang
审 定 人 Reviewer(s)	胡慧建、杨维康 / Huijian Hu, Weikang Yang
其他贡献人 Other Contributor(s)	李立立、丁晨晨 / Lili Li, Chenchen Ding

理由 Justification: 塔尔羊在中国的分布区狭窄、种群小。据估计，目前塔尔羊种群数量仅为400～500只。由于人类活动，其栖息地面积减少，种群数量下降，严重威胁着该种的生存。因此，基于其种群与栖息地现状，将塔尔羊列为极危等级 / Himalayan Tahr has a limited range and small population in China. It is estimated that the current population size of Himalayan Tahr is around 400~500 individuals. The areas of its habitats and population sizes are declining due to human encroachment. The survival of this species is severely threatened. Thus, based on its habitat and population status, Himalayan Tahr is listed as Critically Endangered

地理分布 Geographical Distribution

国内分布 Domestic Distribution
西藏 / Tibet (Xizang)
世界分布 World Distribution
中国、印度、尼泊尔 / China, India, Nepal
分布标注 Distribution Note
非特有种 / Non-endemic

国内分布图 Map of Domestic Distribution

种群 Population

种群数量 Population Size	400～500 只 / 400~500 individuals
种群趋势 Population Trend	下降 / Decreasing

生境与生态系统 Habitat (s) and Ecosystem (s)

生境 Habitat(s)	内陆岩石区域、灌丛 / Inland Rocky Area, Shrubland
生态系统 Ecosystem(s)	灌丛生态系统 / Shrubland Ecosystem

威胁 Threat (s)

主要威胁 Major Threat(s)	人类干扰、狩猎 / Human Disturbance, Hunting

保护级别与保护行动 Protection Category and Conservation Action (s)

国家重点保护野生动物等级 (2021) Category of National Key Protected Wild Animals (2021)	一级 / Category I
IUCN 红色名录 (2020-2) IUCN Red List (2020-2)	近危 / NT
CITES 附录 (2019) CITES Appendix (2019)	无 / NA
保护行动 Conservation Action(s)	无 / None

相关文献 Relevant References

Burgin *et al.*, 2018; Jiang *et al.* (蒋志刚等), 2017; Castelló, 2016; Groves and Grubb, 2011; Wilson and Mittermeier, 2011; Smith *et al.* (史密斯等), 2009; Pan *et al.* (潘清华等), 2007

塔尔羊 *Hemitragus jemlahicus*   蒋志刚 摄   By Zhigang Jiang

中国生物多样性 红色名录

阿尔泰盘羊
Ovis ammon

极危 CR B1ab(i, ii, iii)

| 数据缺乏 DD | 无危 LC | 近危 NT | 易危 VU | 濒危 EN | 极危 CR | 区域灭绝 RE | 野外灭绝 EW | 灭绝 EX |

分类地位 Taxonomic Status

动物界 Animalia	脊索动物门 Chordata	哺乳纲 Mammalia	偶蹄目 Artiodactyla	牛科 Bovidae

学 名 Scientific Name	*Ovis ammon*
命名人 Species Authority	Linnaeus, 1758
英文名 English Name(s)	Argali
同物异名 Synonym(s)	无 / None
种下单元评估 Infra-specific Taxa Assessed	无 / None

评估信息 Assessment Information

评估年份 Year Assessed	2020
评定人 Assessor(s)	蒋志刚 / Zhigang Jiang
审定人 Reviewer(s)	初红军、杨维康 / Hongjun Chu, Weikang Yang
其他贡献人 Other Contributor(s)	李立立、丁晨晨 / Lili Li, Chenchen Ding

理由 Justification: 阿尔泰盘羊分布在新疆阿勒泰中蒙边境地区的阿尔泰山区域，尚缺乏野外调查数据。种群数量估计为数百只。然而，目前仅在青河县特克尼特山发现4只（1雄3雌）的盘羊小群（邸杰，个人通讯）。其分布区由于人类活动而缩小，种群数量下降。其迁徙通道被边境围栏所阻断。根据其发生区面积，阿尔泰盘羊定为极危等级 / Argali is distributed in the Altay Shan and along the China-Mongolia border in northern Xinjiang, but its field data is deficient. Population size is about several hundreds according to estimation. Only 4 Argali (1 male and 3 females) was spotted in Tekenite Shan of Qinghe County (Jie Di, Personal communications). The range of Argali is encroached by human activities; its population is declining rapidly. Furthermore, its migration route is cut off by the border barred wired fence. Based on its population decline rate and habitat, according to its extent of occurrence, Argali is listed as Critically Endangered

地理分布 Geographical Distribution

国内分布 Domestic Distribution
新疆 / Xinjiang
世界分布 World Distribution
中国、蒙古、哈萨克斯坦 / China, Mongolia, Kazakhstan
分布标注 Distribution Note
非特有种 / Non-endemic

国内分布图
Map of Domestic Distribution

种群 Population

种群数量 Population Size	估计为数百只 / Several hundreds according to estimation
种群趋势 Population Trend	未知 / Unknown

生境与生态系统 Habitat (s) and Ecosystem (s)

生境 Habitat(s)	内陆岩石区域、灌丛 / Inland Rocky Area, Shrubland
生态系统 Ecosystem(s)	荒漠生态系统 / Desert Ecosystem

威胁 Threat (s)

主要威胁 Major Threat(s)	人类干扰、狩猎、迁徙通道阻断 / Human Disturbance, Hunting, Migration Route Interruption

保护级别与保护行动 Protection Category and Conservation Action (s)

国家重点保护野生动物等级 (2021) Category of National Key Protected Wild Animals (2021)	二级 / Category II
IUCN红色名录(2020-2) IUCN Red List (2020-2)	近危 / NT
CITES 附录 (2019) CITES Appendix (2019)	II
保护行动 Conservation Action(s)	自然保护区内种群得到保护；列入《迁徙物种保护公约》的"迁徙物种行动计划" / Populations in nature reserves are protected; included in the *Convention on Migratory Species* (CMS) － Species Action Plan

相关文献 Relevant References

Jiang *et al*. (蒋志刚等), 2017; Wilson and Mittermeier, 2012; Groves and Grubb, 2011

阿尔泰盘羊 *Ovis ammon*　　　　蒋志刚 摄　By Zhigang Jiang

戈壁盘羊
Ovis darwini

极危 CR B1ab(i, ii, iii)

| 数据缺乏 DD | 无危 LC | 近危 NT | 易危 VU | 濒危 EN | 极危 CR | 区域灭绝 RE | 野外灭绝 EW | 灭绝 EX |

分类地位 Taxonomic Status

动物界 Animalia	脊索动物门 Chordata	哺乳纲 Mammalia	偶蹄目 Artiodactyla	牛科 Bovidae
学名 Scientific Name		*Ovis darwini*		
命名人 Species Authority		Przewalski, 1883		
英文名 English Name(s)		Gobi Argali		
同物异名 Synonym(s)		无 / None		
种下单元评估 Infra-specific Taxa Assessed		无 / None		

评估信息 Assessment Information

评估年份 Year Assessed	2020
评定人 Assessor(s)	蒋志刚 / Zhigang Jiang
审定人 Reviewer(s)	初红军、杨维康 / Hongjun Chu, Weikang Yang
其他贡献人 Other Contributor(s)	李立立、丁晨晨 / Lili Li, Chenchen Ding

理由 Justification: 戈壁盘羊分布在中蒙边境。杨维康等在北塔山区域多次调查，仅见到少数戈壁盘羊个体。2008年和2010年在大小哈甫提克山调查，仅遇见2群戈壁盘羊。其分布区由于人类活动而缩小，种群数量下降。其迁徙通道被中蒙边境围栏所阻断。基于其种群下降速率与栖息地现状，将戈壁盘羊定为极危等级 / Gobi Argali migrate across the China-Mongolia border. During the field surveys in Baytik Shan of China-Mongolia border region, the Gobi Argali was rare. In 2008 and 2010, in Major and Minor Hafutike Shan, Weikang Yang only sighted 2 groups of Gobi Argali. The range of Gobi Argali is encroached by human activities; its population is declining rapidly. Furthermore, its migration route is cut off by the border barred wired fence. Based on its rate of population decline and habitat status, Gobi Argali is listed as Critically Endangered

地理分布 Geographical Distribution

国内分布 Domestic Distribution	
内蒙古 / Inner Mongolia (Nei Mongol)	
世界分布 World Distribution	
中国、蒙古 / China, Mongolia	
分布标注 Distribution Note	
非特有种 / Non-endemic	

国内分布图
Map of Domestic Distribution

种群 Population

种群数量 Population Size	稀少 / Rare
种群趋势 Population Trend	未知 / Unknown

生境与生态系统 Habitat (s) and Ecosystem (s)

生　境 Habitat(s)	荒漠 / Desert
生态系统 Ecosystem(s)	荒漠生态系统 / Desert Ecosystem

威胁 Threat (s)

主要威胁 Major Threat(s)	未知 / Unknown

保护级别与保护行动 Protection Category and Conservation Action (s)

国家重点保护野生动物等级 (2021) Category of National Key Protected Wild Animals (2021)	二级 / Category II
IUCN红色名录(2020-2) IUCN Red List (2020-2)	未列入 / NA
CITES 附录 (2019) CITES Appendix (2019)	II
保护行动 Conservation Action(s)	列入《迁徙物种保护公约》的"迁徙物种行动计划" / Included in the *Convention on Migratory Species* (CMS)－Species Action Plan

相关文献 Relevant References

Jiang *et al.*（蒋志刚等），2017; Castelló, 2016; Groves and Grubb, 2011; Wilson and Mittermeier, 2011

戈壁盘羊 *Ovis darwini*　　　　　　　　　　　　　　　　邸杰 摄　By Jie Di

白鱀豚
Lipotes vexillifer

极危　CR A1d ver.3.1

| 数据缺乏 DD | 无危 LC | 近危 NT | 易危 VU | 濒危 EN | **极危 CR** | 区域灭绝 RE | 野外灭绝 EW | 灭绝 EX |

分类地位 Taxonomic Status

| 动物界 Animalia | 脊索动物门 Chordata | 哺乳纲 Mammalia | 鲸目 Cetacea | 白鱀豚科 Lipotidae |

学　　名 Scientific Name	*Lipotes vexillifer*
命 名 人 Species Authority	Miller, 1918
英 文 名 English Name(s)	Baiji
同物异名 Synonym(s)	Chinese River Dolphin, Yangtze River Dolphin, Yangtze Dolphin, White-fin Dolphin
种下单元评估 Infra-specific Taxa Assessed	无 / None

评估信息 Assessment Information

评 估 年 份 Year Assessed	2020
评　定　人 Assessor(s)	周开亚 / Kaiya Zhou
审　定　人 Reviewer(s)	张先锋、王克雄、王丁、祝茜、蒋志刚 / Xianfeng Zhang, Kexiong Wang, Ding Wang, Qian Zhu, Zhigang Jiang
其他贡献人 Other Contributor(s)	李立立、丁晨晨 / Lili Li, Chenchen Ding

理由 Justification: 周开亚(1982)估计 1979～1981 年在枝城市（现称宜都市）以下的长江中约有 400 头白鱀豚。Chen 和 Hua(1989) 估测 1985～1986 年有约 300 头白鱀豚。Zhou 和 Li(1989)1982～1986 年调查发现从江西湖口县到长江口约有 100 头白鱀豚。Zhang *et al.* (2003) 报道 1997～1999 年 3 次长江中下游全江段考察中，只发现 13 头白鱀豚。Turvey *et al.* (2007) 报道 2006 年 11～12 月在白鱀豚考察中未发现一头白鱀豚。近年有一些目击白鱀豚的报道，但没有照片或视频来支持这些报道。2017 年 11～12 月的长江江豚考察中，仍未发现一头白鱀豚。因此，白鱀豚的红色名录等级定为极危（可能灭绝）/ Zhou (1982) estimated there were about 400 Baijis in the Changjiang River below Zhicheng City (Newly called Yidu City) between 1979 to 1981. Chen and Hua (1989) estimated there were 300 Baijis between 1985 and 1986. Surveys conducted by Zhou and Li (1989) from 1982 to 1986 showed that there were about 100 Baijis in the lower Changjiang River from Hukou County of Jiangxi Province to Changjiang River estuary. Between 1997 and 1999, three surveys were conducted in the middle and lower reaches of the Changjiang River, only about 13 Baijis were discovered (Zhang *et al.*, 2003). Turvey *et al.* (2007) reported that no Baiji was found during a survey conducted in the Changjiang River from November to December 2006. In recent years, there have been several reports from volunteers and fishermen who witnessed Baiji, however no photographic or video evidence has been available to support those reports. No Baiji was observed during the Survey of Yangtze Finless Porpoise from November to December, 2017. Based on these evidences, Baiji is listed as CR (possibly Extinct)

地理分布 Geographical Distribution

国内分布 Domestic Distribution	
长江、钱塘江 / Changjiang River, Qiantang River	
世界分布 World Distribution	
中国 / China	
分布标注 Distribution Note	
特有种 / Endemic	

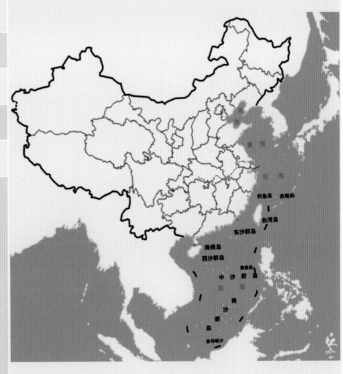

国内分布图
Map of Domestic Distribution

种群 Population

种群数量 Population Size	2001 年 11 月在镇江发现的雌性白鱀豚是最后搁浅记录。2002 年 5 月铜陵保护区江段内的 1 头白鱀豚的照片是最后 1 头在野外见到的白鱀豚 (Turvey et al., 2007) / The last confirmed beaching was in November 2001 in Zhenjiang and the last documented sighting (with photographic evidence) was in May 2002 in Tongling Reserve (Turvey et al., 2007)
种群趋势 Population Trend	下降 / Decreasing

生境与生态系统 Habitat (s) and Ecosystem (s)

生　　境 Habitat(s)	江河 / River
生态系统 Ecosystem(s)	湖泊河流生态系统 / Lake and River Ecosystem

威胁 Threat (s)

主要威胁 Major Threat(s)	渔业、航道、污染、人类活动干扰 / Fishing, Shipping Lane, Pollution, Human Disturbance

保护级别与保护行动 Protection Category and Conservation Action (s)

国家重点保护野生动物等级 (2021) Category of National Key Protected Wild Animals (2021)	一级 / Category I
IUCN 红色名录 (2020-2) IUCN Red List (2020-2)	极危 / CR
CITES 附录 (2019) CITES Appendix (2019)	I
保护行动 Conservation Action(s)	法律保护物种 / Legally protected species

相关文献 Relevant References

Burgin *et al*., 2018; Zhou, 2018; Jiang *et al*. (蒋志刚等), 2015a; Turvey *et al*., 2007; Zhang *et al*., 2003; Zhou *et al*. (周开亚等), 1998; Chen and Hua, 1989; Zhou and Li, 1989

白鱀豚 *Lipotes vexillifer*　　　中国科学院水生生物研究所 提供　By Institute of Hydrobiology, Chinese Academy of Sciences

长江江豚
Neophocaena asiaeorientalis
极危 CR A1acd

| 数据缺乏 DD | 无危 LC | 近危 NT | 易危 VU | 濒危 EN | **极危 CR** | 区域灭绝 RE | 野外灭绝 EW | 灭绝 EX |

分类地位 Taxonomic Status

动物界 Animalia	脊索动物门 Chordata	哺乳纲 Mammalia	鲸目 Cetacea	鼠海豚科 Phocoenidae
学 名 Scientific Name		*Neophocaena asiaeorientalis*		
命 名 人 Species Authority		Pilleri and Gihr, 1972		
英 文 名 English Name(s)		Yangtze Finless Porpoise		
同物异名 Synonym(s)		窄脊江豚		
种下单元评估 Infra-specific Taxa Assessed		无 / None		

评估信息 Assessment Information

评 估 年 份 Year Assessed	2020
评 定 人 Assessor(s)	周开亚 / Kaiya Zhou
审 定 人 Reviewer(s)	张先锋、王克雄、王丁、祝茜、蒋志刚 / Xianfeng Zhang, Kexiong Wang, Ding Wang, Qian Zhu, Zhigang Jiang
其他贡献人 Other Contributor(s)	李立立、丁晨晨 / Lili Li, Chenchen Ding

理由 Justification: Zhou *et al.* (2018) 通过全基因组的研究，将长江江豚提升为种，为中国特有物种。张先锋等 (1993) 首次报道长江江豚种群数量。基于 1984～1991 年的调查，他们估计长江中下游约有 2,700 头长江江豚。周开亚等 (1998) 基于 1989～1992 年的调查，估计南京至湖口间 421km 江段中长江江豚的种群数量为 697 头。Zhao *et al.* (2008) 根据 2006 年 11～12 月全江段考察获得的数据，估算长江中长江江豚的数量约为 1,225 头，另加洞庭湖和鄱阳湖的数据，估计长江江豚总数约 1,800 头。Mei *et al.* (2014) 于 2012 年在长江中下游对长江江豚进行考察，发现只有 505 头长江江豚残留在长江干流中，长江江豚种群加速下降。因此，长江江豚被列为极危等级 / Through genomic studies, Zhou *et al.* (2018) have suggested that the Yangtze Finless Porpoise has an independent species status and is endemic to China. Zhang *et al.* (1993) first reported on the population of the Yangtze Finless Porpoise based on surveys conducted between 1984 and 1991, where they estimated that there were about 2,700 Yangtze Finless Porpoises in the middle and lower reaches of the Changjiang River. Zhou *et al.* (1998) estimated that there were 697 individuals in the 421km section between Nanjing and Hukou based on surveys conducted from 1989 to 1992. Zhao *et al.* (2008) estimated numbers in the Changjiang River to be 1,225 individuals, based on data obtained during the survey of the entire river section from November to December 2006. In addition to the data of Dongting Lake and Poyang Lake, it is estimated that the total number of Yangtze Finless Porpoise is about 1,800. Mei *et al.* (2014) surveyed the Yangtze Finless Porpoise in the middle and lower reaches of the Changjiang River in 2012 and only 505 individuals remained in the main stream of the Changjiang River, suggesting that the Yangtze Finless Porpoise population is decreasing rapidly. Thus, Yangtze Finless Porpoise is listed as Critically Endangered

地理分布 Geographical Distribution

国内分布 Domestic Distribution	长江、洞庭湖、鄱阳湖 / Changjiang River, Dongting Lake, Poyang Lake
世界分布 World Distribution	中国 / China
分布标注 Distribution Note	特有种 / Endemic

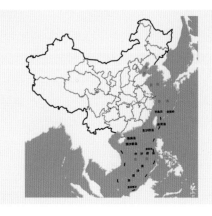

国内分布图
Map of Domestic Distribution

🦌 种群 Population

种群数量 Population Size	1,000 头 / 1,000 individuals
种群趋势 Population Trend	下降 / Decreasing

🦌 生境与生态系统 Habitat (s) and Ecosystem (s)

生　　境 Habitat(s)	江河、湖泊 / River and lake
生态系统 Ecosystem(s)	湖泊与河流生态系统 / Lake and River Ecosystem

🦌 威胁 Threat (s)

主要威胁 Major Threat(s)	船只撞击、渔业误捕、栖息地丧失和退化 / Collision with Ship, Fishery Incidental Capture, Habitat Loss and Degradation

🦌 保护级别与保护行动 Protection Category and Conservation Action (s)

国家重点保护野生动物等级 (2021) Category of National Key Protected Wild Animals (2021)	一级 / Category I
IUCN 红色名录 (2020-2) IUCN Red List (2020-2)	濒危 / EN
CITES 附录 (2019) CITES Appendix (2019)	无 / NA
保护行动 Conservation Action(s)	部分种群位于自然保护区内 / Part of populations are covered by nature reserves

🦌 相关文献 Relevant References

Burgin *et al.*, 2018; Zhou *et al.*, 2018; Jiang *et al.* (蒋志刚 等), 2017, 2015a; Mei *et al.*, 2014; Zhao *et al.*, 2013b; Zhou *et al.* (周开亚等), 1998; Zhang *et al.* (张先锋等), 1993

长江江豚 *Neophocaena asiaeorientalis*　　　　张先锋 摄　By Xianfeng Zhang

比氏鼯鼠
Biswamoyopterus biswasi

极危　CR B1ab(iii)

| 数据缺乏 DD | 无危 LC | 近危 NT | 易危 VU | 濒危 EN | **极危 CR** | 区域灭绝 RE | 野外灭绝 EW | 灭绝 EX |

分类地位 Taxonomic Status

| 动物界 Animalia | 脊索动物门 Chordata | 哺乳纲 Mammalia | 啮齿目 Rodentia | 松鼠科 Sciuridae |

学　名 Scientific Name	*Biswamoyopterus biswasi*
命名人 Species Authority	Saha, 1981
英文名 English Name(s)	Namdapha Flying Squirrel
同物异名 Synonym(s)	Namdapha Giant Flying Squirrel
种下单元评估 Infra-specific Taxa Assessed	无 / None

评估信息 Assessment Information

评估年份 Year Assessed	2020
评定人 Assessor(s)	刘少英、蒋志刚 / Shaoying Liu, Zhigang Jiang
审定人 Reviewer(s)	马勇、鲍毅新 / Yong Ma, Yixin Bao
其他贡献人 Other Contributor(s)	李立立、丁晨晨 / Lili Li, Chenchen Ding

理由 Justification: 比氏鼯鼠是一个稀有种。目前对其种群趋势尚不了解。仅从 Saha (1981) 采集的正模标本得知该种存在。估计占有区为 10～100km²。基于其占有区面积，将比氏鼯鼠濒危等级列为极危 / Namdapha Flying Squirrel appears to be a rare species, it may have a very small population. Nothing is known about its population trends. The existence of the species was only known from the holotype specimen collected by Saha (1981). Estimated area of occupancy is 10~100km². Thus, according to its area of occupancy, Namdapha Flying Squirrel is listed as Critically Endangered

地理分布 Geographical Distribution

国内分布 Domestic Distribution	西藏 / Tibet (Xizang)
世界分布 World Distribution	中国、印度 / China, India
分布标注 Distribution Note	非特有种 / Non-endemic

国内分布图
Map of Domestic Distribution

种群 Population

种群数量 Population Size	未知 / Unknown
种群趋势 Population Trend	下降 / Decreasing

生境与生态系统 Habitat (s) and Ecosystem (s)

生　　境 Habitat(s)	在落叶林中沿溪流的潮湿林间小径活动 / Occurs in deciduous montane forest, occupying moist forest tract along stream
生态系统 Ecosystem(s)	落叶林生态系统 / Deciduous Forest Ecosystem

威胁 Threat (s)

主要威胁 Major Threat(s)	未知 / Unknown

保护级别与保护行动 Protection Category and Conservation Action (s)

国家重点保护野生动物等级 (2021) Category of National Key Protected Wild Animals (2021)	无 / NA
IUCN红色名录(2020-2) IUCN Red List (2020-2)	极危 / CR
CITES 附录 (2019) CITES Appendix (2019)	无 / NA
保护行动 Conservation Action(s)	无 / None

相关文献 Relevant References

Burgin *et al*., 2018; Jiang *et al*. (蒋志刚等), 2017; Zhang (张荣祖), 1997

比氏鼯鼠 *Biswamoyopterus biswasi*

河狸 *Castor fiber*

极危 CR B1ab(i, ii, iii)+2ab(i, ii, iii); C2a(ii)

| 数据缺乏 DD | 无危 LC | 近危 NT | 易危 VU | 濒危 EN | **极危 CR** | 区域灭绝 RE | 野外灭绝 EW | 灭绝 EX |

分类地位 Taxonomic Status

动物界 Animalia	脊索动物门 Chordata	哺乳纲 Mammalia	啮齿目 Rodentia	河狸科 Castoridae
学　　名 Scientific Name		*Castor fiber*		
命 名 人 Species Authority		Linnaeus, 1758		
英 文 名 English Name(s)		Eurasian Beaver		
同物异名 Synonym(s)		蒙新河狸；*Castor albicus* (Matschie, 1907); *albus* (Kerr, 1792); *balticus* (Matschie, 1907); *belarusicus* (Lavrov, 1974); *belorussicus* (Lavrov, 1981); *bielorussieus* （转下页）		
种下单元评估 Infra-specific Taxa Assessed		无 / None		

评估信息 Assessment Information

评 估 年 份 Year Assessed	2020
评 　定　 人 Assessor(s)	蒋志刚、刘少英 / Zhigang Jiang, Shaoying Liu
审 　定　 人 Reviewer(s)	初红军 / Hongjun Chu
其他贡献人 Other Contributor(s)	李立立、丁晨晨 / Lili Li, Chenchen Ding

理由 Justification: 河狸在中国仅分布在新疆北部，为河狸蒙新亚种 (*C. f. birulai*)。由于其毛皮和毛囊分泌物的经济价值，河狸曾遭到大量捕杀。其占有面积由于受到人类开发活动影响而日益减少。经过20余年的保护，如今在中国分布的河狸数量仍不足700只。因此，根据该种的种群大小和占有面积，河狸列为极危等级 / Eurasian Beaver is marginally distributed in Xinjiang, China, is the subspecies of *C. f. birulai*. Its areas of occupancy are decreasing due to human activities. In addition, it was over-hunted for its fur and musk. It is estimated that there is no more than 700 individuals left in China after more than 20 years of protection. Thus, according to its population size and area of occupancy, Eurasian Beaver is listed as Critically Endangered

地理分布 Geographical Distribution

国内分布 Domestic Distribution
新疆 / Xinjiang
世界分布 World Distribution
欧亚大陆 / Eurasian Continent
分布标注 Distribution Note
非特有种 / Non-endemic

国内分布图
Map of Domestic Distribution

种群 Population

种群数量 Population Size	700 只 / 700 individuals
种群趋势 Population Trend	上升 / Increasing

生境与生态系统 Habitat(s) and Ecosystem(s)

生境 Habitat(s)	淡水湖、江河 / Freshwater Lake, River
生态系统 Ecosystem(s)	湖泊河流生态系统 / Lake and River Ecosystem

威胁 Threat(s)

主要威胁 Major Threat(s)	人类活动干扰、家畜放牧 / Human Disturbance, Livestock Ranching

保护级别与保护行动 Protection Category and Conservation Action(s)

国家重点保护野生动物等级 (2021) Category of National Key Protected Wild Animals (2021)	一级 / Category I
IUCN 红色名录 (2020-2) IUCN Red List (2020-2)	无危 / LC
CITES 附录 (2019) CITES Appendix (2019)	无 / NA
保护行动 Conservation Action(s)	已经建立国家级自然保护区 / National nature reserve established

相关文献 Relevant References

Burgin *et al.*, 2018; Jiang *et al.* (蒋志刚等), 2017; Zheng *et al.* (郑智民等), 2012; Chu and Jiang, 2009; Smith *et al.* (史密斯等), 2009; Zhao *et al.* (赵景辉等), 2005; Chen *et al.* (陈道富等), 2003

(接上页)
(Lavrov, 1983); *birulai* (Serebrennikov, 1929); *flavus* (Desmarest, 1822); *fulvus* (Bechstein, 1801); *galliae* (É. Geoffroy, 1803); *gallicus* (Fischer, 1829); *introductus* (Saveljev, 1997); *niger* (Desmarest, 1822); *orientoeuropaeus* (Lavrov, 1981); *osteuropaeus* (Lavrov, 1974); *pohlei* (Serebrennikov, 1929); *proprius* (Billberg, 1833); *solitarius* (Kerr, 1792); *tuvinicus* (Lavrov, 1969); *variegatus* (Bechstein, 1801); *varius* (Desmarest, 1822); *vistulanus* (Matschie, 1907)

河狸 *Castor fiber*

中国生物多样性 红色名录

海南兔
Lepus hainanus
极危 CR A1acd

| 数据缺乏 DD | 无危 LC | 近危 NT | 易危 VU | 濒危 EN | **极危 CR** | 区域灭绝 RE | 野外灭绝 EW | 灭绝 EX |

分类地位 Taxonomic Status

动物界 Animalia	脊索动物门 Chordata	哺乳纲 Mammalia	兔形目 Lagomorpha	兔科 Leporidae

学 名 Scientific Name	*Lepus hainanus*
命 名 人 Species Authority	Swinhoe, 1870
英 文 名 English Name(s)	Hainan Hare
同物异名 Synonym(s)	无 / None
种下单元评估 Infra-specific Taxa Assessed	无 / None

评估信息 Assessment Information

评 估 年 份 Year Assessed	2020
评 定 人 Assessor(s)	江海生、陈辈乐、李飞 / Haisheng Jiang, Bosco P. L. Chan, Fei Li
审 定 人 Reviewer(s)	蒋志刚 / Zhigang Jiang
其他贡献人 Other Contributor(s)	李立立、丁晨晨 / Lili Li, Chenchen Ding

理由 Justification: 海南兔栖息于海南岛丘陵平原地带灌草丛中，很少出现在森林之中，在高山地区从未记录到海南兔。近年的调查与访问显示，由于捕猎及栖息地丧失，海南兔种群数量急剧下降，在多个地区甚至保护区内已经消失（嘉道理农场暨植物园，未发表数据）。基于其种群下降速率，将海南兔列为极危等级 / Hainan Hare inhabits the short grass and shrublands on the hills and plains of Hainan island, it is seldom seen in the forests and it has never been recorded in alpine areas. Surveys and interviews in recent years have shown that the number of Hainan Hare has declined rapidly due to hunting and habitat loss, it disappeared in many areas and even protected areas (Kadoorie Farm and Botanic Garden, unpublished data). Thus, based on its rate of population decline, it is listed as Critically Endangered

地理分布 Geographical Distribution

国内分布 Domestic Distribution
海南 / Hainan
世界分布 World Distribution
中国 / China
分布标注 Distribution Note
特有种 / Endemic

国内分布图 Map of Domestic Distribution

种群 Population

种群数量 Population Size	未知 / Unknown
种群趋势 Population Trend	下降 / Decreasing

生境与生态系统 Habitat (s) and Ecosystem (s)

生　　境 Habitat(s)	灌丛、草地 / Shrubland, Grassland
生态系统 Ecosystem(s)	灌丛生态系统、草地生态系统 / Shrubland Ecosystem, Grassland Ecosystem

威胁 Threat (s)

主要威胁 Major Threat(s)	捕猎，栖息地丧失 / Hunting, Habitat Loss

保护级别与保护行动 Protection Category and Conservation Action (s)

国家重点保护野生动物等级 (2021) Category of National Key Protected Wild Animals (2021)	二级 / Category II
IUCN 红色名录 (2020-2) IUCN Red List (2020-2)	濒危 / EN
CITES 附录 (2019) CITES Appendix (2019)	无 / NA
保护行动 Conservation Action(s)	无 / None

相关文献 Relevant References

Burgin *et al*., 2018; Jiang *et al*.（蒋志刚等），2017; Zheng *et al*.（郑智民等），2012; Smith *et al*.（史密斯等），2009; Pan *et al*.（潘清华等），2007

海南兔 *Lepus hainanus*

海南新毛猬
Neohylomys hainanensis

濒危 EN B1+2ab(i,ii,iii)

| 数据缺乏 DD | 无危 LC | 近危 NT | 易危 VU | **濒危 EN** | 极危 CR | 区域灭绝 RE | 野外灭绝 EW | 灭绝 EX |

分类地位 Taxonomic Status

动物界 Animalia	脊索动物门 Chordata	哺乳纲 Mammalia	劳亚食虫目 Eulipotyphla	猬科 Erinaceidae
学名 Scientific Name		*Neohylomys hainanensis*		
命名人 Species Authority		Shaw and Wang, 1959		
英文名 English Name(s)		Hainan Gymnure		
同物异名 Synonym(s)		Hainan Moonrat; *Hylomys hainanensis* (Shaw and Wang, 1959)		
种下单元评估 Infra-specific Taxa Assessed		无 / None		

评估信息 Assessment Information

评估年份 Year Assessed	2020
评定人 Assessor(s)	蒋志刚 / Zhigang Jiang
审定人 Reviewer(s)	陈辈乐、李飞 / Bosco P. L. Chan, Fei Li
其他贡献人 Other Contributor(s)	李立立、丁晨晨 / Lili Li, Chenchen Ding

理由 Justification: 海南新毛猬为中国特有种，仅分布在海南。迄今所采的海南新毛猬标本数量极少。对该种研究缺乏。海南森林砍伐严重，随着农业发展，其栖息地人为干扰严重，导致该种占有区下降。因此，根据其栖息地现状和发生区数据，将该种列为濒危等级 / Hainan Gymnure is an endemic species and only distributed in the Hainan Province. Only few specimens of the species had been collected and researches on the species are limited. However, the deforestation of natural forest in Hainan Province is severe and the human encroachment, mainly agricultural development, bring significant pressure on the species' habitat, causing its areas of occupancy decline. Thus, based on its habitat status and data for area of occurrence, Hainan Gymnure is listed as Endangered

地理分布 Geographical Distribution

国内分布 Domestic Distribution
海南 / Hainan
世界分布 World Distribution
中国 / China
分布标注 Distribution Note
特有种 / Endemic

国内分布图
Map of Domestic Distribution

种群 Population

种群数量 Population Size	未知 / Unknown
种群趋势 Population Trend	下降 / Decreasing

生境与生态系统 Habitat (s) and Ecosystem (s)

生　　境 Habitat(s)	海拔 250m 的次生林以及橡胶林、海拔 600～1000m 的山地森林 / Secondary forest and rubber forest around 250m a.s.l., mountain forest around 600~1,000m a.s.l.
生态系统 Ecosystem(s)	森林生态系统 / Forest Ecosystem

威胁 Threat (s)

主要威胁 Major Threat(s)	伐木、耕种 / Logging, Plantation

保护级别与保护行动 Protection Category and Conservation Action (s)

国家重点保护野生动物等级 (2021) Category of National Key Protected Wild Animals (2021)	无 / NA
IUCN 红色名录 (2020-2) IUCN Red List (2020-2)	濒危 / EN
CITES 附录 (2019) CITES Appendix (2019)	无 / NA
保护行动 Conservation Action(s)	位于自然保护区内的种群与生境得到保护，宣传教育、生境及栖息地保护 / Populations and habitats located in nature reserves are protected; conservation communication and education, habitat protection

相关文献 Relevant References

Burgin *et al*., 2018; Jiang *et al*. (蒋志刚等), 2017; Stone, 1995; Shou *et al*. (寿振黄等), 1966; Shou and Wang (寿振黄和汪松), 1959

海南新毛猬 *Neohylomys hainanensis*

琉球狐蝠
Pteropus dasymallus

濒危 EN B1+ 2a(i)

| 数据缺乏 DD | 无危 LC | 近危 NT | 易危 VU | 濒危 EN | 极危 CR | 区域灭绝 RE | 野外灭绝 EW | 灭绝 EX |

分类地位 Taxonomic Status

动物界 Animalia	脊索动物门 Chordata	哺乳纲 Mammalia	翼手目 Chiroptera	狐蝠科 Pteropodidae
学名 Scientific Name		*Pteropus dasymallus*		
命名人 Species Authority		Temminck, 1825		
英文名 English Name(s)		Ryukyu Flying Fox		
同物异名 Synonym(s)		Ryukyu Fruit Bat		
种下单元评估 Infra-specific Taxa Assessed		无 / None		

评估信息 Assessment Information

评估年份 Year Assessed	2020
评定人 Assessor(s)	吴毅、蒋志刚 / Yi Wu, Zhigang Jiang
审定人 Reviewer(s)	张礼标、毛秀光、张树义 / Libiao Zhang, Xiuguang Mao, Shuyi Zhang
其他贡献人 Other Contributor(s)	李立立、丁晨晨 / Lili Li, Chenchen Ding

理由 Justification: 在中国，琉球狐蝠仅分布在台湾，由于其栖息地面积缩小，且受到严重的人为干扰，种群数量下降。基于种群与生境现状，将琉球狐蝠列为濒危等级 / In China, Ryukyu Flying Fox is only distributed in Taiwan Province. Its populations are declining because of severe human disturbance and habitat loss. Thus, based on its population and habitat status, Ryukyu Flying Fox is listed as Endangered

地理分布 Geographical Distribution

国内分布 Domestic Distribution
台湾 / Taiwan
世界分布 World Distribution
中国、日本、菲律宾 / China, Japan, Philippines
分布标注 Distribution Note
非特有种 / Non-endemic

国内分布图
Map of Domestic Distribution

种群 Population

种群数量 Population Size	稀少 / Rare
种群趋势 Population Trend	下降 / Decreasing

生境与生态系统 Habitat(s) and Ecosystem(s)

生　　境 Habitat(s)	未知 / Unknown
生态系统 Ecosystem(s)	森林生态系统 / Forest Ecosystem

威胁 Threat(s)

主要威胁 Major Threat(s)	生境丧失 / Habitat Loss

保护级别与保护行动 Protection Category and Conservation Action(s)

国家重点保护野生动物等级 (2021) Category of National Key Protected Wild Animals (2021)	无 / NA
IUCN 红色名录 (2020-2) IUCN Red List (2020-2)	易危 / VU
CITES 附录 (2019) CITES Appendix (2019)	II
保护行动 Conservation Action(s)	政策性措施和立法保护、栖息地保护 / Policy-based actions and legistation, habitat protection

相关文献 Relevant References

Burgin *et al*., 2018; Jiang *et al*.（蒋志刚等）, 2017; Lin（林良恭）, 2002; Lin *et al*.（林良恭等）, 1997; Chen（陈兼善）, 1969

琉球狐蝠 *Pteropus dasymallus*　　　　　　　　　　　　　　　　　　　By Koolah, CC BY-SA 3.0

安氏长舌果蝠
Macroglossus sobrinus

濒危 EN B1+2a(i)

| DD | LC | NT | VU | **EN** | CR | RE | EW | EX |

分类地位 Taxonomic Status

| 动物界 Animalia | 脊索动物门 Chordata | 哺乳纲 Mammalia | 翼手目 Chiroptera | 狐蝠科 Pteropodidae |

学名 Scientific Name	*Macroglossus sobrinus*
命名人 Species Authority	K. Andersen, 1911
英文名 English Name(s)	Hill Long-tongued Fruit Bat
同物异名 Synonym(s)	Long-tongued Fruit Bat; *Macroglossus fraternus* (Chasen and Kloss, 1928); *minimus* (Andersen, 1911) subsp. *sobrinus*
种下单元评估 Infra-specific Taxa Assessed	无 / None

评估信息 Assessment Information

评估年份 Year Assessed	2020
评定人 Assessor(s)	吴毅、蒋志刚 / Yi Wu, Zhigang Jiang
审定人 Reviewer(s)	张礼标、毛秀光、张树义 / Libiao Zhang, Xiuguang Mao, Shuyi Zhang
其他贡献人 Other Contributor(s)	李立立、丁晨晨 / Lili Li, Chenchen Ding

理由 Justification: 在中国，安氏长舌果蝠仅分布在云南，由于其栖息地面积缩小，且受到严重的人为干扰，种群数量不断下降。因此，基于其种群与生境现状，安氏长舌果蝠列为濒危等级 / In China, Hill Long-tongued Fruit Bat is only distributed in Yunnan Province. Its populations are declining due to the shrinking of habitat area and severe human disturbance. Thus, Hill Long-tongued Fruit Bat is listed as Endangered based on its population and habitat status

地理分布 Geographical Distribution

国内分布 Domestic Distribution
云南 / Yunnan
世界分布 World Distribution
柬埔寨、中国、印度、印度尼西亚、老挝、马来西亚、缅甸、泰国、越南 / Cambodia, China, India, Indonesia, Laos, Malaysia, Myanmar, Thailand, Viet Nam
分布标注 Distribution Note
非特有种 / Non-endemic

国内分布图
Map of Domestic Distribution

种群 Population

种群数量 Population Size	稀少 / Rare
种群趋势 Population Trend	持平 / Stable

生境与生态系统 Habitat (s) and Ecosystem (s)

生境 Habitat(s)	森林、次生林、种植园、人造建筑 / Forest, Secondary Forest, Plantation, Building
生态系统 Ecosystem(s)	森林生态系统、人类聚落生态系统 / Forest Ecosystem, Human Settlement Ecosystem

威胁 Threat (s)

主要威胁 Major Threat(s)	伐木、耕种、狩猎及采集 / Logging, Plantation, Hunting and Collection

保护级别与保护行动 Protection Category and Conservation Action (s)

国家重点保护野生动物等级 (2021) Category of National Key Protected Wild Animals (2021)	无 / NA
IUCN 红色名录 (2020-2) IUCN Red List (2020-2)	无危 / LC
CITES 附录 (2019) CITES Appendix (2019)	无 / NA
保护行动 Conservation Action(s)	无 / None

相关文献 Relevant References

Burgin *et al.*, 2018; Jiang *et al.* (蒋志刚等), 2017; Zhang *et al.*, 2010; Feng *et al.* (冯庆等), 2007; Lekagul and McNeely, 1977

安氏长舌果蝠 *Macroglossus sobrinus*   By Alice Hughes

琉球长翼蝠
Miniopterus fuscus

濒危 EN B1ab(iii)

| 数据缺乏 DD | 无危 LC | 近危 NT | 易危 VU | **濒危 EN** | 极危 CR | 区域灭绝 RE | 野外灭绝 EW | 灭绝 EX |

分类地位 Taxonomic Status

动物界 Animalia	脊索动物门 Chordata	哺乳纲 Mammalia	翼手目 Chiroptera	蝙蝠科 Vespertilionidae

学 名 Scientific Name	*Miniopterus fuscus*
命 名 人 Species Authority	Bonhote, 1902
英 文 名 English Name(s)	Southeast Asian Long-fingered Bat
同物异名 Synonym(s)	无 / None
种下单元评估 Infra-specific Taxa Assessed	无 / None

评估信息 Assessment Information

评估年份 Year Assessed	2020
评 定 人 Assessor(s)	吴毅、蒋志刚 / Yi Wu, Zhigang Jiang
审 定 人 Reviewer(s)	张礼标、毛秀光、张树义 / Libiao Zhang, Xiuguang Mao, Shuyi Zhang
其他贡献人 Other Contributor(s)	李立立、丁晨晨 / Lili Li, Chenchen Ding

理由 Justification: 琉球长翼蝠在中国的分布区狭窄，其种群数量极少，且栖息地的面积因人类活动而不断减少。因此，基于其种群与生境现状，琉球长翼蝠列为濒危等级 / Southeast Asian Long-fingered Bat is narrowly distributed in China, and its populations are small. Furthermore, the areas of its habitats are declining due to human encroachment. Thus, based on its population and habitat status, Southeast Asian Long-fingered Bat is listed as Endangered

地理分布 Geographical Distribution

国内分布 Domestic Distribution	福建、台湾 / Fujian, Taiwan
世界分布 World Distribution	中国、日本 / China, Japan
分布标注 Distribution Note	非特有种 / Non-endemic

国内分布图
Map of Domestic Distribution

种群 Population

种群数量 Population Size	未知 / Unknown
种群趋势 Population Trend	下降 / Decreasing

生境与生态系统 Habitat (s) and Ecosystem (s)

生境 Habitat(s)	洞穴、森林 / Cave, Forest
生态系统 Ecosystem(s)	森林生态系统 / Forest Ecosystem

威胁 Threat (s)

主要威胁 Major Threat(s)	砍伐、旅游 / Logging, Tourism

保护级别与保护行动 Protection Category and Conservation Action (s)

国家重点保护野生动物等级 (2021) Category of National Key Protected Wild Animals (2021)	无 / NA
IUCN 红色名录 (2020-2) IUCN Red List (2020-2)	濒危 / EN
CITES 附录 (2019) CITES Appendix (2019)	无 / NA
保护行动 Conservation Action(s)	无 / None

相关文献 Relevant References

Burgin *et al.*, 2018; Jiang *et al.* (蒋志刚等), 2015a; Wang (王应祥), 2003; Tan (谭邦杰), 1992

琉球长翼蝠 *Miniopterus fuscus*

彩蝠 *Kerivoula picta*

濒危 EN B1; C2a(i)

| 数据缺乏 DD | 无危 LC | 近危 NT | 易危 VU | **濒危 EN** | 极危 CR | 区域灭绝 RE | 野外灭绝 EW | 灭绝 EX |

分类地位 Taxonomic Status

动物界 Animalia	脊索动物门 Chordata	哺乳纲 Mammalia	翼手目 Chiroptera	蝙蝠科 Vespertilionidae
学名 Scientific Name		*Kerivoula picta*		
命名人 Species Authority		Pallas, 1767		
英文名 English Name(s)		Painted Woolly Bat		
同物异名 Synonym(s)		Painted Bat; *Vespertilio kirivoula* (Cuvier, 1832); *pictus* (Pallas, 1767)		
种下单元评估 Infra-specific Taxa Assessed		无 / None		

评估信息 Assessment Information

评估年份 Year Assessed	2020
评定人 Assessor(s)	吴毅、蒋志刚 / Yi Wu, Zhigang Jiang
审定人 Reviewer(s)	张礼标、毛秀光、张树义 / Libiao Zhang, Xiuguang Mao, Shuyi Zhang
其他贡献人 Other Contributor(s)	李立立、丁晨晨 / Lili Li, Chenchen Ding

理由 Justification: 彩蝠种群数量非常少，且其所在的栖息地面积锐减。基于其种群与生境现状，彩蝠列为濒危等级 / The population sizes of Painted Woolly Bat are extremely small and the areas of its habitats are declining rapidly. Thus, based on its population and habitat status, Painted Woolly Bat is listed as Endangered

地理分布 Geographical Distribution

国内分布 Domestic Distribution	广东、广西、海南、福建、贵州 / Guangdong, Guangxi, Hainan, Fujian, Guizhou
世界分布 World Distribution	中国；南亚、东南亚 / China; South Asia, Southeast Asia
分布标注 Distribution Note	非特有种 / Non-endemic

国内分布图
Map of Domestic Distribution

种群 Population

种群数量 Population Size	未知 / Unknown
种群趋势 Population Trend	未知 / Unknown

生境与生态系统 Habitat (s) and Ecosystem (s)

生　　境 Habitat(s)	未知 / Unknown
生态系统 Ecosystem(s)	森林生态系统、农田生态系统 / Forest Ecosystem, Cropland Ecosystem

威胁 Threat (s)

主要威胁 Major Threat(s)	伐木、耕种 / Logging, Plantation

保护级别与保护行动 Protection Category and Conservation Action (s)

国家重点保护野生动物等级 (2021) Category of National Key Protected Wild Animals (2021)	无 / NA
IUCN 红色名录 (2020-2) IUCN Red List (2020-2)	近危 / NT
CITES 附录 (2019) CITES Appendix (2019)	无 / NA
保护行动 Conservation Action(s)	无 / None

相关文献 Relevant References

Burgin *et al.*, 2018; Jiang *et al.*（蒋志刚等）, 2017; Li *et al.*（李德伟等）, 2010; Smith *et al.*（史密斯等）, 2009; Wu *et al.*（吴毅等）, 2007

彩蝠 *Kerivoula picta*　　　　吴毅 摄　By Yi Wu

中国生物多样性红色名录

蜂猴
Nycticebus bengalensis

濒危 EN A2bd

| 数据缺乏 DD | 无危 LC | 近危 NT | 易危 VU | **濒危 EN** | 极危 CR | 区域灭绝 RE | 野外灭绝 EW | 灭绝 EX |

分类地位 Taxonomic Status

动物界 Animalia	脊索动物门 Chordata	哺乳纲 Mammalia	灵长目 Primates	懒猴科 Lorisidae

学 名 Scientific Name	*Nycticebus bengalensis*
命 名 人 Species Authority	Lacépède, 1800
英 文 名 English Name(s)	Bengal Slow Loris
同物异名 Synonym(s)	Northern Slow Loris; *Nycticebus cinereus* (Milne-Edwards, 1867); *incanus* (Thomas, 1921); *tenasserimensis* (Elliot, 1913)
种下单元评估 Infra-specific Taxa Assessed	无 / None

评估信息 Assessment Information

评 估 年 份 Year Assessed	2020
评 定 人 Assessor(s)	蒋志刚、蒋学龙 / Zhigang Jiang, Xuelong Jiang
审 定 人 Reviewer(s)	黄乘明、蒋学龙、范朋飞 / Chengming Huang, Xuelong Jiang, Pengfei Fan
其他贡献人 Other Contributor(s)	李立立、丁晨晨 / Lili Li, Chenchen Ding

理由 Justification: 蜂猴栖息地中的森林被砍伐，使该种的栖息地丧失。此外，宠物市场常见蜂猴的非法贸易。非法捕猎和交易严重威胁着该物种的生存。因此，基于其种群大小、种群与栖息地下降速率，将该种列为濒危等级 / The forest inhabited by Bengal Slow Loris are suffering from logging, leading to habitat loss. Furthermore, this species is often illegally hunted and sold in pet markets; poaching also threatens the survival of this species. Thus, based on the population size, population declining and habitat, Bangal Slow Loris is listed as Endangered

地理分布 Geographical Distribution

国内分布 Domestic Distribution
广西、云南 / Guangxi, Yunnan
世界分布 World Distribution
中国；南亚、东南亚 / China; South Asia, Southeast Asia
分布标注 Distribution Note
非特有种 / Non-endemic

国内分布图
Map of Domestic Distribution

种群 Population

种群数量 Population Size	630 只 / 630 individuals
种群趋势 Population Trend	下降 / Decreasing

生境与生态系统 Habitat (s) and Ecosystem (s)

生境 Habitat(s)	森林、灌丛、竹丛、种植园 / Forest, Shrubland, Bamboo Grove, Plantation
生态系统 Ecosystem(s)	森林生态系统、灌丛生态系统、农田生态系统 / Forest Ecosystem, Shrubland Ecosystem, Cropland Ecosystem

威胁 Threat (s)

主要威胁 Major Threat(s)	耕种、伐木、公路和铁路、堤坝及水道建设、服务运输线路、火灾、狩猎、公路碾压事件 / Plantation, Logging, Road and Railroad, Dam and Cannal Construction, Linear Infrastructure, Fire, Hunting, Road Kill

保护级别与保护行动 Protection Category and Conservation Action (s)

国家重点保护野生动物等级 (2021) Category of National Key Protected Wild Animals (2021)	一级 / Category I
IUCN 红色名录 (2020-2) IUCN Red List (2020-2)	易危 / VU
CITES 附录 (2019) CITES Appendix (2019)	I
保护行动 Conservation Action(s)	在自然保护区内的种群及栖息地得到保护 / Populations and habitats in nature reserves are protected

相关文献 Relevant References

Burgin *et al*., 2018; Jiang *et al*. (蒋志刚等), 2017; Yu *et al*. (余梁哥等), 2013; Duan *et al*. (段艳芳等), 2012; Wilson and Mittermeier, 2012; Pan *et al*., 2007

蜂猴 *Nycticebus bengalensis* 　　　　关翔宇 摄（西南山地供图）　　By Xiangyu Guan (Swild.cn)

台湾猴
Macaca cyclopis

濒危 EN A2bd; B2

| 数据缺乏 DD | 无危 LC | 近危 NT | 易危 VU | **濒危 EN** | 极危 CR | 区域灭绝 RE | 野外灭绝 EW | 灭绝 EX |

分类地位 Taxonomic Status

动物界 Animalia	脊索动物门 Chordata	哺乳纲 Mammalia	灵长目 Primates	猴科 Cercopithecidae

学 名 Scientific Name	*Macaca cyclopis*
命 名 人 Species Authority	Swinhoe, 1863
英 文 名 English Name(s)	Formosan Rock Macaque
同物异名 Synonym(s)	Formosan Rock Monkey; Taiwan Macaque; *Macaca affinis* (Blyth, 1863)
种下单元评估 Infra-specific Taxa Assessed	无 / None

评估信息 Assessment Information

评 估 年 份 Year Assessed	2020
评 定 人 Assessor(s)	蒋志刚、蒋学龙 / Zhigang Jiang, Xuelong Jiang
审 定 人 Reviewer(s)	黄乘明、蒋学龙、范朋飞 / Chengming Huang, Xuelong Jiang, Pengfei Fan
其他贡献人 Other Contributor(s)	李立立、丁晨晨 / Lili Li, Chenchen Ding

理由 Justification: 台湾猴为岛屿性灵长类，其分布区狭窄，且目前其栖息地受到人类活动的影响。因此，基于其栖息地现状，台湾猴列为濒危等级 / Formosan Rock Macaque is an island species with small distribution range. The habitats of this species are affected by human encroachment. Thus, based on the status of habitat, Formosan Rock Macaque is listed as Endangered

地理分布 Geographical Distribution

国内分布 Domestic Distribution
台湾 / Taiwan
世界分布 World Distribution
中国、日本 / China, Japan
分布标注 Distribution Note
非特有种 / Non-endemic

国内分布图
Map of Domestic Distribution

种群 Population

种群数量 Population Size	未知 / Unknown
种群趋势 Population Trend	持平 / Stable

生境与生态系统 Habitat(s) and Ecosystem(s)

生境 Habitat(s)	亚热带湿润山地森林、竹林、次生林、耕地 / Subtropical Moist Montane Forest, Bamboo Grove, Secondary Forest, Arable Land
生态系统 Ecosystem(s)	森林生态系统、农田生态系统 / Forest Ecosystem, Cropland Ecosystem

威胁 Threat(s)

主要威胁 Major Threat(s)	狩猎 / Hunting

保护级别与保护行动 Protection Category and Conservation Action(s)

国家重点保护野生动物等级 (2021) Category of National Key Protected Wild Animals (2021)	一级 / Category I
IUCN 红色名录 (2020-2) IUCN Red List (2020-2)	无危 / LC
CITES 附录 (2019) CITES Appendix (2019)	II
保护行动 Conservation Action(s)	已建立自然保护区 / Nature reserve established

相关文献 Relevant References

Burgin *et al*., 2018; Jiang *et al*.（蒋志刚等）, 2017; Wilson and Mittermeier, 2012; Smith *et al*.（史密斯等）, 2009; Wang（王应祥）, 2003; Masui *et al*., 1986; Poirier, 1986

台湾猴 *Macaca cyclopis*

中国生物多样性红色名录 China's Red List of Biodiversity

达旺猴
Macaca munzala

濒危 EN A2bd; B2; C2a(ii)

| 数据缺乏 DD | 无危 LC | 近危 NT | 易危 VU | **濒危 EN** | 极危 CR | 区域灭绝 RE | 野外灭绝 EW | 灭绝 EX |

分类地位 Taxonomic Status

动物界 Animalia	脊索动物门 Chordata	哺乳纲 Mammalia	灵长目 Primates	猴科 Cercopithecidae
学 名 Scientific Name		*Macaca munzala*		
命 名 人 Species Authority		Madhusudan and Mishra, 2005		
英 文 名 English Name(s)		Arunachal Macaque		
同物异名 Synonym(s)		藏南猕猴		
种下单元评估 Infra-specific Taxa Assessed		无 / None		

评估信息 Assessment Information

评估年份 Year Assessed	2020
评定人 Assessor(s)	蒋志刚、蒋学龙 / Zhigang Jiang, Xuelong Jiang
审定人 Reviewer(s)	黄乘明、蒋学龙、龙勇诚、范朋飞 / Chengming Huang, Xuelong Jiang, Yongcheng Long, Pengfei Fan
其他贡献人 Other Contributor(s)	李立立、丁晨晨 / Lili Li, Chenchen Ding

理由 Justification: 达旺猴是一个21世纪初基于照片报道的新种，分布在中国藏南，其记录较少，分布区小。因此，基于其发生区面积，将达旺猴列为濒危等级 / Arunachal Macaque is a new species discovered based on photos in China's Zangnan area in the early 21st century. Little information about this species is recorded, its range is small. Thus, based on the area of occurrence, Arunachal Macaque is listed as Endangered

地理分布 Geographical Distribution

国内分布 Domestic Distribution
西藏 / Tibet (Xizang)
世界分布 World Distribution
中国、印度、不丹 / China, India, Bhutan
分布标注 Distribution Note
非特有种 / Non-endemic

国内分布图
Map of Domestic Distribution

种群 Population

种群数量 Population Size	未知 / Unknown
种群趋势 Population Trend	下降 / Decreasing

生境与生态系统 Habitat (s) and Ecosystem (s)

生　　境 Habitat(s)	阔叶林、针叶林、灌丛、耕地 / Broad-leaved Forest, Coniferous Forest, Shrubland, Arable Land
生态系统 Ecosystem(s)	森林生态系统、灌丛生态系统、农田生态系统 / Forest Ecosystem, Shrubland Ecosystem, Cropland Ecosystem

威胁 Threat (s)

主要威胁 Major Threat(s)	狩猎 / Hunting

保护级别与保护行动 Protection Category and Conservation Action (s)

国家重点保护野生动物等级 (2021) Category of National Key Protected Wild Animals (2021)	二级 / Category II
IUCN 红色名录 (2020-2) IUCN Red List (2020-2)	濒危 / EN
CITES 附录 (2019) CITES Appendix (2019)	II
保护行动 Conservation Action(s)	无 / None

相关文献 Relevant References

Burgin *et al*., 2018; Chang *et al*. (常勇斌等), 2018; Jiang *et al*. (蒋志刚等), 2017; Sinha *et al*., 2005

达旺猴 *Macaca munzala*　　齐硕 摄（西南山地供图）　By Shuo Qi (Swild.cn)

印支灰叶猴
Trachypithecus crepusculus

濒危 EN A2bd; B2; C2a(ii)

| 数据缺乏 DD | 无危 LC | 近危 NT | 易危 VU | 濒危 EN | 极危 CR | 区域灭绝 RE | 野外灭绝 EW | 灭绝 EX |

分类地位 Taxonomic Status

动物界 Animalia	脊索动物门 Chordata	哺乳纲 Mammalia	灵长目 Primates	猴科 Cercopithecidae
学名 Scientific Name		*Trachypithecus crepusculus*		
命名人 Species Authority		Elliot, 1909		
英文名 English Name(s)		Indochinese Gray Langur		
同物异名 Synonym(s)		无 / None		
种下单元评估 Infra-specific Taxa Assessed		无 / None		

评估信息 Assessment Information

评估年份 Year Assessed	2020
评定人 Assessor(s)	蒋志刚、蒋学龙 / Zhigang Jiang, Xuelong Jiang
审定人 Reviewer(s)	黄乘明、蒋学龙、龙勇诚、范朋飞 / Chengming Huang, Xuelong Jiang, Yongcheng Long, Pengfei Fan
其他贡献人 Other Contributor(s)	李立立、丁晨晨 / Lili Li, Chenchen Ding

理由 Justification: 目前在云南无量山分布着43群、约2,000只印支灰叶猴。这些印支灰叶猴容易被观察到。考虑到印支灰叶猴的占有区面积,将其列为濒危等级 / Presently, there are 43 groups of Indochinese Gray Langur with around 2,000 individuals in Wuliang Shan, Yunnan Province. These langurs are easily observed. Based on the area of occupancy of Indochinese Gray Langurs, it is listed as Endangered

地理分布 Geographical Distribution

国内分布 Domestic Distribution
云南 / Yunnan
世界分布 World Distribution
孟加拉国、中国、老挝、泰国、越南 / Bangladesh, China, Laos, Thailand, Viet Nam
分布标注 Distribution Note
非特有种 / Non-endemic

国内分布图
Map of Domestic Distribution

种群 Population

种群数量 Population Size	未知 / Unknown
种群趋势 Population Trend	稳定 / Stable

生境与生态系统 Habitat (s) and Ecosystem (s)

生境 Habitat(s)	森林 / Forest
生态系统 Ecosystem(s)	森林生态系统 / Forest Ecosystem

威胁 Threat (s)

主要威胁 Major Threat(s)	狩猎、耕种、人类活动干扰、火灾 / Hunting, Plantation, Human Disturbance, Fire

保护级别与保护行动 Protection Category and Conservation Action (s)

国家重点保护野生动物等级 (2021) Category of National Key Protected Wild Animals (2021)	一级 / Category I
IUCN 红色名录(2020-2) IUCN Red List (2020-2)	濒危 / EN
CITES 附录 (2019) CITES Appendix (2019)	II
保护行动 Conservation Action(s)	在自然保护区内的种群及栖息地得到保护 / Populations and habitats in nature reserves are protected

相关文献 Relevant References

Burgin *et al.*, 2018; Jiang *et al.* (蒋志刚等), 2017; He *et al.*, 2012b; Wilson and Mittermeier, 2012

印支灰叶猴 *Trachypithecus crepusculus* 　　　　　　　　董磊 摄（西南山地供图） By Lei Dong (Swild.cn)

黑叶猴
Trachypithecus francoisi

濒危 EN A2bd; B2; C2a(ii)

| 数据缺乏 DD | 无危 LC | 近危 NT | 易危 VU | 濒危 EN | 极危 CR | 区域灭绝 RE | 野外灭绝 EW | 灭绝 EX |

分类地位 Taxonomic Status

动物界 Animalia	脊索动物门 Chordata	哺乳纲 Mammalia	灵长目 Primates	猴科 Cercopithecidae
学　　名 Scientific Name		*Trachypithecus francoisi*		
命 名 人 Species Authority		Pousargues, 1898		
英 文 名 English Name(s)		François' Langur		
同物异名 Synonym(s)		François' Leaf Monkey; Tonkin Leaf Monkey; White Side-burned Black Langur; *Trachypithecus francoisi* (Pousargues, 1858) subsp. *francoisi*		
种下单元评估 Infra-specific Taxa Assessed		无 / None		

评估信息 Assessment Information

评 估 年 份 Year Assessed	2020
评 定 人 Assessor(s)	蒋志刚、蒋学龙 / Zhigang Jiang, Xuelong Jiang
审 定 人 Reviewer(s)	黄乘明、蒋学龙、龙勇诚、范朋飞 / Chengming Huang, Xuelong Jiang, Yongcheng Long, Pengfei Fan
其他贡献人 Other Contributor(s)	李立立、丁晨晨 / Lili Li, Chenchen Ding

理由 Justification: 20 世纪 70 年代，广西有 4,000～5,000 只黑叶猴。由于狩猎，广西黑叶猴数量下降了 90%。对黑叶猴的另一个威胁是对其栖息地的破坏。黑叶猴生活在石灰岩悬崖上，而当地农民仍在山坡上"刀耕火种"，这种耕作方式不仅破坏了黑叶猴栖息地，也造成了黑叶猴食物短缺。尽管黑叶猴种群出现过持续下降，但目前该种种群正在恢复。但是，该种仍处于濒危状态。因此，该种列为濒危等级 / Around 4,000~5,000 François' Langur were reported in the 1970s. In Guangxi, François' Langur has declined in numbers by an estimated of 90% since the 1970s due to hunting. Another threat to François' Langur is the destruction of its habitat. The langur lives on limestone cliffs, near to where farmers still practice a slash and burn farming regime on the lower slopes of karst hills. Limestone is particularly susceptible to fire; therefore, this practice not only destroys the langurs' habitats, but also causes significant food shortages because their diet is primarily foliage. Despite the continuing extreme decline in the Francois' Langur's population, actions are being taken toward the conservation of this species and its population size is on a path to recovery. But the species is still endangered. Thus, it is listed as Endangered

地理分布 Geographical Distribution

国内分布 Domestic Distribution
广西、贵州、重庆 / Guangxi, Guizhou, Chongqing
世界分布 World Distribution
中国、越南 / China, Viet Nam
分布标注 Distribution Note
非特有种 / Non-endemic

国内分布图
Map of Domestic Distribution

种群 Population

种群数量 Population Size	1,450～1,850 只 / 1,450~1,850 individuals
种群趋势 Population Trend	恢复 / Recovering

生境与生态系统 Habitat(s) and Ecosystem(s)

生境 Habitat(s)	热带和亚热带湿润低地森林、喀斯特地貌 / Tropical and Subtropical Moist Lowland Forest, Karst Landscape
生态系统 Ecosystem(s)	森林生态系统、喀斯特生态系统 / Forest Ecosystem, Karst Ecosystem

威胁 Threat(s)

主要威胁 Major Threat(s)	狩猎、耕种、人类活动干扰、火灾 / Hunting, Plantation, Human Disturbance, Fire

保护级别与保护行动 Protection Category and Conservation Action(s)

国家重点保护野生动物等级 (2021) Category of National Key Protected Wild Animals (2021)	一级 / Category I
IUCN 红色名录 (2020-2) IUCN Red List (2020-2)	濒危 / EN
CITES 附录 (2019) CITES Appendix (2019)	II
保护行动 Conservation Action(s)	在自然保护区内的种群及栖息地得到保护 / Populations and habitats in nature reserves are protected

相关文献 Relevant References

Burgin *et al*., 2018; Jiang *et al*. (蒋志刚等), 2017; Li and Wei (李友邦和韦振逸), 2012; Wilson and Mittermeier, 2012; Hu *et al*. (胡刚等), 2011; Tang *et al*. (唐华兴等), 2011; Ma and Su (马强和苏化龙), 2004

黑叶猴 *Trachypithecus francoisi* 　　　　　蒋志刚 摄　By Zhigang Jiang

戴帽叶猴
Trachypithecus pileatus

濒危 EN A2bd; B2; C2a(ii)

| 数据缺乏 DD | 无危 LC | 近危 NT | 易危 VU | 濒危 EN | 极危 CR | 区域灭绝 RE | 野外灭绝 EW | 灭绝 EX |

分类地位 Taxonomic Status

动物界 Animalia	脊索动物门 Chordata	哺乳纲 Mammalia	灵长目 Primates	猴科 Cercopithecidae
学 名 Scientific Name		*Trachypithecus pileatus*		
命 名 人 Species Authority		Blyth, 1843		
英 文 名 English Name(s)		Capped Langur		
同物异名 Synonym(s)		Bonneted Langur; Capped Leaf Monkey; Capped Monkey; *Trachypithecus argentatus* (Horsfield, 1851); *saturatus* (Hinton, 1923)		
种下单元评估 Infra-specific Taxa Assessed		无 / None		

评估信息 Assessment Information

评估年份 Year Assessed	2020
评 定 人 Assessor(s)	蒋志刚、蒋学龙 / Zhigang Jiang, Xuelong Jiang
审 定 人 Reviewer(s)	黄乘明 / Chengming Huang
其他贡献人 Other Contributor(s)	李立立、丁晨晨 / Lili Li, Chenchen Ding

理由 Justification: Hu *et al.* (2017) 在西藏山南地区考察时记录到叶猴 (*Trachypithecus pileatus*) 在中国的新分布。而在《中国哺乳动物多样性》(蒋志刚等, 2015a) 的编目中仅收录了菲氏叶猴 (*T. phayrei*) 和 Shortridge's Langur (*T. shortridgei*, 中文名翻译为"戴帽叶猴")。于是,建议将 Shortridge's Langur (*T. shortirdgei*) 更名为"灰叶猴",将 *T. pileatus* 的中文名翻译为"戴帽叶猴"。戴帽叶猴在中国为边缘分布,数量很少。因此,将其列为濒危等级 / Hu *et al.* (2017) recorded a new distribution of *Trachypithecus pileatus* in China during a survey to the Shannan area of Tibet (Xizang). However, in the inventory of *China's Mammal Diversity* (Jiang *et al.*, 2015a), only the *T. phayrei* and Shortridge's Langur (*T. shortridgei*) were included. It was suggested that Shortridge's Langur (*T. shortirdgei*) be renamed as "Gray Langurs", and that *T. pileatus*'s Chinese name should be translated as "a hooded langur" (戴帽叶猴). The range of *T. pileatus* in the country is its peripheral range, its population in the country is rather small. Thus, it is listed as Endangered

地理分布 Geographical Distribution

国内分布 Domestic Distribution
西藏 / Tibet (Xizang)
世界分布 World Distribution
中国、印度 / China, India
分布标注 Distribution Note
非特有种 / Non-endemic

国内分布图
Map of Domestic Distribution

种群 Population

种群数量 Population Size	未知 / Unknown
种群趋势 Population Trend	未知 / Unknown

生境与生态系统 Habitat (s) and Ecosystem (s)

生境 Habitat(s)	灌丛、森林、喀斯特地貌 / Shrubland, Forest, Karst Landscape
生态系统 Ecosystem(s)	森林生态系统、喀斯特生态系统 / Forest Ecosystem, Karst Ecosystem

威胁 Threat (s)

主要威胁 Major Threat(s)	伐木、火灾、耕种 / Logging, Fire, Plantation

保护级别与保护行动 Protection Category and Conservation Action (s)

国家重点保护野生动物等级 (2021) Category of National Key Protected Wild Animals (2021)	一级 / Category I
IUCN 红色名录 (2020-2) IUCN Red List (2020-2)	易危 / VU
CITES 附录 (2019) CITES Appendix (2019)	I
保护行动 Conservation Action(s)	在自然保护区内的种群及栖息地得到保护 / Populations and habitats in nature reserves are protected

相关文献 Relevant References

Burgin *et al*., 2018; Hu *et al*., 2017; Jiang *et al*. (蒋志刚等), 2017; Wilson and Mittermeier, 2012

戴帽叶猴 *Trachypithecus pileatus*　　　　董磊 摄（西南山地供图）　By Lei Dong (Swild.cn)

萧氏叶猴
Trachypithecus shortridgei

濒危 EN A2bd; B1ab(i, ii, iii); C2a(i)

| 数据缺乏 DD | 无危 LC | 近危 NT | 易危 VU | 濒危 EN | 极危 CR | 区域灭绝 RE | 野外灭绝 EW | 灭绝 EX |

分类地位 Taxonomic Status

动物界 Animalia	脊索动物门 Chordata	哺乳纲 Mammalia	灵长目 Primates	猴科 Cercopithecidae
学名 Scientific Name		*Trachypithecus shortridgei*		
命名人 Species Authority		Wroughton, 1915		
英文名 English Name(s)		Shortridge's Langur		
同物异名 Synonym(s)		肖氏乌叶猴；*Trachypithecus belliger* (Wroughton, 1915)		
种下单元评估 Infra-specific Taxa Assessed		无 / None		

评估信息 Assessment Information

评估年份 Year Assessed	2020
评定人 Assessor(s)	蒋志刚、蒋学龙 / Zhigang Jiang, Xuelong Jiang
审定人 Reviewer(s)	黄乘明、蒋学龙、龙勇诚、范朋飞 / Chengming Huang, Xuelong Jiang, Yongcheng Long, Pengfei Fan
其他贡献人 Other Contributor(s)	李立立、丁晨晨 / Lili Li, Chenchen Ding

理由 Justification: 萧氏叶猴仅分布在我国怒江地区，种群数量极为稀少。其栖息地受到人类活动的影响，种群数量仍在下降。因此，基于其种群与占有区面积，将萧氏叶猴列为濒危等级 / In China, Shortridge's Langur is only distributed in Nujiang River area, China and its population densities are low. Furthermore, its habitats are disturbed by human, leading to a declining population. Thus, based on the population and area of occupancy, Shortridge's Langur is listed as Endangered

地理分布 Geographical Distribution

国内分布 Domestic Distribution
云南、西藏 / Yunnan, Tibet (Xizang)
世界分布 World Distribution
中国、缅甸 / China, Myanmar
分布标注 Distribution Note
非特有种 / Non-endemic

国内分布图
MAP of Domestic Distribution

种群 Population

种群数量 Population Size	250 只 / 250 individuals
种群趋势 Population Trend	下降 / Decreasing

生境与生态系统 Habitat (s) and Ecosystem (s)

生境 Habitat(s)	亚热带湿润低地森林 / Subtropical Moist Lowland Forest
生态系统 Ecosystem(s)	森林生态系统 / Forest Ecosystem

威胁 Threat (s)

主要威胁 Major Threat(s)	狩猎、耕种、伐木、堤坝及水利建设 / Hunting, Plantation, Logging, Dam and Water Management Construction

保护级别与保护行动 Protection Category and Conservation Action (s)

国家重点保护野生动物等级 (2021) Category of National Key Protected Wild Animals (2021)	一级 / Category I
IUCN 红色名录 (2020-2) IUCN Red List (2020-2)	濒危 / EN
CITES 附录 (2019) CITES Appendix (2019)	I
保护行动 Conservation Action(s)	在自然保护区内的种群及栖息地得到保护 / Populations and habitats in nature reserves are protected

相关文献 Relevant References

Burgin *et al*., 2018; Jiang *et al*. (蒋志刚等), 2017; Wilson and Mittermeier, 2012; Smith *et al*. (史密斯等), 2009; Wang *et al*. (王应祥等), 1999; Li and Lin (李致祥和林正玉), 1983

萧氏叶猴 *Trachypithecus shortridgei*

滇金丝猴
Rhinopithecus bieti

濒危 EN A2bd; B1ab(i, ii, iii)+2bd

| 数据缺乏 DD | 无危 LC | 近危 NT | 易危 VU | **濒危 EN** | 极危 CR | 区域灭绝 RE | 野外灭绝 EW | 灭绝 EX |

分类地位 Taxonomic Status

动物界 Animalia	脊索动物门 Chordata	哺乳纲 Mammalia	灵长目 Primates	猴科 Cercopithecidae
学 名 Scientific Name		*Rhinopithecus bieti*		
命 名 人 Species Authority		Milne-Edwards, 1897		
英 文 名 English Name(s)		Black Snub-nosed Monkey		
同物异名 Synonym(s)		Yunnan Snub-nosed Monkey; *Pygathrix roxellana* subsp. *bieti* (Milne-Edwards, 1897)		
种下单元评估 Infra-specific Taxa Assessed		无 / None		

评估信息 Assessment Information

评估年份 Year Assessed	2020
评定人 Assessor(s)	蒋志刚、蒋学龙 / Zhigang Jiang, Xuelong Jiang
审定人 Reviewer(s)	黄乘明、蒋学龙、龙勇诚、范朋飞 / Chengming Huang, Xuelong Jiang, Yongcheng Long, Pengfei Fan
其他贡献人 Other Contributor(s)	李立立、丁晨晨 / Lili Li, Chenchen Ding

理由 Justification: 滇金丝猴目前仅分布在云南省德钦、维西、芒康县境内，所在的生境部分由于人类活动而退化或丧失。滇金丝猴种群数量仅有 2,000 只左右。因此，基于其占有区现状，将滇金丝猴列为濒危等级 / To date, Black Snub-nosed Monkey is known only from Deqin, Weixi and Mangkang counties, and its habitats are degrading or lost due to human activities. The number of this species is about 2,000 individuals. Thus, based on the status of its area of occupancy, Black Snub-nosed Monkey is listed as Endangered

地理分布 Geographical Distribution

国内分布 Domestic Distribution
云南、西藏 / Yunnan, Tibet (Xizang)
世界分布 World Distribution
中国 / China
分布标注 Distribution Note
特有种 / Endemic

国内分布图
Map of Domestic Distribution

种群 Population

种群数量 Population Size	2,150 只 / 2,150 individuals
种群趋势 Population Trend	下降 / Decreasing

生境与生态系统 Habitat (s) and Ecosystem (s)

生　　境 Habitat(s)	针叶林 / Coniferous Forest
生态系统 Ecosystem(s)	森林生态系统 / Forest Ecosystem

威胁 Threat (s)

主要威胁 Major Threat(s)	伐木、耕种、农业或林业污染 / Logging, Plantation, Agricultural or Forestry Effluent

保护级别与保护行动 Protection Category and Conservation Action (s)

国家重点保护野生动物等级 (2021) Category of National Key Protected Wild Animals (2021)	一级 / Category I
IUCN 红色名录 (2020-2) IUCN Red List (2020-2)	濒危 / EN
CITES 附录 (2019) CITES Appendix (2019)	I
保护行动 Conservation Action(s)	在自然保护区内的种群及栖息地得到保护 / Populations and habitats in nature reserves are protected

相关文献 Relevant References

Burgin *et al*., 2018; Jiang *et al*.（蒋志刚等），2017; Grueter *et al*., 2012; Wilson and Mittermeier, 2012; Wang *et al*.（王亚明等），2011; Wu and Lu（吴建国和吕佳佳），2009; Ren *et al*.（任宝平等），2004

滇金丝猴 *Rhinopithecus bieti*　　　　　蒋志刚 摄　By Zhigang Jiang

豺
Cuon alpinus

濒危 EN A3d; B1ab(i, ii, iii)

| 数据缺乏 DD | 无危 LC | 近危 NT | 易危 VU | **濒危 EN** | 极危 CR | 区域灭绝 RE | 野外灭绝 EW | 灭绝 EX |

分类地位 Taxonomic Status

动物界 Animalia	脊索动物门 Chordata	哺乳纲 Mammalia	食肉目 Carnivora	犬科 Canidae
学 名 Scientific Name		*Cuon alpinus*		
命 名 人 Species Authority		Pallas, 1811		
英 文 名 English Name(s)		Dhole		
同物异名 Synonym(s)		Asian Wild Dog, Asiatic Wild Dog, Indian Wild Dog, Whistling Dog, Red Dog, Mountain Wolf		
种下单元评估 Infra-specific Taxa Assessed		无 / None		

评估信息 Assessment Information

评估年份 Year Assessed	2020
评定人 Assessor(s)	蒋志刚 / Zhigang Jiang
审定人 Reviewer(s)	马建章、鲍伟东、姜广顺、胡慧建、徐爱春 / Jianzhang Ma, Weidong Bao, Guangshun Jiang, Huijian Hu, Aichun Xu
其他贡献人 Other Contributor(s)	李立立、丁晨晨 / Lili Li, Chenchen Ding

理由 Justification: 豺原在中国分布广泛，由于其栖息地受到严重破坏，人类的过度猎杀及生态系统中食物链的断裂，其分布区与种群数量急剧下降，因此，基于其种群与栖息地下降速率，将豺列为濒危等级 / Dhole was once widely distributed in China. However, due to habitat destruction, over-hunting and a breakdown in the food chain in its habitats and ecosystems, its distribution range and populations are dramatically declining. Thus, based on the rates of population and habitat decline, Dhole is listed as Endangered

地理分布 Geographical Distribution

国内分布 Domestic Distribution

新疆、青海、黑龙江、西藏、浙江、山西、内蒙古、吉林、辽宁、江苏、安徽、福建、江西、湖北、湖南、河南、广东、广西、四川、贵州、云南、陕西、甘肃、宁夏、重庆 / Xinjiang, Qinghai, Heilongjiang, Tibet (Xizang), Zhejiang, Shanxi, Inner Mongolia (Nei Mongol), Jilin, Liaoning, Jiangsu, Anhui, Fujian, Jiangxi, Hubei, Hunan, Henan, Guangdong, Guangxi, Sichuan, Guizhou, Yunnan, Shaanxi, Gansu, Ningxia, Chongqing

世界分布 World Distribution

中国；南亚、中亚、东南亚 / China; South Asia, Central Asia, Southeast Asia

分布标注 Distribution Note

非特有种 / Non-endemic

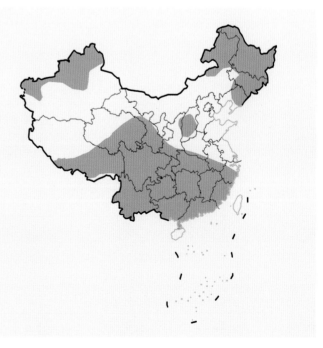

国内分布图
Map of Domestic Distribution

种群 Population

种群数量 Population Size	32,000 只 / 32,000 individuals
种群趋势 Population Trend	下降 / Decreasing

生境与生态系统 Habitat (s) and Ecosystem (s)

生境 Habitat(s)	森林、灌丛 / Forest, Shrubland
生态系统 Ecosystem(s)	森林生态系统、灌丛生态系统 / Forest Ecosystem, Shrubland Ecosystem

威胁 Threat (s)

主要威胁 Major Threat(s)	食物缺乏、投毒、栖息地变化 / Food Scarcity, Poisoning, Habitat Shifting or Alteration

保护级别与保护行动 Protection Category and Conservation Action (s)

国家重点保护野生动物等级 (2021) Category of National Key Protected Wild Animals (2021)	一级 / Category I
IUCN红色名录(2020-2) IUCN Red List (2020-2)	濒危 / EN
CITES 附录 (2019) CITES Appendix (2019)	II
保护行动 Conservation Action(s)	无 / None

相关文献 Relevant References

Burgin *et al.*, 2018; Jiang et al. (蒋志刚等), 2017; Zhang (张进), 2014; Hunter and Barrett, 2011; Smith *et al.* (史密斯等), 2009; Pan *et al.* (潘清华等), 2007; Wang (王应祥), 2003; Gao *et al.* (高耀亭等), 1987

豺 *Cuon alpinus*

懒熊
Melursus ursinus

濒危 EN A3d; B1ab(i, ii, iii)

| 数据缺乏 DD | 无危 LC | 近危 NT | 易危 VU | 濒危 EN | 极危 CR | 区域灭绝 RE | 野外灭绝 EW | 灭绝 EX |

分类地位 Taxonomic Status

动物界 Animalia	脊索动物门 Chordata	哺乳纲 Mammalia	食肉目 Carnivora	熊科 Ursidae
学名 Scientific Name		*Melursus ursinus*		
命名人 Species Authority		Shaw, 1791		
英文名 English Name(s)		Sloth Bear		
同物异名 Synonym(s)		*Bradypus ursinus* (Shaw, 1791)		
种下单元评估 Infra-specific Taxa Assessed		无 / None		

评估信息 Assessment Information

评估年份 Year Assessed	2020
评定人 Assessor(s)	蒋志刚 / Zhigang Jiang
审定人 Reviewer(s)	胡慧建、徐爱春 / Huijian Hu, Aichun Xu
其他贡献人 Other Contributor(s)	李立立、丁晨晨 / Lili Li, Chenchen Ding

理由 Justification: 懒熊分布在西藏旺东地区和西卡门县 (Choudhury, 2003; Garshelis *et al.*, 1999)。由于当地门巴族人狩猎，其种群数量下降，被列入了 CITES 附录 I 和《IUCN 熊类保育行动计划》(Garshelis *et al.*, 1999)。因此，基于其种群与栖息地下降速率，将懒熊列为濒危等级 / Sloth Bear is distributed in Wangdong area and Xikamen County of Tibet (Xizang) (Choudhury, 2003; Garshelis *et al.*, 1999). As local people hunt the bear, its population is in decline, and is included in CITES Appendix I and the *IUCN Bear Conservation Action Plan* (Garshelis *et al.*, 1999). Thus, based on its population and habitat decline rate, it is listed as Endangered

地理分布 Geographical Distribution

国内分布 Domestic Distribution
西藏 / Tibet (Xizang)
世界分布 World Distribution
中国、印度 / China, India
分布标注 Distribution Note
非特有种 / Non-endemic

国内分布图
Map of Domestic Distribution

种群 Population

种群数量 Population Size	未知 / Unknown
种群趋势 Population Trend	下降 / Decreasing

生境与生态系统 Habitat (s) and Ecosystem (s)

生　　境 Habitat(s)	森林、灌丛 / Forest, Shrubland
生态系统 Ecosystem(s)	森林生态系统、灌丛生态系统 / Forest Ecosystem, Shrubland Ecosystem

威胁 Threat (s)

主要威胁 Major Threat(s)	狩猎 / Hunting

保护级别与保护行动 Protection Category and Conservation Action (s)

国家重点保护野生动物等级 (2021) Category of National Key Protected Wild Animals (2021)	二级 / Category II
IUCN 红色名录 (2020-2) IUCN Red List (2020-2)	易危 / VU
CITES 附录 (2019) CITES Appendix (2019)	I
保护行动 Conservation Action(s)	禁止国际贸易，并制定了种群恢复计划 / International trade is banned and is under population recovery plan

相关文献 Relevant References

Burgin *et al*., 2018; Jiang *et al*. (蒋志刚等), 2017; Hunter and Barrett, 2011; Choudhury, 2003; Garshelis *et al*., 1999; Gao *et al*. (高耀亭等), 1987

懒熊 *Melursus ursinus*

石貂
Martes foina

濒危　EN A3d; B1ab(i, ii, iii)+2ab(i, ii, iii); C2a(i)

分类地位 Taxonomic Status

动物界 Animalia	脊索动物门 Chordata	哺乳纲 Mammalia	食肉目 Carnivora	鼬科 Mustelidae
学名 Scientific Name		*Martes foina*		
命名人 Species Authority		Erxleben, 1777		
英文名 English Name(s)		Stone Marten		
同物异名 Synonym(s)		Beech Marten, House Marten, White Breasted Marten		
种下单元评估 Infra-specific Taxa Assessed		无 / None		

评估信息 Assessment Information

评估年份 Year Assessed	2020
评定人 Assessor(s)	蒋志刚 / Zhigang Jiang
审定人 Reviewer(s)	李晟、王昊、周友兵 / Sheng Li, Hao Wang, Youbing Zhou
其他贡献人 Other Contributor(s)	李立立、丁晨晨 / Lili Li, Chenchen Ding

理由 Justification: 石貂在中国西部分布较广，但是石貂历来是重要狩猎对象，捕猎是为了毛皮利用，捕猎导致了该种种群的快速下降。石貂栖息地也受到人类活动的破坏。依据其种群下降与栖息地破坏危及石貂的生存这一事实，将石貂列为濒危等级 / Stone Marten is widely distributed in western China. Stone Marten used to be an important hunting species for its pelts; over-hunting leads to dramatic decrease of its populations. Furthermore, the habitats of Stone Marten have been greatly destroyed due to human activities. Both population and habitat decline threat the survival of Stone Marten. Thus, it is listed as Endangered

地理分布 Geographical Distribution

国内分布 Domestic Distribution	山西、青海、新疆、陕西、甘肃、河北、内蒙古、四川、云南、西藏、宁夏 / Shanxi, Qinghai, Xinjiang, Shaanxi, Gansu, Hebei, Inner Mongolia (Nei Mongol), Sichuan, Yunnan, Tibet (Xizang), Ningxia
世界分布 World Distribution	中国；中亚、西亚和欧洲各国 / China; Central Asia, West Asia and European Countries
分布标注 Distribution Note	非特有种 / Non-endemic

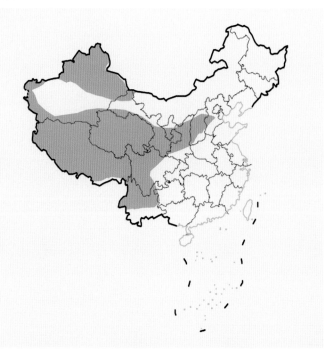

国内分布图
Map of Domestic Distribution

种群 Population

种群数量 Population Size	未知 / Unknown
种群趋势 Population Trend	稳定 / Stable

生境与生态系统 Habitat (s) and Ecosystem (s)

生境 Habitat(s)	内陆岩石区域、森林、灌丛、耕地 / Inland Rocky Area, Forest, Shrubland, Arable Land
生态系统 Ecosystem(s)	森林生态系统、灌丛生态系统、农田生态系统 / Forest Ecosystem, Shrubland Ecosystem, Cropland Ecosystem

威胁 Threat (s)

主要威胁 Major Threat(s)	狩猎、疾病 / Hunting, Disease

保护级别与保护行动 Protection Category and Conservation Action (s)

国家重点保护野生动物等级 (2021) Category of National Key Protected Wild Animals (2021)	二级 / Category II
IUCN 红色名录 (2020-2) IUCN Red List (2020-2)	无危 / LC
CITES 附录 (2019) CITES Appendix (2019)	III
保护行动 Conservation Action(s)	部分种群位于自然保护区内 / Part of its populations are covered by nature reserves

相关文献 Relevant References

Burgin *et al.*, 2018; Jiang *et al.*（蒋志刚等）, 2017; Hunter and Barrett, 2011; Zhu *et al.*（朱红艳等）, 2010; Yu *et al.*（于晓东等）, 2006; Gao *et al.*（高耀亭等）, 1987

石貂 *Martes foina*

貂熊
Gulo gulo

濒危　EN A3d; B1ab(i, ii, iii) + 2ab(i, ii, iii); C2a(i)

| 数据缺乏 DD | 无危 LC | 近危 NT | 易危 VU | 濒危 EN | 极危 CR | 区域灭绝 RE | 野外灭绝 EW | 灭绝 EX |

分类地位 Taxonomic Status

动物界 Animalia	脊索动物门 Chordata	哺乳纲 Mammalia	食肉目 Carnivora	鼬科 Mustelidae
学　名 Scientific Name		*Gulo gulo*		
命名人 Species Authority		Linnaeus, 1758		
英文名 English Name(s)		Wolverine		
同物异名 Synonym(s)		无 / None		
种下单元评估 Infra-specific Taxa Assessed		无 / None		

评估信息 Assessment Information

评估年份 Year Assessed	2020
评定人 Assessor(s)	蒋志刚、李峰 / Zhigang Jiang, Feng Li
审定人 Reviewer(s)	朱欢兵 / Huanbing Zhu
其他贡献人 Other Contributor(s)	李立立、丁晨晨 / Lili Li, Chenchen Ding

理由 Justification: 貂熊在中国阿勒泰地区与大小兴安岭地区的分布区为其边缘分布区，其分布区小、种群小。鉴于其种群数量与占有区面积，将貂熊列为濒危等级 / Wolverine is in its peripheral ranges in the Altay area and the Da and Xiao Hinggan Ling. Its range and population are both small. Based on its population size and area of occupancy, Wolverine is listed as Endangered

地理分布 Geographical Distribution

国内分布 Domestic Distribution
内蒙古、黑龙江、新疆 / Inner Mongolia (Nei Mongol), Heilongjiang, Xinjiang
世界分布 World Distribution
中国；环北极圈 / China; Circum-Arctic
分布标注 Distribution Note
非特有种 / Non-endemic

国内分布图
Map of Domestic Distribution

种群 Population

种群数量 Population Size	180 只 / 180 individuals
种群趋势 Population Trend	下降 / Decreasing

生境与生态系统 Habitat (s) and Ecosystem (s)

生　　境 Habitat(s)	针阔混交林、泰加林 / Coniferous and Broad-leaved Mixed Forest, Taiga Forest
生态系统 Ecosystem(s)	森林生态系统 / Forest Ecosystem

威胁 Threat (s)

主要威胁 Major Threat(s)	狩猎、栖息地改变 / Hunting, Habitat Shifting or Alteration

保护级别与保护行动 Protection Category and Conservation Action (s)

国家重点保护野生动物等级 (2021) Category of National Key Protected Wild Animals (2021)	一级 / Category I
IUCN 红色名录 (2020-2) IUCN Red List (2020-2)	无危 / LC
CITES 附录 (2019) CITES Appendix (2019)	无 / NA
保护行动 Conservation Action(s)	无 / None

相关文献 Relevant References

Burgin *et al.*, 2018; Jiang *et al.* (蒋志刚等), 2017; Hunter and Barrett, 2011; Zhao (赵正阶), 1999; Zheng (郑生武), 1994; Gao *et al.* (高耀亭等), 1987

貂熊 *Gulo gulo*

白鼬
Mustela erminea

濒危　EN A3d; C2a(i)

| 数据缺乏 DD | 无危 LC | 近危 NT | 易危 VU | **濒危 EN** | 极危 CR | 区域灭绝 RE | 野外灭绝 EW | 灭绝 EX |

分类地位 Taxonomic Status

| 动物界 Animalia | 脊索动物门 Chordata | 哺乳纲 Mammalia | 食肉目 Carnivora | 鼬科 Mustelidae |

学名 Scientific Name	*Mustela erminea*
命名人 Species Authority	Linnaeus, 1758
英文名 English Name(s)	Ermine
同物异名 Synonym(s)	Stoat, Short-tailed Weasel
种下单元评估 Infra-specific Taxa Assessed	无 / None

评估信息 Assessment Information

评估年份 Year Assessed	2020
评定人 Assessor(s)	蒋志刚 / Zhigang Jiang
审定人 Reviewer(s)	李晟、胡慧建、王昊、周友兵 / Sheng Li, Huijian Hu, Hao Wang, Youbing Zhou
其他贡献人 Other Contributor(s)	李立立、丁晨晨 / Lili Li, Chenchen Ding

理由 Justification: 白鼬生存受到人类活动的影响。其栖息地面积由于人类活动而减少，农业污染和杀虫剂的使用也影响了白鼬的生存，因为其猎物被毒杀，其由于食物链断裂或二次中毒而濒危。此外，捕猎也是白鼬种群数量不断下降的原因。因此，白鼬列为濒危等级 / Ermine is threatened by human activities. The habitats of Ermine is declining due to human encroachment. Both the agricultural pollution and the using of pesticides have impacted the survival of Ermine due to the poisoning of its preys, thus Ermine is threatened by a breakdown in its food chain and through secondary poison. Furthermore, over-hunting was also a cause of its population decline. Thus, Ermine is listed as Endangered

地理分布 Geographical Distribution

国内分布 Domestic Distribution
黑龙江、内蒙古、新疆、河北、山西、陕西 / Heilongjiang, Inner Mongolia (Nei Mongol), Xinjiang, Hebei, Shanxi, Shaanxi
世界分布 World Distribution
中国；中亚、西亚和欧洲各国 / China; Central Asia, Western Asia and European Countries
分布标注 Distribution Note
非特有种 / Non-endemic

国内分布图
Map of Domestic Distribution

种群 Population

种群数量 Population Size	未知 / Unknown
种群趋势 Population Trend	稳定 / Stable

生境与生态系统 Habitat (s) and Ecosystem (s)

生 境 Habitat(s)	苔原、草甸、森林、沼泽、农田、灌丛 / Tundra, Meadow, Forest, Swamp, Cropland, Shrubland
生态系统 Ecosystem(s)	森林生态系统、灌丛生态系统、草地生态系统、农田生态系统、荒漠生态系统 / Forest Ecosystem, Shrubland Ecosystem, Grassland Ecosystem, Cropland Ecosystem, Desert Ecosystem

威胁 Threat (s)

主要威胁 Major Threat(s)	猎捕、伐木、人类活动干扰 / Hunting, Logging, Human Disturbance

保护级别与保护行动 Protection Category and Conservation Action (s)

国家重点保护野生动物等级 (2021) Category of National Key Protected Wild Animals (2021)	无 / NA
IUCN红色名录(2020-2) IUCN Red List (2020-2)	无危 / LC
CITES 附录 (2019) CITES Appendix (2019)	III
保护行动 Conservation Action(s)	无 / None

相关文献 Relevant References

Burgin *et al*., 2018; Jiang *et al*. (蒋志刚等), 2017; Hunter and Barrett, 2011; Xiao *et al*. (肖红等), 2003; Guo and Liang (郭世芳和梁栓柱), 1997; Gao *et al*. (高耀亭等), 1987; Xu (徐学良), 1975

白鼬 *Mustela erminea*

纹鼬
Mustela strigidorsa

濒危 EN A3d; C2a(i)

| 数据缺乏 DD | 无危 LC | 近危 NT | 易危 VU | **濒危 EN** | 极危 CR | 区域灭绝 RE | 野外灭绝 EW | 灭绝 EX |

分类地位 Taxonomic Status

动物界 Animalia	脊索动物门 Chordata	哺乳纲 Mammalia	食肉目 Carnivora	鼬科 Mustelidae

学名 Scientific Name	*Mustela strigidorsa*
命名人 Species Authority	Gray, 1853
英文名 English Name(s)	Stripe-backed Weasel
同物异名 Synonym(s)	Back-striped Weasel
种下单元评估 Infra-specific Taxa Assessed	无 / None

评估信息 Assessment Information

评估年份 Year Assessed	2020
评定人 Assessor(s)	蒋志刚 / Zhigang Jiang
审定人 Reviewer(s)	马建章、徐爱春、李晟、胡慧建、王昊、张明海、周友兵 / Jianzhang Ma, Aichun Xu, Sheng Li, Huijian Hu, Hao Wang, Minghai Zhang, Youbing Zhou
其他贡献人 Other Contributor(s)	李立立、丁晨晨 / Lili Li, Chenchen Ding

理由 Justification: 纹鼬的栖息地由于人类活动而不断减少，且农业污染和杀虫剂的使用均影响到了该物种的生存。另外，人类捕猎也是纹鼬种群数量不断下降的重要原因。因此，纹鼬列为濒危等级 / The habitats of Stripe-backed Weasel is declining due to human encroachment. Both the agricultural pollution and the use of pesticides affect the survival of Stripe-backed Weasel. Furthermore, over-hunting was also a main cause of its decreasing population. Thus, Stripe-backed Weasel is listed as Endangered

地理分布 Geographical Distribution

国内分布 Domestic Distribution	广西、贵州、云南、西藏 / Guangxi, Guizhou, Yunnan, Tibet (Xizang)
世界分布 World Distribution	中国、印度、老挝、缅甸、泰国、越南 / China, India, Laos, Myanmar, Thailand, Viet Nam
分布标注 Distribution Note	非特有种 / Non-endemic

国内分布图
Map of Domestic Distribution

种群 Population

种群数量 Population Size	未知 / Unknown
种群趋势 Population Trend	未知 / Unknown

生境与生态系统 Habitat (s) and Ecosystem (s)

生境 Habitat(s)	农田、森林 / Cropland, Forest
生态系统 Ecosystem(s)	森林生态系统、农田生态系统 / Forest Ecosystem, Cropland Ecosystem

威胁 Threat (s)

主要威胁 Major Threat(s)	猎捕、农业污染和农药使用 / Trapping, Agricultural Pollution and Use of Pesticide

保护级别与保护行动 Protection Category and Conservation Action (s)

国家重点保护野生动物等级 (2021) Category of National Key Protected Wild Animals (2021)	无 / NA
IUCN红色名录(2020-2) IUCN Red List (2020-2)	无危 / LC
CITES 附录 (2019) CITES Appendix (2019)	无 / NA
保护行动 Conservation Action(s)	无 / None

相关文献 Relevant References

Burgin *et al.*, 2018; Jiang *et al.* (蒋志刚等), 2017; Hunter and Barrett, 2011; Smith *et al.* (史密斯等), 2009; Abramov *et al.*, 2008; Wilson and Reeder, 2005; Wang (王应祥), 2003; Gao *et al.* (高耀亭等), 1987

纹鼬 *Mustela strigidorsa*

虎鼬
Vormela peregusna

濒危　EN A3d; C2a(i)

| 数据缺乏 DD | 无危 LC | 近危 NT | 易危 VU | 濒危 EN | 极危 CR | 区域灭绝 RE | 野外灭绝 EW | 灭绝 EX |

分类地位 Taxonomic Status

动物界 Animalia	脊索动物门 Chordata	哺乳纲 Mammalia	食肉目 Carnivora	鼬科 Mustelidae

学名 Scientific Name	*Vormela peregusna*
命名人 Species Authority	Güldenstädt, 1770
英文名 English Name(s)	Marbled Polecat
同物异名 Synonym(s)	European Marbled Polecat
种下单元评估 Infra-specific Taxa Assessed	无 / None

评估信息 Assessment Information

评估年份 Year Assessed	2020
评定人 Assessor(s)	蒋志刚 / Zhigang Jiang
审定人 Reviewer(s)	马建章、徐爱春、李晟、胡慧建、王昊、张明海、周友兵 / Jianzhang Ma, Aichun Xu, Sheng Li, Huijian Hu, Hao Wang, Minghai Zhang, Youbing Zhou
其他贡献人 Other Contributor(s)	李立立、丁晨晨 / Lili Li, Chenchen Ding

理由 Justification: 虎鼬的栖息地由于人类活动而减少，且农业污染和杀虫剂的使用都影响了该物种的生存。另外，人类捕猎也是虎鼬种群数量不断下降的重要原因。因此，虎鼬列为濒危等级 / The habitats of Marbled Polecat are declining due to human encroachment. Both the agricultural pollution and the use of pesticides affect the survival of Marbled Polecat. Furthermore, over-hunting is also a main cause of its population decline. Thus, Marbled Polecat is listed as Endangered

地理分布 Geographical Distribution

国内分布 Domestic Distribution	山西、内蒙古、陕西、甘肃、宁夏、新疆 / Shanxi, Inner Mongolia (Nei Mongol), Shaanxi, Gansu, Ningxia, Xinjiang
世界分布 World Distribution	亚洲、欧洲 / Asia, Europe
分布标注 Distribution Note	非特有种 / Non-endemic

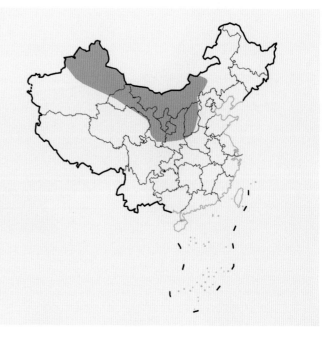

国内分布图
Map of Domestic Distribution

种群 Population

种群数量 Population Size	未知 / Unknown
种群趋势 Population Trend	下降 / Decreasing

生境与生态系统 Habitat (s) and Ecosystem (s)

生境 Habitat(s)	草地、荒漠、半荒漠 / Grassland, Desert, Semi-desert
生态系统 Ecosystem(s)	草地生态系统、荒漠生态系统 / Grassland Ecosystem, Desert Ecosystem

威胁 Threat (s)

主要威胁 Major Threat(s)	栖息地变化、猎捕、农业污染和农药使用 / Habitat Shifting or Alteration, Trapping, Agricultural Pollution and Use of Pesticide

保护级别与保护行动 Protection Category and Conservation Action (s)

国家重点保护野生动物等级 (2021) Category of National Key Protected Wild Animals (2021)	无 / NA
IUCN 红色名录 (2020-2) IUCN Red List (2020-2)	易危 / VU
CITES 附录 (2019) CITES Appendix (2019)	无 / NA
保护行动 Conservation Action(s)	无 / None

相关文献 Relevant References

Burgin *et al*., 2018; Jiang *et al*.（蒋志刚等）, 2017; Hunter and Barrett, 2011; Huang *et al*.（黄薇等）, 2008; Wang（王应祥）, 2003; Hou（侯兰新）, 2000; Gao *et al*.（高耀亭等）, 1987

虎鼬 *Vormela peregusna*

缅甸鼬獾
Melogale personata

濒危 EN A3d; C2a(i)

| 数据缺乏 DD | 无危 LC | 近危 NT | 易危 VU | **濒危 EN** | 极危 CR | 区域灭绝 RE | 野外灭绝 EW | 灭绝 EX |

分类地位 Taxonomic Status

动物界 Animalia	脊索动物门 Chordata	哺乳纲 Mammalia	食肉目 Carnivora	鼬科 Mustelidae
学名 Scientific Name		*Melogale personata*		
命名人 Species Authority		I. Geoffroy Saint-Hilaire, 1831		
英文名 English Name(s)		Large-toothed Ferret Badger		
同物异名 Synonym(s)		Burmese Ferret Badger		
种下单元评估 Infra-specific Taxa Assessed		无 / None		

评估信息 Assessment Information

评估年份 Year Assessed	2020
评定人 Assessor(s)	蒋志刚 / Zhigang Jiang
审定人 Reviewer(s)	李晟、胡慧建、王昊、张明海、周友兵 / Sheng Li, Huijian Hu, Hao Wang, Minghai Zhang, Youbing Zhou
其他贡献人 Other Contributor(s)	李立立、丁晨晨 / Lili Li, Chenchen Ding

理由 Justification: 缅甸鼬獾在中国的分布区为其边缘分布区，占有区小，种群数量较少，在近期的野外考察鲜有报道。因此，缅甸鼬獾列为濒危等级 / Distribution range of Large-toothed Ferret Badger in China is its peripheral range, its area of occupancy and population size are both small. In recent field surveys, reports about Large-toothed Ferret Badger are scarce. Thus, Large-toothed Ferret Badger is listed as Endangered

地理分布 Geographical Distribution

国内分布 Domestic Distribution
广东 / Guangdong
世界分布 World Distribution
中国；南亚、东南亚 / China; South Asia, Southeast Asia
分布标注 Distribution Note
非特有种 / Non-endemic

国内分布图
Map of Domestic Distribution

种群 Population

种群数量 Population Size	未知 / Unknown
种群趋势 Population Trend	未知 / Unknown

生境与生态系统 Habitat (s) and Ecosystem (s)

生境 Habitat(s)	森林、草地、耕地 / Forest, Grassland, Arable Land
生态系统 Ecosystem(s)	森林生态系统、农田生态系统、草地生态系统 / Forest Ecosystem, Cropland Ecosystem, Grassland Ecosystem

威胁 Threat (s)

主要威胁 Major Threat(s)	栖息地变化、猎捕 / Habitat Shifting or Alteration, Trapping

保护级别与保护行动 Protection Category and Conservation Action (s)

国家重点保护野生动物等级 (2021) Category of National Key Protected Wild Animals (2021)	无 / NA
IUCN红色名录(2020-2) IUCN Red List (2020-2)	无危 / LC
CITES 附录 (2019) CITES Appendix (2019)	无 / NA
保护行动 Conservation Action(s)	无 / None

相关文献 Relevant References

Burgin *et al.*, 2018; Jiang *et al.*（蒋志刚等），2017; Hunter and Barrett, 2011; Liu *et al.*（刘少英等），2005; Gao *et al.*（高耀亭等），1987; Zheng（郑永烈），1981

缅甸鼬獾 *Melogale personata*

水獭
Lutra lutra

濒危 EN B1ab(i, ii, iii)+2ab(i, ii, iii); C2a(i)

| 数据缺乏 DD | 无危 LC | 近危 NT | 易危 VU | **濒危 EN** | 极危 CR | 区域灭绝 RE | 野外灭绝 EW | 灭绝 EX |

分类地位 Taxonomic Status

动物界 Animalia	脊索动物门 Chordata	哺乳纲 Mammalia	食肉目 Carnivora	鼬科 Mustelidae
学　名 Scientific Name		*Lutra lutra*		
命名人 Species Authority		Linnaeus, 1758		
英文名 English Name(s)		Eurasian Otter		
同物异名 Synonym(s)		Eurasian River Otter; Common Otter; Old World Otter; *Lutra nippon* (Imaizumi and Yoshiyuki, 1989); *Viverra lutra* (Linnaeus, 1758)		
种下单元评估 Infra-specific Taxa Assessed		无 / None		

评估信息 Assessment Information

评估年份 Year Assessed	2020
评定人 Assessor(s)	蒋志刚 / Zhigang Jiang
审定人 Reviewer(s)	李晟、胡慧建、王昊、张明海、周友兵 / Sheng Li, Huijian Hu, Hao Wang, Minghai Zhang, Youbing Zhou
其他贡献人 Other Contributor(s)	李立立、丁晨晨 / Lili Li, Chenchen Ding

理由 Justification: 水獭曾经广泛分布。水獭的栖息地因为人类活动正遭受持续退化。工业废水、农药使用直接毒害着水獭。此外，电击、炸药等毁灭性捕鱼方式，造成了其食物资源匮乏。这些原因造成水獭野外种群数量的严重下降。因此，水獭被列为濒危等级 / Eurasian Otter once widely distributed. The habitats of Eurasian Otter are constantly being degraded due to human encroachment. Industrial wastewater and agricultural chemicals can directly poison the Eurasian Otter. Additionally, destructive fishing techniques such as electrocution and explosions can lead to food shortages for the species. All of these reasons are leading toward the dramatic decrease of this species in wild. Thus, it is listed as Endangered

地理分布 Geographical Distribution

国内分布 Domestic Distribution	山西、湖南、海南、江苏、浙江、河南、内蒙古、辽宁、吉林、黑龙江、上海、安徽、福建、江西、湖北、广东、香港、广西、四川、贵州、云南、西藏、陕西、甘肃、青海、新疆、台湾、重庆 / Shanxi, Hunan, Hainan, Jiangsu, Zhejiang, Henan, Inner Mongolia (Nei Mongol), Liaoning, Jilin, Heilongjiang, Shanghai, Anhui, Fujian, Jiangxi, Hubei, Guangdong, Hong Kong, Guangxi, Sichuan, Guizhou, Yunnan, Tibet (Xizang), Shaanxi, Gansu, Qinghai, Xinjiang, Taiwan, Chongqing
世界分布 World Distribution	欧亚大陆及不列颠群岛 / Eurasian Continent and British Isles
分布标注 Distribution Note	非特有种 / Non-endemic

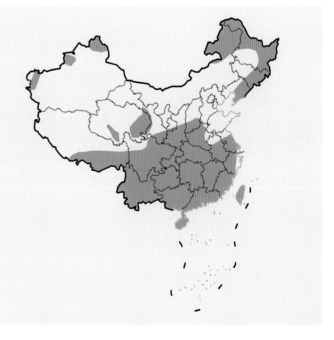

国内分布图 Map of Domestic Distribution

🦌 种群 Population

种群数量 Population Size	未知 / Unknown
种群趋势 Population Trend	下降 / Decreasing

🦌 生境与生态系统 Habitat (s) and Ecosystem (s)

生　　境 Habitat(s)	池塘、淡水湖、沼泽、溪流边、江河、农田 / Pond, Freshwater Lake, Swamp, Stream, River, Cropland
生态系统 Ecosystem(s)	湿地生态系统、湖泊河流生态系统、农田生态系统 / Wetland Ecosystem, Lake and River Ecosystem, Cropland Ecosystem

🦌 威胁 Threat (s)

主要威胁 Major Threat(s)	堤坝及水道改变、狩猎、污染、公路碾压 / Dam and Water Management Shift or Alternation, Hunting, Pollution, Road Killing

🦌 保护级别与保护行动 Protection Category and Conservation Action (s)

国家重点保护野生动物等级 (2021) Category of National Key Protected Wild Animals (2021)	二级 / Category II
IUCN 红色名录 (2020-2) IUCN Red List (2020-2)	近危 / NT
CITES 附录 (2019) CITES Appendix (2019)	I
保护行动 Conservation Action(s)	部分种群位于自然保护区内 / Part of populations are covered by nature reserves

🦌 相关文献 Relevant References

Burgin *et al.*, 2018; Jiang *et al.*（蒋志刚等）, 2017; Li and Chan（李飞和陈辈乐）, 2017; Li *et al.*（李飞等）, 2017; Hu *et al.*（胡一鸣等）, 2014; Hunter and Barrett, 2011; Lau *et al.*, 2010; Lei and Li（雷伟和李玉春）, 2008; Gao *et al.*（高耀亭等）, 1987

水獭 *Lutra lutra*

椰子狸
Paradoxurus hermaphroditus

濒危 EN A3d; C2a(i)

| 数据缺乏 DD | 无危 LC | 近危 NT | 易危 VU | **濒危 EN** | 极危 CR | 区域灭绝 RE | 野外灭绝 EW | 灭绝 EX |

分类地位 Taxonomic Status

动物界 Animalia	脊索动物门 Chordata	哺乳纲 Mammalia	食肉目 Carnivora	灵猫科 Viverridae
学 名 Scientific Name		*Paradoxurus hermaphroditus*		
命 名 人 Species Authority		Pallas, 1777		
英 文 名 English Name(s)		Common Palm Civet		
同物异名 Synonym(s)		Asian Palm Civet		
种下单元评估 Infra-specific Taxa Assessed		无 / None		

评估信息 Assessment Information

评估年份 Year Assessed	2020
评定人 Assessor(s)	蒋志刚 / Zhigang Jiang
审定人 Reviewer(s)	陈辈乐、李飞、李晟、胡慧建、王大军、王昊、周友兵 / Bosco P. L. Chan, Fei Li, Sheng Li, Huijian Hu, Dajun Wang, Hao Wang, Youbing Zhou
其他贡献人 Other Contributor(s)	李立立、丁晨晨 / Lili Li, Chenchen Ding

理由 Justification: 椰子狸曾分布于四川、贵州、云南、广西、广东及海南。近年调查发现其仅在云南腾冲高黎贡山以南、屏边大围山以西的边境热带地区及海南岛中部山区有分布。由于椰子狸分布区缩减，故将其列为濒危等级 / Common Palm Civet was once distributed in Sichuan, Guizhou, Yunnan, Guangxi, Guangdong and Hainan. In recent years, it has been found in tropical border areas of south of Gaoligong Shan in Tengchong, Yunnan, to west of Dawei Shan in Pingbian, Yunnan and the central mountainous area of Hainan Island. Thus, due to the severe shrinkage of its range in China, Common Palm Civet is listed as Endangered

地理分布 Geographical Distribution

国内分布 Domestic Distribution	海南、广东、广西、四川、贵州、云南 / Hainan, Guangdong, Guangxi, Sichuan, Guizhou, Yunnan
世界分布 World Distribution	中国；南亚、东南亚 / China; South Asia, Southeast Asia
分布标注 Distribution Note	非特有种 / Non-endemic

国内分布图
Map of Domestic Distribution

种群 Population

种群数量 Population Size	未知 / Unknown
种群趋势 Population Trend	稳定 / Stable

生境与生态系统 Habitat (s) and Ecosystem (s)

生　　境 Habitat(s)	热带和亚热带湿润山地森林、种植园、次生林 / Tropical and Subtropical Moist Montane Forest, Plantation, Secondary Forest
生态系统 Ecosystem(s)	森林生态系统、农田生态系统 / Forest Ecosystem, Cropland Ecosystem

威胁 Threat (s)

主要威胁 Major Threat(s)	狩猎 / Hunting

保护级别与保护行动 Protection Category and Conservation Action (s)

国家重点保护野生动物等级 (2021) Category of National Key Protected Wild Animals (2021)	二级 / Category II
IUCN 红色名录 (2020-2) IUCN Red List (2020-2)	无危 / LC
CITES 附录 (2019) CITES Appendix (2019)	III
保护行动 Conservation Action(s)	部分种群位于自然保护区内 / Part of populations are covered by nature reserves

相关文献 Relevant References

Burgin *et al*., 2018; Jiang *et al*.（蒋志刚等），2017; Hunter and Barrett, 2011; Lau *et al*., 2010; Zeng *et al*.（曾国仕等），2010; Zhu *et al*.（朱红艳等），2010; Wang *et al*.（王健等），2009; Gao *et al*.（高耀亭等），1987

椰子狸 *Paradoxurus hermaphroditus*

野猫
Felis silvestris

濒危 EN A2ab

| 数据缺乏 DD | 无危 LC | 近危 NT | 易危 VU | 濒危 EN | 极危 CR | 区域灭绝 RE | 野外灭绝 EW | 灭绝 EX |

分类地位 Taxonomic Status

动物界 Animalia	脊索动物门 Chordata	哺乳纲 Mammalia	食肉目 Carnivora	猫科 Felidae

学名 Scientific Name	*Felis silvestris*
命名人 Species Authority	Schreber, 1777
英文名 English Name(s)	Wild Cat
同物异名 Synonym(s)	草原斑猫，欧林猫
种下单元评估 Infra-specific Taxa Assessed	无 / None

评估信息 Assessment Information

评估年份 Year Assessed	2020
评定人 Assessor(s)	蒋志刚 / Zhigang Jiang
审定人 Reviewer(s)	胡慧建、王大军、王昊、周友兵 / Huijian Hu, Dajun Wang, Hao Wang, Youbing Zhou
其他贡献人 Other Contributor(s)	李立立、丁晨晨 / Lili Li, Chenchen Ding

理由 Justification: 野猫的栖息地面积由于人类活动，如伐木、采矿等而不断减少。野猫在国内的标本数量少于10具。且过度捕猎、农业污染等使该种种群急速下降。基于其种群下降速率，将野猫列为濒危 / The habitats of Wild Cat are decreasing in area due to human activities such as logging and quarrying. The number of Wild Cat specimen in China is fewer than 10. Over-hunting, agricultural pollution and other threats have driven Wild Cat populations down. Thus, based on the rate of population decline, Wild Cat has now reached the Endangered category

地理分布 Geographical Distribution

国内分布 Domestic Distribution	新疆、青海、陕西、内蒙古、四川、西藏、甘肃、宁夏 / Xinjiang, Qinghai, Shaanxi, Inner Mongolia (Nei Mongol), Sichuan, Tibet (Xizang), Gansu, Ningxia
世界分布 World Distribution	欧亚大陆及非洲大陆 / Eurasian and African Continents
分布标注 Distribution Note	非特有种 / Non-endemic

国内分布图
Map of Domestic Distribution

种群 Population

种群数量 Population Size	未知 / Unknown
种群趋势 Population Trend	下降 / Decreasing

生境与生态系统 Habitat (s) and Ecosystem (s)

生　　境 Habitat(s)	草地、荒漠、半荒漠、灌丛、沙漠、绿洲、乡村庭院 / Grassland, Wilderness, Semi-desert, Shrubland, Desert, Oasis, Rural Garden
生态系统 Ecosystem(s)	灌丛生态系统、草地生态系统、农田生态系统、荒漠生态系统 / Shrubland Ecosystem, Grassland Ecosystem, Cropland Ecosystem, Desert Ecosystem

威胁 Threat (s)

主要威胁 Major Threat(s)	杂交、投毒 / Hybridization, Poison

保护级别与保护行动 Protection Category and Conservation Action (s)

国家重点保护野生动物等级 (2021) Category of National Key Protected Wild Animals (2021)	二级 / Category II
IUCN 红色名录(2020-2) IUCN Red List (2020-2)	无危 / LC
CITES 附录 (2019) CITES Appendix (2019)	II
保护行动 Conservation Action(s)	无 / None

相关文献 Relevant References

Burgin *et al.*, 2018; Jiang *et al.* (蒋志刚等), 2017; Abdukadir and Khan, 2013; Hunter and Barrett, 2011; Abdukadir *et al.*, 2010; Smith *et al.* (史密斯等), 2009; Wang (王应祥), 2003; Gao *et al.* (高耀亭等), 1987

野猫 *Felis silvestris*

兔狲
Otocolobus manul

濒危　EN A2ab; B1ab(i, ii, iii)

| 数据缺乏 DD | 无危 LC | 近危 NT | 易危 VU | **濒危 EN** | 极危 CR | 区域灭绝 RE | 野外灭绝 EW | 灭绝 EX |

分类地位 Taxonomic Status

动物界 Animalia	脊索动物门 Chordata	哺乳纲 Mammalia	食肉目 Carnivora	猫科 Felidae
学　名 Scientific Name		*Otocolobus manul*		
命 名 人 Species Authority		Pallas, 1776		
英 文 名 English Name(s)		Pallas' Cat		
同物异名 Synonym(s)		Manul		
种下单元评估 Infra-specific Taxa Assessed		无 / None		

评估信息 Assessment Information

评 估 年 份 Year Assessed	2020
评　定　人 Assessor(s)	蒋志刚 / Zhigang Jiang
审　定　人 Reviewer(s)	李晟、胡慧建、王大军、王昊、周友兵 / Sheng Li, Huijian Hu, Dajun Wang, Hao Wang, Youbing Zhou
其他贡献人 Other Contributor(s)	李立立、丁晨晨 / Lili Li, Chenchen Ding

理由 Justification: 兔狲分布于中国西北地区。由于毛绒质量好，在20世纪50年代兔狲被大量捕杀。过度猎捕，加之灭鼠造成的二次中毒导致兔狲种群数量下降。此外，人类活动导致其栖息地萎缩。基于其种群数量与栖息地状况，将兔狲列为濒危等级 / Pallas' Cat distributes in the northwest China. As the fur of this species was of high quality and greatly valued, it was heavily hunted in 1950s. Over-hunting and secondary poison in deratization drove its population down. Furthermore, its habitats are declining due to human activities. Based on populations size and habitat status, it is listed as Endangered

地理分布 Geographical Distribution

国内分布 Domestic Distribution
新疆、青海、甘肃、山西、内蒙古、四川、西藏、宁夏 / Xinjiang, Qinghai, Gansu, Shanxi, Inner Mongolia (Nei Mongol), Sichuan, Tibet (Xizang), Ningxia
世界分布 World Distribution
中国；南亚、中亚、西亚 / China; South Asia, Central Asia, West Asia
分布标注 Distribution Note
非特有种 / Non-endemic

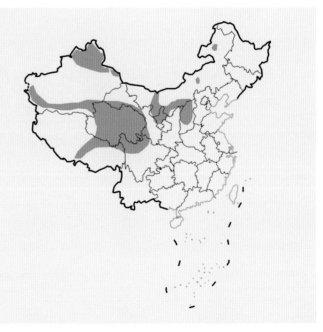

国内分布图
Map of Domestic Distribution

种群 Population

种群数量 Population Size	未知 / Unknown
种群趋势 Population Trend	下降 / Decreasing

生境与生态系统 Habitat (s) and Ecosystem (s)

生　　境 Habitat(s)	荒漠、草地 / Desert, Grassland
生态系统 Ecosystem(s)	草地生态系统、荒漠生态系统 / Grassland Ecosystem, Desert Ecosystem

威胁 Threat (s)

主要威胁 Major Threat(s)	狩猎、灭鼠、食物缺乏、耕种、家畜放牧 / Hunting, Deratization, Food Scarcity, Plantation or Ranching

保护级别与保护行动 Protection Category and Conservation Action (s)

国家重点保护野生动物等级 (2021) Category of National Key Protected Wild Animals (2021)	二级 / Category II
IUCN 红色名录(2020-2) IUCN Red List (2020-2)	近危 / NT
CITES 附录 (2019) CITES Appendix (2019)	II
保护行动 Conservation Action(s)	在自然保护区内的种群得到保护 / Populations in nature reserves are under protection

相关文献 Relevant References

Burgin *et al*., 2018; Jiang *et al*.（蒋志刚等）, 2017; Jiang *et al*.（姜雪松等）, 2013; Hunter and Barrett, 2011; Smith *et al*.（史密斯等）, 2009; Pan *et al*.（潘清华等）, 2007; Wang（王应祥）, 2003; Gao *et al*.（高耀亭等）, 1987

兔狲 *Otocolobus manul*

猞猁
Lynx lynx

濒危　EN A2ab

| 数据缺乏 DD | 无危 LC | 近危 NT | 易危 VU | 濒危 EN | 极危 CR | 区域灭绝 RE | 野外灭绝 EW | 灭绝 EX |

分类地位 Taxonomic Status

| 动物界 Animalia | 脊索动物门 Chordata | 哺乳纲 Mammalia | 食肉目 Carnivora | 猫科 Felidae |

学　名 Scientific Name	*Lynx lynx*
命 名 人 Species Authority	Linnaeus, 1758
英 文 名 English Name(s)	Eurasian Lynx
同物异名 Synonym(s)	无 / None
种下单元评估 Infra-specific Taxa Assessed	无 / None

评估信息 Assessment Information

评 估 年 份 Year Assessed	2020
评　定　人 Assessor(s)	蒋志刚 / Zhigang Jiang
审　定　人 Reviewer(s)	李晟、胡慧建、王大军、王昊、周友兵 / Sheng Li, Huijian Hu, Dajun Wang, Hao Wang, Youbing Zhou
其他贡献人 Other Contributor(s)	李立立、丁晨晨 / Lili Li, Chenchen Ding

理由 Justification: 猞猁曾广泛分布在中国。其种群数量急剧下降的主要原因是其皮毛质量高引起过度捕猎。另外，其栖息地由于人类活动而不断减少。基于其种群现状，将猞猁列为濒危等级 / Eurasian Lynx was widely distributed across China in the past. The main reason for its rapid population decline is over-hunting due to the high quality of its fur. Furthermore, its habitats are declining due to human activities. Based on its population status, Eurasian Lynx is listed as Endangered

地理分布 Geographical Distribution

国内分布 Domestic Distribution	西藏、吉林、黑龙江、内蒙古、新疆、青海、甘肃、河北、山西、辽宁、四川、云南、陕西 / Tibet (Xizang), Jilin, Heilongjiang, Inner Mongolia (Nei Mongol), Xinjiang, Qinghai, Gansu, Hebei, Shanxi, Liaoning, Sichuan, Yunnan, Shaanxi
世界分布 World Distribution	欧亚大陆 / Eurasian Continent
分布标注 Distribution Note	非特有种 / Non-endemic

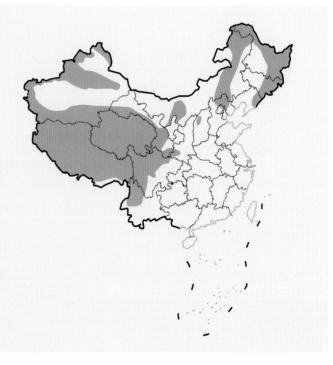

国内分布图
Map of Domestic Distribution

种群 Population

种群数量 Population Size	27,000 只 / 27,000 individuals
种群趋势 Population Trend	稳定 / Stable

生境与生态系统 Habitat (s) and Ecosystem (s)

生境 Habitat(s)	森林、草地、内陆岩石区域、泰加林 / Forest, Grassland, Inland Rocky Area, Taiga Forest
生态系统 Ecosystem(s)	森林生态系统、草地生态系统 / Forest Ecosystem, Grassland Ecosystem

威胁 Threat (s)

主要威胁 Major Threat(s)	狩猎、食物缺乏 / Hunting, Food Scarcity

保护级别与保护行动 Protection Category and Conservation Action (s)

国家重点保护野生动物等级 (2021) Category of National Key Protected Wild Animals (2021)	二级 / Category II
IUCN 红色名录 (2020-2) IUCN Red List (2020-2)	无危 / LC
CITES 附录 (2019) CITES Appendix (2019)	II
保护行动 Conservation Action(s)	在自然保护区内的种群得到保护 / Populations in nature reserves are under protection

相关文献 Relevant References

Burgin *et al.*, 2018; Jiang *et al.* （蒋志刚等）, 2017; Ju *et al.* （鞠丹等）, 2013; Hunter and Barrett, 2011; Piao *et al.* （朴正吉等）, 2011; Bao, 2010; Smith *et al.* （史密斯等）, 2009; Wang（王应祥）, 2003; Gao *et al.* （高耀亭等）, 1987

猞猁 *Lynx lynx*

中国生物多样性红色名录
云猫
Pardofelis marmorata

濒危 EN A2ab

| 数据缺乏 DD | 无危 LC | 近危 NT | 易危 VU | 濒危 EN | 极危 CR | 区域灭绝 RE | 野外灭绝 EW | 灭绝 EX |

分类地位 Taxonomic Status

动物界 Animalia	脊索动物门 Chordata	哺乳纲 Mammalia	食肉目 Carnivora	猫科 Felidae
学名 Scientific Name		*Pardofelis marmorata*		
命名人 Species Authority		Martin, 1837		
英文名 English Name(s)		Marbled Cat		
同物异名 Synonym(s)		无 / None		
种下单元评估 Infra-specific Taxa Assessed		无 / None		

评估信息 Assessment Information

评估年份 Year Assessed	2020
评定人 Assessor(s)	蒋志刚 / Zhigang Jiang
审定人 Reviewer(s)	陈辈乐、李飞、李晟、胡慧建、王大军、王昊、周友兵 / Bosco P. L. Chan, Fei Li, Sheng Li, Huijian Hu, Dajun Wang, Hao Wang, Youbing Zhou
其他贡献人 Other Contributor(s)	李立立、丁晨晨 / Lili Li, Chenchen Ding

理由 Justification: 以往云猫仅在云南有记录，但近年的红外相机调查发现其在西藏东南部亦有分布。目前云猫仍然受到盗猎、栖息地减少等威胁。基于其种群与栖息地现状，故将其列为濒危 / In the past, Marbled Cat was only recorded in Yunnan, but in recent years, infrared camera surveys have found that the Marbled Cat is also distributed in southeast Tibet (Xizang). The species is still threatened by poaching and habitat loss. Based on its populations and habitat status, Marbled Cat is listed as Endangered

地理分布 Geographical Distribution

国内分布 Domestic Distribution
云南、西藏 / Yunnan, Tibet (Xizang)
世界分布 World Distribution
中国；南亚、东南亚 / China; South Asia, Southeast Asia
分布标注 Distribution Note
非特有种 / Non-endemic

国内分布图
Map of Domestic Distribution

种群 Population

种群数量 Population Size	未知 / Unknown
种群趋势 Population Trend	下降 / Decreasing

生境与生态系统 Habitat (s) and Ecosystem (s)

生境 Habitat(s)	热带湿润低地森林、灌丛 / Tropical Moist Lowland Forest, Shrubland
生态系统 Ecosystem(s)	森林生态系统、灌丛生态系统 / Forest Ecosystem, Shrubland Ecosystem

威胁 Threat (s)

主要威胁 Major Threat(s)	耕种、伐木、树木种植园、狩猎 / Plantation, Logging, Wood Plantation, Hunting

保护级别与保护行动 Protection Category and Conservation Action (s)

国家重点保护野生动物等级 (2021) Category of National Key Protected Wild Animals (2021)	二级 / Category II
IUCN 红色名录 (2020-2) IUCN Red List (2020-2)	易危 / VU
CITES 附录 (2019) CITES Appendix (2019)	I
保护行动 Conservation Action(s)	在自然保护区内的种群得到保护 / Populations in nature reserves are under protection

相关文献 Relevant References

Burgin *et al.*, 2018; Jiang *et al.*（蒋志刚等）, 2017; Shi *et al.*（师蕾等）, 2013; Hunter and Barrett, 2011; Smith *et al.*（史密斯等）, 2009; Wang（王应祥）, 2003; Gao *et al.*（高耀亭等）, 1987; Sun and Gao（孙崇烁和高耀亭）, 1976

云猫 *Pardofelis marmorata*

中国生物多样性 红色名录
China's Red List of Biodiversity

金猫
Pardofelis temminckii

濒危　EN A2d; B1ab(i, ii, iii); C2(i)

| 数据缺乏 DD | 无危 LC | 近危 NT | 易危 VU | **濒危 EN** | 极危 CR | 区域灭绝 RE | 野外灭绝 EW | 灭绝 EX |

分类地位 Taxonomic Status

动物界 Animalia	脊索动物门 Chordata	哺乳纲 Mammalia	食肉目 Carnivora	猫科 Felidae

学名 Scientific Name	*Pardofelis temminckii*
命名人 Species Authority	Vigors and Horsfield, 1827
英文名 English Name(s)	Asiatic Golden Cat
同物异名 Synonym(s)	Asian Golden Cat; *Catopuma temminckii* (Vigors and Horsfield, 1827)
种下单元评估 Infra-specific Taxa Assessed	无 / None

评估信息 Assessment Information

评估年份 Year Assessed	2020
评定人 Assessor(s)	蒋志刚 / Zhigang Jiang
审定人 Reviewer(s)	陈辈乐、李飞 / Bosco P. L. Chan, Fei Li
其他贡献人 Other Contributor(s)	李立立、丁晨晨 / Lili Li, Chenchen Ding

理由 Justification: 金猫曾广泛分布于中国的中部和南部。最近的调查发现金猫仍存在于中国中部及西南部的一些保护区中，但分布于中国东南部的亚种 *P. t. dominicanorum* 却多年杳无音信，濒临灭绝。金猫在中国的种群数量及分布区严重缩减，且金猫仍然受到盗猎、栖息地减少等威胁，故将其列为濒危 / The Asiatic Golden Cat was once widespread in central and southern China. Recent surveys found that this cat still lives in some protected areas in central and southwest China, however the subspecies *P. t. dominicanorum* in previously known area in southeast China has not been seen for many years and is on the brink of extinction. Due to a severe decline in its population and shrinking distribution, the Asiatic Golden Cat is threatened by poaching and habitat loss, and thus it is listed as Endangered

地理分布 Geographical Distribution

国内分布 Domestic Distribution	浙江、河南、安徽、福建、江西、湖北、湖南、广东、广西、四川、贵州、云南、西藏、陕西、甘肃、重庆 / Zhejiang, Henan, Anhui, Fujian, Jiangxi, Hubei, Hunan, Guangdong, Guangxi, Sichuan, Guizhou, Yunnan, Tibet (Xizang), Shaanxi, Gansu, Chongqing
世界分布 World Distribution	中国；南亚、东南亚 / China; South Asia, Southeast Asia
分布标注 Distribution Note	非特有种 / Non-endemic

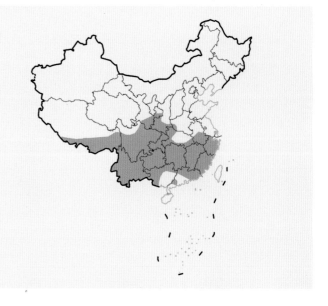

国内分布图
Map of Domestic Distribution

种群 Population

种群数量 Population Size	7,300 只 /7,300 individuals
种群趋势 Population Trend	下降 / Decreasing

生境与生态系统 Habitat (s) and Ecosystem (s)

生境 Habitat(s)	森林、草地、灌丛 / Forest, Grassland, Shrubland
生态系统 Ecosystem(s)	森林生态系统、灌丛生态系统、草地生态系统 / Forest Ecosystem, Shrubland Ecosystem, Grassland Ecosystem

威胁 Threat (s)

主要威胁 Major Threat(s)	狩猎、食物缺乏 / Hunting, Food Scarcity

保护级别与保护行动 Protection Category and Conservation Action (s)

国家重点保护野生动物等级 (2021) Category of National Key Protected Wild Animals (2021)	一级 / Category I
IUCN红色名录(2020-2) IUCN Red List (2020-2)	近危 / NT
CITES 附录 (2019) CITES Appendix (2019)	无 / NA
保护行动 Conservation Action(s)	在自然保护区内的种群得到保护 / Populations in nature reserves are under protection

相关文献 Relevant References

Burgin *et al.*, 2018; Jiang *et al.* (蒋志刚等), 2017; Chen and Shi (陈鹏和师杜鹃), 2013; Deng *et al.* (邓可等), 2013; Hunter and Barrett, 2011; Smith *et al.* (史密斯等), 2009; Gao *et al.* (高耀亭等), 1987

金猫 *Pardofelis temminckii*

金钱豹
Panthera pardus

中国生物多样性红色名录

濒危 EN A2ab; B1ab(i, ii, iii)

| 数据缺乏 DD | 无危 LC | 近危 NT | 易危 VU | **濒危 EN** | 极危 CR | 区域灭绝 RE | 野外灭绝 EW | 灭绝 EX |

分类地位 Taxonomic Status

动物界 Animalia	脊索动物门 Chordata	哺乳纲 Mammalia	食肉目 Carnivora	猫科 Felidae
学 名 Scientific Name		*Panthera pardus*		
命 名 人 Species Authority		Linnaeus, 1758		
英 文 名 English Name(s)		Leopard		
同物异名 Synonym(s)		豹		
种下单元评估 Infra-specific Taxa Assessed		远东豹 (*P. p. orientalis*)：近年来数量有所增加，但是在中国东北仅发现 40 余只个体；故其濒危等级定为极危 (CR)。华北豹 (*P. p. fontanieri*)：近年来 (转后第 2 页)		

评估信息 Assessment Information

评 估 年 份 Year Assessed	2020
评 定 人 Assessor(s)	蒋志刚 / Zhigang Jiang
审 定 人 Reviewer(s)	冯利民、姜广顺、李晟、胡慧建 / Limin Feng, Guangshun Jiang, Sheng Li, Huijian Hu
其他贡献人 Other Contributor(s)	李立立、丁晨晨 / Lili Li, Chenchen Ding

理由 Justification: 金钱豹是大型猫科动物中在全球分布最广的一种，但是其毛皮质量高，且其骨常代替虎骨入传统中药，因此对金钱豹的过度捕猎和非法贸易严重威胁着该物种的生存。20 世纪 50 年代，金钱豹曾被认为是害兽而被加以杀害。加上人类活动造成的栖息地退化和丧失，导致在中国曾经广布的金钱豹的种群越来越小。因此，将金钱豹列为濒危等级 / Of all the big cats, the Leopard has the largest range in the world. However, with its fur being of high quality and its bones are able to be used as substitute for tiger bones in traditional Chinese medicine, over-hunting and illegal trade are severely threatening the survival of this species. In the 1950s, Leopard was considered as a vermin and large numbers were killed. Furthermore, degradation and loss of its habitats due to human activities continue to lead to a rapid decline in the population of leopards, which were once widely distributed across China. Thus, Leopard is listed as Endangered

地理分布 Geographical Distribution

国内分布 Domestic Distribution

山西、河南、黑龙江、浙江、北京、河北、内蒙古、吉林、江苏、安徽、福建、江西、湖北、湖南、广东、广西、四川、贵州、云南、西藏、陕西、甘肃、青海、宁夏、天津、重庆 / Shanxi, Henan, Heilongjiang, Zhejiang, Beijing, Hebei, Inner Mongolia (Nei Mongol), Jilin, Jiangsu, Anhui, Fujian, Jiangxi, Hubei, Hunan, Guangdong, Guangxi, Sichuan, Guizhou, Yunnan, Tibet (Xizang), Shaanxi, Gansu, Qinghai, Ningxia, Tianjin, Chongqing

世界分布 World Distribution

阿富汗、阿尔及利亚、安哥拉、亚美尼亚、阿塞拜疆、孟加拉国、贝宁、不丹、博茨瓦纳、布基纳法索、布隆迪、柬埔寨、喀麦隆、中非、乍得、中国、刚果、刚果民主共和国、科特迪瓦、吉布提、埃及、赤道几内亚、厄立特里亚、埃塞俄比亚、加蓬、冈比亚、格鲁吉亚、加纳、几内亚、几内亚比绍、印度、印度尼西亚、伊朗、以色列、约旦、肯尼亚、朝鲜、老挝、利比里亚、马拉维、马来西亚、马里、摩洛哥、莫桑比克、缅甸、纳米比亚、尼泊尔、尼日尔、尼日利亚、阿曼、巴基斯坦、俄罗斯、卢旺达、沙特阿拉伯、塞内加尔、塞拉利昂、索马里、南非、斯里兰卡、

世界分布 World Distribution

苏丹、斯威士兰、塔吉克斯坦、坦桑尼亚、泰国、多哥、土耳其、土库曼斯坦、乌干达、阿拉伯联合酋长国、乌兹别克斯坦、越南、也门、赞比亚、津巴布韦 / Afghanistan, Algeria, Angola, Armenia, Azerbaijan, Bangladesh, Benin, Bhutan, Botswana, Burkina Faso, Burundi, Cambodia, Cameroon, Central African Republic, Chad, China, Congo, Congo(the Democratic Republic of the), Côte d'Ivoire, Djibouti, Egypt, Equatorial Guinea, Eritrea, Ethiopia, Gabon, Gambia, Georgia, Ghana, Guinea, Guinea-Bissau, India, Indonesia, Iran, Israel, Jordan, Kenya, Korea (the Democratic People's Republic of), Laos, Liberia, Malawi, Malaysia, Mali, Morocco, Mozambique, Myanmar, Namibia, Nepal, Niger, Nigeria, Oman, Pakistan, Russia, Rwanda, Saudi Arabia, Senegal, Sierra Leone, Somalia, South Africa, Sri Lanka, Sudan, Swaziland, Tajikistan, Tanzania, Thailand, Togo, Turkey, Turkmenistan, Uganda, United Arab Emirates, Uzbekistan, Viet Nam, Yemen, Zambia, Zimbabwe

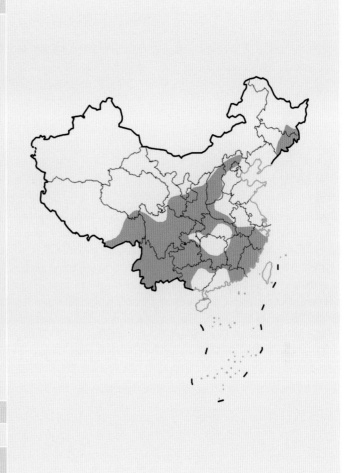

国内分布图
Map of Domestic Distribution

分布标注 Distribution Note

非特有种 / Non-endemic

种群 Population

种群数量 Population Size	3,310 只 / 3,310 individuals
种群趋势 Population Trend	上升 / Increasing

生境与生态系统 Habitat (s) and Ecosystem (s)

生　　境 Habitat(s)	灌丛、森林 / Shrubland, Forest
生态系统 Ecosystem(s)	森林生态系统、灌丛生态系统 / Forest Ecosystem, Shrubland Ecosystem

威胁 Threat (s)

主要威胁 Major Threat(s)	狩猎、人兽冲突、栖息地改变 / Hunting, Human-Animal Conflict, Habitat Shifting or Alteration

保护级别与保护行动 Protection Category and Conservation Action (s)

国家重点保护野生动物等级 (2021) Category of National Key Protected Wild Animals (2021)	一级 / Category I
IUCN 红色名录 (2020-2) IUCN Red List (2020-2)	易危 / VU
CITES 附录 (2019) CITES Appendix (2019)	I
保护行动 Conservation Action(s)	在自然保护区内的种群得到保护 / Populations in nature reserves are under protection

相关文献 Relevant References

Burgin *et al*., 2018; Jiang *et al*. (蒋志刚等), 2017; Hunter and Barrett, 2011; Jutzeler *et al*., 2010a; Liu *et al*. (刘长乐等), 2009; Smith *et al*. (史密斯等), 2009; Wang *et al*. (王好峰等), 2008; Liu *et al*. (刘伟石等), 2007; Gao *et al*. (高耀亭等), 1987

(续前)
曾在山西、陕西、河北、河南、甘肃等地被红外相机拍摄到，在野外仍有可生存种群；故其濒危等级定为濒危 (EN)。印支豹 (*P. p. delacouri*)：现在中国南方野外稀少，极少见到；故其濒危等级定为极危 (CR)。印度豹 (*P. p. fusca*)：仅分布于喜马拉雅山脉局部低海拔区域，在西藏吉隆曾被发现和记录；故其濒危等级定为极危 (CR)。此外，目前分布在青海及青藏高原东缘西藏昌都、四川甘孜的金钱豹，其分类地位未定，可能是介于华北豹和印支豹之间的中间类型 / Amur Leopard (*P. p. orientalis*): Though more individuals have been camera trapped, only about four dozens individuals of Amur Leopard were found in northeast China; it is classified as Critically Endangered (CR). North China Leopard (*P. p. fontanieri*): It has been trapped by infrared cameras in Shanxi, Shaanxi, Hebei, Henan, Gansu and some other places in the country in recent years, there are still viable populations of the North China Leopard in the wild; the North China Leopard is classified as Endangered (EN). Indochinese Leopard (*P. p. delacouri*): Now, Indochinese Leopard has seldom been seen in the wild in southern China due to its extremely low density; it is classified as Critically Endangered (CR). Indian Leopard (*P. p. fusca*): It was only found in low altitude areas of Himalayas, such as Jilong, Tibet (Xizang); Indian Leopard is classified as Critically Endangered (CR). Additionally, the taxonomic position of the leopards found in Qinghai, and eastern fringe of the Qinghai-Tibet (Xizang) Plateau, Qamdo of Tibet (Xizang) and Ganzi of Sichuan, is not clear at present, those leopards may be an intermediate type between the North China Leopard and the Indochinese Leopard

蒋志刚 摄 By Zhigang Jiang

中国生物多样性红色名录 China's Red List of Biodiversity

雪豹
Panthera uncia

濒危 EN A2ab; B1ab(i, ii, iii)

| 数据缺乏 DD | 无危 LC | 近危 NT | 易危 VU | **濒危 EN** | 极危 CR | 区域灭绝 RE | 野外灭绝 EW | 灭绝 EX |

分类地位 Taxonomic Status

动物界 Animalia	脊索动物门 Chordata	哺乳纲 Mammalia	食肉目 Carnivora	猫科 Felidae
学　　名 Scientific Name		*Panthera uncia*		
命　名　人 Species Authority		Schreber, 1775		
英　文　名 English Name(s)		Snow Leopard		
同物异名 Synonym(s)		*Felis uncia* (Schreber, 1775); *Uncia uncia* (Schreber, 1775)		
种下单元评估 Infra-specific Taxa Assessed		无 / None		

评估信息 Assessment Information

评估年份 Year Assessed	2020
评　定　人 Assessor(s)	蒋志刚 / Zhigang Jiang
审　定　人 Reviewer(s)	时坤、姜广顺、胡慧建 / Kun Shi, Guangshun Jiang, Huijian Hu
其他贡献人 Other Contributor(s)	李立立、丁晨晨 / Lili Li, Chenchen Ding

理由 Justification: 雪豹主要分布在中国西部。为了获取雪豹的皮毛和骨，过度捕猎和非法贸易曾一度威胁着雪豹的生存。还有一些放牧者因为雪豹捕杀了羊群，造成了损失而报复性捕杀雪豹。人类的过度放牧使其脆弱的生境逐渐退化、丧失。近年来，通过红外相机等技术，发现了越来越多的雪豹种群，然而，由于全球环境变化，雪豹仍面临灭绝风险。因此，雪豹列为濒危等级 / Snow Leopard is mainly distributed in west China. For obtaining its fur and bone, over-hunting and illegal trades once threatened the survival of this species. Some herdsmen retaliate killing the Snow Leopard in response to livestock depredation. Furthermore, the fragile habitats of Snow Leopards are degraded or lost due to human encroachment. Though, more and more populations of Snow Leopard are discovered by infrared camera in recent years, with its small populations, the Snow Leopard is still facing the extinction risk due to global changes. Thus, Snow Leopard is listed as Endangered

地理分布 Geographical Distribution

国内分布 Domestic Distribution	青海、内蒙古、新疆、四川、云南、西藏、甘肃 / Qinghai, Inner Mongolia (Nei Mongol), Xinjiang, Sichuan, Yunnan, Tibet (Xizang), Gansu
世界分布 World Distribution	中国；中亚、南亚 / China; Central Asia, South Asia
分布标注 Distribution Note	非特有种 / Non-endemic

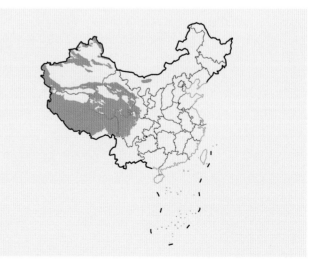

国内分布图 Map of Domestic Distribution

种群 Population

种群数量 Population Size	4,100 只 / 4,100 individuals
种群趋势 Population Trend	下降 / Decreasing

生境与生态系统 Habitat (s) and Ecosystem (s)

生境 Habitat(s)	内陆岩石区域、草地、灌丛 / Inland Rocky Area, Grassland, Shrubland
生态系统 Ecosystem(s)	灌丛生态系统、草地生态系统 / Shrubland Ecosystem, Grassland Ecosystem

威胁 Threat (s)

主要威胁 Major Threat(s)	狩猎、食物缺乏、人兽冲突、人类活动干扰 / Hunting, Food Scarcity, Human-Animal Conflict, Human Disturbance

保护级别与保护行动 Protection Category and Conservation Action (s)

国家重点保护野生动物等级 (2021) Category of National Key Protected Wild Animals (2021)	一级 / Category I
IUCN红色名录(2020-2) IUCN Red List (2020-2)	易危 / VU
CITES 附录 (2019) CITES Appendix (2019)	I
保护行动 Conservation Action(s)	在自然保护区内的种群得到保护；列入《迁徙物种保护公约》的"迁徙物种行动计划" / Populations in nature reserves are under protection; included in the *Convention on Migratory Species* (CMS)－Species Action Plan

相关文献 Relevant References

Burgin *et al*., 2018; Jiang *et al*. (蒋志刚等), 2017, 2015a; McCarthy *et al*., 2017; Hunter and Barrett, 2011; Xu *et al*. (徐峰等), 2011; Wei *et al*., 2009; Zhang *et al*. (张于光等), 2009; Xu *et al*., 2008; Gao *et al*. (高耀亭等), 1987

雪豹 *Panthera uncia*

威氏鼷鹿
Tragulus williamsoni

濒危 EN B1ab(i, ii, iii)+2ab(i, ii, iii)

| 数据缺乏 DD | 无危 LC | 近危 NT | 易危 VU | **濒危 EN** | 极危 CR | 区域灭绝 RE | 野外灭绝 EW | 灭绝 EX |

分类地位 Taxonomic Status

动物界 Animalia	脊索动物门 Chordata	哺乳纲 Mammalia	偶蹄目 Artiodactyla	鼷鹿科 Tragulidae

学名 Scientific Name	*Tragulus williamsoni*
命名人 Species Authority	Kloss, 1916
英文名 English Name(s)	鼷鹿，Williamson's Mouse Deer
同物异名 Synonym(s)	无 / None
种下单元评估 Infra-specific Taxa Assessed	无 / None

评估信息 Assessment Information

评估年份 Year Assessed	2020
评定人 Assessor(s)	蒋志刚 / Zhigang Jiang
审定人 Reviewer(s)	胡慧建、胡一鸣 / Huijian Hu, Yiming Hu
其他贡献人 Other Contributor(s)	李立立、丁晨晨 / Lili Li, Chenchen Ding

理由 Justification: 威氏鼷鹿在中国的分布为其边缘分布，仅分布于靠近中国与老挝边境的云南省勐腊县。据蒋志刚团队2015年的野外考察与实验室研究，中国鼷鹿并不是爪哇鼷鹿 (*Tragulus javanicus*)，尚不能确认中国鼷鹿是威氏鼷鹿 (*T. williamsoni*) 还是小鼷鹿 (*T. kanchil*)。Meijaard et al. (2017) 通过研究中国科学院昆明动物研究所的鼷鹿标本，解开了这一疑团：勐腊的鼷鹿是威氏鼷鹿 (*Tragulus williamsoni*)。威氏鼷鹿分布在云南勐腊临近老挝边境地区，分布区面积小于128km^2，种群数量小于3,000只。因此，威氏鼷鹿列为濒危等级 / In China, Williamson's Mouse Deer is only marginally distributed in Mengla County in Yunnan Province, near the China-Laos border. According to Zhigang Jiang *et al.*'s field and laboratory study in 2015, the Mouse Deer in China is not *Tragulus javanicus* (Java Mouse Deer), however it is not certain whether it is *T. williamsoni* (Williamson's Mouse Deer) or *T. kanchil* (Lesser Mouse Deer) as no DNA series is available for comparison. However, Meijaard *et al.* (2017) resolved the riddle, they found the Mouse Deer in Mengla, Yunan is *T. williamsoni* according to the study of specimens in the Kunming Institute of Zoology, Chinese Academy of Sciences. The distribution of Williamson's Mouse Deer in China covers less than 128km^2 and its population size is under 3,000 individuals. Thus, Williamson's Mouse Deer is listed as Endangered

地理分布 Geographical Distribution

国内分布 Domestic Distribution	云南 / Yunnan
世界分布 World Distribution	中国、泰国、老挝 / China, Thailand, Laos
分布标注 Distribution Note	非特有种 / Non-endemic

国内分布图
Map of Domestic Distribution

种群 Population

种群数量 Population Size	1,681 ~ 2,555 只 / 1,681~2,555 individuals
种群趋势 Population Trend	下降 / Decreasing

生境与生态系统 Habitat (s) and Ecosystem (s)

生　境 Habitat(s)	栖息生境为人为干扰少的原始热带雨林，林内多高大乔木，树干多附生植物及藤本植物攀缘，野生浆果丰富 / Its habitat is pristine tropical rainforest where there is little human disturbance. This forest type has tall trees with many epiphytes and vines on the trunk and wild berries are abundant
生态系统 Ecosystem(s)	森林生态系统 / Forest Ecosystem

威胁 Threat (s)

主要威胁 Major Threat(s)	未知 / Unknown

保护级别与保护行动 Protection Category and Conservation Action (s)

国家重点保护野生动物等级 (2021) Category of National Key Protected Wild Animals (2021)	一级 / Category I
IUCN 红色名录 (2020-2) IUCN Red List (2020-2)	数据缺乏 / DD
CITES 附录 (2019) CITES Appendix (2019)	无 / NA
保护行动 Conservation Action(s)	自然保护区内种群得到保护 / Populations in nature reserves are protected

相关文献 Relevant References

Burgin *et al*., 2018; Jiang *et al*.（蒋志刚等），2017; Meijaard *et al*., 2017; Wilson and Mittermeier, 2012; Cao *et al*.（曹明等），2010; Meijaard and Groves, 2004; Shi and Chen（施立明和陈玉泽），1989

威氏鼷鹿 *Tragulus williamsoni*　　胡一鸣 摄　By Yiming Hu

安徽麝
Moschus anhuiensis

濒危　EN B2ab(i, ii, iii)

| 数据缺乏 DD | 无危 LC | 近危 NT | 易危 VU | **濒危 EN** | 极危 CR | 区域灭绝 RE | 野外灭绝 EW | 灭绝 EX |

分类地位 Taxonomic Status

动物界 Animalia	脊索动物门 Chordata	哺乳纲 Mammalia	偶蹄目 Artiodactyla	麝科 Moschidae

学名 Scientific Name	*Moschus anhuiensis*
命名人 Species Authority	Wang, Hu and Yan, 1982
英文名 English Name(s)	Anhui Musk Deer
同物异名 Synonym(s)	无 / None
种下单元评估 Infra-specific Taxa Assessed	无 / None

评估信息 Assessment Information

评估年份 Year Assessed	2020
评定人 Assessor(s)	蒋志刚 / Zhigang Jiang
审定人 Reviewer(s)	鲍毅新、李春林 / Yixin Bao, Chunlin Li
其他贡献人 Other Contributor(s)	李立立、丁晨晨 / Lili Li, Chenchen Ding

理由 Justification: 安徽麝是中国特有种，分布区狭窄，但近年来红外相机监测发现安徽麝在安徽省金寨天马国家级自然保护区仍有一定种群数量。因此，基于其占有区面积，安徽麝的濒危等级定为濒危 / Anhui Musk Deer is endemic to China and has a narrow range. However, recently infrared camera survey revealed that Anhui Musk Deer still has a certain population size in the Jinzhai Tianma National Nature Reserve, Anhui Province. Thus, Based on its area of occupancy, Anhui Musk Deer is listed as Endangered

地理分布 Geographical Distribution

国内分布 Domestic Distribution
安徽 / Anhui
世界分布 World Distribution
中国 / China
分布标注 Distribution Note
特有种 / Endemic

国内分布图
Map of Domestic Distribution

种群 Population

种群数量 Population Size	1,230 只 / 1,230 individuals
种群趋势 Population Trend	稳定 / Stable

生境与生态系统 Habitat(s) and Ecosystem(s)

生境 Habitat(s)	阔叶林、针阔混交林 / Broad-leaved Forest, Coniferous and Broad-leaved Mixed Forest
生态系统 Ecosystem(s)	森林生态系统 / Forest Ecosystem

威胁 Threat(s)

主要威胁 Major Threat(s)	狩猎、生境破碎、伐木 / Hunting, Habitat Fragmentation, Logging

保护级别与保护行动 Protection Category and Conservation Action(s)

国家重点保护野生动物等级 (2021) Category of National Key Protected Wild Animals (2021)	一级 / Category I
IUCN 红色名录 (2020-2) IUCN Red List (2020-2)	濒危 / EN
CITES 附录 (2019) CITES Appendix (2019)	II
保护行动 Conservation Action(s)	自然保护区内种群得到保护 / Populations in nature reserves are protected

相关文献 Relevant References

Burgin *et al.*, 2018; Jiang *et al.*（蒋志刚等）, 2017; Zhu *et al.*, 2013; Peng *et al.*（彭红元等）, 2010; Wu and Wang（吴家炎和王伟）, 2006; Liu and Tong（刘文华和佟建明）, 2005

安徽麝 *Moschus anhuiensis*

喜马拉雅麝
Moschus leucogaster

濒危 EN A1acd; B1ab(i, ii, iii)

| DD 数据缺乏 | LC 无危 | NT 近危 | VU 易危 | **EN 濒危** | CR 极危 | RE 区域灭绝 | EW 野外灭绝 | EX 灭绝 |

分类地位 Taxonomic Status

动物界 Animalia	脊索动物门 Chordata	哺乳纲 Mammalia	偶蹄目 Artiodactyla	麝科 Moschidae

学名 Scientific Name	*Moschus leucogaster*
命名人 Species Authority	Hodgson, 1839
英文名 English Name(s)	Himalayan Musk Deer
同物异名 Synonym(s)	White-bellied Musk Deer; *Moschus chrysogaster* (Hodgson, 1839) subsp. *leucogaster*
种下单元评估 Infra-specific Taxa Assessed	无 / None

评估信息 Assessment Information

评估年份 Year Assessed	2020
评定人 Assessor(s)	蒋志刚 / Zhigang Jiang
审定人 Reviewer(s)	孟秀祥、胡德夫、鲍毅新 / Xiuxiang Meng, Defu Hu, Yixin Bao
其他贡献人 Other Contributor(s)	李立立、丁晨晨 / Lili Li, Chenchen Ding

理由 Justification: 喜马拉雅麝分布区狭窄，面临偷猎压力，野外种群数量低。故将喜马拉雅麝列为濒危等级 / Himalayan Musk Deer has a narrow range in the country. It is facing the pressure of poaching and its population density is low in field. Thus, Himalayan Musk Deer is listed as Endangered

地理分布 Geographical Distribution

国内分布 Domestic Distribution
西藏 / Tibet (Xizang)

世界分布 World Distribution
中国；南亚 / China; South Asia

分布标注 Distribution Note
非特有种 / Non-endemic

国内分布图
Map of Domestic Distribution

种群 Population

种群数量 Population Size	3,000 只 /3,000 individuals
种群趋势 Population Trend	下降 / Decreasing

生境与生态系统 Habitat (s) and Ecosystem (s)

生　境 Habitat(s)	高山生境 / Alpine Habitat
生态系统 Ecosystem(s)	森林生态系统 / Forest Ecosystem

威胁 Threat (s)

主要威胁 Major Threat(s)	狩猎、家畜放牧、耕种、伐木 / Hunting, Livestock Ranching, Plantation, Logging

保护级别与保护行动 Protection Category and Conservation Action (s)

国家重点保护野生动物等级 (2004) Category of National Key Protected Wild Animals (2004)	一级 / Category I
IUCN 红色名录 (2020-2) IUCN Red List (2020-2)	濒危 / EN
CITES 附录 (2019) CITES Appendix (2019)	I
保护行动 Conservation Action(s)	自然保护区内种群得到保护/ Populations in nature reserves are protected

相关文献 Relevant References

Burgin *et al.*, 2018; Jiang *et al.* (蒋志刚等), 2017; Wu and Wang (吴家炎和王伟), 2006; Liu and Tong (刘文华和佟建明), 2005; Cai and Feng (蔡桂全和冯祚建), 1982

喜马拉雅麝 *Moschus leucogaster*

黑麂
Muntiacus crinifrons

濒危 EN A1acd; B1ab(i,ii,iii)

| 数据缺乏 DD | 无危 LC | 近危 NT | 易危 VU | 濒危 EN | 极危 CR | 区域灭绝 RE | 野外灭绝 EW | 灭绝 EX |

分类地位 Taxonomic Status

动物界 Animalia	脊索动物门 Chordata	哺乳纲 Mammalia	偶蹄目 Artiodactyla	鹿科 Cervidae

学 名 Scientific Name	*Muntiacus crinifrons*
命 名 人 Species Authority	Sclater, 1885
英 文 名 English Name(s)	Black Muntjac
同物异名 Synonym(s)	Hairy-fronted Muntjac; *Cervulus crinifrons* (Sclater, 1885)
种下单元评估 Infra-specific Taxa Assessed	无 / None

评估信息 Assessment Information

评 估 年 份 Year Assessed	2020
评 定 人 Assessor(s)	蒋志刚 / Zhigang Jiang
审 定 人 Reviewer(s)	鲍伟东、鲍毅新、李言阔、丁平、蒋学龙 / Weidong Bao, Yixin Bao, Yankuo Li, Ping Ding, Xuelong Jiang
其他贡献人 Other Contributor(s)	李立立、丁晨晨 / Lili Li, Chenchen Ding

理由 Justification: 黑麂是中国的特有种。其栖息地由于人类活动日益缩减。为获得黑麂皮和黑麂肉，人类过度捕猎，严重威胁着该物种的生存。其种群数量呈现下降的趋势。因此，将其列为濒危等级 / Black Muntjac is an endemic species in China. Its habitats are declining due to human encroachment. People used to hunt Black Muntjac for its skin and meat, which severely threatened its survival. Its populations are decreasing. Thus, Black Muntjac is listed as Endangered

地理分布 Geographical Distribution

国内分布 Domestic Distribution	江西、浙江、安徽、福建 / Jiangxi, Zhejiang, Anhui, Fujian
世界分布 World Distribution	中国 / China
分布标注 Distribution Note	特有种 / Endemic

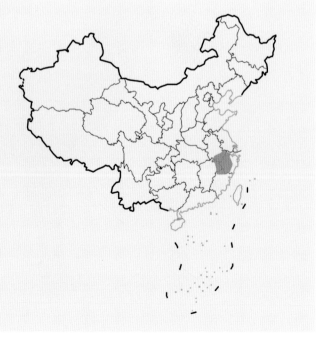

国内分布图
Map of Domestic Distribution

种群 Population

种群数量 Population Size	8,800 只 / 8,800 individuals
种群趋势 Population Trend	下降 / Decreasing

生境与生态系统 Habitat (s) and Ecosystem (s)

生　　境 Habitat(s)	森林 / Forest
生态系统 Ecosystem(s)	阔叶林生态系统 / Broad-leaved Forest Ecosystem

威胁 Threat (s)

主要威胁 Major Threat(s)	生境破碎、伐木、狩猎 / Habitat Fragmentation, Logging, Hunting

保护级别与保护行动 Protection Category and Conservation Action (s)

国家重点保护野生动物等级 (2021) Category of National Key Protected Wild Animals (2021)	一级 / Category I
IUCN 红色名录 (2020-2) IUCN Red List (2020-2)	易危 / VU
CITES 附录 (2019) CITES Appendix (2019)	I
保护行动 Conservation Action(s)	自然保护区内种群得到保护 / Populations in nature reserves are protected

相关文献 Relevant References

Burgin *et al*., 2018; Jiang *et al*. (蒋志刚等), 2017; Zheng *et al*. (郑伟成等), 2012; Chen *et al*. (陈良等), 2010; Smith *et al*. (史密斯等), 2009; Cheng *et al*. (程宏毅等), 2008

黑麂 *Muntiacus crinifrons*

贡山麂
Muntiacus gongshanensis

濒危　EN B1ab(i, ii, iii)

| 数据缺乏 DD | 无危 LC | 近危 NT | 易危 VU | 濒危 EN | 极危 CR | 区域灭绝 RE | 野外灭绝 EW | 灭绝 EX |

分类地位 Taxonomic Status

动物界 Animalia	脊索动物门 Chordata	哺乳纲 Mammalia	偶蹄目 Artiodactyla	鹿科 Cervidae

学　　名 Scientific Name	*Muntiacus gongshanensis*
命　名　人 Species Authority	Ma, Wang and Shi, 1990
英　文　名 English Name(s)	Gongshan Muntjac
同物异名 Synonym(s)	无 / None
种下单元评估 Infra-specific Taxa Assessed	无 / None

评估信息 Assessment Information

评估年份 Year Assessed	2020
评　定　人 Assessor(s)	蒋志刚 / Zhigang Jiang
审　定　人 Reviewer(s)	蒋学龙、陈辈乐、李飞 / Xuelong Jiang, Bosco P. L. Chan, Fei Li
其他贡献人 Other Contributor(s)	李立立、丁晨晨 / Lili Li, Chenchen Ding

理由 Justification: 贡山麂是中国特有种，分布在云南高黎贡山地区，藏东南也可能有分布。近年来西藏墨脱（李成，私人通信）、云南腾冲（嘉道理农场暨植物园，未发表数据）以及盈江（猫科动物保护联盟，私人通信）的红外相机调查皆拍摄到与贡山麂外形极为相似的麂。贡山麂常遭偷猎，种群数量下降。因此，将其列为濒危等级 / Gongshan Muntjac is an endemic species in China. It is distributed in Gaoligong Shan region of Yunnan Province and may also in southeast Tibet (Xizang). In recent years, infrared camera surveys in Mêdog, Tibet (Xizang)(Cheng Li, Personal communications) and Tengchong in Yunnan (Kadoorie Farm and Botanic Garden, unpublished data) and Yingjiang (Cat Conservation Alliance, Personal communications) have captured the images of Muntjac that closely resembled to Gongshan Muntjac. Its population declines due to the frequent occurring poaching. Thus, Gongshan Muntjac is listed as Endangered

地理分布 Geographical Distribution

国内分布 Domestic Distribution
云南、西藏 / Yunnan, Tibet (Xizang)
世界分布 World Distribution
中国 / China
分布标注 Distribution Note
特有种 / Endemic

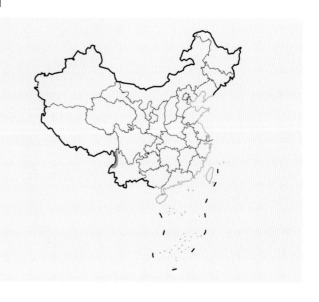

国内分布图
Map of Domestic Distribution

种群 Population

种群数量 Population Size	未知 / Unknown
种群趋势 Population Trend	下降 / Decreasing

生境与生态系统 Habitat(s) and Ecosystem(s)

生　　境 Habitat(s)	森林 / Forest
生态系统 Ecosystem(s)	森林生态系统 / Forest Ecosystem

威胁 Threat(s)

主要威胁 Major Threat(s)	狩猎、开垦 / Hunting, Land Reclaimed for Farming

保护级别与保护行动 Protection Category and Conservation Action(s)

国家重点保护野生动物等级 (2021) Category of National Key Protected Wild Animals (2021)	二级 / Category II
IUCN 红色名录 (2020-2) IUCN Red List (2020-2)	数据缺乏 / DD
CITES 附录 (2019) CITES Appendix (2019)	无 / NA
保护行动 Conservation Action(s)	未知 / Unknown

相关文献 Relevant References

Burgin *et al.*, 2018; Jiang *et al.*（蒋志刚等）, 2017; Smith *et al.*（史密斯等）, 2009; Wang（王应祥）, 2003; Ma *et al.*（马世来等）, 1990

贡山麂 *Muntiacus gongshanensis*

马鹿
Cervus canadensis

濒危 EN B1ab(i, ii, iii)+2ab(i, ii, iii)

| 数据缺乏 DD | 无危 LC | 近危 NT | 易危 VU | 濒危 EN | 极危 CR | 区域灭绝 RE | 野外灭绝 EW | 灭绝 EX |

分类地位 Taxonomic Status

动物界 Animalia	脊索动物门 Chordata	哺乳纲 Mammalia	偶蹄目 Artiodactyla	鹿科 Cervidae

学 名 Scientific Name	*Cervus canadensis*
命 名 人 Species Authority	Milne-Edwards, 1867
英 文 名 English Name(s)	Elk
同物异名 Synonym(s)	东北马鹿，加拿大马鹿，Canadian Elk, Manchurian Wapiti
种下单元评估 Infra-specific Taxa Assessed	马鹿在东亚有分布。在中国，马鹿的亚种包括满洲里马鹿 (*C. c. xanthopygus*)、阿尔泰马鹿 (*C. c. sibiricus*)、天山马鹿 (*C. c. songaricus*) 和分布在阿拉善、陕西和（转下页）

评估信息 Assessment Information

评 估 年 份 Year Assessed	2020
评 定 人 Assessor(s)	蒋志刚 / Zhigang Jiang
审 定 人 Reviewer(s)	张明海 / Minghai Zhang
其他贡献人 Other Contributor(s)	李立立、丁晨晨 / Lili Li, Chenchen Ding

理由 Justification: 由于偷猎与生境丧失，马鹿在原分布区多数区域种群数量锐减，在一些原来分布的区域已经绝迹。基于其种群下降和栖息地状态，马鹿被列为濒危等级 / Due to poaching and habitat loss, Elk populations in the wild are decimated in most areas and are now extinct in some areas of its original range. Based on its population decline and habitat status, it is listed as Endangered

地理分布 Geographical Distribution

国内分布 Domestic Distribution	内蒙古、黑龙江、吉林、新疆 / Inner Mongolia (Nei Mongol), Heilongjiang, Jilin, Xinjiang
世界分布 World Distribution	中国、俄罗斯 / China, Russia
分布标注 Distribution Note	非特有种 / Non-endemic

国内分布图
Map of Domestic Distribution

种群 Population

种群数量 Population Size	未知 / Unknown
种群趋势 Population Trend	下降 / Decreasing

生境与生态系统 Habitat (s) and Ecosystem (s)

生　　境 Habitat(s)	温带阔叶林和针叶林 / Temperate Broad-leaved Forest and Coniferous Forest
生态系统 Ecosystem(s)	森林生态系统 / Forest Ecosystem

威胁 Threat (s)

主要威胁 Major Threat(s)	猎杀、生境丧失 / Hunting, Habitat Loss

保护级别与保护行动 Protection Category and Conservation Action (s)

国家重点保护野生动物等级 (2021) Category of National Key Protected Wild Animals (2021)	二级 (仅限野外种群) / Category II (Only Wild Population)
IUCN红色名录(2020-2) IUCN Red List (2020-2)	未列入 / NA
CITES 附录 (2019) CITES Appendix (2019)	无 / NA
保护行动 Conservation Action(s)	自然保护区内种群得到保护 / Populations in nature reserves are protected

相关文献 Relevant References

Burgin *et al*., 2018; Jiang *et al*. (蒋志刚等), 2017, 2015a; Groves and Grubb, 2011; Wilson and Mittermeier, 2011

(接上页)
内蒙古南部的阿拉善马鹿 (*C. c. alashanicus*) (Ludt, 2004) / Wapiti is found in eastern Asia. In China, its subspecies include Manchurian Wapiti (*C. c. xanthopygus*), Altai Wapiti (*C. c. sibiricus*), Tian Shan Wapiti (*C. c. songaricus*) and the Alashan Wapiti (*C. c. alashanicus*) that distributes in Gansu, Shaanxi and southern Inner Mongolia (Nei Mongol) (Ludt, 2004)

马鹿 *Cervus canadensis* 　　蒋志刚 摄　By Zhigang Jiang

梅花鹿
Cervus nippon

濒危 EN A1acd; B1ab(i, ii, iii)+2ab(i, ii, iii)

| 数据缺乏 DD | 无危 LC | 近危 NT | 易危 VU | **濒危 EN** | 极危 CR | 区域灭绝 RE | 野外灭绝 EW | 灭绝 EX |

分类地位 Taxonomic Status

动物界 Animalia	脊索动物门 Chordata	哺乳纲 Mammalia	偶蹄目 Artiodactyla	鹿科 Cervidae

学 名 Scientific Name	*Cervus nippon*
命 名 人 Species Authority	Swinhoe, 1864
英 文 名 English Name(s)	Sika Deer
同物异名 Synonym(s)	Japanese Deer, Spotted Deer, *Cervus hortulorum*
种下单元评估 Infra-specific Taxa Assessed	中国有6个梅花鹿亚种。其中，华北亚种 (*Cervus nippon mandarins*)、山西亚种 (*C. n. grassianua*) 已经灭绝，台湾亚种 (*C. n. taiouanns*) 野外种群已经灭绝，（转下页）

评估信息 Assessment Information

评估年份 Year Assessed	2020
评 定 人 Assessor(s)	蒋志刚 / Zhigang Jiang
审 定 人 Reviewer(s)	王小明、蒋学龙、李言阔、李玉春 / Xiaoming Wang, Xuelong Jiang, Yankuo Li, Yuchun Li
其他贡献人 Other Contributor(s)	李立立、丁晨晨 / Lili Li, Chenchen Ding

理由 Justification: 尽管梅花鹿养殖业在中国蓬勃发展，但是野生梅花鹿却陷入了"驯化的悲剧"。由于栖息地破坏、猎捕和人类干扰，东北、西南和江南的梅花鹿种群小而分散。基于野生梅花鹿种群及其栖息地状况，梅花鹿列为濒危等级 / Although Sika Deer farming is booming in China, Sika Deer is still trapped in "Tragedy of Domestication". Wild Sika Deer populations live in small, scattered and isolated habitats, including in north-eastern, southern and south-western China regions. Thus, based on its population and habitat status, Sika Deer is classified as Endangered

地理分布 Geographical Distribution

国内分布 Domestic Distribution	江西、浙江、安徽、吉林、黑龙江、四川、甘肃、台湾 / Jiangxi, Zhejiang, Anhui, Jilin, Heilongjiang, Sichuan, Gansu, Taiwan
世界分布 World Distribution	中国、俄罗斯 / China, Russia
分布标注 Distribution Note	非特有种 / Non-endemic

国内分布图
Map of Domestic Distribution

种群 Population

种群数量 Population Size	7,000 只 / 7,000 individuals
种群趋势 Population Trend	增长 / Increasing

生境与生态系统 Habitat(s) and Ecosystem(s)

生　　境 Habitat(s)	落叶阔叶林 / Deciduous and Broad-leaved Forest
生态系统 Ecosystem(s)	森林生态系统 / Forest Ecosystem

威胁 Threat(s)

主要威胁 Major Threat(s)	生境丧失 / Habitat Loss

保护级别与保护行动 Protection Category and Conservation Action(s)

国家重点保护野生动物等级 (2021) Category of National Key Protected Wild Animals (2021)	一级 (仅限野外种群) / Category I (Only Wild Population)
IUCN 红色名录 (2020-2) IUCN Red List (2020-2)	无危 / LC
CITES 附录 (2019) CITES Appendix (2019)	无 / NA
保护行动 Conservation Action(s)	自然保护区内种群得到保护 / Populations in nature reserves are protected

相关文献 Relevant References

Burgin *et al.*, 2018; Jiang *et al.* (蒋志刚等), 2017; Smith and Xie, 2013; Wilson and Mittermeier, 2012; Guo (郭延蜀), 2000

(接上页)
但正在垦丁公园恢复野生种群。东北亚种 (*C. n. hortulorum*)、四川亚种 (*C. n. sichuanicus*) 和华南亚种 (*C. n. kopschi*) 存在野外种群，但 3 个亚种相互隔离，且分布范围不断缩小，已经到达了灭绝的边缘 (Smith and Xie, 1992; Ohitaishi and Gao, 1990)。梅花鹿华南亚种，是我国地理分布最东和最南的野生种群，具有重要的科研价值 / There are six subspecies of Sika Deer in China. Among them, the northern China subspecies (*Cervus nippon mandarinus*) and Shanxi subspecies (*C. n. grassianua*) have gone extinct, and the Taiwan subspecies (*C. n. taiouanus*) is extinct in the wild but is being restored in Kenting Park. On the other hand, the northeastern subspecies (*C. n. hortulorum*), Sichuan subspecies (*C. n. sichuanicus*) and southern China subspecies (*C. n. kopschi*) are still exist in the wild, however the three subspecies are isolated and with distribution ranges shrinking, bringing them close to the brink of extinction (Smith and Xie, 1992; Ohitaishi and Gao, 1990). The southern China subspecies of Sika Deer is the easternmost and southernmost wild population in China and has important scientific research value

梅花鹿 *Cervus nippon*

西藏马鹿
Cervus wallichii

濒危　EN B1ab(i, ii, iii)+2ab(i, ii, iii)

| 数据缺乏 DD | 无危 LC | 近危 NT | 易危 VU | **濒危 EN** | 极危 CR | 区域灭绝 RE | 野外灭绝 EW | 灭绝 EX |

分类地位 Taxonomic Status

动物界 Animalia	脊索动物门 Chordata	哺乳纲 Mammalia	偶蹄目 Artiodactyla	鹿科 Cervidae

学　　名 Scientific Name	*Cervus wallichii*
命　名　人 Species Authority	G. Cuvier, 1823
英　文　名 English Name(s)	Tibetan Red Deer
同物异名 Synonym(s)	Central Asian Red Deer, *Cervus canadensis wallichi*
种下单元评估 Infra-specific Taxa Assessed	西藏马鹿有4个亚种，其中西藏亚种（*C. w. wallichii*）、川西亚种（*C. w. maceilli*）和甘肃亚种（*C. w. kansuensis*）在中国有分布（Mattioli, 2011）/ The Tibetan（转下页）

评估信息 Assessment Information

评估年份 Year Assessed	2020
评　定　人 Assessor(s)	蒋志刚 / Zhigang Jiang
审　定　人 Reviewer(s)	胡慧建、刘务林、邹二虎 / Huijian Hu, Wulin Liu, Erhu Gao
其他贡献人 Other Contributor(s)	李立立、丁晨晨 / Lili Li, Chenchen Ding

理由 Justification: 野生西藏马鹿种群在其分布区多数区域数量下降。因此，将其列为濒危 / The population of Tibetan Red Deer in the wild has been reduced in most areas compare to its original range. Thus, it is listed as Endangered

地理分布 Geographical Distribution

国内分布 Domestic Distribution	西藏、青海、四川、甘肃 / Tibet (Xizang), Qinghai, Sichuan, Gansu
世界分布 World Distribution	中国、印度、巴基斯坦 / China, India, Pakistan
分布标注 Distribution Note	非特有种 / Non-endemic

国内分布图
Map of Domestic Distribution

种群 Population

种群数量 Population Size	未知 / Unknown
种群趋势 Population Trend	下降 / Decreasing

生境与生态系统 Habitat (s) and Ecosystem (s)

生境 Habitat(s)	森林 / Forest
生态系统 Ecosystem(s)	森林生态系统、草地生态系统 / Forest Ecosystem, Grassland Ecosystem

威胁 Threat (s)

主要威胁 Major Threat(s)	猎杀、生境丧失 / Hunting, Habitat Loss

保护级别与保护行动 Protection Category and Conservation Action (s)

国家重点保护野生动物等级 (2021) Category of National Key Protected Wild Animals (2021)	一级 / Category I
IUCN 红色名录 (2020-2) IUCN Red List (2020-2)	未列入 / NA
CITES 附录 (2019) CITES Appendix (2019)	无 / NA
保护行动 Conservation Action(s)	自然保护区内种群得到保护 / Populations in nature reserves are protected

相关文献 Relevant References

Burgin *et al.*, 2018; Jiang *et al.* (蒋志刚等), 2017, 2015a; Wilson and Mittermeier, 2012; Groves and Grubb, 2011; Mattioli, 2011

(接上页)
Red Deer comprises 4 subspecies, *C. w. wallichii*, *C. w. maceilli* and *C. w. kansuensis* are found in China (Mattioli, 2011)

西藏马鹿 *Cervus wallichii*

白唇鹿
Przewalskium albirostris

中国生物多样性 红色名录
China's Red List of Biodiversity

濒危 EN B1ab(i, ii, iii)+2ab(i, ii, iii)

| 数据缺乏 DD | 无危 LC | 近危 NT | 易危 VU | **濒危 EN** | 极危 CR | 区域灭绝 RE | 野外灭绝 EW | 灭绝 EX |

分类地位 Taxonomic Status

动物界 Animalia	脊索动物门 Chordata	哺乳纲 Mammalia	偶蹄目 Artiodactyla	鹿科 Cervidae

学名 Scientific Name	*Przewalskium albirostris*
命名人 Species Authority	Przewalski, 1883
英文名 English Name(s)	White-lipped Deer
同物异名 Synonym(s)	Thorold's deer; *albirostris* (Przewalski, 1883); *sellatus* (Przewalski, 1883); *thoroldi* (Blanford, 1893)
种下单元评估 Infra-specific Taxa Assessed	无 / None

评估信息 Assessment Information

评估年份 Year Assessed	2020
评定人 Assessor(s)	蒋志刚 / Zhigang Jiang
审定人 Reviewer(s)	胡慧建 / Huijian Hu
其他贡献人 Other Contributor(s)	李立立、丁晨晨 / Lili Li, Chenchen Ding

理由 Justification: 白唇鹿是青藏高原特有种，近年种群数量趋于稳定，但基于其生境的破碎程度，白唇鹿评定为濒危等级 / The White-lipped Deer is endemic to the Qinghai-Tibet (Xizang) Plateau. Its populations has been relatively stable in recent years. Due to the fragmentation of its habitat, White-lipped Deer is listed as Endangered

地理分布 Geographical Distribution

国内分布 Domestic Distribution	青海、四川、云南、西藏、甘肃 / Qinghai, Sichuan, Yunnan, Tibet (Xizang), Gansu
世界分布 World Distribution	中国 / China
分布标注 Distribution Note	特有种 / Endemic

国内分布图
Map of Domestic Distribution

种群 Population

种群数量 Population Size	37,000 只 / 37,000 individuals
种群趋势 Population Trend	稳定 / Stable

生境与生态系统 Habitat(s) and Ecosystem(s)

生境 Habitat(s)	泰加林、灌丛、草甸 / Taiga Forest, Shrubland, Meadow
生态系统 Ecosystem(s)	森林生态系统、灌丛生态系统、草地生态系统 / Forest Ecosystem, Shrubland Ecosystem, Grassland Ecosystem

威胁 Threat(s)

主要威胁 Major Threat(s)	家畜放牧、狩猎 / Livestock Ranching, Hunting

保护级别与保护行动 Protection Category and Conservation Action(s)

国家重点保护野生动物等级 (2021) Category of National Key Protected Wild Animals (2021)	一级 / Category I
IUCN 红色名录 (2020-2) IUCN Red List (2020-2)	易危 / VU
CITES 附录 (2019) CITES Appendix (2019)	无 / NA
保护行动 Conservation Action(s)	自然保护区内种群得到保护 / Populations in nature reserves are protected

相关文献 Relevant References

Burgin *et al.*, 2018; Cui *et al.* (崔绍朋等), 2018; Jiang *et al.* (蒋志刚等), 2017; You *et al.* (游章强等), 2014; Smith *et al.* (史密斯等), 2009; Wu and Pei (吴家炎和裴俊峰), 2007

白唇鹿 *Przewalskium albirostris*　　　　蒋志刚 摄　By Zhigang Jiang

普氏原羚
Procapra przewalskii

濒危 EN A1acd; B1ab(i, ii, iii)

| 数据缺乏 DD | 无危 LC | 近危 NT | 易危 VU | **濒危 EN** | 极危 CR | 区域灭绝 RE | 野外灭绝 EW | 灭绝 EX |

分类地位 Taxonomic Status

动物界 Animalia	脊索动物门 Chordata	哺乳纲 Mammalia	偶蹄目 Artiodactyla	牛科 Bovidae

学 名 Scientific Name	*Procapra przewalskii*
命 名 人 Species Authority	Büchner, 1891
英 文 名 English Name(s)	Przewalski's Gazelle
同物异名 Synonym(s)	无 / None
种下单元评估 Infra-specific Taxa Assessed	普氏原羚有 3 个亚种 *P. p. przewalskii*，*P. p. diversicornis* 和 *P. p. wayu*。亚种 *P. p. diversicornis* 已经在其历史分布区——青海湖流域灭绝，*P. p. przewalskii* 扩散至 (转下页)

评估信息 Assessment Information

评 估 年 份 Year Assessed	2020
评 定 人 Assessor(s)	蒋志刚 / Zhigang Jiang
审 定 人 Reviewer(s)	胡慧建、李忠秋、李春旺 / Huijian Hu, Zhongqiu Li, Chunwang Li
其他贡献人 Other Contributor(s)	李立立、丁晨晨 / Lili Li, Chenchen Ding

理由 Justification: 普氏原羚有三个亚种：*P. p. przewalskii*，*P. p. wayu* 和 *P. p. diversicornis*。其中 *P. p. diversicornis* 已经灭绝。*P. p. wayu* 的发生区面积不到 $100 km^2$。*P. p. przewalskii* 仅分布在青海湖周边地区的 11 个种群之中，近年来种群数量上升，但其面临的威胁因子尚未消除。基于其种群与栖息地现状，将普氏原羚列为濒危等级 / Przewalski's Gazelle has three subspecies, namely: *P. p. przewalskii*, *P. p. wayu* and *P. p. diversicornis*. *P. p. diversicornis* is extinct. *P. p. wayu* has an area of occurrence less than $100 km^2$. *P. p. przewalskii* has 11 small populations in fragmented habitat. Though its populations are increasing, but the threats to its survival still remain. Thus, based on its population and habitat status, Przewalski's Gazelle is listed as Endangered

地理分布 Geographical Distribution

国内分布 Domestic Distribution
青海 / Qinghai
世界分布 World Distribution
中国 / China
分布标注 Distribution Note
特有种 / Endemic

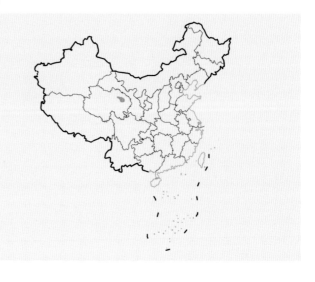

国内分布图
Map of Domestic Distribution

种群 Population

种群数量 Population Size	1,400 只 / 1,400 individuals
种群趋势 Population Trend	稳定 / Stable

生境与生态系统 Habitat (s) and Ecosystem (s)

生　　境 Habitat(s)	草地 / Steppe
生态系统 Ecosystem(s)	高寒草地生态系统 / Apline Grassland Ecosystem

威胁 Threat (s)

主要威胁 Major Threat(s)	栖息地改变、草原围栏 / Habitat Alteration, Fencing

保护级别与保护行动 Protection Category and Conservation Action (s)

国家重点保护野生动物等级 (2021) Category of National Key Protected Wild Animals (2021)	一级 / Category I
IUCN 红色名录 (2020-2) IUCN Red List (2020-2)	濒危 / EN
CITES 附录 (2019) CITES Appendix (2019)	无 / NA
保护行动 Conservation Action(s)	自然保护区内种群得到保护 / Populations in nature reserves are protected

相关文献 Relevant References

Burgin *et al.*, 2018; Ping *et al.*（平晓鸽等）, 2018; Jiang *et al.*（蒋志刚等）, 2015a; Zhang *et al.*, 2014, 2013b; Hu *et al.*, 2013; Turghan *et al.*, 2013; Li *et al.*, 2012; Yang and Jiang, 2011

（接上页）
青海湖地区，分布在几个彼此隔离的栖息地斑块之中，处境濒危。*P. p. wayu* 是买尔旦·吐尔干等于 2013 年发现的新亚种，其占有区面积狭小，极危 / Przewalski's Gazelle has three subspecies, namely: *P. p. przewalskii*, *P. p. diversicornis* and *P. p. wayu*. *P. p. diversicornis* is extinct in its historical range—the Qinghai Lake Basin. *P. p. przewalskii* dispersed into the Qinghai Lake Basin and live in several habitat patches; *P. p. przewalskii* is Endangered. Turgan *et al.* discovered a new subspecies—*P. p. wayu* in 2013, its area of occupancy is rather small, it is critically Endangered

普氏原羚 *Procapra przewalskii*　　　　　　　　　　　　　　蒋志刚 摄　By Zhigang Jiang

中国生物多样性红色名录

赤斑羚
Naemorhedus baileyi

濒危 EN B1ab(i, ii, iii)

| 数据缺乏 DD | 无危 LC | 近危 NT | 易危 VU | **濒危 EN** | 极危 CR | 区域灭绝 RE | 野外灭绝 EW | 灭绝 EX |

分类地位 Taxonomic Status

动物界 Animalia	脊索动物门 Chordata	哺乳纲 Mammalia	偶蹄目 Artiodactyla	牛科 Bovidae

学名 Scientific Name	*Naemorhedus baileyi*
命名人 Species Authority	Pocock, 1914
英文名 English Name(s)	Red Goral
同物异名 Synonym(s)	*Nemorhaedus baileyi*
种下单元评估 Infra-specific Taxa Assessed	无 / None

评估信息 Assessment Information

评估年份 Year Assessed	2020
评定人 Assessor(s)	蒋志刚 / Zhigang Jiang
审定人 Reviewer(s)	蒋学龙 / Xuelong Jiang
其他贡献人 Other Contributor(s)	李立立、丁晨晨 / Lili Li, Chenchen Ding

理由 Justification: 赤斑羚分布于云南西部高黎贡山北段及西藏东南部地区，有一定的种群数量，但常遭猎杀，还受到当地居民活动的干扰，生存受到威胁。因此，基于其种群与栖息地现状，将赤斑羚列为濒危等级 / The Red Goral is distributed in Gaoligong Shan of west Yunnan and in southeast Tibet (Xizang), and has certain populations. Local residents hunted the Red Goral, causing its population decline. In addition, due to human encroachment, the habitat area is also deteriorating, which threatens the survival of this species. Thus, based on its population and habitat status, it is listed as Endangered

地理分布 Geographical Distribution

国内分布 Domestic Distribution
云南、西藏 / Yunnan, Tibet (Xizang)
世界分布 World Distribution
中国、印度、缅甸 / China, India, Myanmar
分布标注 Distribution Note
非特有种 / Non-endemic

国内分布图
Map of Domestic Distribution

种群 Population

种群数量 Population Size	< 1,500 只 / < 1,500 individuals
种群趋势 Population Trend	下降 / Decreasing

生境与生态系统 Habitat (s) and Ecosystem (s)

生　　境 Habitat(s)	森林、峭壁、灌丛、草甸 / Forest, Cliff, Shrubland, Meadow
生态系统 Ecosystem(s)	森林生态系统、灌丛生态系统、草地生态系统 / Forest Ecosystem, Shrubland Ecosystem, Grassland Ecosystem

威胁 Threat (s)

主要威胁 Major Threat(s)	狩猎、栖息地丧失 / Hunting, Habitat Loss

保护级别与保护行动 Protection Category and Conservation Action (s)

国家重点保护野生动物等级 (2021) Category of National Key Protected Wild Animals (2021)	一级 / Category I
IUCN 红色名录 (2020-2) IUCN Red List (2020-2)	易危 / VU
CITES 附录 (2019) CITES Appendix (2019)	I
保护行动 Conservation Action(s)	无 / None

相关文献 Relevant References

Burgin *et al.*, 2018; Jiang *et al.*（蒋志刚等），2017; Hu *et al.*（胡一鸣等），2014; Xiong *et al.*, 2013; Wilson and Mittermeier, 2012; Smith *et al.*（史密斯等），2009; Huang *et al.*（黄薇等），2008

赤斑羚 *Naemorhedus baileyi*

喜马拉雅斑羚
Naemorhedus goral

濒危 EN B1ab(i, ii, iii)

| 数据缺乏 DD | 无危 LC | 近危 NT | 易危 VU | **濒危 EN** | 极危 CR | 区域灭绝 RE | 野外灭绝 EW | 灭绝 EX |

分类地位 Taxonomic Status

动物界 Animalia	脊索动物门 Chordata	哺乳纲 Mammalia	偶蹄目 Artiodactyla	牛科 Bovidae

学　名 Scientific Name	*Naemorhedus goral*
命名人 Species Authority	Hardwicke, 1825
英文名 English Name(s)	Himalayan Goral
同物异名 Synonym(s)	*Nemorhaedus goral*
种下单元评估 Infra-specific Taxa Assessed	无 / None

评估信息 Assessment Information

评估年份 Year Assessed	2020
评定人 Assessor(s)	蒋志刚 / Zhigang Jiang
审定人 Reviewer(s)	蒋学龙、陈辈乐、李飞 / Xuelong Jiang, Bosco P. L. Chan, Fei Li
其他贡献人 Other Contributor(s)	李立立、丁晨晨 / Lili Li, Chenchen Ding

理由 Justification: 喜马拉雅斑羚分布在喜马拉雅山区，分布区狭窄，受到当地居民的猎杀，导致其种群数量下降。另外，人类活动干扰了其栖息地，威胁着该种的生存。基于栖息地与种群数量，将喜马拉雅斑羚列为濒危 / Himalayan Goral is found only within a narrow range in the Himalaya Mountains. Local people have long hunted this species, leading its populations to decline. Furthermore, its habitat has been influenced due to human encroachment, and the survival of this species is threatened. Thus, based on its population and habitat status, Himalayan Goral is listed as Endangered

地理分布 Geographical Distribution

国内分布 Domestic Distribution
西藏 / Tibet (Xizang)
世界分布 World Distribution
中国；南亚 / China; South Asia
分布标注 Distribution Note
非特有种 / Non-endemic

国内分布图
Map of Domestic Distribution

种群 Population

种群数量 Population Size	未知 / Unknown
种群趋势 Population Trend	下降 / Decreasing

生境与生态系统 Habitat (s) and Ecosystem (s)

生　　境 Habitat(s)	峭壁、森林 / Cliff, Forest
生态系统 Ecosystem(s)	森林生态系统 / Forest Ecosystem

威胁 Threat (s)

主要威胁 Major Threat(s)	人类干扰、狩猎 / Human Disturbance, Hunting

保护级别与保护行动 Protection Category and Conservation Action (s)

国家重点保护野生动物等级 (2021) Category of National Key Protected Wild Animals (2021)	一级 / Category I
IUCN 红色名录 (2020-2) IUCN Red List (2020-2)	近危 / NT
CITES 附录 (2019) CITES Appendix (2019)	I
保护行动 Conservation Action(s)	自然保护区内种群得到保护 / Populations in nature reserves are protected

相关文献 Relevant References

Burgin *et al.*, 2018; Jiang *et al.*（蒋志刚等）, 2017; Wilson and Mittermeier, 2012; Groves and Grubb, 2011; Smith *et al.*（史密斯等）, 2009; Pan *et al.*（潘清华等）, 2007; Zhang（张荣祖）, 1997

喜马拉雅斑羚 *Naemorhedus goral*

天山盘羊
Ovis karelini

濒危　EN B1ab(i, ii, iii)

分类地位 Taxonomic Status

动物界 Animalia	脊索动物门 Chordata	哺乳纲 Mammalia	偶蹄目 Artiodactyla	牛科 Bovidae
学　　名 Scientific Name		*Ovis karelini*		
命 名 人 Species Authority		Severtzov, 1873		
英 文 名 English Name(s)		Tianshan Argali		
同物异名 Synonym(s)		*Ovis ammon karelini*		
种下单元评估 Infra-specific Taxa Assessed		无 / None		

评估信息 Assessment Information

评 估 年 份 Year Assessed	2020
评 定 人 Assessor(s)	蒋志刚 / Zhigang Jiang
审 定 人 Reviewer(s)	初红军、杨维康 / Hongjun Chu, Weikang Yang
其他贡献人 Other Contributor(s)	李立立、丁晨晨 / Lili Li, Chenchen Ding

理由 Justification: 据杨维康等在新疆北部天山一带的调查，由于偷猎、过度放牧、开矿和草原围栏建设，以前天山盘羊和北山羊分布的区域，目前只见北山羊，而天山盘羊消失了。自2008年起，随着偷猎加剧，在东天山木垒盘羊国际狩猎场盘羊遇见率锐减。基于其种群下降速率，将天山盘羊定为濒危等级 / According to survey of Weikang Yang *et al.* in the Tian Shan area, northern Xinjiang, in the range where Tianshan Argali lived sympatrically with Ibex, now, only Ibex is found while Tianshan Argali disappeared, due to poaching, overgrazing, mining and the construction of grassland fences. Since 2008, as poaching increased, the encounter rate of Tianshan Argali in the eastern Tian Shan mountains has decreased sharply. Based on its population decline rate, Tianshan Argali is listed as Endangered

地理分布 Geographical Distribution

国内分布 Domestic Distribution
新疆 / Xinjiang
世界分布 World Distribution
中国；中亚 / China; Central Asia
分布标注 Distribution Note
非特有种 / Non-endemic

国内分布图
Map of Domestic Distribution

种群 Population

种群数量 Population Size	未知 / Unknown
种群趋势 Population Trend	下降 / Decreasing

生境与生态系统 Habitat (s) and Ecosystem (s)

生　　境 Habitat(s)	草地 / Grassland
生态系统 Ecosystem(s)	草地生态系统 / Grassland Ecosystem

威胁 Threat (s)

主要威胁 Major Threat(s)	家畜放牧压力增加、人类干扰、围栏和偷猎 / Increasing Livestock Grazing Pressure, Human Disturbance, Fencing, Poaching

保护级别与保护行动 Protection Category and Conservation Action (s)

国家重点保护野生动物等级 (2021) Category of National Key Protected Wild Animals (2021)	二级 / Category II
IUCN 红色名录 (2020-2) IUCN Red List (2020-2)	未列入 / NA
CITES 附录 (2019) CITES Appendix (2019)	II
保护行动 Conservation Action(s)	在自然保护区内的种群得到保护；列入《迁徙物种保护公约》的"迁徙物种行动计划" / Populations in nature reserves are under protection; included in the *Convention on Migratory Species* (CMS)—Species Action Plan

相关文献 Relevant References

Jiang *et al*. (蒋志刚等), 2017; Wilson and Mittermeier, 2012; Groves and Grubb, 2011

天山盘羊 *Ovis karelini*　　　　　　　　　　　　　　　　　　邱杰 摄　By Jie Di

喜马拉雅鬣羚
Capricornis thar

濒危 EN A1acd; B1ab(i, ii, iii)

| 数据缺乏 DD | 无危 LC | 近危 NT | 易危 VU | 濒危 EN | 极危 CR | 区域灭绝 RE | 野外灭绝 EW | 灭绝 EX |

分类地位 Taxonomic Status

动物界 Animalia	脊索动物门 Chordata	哺乳纲 Mammalia	偶蹄目 Artiodactyla	牛科 Bovidae
学 名 Scientific Name		*Capricornis thar*		
命 名 人 Species Authority		Hodgson, 1831		
英 文 名 English Name(s)		Himalayan Serow		
同物异名 Synonym(s)		*Capricornis sumatraensis* (Hodgson, 1831) subsp. *thar*		
种下单元评估 Infra-specific Taxa Assessed		无 / None		

评估信息 Assessment Information

评估年份 Year Assessed	2020
评 定 人 Assessor(s)	蒋志刚 / Zhigang Jiang
审 定 人 Reviewer(s)	蒋学龙、陈辈乐、李飞 / Xuelong Jiang, Bosco P. L. Chan, Fei Li
其他贡献人 Other Contributor(s)	李立立、丁晨晨 / Lili Li, Chenchen Ding

理由 Justification: 喜马拉雅鬣羚分布区狭窄，种群数量较少，面临放牧、偷猎和生境破坏的压力。基于其种群大小与栖息地状态，喜马拉雅鬣羚被列为濒危等级 / The distribution area of the Himalayan Serow is narrow and its population density is sparse. Its survival is threatened by livestock grazing, poaching and habitat destruction. Thus, based on its population size and habitat status, Himalayan Serow is listed as Endangered

地理分布 Geographical Distribution

国内分布 Domestic Distribution
西藏 / Tibet (Xizang)
世界分布 World Distribution
中国；南亚 / China; South Asia
分布标注 Distribution Note
非特有种 / Non-endemic

国内分布图
Map of Domestic Distribution

种群 Population

种群数量 Population Size	未知 / Unknown
种群趋势 Population Trend	下降 / Decreasing

生境与生态系统 Habitat (s) and Ecosystem (s)

生 境 Habitat(s)	森林 / Forest
生态系统 Ecosystem(s)	森林生态系统 / Forest Ecosystem

威胁 Threat (s)

主要威胁 Major Threat(s)	狩猎、耕种、伐木、火灾 / Hunting, Plantation, Logging, Fire

保护级别与保护行动 Protection Category and Conservation Action (s)

国家重点保护野生动物等级 (2021) Category of National Key Protected Wild Animals (2021)	一级 / Category I
IUCN 红色名录 (2020-2) IUCN Red List (2020-2)	近危 / NT
CITES 附录 (2019) CITES Appendix (2019)	I
保护行动 Conservation Action(s)	无 / None

相关文献 Relevant References

Burgin *et al*., 2018; Jiang *et al*. (蒋志刚等), 2017; Wilson and Mittermeier, 2012; Groves and Grubb, 2011; Smith *et al*. (史密斯等), 2009

喜马拉雅鬣羚 *Capricornis thar*

北太平洋露脊鲸
Eubalaena japonica

濒危 EN D

| 数据缺乏 DD | 无危 LC | 近危 NT | 易危 VU | **濒危 EN** | 极危 CR | 区域灭绝 RE | 野外灭绝 EW | 灭绝 EX |

分类地位 Taxonomic Status

动物界 Animalia	脊索动物门 Chordata	哺乳纲 Mammalia	鲸目 Cetacea	露脊鲸科 Balaenidae
学名 Scientific Name		*Eubalaena japonica*		
命名人 Species Authority		Lacépède, 1818		
英文名 English Name(s)		North Pacific Right Whale		
同物异名 Synonym(s)		*Eubalaena glacialis* (P. L. S. Müller, 1776)		
种下单元评估 Infra-specific Taxa Assessed		无 / None		

评估信息 Assessment Information

评估年份 Year Assessed	2020
评定人 Assessor(s)	周开亚 / Kaiya Zhou
审定人 Reviewer(s)	张先锋、王克雄、王丁、祝茜、蒋志刚 / Xianfeng Zhang, Kexiong Wang, Ding Wang, Qian Zhu, Zhigang Jiang
其他贡献人 Other Contributor(s)	李立立、丁晨晨 / Lili Li, Chenchen Ding

理由 Justification: 在19世纪和20世纪被捕鲸船捕杀的北太平洋露脊鲸超过15,000头，种群接近灭绝。因此，基于其种群下降速率，北太平洋露脊鲸被列为濒危等级 / In the 19th and 20th centuries, more than 15,000 North Pacific Right Whales were killed by whaling ships. The population is near extinction. Based on its population decline rate, North Pacific Right Whale is listed as Endangered

地理分布 Geographical Distribution

国内分布 Domestic Distribution
黄海、台湾海峡 / Yellow Sea, Taiwan Strait
世界分布 World Distribution
北太平洋 / North Pacific Ocean
分布标注 Distribution Note
非特有种 / Non-endemic

国内分布图
Map of Domestic Distribution

种群 Population

种群数量 Population Size	未知 / Unknown
种群趋势 Population Trend	下降 / Decreasing

生境与生态系统 Habitat (s) and Ecosystem (s)

生　　境 Habitat(s)	海洋 / Ocean
生态系统 Ecosystem(s)	海洋生态系统 / Ocean Ecosystem

威胁 Threat (s)

主要威胁 Major Threat(s)	人类活动干扰 / Human Disturbance

保护级别与保护行动 Protection Category and Conservation Action (s)

国家重点保护野生动物等级 (2021) Category of National Key Protected Wild Animals (2021)	一级 / Category I
IUCN 红色名录 (2020-2) IUCN Red List (2020-2)	濒危 / EN
CITES 附录 (2019) CITES Appendix (2019)	I
保护行动 Conservation Action(s)	无 / None

相关文献 Relevant References

Burgin *et al.*, 2018; Kenney, 2018; Jiang *et al.* (蒋志刚等), 2017; Zhou (周开亚), 2008, 2004; Shi and Wang (施友仁和王秀玉), 1978

北太平洋露脊鲸 *Eubalaena japonica*

塞鲸
Balaenoptera borealis

濒危 EN

| 数据缺乏 DD | 无危 LC | 近危 NT | 易危 VU | 濒危 EN | 极危 CR | 区域灭绝 RE | 野外灭绝 EW | 灭绝 EX |

分类地位 Taxonomic Status

动物界 Animalia	脊索动物门 Chordata	哺乳纲 Mammalia	鲸目 Cetacea	须鲸科 Balaenopteridae

学名 Scientific Name	*Balaenoptera borealis*
命名人 Species Authority	Lesson, 1828
英文名 English Name(s)	Sei Whale
同物异名 Synonym(s)	Lesser Fin Whale
种下单元评估 Infra-specific Taxa Assessed	无 / None

评估信息 Assessment Information

评估年份 Year Assessed	2020
评定人 Assessor(s)	周开亚 / Kaiya Zhou
审定人 Reviewer(s)	张先锋、王克雄、王丁、祝茜、蒋志刚 / Xianfeng Zhang, Kexiong Wang, Ding Wang, Qian Zhu, Zhigang Jiang
其他贡献人 Other Contributor(s)	李立立、丁晨晨 / Lili Li, Chenchen Ding

理由 Justification: 在 19 世纪和 20 世纪，被捕鲸船捕杀的塞鲸超过 110,000 头，塞鲸种群下降了 80%。因此，基于其种群下降速率，塞鲸被列为濒危等级 / In the 19th and 20th centuries, the number of Sei Whale captured and killed by whaling ships exceeded 110,000, and the population of Sei Whale declined 80%. Based on its population decline rate, Sei Whale is therefore listed as Endangered

地理分布 Geographical Distribution

国内分布 Domestic Distribution
黄海、东海、台湾海峡、南海 / Yellow Sea, East China Sea, Taiwan Strait, South China Sea
世界分布 World Distribution
太平洋、印度洋、大西洋、地中海 / Pacific Ocean, Indian Ocean, Atlantic Ocean, Mediterranean Sea
分布标注 Distribution Note
非特有种 / Non-endemic

国内分布图
Map of Domestic Distribution

种群 Population

种群数量 Population Size	南半球和北半球现存的塞鲸 70,000 头 / There are approximately 70,000 Sei Whales in the southern and northern hemispheres
种群趋势 Population Trend	下降 / Decreasing

生境与生态系统 Habitat (s) and Ecosystem (s)

生　　境 Habitat(s)	海洋 / Ocean
生态系统 Ecosystem(s)	海洋生态系统 / Ocean Ecosystem

威胁 Threat (s)

主要威胁 Major Threat(s)	海洋污染、渔业纠缠 / Ocean Pollution, Fishery Entanglement

保护级别与保护行动 Protection Category and Conservation Action (s)

国家重点保护野生动物等级 (2021) Category of National Key Protected Wild Animals (2021)	一级 / Category I
IUCN 红色名录 (2020-2) IUCN Red List (2020-2)	濒危 / EN
CITES 附录 (2019) CITES Appendix (2019)	I
保护行动 Conservation Action(s)	法律保护物种 / Legally protected species

相关文献 Relevant References

Burgin *et al.*, 2018; Horwood, 2018; Jiang *et al.* (蒋志刚等), 2015a; Wang (王丕烈), 2011; Zhou (周开亚), 2008, 2004; Zhou *et al.* (周开亚等), 2001

塞鲸 *Balaenoptera borealis*

中国生物多样性红色名录

蓝鲸
Balaenoptera musculus

濒危　EN A1abd

| 数据缺乏 DD | 无危 LC | 近危 NT | 易危 VU | **濒危 EN** | 极危 CR | 区域灭绝 RE | 野外灭绝 EW | 灭绝 EX |

分类地位 Taxonomic Status

动物界 Animalia	脊索动物门 Chordata	哺乳纲 Mammalia	鲸目 Cetacea	须鲸科 Balaenopteridae

学名 Scientific Name	*Balaenoptera musculus*
命名人 Species Authority	Linnaeus, 1758
英文名 English Name(s)	Blue Whale
同物异名 Synonym(s)	无 / None
种下单元评估 Infra-specific Taxa Assessed	无 / None

评估信息 Assessment Information

评估年份 Year Assessed	2020
评定人 Assessor(s)	周开亚 / Kaiya Zhou
审定人 Reviewer(s)	张先锋、王克雄、王丁、祝茜、蒋志刚 / Xianfeng Zhang, Kexiong Wang, Ding Wang, Qian Zhu, Zhigang Jiang
其他贡献人 Other Contributor(s)	李立立、丁晨晨 / Lili Li, Chenchen Ding

理由 Justification: 由于体型巨大，蓝鲸在19世纪成为捕猎的目标。20世纪上半叶在南极捕杀的蓝鲸超过300,000头，使蓝鲸几乎灭绝。基于其种群下降速率，将蓝鲸列为濒危 / Due to its large size, the Blue Whale was targeted by hunting in the 19th century. In the first half of the 20th century, more than 300,000 Blue Whales were killed in the Antarctic, leaving the Blue Whale almost extinct. Based on its population decline rate, it is listed as Endangered

地理分布 Geographical Distribution

国内分布 Domestic Distribution	黄海、东海、台湾海峡 / Yellow Sea, East China Sea, Taiwan Strait
世界分布 World Distribution	太平洋、印度洋、大西洋、地中海 / Pacific Ocean, Indian Ocean, Atlantic Ocean, Mediterranean Sea
分布标注 Distribution Note	非特有种 / Non-endemic

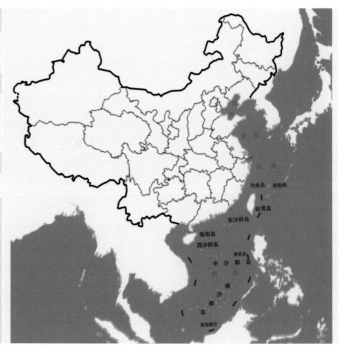

国内分布图
Map of Domestic Distribution

🦌 种群 Population

种群数量 Population Size	少于 10,000 头 / Fewer than 10,000 individuals
种群趋势 Population Trend	上升 / Increasing

🦌 生境与生态系统 Habitat (s) and Ecosystem (s)

生　　境 Habitat(s)	海洋 / Ocean
生态系统 Ecosystem(s)	海洋生态系统 / Ocean Ecosystem

🦌 威胁 Threat (s)

主要威胁 Major Threat(s)	海洋污染、气候变化 / Ocean Pollution, Climate Change

🦌 保护级别与保护行动 Protection Category and Conservation Action (s)

国家重点保护野生动物等级 (2021) Category of National Key Protected Wild Animals (2021)	一级 / Category I
IUCN 红色名录(2020-2) IUCN Red List (2020-2)	濒危 / EN
CITES 附录 (2019) CITES Appendix (2019)	I
保护行动 Conservation Action(s)	无 / None

🦌 相关文献 Relevant References

Burgin *et al*., 2018; Jiang *et al*. (蒋志刚等), 2015a; Wang (王丕烈), 2011; Zhou (周开亚), 2008, 2004

蓝鲸 *Balaenoptera musculus*

中国生物多样性红色名录

长须鲸
Balaenoptera physalus

濒危 EN A1d

| 数据缺乏 DD | 无危 LC | 近危 NT | 易危 VU | 濒危 EN | 极危 CR | 区域灭绝 RE | 野外灭绝 EW | 灭绝 EX |

分类地位 Taxonomic Status

动物界 Animalia	脊索动物门 Chordata	哺乳纲 Mammalia	鲸目 Cetacea	须鲸科 Balaenopteridae
学名 Scientific Name		*Balaenoptera physalus*		
命名人 Species Authority		Linnaeus, 1758		
英文名 English Name(s)		Fin Whale		
同物异名 Synonym(s)		Finback Whale, Common Rorqual (formerly also known as Herring Whale and Razorback Whale)		
种下单元评估 Infra-specific Taxa Assessed		无 / None		

评估信息 Assessment Information

评估年份 Year Assessed	2020
评定人 Assessor(s)	周开亚 / Kaiya Zhou
审定人 Reviewer(s)	张先锋、王克雄、王丁、祝茜、蒋志刚 / Xianfeng Zhang, Kexiong Wang, Ding Wang, Qian Zhu, Zhigang Jiang
其他贡献人 Other Contributor(s)	李立立、丁晨晨 / Lili Li, Chenchen Ding

理由 Justification: 全球长须鲸的数量在1929～2007年的三个世代中下降了70%以上。因此，基于其种群下降速率，将长须鲸列为濒危等级 / The global population of Fin Whale has declined by more than 70% over the three generations from 1929 to 2007. Thus, based on its rate of population decline, Fin Whale is listed as Endangered

地理分布 Geographical Distribution

国内分布 Domestic Distribution
渤海、黄海、东海、台湾海峡、南海 / Bohai Sea, Yellow Sea, East China Sea, Taiwan Strait, South China Sea
世界分布 World Distribution
太平洋、印度洋、大西洋、地中海 / Pacific Ocean, Indian Ocean, Atlantic Ocean, Mediterranean Sea
分布标注 Distribution Note
非特有种 / Non-endemic

国内分布图
Map of Domestic Distribution

种群 Population

种群数量 Population Size	未知 / Unknown
种群趋势 Population Trend	下降 / Decreasing

生境与生态系统 Habitat (s) and Ecosystem (s)

生境 Habitat(s)	海洋 / Ocean
生态系统 Ecosystem(s)	海洋生态系统 / Ocean Ecosystem

威胁 Threat (s)

主要威胁 Major Threat(s)	船只撞击、海洋污染、渔具误捕 / Vessel Collision, Ocean Pollution, Incidental Catch in Fishing Gear

保护级别与保护行动 Protection Category and Conservation Action (s)

国家重点保护野生动物等级 (2021) Category of National Key Protected Wild Animals (2021)	一级 / Category I
IUCN 红色名录 (2020-2) IUCN Red List (2020-2)	易危 / VU
CITES 附录 (2019) CITES Appendix (2019)	I
保护行动 Conservation Action(s)	法律保护物种 / Legally protected species

相关文献 Relevant References

Burgin *et al.*, 2018; Jiang *et al.* (蒋志刚等), 2015a; Wang (王丕烈), 2011; Zhou (周开亚), 2008, 2004; Cai *et al.* (蔡仁逵等), 1959

长须鲸 *Balaenoptera physalus*

东亚江豚
Neophocaena sunameri
濒危　EN A1d

| 数据缺乏 DD | 无危 LC | 近危 NT | 易危 VU | 濒危 EN | 极危 CR | 区域灭绝 RE | 野外灭绝 EW | 灭绝 EX |

分类地位 Taxonomic Status

| 动物界 Animalia | 脊索动物门 Chordata | 哺乳纲 Mammalia | 鲸目 Cetacea | 鼠海豚科 Phocoenidae |

学　名 Scientific Name	*Neophocaena sunameri*
命名人 Species Authority	Pilleri and Gihr, 1975
英文名 English Name(s)	East Asian Finless Porpoise
同物异名 Synonym(s)	*Neophocaena phocaenoides sunameri*
种下单元评估 Infra-specific Taxa Assessed	无 / None

评估信息 Assessment Information

评估年份 Year Assessed	2020
评定人 Assessor(s)	周开亚 / Kaiya Zhou
审定人 Reviewer(s)	张先锋、王克雄、王丁、祝茜、蒋志刚 / Xianfeng Zhang, Kexiong Wang, Ding Wang, Qian Zhu, Zhigang Jiang
其他贡献人 Other Contributor(s)	李立立、丁晨晨 / Lili Li, Chenchen Ding

理由 Justification: 在日本和朝鲜半岛海域的东亚江豚估计分别约有 19,000 头和 13,000 头。由于其栖息地近岸，东亚江豚往往受到许多人类活动的威胁。韩国西海岸和日本濑户内海的东亚江豚呈下降趋势。生活在中国东海、黄海、渤海沿岸海域的东亚江豚种群数量约 60,000 头。根据 20 世纪 80 年代在江苏沿岸的调查，东亚江豚易被刺网、张网、插网和陷阱等被动渔具误捕。每年都有数十头甚至数百头东亚江豚遭被动渔具误捕。因此，根据其种群下降速率，将东亚江豚列为濒危等级 / The estimated number of East Asian Finless Porpoise in the waters surrounding Japan and the Korean Peninsula is approximately 19,000 and 13,000, respectively. East Asian Finless Porpoises are threatened by many human activities because of their near-shore habitat. A decreasing population trend was reported off the west coast of Korea (the Republic of) and in the Seto Inland Sea, Japan. The population number of East Asian Finless Porpoise living in coastal waters of the East China Sea, Yellow Sea and Bohai Sea is about 60,000 individuals. In surveys undertaken along the coast of Jiangsu Province in the 1980s, the species was found to be vulnerable to incidental mortality from drift gillnets, set gillnets, stow nets and traps. Dozens and even hundreds of East Asian Finless Porpoises were entangled by passive fishing gear each year. Based on its rate of population decline, East Asian Finless Porpoise is listed as Endangered

地理分布 Geographical Distribution

国内分布 Domestic Distribution	东海、渤海、黄海 / East China Sea, Bohai Sea, Yellow Sea
世界分布 World Distribution	中国（东海、渤海、黄海）；日本海 / China (East China Sea, Bohai Sea, Yellow Sea); Sea of Japan
分布标注 Distribution Note	非特有种 / Non-endemic

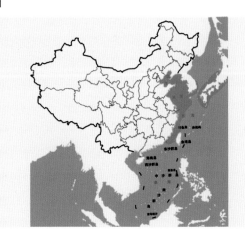

国内分布图
Map of Domestic Distribution

种群 Population

种群数量 Population Size	60,000 头 / 60,000 individuals
种群趋势 Population Trend	下降 / Decreasing

生境与生态系统 Habitat(s) and Ecosystem(s)

生境 Habitat(s)	海洋 / Ocean
生态系统 Ecosystem(s)	海洋生态系统 / Ocean Ecosystem

威胁 Threat(s)

主要威胁 Major Threat(s)	渔具误捕致死，沿海开发和工业化、污染、船只交通导致的栖息地退化和丧失 / Incidental Mortality Caused by Fishing Gear, Habitat Degradation and Loss Caused by Coastal Development and Industrialization, Pollution, and High Level of Vessel Traffic

保护级别与保护行动 Protection Category and Conservation Action(s)

国家重点保护野生动物等级 (2021) Category of National Key Protected Wild Animals (2021)	二级 / Category II
IUCN 红色名录 (2020-2) IUCN Red List (2020-2)	未列入 / NA
CITES 附录 (2019) CITES Appendix (2019)	无 / NA
保护行动 Conservation Action(s)	法律保护物种 / Legally protected species

相关文献 Relevant References

Amano, 2018; Burgin *et al*., 2018; Zhou *et al*., 2018; Yang and Zhou（杨光和周开亚), 1996; Zhou and Wang, 1994

东亚江豚 *Neophocaena sunameri* 　　中国科学院水生生物研究所 提供　By Institute of Hydrobiology, Chinese Academy of Sciences

恒河豚
Platanista gangetica

濒危 EN A2abcde+3bcde+4abcde

| 数据缺乏 DD | 无危 LC | 近危 NT | 易危 VU | 濒危 EN | 极危 CR | 区域灭绝 RE | 野外灭绝 EW | 灭绝 EX |

分类地位 Taxonomic Status

动物界 Animalia	脊索动物门 Chordata	哺乳纲 Mammalia	鲸目 Cetacea	恒河豚科 Platanistidae
学名 Scientific Name		*Platanista gangetica*		
命名人 Species Authority		Lebeck, 1801		
英文名 English Name(s)		South Asian River Dolphin		
同物异名 Synonym(s)		*Delphinus gangetica* (Roxburgh, 1801); *Platanista gangetica* (Roxburgh, 1801); *P. indi* (Blyth, 1859); *P. minor* (Owen, 1853)		
种下单元评估 Infra-specific Taxa Assessed		无 / None		

评估信息 Assessment Information

评估年份 Year Assessed	2020
评定人 Assessor(s)	周开亚 / Kaiya Zhou
审定人 Reviewer(s)	张先锋、王克雄、王丁、祝茜、蒋志刚 / Xianfeng Zhang, Kexiong Wang, Ding Wang, Qian Zhu, Zhigang Jiang
其他贡献人 Other Contributor(s)	李立立、丁晨晨 / Lili Li, Chenchen Ding

理由 Justification: 恒河豚面临的主要威胁是大坝和灌溉项目建设、污染、渔业兼捕造成的栖息地退化。近年来，恒河豚在恒河上游几乎绝迹。因此，基于其种群下降速率，将恒河豚列为濒危等级 / South Asian River Dolphin is threatened by habitat degradation resulting from construction of dams and irrigation projects, pollution and fisheries by catch. It is rarely seen in recent years in the upper reach of the Ganges River, Thus, based on its rate of population decline, South Asian River Dolphin is listed as Endangered

地理分布 Geographical Distribution

国内分布 Domestic Distribution
西藏 / Tibet (Xizang)
世界分布 World Distribution
孟加拉国、中国、印度、尼泊尔、巴基斯坦 / Bangladesh, China, India, Nepal, Pakistan
分布标注 Distribution Note
非特有种 / Non-endemic

国内分布图
Map of Domestic Distribution

种群 Population

种群数量 Population Size	未知 / Unknown
种群趋势 Population Trend	下降 / Decreasing

生境与生态系统 Habitat (s) and Ecosystem (s)

生境 Habitat(s)	江河 / River
生态系统 Ecosystem(s)	湖泊河流生态系统、海洋生态系统 / Lake and River Ecosystem, Ocean Ecosystem

威胁 Threat (s)

主要威胁 Major Threat(s)	直接捕杀、误捕、污染、栖息地退化 / Direct Catch, Incidental Catch, Pollution, Habitat Degradation

保护级别与保护行动 Protection Category and Conservation Action (s)

国家重点保护野生动物等级 (2021) Category of National Key Protected Wild Animals (2021)	一级 / Category I
IUCN红色名录(2020-2) IUCN Red List (2020-2)	濒危 / EN
CITES 附录 (2019) CITES Appendix (2019)	I
保护行动 Conservation Action(s)	法律保护物种 / Legally protected species

相关文献 Relevant References

Burgin *et al*., 2018; Jiang *et al*. (蒋志刚等), 2017; Culik, 2011; Choudhury, 2003

恒河豚 *Platanista gangetica*

中华白海豚 *Sousa chinensis*

濒危 EN C1

| 数据缺乏 DD | 无危 LC | 近危 NT | 易危 VU | **濒危 EN** | 极危 CR | 区域灭绝 RE | 野外灭绝 EW | 灭绝 EX |

分类地位 Taxonomic Status

动物界 Animalia	脊索动物门 Chordata	哺乳纲 Mammalia	鲸目 Cetacea	海豚科 Delphinidae

学名 Scientific Name	*Sousa chinensis*
命名人 Species Authority	Osbeck, 1765
英文名 English Name(s)	Chinese White Dolphin
同物异名 Synonym(s)	Indo-Pacific Humpback Dolphin; *Sousa borneensis* (Lydekker, 1901)
种下单元评估 Infra-specific Taxa Assessed	无 / None

评估信息 Assessment Information

评估年份 Year Assessed	2020
评定人 Assessor(s)	周开亚 / Kaiya Zhou
审定人 Reviewer(s)	张先锋、王克雄、王丁、祝茜、蒋志刚 / Xianfeng Zhang, Kexiong Wang, Ding Wang, Qian Zhu, Zhigang Jiang
其他贡献人 Other Contributor(s)	李立立、丁晨晨 / Lili Li, Chenchen Ding

理由 Justification: 在最近几十年中国经济迅速崛起过程中,中国沿岸海域有大规模的人类活动。自20世纪80年代以来,历史上分布在长江以南沿岸水域的中华白海豚种群由于人为因素而急剧下降。香港-珠江口种群的年下降率为2.46%。因此,中国水域的中华白海豚被列为濒危等级 / There have been massive human activities along the Chinese coast for several decades during China's rapid economic development. Historically distributed in near shore coastal waters south of the Changjiang River, the *Sousa chinensis* populations in Chinese waters have declined drastically due to anthropogenic factors since the 1980s. The estimated annual decline rate of the population in Hong Kong and the Zhujiang River Estuary is 2.46%. Thus, Chinese White Dolphin in Chinese waters is listed as Endangered

地理分布 Geographical Distribution

国内分布 Domestic Distribution	黄海、东海、台湾海峡、南海 / Yellow Sea, East China Sea, Taiwan Strait, South China Sea
世界分布 World Distribution	太平洋、印度洋、大西洋、地中海 / Pacific Ocean, Indian Ocean, Atlantic Ocean, Mediterranean Sea
分布标注 Distribution Note	非特有种 / Non-endemic

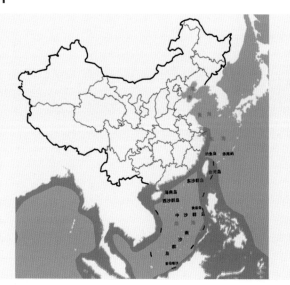

国内分布图
Map of Domestic Distribution

🦌 种群 Population

种群数量 Population Size	中国沿海 5 个中华白海豚种群合计约 4,368 头 / Up to 4,368 individuals in five populations along China's coast
种群趋势 Population Trend	下降 / Decreasing

🦌 生境与生态系统 Habitat (s) and Ecosystem (s)

生　　境 Habitat(s)	海洋 / Ocean
生态系统 Ecosystem(s)	湖泊河流生态系统、海洋生态系统 / Lake and River Ecosystem, Ocean Ecosystem

🦌 威胁 Threat (s)

主要威胁 Major Threat(s)	港口建设、填海、繁忙的海上交通、过度捕捞、水污染、栖息地退化 / Port Construction, Land Reclamation, Busy Sea Traffic, Over Fishing, Water Pollution, Habitat Degradation

🦌 保护级别与保护行动 Protection Category and Conservation Action (s)

国家重点保护野生动物等级 (2021) Category of National Key Protected Wild Animals (2021)	一级 / Category I
IUCN 红色名录(2020-2) IUCN Red List (2020-2)	易危 / VU
CITES 附录 (2019) CITES Appendix (2019)	I
保护行动 Conservation Action(s)	部分种群位于自然保护区内 / Part of populations are covered by nature reserves

🦌 相关文献 Relevant References

Burgin *et al*., 2018; Jiang *et al*. (蒋志刚等), 2015a; Xu *et al*., 2015; Huang *et al*., 2012; Zhou (周开亚), 2008, 2004

中华白海豚 *Sousa chinensis*

滇攀鼠
Vernaya fulva

濒危 EN B1ab(i, ii, iii)+2ab(i, ii, iii); C1

| 数据缺乏 DD | 无危 LC | 近危 NT | 易危 VU | 濒危 EN | 极危 CR | 区域灭绝 RE | 野外灭绝 EW | 灭绝 EX |

分类地位 Taxonomic Status

动物界 Animalia	脊索动物门 Chordata	哺乳纲 Mammalia	啮齿目 Rodentia	鼠科 Muridae
学名 Scientific Name		*Vernaya fulva*		
命名人 Species Authority		G. M. Allen, 1927		
英文名 English Name(s)		Vernay's Climbing Mouse		
同物异名 Synonym(s)		Red Climbing Mouse; *Vernaya foramena* (Wang, Hu, and Chen, 1980)		
种下单元评估 Infra-specific Taxa Assessed		无 / None		

评估信息 Assessment Information

评估年份 Year Assessed	2020
评定人 Assessor(s)	刘少英、蒋志刚 / Shaoying Liu, Zhigang Jiang
审定人 Reviewer(s)	马勇、鲍毅新 / Yong Ma, Yixin Bao
其他贡献人 Other Contributor(s)	李立立、丁晨晨 / Lili Li, Chenchen Ding

理由 Justification: 滇攀鼠种群数量少，其栖息地的面积因人类伐木、放牧等活动而减少。因此，基于其种群数量和栖息地下降速率，将滇攀鼠列为濒危等级 / The populations of Vernay's Climbing Mouse are small. Furthermore, the areas of its habitats are declining due to human encroachment such as logging and grazing by livestock. Thus, based on its population size and the rate of habitat decline, Vernay's Climbing Mouse is listed as Endangered

地理分布 Geographical Distribution

国内分布 Domestic Distribution
云南、四川、甘肃、陕西 / Yunnan, Sichuan, Gansu, Shaanxi
世界分布 World Distribution
中国、缅甸 / China, Myanmar
分布标注 Distribution Note
非特有种 / Non-endemic

国内分布图
Map of Domestic Distribution

种群 Population

种群数量 Population Size	未知 / Unknown
种群趋势 Population Trend	未知 / Unknown

生境与生态系统 Habitat(s) and Ecosystem(s)

生境 Habitat(s)	森林 / Forest
生态系统 Ecosystem(s)	森林生态系统 / Forest Ecosystem

威胁 Threat(s)

主要威胁 Major Threat(s)	伐木、家畜放牧 / Logging, Livestock Ranching

保护级别与保护行动 Protection Category and Conservation Action(s)

国家重点保护野生动物等级 (2021) Category of National Key Protected Wild Animals (2021)	无 / NA
IUCN 红色名录 (2020-2) IUCN Red List (2020-2)	无危 / LC
CITES 附录 (2019) CITES Appendix (2019)	无 / NA
保护行动 Conservation Action(s)	无 / None

相关文献 Relevant References

Burgin *et al.*, 2018; Jiang *et al.* (蒋志刚等), 2017; Zheng *et al.* (郑智民等), 2012; Smith *et al.* (史密斯等), 2009; Li and Wang (李晓晨和王廷正), 1995

滇攀鼠 *Vernaya fulva*

中国生物多样性红色名录

耐氏大鼠
Leopoldamys neilli

濒危 EN B2a

| 数据缺乏 DD | 无危 LC | 近危 NT | 易危 VU | 濒危 EN | 极危 CR | 区域灭绝 RE | 野外灭绝 EW | 灭绝 EX |

分类地位 Taxonomic Status

| 动物界 Animalia | 脊索动物门 Chordata | 哺乳纲 Mammalia | 啮齿目 Rodentia | 鼠科 Muridae |

学名 Scientific Name	*Leopoldamys neilli*
命名人 Species Authority	J. T. Marshall Jr., 1976
英文名 English Name(s)	Neill's Long-tailed Giant Rat
同物异名 Synonym(s)	无 / None
种下单元评估 Infra-specific Taxa Assessed	无 / None

评估信息 Assessment Information

评估年份 Year Assessed	2020
评定人 Assessor(s)	刘少英、蒋志刚 / Shaoying Liu, Zhigang Jiang
审定人 Reviewer(s)	马勇、鲍毅新 / Yong Ma, Yixin Bao
其他贡献人 Other Contributor(s)	李立立、丁晨晨 / Lili Li, Chenchen Ding

理由 Justification: 耐氏大鼠种群数量少，发生区面积小，且栖息地面积因人类伐木、放牧等活动而减少。因此，基于其种群数量和发生区面积，将耐氏大鼠列为濒危等级 / The populations of Neill's Long-tailed Giant Rat are small. Furthermore, its area of occurance is small and its habitat is declining due to human encroachment such as logging and grazing by livestock. Thus, based on its population size and the area of occurrence, Neill's Long-tailed Giant Rat is listed as Endangered

地理分布 Geographical Distribution

国内分布 Domestic Distribution	云南 / Yunnan
世界分布 World Distribution	中国、泰国 / China, Thailand
分布标注 Distribution Note	非特有种 / Non-endemic

国内分布图
Map of Domestic Distribution

种群 Population

种群数量 Population Size	未知 / Unknown
种群趋势 Population Trend	未知 / Unknown

生境与生态系统 Habitat (s) and Ecosystem (s)

生境 Habitat(s)	森林、内陆岩石区域 / Forest, Inland Rocky Area
生态系统 Ecosystem(s)	森林生态系统 / Forest Ecosystem

威胁 Threat (s)

主要威胁 Major Threat(s)	伐木、放牧 / Logging, Livestock Ranching

保护级别与保护行动 Protection Category and Conservation Action (s)

国家重点保护野生动物等级 (2021) Category of National Key Protected Wild Animals (2021)	无 / NA
IUCN 红色名录(2020-2) IUCN Red List (2020-2)	无危 / LC
CITES 附录 (2019) CITES Appendix (2019)	无 / NA
保护行动 Conservation Action(s)	无 / None

相关文献 Relevant References

Burgin *et al.*, 2018; Jiang *et al.* (蒋志刚等), 2017; Chen *et al.* (陈鹏等), 2014; Wilson and Reeder, 2005

耐氏大鼠 *Leopoldamys neilli*

中国生物多样性红色名录

四川毛尾睡鼠
Chaetocauda sichuanensis

濒危 EN B1ab(i, ii, iii)+2ab(i, ii, iii); C1

| 数据缺乏 DD | 无危 LC | 近危 NT | 易危 VU | 濒危 EN | 极危 CR | 区域灭绝 RE | 野外灭绝 EW | 灭绝 EX |

分类地位 Taxonomic Status

动物界 Animalia	脊索动物门 Chordata	哺乳纲 Mammalia	啮齿目 Rodentia	睡鼠科 Gliridae

学名 Scientific Name	*Chaetocauda sichuanensis*
命名人 Species Authority	Wang, 1985
英文名 English Name(s)	Sichuan Dormouse
同物异名 Synonym(s)	Chinese Dormouse; *Dryomys sichuanensis* (Wang, 1985)
种下单元评估 Infra-specific Taxa Assessed	无 / None

评估信息 Assessment Information

评估年份 Year Assessed	2020
评定人 Assessor(s)	蒋志刚、刘少英 / Zhigang Jiang, Shaoying Liu
审定人 Reviewer(s)	马勇、刘伟 / Yong Ma, Wei Liu
其他贡献人 Other Contributor(s)	李立立、丁晨晨 / Lili Li, Chenchen Ding

理由 Justification: 四川毛尾睡鼠种群小，发生区面积小，且栖息地面积因人类伐木、放牧等活动而减少。因此，基于其种群数量和发生区面积下降速率，将四川毛尾睡鼠列为濒危等级 / The populations of Sichuan Dormouse are small. Furthermore, its area of occurance is small and its habitat is declining due to human encroachment such as logging and grazing by livestock. Thus, based on its population size and the rate of habitat decline, Sichuan Dormouse is listed as Endangered

地理分布 Geographical Distribution

国内分布 Domestic Distribution	四川 / Sichuan
世界分布 World Distribution	中国 / China
分布标注 Distribution Note	特有种 / Endemic

国内分布图
Map of Domestic Distribution

种群 Population

种群数量 Population Size	未知 / Unknown
种群趋势 Population Trend	未知 / Unknown

生境与生态系统 Habitat (s) and Ecosystem (s)

生　　境 Habitat(s)	针阔混交林 / Coniferous and Broad-leaved Mixed Forest
生态系统 Ecosystem(s)	森林生态系统 / Forest Ecosystem

威胁 Threat (s)

主要威胁 Major Threat(s)	伐木、放牧 / Logging, Livestock Ranching

保护级别与保护行动 Protection Category and Conservation Action (s)

国家重点保护野生动物等级 (2021) Category of National Key Protected Wild Animals (2021)	无 / NA
IUCN 红色名录 (2020-2) IUCN Red List (2020-2)	无危 / LC
CITES 附录 (2019) CITES Appendix (2019)	无 / NA
保护行动 Conservation Action(s)	无 / None

相关文献 Relevant References

Burgin *et al.*, 2018; Jiang *et al.* (蒋志刚等), 2017; Zheng *et al.* (郑智民等), 2012; Liu *et al.* (刘少英等), 2005; Wang and Xie (汪松和解焱), 2004; Wang (王酉之), 1985

四川毛尾睡鼠 *Chaetocauda sichuanensis*

扁颅鼠兔
Ochotona flatcalvariam

濒危 EN B1+2a(i); D2

| 数据缺乏 DD | 无危 LC | 近危 NT | 易危 VU | **濒危 EN** | 极危 CR | 区域灭绝 RE | 野外灭绝 EW | 灭绝 EX |

分类地位 Taxonomic Status

动物界 Animalia	脊索动物门 Chordata		哺乳纲 Mammalia	兔形目 Lagomorpha	鼠兔科 Ochotonidae
学　　名 Scientific Name			*Ochotona flatcalvariam*		
命 名 人 Species Authority			Liu *et al.*, 2017		
英 文 名 English Name(s)			Flat-cranium Pika		
同物异名 Synonym(s)			无 / None		
种下单元评估 Infra-specific Taxa Assessed			无 / None		

评估信息 Assessment Information

评估年份 Year Assessed	2020
评　定　人 Assessor(s)	刘少英、蒋志刚 / Shaoying Liu, Zhigang Jiang
审　定　人 Reviewer(s)	苏建平、冯祚建、宗浩、廖继承 / Jianping Su, Zuojian Feng, Hao Zong, Jicheng Liao
其他贡献人 Other Contributor(s)	李立立、丁晨晨 / Lili Li, Chenchen Ding

理由 Justification: 扁颅鼠兔是一个新种，其发生区极为狭窄，种群数量小。基于其发生区面积与种群数量，故将扁颅鼠兔定为濒危 / Flat-cranium Pika is a new species, it has very narrow area of occurrence and small populations. Thus, based on its area of occurrence and population size, Flat-cranium Pika is listed as Endangered

地理分布 Geographical Distribution

国内分布 Domestic Distribution
重庆、四川 / Chongqing, Sichuan
世界分布 World Distribution
中国 / China
分布标注 Distribution Note
特有种 / Endemic

国内分布图
Map of Domestic Distribution

种群 Population

种群数量 Population Size	未知 / Unknown
种群趋势 Population Trend	未知 / Unknown

生境与生态系统 Habitat (s) and Ecosystem (s)

生　　境 Habitat(s)	草原岩石区域 / Grassland and Rocky Areas
生态系统 Ecosystem(s)	草地生态系统 / Grassland Ecosystem

威胁 Threat (s)

主要威胁 Major Threat(s)	未知 / Unknown

保护级别与保护行动 Protection Category and Conservation Action (s)

国家重点保护野生动物等级 (2021) Category of National Key Protected Wild Animals (2021)	无 / NA
IUCN 红色名录 (2020-2) IUCN Red List (2020-2)	未列入 / NA
CITES 附录 (2019) CITES Appendix (2019)	无 / NA
保护行动 Conservation Action(s)	无 / None

相关文献 Relevant References

Burgin *et al.*, 2018; Jiang *et al.* (蒋志刚等), 2017; Liu *et al.* (刘少英等), 2017

扁颅鼠兔 *Ochotona flatcalvariam*　　　　方红霞 绘　By Hongxia Fang

伊犁鼠兔
Ochotona iliensis

濒危 EN A2c

| 数据缺乏 DD | 无危 LC | 近危 NT | 易危 VU | 濒危 EN | 极危 CR | 区域灭绝 RE | 野外灭绝 EW | 灭绝 EX |

分类地位 Taxonomic Status

动物界 Animalia	脊索动物门 Chordata	哺乳纲 Mammalia	兔形目 Lagomorpha	鼠兔科 Ochotonidae
学名 Scientific Name		*Ochotona iliensis*		
命名人 Species Authority		Li and Ma, 1986		
英文名 English Name(s)		Ili Pika		
同物异名 Synonym(s)		无 / None		
种下单元评估 Infra-specific Taxa Assessed		无 / None		

评估信息 Assessment Information

评估年份 Year Assessed	2020
评定人 Assessor(s)	刘少英、蒋志刚 / Shaoying Liu, Zhigang Jiang
审定人 Reviewer(s)	苏建平、冯祚建、宗浩、廖继承 / Jianping Su, Zuojian Feng, Hao Zong, Jicheng Liao
其他贡献人 Other Contributor(s)	李立立、丁晨晨 / Lili Li, Chenchen Ding

理由 Justification: 伊犁鼠兔是中国特有种，仅分布在新疆。由于栖息地的丧失，近10年间伊犁鼠兔的种群数量减少了55%以上。因此，基于其种群下降速率，将伊犁鼠兔列为濒危等级 / Ili Pika is an endemic species in China, only found in Xinjiang. Due to habitat loss, its populations have decreased more than 55% in recent 10 years. Thus, based on its rate of population decline, Ili Pika is listed as Endangered

地理分布 Geographical Distribution

国内分布 Domestic Distribution
新疆 / Xinjiang
世界分布 World Distribution
中国 / China
分布标注 Distribution Note
特有种 / Endemic

国内分布图
Map of Domestic Distribution

种群 Population

种群数量 Population Size	4,000 只左右 / About 4,000 individuals
种群趋势 Population Trend	下降 / Decreasing

生境与生态系统 Habitat(s) and Ecosystem(s)

生境 Habitat(s)	草原、内陆岩石区域 / Grassland, Inland Rocky Area
生态系统 Ecosystem(s)	草地生态系统 / Grassland Ecosystem

威胁 Threat(s)

主要威胁 Major Threat(s)	家畜放牧、气候变化、繁殖能力低 / Livestock Ranching, Climate Change, Low Reproductive Ability

保护级别与保护行动 Protection Category and Conservation Action(s)

国家重点保护野生动物等级 (2021) Category of National Key Protected Wild Animals (2021)	二级 / Category II
IUCN 红色名录 (2020-2) IUCN Red List (2020-2)	濒危 / EN
CITES 附录 (2019) CITES Appendix (2019)	无 / NA
保护行动 Conservation Action(s)	无 / None

相关文献 Relevant References

Burgin *et al.*, 2018; Jiang *et al.* (蒋志刚等), 2017; Cui *et al.* (崔鹏等), 2014; Zheng *et al.* (郑智民等), 2012; Li and Smith, 2005; Li, 2003

伊犁鼠兔 *Ochotona iliensis* 　　　　李维东 摄　By Weidong Li

柯氏鼠兔
Ochotona koslowi

濒危　EN A2c

| 数据缺乏 DD | 无危 LC | 近危 NT | 易危 VU | 濒危 EN | 极危 CR | 区域灭绝 RE | 野外灭绝 EW | 灭绝 EX |

分类地位 Taxonomic Status

动物界 Animalia	脊索动物门 Chordata	哺乳纲 Mammalia	兔形目 Lagomorpha	鼠兔科 Ochotonidae

学名 Scientific Name	*Ochotona koslowi*
命名人 Species Authority	Büchner, 1894
英文名 English Name(s)	Kozlov's Pika
同物异名 Synonym(s)	Koslov's Pika
种下单元评估 Infra-specific Taxa Assessed	无 / None

评估信息 Assessment Information

评估年份 Year Assessed	2020
评定人 Assessor(s)	刘少英、蒋志刚 / Shaoying Liu, Zhigang Jiang
审定人 Reviewer(s)	苏建平、冯祚建、宗浩、廖继承 / Jianping Su, Zuojian Feng, Hao Zong, Jicheng Liao
其他贡献人 Other Contributor(s)	李立立、丁晨晨 / Lili Li, Chenchen Ding

理由 Justification: 柯氏鼠兔发生区狭窄，种群小，种群数量下降。因此，基于其发生区面积与种群下降速率，将柯氏鼠兔列为濒危等级 / Kozlov's Pika has a narrow range of occurrence and its population is declining, Thus, based on its area of occurrence and its rate of population decline, Kozlov's Pika is listed as Endangered

地理分布 Geographical Distribution

国内分布 Domestic Distribution	
新疆 / Xinjiang	
世界分布 World Distribution	
中国 / China	
分布标注 Distribution Note	
特有种 / Endemic	

国内分布图
Map of Domestic Distribution

种群 Population

种群数量 Population Size	未知 / Unknown
种群趋势 Population Trend	下降 / Decreasing

生境与生态系统 Habitat (s) and Ecosystem (s)

生境 Habitat(s)	亚寒带草地 / Subarctic Grassland
生态系统 Ecosystem(s)	草地生态系统 / Grassland Ecosystem

威胁 Threat (s)

主要威胁 Major Threat(s)	无 / None

保护级别与保护行动 Protection Category and Conservation Action (s)

国家重点保护野生动物等级 (2021) Category of National Key Protected Wild Animals (2021)	无 / NA
IUCN 红色名录 (2020-2) IUCN Red List (2020-2)	濒危 / EN
CITES 附录 (2019) CITES Appendix (2019)	无 / NA
保护行动 Conservation Action(s)	无 / None

相关文献 Relevant References

Burgin *et al*., 2018; Jiang *et al*. (蒋志刚等), 2017; Zheng *et al*. (郑智民等), 2012; Cao *et al*. (曹伊凡等), 2009; Smith *et al*. (史密斯等), 2009; Li *et al*., 2006; Wang (王应祥), 2003

柯氏鼠兔 *Ochotona koslowi*　　　　李维东 摄　By Weidong Li

原仓鼠
Cricetus cricetus

濒危　EN A3c

| 数据缺乏 DD | 无危 LC | 近危 NT | 易危 VU | **濒危 EN** | 极危 CR | 区域灭绝 RE | 野外灭绝 EW | 灭绝 EX |

分类地位 Taxonomic Status

动物界 Animalia	脊索动物门 Chordata	哺乳纲 Mammalia	啮齿目 Rodentia	仓鼠科 Cricetidae

学　名 Scientific Name	*Cricetus cricetus*
命名人 Species Authority	Linnaeus, 1758
英文名 English Name(s)	Black-bellied Hamster
同物异名 Synonym(s)	Common Hamster; Eurasian Hamster; European Hamster; *Cricetus albus* (Fitzinger, 1867); *babylonicus* (Nehring, 1903); *canescens* (Nehring, 1899); *frumentarius* (转下页)
种下单元评估 Infra-specific Taxa Assessed	无 / None

评估信息 Assessment Information

评估年份 Year Assessed	2020
评定人 Assessor(s)	刘少英、蒋志刚 / Shaoying Liu, Zhigang Jiang
审定人 Reviewer(s)	马勇、鲍毅新 / Yong Ma, Yixin Bao
其他贡献人 Other Contributor(s)	李立立、丁晨晨 / Lili Li, Chenchen Ding

理由 Justification: 原仓鼠在中国分布区狭窄，种群数量很少，占有的栖息地面积小，近年来，原仓鼠在其主要分布区的种群数量急剧下降。刘少英团队在新疆的3次野外考察中均未采集到标本。因此，原仓鼠列为濒危等级 / Black-bellied Hamster has a narrow range and its population size is small, its area of occupancy is small in the country. The populations of Black-bellied Hamster are experiencing rapid declines in its major ranges. Shaoying Liu's team failed to acquire any specimen of the species during their 3 field expeditions in Xinjiang, China in recent years. Thus, Black-bellied Hamster is listed as Endangered

地理分布 Geographical Distribution

国内分布 Domestic Distribution
新疆 / Xinjiang
世界分布 World Distribution
欧亚大陆 / Eurasian Continent
分布标注 Distribution Note
非特有种 / Non-endemic

国内分布示意图
Schematic Map of Domestic Distribution

种群 Population

种群数量 Population Size	未知 / Unknown
种群趋势 Population Trend	下降 / Decreasing

生境与生态系统 Habitat (s) and Ecosystem (s)

生　　境 Habitat(s)	干旱低地草地、草甸 / Dry Lowland Grassland, Meadow
生态系统 Ecosystem(s)	草地生态系统 / Grassland Ecosystem

威胁 Threat (s)

主要威胁 Major Threat(s)	毒杀、耕种、农林业排放物污染 / Poisoning, Plantation, Agricultural and Forestry Effluent

保护级别与保护行动 Protection Category and Conservation Action (s)

国家重点保护野生动物等级 (2021) Category of National Key Protected Wild Animals (2021)	无 / NA
IUCN红色名录(2020-2) IUCN Red List (2020-2)	极危 / CR
CITES 附录 (2019) CITES Appendix (2019)	无 / NA
保护行动 Conservation Action(s)	无 / None

相关文献 Relevant References

Banaszek *et al.*, 2020; Burgin *et al.*, 2018; Jiang *et al.*（蒋志刚等）, 2017; Wilson *et al.*, 2017; Zheng *et al.*（郑智民等）, 2012; Smith *et al.*（史密斯等）, 2009; Wang（王应祥）, 2003; Luo（罗泽珣）, 2000

（接上页）
(Pallas, 1811); *fulvus* (Bechstein, 1801); *fuscidorsis* (Argyropulo, 1932); *fuscidorsis* (Argyropulo, 1936); *germanicus* (Kerr, 1792); *jeudii* (Gray, 1873); *latycranius* (Ognev, 1923); *nehringi* (Matschie, 1901); *niger* (Fitzinger, 1867); *niger* (Bogdanov, 1871); *niger* (Simroth, 1906); *nigricans* (Lacépède, 1799); *polychroma* (Krulikovski, 1916); *rufescens* (Nehring, 1899); *stavropolicus* (Satunin, 1907); *tauricus* (Ognev, 1924); *tomensis* (Ognev, 1924); *varius* (Fitzinger, 1867); *vulgaris* (Geoffroy, 1803)

原仓鼠 *Cricetus cricetus*

粗毛兔
Caprolagus hispidus

濒危 EN B2ab (ii, iii, v)

| 数据缺乏 DD | 无危 LC | 近危 NT | 易危 VU | **濒危 EN** | 极危 CR | 区域灭绝 RE | 野外灭绝 EW | 灭绝 EX |

分类地位 Taxonomic Status

动物界 Animalia	脊索动物门 Chordata	哺乳纲 Mammalia	兔形目 Lagomorpha	兔科 Leporidae

学名 Scientific Name	*Caprolagus hispidus*
命名人 Species Authority	Pearson, 1839
英文名 English Name(s)	Hispid Hare
同物异名 Synonym(s)	Assam Rabbit, Bristly Rabbit
种下单元评估 Infra-specific Taxa Assessed	无 / None

评估信息 Assessment Information

评估年份 Year Assessed	2020
评定人 Assessor(s)	蒋志刚 / Zhigang Jiang
审定人 Reviewer(s)	胡慧建、胡一鸣 / Huijian Hu, Yiming Hu
其他贡献人 Other Contributor(s)	李立立、丁晨晨 / Lili Li, Chenchen Ding

理由 Justification: 粗毛兔历史上一直分布到喜马拉雅山脉山麓地带。据估计，分布区在 5,000～20,000km^2，在人口稀少的生境中，粗毛兔占有区面积为 11～500km^2。它发生在海拔 100～2,500m，其种群破碎化。根据其占有区面积，将粗毛兔列为濒危等级 / The historic range of the species extended along the foothill region of the southern Himalayas. The geographic extent of the Hispid Hare is estimated at between 5,000 and 20,000km^2. Its area of occupancy is estimated between 11 and 500km^2, in highly fragmented populations. Hispid Hare occurs at elevations ranging from 100 to 2,500m, in highly fragmented populations. Based on its area of occupancy, Hispid Hare is listed as Endangered

地理分布 Geographical Distribution

国内分布 Domestic Distribution
西藏 / Tibet (Xizang)
世界分布 World Distribution
中国；南亚 / China; South Asia
分布标注 Distribution Note
非特有种 / Non-endemic

国内分布图
Map of Domestic Distribution

种群 Population

种群数量 Population Size	未知 / Unknown
种群趋势 Population Trend	下降 / Decreasing

生境与生态系统 Habitat (s) and Ecosystem (s)

生　　境 Habitat(s)	生活在演替早期的高草草原 / Inhabits the Tall Grassland of early successional stage
生态系统 Ecosystem(s)	草地生态系统 / Grassland Ecosystem

威胁 Threat (s)

主要威胁 Major Threat(s)	无 / None

保护级别与保护行动 Protection Category and Conservation Action (s)

国家重点保护野生动物等级 (2021) Category of National Key Protected Wild Animals (2021)	二级 / Category II
IUCN 红色名录 (2020-2) IUCN Red List (2020-2)	濒危 / EN
CITES 附录 (2019) CITES Appendix (2019)	无 / NA
保护行动 Conservation Action(s)	无 / None

相关文献 Relevant References

Burgin *et al.*, 2018; Jiang *et al.* (蒋志刚等), 2017; Choudhury, 2003

粗毛兔 *Caprolagus hispidus*

峨眉鼩鼹
Uropsilus andersoni

易危 VU B1+2a, b(i, ii, iii)

| 数据缺乏 DD | 无危 LC | 近危 NT | 易危 VU | 濒危 EN | 极危 CR | 区域灭绝 RE | 野外灭绝 EW | 灭绝 EX |

分类地位 Taxonomic Status

动物界 Animalia	脊索动物门 Chordata	哺乳纲 Mammalia	劳亚食虫目 Eulipotyphla	鼹科 Talpidae
学　　名 Scientific Name		*Uropsilus andersoni*		
命　名　人 Species Authority		Thomas, 1911		
英　文　名 English Name(s)		Anderson's Shrew Mole		
同物异名 Synonym(s)		无 / None		
种下单元评估 Infra-specific Taxa Assessed		无 / None		

评估信息 Assessment Information

评估年份 Year Assessed	2020
评　定　人 Assessor(s)	蒋志刚 / Zhigang Jiang
审　定　人 Reviewer(s)	蒋学龙、冯祚建 / Xuelong Jiang, Zuojian Feng
其他贡献人 Other Contributor(s)	李立立、丁晨晨 / Lili Li, Chenchen Ding

理由 Justification: 峨眉鼩鼹为中国特有种，分布区狭窄，仅分布在四川。未见种群数量和分布现状的研究，然而，四川许多地方正在修建水电站，威胁到峨眉鼩鼹的生存。因此，峨眉鼩鼹列为易危等级 / Anderson's Shrew Mole is an endemic species narrowly distributed in Sichuan Province. Little information is available on the population and habitat status of this species, apparently the construction of hydropower stations in Sichuan Province threatens the Anderson's Shrew Mole. Thus, based on its habitat status, Anderson's Shrew Mole is listed as Vulnerable

地理分布 Geographical Distribution

国内分布 Domestic Distribution
四川 / Sichuan
世界分布 World Distribution
中国 / China
分布标注 Distribution Note
特有种 / Endemic

国内分布图
Map of Domestic Distribution

种群 Population

种群数量 Population Size	未知 / Unknown
种群趋势 Population Trend	未知 / Unknown

生境与生态系统 Habitat(s) and Ecosystem(s)

生境 Habitat(s)	常绿阔叶林 / Evergreen Broad-leaved Forest
生态系统 Ecosystem(s)	森林生态系统 / Forest Ecosystem

威胁 Threat(s)

主要威胁 Major Threat(s)	未知 / Unknown

保护级别与保护行动 Protection Category and Conservation Action(s)

国家重点保护野生动物等级 (2021) Category of National Key Protected Wild Animals (2021)	无 / NA
IUCN 红色名录 (2020-2) IUCN Red List (2020-2)	数据缺乏 / DD
CITES 附录 (2019) CITES Appendix (2019)	无 / NA
保护行动 Conservation Action(s)	无 / None

相关文献 Relevant References

Burgin *et al.*, 2018; Jiang *et al.* (蒋志刚等), 2017; Zhang *et al.* (张君等), 2010; Zhu *et al.* (朱红艳等), 2010; Liang *et al.* (梁艺于等), 2009; Li *et al.* (李艳红等), 2007; Hoffmann, 1987

峨眉鼩鼹 *Uropsilus andersoni*

中国生物多样性 红色名录

宽齿鼹
Euroscaptor grandis

易危　VU B2a, b(i, ii, iii)

| 数据缺乏 DD | 无危 LC | 近危 NT | 易危 VU | 濒危 EN | 极危 CR | 区域灭绝 RE | 野外灭绝 EW | 灭绝 EX |

分类地位 Taxonomic Status

动物界 Animalia	脊索动物门 Chordata	哺乳纲 Mammalia	劳亚食虫目 Eulipotyphla	鼹科 Talpidae
学　　名 Scientific Name		*Euroscaptor grandis*		
命　名　人 Species Authority		Miller, 1940		
英　文　名 English Name(s)		Greater Chinese Mole		
同物异名 Synonym(s)		无 / None		
种下单元评估 Infra-specific Taxa Assessed		无 / None		

评估信息 Assessment Information

评估年份 Year Assessed	2020
评　定　人 Assessor(s)	蒋志刚 / Zhigang Jiang
审　定　人 Reviewer(s)	蒋学龙、冯祚建 / Xuelong Jiang, Zuojian Feng
其他贡献人 Other Contributor(s)	李立立、丁晨晨 / Lili Li, Chenchen Ding

理由 Justification: 宽齿鼹为中国特有种，其分布区狭窄，数量很少。人类活动和基础设施建设，如交通道路建设对其栖息地影响大。因此，将宽齿鼹列为易危等级 / Greater Chinese Mole is endemic to China, with a narrow distribution and small populations. Human disturbance and infrastructure like transportation have significant negative effect on the habitat of Greater Chinese Mole. Thus, it is listed as Vulnerable

地理分布 Geographical Distribution

国内分布 Domestic Distribution
四川、云南 / Sichuan, Yunnan
世界分布 World Distribution
中国 / China
分布标注 Distribution Note
特有种 / Endemic

国内分布图
Map of Domestic Distribution

种群 Population

种群数量 Population Size	未知 / Unknown
种群趋势 Population Trend	未知 / Unknown

生境与生态系统 Habitat (s) and Ecosystem (s)

生　　境 Habitat(s)	森林 / Forest
生态系统 Ecosystem(s)	森林生态系统 / Forest Ecosystem

威胁 Threat (s)

主要威胁 Major Threat(s)	道路建设 / Road Construction

保护级别与保护行动 Protection Category and Conservation Action (s)

国家重点保护野生动物等级 (2021) Category of National Key Protected Wild Animals (2021)	无 / NA
IUCN 红色名录 (2020-2) IUCN Red List (2020-2)	无危 / LC
CITES 附录 (2019) CITES Appendix (2019)	无 / NA
保护行动 Conservation Action(s)	位于自然保护区内的种群与生境得到保护 / Populations and habitats located in nature reserves are protected

相关文献 Relevant References

Burgin *et al*., 2018; Jiang *et al*. (蒋志刚等), 2017; Smith *et al*. (史密斯等), 2009; Wang (王应祥), 2003; Miller, 1940

宽齿鼹 *Euroscaptor grandis*

短尾鼹
Euroscaptor micrura
易危 VU B1

| 数据缺乏 DD | 无危 LC | 近危 NT | 易危 VU | 濒危 EN | 极危 CR | 区域灭绝 RE | 野外灭绝 EW | 灭绝 EX |

分类地位 Taxonomic Status

动物界 Animalia	脊索动物门 Chordata	哺乳纲 Mammalia	劳亚食虫目 Eulipotyphla	鼹科 Talpidae
学名 Scientific Name		*Euroscaptor micrura*		
命名人 Species Authority		Hodgson, 1841		
英文名 English Name(s)		Himalayan Mole		
同物异名 Synonym(s)		无 / None		
种下单元评估 Infra-specific Taxa Assessed		无 / None		

评估信息 Assessment Information

评估年份 Year Assessed	2020
评定人 Assessor(s)	蒋志刚 / Zhigang Jiang
审定人 Reviewer(s)	蒋学龙、冯祚建 / Xuelong Jiang, Zuojian Feng
其他贡献人 Other Contributor(s)	李立立、丁晨晨 / Lili Li, Chenchen Ding

理由 Justification: 短尾鼹在中国仅分布于云南省，分布区狭窄，部分种群分布在保护区内，但其种群密度低。因此，将短尾鼹列为易危等级 / In China, Himalayan Mole is only narrowly distributed in Yunnan Province, only parts of its populations are distributed in nature reserves, and its population densities are low. Thus, based on its population size, Himalayan Mole is listed as Vulnerable

地理分布 Geographical Distribution

国内分布 Domestic Distribution
云南 / Yunnan
世界分布 World Distribution
中国；南亚、东南亚 / China; South Asia, Southeast Asia
分布标注 Distribution Note
非特有种 / Non-endemic

国内分布图
Map of Domestic Distribution

种群 Population

种群数量 Population Size	未知 / Unknown
种群趋势 Population Trend	未知 / Unknown

生境与生态系统 Habitat (s) and Ecosystem (s)

生　　境 Habitat(s)	森林 / Forest
生态系统 Ecosystem(s)	森林生态系统 / Forest Ecosystem

威胁 Threat (s)

主要威胁 Major Threat(s)	无 / None

保护级别与保护行动 Protection Category and Conservation Action (s)

国家重点保护野生动物等级 (2021) Category of National Key Protected Wild Animals (2021)	无 / NA
IUCN红色名录(2020-2) IUCN Red List (2020-2)	无危 / LC
CITES 附录 (2019) CITES Appendix (2019)	无 / NA
保护行动 Conservation Action(s)	位于自然保护区内的种群与生境得到保护 / Populations and habitats located in nature reserves are protected

相关文献 Relevant References

Burgin *et al*., 2018; Jiang *et al*. (蒋志刚等), 2017; Smith *et al*. (史密斯等), 2009; Kawada *et al*., 2003; Wang (王应祥), 2003

短尾鼹 *Euroscaptor micrura*　　　　方红霞 绘　By Hongxia Fang

中国生物多样性红色名录

小齿鼹
Euroscaptor parvidens

易危 VU B2a, b(i, ii, iii)

| 数据缺乏 DD | 无危 LC | 近危 NT | 易危 VU | 濒危 EN | 极危 CR | 区域灭绝 RE | 野外灭绝 EW | 灭绝 EX |

分类地位 Taxonomic Status

动物界 Animalia	脊索动物门 Chordata	哺乳纲 Mammalia	劳亚食虫目 Eulipotyphla	鼹科 Talpidae

学 名 Scientific Name	*Euroscaptor parvidens*
命名人 Species Authority	Miller, 1940
英文名 English Name(s)	Small-toothed Mole
同物异名 Synonym(s)	无 / None
种下单元评估 Infra-specific Taxa Assessed	无 / None

评估信息 Assessment Information

评估年份 Year Assessed	2020
评定人 Assessor(s)	蒋志刚 / Zhigang Jiang
审定人 Reviewer(s)	蒋学龙、冯祚建 / Xuelong Jiang, Zuojian Feng
其他贡献人 Other Contributor(s)	李立立、丁晨晨 / Lili Li, Chenchen Ding

理由 Justification: 小齿鼹在中国仅发现于云南省南部与越南边界地区，种群分布在保护区内，但其分布区狭窄，数量很少。因此，将小齿鼹列入易危等级 / Small-toothed Mole is only found near the border with Viet Nam in southern Yunnan Province. The populations of the species are very small and narrowly distributed and only some of its populations are located in nature reserves. Thus, Small-toothed Mole is listed as Vulnerable

地理分布 Geographical Distribution

国内分布 Domestic Distribution	
云南 / Yunnan	
世界分布 World Distribution	
中国、越南 / China, Viet Nam	
分布标注 Distribution Note	
非特有种 / Non-endemic	

国内分布图
Map of Domestic Distribution

种群 Population

种群数量 Population Size	未知 / Unknown
种群趋势 Population Trend	未知 / Unknown

生境与生态系统 Habitat (s) and Ecosystem (s)

生境 Habitat(s)	森林 / Forest
生态系统 Ecosystem(s)	森林生态系统 / Forest Ecosystem

威胁 Threat (s)

主要威胁 Major Threat(s)	无 / None

保护级别与保护行动 Protection Category and Conservation Action (s)

国家重点保护野生动物等级 (2021) Category of National Key Protected Wild Animals (2021)	无 / NA
IUCN 红色名录 (2020-2) IUCN Red List (2020-2)	数据缺乏 / DD
CITES 附录 (2019) CITES Appendix (2019)	无 / NA
保护行动 Conservation Action(s)	位于自然保护区内的种群与生境得到保护 / Populations and habitats in nature reserves are protected

相关文献 Relevant References

Burgin *et al.*, 2018; Jiang *et al.* (蒋志刚等), 2017; Smith *et al.* (史密斯等), 2009; Wang (王应祥), 2003

小齿鼹 *Euroscaptor parvidens*

中国生物多样性红色名录

大鼩鼱
Sorex mirabilis

易危 VU B1+2a, b(i, ii, iii)

| 数据缺乏 DD | 无危 LC | 近危 NT | **易危 VU** | 濒危 EN | 极危 CR | 区域灭绝 RE | 野外灭绝 EW | 灭绝 EX |

分类地位 Taxonomic Status

动物界 Animalia	脊索动物门 Chordata	哺乳纲 Mammalia	劳亚食虫目 Eulipotyphla	鼩鼱科 Soricidae

学名 Scientific Name	*Sorex mirabilis*
命名人 Species Authority	Ognev, 1937
英文名 English Name(s)	Ussuri Shrew
同物异名 Synonym(s)	Giant Shrew
种下单元评估 Infra-specific Taxa Assessed	无 / None

评估信息 Assessment Information

评估年份 Year Assessed	2020
评定人 Assessor(s)	蒋志刚 / Zhigang Jiang
审定人 Reviewer(s)	蒋学龙、冯祚建 / Xuelong Jiang, Zuojian Feng
其他贡献人 Other Contributor(s)	李立立、丁晨晨 / Lili Li, Chenchen Ding

理由 Justification: 大鼩鼱的分布区较小，分布点少，种群数量较少，分布地区生境遭到破坏。因此，大鼩鼱列为易危等级 / Ussuri Shrew is narrowly distributed in small populations in limited localities. Furthermore, its habitats are under destruction. Thus, Ussuri Shrew is listed as Vulnerable

地理分布 Geographical Distribution

国内分布 Domestic Distribution	黑龙江、吉林 / Heilongjiang, Jilin
世界分布 World Distribution	中国、朝鲜、俄罗斯 / China, Korea (the Democratic People's Republic of), Russia
分布标注 Distribution Note	非特有种 / Non-endemic

国内分布图
Map of Domestic Distribution

种群 Population

种群数量 Population Size	未知 / Unknown
种群趋势 Population Trend	未知 / Unknown

生境与生态系统 Habitat (s) and Ecosystem (s)

生　　　境 Habitat(s)	针阔混交林、温带森林地带的岩石山坡 / Live on Rocky Slope in Coniferous and Broad-leaved Mixed Forest, Temperate Forest
生态系统 Ecosystem(s)	森林生态系统 / Forest Ecosystem

威胁 Threat (s)

主要威胁 Major Threat(s)	未知 / Unknown

保护级别与保护行动 Protection Category and Conservation Action (s)

国家重点保护野生动物等级 (2021) Category of National Key Protected Wild Animals (2021)	无 / NA
IUCN红色名录(2020-2) IUCN Red List (2020-2)	数据缺乏 / DD
CITES 附录 (2019) CITES Appendix (2019)	无 / NA
保护行动 Conservation Action(s)	无 / None

相关文献 Relevant References

Liu *et al*.（刘铸等）, 2019; Burgin *et al*., 2018; Jiang *et al*.（蒋志刚等）, 2017; Pan *et al*.（潘清华等）, 2007; Wang（王应祥）, 2003; Wilson and Reeder, 1993; Mammal Research Group, Institute of Zoology, Chinese Academy of Sciences（中国科学院动物研究所兽类研究组）, 1958

大䶄鼩 *Sorex mirabilis*

中国生物多样性 红色名录

扁颅鼩鼱
Sorex roboratus

易危 VU B1+2a, b(i, ii, iii)

| 数据缺乏 DD | 无危 LC | 近危 NT | 易危 VU | 濒危 EN | 极危 CR | 区域灭绝 RE | 野外灭绝 EW | 灭绝 EX |

分类地位 Taxonomic Status

动物界 Animalia	脊索动物门 Chordata	哺乳纲 Mammalia	劳亚食虫目 Eulipotyphla	鼩鼱科 Soricidae
学 名 Scientific Name		*Sorex roboratus*		
命 名 人 Species Authority		Hollister, 1913		
英 文 名 English Name(s)		Flat-skulled Shrew		
同物异名 Synonym(s)		*Sorex araneus* (Dukelsky, 1928) subsp. *jakutensis*; *dukelskiae* (Ognev, 1933); *macropygmaeus* (Ognev, 1922) subsp. *araneoides*; *platycranius* (Ognev, 1922); *thomasi* (转下页)		
种下单元评估 Infra-specific Taxa Assessed		无 / None		

评估信息 Assessment Information

评 估 年 份 Year Assessed	2020
评 定 人 Assessor(s)	蒋志刚 / Zhigang Jiang
审 定 人 Reviewer(s)	蒋学龙、冯祚建 / Xuelong Jiang, Zuojian Feng
其他贡献人 Other Contributor(s)	李立立、丁晨晨 / Lili Li, Chenchen Ding

理由 Justification: 扁颅鼩鼱分布区较狭窄，分布点少，且种群数量少。因此，基于其占有区与种群现状，将扁颅鼩鼱列为易危等级 / Flat-skulled Shrew has narrow distribution range at limited localities, and its population density is low. Thus, based on its area of occupation, Flat-skulled Shrew is listed as Vulnerable

地理分布 Geographical Distribution

国内分布 Domestic Distribution
新疆、黑龙江 / Xinjiang, Heilongjiang
世界分布 World Distribution
中国、蒙古、俄罗斯 / China, Mongolia, Russia
分布标注 Distribution Note
非特有种 / Non-endemic

国内分布图
Map of Domestic Distribution

种群 Population

种群数量 Population Size	未知 / Unknown
种群趋势 Population Trend	未知 / Unknown

生境与生态系统 Habitat (s) and Ecosystem (s)

生　　境 Habitat(s)	针阔混交林、温带森林地带的岩石山坡 / Live on Rocky Slope in Coniferous and Broad-leaved Mixed Forest, Temperate Forest
生态系统 Ecosystem(s)	森林生态系统、荒漠生态系统 / Forest Ecosystem, Desert Ecosystem

威胁 Threat (s)

主要威胁 Major Threat(s)	伐木、火灾、干旱 / Logging, Fire, Drought

保护级别与保护行动 Protection Category and Conservation Action (s)

国家重点保护野生动物等级 (2021) Category of National Key Protected Wild Animals (2021)	无 / NA
IUCN 红色名录 (2020-2) IUCN Red List (2020-2)	无危 / LC
CITES 附录 (2019) CITES Appendix (2019)	无 / NA
保护行动 Conservation Action(s)	无 / None

相关文献 Relevant References

Burgin *et al.*, 2018; Jiang *et al.* (蒋志刚等), 2017; Liu *et al.*(刘铸等), 2016; Smith *et al.* (史密斯等), 2009

(接上页)
(Ognev, 1922); *vir* (G. Allen, 1914)

扁颅鼩鼱 *Sorex roboratus*

水駒鼱
Neomys fodiens

易危 VU B1+2a, b(i, ii, iii)

| 数据缺乏 DD | 无危 LC | 近危 NT | 易危 VU | 濒危 EN | 极危 CR | 区域灭绝 RE | 野外灭绝 EW | 灭绝 EX |

分类地位 Taxonomic Status

动物界 Animalia	脊索动物门 Chordata	哺乳纲 Mammalia	劳亚食虫目 Eulipotyphla	駒鼱科 Soricidae
学 名 Scientific Name		*Neomys fodiens*		
命 名 人 Species Authority		Pennant, 1771		
英 文 名 English Name(s)		Eurasian Water Shrew		
同物异名 Synonym(s)		无 / None		
种下单元评估 Infra-specific Taxa Assessed		无 / None		

评估信息 Assessment Information

评 估 年 份 Year Assessed	2020
评 定 人 Assessor(s)	蒋志刚 / Zhigang Jiang
审 定 人 Reviewer(s)	蒋学龙 / Xuelong Jiang
其他贡献人 Other Contributor(s)	李立立、丁晨晨 / Lili Li, Chenchen Ding

理由 Justification: 水駒鼱首次发现是在吉林省，后在长白山被发现，其他记载包括黑龙江及新疆。水駒鼱数量稀少，迄今所采的标本也极少，也未有研究其种群动态的报道（汪松，1958；朴仁峰和俞曙林，1990；李永项，2012）。而分布在长白山地区的水駒鼱，已经受到当地水资源利用的影响（黄乃伟等，2012）。因此，将其列为易危等级 / Eurasian Water Shrew was first discovered in Jilin Province, and later found in Changbai Shan. Other records of the species are in Heilongjiang and Xinjiang. The Eurasian Water Shrew is extremely rare. So far, only a few specimens have been collected and reports on its population dynamics are lacking (Wang, 1958; Piao and Yu, 1990; Li, 2012). Furthermore, its population in the Changbai Shan area is threatened by different forms of water resources utilization (Huang *et al.*, 2012). Thus, Eurasian Water Shrew is listed as Vulnerable

地理分布 Geographical Distribution

国内分布 Domestic Distribution
黑龙江、吉林、新疆 / Heilongjiang, Jilin, Xinjiang
世界分布 World Distribution
欧亚大陆各国 / The Countries on Eurasian Continent
分布标注 Distribution Note
非特有种 / Non-endemic

国内分布图
Map of Domestic Distribution

种群 Population

种群数量 Population Size	未知 / Unknown
种群趋势 Population Trend	持平 / Stable

生境与生态系统 Habitat (s) and Ecosystem (s)

生　　境 Habitat(s)	森林、草地、溪流边 / Forest, Grassland, Near Stream
生态系统 Ecosystem(s)	森林生态系统、草地生态系统、湖泊河流生态系统 / Forest Ecosystem, Grassland Ecosystem, Lake and River Ecosystem

威胁 Threat (s)

主要威胁 Major Threat(s)	耕地、农业或林业污染，湿地排水 / Arable Land, Agricultural or Forestry Effluent; Wetland Drainage

保护级别与保护行动 Protection Category and Conservation Action (s)

国家重点保护野生动物等级 (2021) Category of National Key Protected Wild Animals (2021)	无 / NA
IUCN 红色名录 (2020-2) IUCN Red List (2020-2)	无危 / LC
CITES 附录 (2019) CITES Appendix (2019)	无 / NA
保护行动 Conservation Action(s)	无 / None

相关文献 Relevant References

Burgin *et al.*, 2018; Jiang *et al.* (蒋志刚等), 2017; Huang *et al.* (黄乃伟等), 2012; Li (李永项), 2012; Wang (王东风), 1993; Piao and Yu (朴仁峰和俞曙林), 1990; Wang (汪松), 1958

水鼩鼱 *Neomys fodiens*

灰腹水鼩
Chimarrogale styani

易危 VU B1+2a(i)

| 数据缺乏 DD | 无危 LC | 近危 NT | 易危 VU | 濒危 EN | 极危 CR | 区域灭绝 RE | 野外灭绝 EW | 灭绝 EX |

分类地位 Taxonomic Status

动物界 Animalia	脊索动物门 Chordata	哺乳纲 Mammalia	劳亚食虫目 Eulipotyphla	鼩鼱科 Soricidae
学名 Scientific Name		*Chimarrogale styani*		
命名人 Species Authority		de Winton, 1899		
英文名 English Name(s)		Chinese Water Shrew		
同物异名 Synonym(s)		无 / None		
种下单元评估 Infra-specific Taxa Assessed		无 / None		

评估信息 Assessment Information

评估年份 Year Assessed	2020
评定人 Assessor(s)	蒋志刚 / Zhigang Jiang
审定人 Reviewer(s)	蒋学龙 / Xuelong Jiang
其他贡献人 Other Contributor(s)	李立立、丁晨晨 / Lili Li, Chenchen Ding

理由 Justification: 在中国分布的灰腹水鼩数量较少，至今所采集到的标本也很少。其生境特殊且受到水体污染、森林破坏的影响。因此，灰腹水鼩列为易危等级 / The populations of Chinese Water Shrew in China are small and few specimens of the species have been collected so far. As the species requires specialized habitats, it is confronting threats from water pollution and deforestation. Thus, based on its habitat status, Chinese Water Shrew is listed as Vulnerable

地理分布 Geographical Distribution

国内分布 Domestic Distribution
四川、云南、西藏、青海、甘肃 / Sichuan, Yunnan, Tibet (Xizang), Qinghai, Gansu
世界分布 World Distribution
中国、缅甸 / China, Myanmar
分布标注 Distribution Note
非特有种 / Non-endemic

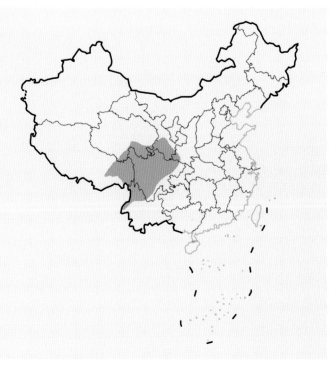

国内分布图
Map of Domestic Distribution

种群 Population

种群数量 Population Size	未知 / Unknown
种群趋势 Population Trend	未知 / Unknown

生境与生态系统 Habitat (s) and Ecosystem (s)

生　　境 Habitat(s)	亚热带湿润山地森林、溪流边 / Subtropical Moist Montane Forest, Near Stream
生态系统 Ecosystem(s)	森林生态系统、湖泊河流生态系统 / Forest Ecosystem, Lake and River Ecosystem

威胁 Threat (s)

主要威胁 Major Threat(s)	未知 / Unknown

保护级别与保护行动 Protection Category and Conservation Action (s)

国家重点保护野生动物等级 (2021) Category of National Key Protected Wild Animals (2021)	无 / NA
IUCN红色名录(2020-2) IUCN Red List (2020-2)	无危 / LC
CITES 附录 (2019) CITES Appendix (2019)	无 / NA
保护行动 Conservation Action(s)	位于自然保护区内的种群与生境得到保护 / Populations and habitats located in nature reserves are protected

相关文献 Relevant References

Burgin *et al*., 2018; Jiang *et al.* (蒋志刚等), 2017; Zha *et al.* (扎史其等), 2014; Jiang *et al.* (姜雪松等), 2013; Hoffmann, 1987

灰腹水鼩 *Chimarrogale styani*

小臭鼩
Suncus etruscus

易危　VU B1+2a, b(i, ii, iii)

| DD | LC | NT | VU | EN | CR | RE | EW | EX |

分类地位 Taxonomic Status

| 动物界 Animalia | 脊索动物门 Chordata | 哺乳纲 Mammalia | 劳亚食虫目 Eulipotyphla | 鼩鼱科 Soricidae |

学名 Scientific Name	*Suncus etruscus*
命名人 Species Authority	Savi, 1822
英文名 English Name(s)	Etruscan Shrew
同物异名 Synonym(s)	无 / None
种下单元评估 Infra-specific Taxa Assessed	无 / None

评估信息 Assessment Information

评估年份 Year Assessed	2020
评定人 Assessor(s)	蒋志刚 / Zhigang Jiang
审定人 Reviewer(s)	蒋学龙 / Xuelong Jiang
其他贡献人 Other Contributor(s)	李立立、丁晨晨 / Lili Li, Chenchen Ding

理由 Justification: 小臭鼩分布区较狭窄，分布点少，种群数量较少，且其所在的热带雨林受到破坏，目前仅知两个采集点，且仅采集到两号标本。因此，小臭鼩列为易危等级 / Etruscan Shrew is narrowly distributed and has only fairly small populations, with a few distribution locations, and only two specimens are known at present. The tropical rainforests where it lives are now being destroyed. Thus, Etruscan Shrew is listed as Vulnerable

地理分布 Geographical Distribution

| 国内分布 Domestic Distribution |
| 云南 / Yunnan |
| 世界分布 World Distribution |
| 欧亚大陆各国 / The Countries on Eurasian Continent |
| 分布标注 Distribution Note |
| 非特有种 / Non-endemic |

国内分布图
Map of Domestic Distribution

种群 Population

种群数量 Population Size	未知 / Unknown
种群趋势 Population Trend	未知 / Unknown

生境与生态系统 Habitat (s) and Ecosystem (s)

生 境 Habitat(s)	草地、灌丛、森林、人工环境 / Grassland, Shrubland, Forest, Artificial Environment
生态系统 Ecosystem(s)	森林生态系统、灌丛生态系统、草地生态系统、农田生态系统、人类聚落生态系统 / Forest Ecosystem, Shrubland Ecosystem, Grassland Ecosystem, Cropland Ecosystem, Human Settlement Ecosystem

威胁 Threat (s)

主要威胁 Major Threat(s)	未知 / Unknown

保护级别与保护行动 Protection Category and Conservation Action (s)

国家重点保护野生动物等级 (2021) Category of National Key Protected Wild Animals (2021)	无 / NA
IUCN 红色名录 (2020-2) IUCN Red List (2020-2)	无危 / LC
CITES 附录 (2019) CITES Appendix (2019)	无 / NA
保护行动 Conservation Action(s)	无 / None

相关文献 Relevant References

Burgin *et al*., 2018; Jiang *et al*. (蒋志刚等), 2017; Smith *et al*. (史密斯等), 2009; Pan *et al*. (潘清华等), 2007; Wang (王应祥), 2003; Wilson and Reeder, 1993

小臭鼩 *Suncus etruscus*

抱尾果蝠
Rousettus amplexicaudatus

易危 VU B1; C2a(i)

DD	LC	NT	VU	EN	CR	RE	EW	EX
数据缺乏	无危	近危	易危	濒危	极危	区域灭绝	野外灭绝	灭绝

分类地位 Taxonomic Status

动物界 Animalia	脊索动物门 Chordata	哺乳纲 Mammalia	翼手目 Chiroptera	狐蝠科 Pteropodidae
学　　名 Scientific Name		*Rousettus amplexicaudatus*		
命　名　人 Species Authority		É. Geoffroy, 1810		
英　文　名 English Name(s)		Geoffroy's Rousette		
同物异名 Synonym(s)		无 / None		
种下单元评估 Infra-specific Taxa Assessed		无 / None		

评估信息 Assessment Information

评估年份 Year Assessed	2020
评　定　人 Assessor(s)	吴毅、蒋志刚 / Yi Wu, Zhigang Jiang
审　定　人 Reviewer(s)	张礼标、毛秀光、张树义 / Libiao Zhang, Xiuguang Mao, Shuyi Zhang
其他贡献人 Other Contributor(s)	李立立、丁晨晨 / Lili Li, Chenchen Ding

理由 Justification: 抱尾果蝠分布狭窄，虽然部分抱尾果蝠种群分布在保护区内，但抱尾果蝠种群数量少，且受关注度较低。因此，基于其占有区面积，将抱尾果蝠列为易危等级 / The distribution range of Geoffroy's Rousette is restricted. Although some of its populations are located in nature reserves, its population sizes are small and little attention is paid to the species. Thus, based on its area of occupancy, Geoffroy's Rousette is listed as Vulnerable

地理分布 Geographical Distribution

国内分布 Domestic Distribution
云南 / Yunnan
世界分布 World Distribution
中国；东南亚 / China; Southeast Asia
分布标注 Distribution Note
非特有种 / Non-endemic

国内分布图
Map of Domestic Distribution

种群 Population

种群数量 Population Size	稀少 / Rare
种群趋势 Population Trend	未知 / Unknown

生境与生态系统 Habitat (s) and Ecosystem (s)

生　　境 Habitat(s)	森林、洞穴、内陆岩石区域 / Forest, Cave, Inland Rocky Area
生态系统 Ecosystem(s)	森林生态系统、喀斯特生态系统 / Forest Ecosystem, Karst Ecosystem

威胁 Threat (s)

主要威胁 Major Threat(s)	未知 / Unknown

保护级别与保护行动 Protection Category and Conservation Action (s)

国家重点保护野生动物等级 (2021) Category of National Key Protected Wild Animals (2021)	无 / NA
IUCN 红色名录 (2020-2) IUCN Red List (2020-2)	无危 / LC
CITES 附录 (2019) CITES Appendix (2019)	无 / NA
保护行动 Conservation Action(s)	无 / None

相关文献 Relevant References

Burgin *et al.*, 2018; Jiang *et al.* (蒋志刚等), 2017; Smith *et al.* (史密斯等), 2009; Pan *et al.* (潘清华等), 2007; Wang (王应祥), 2003

抱尾果蝠 *Rousettus amplexicaudatus*

By Flickr / Sexecutioner

短耳犬蝠
Cynopterus brachyotis
易危 VU B1; C2a(i)

| DD | LC | NT | **VU** | EN | CR | RE | EW | EX |

分类地位 Taxonomic Status

| 动物界 Animalia | 脊索动物门 Chordata | 哺乳纲 Mammalia | 翼手目 Chiroptera | 狐蝠科 Pteropodidae |

学名 Scientific Name	*Cynopterus brachyotis*
命名人 Species Authority	Müller, 1838
英文名 English Name(s)	Lesser Dog-faced Fruit Bat
同物异名 Synonym(s)	Lesser Short-nosed Fruit Bat; *Cynopterus brachysoma* (Dobson, 1871); *C. marginatus* (Dobson, 1873) var. *andamanensis*; *C. marginatus* (Gray, 1871) var. (转下页)
种下单元评估 Infra-specific Taxa Assessed	无 / None

评估信息 Assessment Information

评估年份 Year Assessed	2020
评定人 Assessor(s)	吴毅、蒋志刚 / Yi Wu, Zhigang Jiang
审定人 Reviewer(s)	张礼标、毛秀光、张树义 / Libiao Zhang, Xiuguang Mao, Shuyi Zhang
其他贡献人 Other Contributor(s)	李立立、丁晨晨 / Lili Li, Chenchen Ding

理由 Justification: 短耳犬蝠分布狭窄，其种群数量少，且栖息地的面积因人类活动而减少。因此，短耳犬蝠列为易危等级 / Lesser Dog-faced Fruit Bat is narrowly distributed in China, and its populations are small. Furthermore, the areas of its habitats are declining due to human encroachment. Thus, Lesser Dog-faced Fruit Bat is listed as Vulnerable

地理分布 Geographical Distribution

国内分布 Domestic Distribution	云南、西藏 / Yunnan, Tibet (Xizang)
世界分布 World Distribution	中国；东南亚 / China; Southeast Asia
分布标注 Distribution Note	非特有种 / Non-endemic

国内分布图
Map of Domestic Distribution

种群 Population

种群数量 Population Size	罕见 / Very rare
种群趋势 Population Trend	未知 / Unknown

生境与生态系统 Habitat (s) and Ecosystem (s)

生　　境 Habitat(s)	森林、乡村花园、种植园、城市 / Forest, Rural Garden, Plantation, Urban Area
生态系统 Ecosystem(s)	森林生态系统、人类聚落生态系统 / Forest Ecosystem, Human Settlement Ecosystem

威胁 Threat (s)

主要威胁 Major Threat(s)	未知 / Unknown

保护级别与保护行动 Protection Category and Conservation Action (s)

国家重点保护野生动物等级 (2021) Category of National Key Protected Wild Animals (2021)	无 / NA
IUCN红色名录 (2020-2) IUCN Red List (2020-2)	无危 / LC
CITES 附录 (2019) CITES Appendix (2019)	无 / NA
保护行动 Conservation Action(s)	无 / None

相关文献 Relevant References

Burgin *et al.*, 2018; Jiang *et al.* (蒋志刚等), 2017; Huang *et al.* (黄继展等), 2013; Wang (王应祥), 2003; Corbet and Hill, 1992; Heideman and Heaney, 1989

(接上页)
ceylonensis; *Pachysoma brachyotis* (Muller, 1838)

短耳犬蝠 *Cynopterus brachyotis*

球果蝠
Sphaerias blanfordi

易危 VU B1; C2a(i)

| 数据缺乏 DD | 无危 LC | 近危 NT | 易危 VU | 濒危 EN | 极危 CR | 区域灭绝 RE | 野外灭绝 EW | 灭绝 EX |

分类地位 Taxonomic Status

动物界 Animalia	脊索动物门 Chordata	哺乳纲 Mammalia	翼手目 Chiroptera	狐蝠科 Pteropodidae

学名 Scientific Name	*Sphaerias blanfordi*
命名人 Species Authority	Thomas, 1891
英文名 English Name(s)	Blandford's Fruit Bat
同物异名 Synonym(s)	*Cynopterus blanfordi* (Thomas, 1891)
种下单元评估 Infra-specific Taxa Assessed	无 / None

评估信息 Assessment Information

评估年份 Year Assessed	2020
评定人 Assessor(s)	吴毅、蒋志刚 / Yi Wu, Zhigang Jiang
审定人 Reviewer(s)	张礼标、毛秀光、张树义 / Libiao Zhang, Xiuguang Mao, Shuyi Zhang
其他贡献人 Other Contributor(s)	李立立、丁晨晨 / Lili Li, Chenchen Ding

理由 Justification: 球果蝠分布狭窄，其种群数量少，其占有区面积因人类活动而减少。因此，基于其种群与占有区面积，将球果蝠列入易危等级 / Blandford's Fruit Bat is narrowly distributed in China, and its population is small. Furthermore, the areas of its habitats are declining due to human activities. Thus, based on its population and area of occupancy, Blandford's Fruit Bat is listed as Vulnerable

地理分布 Geographical Distribution

国内分布 Domestic Distribution	西藏、云南 / Tibet (Xizang), Yunnan
世界分布 World Distribution	中国；南亚、东南亚 / China; South Asia, Southeast Asia
分布标注 Distribution Note	非特有种 / Non-endemic

国内分布图 Map of Domestic Distribution

种群 Population

种群数量 Population Size	稀少 / Rare
种群趋势 Population Trend	未知 / Unknown

生境与生态系统 Habitat (s) and Ecosystem (s)

生　　境 Habitat(s)	热带和亚热带湿润山地森林 / Tropical and Subtropical Moist Mountain Forest
生态系统 Ecosystem(s)	森林生态系统 / Forest Ecosystem

威胁 Threat (s)

主要威胁 Major Threat(s)	未知 / Unknown

保护级别与保护行动 Protection Category and Conservation Action (s)

国家重点保护野生动物等级 (2021) Category of National Key Protected Wild Animals (2021)	无 / NA
IUCN 红色名录 (2020-2) IUCN Red List (2020-2)	无危 / LC
CITES 附录 (2019) CITES Appendix (2019)	无 / NA
保护行动 Conservation Action(s)	位于自然保护区内的种群与生境得到保护 / Populations and habitats in nature reserves are protected

相关文献 Relevant References

Burgin *et al*., 2018; Jiang *et al*. (蒋志刚等), 2017; Cheng *et al*. (程志营等), 2011; Feng *et al*. (冯庆等), 2008; Li (李思华), 1989; Feng *et al*. (冯祚建等), 1986

球果蝠 *Sphaerias blanfordi*

长舌果蝠
Eonycteris spelaea

易危 VU B1; C2a(i)

| 数据缺乏 DD | 无危 LC | 近危 NT | **易危 VU** | 濒危 EN | 极危 CR | 区域灭绝 RE | 野外灭绝 EW | 灭绝 EX |

分类地位 Taxonomic Status

动物界 Animalia	脊索动物门 Chordata	哺乳纲 Mammalia	翼手目 Chiroptera	狐蝠科 Pteropodidae

学名 Scientific Name	*Eonycteris spelaea*
命名人 Species Authority	Dobson, 1871
英文名 English Name(s)	Dawn Bat
同物异名 Synonym(s)	Common Dawn Bat; Lesser Dawn Bat; Common Nectar Bat; Cave Nectar Bat; *Eonycteris bernsteini* (Tate, 1942); *E. spelaea* (Lawrence, 1939) subsp. *glandifera*;（转下页）
种下单元评估 Infra-specific Taxa Assessed	无 / None

评估信息 Assessment Information

评估年份 Year Assessed	2020
评定人 Assessor(s)	吴毅、蒋志刚 / Yi Wu, Zhigang Jiang
审定人 Reviewer(s)	张礼标、毛秀光、张树义 / Libiao Zhang, Xiuguang Mao, Shuyi Zhang
其他贡献人 Other Contributor(s)	李立立、丁晨晨 / Lili Li, Chenchen Ding

理由 Justification: 长舌果蝠分布区狭窄，其种群数量少，且栖息地面积因人类活动而减少。因此，将长舌果蝠列入易危等级 / Dawn Bat is narrowly distributed in China with small populations. Furthermore, its habitat areas are declining due to human encroachment. Thus, Dawn Bat is listed as Vulnerable

地理分布 Geographical Distribution

国内分布 Domestic Distribution
广西、云南 / Guangxi, Yunnan
世界分布 World Distribution
中国；南亚、东南亚 / China; South Asia, Southeast Asia
分布标注 Distribution Note
非特有种 / Non-endemic

国内分布图
Map of Domestic Distribution

种群 Population

种群数量 Population Size	稀少 / Rare
种群趋势 Population Trend	未知 / Unknown

生境与生态系统 Habitat (s) and Ecosystem (s)

生　　境 Habitat(s)	洞穴、森林、种植园 / Cave, Forest, Plantation
生态系统 Ecosystem(s)	森林生态系统、人类聚落生态系统 / Forest Ecosystem, Human Settlement Ecosystem

威胁 Threat (s)

主要威胁 Major Threat(s)	伐木、耕种、狩猎及采集 / Logging, Plantation, Hunting and Collection

保护级别与保护行动 Protection Category and Conservation Action (s)

国家重点保护野生动物等级 (2021) Category of National Key Protected Wild Animals (2021)	无 / NA
IUCN 红色名录 (2020-2) IUCN Red List (2020-2)	无危 / LC
CITES 附录 (2019) CITES Appendix (2019)	无 / NA
保护行动 Conservation Action(s)	无 / None

相关文献 Relevant References

Burgin *et al*., 2018; Jiang *et al*.（蒋志刚等），2017; Smith *et al*.（史密斯等），2009; Wang（王应祥），2003; Yang *et al*.（杨德华等），1993

（接上页）
E. spelaea (Jentink, 1889) subsp. *rosenbergii*; *E. spelaea* (Maharadatunkamsi and Kitchener, 1997) subsp. *winnyae*; *E. bernsteini* (Tate, 1942); *E. spelaea* (Lawrence, 1939) subsp. *glandifera*; *E. spelaea* (Jentink, 1889) subsp. *rosenbergii*; *Macroglossus spelaeus* (Dobson, 1871)

长舌果蝠 *Eonycteris spelaea*

By Alice Hughe

印度假吸血蝠
Megaderma lyra

易危 VU B1; C2a(i)

| 数据缺乏 DD | 无危 LC | 近危 NT | **易危 VU** | 濒危 EN | 极危 CR | 区域灭绝 RE | 野外灭绝 EW | 灭绝 EX |

分类地位 Taxonomic Status

动物界 Animalia	脊索动物门 Chordata	哺乳纲 Mammalia	翼手目 Chiroptera	假吸血蝠科 Megadermatidae

学 名 Scientific Name	*Megaderma lyra*
命 名 人 Species Authority	É. Geoffroy, 1810
英 文 名 English Name(s)	Greater False Vampire Bat
同物异名 Synonym(s)	Greater False-vampire
种下单元评估 Infra-specific Taxa Assessed	无 / None

评估信息 Assessment Information

评 估 年 份 Year Assessed	2020
评 定 人 Assessor(s)	吴毅、蒋志刚 / Yi Wu, Zhigang Jiang
审 定 人 Reviewer(s)	张礼标、毛秀光、张树义 / Libiao Zhang, Xiuguang Mao, Shuyi Zhang
其他贡献人 Other Contributor(s)	李立立、丁晨晨 / Lili Li, Chenchen Ding

理由 Justification: 印度假吸血蝠分布虽广，但人类活动，如捕猎、旅游和采石等已经严重影响到其生存，导致其种群数量下降。因此，印度假吸血蝠列为易危等级 / Although the Greater False Vampire Bat is widely distributed, the survival of the species has been threatened by human encroachment such as hunting, travel, and quarrying. Its populations continue to decline. Thus, Greater False Vampire Bat is listed as Vulnerable

地理分布 Geographical Distribution

国内分布 Domestic Distribution
湖南、四川、广西、云南、广东、福建、贵州、重庆、海南、西藏 / Hunan, Sichuan, Guangxi, Yunnan, Guangdong, Fujian, Guizhou, Chongqing, Hainan, Tibet (Xizang)
世界分布 World Distribution
中国；南亚、东南亚 / China; South Asia, Southeast Asia
分布标注 Distribution Note
非特有种 / Non-endemic

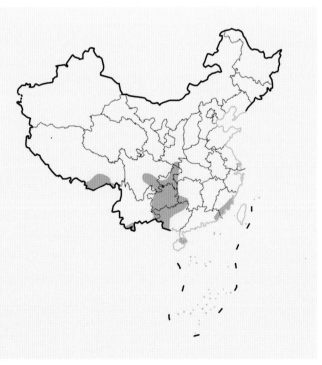

国内分布图
Map of Domestic Distribution

种群 Population

种群数量 Population Size	稀少 / Rare
种群趋势 Population Trend	未知 / Unknown

生境与生态系统 Habitat (s) and Ecosystem (s)

生　　境 Habitat(s)	森林、洞穴、人造建筑 / Forest, Cave, Building
生态系统 Ecosystem(s)	森林生态系统、人类聚落生态系统 / Forest Ecosystem, Human Settlement Ecosystem

威胁 Threat (s)

主要威胁 Major Threat(s)	采石场、伐木、火灾、耕种、狩猎及采集 / Quarrying Field, Logging, Fire, Plantation, Hunting and Collection

保护级别与保护行动 Protection Category and Conservation Action (s)

国家重点保护野生动物等级 (2021) Category of National Key Protected Wild Animals (2021)	无 / NA
IUCN 红色名录 (2020-2) IUCN Red List (2020-2)	无危 / LC
CITES 附录 (2019) CITES Appendix (2019)	无 / NA
保护行动 Conservation Action(s)	无 / None

相关文献 Relevant References

Burgin *et al*., 2018; Jiang *et al*.（蒋志刚等）, 2017; Zhang *et al*.（张礼标等）, 2007; Luo *et al*.（罗蓉等）, 1993; Yang *et al*.（杨德华等）, 1993; Chu（褚新洛）, 1989; Shi and Zhao（施白南和赵尔宓）, 1980

印度假吸血蝠 *Megaderma lyra*　　　　　吴毅 摄　By Yi Wu

单角菊头蝠
Rhinolophus monoceros

易危 VU B1; C2a(i)

| 数据缺乏 DD | 无危 LC | 近危 NT | **易危 VU** | 濒危 EN | 极危 CR | 区域灭绝 RE | 野外灭绝 EW | 灭绝 EX |

分类地位 Taxonomic Status

动物界 Animalia	脊索动物门 Chordata	哺乳纲 Mammalia	翼手目 Chiroptera	菊头蝠科 Rhinolophidae

学 名 Scientific Name	*Rhinolophus monoceros*
命 名 人 Species Authority	K. Andersen, 1905
英 文 名 English Name(s)	Formosan Lesser Horseshoe Bat
同物异名 Synonym(s)	Formosan Least Horseshoe Bat, Formosan Horseshoe Bat
种下单元评估 Infra-specific Taxa Assessed	无 / None

评估信息 Assessment Information

评估年份 Year Assessed	2020
评 定 人 Assessor(s)	吴毅、蒋志刚 / Yi Wu, Zhigang Jiang
审 定 人 Reviewer(s)	张礼标、毛秀光、张树义 / Libiao Zhang, Xiuguang Mao, Shuyi Zhang
其他贡献人 Other Contributor(s)	李立立、丁晨晨 / Lili Li, Chenchen Ding

理由 Justification: 单角菊头蝠是中国特有种，仅分布在台湾。人类活动，如捕猎、旅游和采石等已经严重影响到其生存，导致其种群数量下降。因此，单角菊头蝠列为易危等级 / Formosan Lesser Horseshoe Bat is an endemic species only distributed in Taiwan Province. Due to serious disturbances from human encroachment such as hunting, travel and quarrying, its populations are declining. Thus, Formosan Lesser Horseshoe Bat is listed as Vulnerable

地理分布 Geographical Distribution

国内分布 Domestic Distribution
台湾 / Taiwan
世界分布 World Distribution
中国 / China
分布标注 Distribution Note
特有种 / Endemic

国内分布图
Map of Domestic Distribution

种群 Population

种群数量 Population Size	较常见 / Relatively common
种群趋势 Population Trend	未知 / Unknown

生境与生态系统 Habitat (s) and Ecosystem (s)

生境 Habitat(s)	洞穴、亚热带湿润低地森林 / Cave, Subtropical Moist Lowland Forest
生态系统 Ecosystem(s)	喀斯特生态系统、森林生态系统、人类聚落生态系统 / Karst Ecosystem, Forest Ecosystem, Human Settlement Ecosystem

威胁 Threat (s)

主要威胁 Major Threat(s)	伐木、耕种、采矿、采石场、旅游休闲 / Logging, Plantation, Mining, Quarrying Field, Tourism and Recreation

保护级别与保护行动 Protection Category and Conservation Action (s)

国家重点保护野生动物等级 (2021) Category of National Key Protected Wild Animals (2021)	无 / NA
IUCN 红色名录(2020-2) IUCN Red List (2020-2)	无危 / LC
CITES 附录 (2019) CITES Appendix (2019)	无 / NA
保护行动 Conservation Action(s)	位于自然保护区内的种群与生境得到保护 / Populations and habitats located in nature reserves are protected

相关文献 Relevant References

Burgin *et al*., 2018; Jiang *et al*.（蒋志刚等）, 2017; Smith *et al*.（史密斯等）, 2009; Xu *et al*.（许立杰等）, 2008; Wang（王应祥）, 2003; Lin（林良恭）, 2000

单角菊头蝠 *Rhinolophus monoceros*　　蒋志刚 绘　Drawn by Zhigang Jiang

云南菊头蝠
Rhinolophus yunnanensis

易危　VU B1; C2a(i)

| 数据缺乏 DD | 无危 LC | 近危 NT | 易危 VU | 濒危 EN | 极危 CR | 区域灭绝 RE | 野外灭绝 EW | 灭绝 EX |

分类地位 Taxonomic Status

动物界 Animalia	脊索动物门 Chordata	哺乳纲 Mammalia	翼手目 Chiroptera	菊头蝠科 Rhinolophidae

学　名 Scientific Name	*Rhinolophus yunnanensis*
命 名 人 Species Authority	Dobson, 1872
英 文 名 English Name(s)	Dobson's Horseshoe Bat
同物异名 Synonym(s)	无 / None
种下单元评估 Infra-specific Taxa Assessed	无 / None

评估信息 Assessment Information

评 估 年 份 Year Assessed	2020
评 定 人 Assessor(s)	吴毅、蒋志刚 / Yi Wu, Zhigang Jiang
审 定 人 Reviewer(s)	张礼标、毛秀光、张树义 / Libiao Zhang, Xiuguang Mao, Shuyi Zhang
其他贡献人 Other Contributor(s)	李立立、丁晨晨 / Lili Li, Chenchen Ding

理由 Justification: 云南菊头蝠分布区狭窄，数量少，国内研究机构和博物馆收藏的标本数量极少，且由于人类活动影响，其栖息地面积缩小。因此，云南菊头蝠列为易危等级 / Dobson's Horseshoe Bat is narrowly distributed in China, and its populations are small. Only several specimens have been collected by research institutions and museums nationwide. Furthermore, the areas of its habitats are declining due to human encroachment. Thus, Dobson's Horseshoe Bat is listed as Vulnerable

地理分布 Geographical Distribution

国内分布 Domestic Distribution	云南、四川 / Yunnan, Sichuan
世界分布 World Distribution	中国；南亚、东南亚 / China; South Asia, Southeast Asia
分布标注 Distribution Note	非特有种 / Non-endemic

国内分布图
Map of Domestic Distribution

种群 Population

种群数量 Population Size	稀少 / Rare
种群趋势 Population Trend	未知 / Unknown

生境与生态系统 Habitat(s) and Ecosystem(s)

生境 Habitat(s)	竹林、人造建筑 / Bamboo Grove, Building
生态系统 Ecosystem(s)	竹林生态系统、人类聚落生态系统 / Bamboo Forest Ecosystem, Human Settlement Ecosystem

威胁 Threat(s)

主要威胁 Major Threat(s)	未知 / Unknown

保护级别与保护行动 Protection Category and Conservation Action(s)

国家重点保护野生动物等级 (2021) Category of National Key Protected Wild Animals (2021)	无 / NA
IUCN红色名录 (2020-2) IUCN Red List (2020-2)	无危 / LC
CITES 附录 (2019) CITES Appendix (2019)	无 / NA
保护行动 Conservation Action(s)	无 / None

相关文献 Relevant References

Burgin *et al.*, 2018; Jiang *et al.* (蒋志刚等), 2017; Mao (毛秀光), 2010; Wu *et al.*, 2009a; Chen (陈敏), 2003

云南菊头蝠 *Rhinolophus yunnanensis*　　　　By Alice Hughes

莱氏蹄蝠
Hipposideros lylei

易危 VU B1; C2a(i)

DD	LC	NT	**VU**	EN	CR	RE	EW	EX
数据缺乏	无危	近危	**易危**	濒危	极危	区域灭绝	野外灭绝	灭绝

分类地位 Taxonomic Status

动物界 Animalia	脊索动物门 Chordata	哺乳纲 Mammalia	翼手目 Chiroptera	蹄蝠科 Hipposideridae

学名 Scientific Name	*Hipposideros lylei*
命名人 Species Authority	Thomas, 1913
英文名 English Name(s)	Shield-faced Roundleaf Bat
同物异名 Synonym(s)	Shield-faced Leaf-nosed Bat
种下单元评估 Infra-specific Taxa Assessed	无 / None

评估信息 Assessment Information

评估年份 Year Assessed	2020
评定人 Assessor(s)	吴毅、蒋志刚 / Yi Wu, Zhigang Jiang
审定人 Reviewer(s)	张礼标、毛秀光、张树义 / Libiao Zhang, Xiuguang Mao, Shuyi Zhang
其他贡献人 Other Contributor(s)	李立立、丁晨晨 / Lili Li, Chenchen Ding

理由 Justification: 莱氏蹄蝠分布区较狭窄，种群数量较少，每个栖息地斑块中不超过100只个体。由于人类活动的影响，其栖息地面积缩小。因此，莱氏蹄蝠列为易危等级 / Shield-faced Roundleaf Bat is narrowly distributed in China, and its populations are small, with no more than 100 individuals at each locality. Furthermore, the areas of its habitats are declining due to human encroachment. Thus, Shield-faced Roundleaf Bat is listed as Vulnerable

地理分布 Geographical Distribution

国内分布 Domestic Distribution
云南 / Yunnan
世界分布 World Distribution
中国、马来西亚、缅甸、泰国、越南 / China, Malaysia, Myanmar, Thailand, Viet Nam
分布标注 Distribution Note
非特有种 / Non-endemic

国内分布图
Map of Domestic Distribution

种群 Population

种群数量 Population Size	未知 / Unknown
种群趋势 Population Trend	未知 / Unknown

生境与生态系统 Habitat (s) and Ecosystem (s)

生　　境 Habitat(s)	洞穴、耕地、亚热带和热带退化森林 / Cave, Arable Land, Degraded Subtropical and Tropical Forest
生态系统 Ecosystem(s)	喀斯特生态系统、森林生态系统 / Karst Ecosystem, Forest Ecosystem

威胁 Threat (s)

主要威胁 Major Threat(s)	旅游休闲、采矿、采石场 / Tourism and Recreation, Mining, Quarrying Field

保护级别与保护行动 Protection Category and Conservation Action (s)

国家重点保护野生动物等级 (2021) Category of National Key Protected Wild Animals (2021)	无 / NA
IUCN 红色名录(2020-2) IUCN Red List (2020-2)	无危 / LC
CITES 附录 (2019) CITES Appendix (2019)	无 / NA
保护行动 Conservation Action(s)	无 / None

相关文献 Relevant References

Burgin *et al.*, 2018; Jiang *et al.* (蒋志刚等), 2017; He *et al.* (贺新平等), 2014; Lu *et al.* (陆长坤等), 1965

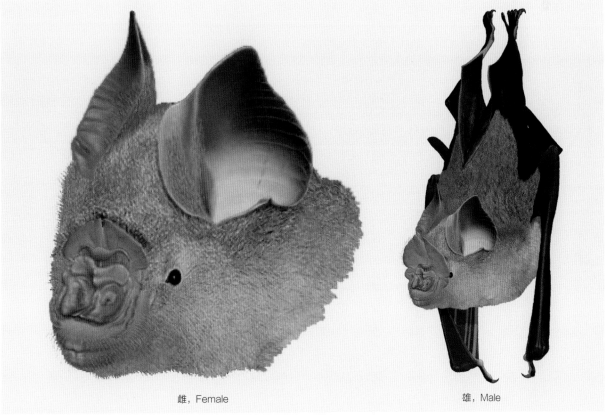

雌, Female　　　　　雄, Male

莱氏蹄蝠 *Hipposideros lylei*　　　　　By Alice Hughes

无尾蹄蝠
Coelops frithii

易危 VU B1; C2a(i)

| 数据缺乏 DD | 无危 LC | 近危 NT | 易危 VU | 濒危 EN | 极危 CR | 区域灭绝 RE | 野外灭绝 EW | 灭绝 EX |

分类地位 Taxonomic Status

动物界 Animalia	脊索动物门 Chordata	哺乳纲 Mammalia	翼手目 Chiroptera	蹄蝠科 Hipposideridae
学　　名 Scientific Name		*Coelops frithii*		
命　名　人 Species Authority		Blyth, 1848		
英　文　名 English Name(s)		Tail-less Leaf-nosed Bat		
同物异名 Synonym(s)		East Asian Tailless Leaf-nosed Bat		
种下单元评估 Infra-specific Taxa Assessed		无 / None		

评估信息 Assessment Information

评估年份 Year Assessed	2020
评　定　人 Assessor(s)	吴毅、蒋志刚 / Yi Wu, Zhigang Jiang
审　定　人 Reviewer(s)	张礼标、毛秀光、张树义 / Libiao Zhang, Xiuguang Mao, Shuyi Zhang
其他贡献人 Other Contributor(s)	李立立、丁晨晨 / Lili Li, Chenchen Ding

理由 Justification: 无尾蹄蝠分布虽广，但旅游和采石等已经影响到其生存，导致其种群数量下降。因此，无尾蹄蝠列为易危等级 / Although the distribution range of Tail-less Leaf-nosed Bat is large, its populations are threatened by human encroachment such as tourism and quarrying, its populations are constantly declining. Thus, Tail-less Leaf-nosed Bat is listed as Vulnerable

地理分布 Geographical Distribution

国内分布 Domestic Distribution	广东、江西、四川、海南、福建、台湾、广西、云南、贵州、重庆 / Guangdong, Jiangxi, Sichuan, Hainan, Fujian, Taiwan, Guangxi, Yunnan, Guizhou, Chongqing
世界分布 World Distribution	中国；南亚、东南亚 / China; South Asia, Southeast Asia
分布标注 Distribution Note	非特有种 / Non-endemic

国内分布图
Map of Domestic Distribution

种群 Population

种群数量 Population Size	稀少 / Rare
种群趋势 Population Trend	未知 / Unknown

生境与生态系统 Habitat (s) and Ecosystem (s)

生境 Habitat(s)	洞穴、耕地、亚热带和热带退化森林 / Cave, Arable Land, Degraded Subtropical and Tropical Forest
生态系统 Ecosystem(s)	喀斯特生态系统、森林生态系统 / Karst Ecosystem, Forest Ecosystem

威胁 Threat (s)

主要威胁 Major Threat(s)	伐木、耕种 / Logging, Plantation

保护级别与保护行动 Protection Category and Conservation Action (s)

国家重点保护野生动物等级 (2021) Category of National Key Protected Wild Animals (2021)	无 / NA
IUCN 红色名录 (2020-2) IUCN Red List (2020-2)	无危 / LC
CITES 附录 (2019) CITES Appendix (2019)	无 / NA
保护行动 Conservation Action(s)	无 / None

相关文献 Relevant References

Burgin *et al*., 2018; Jiang *et al*. (蒋志刚等), 2017; Xu *et al*. (徐忠鲜等), 2013; Liu *et al*. (刘森等), 2008; Zhu (朱斌良), 2008; He (何晓瑞), 1999

无尾蹄蝠 *Coelops frithii*　　　　余文华 摄　By Wenhua Yu

金黄鼠耳蝠
Myotis formosus

易危　VU B1; C2a(i)

| 数据缺乏 DD | 无危 LC | 近危 NT | **易危 VU** | 濒危 EN | 极危 CR | 区域灭绝 RE | 野外灭绝 EW | 灭绝 EX |

分类地位 Taxonomic Status

动物界 Animalia	脊索动物门 Chordata	哺乳纲 Mammalia	翼手目 Chiroptera	蝙蝠科 Vespertilionidae

学名 Scientific Name	*Myotis formosus*
命名人 Species Authority	Hodgson, 1835
英文名 English Name(s)	Hodgson's Bat
同物异名 Synonym(s)	Hodgson's Mouse-eared Bat; Copper-winged Bat; *Kerivoula pallida* (Blyth, 1863); *Myotis formosus* (Touessart, 1897) subsp. *andersoni*; *M. formosus* (Dobson, 1871) (转下页)
种下单元评估 Infra-specific Taxa Assessed	无 / None

评估信息 Assessment Information

评估年份 Year Assessed	2020
评定人 Assessor(s)	吴毅、蒋志刚 / Yi Wu, Zhigang Jiang
审定人 Reviewer(s)	张礼标、毛秀光、张树义 / Libiao Zhang, Xiuguang Mao, Shuyi Zhang
其他贡献人 Other Contributor(s)	李立立、丁晨晨 / Lili Li, Chenchen Ding

理由 Justification: 金黄鼠耳蝠分布虽广，但旅游和采石等影响到其生存，导致其种群数量下降，因此金黄鼠耳蝠列为易危等级 / Although Hodgson's Bat is widely distributed, it is threatened by tourism and quarrying, which lead to a constant decline of its populations. Thus, Hodgson's Bat is listed as Vulnerable

地理分布 Geographical Distribution

国内分布 Domestic Distribution
台湾、江西 / Taiwan, Jiangxi
世界分布 World Distribution
中国；南亚、东南亚 / China; South Asia, Southeast Asia
分布标注 Distribution Note
非特有种 / Non-endemic

国内分布图
Map of Domestic Distribution

种群 Population

种群数量 Population Size	未知 / Unknown
种群趋势 Population Trend	未知 / Unknown

生境与生态系统 Habitat (s) and Ecosystem (s)

生　　境 Habitat(s)	森林、竹林、耕地、牧场 / Forest, Bamboo Grove, Arable Land, Pastureland
生态系统 Ecosystem(s)	森林生态系统、草地生态系统、农田生态系统 / Forest Ecosystem, Grassland Ecosystem, Cropland Ecosystem

威胁 Threat (s)

主要威胁 Major Threat(s)	基础建设及交通、堤坝建设，生境退化、丧失 / Infrastructure and Transportation, Dam Construction, Habitat Degradation and Loss

保护级别与保护行动 Protection Category and Conservation Action (s)

国家重点保护野生动物等级 (2021) Category of National Key Protected Wild Animals (2021)	无 / NA
IUCN 红色名录 (2020-2) IUCN Red List (2020-2)	无危 / LC
CITES 附录 (2019) CITES Appendix (2019)	无 / NA
保护行动 Conservation Action(s)	无 / None

相关文献 Relevant References

Huang *et al*., 2020; Burgin *et al*., 2018; Dang *et al*.（党飞红等）, 2017; Jiang *et al*.（蒋志刚等）, 2017; Luo and Gao（罗键和高红英）, 2006; Lin（林良恭）, 2000; Luo *et al*.（罗蓉等）, 1993; Dong *et al*.（董聿茂等）, 1989; Liang and Dong（梁仁济和董永文）, 1984

（接上页）
subsp. *auratus*; *Vespertilio andersoni* (Trouessart, 1897); *V. auratus* (Dobson, 1871); *V. dobsoni* (Andersen, 1881); *V. formosa* (Hodgson, 1835)

By Wiki CC3.0

蒋志刚 绘 Drawn by Zhigang Jiang

金黄鼠耳蝠 *Myotis formosus*

中国生物多样性 红色名录

小巨足鼠耳蝠
Myotis hasseltii

易危　VU B1; C2a(i)

DD	LC	NT	VU	EN	CR	RE	EW	EX
数据缺乏	无危	近危	**易危**	濒危	极危	区域灭绝	野外灭绝	灭绝

分类地位 Taxonomic Status

动物界 Animalia	脊索动物门 Chordata	哺乳纲 Mammalia	翼手目 Chiroptera	蝙蝠科 Vespertilionidae

学名 Scientific Name	*Myotis hasseltii*
命名人 Species Authority	Temminck, 1840
英文名 English Name(s)	Lesser Large-footed Bat
同物异名 Synonym(s)	Lesser Large-footed Mouse-eared Bat; *Leuconoe hasselti* (Wroughton, 1918); *Vespertilio hasseltii* (Temminck, 1840)
种下单元评估 Infra-specific Taxa Assessed	无 / None

评估信息 Assessment Information

评估年份 Year Assessed	2020
评定人 Assessor(s)	吴毅、蒋志刚 / Yi Wu, Zhigang Jiang
审定人 Reviewer(s)	张礼标、毛秀光、张树义 / Libiao Zhang, Xiuguang Mao, Shuyi Zhang
其他贡献人 Other Contributor(s)	李立立、丁晨晨 / Lili Li, Chenchen Ding

理由 Justification: 小巨足鼠耳蝠在中国仅分布于云南省，且自其被记录后便少有发现，种群数量极少。因此，小巨足鼠耳蝠列为易危等级 / In China, Lesser Large-footed Bat is only distributed in Yunnan Province. Only a few records have been reported since its discovery. Its populations are very small. Thus, Lesser Large-footed Bat is listed as Vulnerable

地理分布 Geographical Distribution

国内分布 Domestic Distribution	云南 / Yunnan
世界分布 World Distribution	中国；南亚、东南亚 / China; South Asia, Southeast Asia
分布标注 Distribution Note	非特有种 / Non-endemic

国内分布图
Map of Domestic Distribution

种群 Population

种群数量 Population Size	未知 / Unknown
种群趋势 Population Trend	未知 / Unknown

生境与生态系统 Habitat (s) and Ecosystem (s)

生境 Habitat(s)	热带和亚热带干旱森林、红树林、竹林、人造建筑、海岸、喀斯特地貌、洞穴 / Tropical and Subtropical Dry Forest, Mangrove, Bamboo Grove, Building, Coast, Karst Landscape, Cave
生态系统 Ecosystem(s)	森林生态系统、海岸生态系统、人类聚落生态系统、喀斯特生态系统 / Forest Ecosystem, Coast Ecosystem, Human Settlement Ecosystem, Karst Ecosystem

威胁 Threat (s)

主要威胁 Major Threat(s)	伐木、耕种 / Logging, Plantation

保护级别与保护行动 Protection Category and Conservation Action (s)

国家重点保护野生动物等级 (2021) Category of National Key Protected Wild Animals (2021)	无 / NA
IUCN 红色名录 (2020-2) IUCN Red List (2020-2)	无危 / LC
CITES 附录 (2019) CITES Appendix (2019)	无 / NA
保护行动 Conservation Action(s)	无 / None

相关文献 Relevant References

Burgin *et al*., 2018; Jiang *et al*. (蒋志刚等), 2017; Zhang *et al*., 2009a; Zhang *et al*. (张礼标等), 2004; Koopman, 1993

小巨足鼠耳蝠 *Myotis hasseltii*   张礼标 摄 By Libiao Zhang

渡濑氏鼠耳蝠
Myotis rufoniger

易危 VU B1+2a(i); D2

| 数据缺乏 DD | 无危 LC | 近危 NT | **易危 VU** | 濒危 EN | 极危 CR | 区域灭绝 RE | 野外灭绝 EW | 灭绝 EX |

分类地位 Taxonomic Status

动物界 Animalia	脊索动物门 Chordata	哺乳纲 Mammalia	翼手目 Chiroptera	蝙蝠科 Vespertilionidae
学 名 Scientific Name		*Myotis rufoniger*		
命 名 人 Species Authority		Tomats, 1858		
英 文 名 English Name(s)		Red and Black Mouse-eared Bat		
同物异名 Synonym(s)		无 / None		
种下单元评估 Infra-specific Taxa Assessed		无 / None		

评估信息 Assessment Information

评估年份 Year Assessed	2020
评 定 人 Assessor(s)	吴毅、蒋志刚 / Yi Wu, Zhigang Jiang
审 定 人 Reviewer(s)	张礼标、毛秀光、张树义 / Libiao Zhang, Xiuguang Mao, Shuyi Zhang
其他贡献人 Other Contributor(s)	李立立、丁晨晨 / Lili Li, Chenchen Ding

理由 Justification: 渡濑氏鼠耳蝠分布区小，种群数量较少。因此，列为易危等级 / The area of occurrence and the populations of Red and Black Mouse-eared Bat are small. Thus, based on its area of occurrence and population, it is listed as Vulnerable

地理分布 Geographical Distribution

国内分布 Domestic Distribution

四川、贵州、广西、广东、福建、浙江、江苏、安徽、上海、湖北、湖南、陕西、吉林、辽宁、重庆、江西、河南 / Sichuan, Guizhou, Guangxi, Guangdong, Fujian, Zhejiang, Jiangsu, Anhui, Shanghai, Hubei, Hunan, Shaanxi, Jilin, Liaoning, Chongqing, Jiangxi, Henan

世界分布 World Distribution

中国；东南亚、东北亚 / China; Southeast Asia, Northeast Asia

分布标注 Distribution Note

非特有种 / Non-endemic

国内分布图
Map of Domestic Distribution

种群 Population

种群数量 Population Size	较常见 / Relatively common
种群趋势 Population Trend	稳定 / Stable

生境与生态系统 Habitat (s) and Ecosystem (s)

生境 Habitat(s)	次生林、溪流边、洞穴、人造建筑 / Secondary Forest, Near Stream, Cave, Building
生态系统 Ecosystem(s)	森林生态系统、湖泊河流生态系统、人类聚落生态系统 / Forest Ecosystem, Lake and River Ecosystem, Human Settlement Ecosystem

威胁 Threat (s)

主要威胁 Major Threat(s)	未知 / Unknown

保护级别与保护行动 Protection Category and Conservation Action (s)

国家重点保护野生动物等级 (2021) Category of National Key Protected Wild Animals (2021)	无 / NA
IUCN 红色名录 (2020-2) IUCN Red List (2020-2)	未列入 / NA
CITES 附录 (2019) CITES Appendix (2019)	无 / NA
保护行动 Conservation Action(s)	无 / None

相关文献 Relevant References

Burgin *et al.*, 2018; Dang *et al.* (党飞红等), 2017; Jiang *et al.* (蒋志刚等), 2017; Jiang *et al.*, 2010c

渡濑氏鼠耳蝠 *Myotis rufoniger* 吴毅 摄 By Yi Wu

大黑伏翼
Arielulus circumdatus
易危 VU B1; C2a(i)

分类地位 Taxonomic Status

动物界 Animalia	脊索动物门 Chordata	哺乳纲 Mammalia	翼手目 Chiroptera	蝙蝠科 Vespertilionidae

学名 Scientific Name	*Arielulus circumdatus*
命名人 Species Authority	Temminck, 1840
英文名 English Name(s)	Bronze Sprite
同物异名 Synonym(s)	Black-gilded Pipistrelle; *Pipistrellus circumdatus* (Temminck, 1840); *Vespertilio circumdatus* (Temminck, 1840)
种下单元评估 Infra-specific Taxa Assessed	无 / None

评估信息 Assessment Information

评估年份 Year Assessed	2020
评定人 Assessor(s)	吴毅、蒋志刚 / Yi Wu, Zhigang Jiang
审定人 Reviewer(s)	张礼标、毛秀光、张树义 / Libiao Zhang, Xiuguang Mao, Shuyi Zhang
其他贡献人 Other Contributor(s)	李立立、丁晨晨 / Lili Li, Chenchen Ding

理由 Justification: 大黑伏翼分布区狭窄，种群数量极少，且栖息地的面积因人类活动而不断减少。因此，大黑伏翼列为易危等级 / Bronze Sprite is narrowly distributed in China, and its populations are small. Furthermore, the areas of its occurence are declining due to human encroachment. Thus, Bronze Sprite is listed as Vulnerable

地理分布 Geographical Distribution

国内分布 Domestic Distribution
云南、广东 / Yunnan, Guangdong
世界分布 World Distribution
中国；南亚、东南亚 / China; South Asia, Southeast Asia
分布标注 Distribution Note
非特有种 / Non-endemic

国内分布图 Map of Domestic Distribution

种群 Population

种群数量 Population Size	稀少 / Rare
种群趋势 Population Trend	未知 / Unknown

生境与生态系统 Habitat(s) and Ecosystem(s)

生境 Habitat(s)	未知 / Unknown
生态系统 Ecosystem(s)	森林生态系统 / Forest Ecosystem

威胁 Threat(s)

主要威胁 Major Threat(s)	未知 / Unknown

保护级别与保护行动 Protection Category and Conservation Action(s)

国家重点保护野生动物等级 (2021) Category of National Key Protected Wild Animals (2021)	无 / NA
IUCN 红色名录 (2020-2) IUCN Red List (2020-2)	无危 / LC
CITES 附录 (2019) CITES Appendix (2019)	无 / NA
保护行动 Conservation Action(s)	无 / None

相关文献 Relevant References

Burgin *et al*., 2018; Jiang *et al*. (蒋志刚等), 2017; Zhang *et al*. (张礼标等), 2014; Hill and Harrison, 1986

大黑伏翼 *Arielulus circumdatus*   张礼标 摄   By Libiao Zhang

黄喉黑伏翼
Arielulus torquatus

易危 VU B1; C2a(i)

| DD | LC | NT | VU | EN | CR | RE | EW | EX |

分类地位 Taxonomic Status

| 动物界 Animalia | 脊索动物门 Chordata | 哺乳纲 Mammalia | 翼手目 Chiroptera | 蝙蝠科 Vespertilionidae |

学名 Scientific Name	*Arielulus torquatus*
命名人 Species Authority	Csorba and Lee, 1999
英文名 English Name(s)	Necklace Sprite
同物异名 Synonym(s)	Necklace Pipistrelle
种下单元评估 Infra-specific Taxa Assessed	无 / None

评估信息 Assessment Information

评估年份 Year Assessed	2020
评定人 Assessor(s)	吴毅、蒋志刚 / Yi Wu, Zhigang Jiang
审定人 Reviewer(s)	张礼标、毛秀光、张树义 / Libiao Zhang, Xiuguang Mao, Shuyi Zhang
其他贡献人 Other Contributor(s)	李立立、丁晨晨 / Lili Li, Chenchen Ding

理由 Justification: 黄喉黑伏翼为中国特有种，仅分布在台湾。其分布狭窄，种群数量极少，且栖息地的面积因人类活动而减少。因此，黄喉黑伏翼列为易危等级 / Necklace Sprite is an endemic species in China, only distributed in Taiwan. Its populations are small. Furthermore, the areas of its habitats are declining due to human encroachment. Thus, it is listed as Vulnerable

地理分布 Geographical Distribution

国内分布 Domestic Distribution
台湾 / Taiwan
世界分布 World Distribution
中国 / China
分布标注 Distribution Note
特有种 / Endemic

国内分布图
Map of Domestic Distribution

种群 Population

种群数量 Population Size	稀少 / Rare
种群趋势 Population Trend	持平 / Stable

生境与生态系统 Habitat(s) and Ecosystem(s)

生境 Habitat(s)	未知 / Unknown
生态系统 Ecosystem(s)	森林生态系统 / Forest Ecosystem

威胁 Threat(s)

主要威胁 Major Threat(s)	住宅及商业开发、公路和铁路建设、耕种 / Residential and Commercial Development, Road and Railroad Construction, Plantation

保护级别与保护行动 Protection Category and Conservation Action(s)

国家重点保护野生动物等级 (2021) Category of National Key Protected Wild Animals (2021)	无 / NA
IUCN 红色名录 (2020-2) IUCN Red List (2020-2)	无危 / LC
CITES 附录 (2019) CITES Appendix (2019)	无 / NA
保护行动 Conservation Action(s)	无 / None

相关文献 Relevant References

Burgin *et al.*, 2018; Jiang *et al.* (蒋志刚等), 2017; Csorba and Lee, 1999; Lin *et al.* (林良恭等), 1997

黄喉黑伏翼 *Arielulus torquatus*

By Alice Hughes

亚洲宽耳蝠
Barbastella leucomelas

易危 VU B1; C2a(i)

| DD | LC | NT | **VU** | EN | CR | RE | EW | EX |

分类地位 Taxonomic Status

动物界 Animalia	脊索动物门 Chordata	哺乳纲 Mammalia	翼手目 Chiroptera	蝙蝠科 Vespertilionidae

学名 Scientific Name	*Barbastella leucomelas*
命名人 Species Authority	Cretzschmar, 1826
英文名 English Name(s)	Asian Barbastelle
同物异名 Synonym(s)	Levant Barbastelle; Eastern Barbastelle; *Barbastella blanfordi* (Bianchi, 1917); *B. caspica* (Satunin, 1908); *B. walteri* (Bianchi, 1916); *Plecotus darjelingensis*
种下单元评估 Infra-specific Taxa Assessed	无 / None

评估信息 Assessment Information

评估年份 Year Assessed	2020
评定人 Assessor(s)	吴毅、蒋志刚 / Yi Wu, Zhigang Jiang
审定人 Reviewer(s)	张礼标、毛秀光、张树义 / Libiao Zhang, Xiuguang Mao, Shuyi Zhang
其他贡献人 Other Contributor(s)	李立立、丁晨晨 / Lili Li, Chenchen Ding

理由 Justification: 亚洲宽耳蝠分布虽广，但其种群数量少，其占有区面积因人类活动而减少。因此，亚洲宽耳蝠列为易危等级 / Although Asian Barbastelle is distributed widely across China, its populations are extremely small. Furthermore, the areas of its habitats are declining due to human encroachment. Thus, Asian Barbastelle is listed as Vulnerable

地理分布 Geographical Distribution

国内分布 Domestic Distribution	内蒙古、新疆、陕西、甘肃、青海、四川、云南、重庆 / Inner Mongolia (Nei Mongol), Xinjiang, Shaanxi, Gansu, Qinghai, Sichuan, Yunnan, Chongqing
世界分布 World Distribution	中国；中亚、西亚、南亚 / China; Central Asia, West Asia, South Asia
分布标注 Distribution Note	非特有种 / Non-endemic

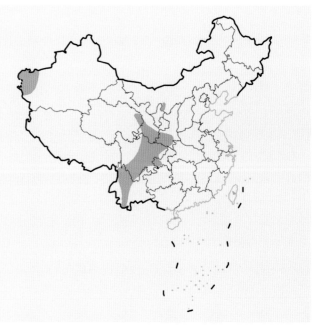

国内分布图
Map of Domestic Distribution

种群 Population

种群数量 Population Size	稀少 / Rare
种群趋势 Population Trend	稳定 / Stable

生境与生态系统 Habitat(s) and Ecosystem(s)

生境 Habitat(s)	热带和亚热带湿润山地森林、洞穴、人造建筑 / Tropical and Subtropical Moist Montane Forest, Cave, Building
生态系统 Ecosystem(s)	森林生态系统、人类聚落生态系统 / Forest Ecosystem, Human Settlement Ecosystem

威胁 Threat(s)

主要威胁 Major Threat(s)	种群密度过低，特殊生境要求 / Low Population Density, Specialized Habitat Requirement

保护级别与保护行动 Protection Category and Conservation Action(s)

国家重点保护野生动物等级 (2021) Category of National Key Protected Wild Animals (2021)	无 / NA
IUCN 红色名录 (2020-2) IUCN Red List (2020-2)	无危 / LC
CITES 附录 (2019) CITES Appendix (2019)	无 / NA
保护行动 Conservation Action(s)	无 / None

相关文献 Relevant References

Burgin *et al.*, 2018; Jiang *et al.*（蒋志刚等），2017; Smith *et al.*（史密斯等），2009; Wang（王应祥），2003; Lin *et al.*, 2002a

亚洲宽耳蝠 *Barbastella leucomelas*　　　　吴毅 摄　By Yi Wu

短尾猴
Macaca arctoides
易危 VU A2bd

| 数据缺乏 DD | 无危 LC | 近危 NT | 易危 VU | 濒危 EN | 极危 CR | 区域灭绝 RE | 野外灭绝 EW | 灭绝 EX |

分类地位 Taxonomic Status

动物界 Animalia	脊索动物门 Chordata	哺乳纲 Mammalia	灵长目 Primates	猴科 Cercopithecidae

学名 Scientific Name	*Macaca arctoides*
命名人 Species Authority	Geoffroy, 1831
英文名 English Name(s)	Stump-tailed Macaque
同物异名 Synonym(s)	Bear Macaque; *Macaca brunneus* (Anderson, 1871); *harmandi* (Trouessart, 1897); *melanotus* (Ogilby, 1839); *melli* (Matschie, 1912); *rufescens* (Anderson, 1872);（转下页）
种下单元评估 Infra-specific Taxa Assessed	无 / None

评估信息 Assessment Information

评估年份 Year Assessed	2020
评定人 Assessor(s)	蒋志刚、蒋学龙 / Zhigang Jiang, Xuelong Jiang
审定人 Reviewer(s)	黄乘明、李进华 / Chengming Huang, Jinhua Li
其他贡献人 Other Contributor(s)	李立立、丁晨晨 / Lili Li, Chenchen Ding

理由 Justification: 短尾猴分布虽广，但其种群数量少，栖息地破碎，栖息地面积因人类活动而减少。因此，将短尾猴列为易危等级 / Although Stump-tailed Macaque is widely distributed in China, its populations are small. Furthermore, its habitats are fragmented, and the areas of its habitats are declining due to human encroachment. Thus, it is listed as Vulnerable

地理分布 Geographical Distribution

国内分布 Domestic Distribution
广西、云南、湖南、广东、贵州、江西 / Guangxi, Yunnan, Hunan, Guangdong, Guizhou, Jiangxi
世界分布 World Distribution
中国；南亚、东南亚 / China; South Asia, Southeast Asia
分布标注 Distribution Note
非特有种 / Non-endemic

国内分布图
Map of Domestic Distribution

种群 Population

种群数量 Population Size	200,000 只 / 200,000 individuals
种群趋势 Population Trend	下降 / Decreasing

生境与生态系统 Habitat (s) and Ecosystem (s)

生　　境 Habitat(s)	热带和亚热带湿润山地森林 / Tropical and Subtropical Moist Montane Forest
生态系统 Ecosystem(s)	森林生态系统 / Forest Ecosystem

威胁 Threat (s)

主要威胁 Major Threat(s)	人类活动，栖息地减少 / Human Encroachment, Habitat Decreasing

保护级别与保护行动 Protection Category and Conservation Action (s)

国家重点保护野生动物等级 (2021) Category of National Key Protected Wild Animals (2021)	二级 / Category II
IUCN 红色名录(2020-2) IUCN Red List (2020-2)	易危 / VU
CITES 附录 (2019) CITES Appendix (2019)	II
保护行动 Conservation Action(s)	在自然保护区内的种群及栖息地得到保护 / Populations and habitats are protected in nature reserves

相关文献 Relevant References

Burgin *et al*., 2018; Jiang *et al*.（蒋志刚等），2017; Wilson and Mittermeier, 2012; Smith *et al*.（史密斯等），2009; Wang（王应祥），2003; Wang *et al*.（王岐山等），1994a, 1994b

(接上页)
speciosus (Murie, 1875); *ursinus* (Gervais, 1854)

短尾猴 *Macaca arctoides*　　　　蒋志刚 摄　By Zhigang Jiang

熊猴
Macaca assamensis

易危 VU A2bd; B2

| DD | LC | NT | VU | EN | CR | RE | EW | EX |

分类地位 Taxonomic Status

| 动物界 Animalia | 脊索动物门 Chordata | 哺乳纲 Mammalia | 灵长目 Primates | 猴科 Cercopithecidae |

学名 Scientific Name	*Macaca assamensis*
命名人 Species Authority	M'Clelland, 1840
英文名 English Name(s)	Assam Macaque
同物异名 Synonym(s)	Assamese Macaque; *Macaca coolidgei* (Osgood, 1932); *macclellandii* (Gray, 1846); *problematicus* (Gray, 1870); *rhesosimilis* (Sclater, 1872); *sikimensis* (Hodgson, 1867)
种下单元评估 Infra-specific Taxa Assessed	无 / None

评估信息 Assessment Information

评估年份 Year Assessed	2020
评定人 Assessor(s)	蒋志刚、蒋学龙 / Zhigang Jiang, Xuelong Jiang
审定人 Reviewer(s)	黄乘明、蒋学龙、范朋飞 / Chengming Huang, Xuelong Jiang, Pengfei Fan
其他贡献人 Other Contributor(s)	李立立、丁晨晨 / Lili Li, Chenchen Ding

理由 Justification: 熊猴在中国分布虽广，但其种群数量少，且占有区面积小。因此，熊猴列入易危等级 / Although Assam Macaque is widely distributed in China, its populations are small. Furthermore, the areas of its occupancy are limited. Thus, Assam Macaque is listed as Vulnerable

地理分布 Geographical Distribution

| 国内分布 Domestic Distribution |
| 广东、云南、广西、贵州、西藏 / Guangdong, Yunnan, Guangxi, Guizhou, Tibet (Xizang) |
| 世界分布 World Distribution |
| 中国；南亚、东南亚 / China; South Asia, Southeast Asia |
| 分布标注 Distribution Note |
| 非特有种 / Non-endemic |

国内分布图
Map of Domestic Distribution

种群 Population

种群数量 Population Size	8,200 只 / 8,200 individuals
种群趋势 Population Trend	下降 / Decreasing

生境与生态系统 Habitat(s) and Ecosystem(s)

生　　境 Habitat(s)	森林 / Forest
生态系统 Ecosystem(s)	森林生态系统 / Forest Ecosystem

威胁 Threat(s)

主要威胁 Major Threat(s)	伐木、牧场、狩猎 / Logging, Ranching, Hunting

保护级别与保护行动 Protection Category and Conservation Action(s)

国家重点保护野生动物等级 (2021) Category of National Key Protected Wild Animals (2021)	二级 / Category II
IUCN 红色名录 (2020-2) IUCN Red List (2020-2)	近危 / NT
CITES 附录 (2019) CITES Appendix (2019)	II
保护行动 Conservation Action(s)	在自然保护区内的种群及栖息地得到保护 / Populations and habitats in nature reserves are protected

相关文献 Relevant References

Burgin *et al*., 2018; Jiang *et al*. (蒋志刚等), 2017; Wilson and Mittermeier, 2012; Zhang (张荣祖), 2002; Groves, 2001; Jiang *et al*. (蒋学龙等), 1993

熊猴 *Macaca assamensis* 　　　　　王昌大 摄（西南山地供图）　By Changda Wang

藏酋猴
Macaca thibetana

易危 VU A2bd; B2

| 数据缺乏 DD | 无危 LC | 近危 NT | **易危 VU** | 濒危 EN | 极危 CR | 区域灭绝 RE | 野外灭绝 EW | 灭绝 EX |

分类地位 Taxonomic Status

动物界 Animalia	脊索动物门 Chordata	哺乳纲 Mammalia	灵长目 Primates	猴科 Cercopithecidae

学 名 Scientific Name	*Macaca thibetana*
命 名 人 Species Authority	Milne-Edwards, 1870
英 文 名 English Name(s)	Tibetan Macaque
同物异名 Synonym(s)	Chinese Stump-tailed Macaque; Milne-Edwards' Macaque; *Macaca pullus* (Howell, 1928)
种下单元评估 Infra-specific Taxa Assessed	无 / None

评估信息 Assessment Information

评估年份 Year Assessed	2020
评 定 人 Assessor(s)	蒋志刚、蒋学龙 / Zhigang Jiang, Xuelong Jiang
审 定 人 Reviewer(s)	黄乘明、蒋学龙、龙勇诚、范朋飞 / Chengming Huang, Xuelong Jiang, Yongcheng Long, Pengfei Fan
其他贡献人 Other Contributor(s)	李立立、丁晨晨 / Lili Li, Chenchen Ding

理由 Justification: 藏酋猴为中国特有种，种群数量少，且栖息地的面积因人类活动而不断减少。因此，藏酋猴列入易危等级 / Tibetan Macaque is endemic to China, and its populations are small. Furthermore, the areas of its habitats are fragmented due to human activities. Thus, Tibetan Macaque is listed as Vulnerable

地理分布 Geographical Distribution

国内分布 Domestic Distribution	浙江、湖南、安徽、福建、江西、湖北、广东、广西、四川、贵州、云南、西藏、甘肃、重庆 / Zhejiang, Hunan, Anhui, Fujian, Jiangxi, Hubei, Guangdong, Guangxi, Sichuan, Guizhou, Yunnan, Tibet (Xizang), Gansu, Chongqing
世界分布 World Distribution	中国 / China
分布标注 Distribution Note	特有种 / Endemic

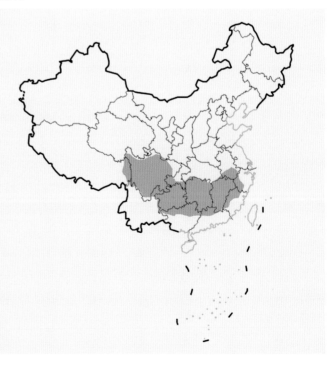

国内分布图
Map of Domestic Distribution

种群 Population

种群数量 Population Size	17,000 只 / 17,000 individuals
种群趋势 Population Trend	稳定 / Stable

生境与生态系统 Habitat(s) and Ecosystem(s)

生境 Habitat(s)	热带和亚热带湿润山地森林，洞穴，次生林 / Tropical and Subtropical Moist Montane Forest, Cave, Secondary Forest
生态系统 Ecosystem(s)	森林生态系统 / Forest Ecosystem

威胁 Threat(s)

主要威胁 Major Threat(s)	狩猎 / Hunting

保护级别与保护行动 Protection Category and Conservation Action(s)

国家重点保护野生动物等级 (2021) Category of National Key Protected Wild Animals (2021)	二级 / Category II
IUCN 红色名录 (2020-2) IUCN Red List (2020-2)	近危 / NT
CITES 附录 (2019) CITES Appendix (2019)	II
保护行动 Conservation Action(s)	在自然保护区内的种群及栖息地得到保护 / Populations and habitats are protected in nature reserves

相关文献 Relevant References

Burgin *et al.*, 2018; Jiang *et al.*（蒋志刚等）, 2017; Wilson and Mittermeier, 2012; Jiang *et al.*（蒋学龙等）, 1996

藏酋猴 *Macaca thibetana* 蒋志刚 摄 By Zhigang Jiang

中国生物多样性红色名录

菲氏叶猴
Trachypithecus phayrei

易危 VU A2bd; B1ab(i, ii, iii)

| 数据缺乏 DD | 无危 LC | 近危 NT | **易危 VU** | 濒危 EN | 极危 CR | 区域灭绝 RE | 野外灭绝 EW | 灭绝 EX |

分类地位 Taxonomic Status

动物界 Animalia	脊索动物门 Chordata	哺乳纲 Mammalia	灵长目 Primates	猴科 Cercopithecidae

学　　名 Scientific Name	*Trachypithecus phayrei*
命　名　人 Species Authority	Blyth, 1847
英　文　名 English Name(s)	Phayre's Leaf Monkey
同物异名 Synonym(s)	Phayre's Langur
种下单元评估 Infra-specific Taxa Assessed	无 / None

评估信息 Assessment Information

评估年份 Year Assessed	2020
评　定　人 Assessor(s)	蒋志刚、蒋学龙 / Zhigang Jiang, Xuelong Jiang
审　定　人 Reviewer(s)	黄乘明、蒋学龙、龙勇诚、范朋飞 / Chengming Huang, Xuelong Jiang, Yongcheng Long, Pengfei Fan
其他贡献人 Other Contributor(s)	丁晨晨 / Chenchen Ding

理由 Justification: 菲氏叶猴在中国仅分布在云南地区，有一定种群数量，但其栖息地面积因人类活动而减少。因此，列为易危等级 / In China, Phayre's Leaf Monkey is narrowly distributed in Yunnan Province, and it has a certain number of population. However, the areas of its habitats are declining due to human activities. Thus, Phayre's Leaf Monkey is listed as Vulnerable

地理分布 Geographical Distribution

国内分布 Domestic Distribution
云南 / Yunnan
世界分布 World Distribution
中国；南亚、东南亚 / China; South Asia, Southeast Asia
分布标注 Distribution Note
非特有种 / Non-endemic

国内分布图
Map of Domestic Distribution

种群 Population

种群数量 Population Size	700 只 / 700 individuals
种群趋势 Population Trend	下降 / Decreasing

生境与生态系统 Habitat(s) and Ecosystem(s)

生境 Habitat(s)	森林、盐碱地 / Forest; Saline, Brackish or Alkaline Land
生态系统 Ecosystem(s)	森林生态系统 / Forest Ecosystem

威胁 Threat(s)

主要威胁 Major Threat(s)	伐木、火灾、耕种 / Logging, Fire, Plantation

保护级别与保护行动 Protection Category and Conservation Action(s)

国家重点保护野生动物等级 (2021) Category of National Key Protected Wild Animals (2021)	一级 / Category I
IUCN 红色名录 (2020-2) IUCN Red List (2020-2)	濒危 / EN
CITES 附录 (2019) CITES Appendix (2019)	II
保护行动 Conservation Action(s)	在自然保护区内的种群及栖息地得到保护 / Populations and habitats in nature reserves are protected

相关文献 Relevant References

Burgin *et al.*, 2018; Jiang *et al.* (蒋志刚等), 2017; He *et al.*, 2012b; Wilson and Mittermeier, 2012; Wang *et al.* (王应祥等), 1999; Li and Lin (李致祥和林正玉), 1983; He and Yang (何晓瑞和杨德华), 1982

菲氏叶猴 *Trachypithecus phayrei*　王斌 摄（西南山地供图）　By Bin Wang (Swild.cn)　　魏行智 摄（西南山地供图）　By Xingzhi Wei (Swild.cn)

中国生物多样性 红色名录
China's Red List of Biodiversity

棕熊
Ursus arctos

易危 VU A3d; B1ab(i, ii, iii)

| 数据缺乏 DD | 无危 LC | 近危 NT | **易危 VU** | 濒危 EN | 极危 CR | 区域灭绝 RE | 野外灭绝 EW | 灭绝 EX |

分类地位 Taxonomic Status

动物界 Animalia	脊索动物门 Chordata	哺乳纲 Mammalia	食肉目 Carnivora	熊科 Ursidae

学 名 Scientific Name	*Ursus arctos*
命 名 人 Species Authority	Linnaeus, 1758
英 文 名 English Name(s)	Brown Bear
同物异名 Synonym(s)	无 / None
种下单元评估 Infra-specific Taxa Assessed	无 / None

评估信息 Assessment Information

评 估 年 份 Year Assessed	2020
评 定 人 Assessor(s)	蒋志刚 / Zhigang Jiang
审 定 人 Reviewer(s)	马建章、徐爱春、张明海 / Jianzhang Ma, Aichun Xu, Minghai Zhang
其他贡献人 Other Contributor(s)	李立立、丁晨晨 / Lili Li, Chenchen Ding

理由 Justification: 棕熊分布范围广，但其种群数量少，其占有区面积因人类活动而减少。因此，将棕熊列为易危等级 / Brown Bear is widely distributed in China, but its populations are small. Furthermore, the areas of its occupancy are declining due to human activities. Thus, Brown Bear is listed as Vulnerable

地理分布 Geographical Distribution

国内分布 Domestic Distribution
吉林、黑龙江、内蒙古、辽宁、四川、云南、西藏、甘肃、青海、新疆 / Jilin, Heilongjiang, Inner Mongolia (Nei Mongol), Liaoning, Sichuan, Yunnan, Tibet (Xizang), Gansu, Qinghai, Xinjiang
世界分布 World Distribution
中国；环北极圈 / China; Circum-Arctic
分布标注 Distribution Note
非特有种 / Non-endemic

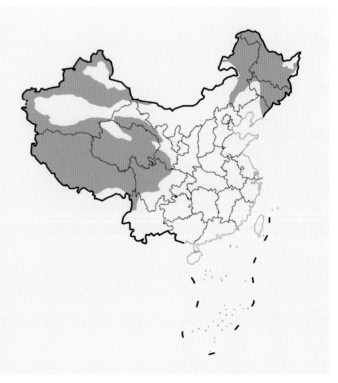

国内分布图
Map of Domestic Distribution

种群 Population

种群数量 Population Size	15,000 只 / 15,000 individuals
种群趋势 Population Trend	稳定 / Stable

生境与生态系统 Habitat (s) and Ecosystem (s)

生　　境 Habitat(s)	森林、青藏高原、苔原 / Forest, Qinghai-Tibet (Xizang) Plateau, Tundra
生态系统 Ecosystem(s)	森林生态系统、高寒草甸生态系统 / Forest Ecosystem, Alpine Meadow Ecosystem

威胁 Threat (s)

主要威胁 Major Threat(s)	狩猎、耕种、树木种植园、人类活动干扰 / Hunting, Plantation, Wood Plantation, Human Disturbance

保护级别与保护行动 Protection Category and Conservation Action (s)

国家重点保护野生动物等级 (2021) Category of National Key Protected Wild Animals (2021)	二级 / Category II
IUCN 红色名录 (2020-2) IUCN Red List (2020-2)	无危 / LC
CITES 附录 (2019) CITES Appendix (2019)	I
保护行动 Conservation Action(s)	部分种群位于自然保护区内 / Part of populations are covered by nature reserve

相关文献 Relevant References

Burgin *et al.*, 2018; Jiang *et al.*（蒋志刚等）, 2017; Malcolm *et al.*, 2014; Xu *et al.*, 2006; Zhang（张明海）, 2002; Ma *et al.*, 1994; Guo *et al.*, 1997; Lei（雷俊宏）, 1991; Gao *et al.*（高耀亭等）, 1987

棕熊 *Ursus arctos*

黑熊
Ursus thibetanus

易危 VU A3d; B1ab(i, ii, iii)

| 数据缺乏 DD | 无危 LC | 近危 NT | **易危 VU** | 濒危 EN | 极危 CR | 区域灭绝 RE | 野外灭绝 EW | 灭绝 EX |

分类地位 Taxonomic Status

动物界 Animalia	脊索动物门 Chordata	哺乳纲 Mammalia	食肉目 Carnivora	熊科 Ursidae

学名 Scientific Name	*Ursus thibetanus*
命名人 Species Authority	G. Cuvier, 1823
英文名 English Name(s)	Asiatic Black Bear
同物异名 Synonym(s)	Asian Black Bear, Moon Bear, White-chested Bear
种下单元评估 Infra-specific Taxa Assessed	无 / None

评估信息 Assessment Information

评估年份 Year Assessed	2020
评定人 Assessor(s)	蒋志刚 / Zhigang Jiang
审定人 Reviewer(s)	马建章、徐爱春 / Jianzhang Ma, Aichun Xu
其他贡献人 Other Contributor(s)	李立立、丁晨晨 / Lili Li, Chenchen Ding

理由 Justification: 黑熊分布范围广，但其生存与人类活动方式冲突，其占有区破碎化。因此，基于其占有区现状，将黑熊列入易危等级 / Asiatic Black Bear is widely distributed in China, but its survival is in conflict with human being, its area of occupancy is fragmented. Thus, based on the status of its area of occupancy, Asiatic Black Bear is listed as Vulnerable

地理分布 Geographical Distribution

国内分布 Domestic Distribution

吉林、黑龙江、西藏、河北、内蒙古、辽宁、浙江、安徽、福建、湖南、广东、广西、海南、四川、贵州、云南、陕西、甘肃、青海、湖北、江西、重庆、台湾 / Jilin, Heilongjiang, Tibet (Xizang), Hebei, Inner Mongolia (Nei Mongol), Liaoning, Zhejiang, Anhui, Fujian, Hunan, Guangdong, Guangxi, Hainan, Sichuan, Guizhou, Yunnan, Shaanxi, Gansu, Qinghai, Hubei, Jiangxi, Chongqing, Taiwan

世界分布 World Distribution

亚洲 / Asia

分布标注 Distribution Note

非特有种 / Non-endemic

国内分布图
Map of Domestic Distribution

🦌 种群 Population

种群数量 Population Size	28,000 只 / 28,000 individuals
种群趋势 Population Trend	下降 / Decreasing

🦌 生境与生态系统 Habitat (s) and Ecosystem (s)

生　　境 Habitat(s)	热带和亚热带湿润山地森林、针阔混交林 / Tropical and Subtropical Moist Montane Forest, Coniferous and Broad-leaved Mixed Forest
生态系统 Ecosystem(s)	森林生态系统 / Forest Ecosystem

🦌 威胁 Threat (s)

主要威胁 Major Threat(s)	狩猎、耕种、伐木、火灾 / Hunting, Plantation, Logging, Fire

🦌 保护级别与保护行动 Protection Category and Conservation Action (s)

国家重点保护野生动物等级 (2021) Category of National Key Protected Wild Animals (2021)	二级 / Category II
IUCN 红色名录 (2020-2) IUCN Red List (2020-2)	易危 / VU
CITES 附录 (2019) CITES Appendix (2019)	I
保护行动 Conservation Action(s)	部分种群位于自然保护区内 / Part of populations are covered by nature reserves

🦌 相关文献 Relevant References

Burgin *et al.*, 2018; Jiang *et al.* (蒋志刚等), 2017; Escobar *et al.*, 2015; Ohnishi and Osawa, 2014; Liu *et al.*, 2011, 2009; Gao *et al.* (高耀亭等), 1987

黑熊 *Ursus thibetanus*

中国生物多样性 红色名录

大熊猫
Ailuropoda melanoleuca

易危 VU C1i

| 数据缺乏 DD | 无危 LC | 近危 NT | **易危 VU** | 濒危 EN | 极危 CR | 区域灭绝 RE | 野外灭绝 EW | 灭绝 EX |

分类地位 Taxonomic Status

动物界 Animalia	脊索动物门 Chordata	哺乳纲 Mammalia	食肉目 Carnivora	大熊猫科 Ailuropodidae
学 名 Scientific Name		*Ailuropoda melanoleuca*		
命名人 Species Authority		David, 1869		
英文名 English Name(s)		Giant Panda		
同物异名 Synonym(s)		Panda, Panda Bear		
种下单元评估 Infra-specific Taxa Assessed		无 / None		

评估信息 Assessment Information

评估年份 Year Assessed	2020
评定人 Assessor(s)	蒋志刚 / Zhigang Jiang
审定人 Reviewer(s)	刘定震、方盛国 / Dingzhen Liu, Shengguo Fang
其他贡献人 Other Contributor(s)	李立立、丁晨晨 / Lili Li, Chenchen Ding

理由 Justification: 第四次全国大熊猫调查发现野生大熊猫种群数量比第三次全国大熊猫调查增长 16.8%，达到 1,864 只，中国现已建设了 67 处大熊猫自然保护区，覆盖了 66.8% 的野生大熊猫种群和 53.8% 的大熊猫栖息地。2019 年，人工繁育的大熊猫数量超过 600 只。启动了野化放归工作，先后将 7 只人工繁育大熊猫放归自然。鉴于大熊猫种群的恢复，2015 年，中国生物多样性红色名录评估已经将大熊猫的濒危等级降为易危 (VU) 等级。本次评估维持易危 (VU) 等级 / The population of Giant Panda reported by the Fourth National Survey on Giant Panda was 1,864 individuals, an increase of 16.8% compared to the Third National Survey on Giant Panda. China has established 67 nature reserves for Giant Panda, encompassing 66.8% of the wild populations of Giant panda and 53.8% of its suitable habitat. In 2019, there were 600 Giant Pandas bred in captivity. Moreover, 7 captively-bred Giant Pandas were released into the wild. Based on the Giant Panda's recovering populations, it has been down-listed to Vulnerable in the Red List of China's Biodiversity (2015). This assessment maintains its status as Vulnerable

地理分布 Geographical Distribution

国内分布 Domestic Distribution
四川、陕西、甘肃 / Sichuan, Shaanxi, Gansu
世界分布 World Distribution
中国 / China
分布标注 Distribution Note
特有种 / Endemic

国内分布图
Map of Domestic Distribution

🦌 种群 Population

种群数量 Population Size	1,864 只 / 1,864 individuals
种群趋势 Population Trend	稳定 / Stable

🦌 生境与生态系统 Habitat (s) and Ecosystem (s)

生　　境 Habitat(s)	亚热带湿润山地森林、竹林 / Subtropical Moist Montane Forest, Bamboo Forest
生态系统 Ecosystem(s)	森林生态系统 / Forest Ecosystem

🦌 威胁 Threat (s)

主要威胁 Major Threat(s)	生境破碎、耕种、伐木、食物 (竹子) 缺乏 / Habitat Fragmentation, Plantation, Logging, Food (bamboo) Scarcity

🦌 保护级别与保护行动 Protection Category and Conservation Action (s)

国家重点保护野生动物等级 (2021) Category of National Key Protected Wild Animals (2021)	一级 / Category I
IUCN 红色名录 (2020-2) IUCN Red List (2020-2)	濒危 / EN
CITES 附录 (2019) CITES Appendix (2019)	I
保护行动 Conservation Action(s)	大部分种群位于自然保护区内 / Most populations are covered by nature reserve

🦌 相关文献 Relevant References

Burgin *et al.*, 2018; Jiang *et al.* (蒋志刚等), 2017; Wang *et al.* (王志学等), 2008; Yan (严旬), 2005; Wang *et al.* (王昊等), 2002; Hu (胡锦矗), 2001; Gao *et al.* (高耀亭等), 1987

大熊猫 *Ailuropoda melanoleuca*

小熊猫
Ailurus fulgens

易危 VU A3d; B1ab(i, ii, iii)

| 数据缺乏 DD | 无危 LC | 近危 NT | 易危 VU | 濒危 EN | 极危 CR | 区域灭绝 RE | 野外灭绝 EW | 灭绝 EX |

分类地位 Taxonomic Status

动物界 Animalia	脊索动物门 Chordata	哺乳纲 Mammalia	食肉目 Carnivora	小熊猫科 Ailuridae
学 名 Scientific Name		*Ailurus fulgens*		
命 名 人 Species Authority		F. G. Cuvier, 1825		
英 文 名 English Name(s)		Red Panda		
同物异名 Synonym(s)		Lesser Panda, Red Bear-cat, Red Cat-bear, Chinese Red Panda, *Ailurus styani*		
种下单元评估 Infra-specific Taxa Assessed		无 / None		

评估信息 Assessment Information

评 估 年 份 Year Assessed	2020
评 定 人 Assessor(s)	蒋志刚 / Zhigang Jiang
审 定 人 Reviewer(s)	李明 / Ming Li
其他贡献人 Other Contributor(s)	李立立、丁晨晨 / Lili Li, Chenchen Ding

理由 Justification: 小熊猫分布范围较广，但其种群数量少，且栖息地面积因人类活动而不断减少。因此，小熊猫列为易危等级 / Red Panda is relatively widely distributed in China, but its population densities are low. Furthermore, the areas of its habitats are declining due to human activities. Thus, Red Panda is listed as Vulnerable

地理分布 Geographical Distribution

国内分布 Domestic Distribution
四川、云南、西藏、青海、甘肃、陕西 / Sichuan, Yunnan, Tibet (Xizang), Qinghai, Gansu, Shaanxi
世界分布 World Distribution
中国；南亚 / China; South Asia
分布标注 Distribution Note
非特有种 / Non-endemic

国内分布图
Map of Domestic Distribution

种群 Population

种群数量 Population Size	8,000 只 / 8,000 individuals
种群趋势 Population Trend	下降 / Decreasing

生境与生态系统 Habitat (s) and Ecosystem (s)

生 境 Habitat(s)	温带森林、针阔混交林、竹林 / Temperate Forest, Coniferous and Broad-leaved Mixed Forest, Bamboo Forest
生态系统 Ecosystem(s)	森林生态系统 / Forest Ecosystem

威胁 Threat (s)

主要威胁 Major Threat(s)	狩猎、伐木、家畜放牧、旅游、公路和铁路建设 / Hunting, Logging, Livestock Ranching, Tourism, Road and Railroad Construction

保护级别与保护行动 Protection Category and Conservation Action (s)

国家重点保护野生动物等级 (2021) Category of National Key Protected Wild Animals (2021)	二级 / Category II
IUCN 红色名录 (2020-2) IUCN Red List (2020-2)	濒危 / EN
CITES 附录 (2019) CITES Appendix (2019)	I
保护行动 Conservation Action(s)	大部分种群位于自然保护区内 / Most populations are covered by nature reserves

相关文献 Relevant References

Burgin *et al*., 2018; Jiang *et al*. (蒋志刚等), 2017; Hu and Du (胡刚和杜勇), 2002; Su *et al*., 2001; Wei *et al*., 1999; Gao *et al*. (高耀亭等), 1987

小熊猫 *Ailurus fulgens*

北海狗
Callorhinus ursinus

易危　VU A3d ver 3.1

| DD 数据缺乏 | LC 无危 | NT 近危 | **VU 易危** | EN 濒危 | CR 极危 | RE 区域灭绝 | EW 野外灭绝 | EX 灭绝 |

分类地位 Taxonomic Status

动物界 Animalia	脊索动物门 Chordata	哺乳纲 Mammalia	食肉目 Carnivora	海狮科 Otariidae

学　　名 Scientific Name	*Callorhinus ursinus*
命　名　人 Species Authority	Linnaeus, 1758
英　文　名 English Name(s)	Northern Fur Seal
同物异名 Synonym(s)	无 / None
种下单元评估 Infra-specific Taxa Assessed	无 / None

评估信息 Assessment Information

评估年份 Year Assessed	2020
评　定　人 Assessor(s)	周开亚 / Kaiya Zhou
审　定　人 Reviewer(s)	张先锋、王克雄、王丁、祝茜、蒋志刚 / Xianfeng Zhang, Kexiong Wang, Ding Wang, Qian Zhu, Zhigang Jiang
其他贡献人 Other Contributor(s)	李立立、丁晨晨 / Lili Li, Chenchen Ding

理由 Justification: 北海狗是一个栖息在北太平洋和白令海的近北极物种。一些游荡个体偶然到达中国黄海、东海和南海沿岸。基于其种群密度，将北海狗列为易危等级 / Northern Fur Seal is a subpolar species that inhabits the North Pacific Ocean and Bering Sea. Some roaming individuals accidentally reached China's Yellow Sea, East China Sea and Coast of South China Sea. Based on its population density, Northern Fur Seal is listed as Vulnerable.

地理分布 Geographical Distribution

国内分布 Domestic Distribution	黄海、东海、台湾海峡、南海 / Yellow Sea, East China Sea, Taiwan Strait, South China Sea
世界分布 World Distribution	北太平洋及白令海 / North Pacific Ocean and Bering Sea
分布标注 Distribution Note	非特有种 / Non-endemic

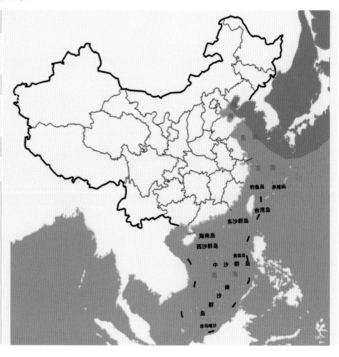

国内分布图
Map of Domestic Distribution

种群 Population

种群数量 Population Size	未知 / Unknown
种群趋势 Population Trend	下降 / Decreasing

生境与生态系统 Habitat (s) and Ecosystem (s)

生　　境 Habitat(s)	海洋 / Ocean
生态系统 Ecosystem(s)	海洋生态系统 / Ocean Ecosystem

威胁 Threat (s)

主要威胁 Major Threat(s)	丢弃渔具、海洋垃圾、漏油 / Discarded Fishing Gear, Marine Debris, Oil Spill

保护级别与保护行动 Protection Category and Conservation Action (s)

国家重点保护野生动物等级 (2021) Category of National Key Protected Wild Animals (2021)	二级 / Category II
IUCN红色名录 (2020-2) IUCN Red List (2020-2)	易危 / VU
CITES 附录 (2019) CITES Appendix (2019)	无 / NA
保护行动 Conservation Action(s)	部分种群位于自然保护区内 / Part of populations are covered by nature reserves

相关文献 Relevant References

Burgin *et al.*, 2018; Jiang *et al.* (蒋志刚等), 2017, 2015a; Smith *et al.* (史密斯等), 2009; Pan *et al.* (潘清华等), 2007; Zhou (周开亚), 2004; Zhou *et al.* (周开亚等), 2001

北海狗 *Callorhinus ursinus*

黄喉貂
Martes flavigula

易危 VU A3d ver 3.1

| 数据缺乏 DD | 无危 LC | 近危 NT | 易危 VU | 濒危 EN | 极危 CR | 区域灭绝 RE | 野外灭绝 EW | 灭绝 EX |

分类地位 Taxonomic Status

| 动物界 Animalia | 脊索动物门 Chordata | 哺乳纲 Mammalia | 食肉目 Carnivora | 鼬科 Mustelidae |

学　　名 Scientific Name	*Martes flavigula*
命　名　人 Species Authority	Boaert, 1785
英　文　名 English Name(s)	Yellow-throated Marten
同物异名 Synonym(s)	无 / None
种下单元评估 Infra-specific Taxa Assessed	无 / None

评估信息 Assessment Information

评估年份 Year Assessed	2020
评　定　人 Assessor(s)	蒋志刚 / Zhigang Jiang
审　定　人 Reviewer(s)	李晟、王昊、周友兵 / Sheng Li, Hao Wang, Youbing Zhou
其他贡献人 Other Contributor(s)	李立立、丁晨晨 / Lili Li, Chenchen Ding

理由 Justification: 黄喉貂在中国的分布区较广，但数量较少，且其所在的栖息地面积不断减小。因此，黄喉貂列为易危等级 / The range of Yellow-throated Marten in China is large, but its population sizes are small. Furthermore, the areas of its habitats are declining. Thus, Yellow-throated Marten is listed as Vulnerable

地理分布 Geographical Distribution

国内分布 Domestic Distribution

山西、湖南、河南、吉林、黑龙江、浙江、江苏、福建、广西、海南、四川、陕西、香港、甘肃、广东、贵州、湖北、江西、辽宁、西藏、云南、福建、重庆、台湾、安徽 / Shanxi, Hunan, Henan, Jilin, Heilongjiang, Zhejiang, Jiangsu, Fujian, Guangxi, Hainan, Sichuan, Shaanxi, Hong Kong, Gansu, Guangdong, Guizhou, Hubei, Jiangxi, Liaoning, Tibet (Xizang), Yunnan, Fujian, Chongqing, Taiwan, Anhui

世界分布 World Distribution

中国；东亚、南亚、东北亚 / China; East Asia, South Asia, Northeast Asia

分布标注 Distribution Note

非特有种 / Non-endemic

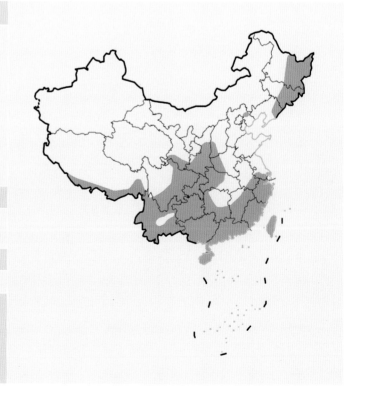

国内分布图
Map of Domestic Distribution

种群 Population

种群数量 Population Size	在华南大部分地区种群数量剧减，多年未见；海南亚种 *M. f. hainana* 自19世纪60年代采集到标本后，再未发现，指名亚种 *M. f. flavigula* 在华南多个省份近年发表的红外相机调查中皆未记录到。但其在中国东北、西南地区及台湾分布较广，有一定数量 / In most areas of south China, the population of Yellow-throated Marten has decreased dramatically and has not been seen for many years. Samples of the Hainan subspecies *M. f. hainana* have not been found since the 1860s, and the type subspecies *M. f. flavigula* has not been recorded in the infrared camera surveys published for several provinces of southern China in recent years. However, the species is still widely distributed in northeast China, southwest China and Taiwan with certain number of population densities
种群趋势 Population Trend	在部分分布区稳定 / Stable in part of its range area

生境与生态系统 Habitat(s) and Ecosystem(s)

生境 Habitat(s)	泰加林，热带和亚热带湿润低地森林 / Taiga Forest, Tropical and Subtropical Moist Lowland Forest
生态系统 Ecosystem(s)	森林生态系统 / Forest Ecosystem

威胁 Threat(s)

主要威胁 Major Threat(s)	狩猎 / Hunting

保护级别与保护行动 Protection Category and Conservation Action(s)

国家重点保护野生动物等级 (2021) Category of National Key Protected Wild Animals (2021)	二级 / Category II
IUCN红色名录 (2020-2) IUCN Red List (2020-2)	无危 / LC
CITES 附录 (2019) CITES Appendix (2019)	III
保护行动 Conservation Action(s)	部分种群位于自然保护区内 / Part of populations are covered by nature reserves

相关文献 Relevant References

Burgin *et al.*, 2018; Jiang *et al.* (蒋志刚等), 2017; Xu *et al.*, 2013; Zhang (张建军), 2000

黄喉貂 *Martes flavigula*

紫貂
Martes zibellina

易危 VU B1; C2a(i)

| 数据缺乏 DD | 无危 LC | 近危 NT | **易危 VU** | 濒危 EN | 极危 CR | 区域灭绝 RE | 野外灭绝 EW | 灭绝 EX |

分类地位 Taxonomic Status

动物界 Animalia	脊索动物门 Chordata	哺乳纲 Mammalia	食肉目 Carnivora	鼬科 Mustelidae

学名 Scientific Name	*Martes zibellina*
命名人 Species Authority	Linnaeus, 1758
英文名 English Name(s)	Sable
同物异名 Synonym(s)	无 / None
种下单元评估 Infra-specific Taxa Assessed	无 / None

评估信息 Assessment Information

评估年份 Year Assessed	2020
评定人 Assessor(s)	蒋志刚 / Zhigang Jiang
审定人 Reviewer(s)	姜广顺、胡慧建、金崑、时坤、李晟、王昊 / Guangshun Jiang, Huijian Hu, Kun Jin, Kun Shi, Sheng Li, Hao Wang
其他贡献人 Other Contributor(s)	李立立、丁晨晨 / Lili Li, Chenchen Ding

理由 Justification: 紫貂分布狭窄，其种群数量少，且栖息地的面积因人类活动而减少。因此，基于其栖息地与种群现状，将紫貂列入易危等级 / Sable is narrowly distributed in China, and its populations are small. Furthermore, the areas of its habitats are declining due to human activities. Thus, based on its habitat and population status, Sable is listed as Vulnerable

地理分布 Geographical Distribution

国内分布 Domestic Distribution	吉林、内蒙古、黑龙江、新疆、辽宁 / Jilin, Inner Mongolia (Nei Mongol), Heilongjiang, Xinjiang, Liaoning
世界分布 World Distribution	中国、芬兰、日本、朝鲜、蒙古、波兰、俄罗斯 / China, Finland, Japan, Korea (the Democratic People's Republic of), Mongolia, Poland, Russia
分布标注 Distribution Note	非特有种 / Non-endemic

国内分布图
Map of Domestic Distribution

种群 Population

种群数量 Population Size	18,000 只 / 18,000 individuals
种群趋势 Population Trend	未知 / Unknown

生境与生态系统 Habitat (s) and Ecosystem (s)

生境 Habitat(s)	泰加林、溪流边 / Taiga Forest, Near Stream
生态系统 Ecosystem(s)	森林生态系统、湖泊河流生态系统 / Forest Ecosystem, Lake and River Ecosystem

威胁 Threat (s)

主要威胁 Major Threat(s)	狩猎、伐木 / Hunting, Logging

保护级别与保护行动 Protection Category and Conservation Action (s)

国家重点保护野生动物等级 (2021) Category of National Key Protected Wild Animals (2021)	一级 / Category I
IUCN 红色名录 (2020-2) IUCN Red List (2020-2)	无危 / LC
CITES 附录 (2019) CITES Appendix (2019)	无 / NA
保护行动 Conservation Action(s)	已经建立自然保护区 / Nature reserves established

相关文献 Relevant References

Burgin *et al.*, 2018; Jiang *et al.* (蒋志刚等), 2017; Zhu *et al.* (朱妍等), 2011; Xu *et al.* (徐纯柱等), 2010; Li *et al.* (李月辉等), 2007; Zhang and Ma (张洪海和马建章), 2000

紫貂 *Martes zibellina*

艾鼬
Mustela eversmanii

易危 VU A3d; C2a(i)

| 数据缺乏 DD | 无危 LC | 近危 NT | **易危 VU** | 濒危 EN | 极危 CR | 区域灭绝 RE | 野外灭绝 EW | 灭绝 EX |

分类地位 Taxonomic Status

动物界 Animalia	脊索动物门 Chordata	哺乳纲 Mammalia	食肉目 Carnivora	鼬科 Mustelidae

学 名 Scientific Name	*Mustela eversmanii*
命 名 人 Species Authority	Lesson, 1827
英 文 名 English Name(s)	Steppe Polecat
同物异名 Synonym(s)	White Polecat; Masked Polecat; *Mustela eversmannii* (Lesson, 1827)
种下单元评估 Infra-specific Taxa Assessed	无 / None

评估信息 Assessment Information

评估年份 Year Assessed	2020
评 定 人 Assessor(s)	蒋志刚 / Zhigang Jiang
审 定 人 Reviewer(s)	徐爱春、李晟、胡慧建、王昊、周友兵 / Aichun Xu, Sheng Li, Huijian Hu, Hao Wang, Youbing Zhou
其他贡献人 Other Contributor(s)	李立立、丁晨晨 / Lili Li, Chenchen Ding

理由 Justification: 艾鼬分布虽较广，但数量稀少，且其栖息地破碎化，此外由于灭鼠导致二次中毒，其种群数量下降。因此，基于其种群与栖息地现状，将艾鼬列为易危等级 / Although Steppe Polecat is widely distributed in China, its habitat is fragmented and its populations are small. Furthermore, its populations are declining due to secondary poisoning in deratization. Thus, based on its population and habitat status, Steppe Polecat is listed as Vulnerable

地理分布 Geographical Distribution

国内分布 Domestic Distribution
黑龙江、吉林、辽宁、内蒙古、西藏、新疆、青海、宁夏、甘肃、四川、陕西、山西、河北、北京、江苏 / Heilongjiang, Jilin, Liaoning, Inner Mongolia (Nei Mongol), Tibet (Xizang), Xinjiang, Qinghai, Ningxia, Gansu, Sichuan, Shaanxi, Shanxi, Hebei, Beijing, Jiangsu
世界分布 World Distribution
欧亚大陆、美洲大陆及环北极国家 / Eurasian Continent, American Continent and Circum-Arctic Countries
分布标注 Distribution Note
非特有种 / Non-endemic

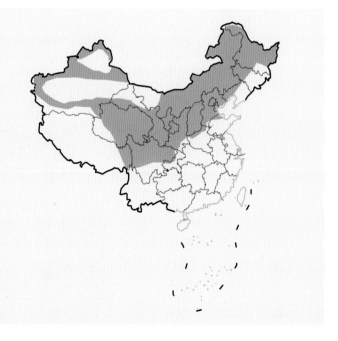

国内分布图
Map of Domestic Distribution

种群 Population

种群数量 Population Size	未知 / Unknown
种群趋势 Population Trend	稳定 / Stable

生境与生态系统 Habitat(s) and Ecosystem(s)

生　　境 Habitat(s)	草地 / Grassland
生态系统 Ecosystem(s)	草地生态系统 / Grassland Ecosystem

威胁 Threat(s)

主要威胁 Major Threat(s)	栖息地丧失、狩猎、二次中毒 / Habitat Loss, Hunting, Secondary Poisoning

保护级别与保护行动 Protection Category and Conservation Action(s)

国家重点保护野生动物等级 (2021) Category of National Key Protected Wild Animals (2021)	无 / NA
IUCN红色名录 (2020-2) IUCN Red List (2020-2)	无危 / LC
CITES 附录 (2019) CITES Appendix (2019)	无 / NA
保护行动 Conservation Action(s)	无 / None

相关文献 Relevant References

Burgin *et al.*, 2018; Jiang *et al.* (蒋志刚等), 2017; Zou *et al.* (邹波等), 2012; Xia (夏亚军), 2011; Huang *et al.* (黄薇等), 2008; Hou (侯兰新), 2000

艾鼬 *Mustela eversmanii*

伶鼬 *Mustela nivalis*

易危 VU A3d; C2a(i)

| DD 数据缺乏 | LC 无危 | NT 近危 | **VU 易危** | EN 濒危 | CR 极危 | RE 区域灭绝 | EW 野外灭绝 | EX 灭绝 |

分类地位 Taxonomic Status

动物界 Animalia	脊索动物门 Chordata	哺乳纲 Mammalia	食肉目 Carnivora	鼬科 Mustelidae

学 名 Scientific Name	*Mustela nivalis*
命 名 人 Species Authority	Linnaeus, 1766
英 文 名 English Name(s)	Least Weasel
同物异名 Synonym(s)	Little Weasel, Common Weasel
种下单元评估 Infra-specific Taxa Assessed	无 / None

评估信息 Assessment Information

评 估 年 份 Year Assessed	2020
评 定 人 Assessor(s)	蒋志刚 / Zhigang Jiang
审 定 人 Reviewer(s)	马建章、徐爱春、张明海 / Jianzhang Ma, Aichun Xu, Minghai Zhang
其他贡献人 Other Contributor(s)	李立立、丁晨晨 / Lili Li, Chenchen Ding

理由 Justification: 伶鼬分布虽较广，但其种群数量少，且栖息地受到人类活动干扰。因此，基于其种群大小，将伶鼬列为易危等级 / Although Least Weasel is widely distributed in China, its populations are very small and its habitats are disturbed by human activities. Thus, based on its population size, Least Weasel is listed as Vulnerable

地理分布 Geographical Distribution

国内分布 Domestic Distribution

陕西、河北、青海、甘肃、四川、内蒙古、辽宁、吉林、黑龙江、新疆 / Shaanxi, Hebei, Qinghai, Gansu, Sichuan, Inner Mongolia (Nei Mongol), Liaoning, Jilin, Heilongjiang, Xinjiang

世界分布 World Distribution

欧亚大陆、美洲大陆及环北极国家 / Eurasian Continent, American Continent and Circum-Arctic Countries

分布标注 Distribution Note

非特有种 / Non-endemic

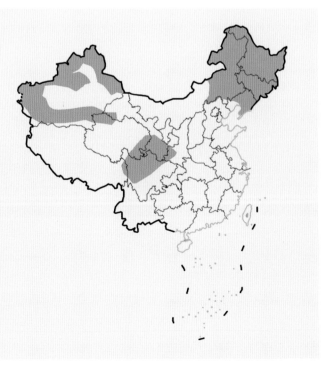

国内分布图
Map of Domestic Distribution

种群 Population

种群数量 Population Size	未知 / Unknown
种群趋势 Population Trend	稳定 / Stable

生境与生态系统 Habitat (s) and Ecosystem (s)

生　　境 Habitat(s)	森林、草地、草甸、乡村花园、农田 / Forest, Grassland, Meadow, Rural Garden, Cropland
生态系统 Ecosystem(s)	森林生态系统、草地生态系统、农田生态系统、人类聚落生态系统 / Forest Ecosystem, Grassland Ecosystem, Cropland Ecosystem, Human Settlement Ecosystem

威胁 Threat (s)

主要威胁 Major Threat(s)	猎捕、毒杀 / Trapping, Poisoning

保护级别与保护行动 Protection Category and Conservation Action (s)

国家重点保护野生动物等级 (2021) Category of National Key Protected Wild Animals (2021)	无 / NA
IUCN 红色名录 (2020-2) IUCN Red List (2020-2)	无危 / LC
CITES 附录 (2019) CITES Appendix (2019)	无 / NA
保护行动 Conservation Action(s)	无 / None

相关文献 Relevant References

Burgin *et al*., 2018; Jiang *et al*. (蒋志刚等), 2017; Smith *et al*. (史密斯等), 2009; Hu and Hu (胡锦矗和胡杰), 2007

伶鼬 *Mustela nivalis*

斑海豹
Phoca largha

易危 VU A3d

| 数据缺乏 DD | 无危 LC | 近危 NT | 易危 VU | 濒危 EN | 极危 CR | 区域灭绝 RE | 野外灭绝 EW | 灭绝 EX |

分类地位 Taxonomic Status

| 动物界 Animalia | 脊索动物门 Chordata | 哺乳纲 Mammalia | 食肉目 Carnivora | 海豹科 Phocidae |

学 名 Scientific Name	*Phoca largha*
命 名 人 Species Authority	Pallas, 1811
英 文 名 English Name(s)	Spotted Seal
同物异名 Synonym(s)	西太平洋斑海豹，Larga Seal, Largha Seal
种下单元评估 Infra-specific Taxa Assessed	无 / None

评估信息 Assessment Information

评 估 年 份 Year Assessed	2020
评 定 人 Assessor(s)	蒋志刚 / Zhigang Jiang
审 定 人 Reviewer(s)	张先锋、王克雄、王丁、祝茜、杨光、蒋志刚 / Xianfeng Zhang, Kexiong Wang, Ding Wang, Qian Zhu, Guang Yang, Zhigang Jiang
其他贡献人 Other Contributor(s)	李立立、丁晨晨 / Lili Li, Chenchen Ding

理由 Justification: 斑海豹分布虽较广，但其种群数量少，易被偷捕，且栖息地受到人类活动干扰。基于其种群小、密度低，将斑海豹列为易危等级 / Although Spotted Seal is relatively widely distributed in China, its populations are small and vulnerable to poaching. Furthermore, its habitats are disturbed by human activities. Based on its population size and density, Spotted Seal is listed as Vulnerable

地理分布 Geographical Distribution

国内分布 Domestic Distribution
渤海、黄海、东海、南海 / Bohai Sea, Yellow Sea, East China Sea, South China Sea
世界分布 World Distribution
北太平洋 / North Pacific Ocean
分布标注 Distribution Note
非特有种 / Non-endemic

国内分布图 Map of Domestic Distribution

种群 Population

种群数量 Population Size	未知 / Unknown
种群趋势 Population Trend	未知 / Unknown

生境与生态系统 Habitat (s) and Ecosystem (s)

生　　境 Habitat(s)	江河、海滩 / River, Beach
生态系统 Ecosystem(s)	海洋生态系统、湖泊河流生态系统 / Ocean Ecosystem, Lake and River Ecosystem

威胁 Threat (s)

主要威胁 Major Threat(s)	狩猎、渔业、工业污染、气候变化 / Hunting, Fishing, Industrial Pollution, Climate Change

保护级别与保护行动 Protection Category and Conservation Action (s)

国家重点保护野生动物等级 (2021) Category of National Key Protected Wild Animals (2021)	一级 / Category I
IUCN红色名录(2020-2) IUCN Red List (2020-2)	无危 / LC
CITES 附录 (2019) CITES Appendix (2019)	无 / NA
保护行动 Conservation Action(s)	已经建立自然保护区 / Nature reserves established

相关文献 Relevant References

Burgin *et al.*, 2018; Jiang *et al.* (蒋志刚等), 2017; Hao *et al.* (郝玉江等), 2011; Smith *et al.* (史密斯等), 2009; Wang (王应祥), 2003

斑海豹 *Phoca largha*

环斑小头海豹
Pusa hispida

易危 VU A3d

| 数据缺乏 DD | 无危 LC | 近危 NT | 易危 VU | 濒危 EN | 极危 CR | 区域灭绝 RE | 野外灭绝 EW | 灭绝 EX |

分类地位 Taxonomic Status

动物界 Animalia	脊索动物门 Chordata	哺乳纲 Mammalia	食肉目 Carnivora	海豹科 Phocidae
学 名 Scientific Name		*Pusa hispida*		
命名人 Species Authority		Schreber, 1775		
英文名 English Name(s)		Ringed Seal		
同物异名 Synonym(s)		环海豹；*Phoca hispida* (Schreber, 1775)		
种下单元评估 Infra-specific Taxa Assessed		无 / None		

评估信息 Assessment Information

评估年份 Year Assessed	2020
评定人 Assessor(s)	周开亚 / Kaiya Zhou
审定人 Reviewer(s)	张先锋、王克雄、王丁、祝茜、杨光、蒋志刚 / Xianfeng Zhang, Kexiong Wang, Ding Wang, Qian Zhu, Guang Yang, Zhigang Jiang
其他贡献人 Other Contributor(s)	李立立、丁晨晨 / Lili Li, Chenchen Ding

理由 Justification: 环斑小头海豹分布虽较广，但其种群数量极少，且栖息地的面积因人类活动而不断减少。因此，环斑小头海豹列为易危等级 / Although Ringed Seal is relatively widely distributed in China, its populations are small. Furthermore, the areas of its habitats are declining due to human encroachment. Thus, Ringed Seal is listed as Vulnerable

地理分布 Geographical Distribution

国内分布 Domestic Distribution

黄海 / Yeallow Sea

世界分布 World Distribution

中国；环极地分布，从北极到北纬35°，分布在北冰洋、北太平洋、白令海南部和南部的鄂霍次克海和日本海 / China; a circumpolar distribution from approximately 35°N to the North Pole, occurring in Arctic Ocean, North Pacific Ocean, Southern Bering Sea and the Seas of Okhotsk and Sea of Japan

分布标注 Distribution Note

非特有种 / Non-endemic

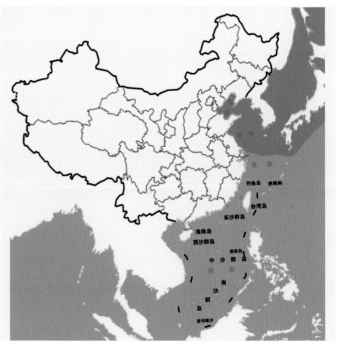

国内分布图
Map of Domestic Distribution

种群 Population

种群数量 Population Size	未知 / Unknown
种群趋势 Population Trend	未知 / Unknown

生境与生态系统 Habitat (s) and Ecosystem (s)

生境 Habitat(s)	冰雪覆盖的海域，适应季节性和永久性的海洋冰面 / Ice-covered Water, Seasonal and Permanent Ice Surface in ocean
生态系统 Ecosystem(s)	海洋生态系统 / Ocean Ecosystem

威胁 Threat (s)

主要威胁 Major Threat(s)	狩猎及采集、工业污染、农业或林业污染、渔业、旅游、气候变化 / Hunting and Collection, Industrial Pollution, Agricultural or Forestry Effluent, Fishing, Tourism, Climate Change

保护级别与保护行动 Protection Category and Conservation Action (s)

国家重点保护野生动物等级 (2021) Category of National Key Protected Wild Animals (2021)	二级 / Category II
IUCN 红色名录(2020-2) IUCN Red List (2020-2)	无危 / LC
CITES 附录 (2019) CITES Appendix (2019)	无 / NA
保护行动 Conservation Action(s)	无 / None

相关文献 Relevant References

Burgin *et al*., 2018; Jiang *et al*. (蒋志刚等), 2017; Mittermeier and Wilson, 2014; Hao *et al*. (郝玉江等), 2011; Wang and Fan (王火根和范忠勇), 2004; Zhu *et al*. (祝茜等), 2000

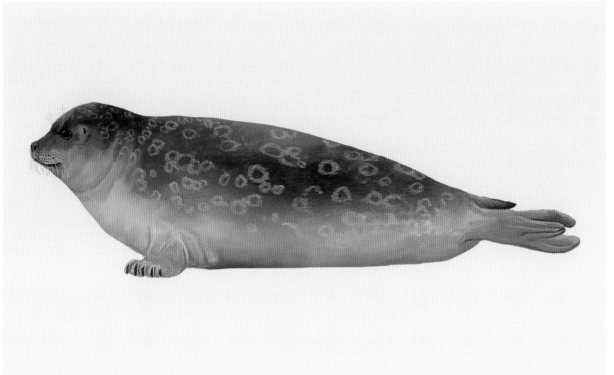

环斑小头海豹 *Pusa hispida*

斑林狸
Prionodon pardicolor

易危 VU A3d; C2a(i)

分类地位 Taxonomic Status

动物界 Animalia	脊索动物门 Chordata	哺乳纲 Mammalia	食肉目 Carnivora	林狸科 Prionodontidae
学名 Scientific Name		*Prionodon pardicolor*		
命名人 Species Authority		Hodgson, 1842		
英文名 English Name(s)		Spotted Linsang		
同物异名 Synonym(s)		无 / None		
种下单元评估 Infra-specific Taxa Assessed		无 / None		

评估信息 Assessment Information

评估年份 Year Assessed	2020
评定人 Assessor(s)	蒋志刚 / Zhigang Jiang
审定人 Reviewer(s)	李晟、胡慧建、王大军、王昊、周友兵 / Sheng Li, Huijian Hu, Dajun Wang, Hao Wang, Youbing Zhou
其他贡献人 Other Contributor(s)	李立立、丁晨晨 / Lili Li, Chenchen Ding

理由 Justification: 斑林狸分布广。近年在多个保护区的红外相机调查中记录到斑林狸，但拍摄到的次数少。由于斑林狸种群密度低，故将其列为易危等级 / Spotted Linsang is widely distributed. In recent years, a few Spotted Linsangs have been trapped in the infrared cameras in several protected areas only at very low frequency. Due to its low density, Spotted Linsang is listed as Vulnerable

地理分布 Geographical Distribution

国内分布 Domestic Distribution	江西、湖南、广东、广西、四川、贵州、云南、西藏 / Jiangxi, Hunan, Guangdong, Guangxi, Sichuan, Guizhou, Yunnan, Tibet (Xizang)
世界分布 World Distribution	中国；东南亚 / China; Southeast Asia
分布标注 Distribution Note	非特有种 / Non-endemic

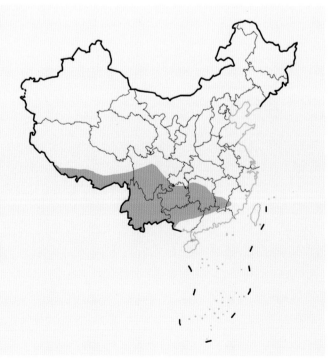

国内分布图
Map of Domestic Distribution

🦌 种群 Population

种群数量 Population Size	未知 / Unknown
种群趋势 Population Trend	未知 / Unknown

🦌 生境与生态系统 Habitat (s) and Ecosystem (s)

生　　境 Habitat(s)	亚热带湿润低地森林 / Subtropical Moist Lowland Forest
生态系统 Ecosystem(s)	森林生态系统 / Forest Ecosystem

🦌 威胁 Threat (s)

主要威胁 Major Threat(s)	狩猎、伐木、耕种 / Hunting, Logging, Plantation

🦌 保护级别与保护行动 Protection Category and Conservation Action (s)

国家重点保护野生动物等级 (2021) Category of National Key Protected Wild Animals (2021)	二级 / Category II
IUCN红色名录(2020-2) IUCN Red List (2020-2)	无危 / LC
CITES 附录 (2019) CITES Appendix (2019)	I
保护行动 Conservation Action(s)	部分种群位于自然保护区内 / Part of populations are covered by nature reserve

🦌 相关文献 Relevant References

Burgin *et al*., 2018; Jiang *et al*. (蒋志刚等), 2017; Lau *et al*., 2010; Patou *et al*., 2010; Wang (王应祥), 2003

斑林狸 *Prionodon pardicolor*

爪哇獴
Herpestes javanicus

易危 VU B1ab(i, ii, iii); C2a(i)

| 数据缺乏 DD | 无危 LC | 近危 NT | 易危 VU | 濒危 EN | 极危 CR | 区域灭绝 RE | 野外灭绝 EW | 灭绝 EX |

分类地位 Taxonomic Status

动物界 Animalia	脊索动物门 Chordata	哺乳纲 Mammalia	食肉目 Carnivora	獴科 Herpestidae

学名 Scientific Name	*Herpestes javanicus*
命名人 Species Authority	I. Geoffroy Saint-Hilaire, 1818
英文名 English Name(s)	Javan Mongoose
同物异名 Synonym(s)	Small Asian Mongoose; Small Indian Mongoose; *Herpestes palustris* (Ghose, 1965)
种下单元评估 Infra-specific Taxa Assessed	无 / None

评估信息 Assessment Information

评估年份 Year Assessed	2020
评定人 Assessor(s)	蒋志刚 / Zhigang Jiang
审定人 Reviewer(s)	李晟、胡慧建、王大军、王昊、周友兵、陈辈乐、李飞 / Sheng Li, Huijian Hu, Dajun Wang, Hao Wang, Youbing Zhou, Bosco P. L. Chan, Fei Li
其他贡献人 Other Contributor(s)	李立立、丁晨晨 / Lili Li, Chenchen Ding

理由 Justification: 爪哇獴分布于中国南部，白天在草地或浓密次生灌丛中活动。爪哇獴栖息区域人类活动较多，在红外相机调查中很少记录到。基于其种群密度，将爪哇獴列为易危 / Javan Mongoose is distributed in grassland or dense secondary shrubs in southern China and it is active during the day. As human is relatively active in its habitat, it is rarely recorded in infrared camera surveys. Based on its population density status, Javan Mongoose is listed as Vulnerable

地理分布 Geographical Distribution

国内分布 Domestic Distribution	海南、广东、广西、贵州、云南、香港 / Hainan, Guangdong, Guangxi, Guizhou, Yunnan, Hong Kong
世界分布 World Distribution	中国；中东地区北部、南亚和东南亚 / China; northern Middle East region, South and Southeast Asia
分布标注 Distribution Note	非特有种 / Non-endemic

国内分布图
Map of Domestic Distribution

种群 Population

种群数量 Population Size	未知 / Unknown
种群趋势 Population Trend	未知 / Unknown

生境与生态系统 Habitat (s) and Ecosystem (s)

生　　境 Habitat(s)	热带和亚热带森林、灌丛、草地、耕地 / Tropical and Subtropical Forest, Shrubland, Grassland, Arable Land
生态系统 Ecosystem(s)	森林生态系统、灌丛生态系统、农田生态系统、草地生态系统 / Forest Ecosystem, Shrubland Ecosystem, Cropland Ecosystem, Grassland Ecosystem

威胁 Threat (s)

主要威胁 Major Threat(s)	未知 / Unknown

保护级别与保护行动 Protection Category and Conservation Action (s)

国家重点保护野生动物等级 (2021) Category of National Key Protected Wild Animals (2021)	无 / NA
IUCN 红色名录(2020-2) IUCN Red List (2020-2)	无危 / LC
CITES 附录 (2019) CITES Appendix (2019)	III
保护行动 Conservation Action(s)	无 / None

相关文献 Relevant References

Burgin *et al.*, 2018; Jiang *et al.* (蒋志刚等), 2017; Lau *et al.*, 2010; Pei *et al.*, 2010; Huo *et al.* (霍晟等), 2003; Wang (王应祥), 2003

爪哇獴 *Herpestes javanicus*

食蟹獴
Herpestes urva

中国生物多样性 红色名录
China's Red List of Biodiversity

易危 VU B1ab(i, ii, iii); C2a(i)

| 数据缺乏 DD | 无危 LC | 近危 NT | 易危 VU | 濒危 EN | 极危 CR | 区域灭绝 RE | 野外灭绝 EW | 灭绝 EX |

分类地位 Taxonomic Status

动物界 Animalia	脊索动物门 Chordata	哺乳纲 Mammalia	食肉目 Carnivora	獴科 Herpestidae
学名 Scientific Name		*Herpestes urva*		
命名人 Species Authority		Hodgson, 1836		
英文名 English Name(s)		Crab-eating Mongoose		
同物异名 Synonym(s)		无 / None		
种下单元评估 Infra-specific Taxa Assessed		无 / None		

评估信息 Assessment Information

评估年份 Year Assessed	2020
评定人 Assessor(s)	蒋志刚 / Zhigang Jiang
审定人 Reviewer(s)	李晟、胡慧建、王大军、王昊、周友兵 / Sheng Li, Huijian Hu, Dajun Wang, Hao Wang, Youbing Zhou
其他贡献人 Other Contributor(s)	李立立、丁晨晨 / Lili Li, Chenchen Ding

理由 Justification: 食蟹獴曾广布于中国南部。近年的红外相机调查显示，食蟹獴在云南南部及西南部、广西弄岗、广东南岭、浙江山区及台湾仍有分布，但一些以往有分布的省份，如四川、湖南、江西、贵州、海南，却再未见到记录。鉴于分布区的缩减，将食蟹獴列为易危 / Crab-eating Mongoose was once widespread in southern China. In recent years, infrared camera surveys show that it still lives in southern and southwestern Yunnan; in Nonggang, Guangxi; in Nanling, Guangdong; and in mountainous areas of Zhejiang and in Taiwan. However, the species is not found in several provinces where it previously was distributed, such as in Sichuan, Hunan, Jiangxi, Guizhou and Hainan. In view of its declining range, Crab-eating Mongoose is listed as Vulnerable

地理分布 Geographical Distribution

国内分布 Domestic Distribution

云南、贵州、四川、重庆、广西、广东、海南、香港、福建、台湾、浙江、江苏、安徽、湖北、湖南、江西 / Yunnan, Guizhou, Sichuan, Chongqing, Guangxi, Guangdong, Hainan, Hong Kong, Fujian, Taiwan, Zhejiang, Jiangsu, Anhui, Hubei, Hunan, Jiangxi

世界分布 World Distribution

中国、印度北部、缅甸北部、泰国、马来半岛、老挝、柬埔寨和越南，在孟加拉国罕见 / China, Northeastern India, Northern Myanmar, Thailand, Malay Peninsular, Laos, Cambodia and Viet Nam. It is rare in Bangladesh

分布标注 Distribution Note

非特有种 / Non-endemic

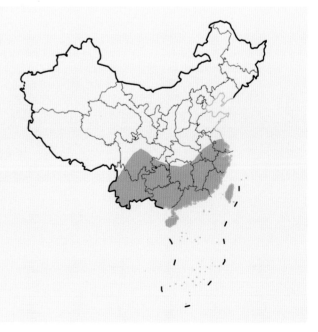

国内分布图
Map of Domestic Distribution

种群 Population

种群数量 Population Size	未知 / Unknown
种群趋势 Population Trend	稳定 / Stable

生境与生态系统 Habitat (s) and Ecosystem (s)

生　　境 Habitat(s)	森林、农田、溪流边 / Forest, Cropland, Near Stream
生态系统 Ecosystem(s)	森林生态系统、农田生态系统、湖泊河流生态系统 / Forest Ecosystem, Cropland Ecosystem, Lake and River Ecosystem

威胁 Threat (s)

主要威胁 Major Threat(s)	狩猎 / Hunting

保护级别与保护行动 Protection Category and Conservation Action (s)

国家重点保护野生动物等级 (2021) Category of National Key Protected Wild Animals (2021)	无 / NA
IUCN 红色名录 (2020-2) IUCN Red List (2020-2)	无危 / LC
CITES 附录 (2019) CITES Appendix (2019)	III
保护行动 Conservation Action(s)	无 / None

相关文献 Relevant References

Burgin *et al*., 2018; Jiang *et al*. (蒋志刚等), 2017; Chen *et al*. (陈小荣等), 2013; Shek *et al*., 2007; Van Rompaey, 2001

食蟹獴 *Herpestes urva*

中国生物多样性红色名录

豹猫
Prionailurus bengalensis

易危 VU A2ab; B1ab(i, ii, iii)

| 数据缺乏 DD | 无危 LC | 近危 NT | **易危 VU** | 濒危 EN | 极危 CR | 区域灭绝 RE | 野外灭绝 EW | 灭绝 EX |

分类地位 Taxonomic Status

动物界 Animalia	脊索动物门 Chordata	哺乳纲 Mammalia	食肉目 Carnivora	猫科 Felidae

学　名 Scientific Name	*Prionailurus bengalensis*
命名人 Species Authority	Kerr, 1792
英文名 English Name(s)	Leopard Cat
同物异名 Synonym(s)	无 / None
种下单元评估 Infra-specific Taxa Assessed	无 / None

评估信息 Assessment Information

评估年份 Year Assessed	2020
评定人 Assessor(s)	蒋志刚 / Zhigang Jiang
审定人 Reviewer(s)	李晟、胡慧建、王大军、王昊、周友兵、陈辈乐、李飞 / Sheng Li, Huijian Hu, Dajun Wang, Hao Wang, Youbing Zhou, Bosco P. L. Chan, Fei Li
其他贡献人 Other Contributor(s)	李立立、丁晨晨 / Lili Li, Chenchen Ding

理由 Justification: 除北部及西部干旱区外，豹猫广布于中国。近年红外相机调查显示，豹猫仍然广泛分布，但捕杀、车辆碾压、宠物贸易及栖息地丧失等因素威胁着豹猫的生存。因此，将豹猫列为易危等级 / Except for the arid regions in northern and western China, the Leopard Cat is widespread in the country. In recent years, infrared camera surveys have shown that Leopard Cat is still widely distributed, but factors such as hunting, road kill, the pet trade and habitat loss threaten its survival. Thus, Leopard Cat is listed as Vulnerable

地理分布 Geographical Distribution

国内分布 Domestic Distribution

山西、河南、湖南、海南、青海、甘肃、北京、河北、内蒙古、辽宁、吉林、黑龙江、江苏、浙江、安徽、福建、江西、山东、湖北、广东、广西、四川、贵州、云南、西藏、陕西、宁夏、台湾、香港、天津、重庆 / Shanxi, Henan, Hunan, Hainan, Qinghai, Gansu, Beijing, Hebei, Inner Mongolia (Nei Mongol), Liaoning, Jilin, Heilongjiang, Jiangsu, Zhejiang, Anhui, Fujian, Jiangxi, Shandong, Hubei, Guangdong, Guangxi, Sichuan, Guizhou, Yunnan, Tibet (Xizang), Shaanxi, Ningxia, Taiwan, Hong Kong, Tianjin, Chongqing

世界分布 World Distribution

中国；中亚、南亚、东北亚、东南亚 / China; Central Asia, South Asia, Northeast Asia, Southeast Asia

分布标注 Distribution Note

非特有种 / Non-endemic

国内分布图
Map of Domestic Distribution

种群 Population

种群数量 Population Size	230,000 只 / 230,000 individuals
种群趋势 Population Trend	稳定 / Stable

生境与生态系统 Habitat (s) and Ecosystem (s)

生　　境 Habitat(s)	森林、泰加林、灌丛、次生林、耕地、种植园 / Forest, Taiga Forest, Shrubland, Secondary Forest, Arable Land, Plantation
生态系统 Ecosystem(s)	森林生态系统、灌丛生态系统、农田生态系统 / Forest Ecosystem, Shrubland Ecosystem, Cropland Ecosystem

威胁 Threat (s)

主要威胁 Major Threat(s)	狩猎、杂交、耕种、伐木 / Hunting, Hybridization, Plantation, Logging

保护级别与保护行动 Protection Category and Conservation Action (s)

国家重点保护野生动物等级 (2021) Category of National Key Protected Wild Animals (2021)	二级 / Category II
IUCN红色名录(2020-2) IUCN Red List (2020-2)	无危 / LC
CITES 附录 (2019) CITES Appendix (2019)	I
保护行动 Conservation Action(s)	在自然保护区内的种群得到保护 / Populations in nature reserves are under protection

相关文献 Relevant References

Burgin *et al*., 2018; Jiang *et al*. (蒋志刚等), 2017; Hu *et al*. (胡一鸣等), 2014; Lau *et al*., 2010; Yu, 2010; Wang *et al*. (王应祥等), 1997; Gao *et al*. (高耀亭等), 1987

豹猫 *Prionailurus bengalensis*　　马文虎 摄　By Wenhu Ma

蒙古野驴
Equus hemionus

易危 VU A1acd; B1ab(i, ii, iii)+2ab(i, ii, iii)

| 数据缺乏 DD | 无危 LC | 近危 NT | **易危 VU** | 濒危 EN | 极危 CR | 区域灭绝 RE | 野外灭绝 EW | 灭绝 EX |

分类地位 Taxonomic Status

动物界 Animalia	脊索动物门 Chordata	哺乳纲 Mammalia	奇蹄目 Perissodactyla	马科 Equidae
学　名 Scientific Name		*Equus hemionus*		
命　名　人 Species Authority		Pallas, 1775		
英　文　名 English Name(s)		Asiatic Wild Ass		
同物异名 Synonym(s)		Kulan; *Equus bahram* (Pocock, 1947); *bedfordi* (Matschie, 1911); *blanfordi* (Pocock, 1947); *castaneus* (Lydekker, 1904); *dzigguetai* (Wood, 1879); *ferus* (Erxleben, 1777);（转下页）		
种下单元评估 Infra-specific Taxa Assessed		无 / None		

评估信息 Assessment Information

评估年份 Year Assessed	2020
评定人 Assessor(s)	蒋志刚 / Zhigang Jiang
审定人 Reviewer(s)	胡德夫、毕俊怀、初红军、杨维康 / Defu Hu, Junhuai Bi, Hongjun Chu, Weikang Yang
其他贡献人 Other Contributor(s)	李立立、丁晨晨 / Lili Li, Chenchen Ding

理由 Justification: 蒙古野驴分布在西北，种群数量少，栖息地受到人类活动挤压。基于其种群与栖息地现状，将蒙古野驴列为易危等级 / Asiatic Wild Ass is distributed in northwest China, its populations are small. Areas of its habitats are declining due to human activities. Based on its population and habitat status, Asiatic Wild Ass is listed as Vulnerable

地理分布 Geographical Distribution

国内分布 Domestic Distribution
甘肃、新疆、内蒙古 / Gansu, Xinjiang, Inner Mongolia (Nei Mongol)
世界分布 World Distribution
中国；中亚 / China; Central Asia
分布标注 Distribution Note
非特有种 / Non-endemic

国内分布图
Map of Domestic Distribution

种群 Population

种群数量 Population Size	14,000 只 / 14,000 individuals
种群趋势 Population Trend	下降 / Decreasing

生境与生态系统 Habitat (s) and Ecosystem (s)

生境 Habitat(s)	荒漠 / Desert
生态系统 Ecosystem(s)	草原生态系统、荒漠生态系统 / Steppe Ecosystem, Desert Ecosystem

威胁 Threat (s)

主要威胁 Major Threat(s)	家畜放牧、人类活动干扰、人兽冲突、耕种、狩猎 / Livestock Ranching, Human Disturbance, Human-Animal Conflict, Plantation, Hunting

保护级别与保护行动 Protection Category and Conservation Action (s)

国家重点保护野生动物等级 (2021) Category of National Key Protected Wild Animals (2021)	一级 / Category I
IUCN 红色名录 (2020-2) IUCN Red List (2020-2)	近危 / NT
CITES 附录 (2019) CITES Appendix (2019)	II
保护行动 Conservation Action(s)	在自然保护区内的种群得到保护；列入《迁徙物种保护公约》的"迁徙物种行动计划" / Populations in nature reserves are under protection; included in the *Convention on Migratory Species* (CMS)－Species Action Plan

相关文献 Relevant References

Burgin *et al*., 2018; Jiang *et al*. (蒋志刚等), 2017; Kaczensky *et al*., 2011; Lin *et al*. (林杰等), 2011; Chu and Jiang, 2009; Li *et al*. (李春旺等), 2002

(接上页)
finschi (Matschie, 1911); *hamar* (C.H. Smith, 1841); *hemionos* (Boaert, 1785); *hemippus* (I. Geoffroy Saint-Hilaire, 1855); *indicus* (George, 1869); *indicus* (Sclater, 1862); *khur* (Lesson, 1827); *kulan* (Groves and Mazák, 1967); *luteus* (Matschie, 1911); *onager* (Boaert, 1785); *onager* (Pallas, 1777); *syriacus* (Milne-Edwards, 1869); *typicus* (Sclater, 1891)

蒙古野驴 *Equus hemionus*

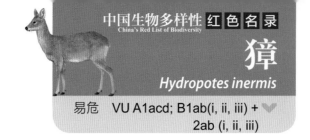

獐
Hydropotes inermis

易危 VU A1acd; B1ab(i, ii, iii) + 2ab (i, ii, iii)

| 数据缺乏 DD | 无危 LC | 近危 NT | 易危 VU | 濒危 EN | 极危 CR | 区域灭绝 RE | 野外灭绝 EW | 灭绝 EX |

分类地位 Taxonomic Status

动物界 Animalia	脊索动物门 Chordata	哺乳纲 Mammalia	偶蹄目 Artiodactyla	鹿科 Cervidae
学　名 Scientific Name		*Hydropotes inermis*		
命名人 Species Authority		Swinhoe, 1870		
英文名 English Name(s)		Water Deer		
同物异名 Synonym(s)		河麂；Chinese Water Deer; *Hydropotes affinis* (Brooke, 1872); *argyropus* (Heude, 1884); *kreyenbergi* (Hilzheimer, 1905)		
种下单元评估 Infra-specific Taxa Assessed		无 / None		

评估信息 Assessment Information

评估年份 Year Assessed	2020
评定人 Assessor(s)	蒋志刚 / Zhigang Jiang
审定人 Reviewer(s)	王小明 / Xiaoming Wang
其他贡献人 Other Contributor(s)	李立立、丁晨晨 / Lili Li, Chenchen Ding

理由 Justification: 獐主要分布于浙江舟山和江苏盐城一带。根据鲍毅新等 2009 年在舟山海岛的调查，种群数量为 2,100～2,600 只，比 1997 年调查的数量 (5,000 只) 下降了近 50%。根据其种群下降速度，将獐列为易危等级 / Water Deer is mainly distributed in China in Zhoushan, Zhejiang and in Yancheng, Jiangsu. According to the survey conducted by Yixin Bao *et al.* in 2009 in the Zhoushan Archipelago, the population of this species is 2,100~2,600, a decline of nearly 50% compared to the 5,000 individuals reported in 1997. Based on the rate of its population decline, Water Deer is listed as Vulnerable

地理分布 Geographical Distribution

国内分布 Domestic Distribution
浙江、上海、江苏、安徽、江西 / Zhejiang, Shanghai, Jiangsu, Anhui, Jiangxi
世界分布 World Distribution
亚洲 (法国、英国有引入的野化种群) / Asia (France and United Kingdom have introduced and rewild its populations)
分布标注 Distribution Note
非特有种 / Non-endemic

国内分布图
Map of Domestic Distribution

🦌 种群 Population

种群数量 Population Size	24,000 只 / 24,000 individuals
种群趋势 Population Trend	下降 / Decreasing

🦌 生境与生态系统 Habitat (s) and Ecosystem (s)

生　　境 Habitat(s)	草地、沼泽、亚热带季节性洪泛低地草原 / Grassland, Swamp, Subtropical Seasonal Flooded Lowland Grassland
生态系统 Ecosystem(s)	草地生态系统、湿地生态系统 / Grassland Ecosystem, Wetland Ecosystem

🦌 威胁 Threat (s)

主要威胁 Major Threat(s)	狩猎、耕种、住宅及商业发展、洪水 / Hunting, Plantation, Residential and Commercial Development, Flooding

🦌 保护级别与保护行动 Protection Category and Conservation Action (s)

国家重点保护野生动物等级 (2021) Category of National Key Protected Wild Animals (2021)	二级 / Category II
IUCN 红色名录 (2020-2) IUCN Red List (2020-2)	易危 / VU
CITES 附录 (2019) CITES Appendix (2019)	无 / NA
保护行动 Conservation Action(s)	自然保护区内种群得到保护 / Populations in nature reserves are protected

🦌 相关文献 Relevant References

Burgin *et al.*, 2018; Jiang *et al.* (蒋志刚等), 2017; Li *et al.* (李言阔等), 2013; Zhu *et al.* (朱曦等), 2010; Zhang and Zhang (张小龙和张恩迪), 2002; Sun and Bao (孙孟军和鲍毅新), 2001

獐 *Hydropotes inermis*

海南麂
Muntiacus nigripes

易危 VU A1acd; B1ab(i, ii, iii)

| 数据缺乏 DD | 无危 LC | 易危 VU | 濒危 EN | 极危 CR | 区域灭绝 RE | 野外灭绝 EW | 灭绝 EX |

分类地位 Taxonomic Status

动物界 Animalia	脊索动物门 Chordata	哺乳纲 Mammalia	偶蹄目 Artiodactyla	鹿科 Cervidae

学　名 Scientific Name	*Muntiacus nigripes*
命名人 Species Authority	G. M. Allen, 1930
英文名 English Name(s)	Black-legged Muntjac
同物异名 Synonym(s)	Black-footed Muntjac, Hainan Muntjac
种下单元评估 Infra-specific Taxa Assessed	无 / None

评估信息 Assessment Information

评估年份 Year Assessed	2020
评　定　人 Assessor(s)	蒋志刚 / Zhigang Jiang
审　定　人 Reviewer(s)	江海生、陈辈乐、李飞 / Haisheng Jiang, Bosco P. L. Chan, Fei Li
其他贡献人 Other Contributor(s)	李立立、丁晨晨 / Lili Li, Chenchen Ding

理由 Justification: 海南麂仅分布于海南岛，曾经遍布全岛。最新的红外相机调查显示，海南麂目前仅分布于鹦哥岭、霸王岭、俄贤岭等中部山区数个保护状况良好的林区中，而低海拔种群数量锐减，如在大田、邦溪等保护区中已经很难见到海南麂。但在海口观澜湖一带仍发现有海南麂活动，说明在北部火山岩地带仍有部分种群残留（嘉道理农场暨植物园，未发表数据）。海南麂分布狭窄，且受到盗猎及栖息地减少的威胁，故列为易危等级 / Black-legged Muntjac is distributed only on Hainan Island in China, and once was found throughout the island. According to the latest infrared camera survey, Black-legged Muntjac is only distributed in several well-protected forest areas in the central mountainous regions, such as Yingeling, Bawangling and Exianling, while the number in low-altitude areas has declined sharply. For example, it is difficult to see Black-legged Muntjac in the protected areas such as Datian and Bangxi reserves. However, some individuals are still found in the Guanlan Lake area of Haikou, indicating that some population may remain in the volcanic rock area in the north part of the island (Kadoorie Farm and Botanic Garden, unpublished data). Black-legged Muntjac has a narrow distribution and is threatened by poaching and habitat loss. Thus, it is listed as Vulnerable

地理分布 Geographical Distribution

国内分布 Domestic Distribution
海南 / Hainan
世界分布 World Distribution
中国、越南 / China, Viet Nam
分布标注 Distribution Note
非特有种 / Non-endemic

国内分布图
Map of Domestic Distribution

种群 Population

种群数量 Population Size	未知 / Unknown
种群趋势 Population Trend	下降 / Decreasing

生境与生态系统 Habitat(s) and Ecosystem(s)

生 境 Habitat(s)	森林 / Forest
生态系统 Ecosystem(s)	森林生态系统 / Forest Ecosystem

威胁 Threat(s)

主要威胁 Major Threat(s)	狩猎、开垦 / Hunting, Land Reclaimed for Farming

保护级别与保护行动 Protection Category and Conservation Action(s)

国家重点保护野生动物等级 (2021) Category of National Key Protected Wild Animals (2021)	二级 / Category II
IUCN红色名录 (2020-2) IUCN Red List (2020-2)	未列入 / NA
CITES 附录 (2019) CITES Appendix (2019)	无 / NA
保护行动 Conservation Action(s)	未知 / Unknown

相关文献 Relevant References

Burgin *et al*., 2018; Groves and Grubb, 2011

海南麂 *Muntiacus nigripes*   嘉道理农场暨植物园 提供   By Kadoorie Farm and Botanic Garden

中国生物多样性红色名录
China's Red List of Biodiversity

野牦牛
Bos mutus

易危 VU A1acd; B1ab(i, ii, iii)

| 数据缺乏 DD | 无危 LC | 近危 NT | 易危 VU | 濒危 EN | 极危 CR | 区域灭绝 RE | 野外灭绝 EW | 灭绝 EX |

分类地位 Taxonomic Status

动物界 Animalia	脊索动物门 Chordata	哺乳纲 Mammalia	偶蹄目 Artiodactyla	牛科 Bovidae

学 名 Scientific Name	*Bos mutus*
命 名 人 Species Authority	Przewalski, 1883
英 文 名 English Name(s)	Wild Yak
同物异名 Synonym(s)	无 / None
种下单元评估 Infra-specific Taxa Assessed	无 / None

评估信息 Assessment Information

评 估 年 份 Year Assessed	2020
评 定 人 Assessor(s)	蒋志刚 / Zhigang Jiang
审 定 人 Reviewer(s)	蒋学龙 / Xuelong Jiang
其他贡献人 Other Contributor(s)	李立立、丁晨晨 / Lili Li, Chenchen Ding

理由 Justification: 野牦牛栖息在青藏高原高海拔地区。野牦牛种群数量在增长，但仍面临来自家畜放牧、捕猎等的威胁。因此，将野牦牛列为易危等级 / Wild Yak inhabits the high altitude areas on the Qinghai-Tibet(Xizang) Plateau. The population of Wild Yak is increasing, however it still faces threats from over-grazing of livestock as well as hunting. Thus, Wild Yak is listed as Vulnerable

地理分布 Geographical Distribution

国内分布 Domestic Distribution
四川、西藏、新疆、青海、甘肃、内蒙古 / Sichuan, Tibet (Xizang), Xinjiang, Qinghai, Gansu, Inner Mongolia (Nei Mongol)
世界分布 World Distribution
中国、印度、尼泊尔 / China, India, Nepal
分布标注 Distribution Note
非特有种 / Non-endemic

国内分布图
Map of Domestic Distribution

种群 Population

种群数量 Population Size	27,220～47,138 头 / 27,220~47,138 individuals
种群趋势 Population Trend	增加 / Increasing

生境与生态系统 Habitat(s) and Ecosystem(s)

生　　境 Habitat(s)	高寒草甸、高寒草地、高寒荒漠 / Alpine Meadow, Alpine Grassland, Alpine Desert
生态系统 Ecosystem(s)	高寒生态系统 / Alpine Ecosystem

威胁 Threat(s)

主要威胁 Major Threat(s)	家畜放牧、杂交、疾病、捕猎 / Livestock Ranching, Hybridization, Disease, Hunting

保护级别与保护行动 Protection Category and Conservation Action(s)

国家重点保护野生动物等级 (2021) Category of National Key Protected Wild Animals (2021)	一级 / Category I
IUCN 红色名录 (2020-2) IUCN Red List (2020-2)	易危 / VU
CITES 附录 (2019) CITES Appendix (2019)	I
保护行动 Conservation Action(s)	在自然保护区内的种群得到保护；列入《迁徙物种保护公约》的"迁徙物种行动计划" / Populations in nature reserves are under protection; included in the *Convention on Migratory Species* (CMS)－Species Action Plan

相关文献 Relevant References

Burgin *et al*., 2018; Hu *et al*.（胡一鸣等）, 2018; Jiang *et al*.（蒋志刚等）, 2015a; Buzzard *et al*., 2010; Guo *et al*.（郭宪等）, 2007; Liu and Schaller（刘务林和乔治·B·夏勒）, 2003

野牦牛 *Bos mutus* 　　　　　　　　　　　　　　　　　　　蒋志刚 摄　By Zhigang Jiang

中国生物多样性红色名录
China's Red List of Biodiversity

鹅喉羚
Gazella subgutturosa

易危 VU A1acd; B1ab(i, ii, iii)

| 数据缺乏 DD | 无危 LC | 近危 NT | 易危 VU | 濒危 EN | 极危 CR | 区域灭绝 RE | 野外灭绝 EW | 灭绝 EX |

分类地位 Taxonomic Status

动物界 Animalia	脊索动物门 Chordata	哺乳纲 Mammalia	偶蹄目 Artiodactyla	牛科 Bovidae
学　　名 Scientific Name		*Gazella subgutturosa*		
命 名 人 Species Authority		Blanford, 1875		
英 文 名 English Name(s)		Yarkand Goitered Gazelle		
同物异名 Synonym(s)		Black-tailed Gazelle, *Gazella subgutturosa yarkandensis*		
种下单元评估 Infra-specific Taxa Assessed		无 / None		

评估信息 Assessment Information

评估年份 Year Assessed	2020
评　定　人 Assessor(s)	蒋志刚 / Zhigang Jiang
审　定　人 Reviewer(s)	杨维康、初红军、毕俊怀、胡德夫 / Weikang Yang, Hongjun Chu, Junhuai Bi, Defu Hu
其他贡献人 Other Contributor(s)	李立立、丁晨晨 / Lili Li, Chenchen Ding

理由 Justification: 鹅喉羚是国家Ⅱ级重点保护野生动物。近半个世纪以来，由于人类活动，如过度放牧、矿业开采、狩猎等的加剧，鹅喉羚栖息地不断恶化，分布区日益缩小。2008～2009年冬季暴风雪后，塔里木鹅喉羚种群在新疆经历了一个快速下降过程，目前种群数量仍未完全恢复。因此，基于其种群数量下降速率，将鹅喉羚列为易危等级 / Yarkand Goitered Gazelle is a species in the Secondary Category of National Key Protected Wild Animals in China. In recent half century, due to intensified human disturbance, such as overgrazing, mining and poaching, its habitats are deteriorating and its distribution range is shrinking. Following the blizzard in winter of 2008~2009, the population of Yarkand Goitered Gazelle declined dramatically and even today it has not yet fully recovered. Thus, based on the rate of its population decline, Yarkand Goitred Gazelle is listed as Vulnerable

地理分布 Geographical Distribution

国内分布 Domestic Distribution
新疆、内蒙古、甘肃、青海 / Xinjiang, Inner Mongolia (Nei Mongol), Gansu, Qinghai
世界分布 World Distribution
阿富汗、中国、伊朗、哈萨克斯坦、吉尔吉斯斯坦、蒙古、巴基斯坦、塔吉克斯坦、土库曼斯坦、乌兹别克斯坦 / Afghanistan, China, Iran, Kazakhstan, Kyrgyzstan, Mongolia, Pakistan, Tajikistan, Turkmenistan, Uzbekistan
分布标注 Distribution Note
非特有种 / Non-endemic

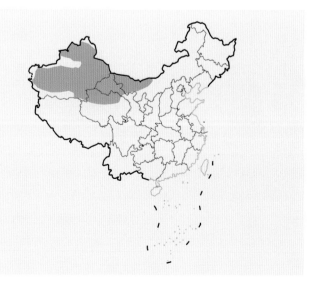

国内分布图
Map of Domestic Distribution

种群 Population

种群数量 Population Size	下降 / Decreasing
种群趋势 Population Trend	稳定 / Stable

生境与生态系统 Habitat (s) and Ecosystem (s)

生境 Habitat(s)	荒漠、半荒漠 / Desert, Semi-desert
生态系统 Ecosystem(s)	荒漠生态系统 / Desert Ecosystem

威胁 Threat (s)

主要威胁 Major Threat(s)	生境丧失、自然灾害、人类干扰 / Habitat Loss, Natural Disaster, Human Disturbance

保护级别与保护行动 Protection Category and Conservation Action (s)

国家重点保护野生动物等级 (2021) Category of National Key Protected Wild Animals (2021)	二级 / Category II
IUCN红色名录 (2020-2) IUCN Red List (2020-2)	易危 / VU
CITES 附录 (2019) CITES Appendix (2019)	无 / NA
保护行动 Conservation Action(s)	在自然保护区内的种群得到保护；列入《迁徙物种保护公约》的"迁徙物种行动计划" / Populations in nature reserves are under protection; included in the *Convention on Migratory Species* (CMS)－Species Action Plan

相关文献 Relevant References

Burgin *et al*., 2018; Jiang *et al*. (蒋志刚等), 2017; Groves and Grubb, 2011; Wilson and Mittermeier, 2011; Chu *et al*., 2009; Xu *et al*. (徐文轩等), 2008; Gao and Yao (高行宜和姚军), 2006; Sun *et al*. (孙铭娟等), 2002

鹅喉羚 *Gazella subgutturosa* 　　　　　　　　　　　　　　蒋志刚 摄　By Zhigang Jiang

秦岭羚牛
Budorcas bedfordi

中国生物多样性红色名录 / China's Red List of Biodiversity

易危 VU A1acd; B1ab(i, ii, iii)

| 数据缺乏 DD | 无危 LC | 近危 NT | 易危 VU | 濒危 EN | 极危 CR | 区域灭绝 RE | 野外灭绝 EW | 灭绝 EX |

分类地位 Taxonomic Status

动物界 Animalia	脊索动物门 Chordata	哺乳纲 Mammalia	偶蹄目 Artiodactyla	牛科 Bovidae

学 名 Scientific Name	*Budorcas bedfordi*
命 名 人 Species Authority	Thomas, 1911
英 文 名 English Name(s)	Golden Takin
同物异名 Synonym(s)	*Budorcas taxicolor bedfordi*
种下单元评估 Infra-specific Taxa Assessed	无 / None

评估信息 Assessment Information

评估年份 Year Assessed	2020
评 定 人 Assessor(s)	蒋志刚 / Zhigang Jiang
审 定 人 Reviewer(s)	宋延龄、曾治高 / Yanling Song, Zhigao Zeng
其他贡献人 Other Contributor(s)	李立立、丁晨晨 / Lili Li, Chenchen Ding

理由 Justification: 秦岭羚牛是中国特有种，分布于秦岭，其分布区狭窄、种群数量少。因此，基于其占有区面积，将其列为易危等级 / Golden Takin is an endemic species in China. It is narrowly distributed in the Qinling Mountains and its populations are small. Thus, based on its area of occupancy, Golden Takin is listed as Vulnerable

地理分布 Geographical Distribution

国内分布 Domestic Distribution
陕西 / Shaanxi
世界分布 World Distribution
中国 / China
分布标注 Distribution Note
特有种 / Endemic

国内分布图
Map of Domestic Distribution

种群 Population

种群数量 Population Size	4,000 头 / 4,000 individuals
种群趋势 Population Trend	稳定 / Stable

生境与生态系统 Habitat (s) and Ecosystem (s)

生　　境 Habitat(s)	草甸、森林 / Meadow, Forest
生态系统 Ecosystem(s)	森林生态系统、草地生态系统 / Forest Ecosystem, Grassland Ecosystem

威胁 Threat (s)

主要威胁 Major Threat(s)	生境丧失 / Habitat Loss

保护级别与保护行动 Protection Category and Conservation Action (s)

国家重点保护野生动物等级 (2021) Category of National Key Protected Wild Animals (2021)	一级 / Category I
IUCN 红色名录 (2020-2) IUCN Red List (2020-2)	未列入 / NA
CITES 附录 (2019) CITES Appendix (2019)	无 / NA
保护行动 Conservation Action(s)	自然保护区内种群得到保护 / Populations in nature reserves are protected

相关文献 Relevant References

Jiang et al. (蒋志刚等), 2017; Castelló, 2016; Groves and Grubb, 2011; Wilson and Mittermeier, 2011; Ma and Wang (麻应太和王西峰), 2008; Zeng and Song (曾治高和宋延龄), 2008

秦岭羚牛 *Budorcas bedfordi*

四川羚牛
Budorcas tibetanus

易危　VU A1acd; B1ab(i, ii, iii)

| 数据缺乏 DD | 无危 LC | 近危 NT | 易危 VU | 濒危 EN | 极危 CR | 区域灭绝 RE | 野外灭绝 EW | 灭绝 EX |

分类地位 Taxonomic Status

动物界 Animalia	脊索动物门 Chordata	哺乳纲 Mammalia	偶蹄目 Artiodactyla	牛科 Bovidae
学名 Scientific Name		*Budorcas tibetanus*		
命名人 Species Authority		Milne-Edwards, 1874		
英文名 English Name(s)		Sichuan Takin		
同物异名 Synonym(s)		无 / None		
种下单元评估 Infra-specific Taxa Assessed		无 / None		

评估信息 Assessment Information

评估年份 Year Assessed	2020
评定人 Assessor(s)	蒋志刚 / Zhigang Jiang
审定人 Reviewer(s)	宋延龄、曾治高 / Yanling Song, Zhigao Zeng
其他贡献人 Other Contributor(s)	李立立、丁晨晨 / Lili Li, Chenchen Ding

理由 Justification: 四川羚牛缺少自然天敌，但是其种群发展受到生境破碎和生境负载量的影响。目前已建成自然保护区网络保护四川羚牛野生种群。基于其栖息地状况，将四川羚牛列为易危等级 / Now, wild populations of Sichuan Takin are stable and instances of predation are rare, but its population growth is restricted by the carrying capacity and fragmented habitat. A well-completed *in situ* protection network has been established to protect Sichuan Takin. Based on its habitat status, Sichuan Takin is listed as Vulnerable

地理分布 Geographical Distribution

国内分布 Domestic Distribution
四川 / Sichuan
世界分布 World Distribution
中国 / China
分布标注 Distribution Note
特有种 / Endemic

国内分布图
Map of Domestic Distribution

种群 Population

种群数量 Population Size	未知 / Unknown
种群趋势 Population Trend	稳定 / Stable

生境与生态系统 Habitat (s) and Ecosystem (s)

生　　境 Habitat(s)	草甸、森林 / Meadow, Forest
生态系统 Ecosystem(s)	森林生态系统、草地生态系统 / Forest Ecosystem, Grassland Ecosystem

威胁 Threat (s)

主要威胁 Major Threat(s)	生境破碎与生境丧失 / Habitat Fragmentation and Loss

保护级别与保护行动 Protection Category and Conservation Action (s)

国家重点保护野生动物等级 (2021) Category of National Key Protected Wild Animals (2021)	一级 / Category I
IUCN红色名录 (2020-2) IUCN Red List (2020-2)	未列入 / NA
CITES 附录 (2019) CITES Appendix (2019)	无 / NA
保护行动 Conservation Action(s)	自然保护区内种群得到保护 / Populations in nature reserves are protected

相关文献 Relevant References

Jiang *et al.*（蒋志刚等）, 2017; Castelló, 2016; Chen *et al.*（陈万里等）, 2013; Groves and Grubb, 2011; Wilson and Mittermeier, 2011

四川羚牛 *Budorcas tibetanus*

不丹羚牛
Budorcas whitei

易危 VU A1acd; B1ab(i, ii, iii)

| 数据缺乏 DD | 无危 LC | 近危 NT | 易危 VU | 濒危 EN | 极危 CR | 区域灭绝 RE | 野外灭绝 EW | 灭绝 EX |

分类地位 Taxonomic Status

动物界 Animalia	脊索动物门 Chordata	哺乳纲 Mammalia	偶蹄目 Artiodactyla	牛科 Bovidae
学 名 Scientific Name		*Budorcas whitei*		
命 名 人 Species Authority		Lydekker, 1907		
英 文 名 English Name(s)		Bhutan Takin		
同物异名 Synonym(s)		无 / None		
种下单元评估 Infra-specific Taxa Assessed		无 / None		

评估信息 Assessment Information

评 估 年 份 Year Assessed	2020
评 定 人 Assessor(s)	蒋志刚 / Zhigang Jiang
审 定 人 Reviewer(s)	蒋学龙 / Xuelong Jiang
其他贡献人 Other Contributor(s)	李立立、丁晨晨 / Lili Li, Chenchen Ding

理由 Justification: 已建成不丹羚牛就地保护网络。然而，不丹羚牛野生种群不大，且其种群发展受到生境破碎和生境负载量的影响。因此，基于其种群小，不丹羚牛列为易危等级 / A complete *in situ* protection network has been developed to protect Bhutan Takin. Nonetheless, wild populations remain small and growth is limited by habitat carrying capacity and habitat fragmentation. Based on its small population size, the Bhutan Takin is listed as Vulnerable

地理分布 Geographical Distribution

国内分布 Domestic Distribution
西藏、云南 / Tibet (Xizang), Yunnan
世界分布 World Distribution
不丹、中国 / Bhutan, China
分布标注 Distribution Note
非特有种 / Non-endemic

国内分布图
Map of Domestic Distribution

种群 Population

种群数量 Population Size	未知 / Unknown
种群趋势 Population Trend	下降 / Decreasing

生境与生态系统 Habitat (s) and Ecosystem (s)

生　　境 Habitat(s)	草甸、森林 / Meadow, Forest
生态系统 Ecosystem(s)	森林生态系统、草地生态系统 / Forest Ecosystem, Grassland Ecosystem

威胁 Threat (s)

主要威胁 Major Threat(s)	生境破碎与生境丧失 / Habitat Fragmentation and Loss

保护级别与保护行动 Protection Category and Conservation Action (s)

国家重点保护野生动物等级 (2021) Category of National Key Protected Wild Animals (2021)	一级 / Category I
IUCN 红色名录 (2020-2) IUCN Red List (2020-2)	未列入 / NA
CITES 附录 (2019) CITES Appendix (2019)	无 / NA
保护行动 Conservation Action(s)	自然保护区内种群得到保护 / Populations in nature reserves are protected

相关文献 Relevant References

Jiang *et al*.（蒋志刚等），2018; Castelló, 2016; Groves and Grubb, 2011; Wilson and Mittermeier, 2011; Wu and Zhang（吴鹏举和张恩迪），2006

不丹羚牛 *Budorcas whitei* 　　　　董磊 摄（西南山地供图）　By Lei Dong (Swild.cn)

中华斑羚
Naemorhedus griseus

易危 VU A2acd

| 数据缺乏 DD | 无危 LC | 近危 NT | **易危 VU** | 濒危 EN | 极危 CR | 区域灭绝 RE | 野外灭绝 EW | 灭绝 EX |

🦌 分类地位 Taxonomic Status

动物界 Animalia	脊索动物门 Chordata	哺乳纲 Mammalia	偶蹄目 Artiodactyla	牛科 Bovidae

学 名 Scientific Name	*Naemorhedus griseus*
命 名 人 Species Authority	Milne-Edwards, 1871
英 文 名 English Name(s)	Chinese Goral
同物异名 Synonym(s)	Grey Long-tailed Goral; *Nemorhaedus griseus*; *caudatus* (Milne-Edwards, 1871) subsp. *griseus*
种下单元评估 Infra-specific Taxa Assessed	无 / None

🦌 评估信息 Assessment Information

评 估 年 份 Year Assessed	2020
评 定 人 Assessor(s)	蒋志刚 / Zhigang Jiang
审 定 人 Reviewer(s)	陈辈乐、李飞 / Bosco P. L. Chan, Fei Li
其他贡献人 Other Contributor(s)	李立立、丁晨晨 / Lili Li, Chenchen Ding

理由 Justification: 中华斑羚广布于中国中部及南部。然而，最近的红外相机调查并未在广东、广西、安徽、贵州等历史分布区记录到中华斑羚，说明其分布区缩减。且受到猎捕的压力，栖息地受人类活动干扰。因此，基于其占有区状况，将中华斑羚列入易危 / Historically, Chinese Goral was widely distributed in central and southern China. However, recent infrared camera surveys did not record Chinese Goral in its former range in Guangdong, Guangxi, Anhui and Guizhou, *etc.*, indicating that its distribution is shrinking. Furthermore, hunting pressure is mounting and its habitats are being disturbed by human encroachment. Thus, based on its area of occupancy, Chinese Goral is listed as Vulnerable

🦌 地理分布 Geographical Distribution

国内分布 Domestic Distribution

内蒙古、河北、北京、河南、山西、陕西、甘肃、宁夏、云南、四川、贵州、重庆、湖北、湖南、广西、广东、江西、福建、浙江、上海、江苏、安徽 / Inner Mongolia (Nei Mongol), Hebei, Beijing, Henan, Shanxi, Shaanxi, Gansu, Ningxia, Yunnan, Sichuan, Guizhou, Chongqing, Hubei, Hunan, Guangxi, Guangdong, Jiangxi, Fujian, Zhejiang, Shanghai, Jiangsu, Anhui

世界分布 World Distribution

中国；南亚、东南亚 / China; South Asia, Southeast Asia

分布标注 Distribution Note

非特有种 / Non-endemic

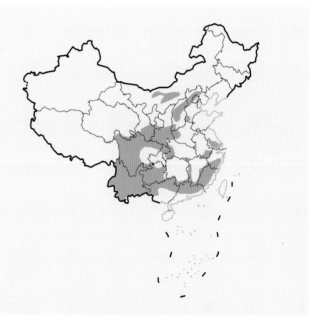

国内分布图
Map of Domestic Distribution

种群 Population

种群数量 Population Size	未知 / Unknown
种群趋势 Population Trend	下降 / Decreasing

生境与生态系统 Habitat (s) and Ecosystem (s)

生　　境 Habitat(s)	森林 / Forest
生态系统 Ecosystem(s)	森林生态系统 / Forest Ecosystem

威胁 Threat (s)

主要威胁 Major Threat(s)	人类干扰、狩猎 / Human Disturbance, Hunting

保护级别与保护行动 Protection Category and Conservation Action (s)

国家重点保护野生动物等级 (2021) Category of National Key Protected Wild Animals (2021)	二级 / Category II
IUCN 红色名录(2020-2) IUCN Red List (2020-2)	易危 / VU
CITES 附录 (2019) CITES Appendix (2019)	I
保护行动 Conservation Action(s)	无 / None

相关文献 Relevant References

Burgin *et al.*, 2018; Jiang *et al.* (蒋 志 刚 等), 2017; Castelló, 2016; Groves and Grubb, 2011; Wilson and Mittermeier, 2011; Smith *et al.* (史密斯等), 2009; Pan *et al.* (潘清华等), 2007

中华斑羚 *Naemorhedus griseus*

帕米尔盘羊
Ovis polii

易危 VU A1acd; B1ab(i, ii, iii)

| DD 数据缺乏 | LC 无危 | NT 近危 | **VU 易危** | EN 濒危 | CR 极危 | RE 区域灭绝 | EW 野外灭绝 | EX 灭绝 |

分类地位 Taxonomic Status

动物界 Animalia	脊索动物门 Chordata	哺乳纲 Mammalia	偶蹄目 Artiodactyla	牛科 Bovidae

学 名 Scientific Name	*Ovis polii*
命名人 Species Authority	Blyth, 1841
英文名 English Name(s)	Pamir Argali
同物异名 Synonym(s)	Marco Polo Sheep, *Ovis ammon polii*
种下单元评估 Infra-specific Taxa Assessed	无 / None

评估信息 Assessment Information

评估年份 Year Assessed	2020
评定人 Assessor(s)	蒋志刚 / Zhigang Jiang
审定人 Reviewer(s)	初红军、杨维康 / Hongjun Chu, Weikang Yang
其他贡献人 Other Contributor(s)	李立立、丁晨晨 / Lili Li, Chenchen Ding

理由 Justification: 帕米尔盘羊又称为马可波罗盘羊，分布在帕米尔高原。估计中国境内有2,500只、吉尔吉斯斯坦有5,000只、塔吉克斯坦有25,000只帕米尔盘羊。帕米尔盘羊面临的主要威胁是家畜放牧与草原火灾。故帕米尔盘羊列为易危等级 / Pamir Argali is also known as Marco Polo Sheep, it is distributed in the Pamir Plateau. Its main population is in Tajikistan (25,000) and Kyrgyzstan (5,000). There are estimated to be around 2,500 Pamir Argalis in China. The main threats to this species are livestock grazing and fires in the grasslands where it resides. Thus, Pamir Argali is listed as Vulnerable

地理分布 Geographical Distribution

国内分布 Domestic Distribution
新疆 / Xinjiang
世界分布 World Distribution
中国、哈萨克斯坦、吉尔吉斯斯坦、塔吉克斯坦 / China, Kazakhstan, Kyrgyzstan, Tajikistan
分布标注 Distribution Note
非特有种 / Non-endemic

国内分布图
Map of Domestic Distribution

种群 Population

种群数量 Population Size	2,500 只 / 2,500 individuals
种群趋势 Population Trend	稳定 / Stable

生境与生态系统 Habitat(s) and Ecosystem(s)

生境 Habitat(s)	开阔草地 / Open Grassland
生态系统 Ecosystem(s)	草地生态系统 / Grassland Ecosystem

威胁 Threat(s)

主要威胁 Major Threat(s)	家畜放牧压力增加 / Increasing Livestock Grazing Pressure

保护级别与保护行动 Protection Category and Conservation Action(s)

国家重点保护野生动物等级 (2021) Category of National Key Protected Wild Animals (2021)	二级 / Category II
IUCN 红色名录 (2020-2) IUCN Red List (2020-2)	未列入 / NA
CITES 附录 (2019) CITES Appendix (2019)	II
保护行动 Conservation Action(s)	在自然保护区内的种群得到保护；列入《迁徙物种保护公约》的"迁徙物种行动计划" / Populations in nature reserves are under protection; included in the *Convention on Migratory Species* (CMS)－Species Action Plan

相关文献 Relevant References

Jiang *et al.* (蒋志刚等), 2017; Castelló, 2016; Wilson and Mittermeier, 2012; Groves and Grubb, 2011

帕米尔盘羊 *Ovis polii*　　　　　阎旭光 摄　By Xuguang Yan

中华鬣羚
Capricornis milneedwardsii

易危 VU A2acd

| 数据缺乏 DD | 无危 LC | 近危 NT | 易危 VU | 濒危 EN | 极危 CR | 区域灭绝 RE | 野外灭绝 EW | 灭绝 EX |

分类地位 Taxonomic Status

动物界 Animalia	脊索动物门 Chordata	哺乳纲 Mammalia	偶蹄目 Artiodactyla	牛科 Bovidae
学名 Scientific Name		*Capricornis milneedwardsii*		
命名人 Species Authority		David, 1869		
英文名 English Name(s)		Chinese Serow		
同物异名 Synonym(s)		Mainland Serow; *Capricornis sumatraensis* (David, 1869) subsp. *milneedwardsii*		
种下单元评估 Infra-specific Taxa Assessed		无 / None		

评估信息 Assessment Information

评估年份 Year Assessed	2020
评定人 Assessor(s)	蒋志刚 / Zhigang Jiang
审定人 Reviewer(s)	陈辈乐、李飞 / Bosco P. L. Chan, Fei Li
其他贡献人 Other Contributor(s)	李立立、丁晨晨 / Lili Li, Chenchen Ding

理由 Justification: 中华鬣羚分布广。最近的红外相机调查显示，中华鬣羚仍然广泛分布于中国中部及南部，但种群数量较少。在一些地区的中华鬣羚仍然受到盗猎及栖息地减少的威胁。故将中华鬣羚列入易危等级 / Chinese Serow has a wide distribution area. Recent infrared camera surveys showed that Chinese Serow is still widely distributed in central and south China, but its populations are small. Chinese Serow in some areas is still threatened by poaching and habitat loss. Thus, it is listed as Vulnerable

地理分布 Geographical Distribution

国内分布 Domestic Distribution
广东、广西、湖南、湖北、四川、云南、贵州、西藏、青海、甘肃、陕西、河南、安徽、浙江、福建、江西 / Guangdong, Guangxi, Hunan, Hubei, Sichuan, Yunnan, Guizhou, Tibet (Xizang), Qinghai, Gansu, Shaanxi, Henan, Anhui, Zhejiang, Fujian, Jiangxi
世界分布 World Distribution
中国；东南亚 / China; Southeast Asia
分布标注 Distribution Note
非特有种 / Non-endemic

国内分布图
Map of Domestic Distribution

种群 Population

种群数量 Population Size	未知 / Unknown
种群趋势 Population Trend	下降 / Decreasing

生境与生态系统 Habitat (s) and Ecosystem (s)

生　　境 Habitat(s)	森林、内陆岩石区域、峭壁 / Forest, Inland Rocky Area, Cliff
生态系统 Ecosystem(s)	森林生态系统 / Forest Ecosystem

威胁 Threat (s)

主要威胁 Major Threat(s)	狩猎、耕种、伐木、火灾 / Hunting, Plantation, Logging, Fire

保护级别与保护行动 Protection Category and Conservation Action (s)

国家重点保护野生动物等级 (2021) Category of National Key Protected Wild Animals (2021)	二级 / Category II
IUCN 红色名录 (2020-2) IUCN Red List (2020-2)	易危 / VU
CITES 附录 (2019) CITES Appendix (2019)	I
保护行动 Conservation Action(s)	部分种群位于自然保护区之内 / Parts of the populations are protected by nature reserves

相关文献 Relevant References

Burgin *et al*., 2018; Jiang *et al*.（蒋志刚等）, 2017; Castelló, 2016; Chen *et al*.（陈永春等）, 2013; Wilson and Mittermeier, 2011; Smith *et al*.（史密斯等）, 2009; Lu *et al*.（陆雪等）, 2007; Wang（王应祥）, 2003

中华鬣羚 *Capricornis milneedwardsii*

中国生物多样性红色名录

抹香鲸
Physeter macrocephalus

易危　VU A2abc

| 数据缺乏 DD | 无危 LC | 近危 NT | **易危 VU** | 濒危 EN | 极危 CR | 区域灭绝 RE | 野外灭绝 EW | 灭绝 EX |

分类地位 Taxonomic Status

动物界 Animalia	脊索动物门 Chordata	哺乳纲 Mammalia	鲸目 Cetacea	抹香鲸科 Physeteridae

学名 Scientific Name	*Physeter macrocephalus*
命名人 Species Authority	Linnaeus, 1758
英文名 English Name(s)	Sperm Whale
同物异名 Synonym(s)	*Physeter catodon* (Linnaeus, 1758)
种下单元评估 Infra-specific Taxa Assessed	无 / None

评估信息 Assessment Information

评估年份 Year Assessed	2020
评定人 Assessor(s)	周开亚 / Kaiya Zhou
审定人 Reviewer(s)	张先锋、王克雄、王丁、祝茜、蒋志刚 / Xianfeng Zhang, Kexiong Wang, Ding Wang, Qian Zhu, Zhigang Jiang
其他贡献人 Other Contributor(s)	李立立、丁晨晨 / Lili Li, Chenchen Ding

理由 Justification: 抹香鲸主要在深海活动，在浅海中不常见。在全世界海洋中，抹香鲸一直被猎捕。现存的约 360,000 头抹香鲸面临着生存威胁，直到 1986 年猎捕暂停令公布。因此，抹香鲸列为易危等级 / Sperm Whale mainly lives in deep seas, Sperm Whale in shallow sea is rare. It was hunted throughout the world's oceans until a 1986 moratorium of hunting released. About 360,000 Sperm Whales in oceans worldwide are facing a variety of threats. Thus, Sperm Whale is listed as Vulnerable

地理分布 Geographical Distribution

国内分布 Domestic Distribution	黄海、东海、台湾海峡、南海 / Yellow Sea, East China Sea, Taiwan Strait, South China Sea
世界分布 World Distribution	世界各地深海海域 / The deep oceans all over the world
分布标注 Distribution Note	非特有种 / Non-endemic

国内分布图
Map of Domestic Distribution

种群 Population

种群数量 Population Size	全球有大约 360,000 头 / Approximately 360,000 individuals world wide
种群趋势 Population Trend	未知 / Unknown

生境与生态系统 Habitat(s) and Ecosystem(s)

生境 Habitat(s)	海洋 / Ocean
生态系统 Ecosystem(s)	海洋生态系统 / Ocean Ecosystem

威胁 Threat(s)

主要威胁 Major Threat(s)	小规模捕鲸、船舶撞击、海洋垃圾、水下噪声 / Small Scale Whaling, Ship Strike, Oceanic Debris, Under Water Noise

保护级别与保护行动 Protection Category and Conservation Action(s)

国家重点保护野生动物等级 (2021) Category of National Key Protected Wild Animals (2021)	一级 / Category I
IUCN 红色名录 (2020-2) IUCN Red List (2020-2)	易危 / VU
CITES 附录 (2019) CITES Appendix (2019)	I
保护行动 Conservation Action(s)	法律保护物种 / Legally protected species

相关文献 Relevant References

Burgin *et al.*, 2018; Jiang *et al.* (蒋志刚等), 2015a; Wang (王丕烈), 2011; Zhou (周开亚), 2008, 2004; Dong *et al.* (董金海等), 1977

抹香鲸 *Physeter macrocephalus*

中国生物多样性红色名录

印太江豚
Neophocaena phocaenoides

易危 VU C2; E

DD	LC	NT	VU	EN	CR	RE	EW	EX
数据缺乏	无危		易危	濒危	极危	区域灭绝	野外灭绝	灭绝

分类地位 Taxonomic Status

动物界 Animalia	脊索动物门 Chordata	哺乳纲 Mammalia	鲸目 Cetacea	鼠海豚科 Phocoenidae

学 名 Scientific Name	*Neophocaena phocaenoides*
命 名 人 Species Authority	G. Cuvier, 1829
英 文 名 English Name(s)	Indo-Pacific Finless Porpoise
同物异名 Synonym(s)	Finless Porpoise
种下单元评估 Infra-specific Taxa Assessed	无 / None

评估信息 Assessment Information

评 估 年 份 Year Assessed	2020
评 定 人 Assessor(s)	周开亚 / Kaiya Zhou
审 定 人 Reviewer(s)	张先锋、王克雄、王丁、祝茜、蒋志刚 / Xianfeng Zhang, Kexiong Wang, Ding Wang, Qian Zhu, Zhigang Jiang
其他贡献人 Other Contributor(s)	李立立、丁晨晨 / Lili Li, Chenchen Ding

理由 Justification: 印太江豚栖息在太平洋和印度洋热带和亚热带沿岸浅水海域，在太平洋北至台湾海峡。种群数量不详。它们靠近多种人类活动，受到栖息地丧失和退化、渔具误捕和船只撞击等的影响。因此，列为易危等级 / Indo-Pacific Finless Porpoise inhabits shallow tropical and subtropical coastal waters of the Pacific Ocean and Indian Ocean, northward to Taiwan Strait in the Pacific Ocean. Total abundance of Indo-Pacific Finless Porpoise is unknown. They may come in close proximity to human activities and therefore are affected by the loss and degradation of habitat, entanglement in fishing gear, and vessel strikes. Thus, it is listed as Vulnerable

地理分布 Geographical Distribution

国内分布 Domestic Distribution	东海、台湾海峡、南海 / East China Sea, Taiwan Strait, South China Sea
世界分布 World Distribution	印度洋、太平洋 / Indian Ocean, Pacific Ocean
分布标注 Distribution Note	非特有种 / Non-endemic

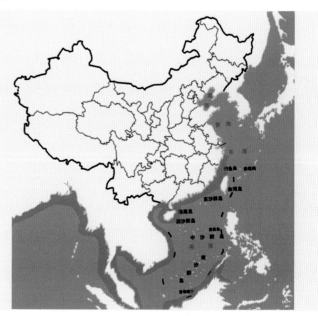

国内分布图
Map of Domestic Distribution

种群 Population

种群数量 Population Size	未知 / Unknown
种群趋势 Population Trend	下降 / Decreasing

生境与生态系统 Habitat (s) and Ecosystem (s)

生　　境 Habitat(s)	海洋 / Ocean
生态系统 Ecosystem(s)	海洋生态系统 / Ocean Ecosystem

威胁 Threat (s)

主要威胁 Major Threat(s)	刺网误捕，沿海污染、船只交通导致的栖息地退化和丧失 / Incidental Mortality in Gillnet; Habitat Degradation and Loss Caused by Pollution and Vessel Traffic

保护级别与保护行动 Protection Category and Conservation Action (s)

国家重点保护野生动物等级 (2021) Category of National Key Protected Wild Animals (2021)	二级 / Category II
IUCN 红色名录 (2020-2) IUCN Red List (2020-2)	易危 / VU
CITES 附录 (2019) CITES Appendix (2019)	I
保护行动 Conservation Action(s)	法律保护物种 / Legally protected species

相关文献 Relevant References

Burgin *et al*., 2018; Jiang *et al*. (蒋志刚等), 2015a; Jefferson and Wang, 2011; Wang (王丕烈), 2011; Zhou (周开亚), 2008, 2004; Gao (高安利), 1991

印太江豚 *Neophocaena phocaenoides*

巨松鼠 *Ratufa bicolor*

中国生物多样性 红色名录

易危 VU A1ac; B1ab(i, ii, iii)+2ab(i, ii, iii)

| 数据缺乏 DD | 无危 LC | 近危 NT | 易危 VU | 濒危 EN | 极危 CR | 区域灭绝 RE | 野外灭绝 EW | 灭绝 EX |

分类地位 Taxonomic Status

动物界 Animalia	脊索动物门 Chordata	哺乳纲 Mammalia	啮齿目 Rodentia	松鼠科 Sciuridae
学名 Scientific Name		*Ratufa bicolor*		
命名人 Species Authority		Sparrman, 1778		
英文名 English Name(s)		Black Giant Squirrel		
同物异名 Synonym(s)		Malayan Giant Squirrel; *Ratufa tennentii* (Layard, in Blyth, 1849)		
种下单元评估 Infra-specific Taxa Assessed		无 / None		

评估信息 Assessment Information

评估年份 Year Assessed	2020
评定人 Assessor(s)	蒋志刚、刘少英 / Zhigang Jiang, Shaoying Liu
审定人 Reviewer(s)	马勇、鲁长虎、鲍毅新 / Yong Ma, Changhu Lu, Yixin Bao
其他贡献人 Other Contributor(s)	李立立、丁晨晨 / Lili Li, Chenchen Ding

理由 Justification: 巨松鼠分布区狭窄，种群数量少，且其占有区因人类活动而不断减少。因此，巨松鼠列为易危等级 / Black Giant Squirrel is narrowly distributed in China. Its populations are small and the areas of its occupancy are declining due to human activities. Thus, Black Giant Squirrel is listed as Vulnerable

地理分布 Geographical Distribution

国内分布 Domestic Distribution	云南、广西、海南、西藏 / Yunnan, Guangxi, Hainan, Tibet (Xizang)
世界分布 World Distribution	中国；南亚、东南亚 / China; South Asia, Southeast Asia
分布标注 Distribution Note	非特有种 / Non-endemic

国内分布图
Map of Domestic Distribution

种群 Population

种群数量 Population Size	未知 / Unknown
种群趋势 Population Trend	下降 / Decreasing

生境与生态系统 Habitat(s) and Ecosystem(s)

生境 Habitat(s)	森林 / Forest
生态系统 Ecosystem(s)	森林生态系统 / Forest Ecosystem

威胁 Threat(s)

主要威胁 Major Threat(s)	耕种、伐木、狩猎、住宅区及商业发展 / Plantation, Logging, Hunting, Residential and Commercial Development

保护级别与保护行动 Protection Category and Conservation Action(s)

国家重点保护野生动物等级 (2021) Category of National Key Protected Wild Animals (2021)	二级 / Category II
IUCN 红色名录 (2020-2) IUCN Red List (2020-2)	近危 / NT
CITES 附录 (2019) CITES Appendix (2019)	II
保护行动 Conservation Action(s)	无 / None

相关文献 Relevant References

Burgin *et al.*, 2018; Jiang *et al.* (蒋志刚等), 2017; Wilson *et al.*, 2017; Zheng *et al.* (郑智民等), 2012; Smith *et al.* (史密斯等), 2009; Li *et al.* (李松等), 2008

巨松鼠 *Ratufa bicolor*

中国生物多样性红色名录
复齿鼯鼠
Trogopterus xanthipes

易危 VU A1acd; B1ab(i, ii, iii)+2ab(i, ii, iii)

| 数据缺乏 DD | 无危 LC | 近危 NT | **易危 VU** | 濒危 EN | 极危 CR | 区域灭绝 RE | 野外灭绝 EW | 灭绝 EX |

分类地位 Taxonomic Status

| 动物界 Animalia | 脊索动物门 Chordata | 哺乳纲 Mammalia | 啮齿目 Rodentia | 松鼠科 Sciuridae |

学 名 Scientific Name	*Trogopterus xanthipes*
命 名 人 Species Authority	Milne-Edwards, 1867
英 文 名 English Name(s)	Complex-toothed Flying Squirrel
同物异名 Synonym(s)	*Trogopterus edithae* (Thomas, 1923); *himalaicus* (Thomas, 1914); *minax* (Thomas, 1923); *mordax* (Thomas, 1914)
种下单元评估 Infra-specific Taxa Assessed	无 / None

评估信息 Assessment Information

评估年份 Year Assessed	2020
评 定 人 Assessor(s)	蒋志刚、刘少英 / Zhigang Jiang, Shaoying Liu
审 定 人 Reviewer(s)	马勇、鲍毅新 / Yong Ma, Yixin Bao
其他贡献人 Other Contributor(s)	李立立、丁晨晨 / Lili Li, Chenchen Ding

理由 Justification: 复齿鼯鼠分布虽较广，但其种群数量极少，且栖息地的面积因人类活动而不断减少。因此，复齿鼯鼠列为易危等级 / Complex-toothed Flying Squirrel is relatively widely distributed in China, its populations are small. Furthermore, the areas of its habitats are declining due to human activities. Thus, Complex-toothed Flying Squirrel is listed as Vulnerable

地理分布 Geographical Distribution

国内分布 Domestic Distribution

北京、河北、辽宁、陕西、山西、河南、四川、青海、贵州、云南、西藏、湖北、湖南、甘肃、重庆 / Beijing, Hebei, Liaoning, Shaanxi, Shanxi, Henan, Sichuan, Qinghai, Guizhou, Yunnan, Tibet (Xizang), Hubei, Hunan, Gansu, Chongqing

世界分布 World Distribution

中国 / China

分布标注 Distribution Note

特有种 / Endemic

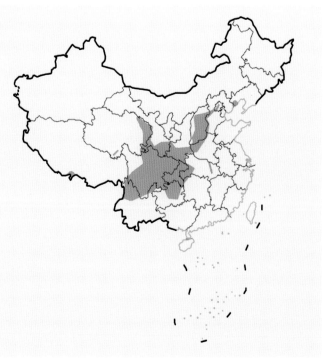

国内分布图
Map of Domestic Distribution

种群 Population

种群数量 Population Size	未知 / Unknown
种群趋势 Population Trend	下降 / Decreasing

生境与生态系统 Habitat (s) and Ecosystem (s)

生境 Habitat(s)	温带森林 / Temperate Forest
生态系统 Ecosystem(s)	森林生态系统 / Forest Ecosystem

威胁 Threat (s)

主要威胁 Major Threat(s)	狩猎、伐木、人工林 / Hunting, Logging, Artificial Forest

保护级别与保护行动 Protection Category and Conservation Action (s)

国家重点保护野生动物等级 (2021) Category of National Key Protected Wild Animals (2021)	无 / NA
IUCN 红色名录(2020-2) IUCN Red List (2020-2)	近危 / NT
CITES 附录 (2019) CITES Appendix (2019)	无 / NA
保护行动 Conservation Action(s)	无 / None

相关文献 Relevant References

Burgin *et al*., 2018; Jiang *et al*. (蒋志刚等), 2017; Wilson *et al*., 2017; Zheng *et al*. (郑智民等), 2012; Smith *et al*. (史密斯等), 2009; Jin *et al*. (金一等), 2007; Wang and Wang (王福麟和王小非), 1995

复齿鼯鼠 *Trogopterus xanthipes*

红背鼯鼠
Petaurista petaurista

易危 VU A1acd; B2b(i, ii, iii)

DD	LC	NT	**VU**	EN	CR	RE	EW	EX
数据缺乏	无危	近危	**易危**	濒危	极危	区域灭绝	野外灭绝	灭绝

分类地位 Taxonomic Status

动物界 Animalia	脊索动物门 Chordata	哺乳纲 Mammalia	啮齿目 Rodentia	松鼠科 Sciuridae

学名 Scientific Name	*Petaurista petaurista*
命名人 Species Authority	Pallas, 1766
英文名 English Name(s)	Red Giant Flying Squirrel
同物异名 Synonym(s)	Common Giant Flying Squirrel
种下单元评估 Infra-specific Taxa Assessed	无 / None

评估信息 Assessment Information

评估年份 Year Assessed	2020
评定人 Assessor(s)	蒋志刚、刘少英 / Zhigang Jiang, Shaoying Liu
审定人 Reviewer(s)	马勇、鲍毅新 / Yong Ma, Yixin Bao
其他贡献人 Other Contributor(s)	李立立、丁晨晨 / Lili Li, Chenchen Ding

理由 Justification: 红背鼯鼠分布虽较广, 但其种群数量下降, 且占有区面积因人类活动而不断减少。因此, 将其列为易危等级 / Although Red Giant Flying Squirrel is relatively widely distributed in China, its populations are small and the areas of its habitats are declining due to human activities. Thus, it is listed as Vulnerable

地理分布 Geographical Distribution

国内分布 Domestic Distribution

湖南、浙江、福建、广东、广西、四川、贵州、云南、西藏、江西 / Hunan, Zhejiang, Fujian, Guangdong, Guangxi, Sichuan, Guizhou, Yunnan, Tibet (Xizang), Jiangxi

世界分布 World Distribution

中国;南亚、东南亚 / China; South Asia, Southeast Asia

分布标注 Distribution Note

非特有种 / Non-endemic

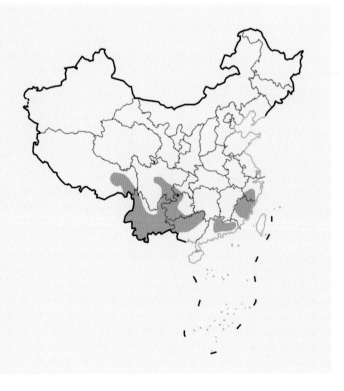

国内分布图
Map of Domestic Distribution

种群 Population

种群数量 Population Size	未知 / Unknown
种群趋势 Population Trend	下降 / Decreasing

生境与生态系统 Habitat (s) and Ecosystem (s)

生　　　境 Habitat(s)	热带和亚热带湿润低地山地森林、泰加林 / Tropical and Subtropical Moist Lowland Montane Forest, Taiga Forest
生态系统 Ecosystem(s)	森林生态系统 / Forest Ecosystem

威胁 Threat (s)

主要威胁 Major Threat(s)	人工林、伐木、火灾、狩猎、住宅区及商业发展 / Artificial Forest, Logging, Fire, Hunting, Residential and Commercial Development

保护级别与保护行动 Protection Category and Conservation Action (s)

国家重点保护野生动物等级 (2021) Category of National Key Protected Wild Animals (2021)	无 / NA
IUCN 红色名录(2020-2) IUCN Red List (2020-2)	无危 / LC
CITES 附录 (2019) CITES Appendix (2019)	无 / NA
保护行动 Conservation Action(s)	无 / None

相关文献 Relevant References

Burgin *et al.*, 2018; Jiang *et al.* (蒋志刚等), 2017; Wilson *et al.*, 2017; Zheng *et al.* (郑智民等), 2012; Smith *et al.* (史密斯等), 2009; Wang (王应祥), 2003

红背鼯鼠 *Petaurista petaurista*

小飞鼠
Pteromys volans

易危 VU A1acd; B2b(i, ii, iii)

| 数据缺乏 DD | 无危 LC | 近危 NT | 易危 VU | 濒危 EN | 极危 CR | 区域灭绝 RE | 野外灭绝 EW | 灭绝 EX |

分类地位 Taxonomic Status

动物界 Animalia	脊索动物门 Chordata	哺乳纲 Mammalia	啮齿目 Rodentia	松鼠科 Sciuridae
学　名 Scientific Name		*Pteromys volans*		
命名人 Species Authority		Linnaeus, 1758		
英文名 English Name(s)		Siberian Flying Squirrel		
同物异名 Synonym(s)		Russian Flying Squirrel		
种下单元评估 Infra-specific Taxa Assessed		无 / None		

评估信息 Assessment Information

评估年份 Year Assessed	2020
评定人 Assessor(s)	刘少英、蒋志刚 / Shaoying Liu, Zhigang Jiang
审定人 Reviewer(s)	马勇、鲍毅新 / Yong Ma, Yixin Bao
其他贡献人 Other Contributor(s)	李立立、丁晨晨 / Lili Li, Chenchen Ding

理由 **Justification**: 小飞鼠分布广，但种群数量少，其栖息地面积因人类活动而减少。因此，将其列为易危等级 / Siberian Flying Squirrel is widely distributed in China, but its populations are small, and its areas of habitats are declining due to human activities. Thus, Siberian Flying Squirrel is listed as Vulnerable

地理分布 Geographical Distribution

国内分布 Domestic Distribution

黑龙江、吉林、辽宁、内蒙古、北京、河北、河南、山西、陕西、甘肃、宁夏、青海、四川、新疆 / Heilongjiang, Jilin, Liaoning, Inner Mongolia (Nei Mongol), Beijing, Hebei, Henan, Shanxi, Shaanxi, Gansu, Ningxia, Qinghai, Sichuan, Xinjiang

世界分布 World Distribution

中国；亚洲北部、欧洲北部 / China; northern Asia, northern Europe

分布标注 Distribution Note

非特有种 / Non-endemic

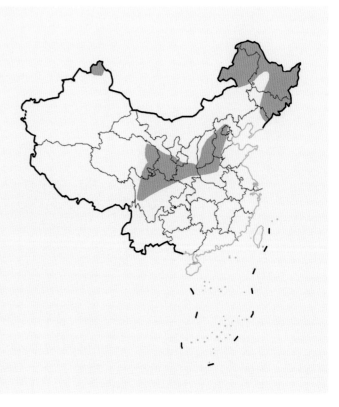

国内分布图
Map of Domestic Distribution

种群 Population

种群数量 Population Size	未知 / Unknown
种群趋势 Population Trend	下降 / Decreasing

生境与生态系统 Habitat (s) and Ecosystem (s)

生境 Habitat(s)	泰加林 / Taiga Forest
生态系统 Ecosystem(s)	森林生态系统 / Forest Ecosystem

威胁 Threat (s)

主要威胁 Major Threat(s)	伐木、狩猎、火灾 / Logging, Hunting, Fire

保护级别与保护行动 Protection Category and Conservation Action (s)

国家重点保护野生动物等级 (2021) Category of National Key Protected Wild Animals (2021)	无 / NA
IUCN 红色名录 (2020-2) IUCN Red List (2020-2)	无危 / LC
CITES 附录 (2019) CITES Appendix (2019)	无 / NA
保护行动 Conservation Action(s)	无 / None

相关文献 Relevant References

Burgin *et al*., 2018; Jiang *et al*. (蒋志刚等), 2017; Wilson *et al*., 2017; Zheng *et al*. (郑智民等), 2012; Smith *et al*. (史密斯等), 2009; Wang (王应祥), 2003

小飞鼠 *Pteromys volans*

沟牙田鼠
Proedromys bedfordi

易危 VU B1ab(i, ii, iii)+2ab(i, ii, iii); C1

| 数据缺乏 DD | 无危 LC | 近危 NT | **易危 VU** | 濒危 EN | 极危 CR | 区域灭绝 RE | 野外灭绝 EW | 灭绝 EX |

分类地位 Taxonomic Status

动物界 Animalia	脊索动物门 Chordata	哺乳纲 Mammalia	啮齿目 Rodentia	仓鼠科 Cricetidae

学名 Scientific Name	*Proedromys bedfordi*
命名人 Species Authority	Thomas, 1911
英文名 English Name(s)	Duke of Bedford's Vole
同物异名 Synonym(s)	*Microtus bedfordi* (Thomas, 1911)
种下单元评估 Infra-specific Taxa Assessed	无 / None

评估信息 Assessment Information

评估年份 Year Assessed	2020
评定人 Assessor(s)	刘少英、蒋志刚 / Shaoying Liu, Zhigang Jiang
审定人 Reviewer(s)	马勇、鲍毅新 / Yong Ma, Yixin Bao
其他贡献人 Other Contributor(s)	李立立、丁晨晨 / Lili Li, Chenchen Ding

理由 Justification: 沟牙田鼠分布区极为狭窄，种群数量极少，且栖息地面积因人类活动而不断减少。因此，将其列为易危等级 / Duke of Bedford's Vole is narrowly distributed in China, and its populations are small. Its areas of habitats are declining due to human activities. Thus, Duke of Bedford's Vole is listed as Vulnerable

地理分布 Geographical Distribution

国内分布 Domestic Distribution
四川、甘肃 / Sichuan, Gansu
世界分布 World Distribution
中国 / China
分布标注 Distribution Note
特有种 / Endemic

国内分布图
Map of Domestic Distribution

种群 Population

种群数量 Population Size	未知 / Unknown
种群趋势 Population Trend	下降 / Decreasing

生境与生态系统 Habitat (s) and Ecosystem (s)

生 境 Habitat(s)	亚热带湿润山地森林 / Subtropical Moist Montane Forest
生态系统 Ecosystem(s)	森林生态系统 / Forest Ecosystem

威胁 Threat (s)

主要威胁 Major Threat(s)	无 / None

保护级别与保护行动 Protection Category and Conservation Action (s)

国家重点保护野生动物等级 (2021) Category of National Key Protected Wild Animals (2021)	无 / NA
IUCN 红色名录 (2020-2) IUCN Red List (2020-2)	易危 / VU
CITES 附录 (2019) CITES Appendix (2019)	无 / NA
保护行动 Conservation Action(s)	无 / None

相关文献 Relevant References

Burgin *et al*., 2018; Jiang *et al*.（蒋志刚等），2017; Wilson *et al*., 2017; Zheng *et al*.（郑智民等），2012; Liu *et al*.（刘少英等），2005; Wang（王应祥），2003

沟牙田鼠 *Proedromys bedfordi*

小狕鼠
Hapalomys delacouri

易危　VU B1; C2a(i)

| 数据缺乏 DD | 无危 LC | 近危 NT | 易危 VU | 濒危 EN | 极危 CR | 区域灭绝 RE | 野外灭绝 EW | 灭绝 EX |

分类地位 Taxonomic Status

动物界 Animalia	脊索动物门 Chordata	哺乳纲 Mammalia	啮齿目 Rodentia	鼠科 Muridae

学　名 Scientific Name	*Hapalomys delacouri*
命名人 Species Authority	Thomas, 1927
英文名 English Name(s)	Lesser Marmoset Rat
同物异名 Synonym(s)	Delacour's Marmoset Rat; *Hapalomys marmosa* (G. M. Allen, 1927); *pasquieri* (Thomas, 1927)
种下单元评估 Infra-specific Taxa Assessed	无 / None

评估信息 Assessment Information

评估年份 Year Assessed	2020
评定人 Assessor(s)	刘少英、蒋志刚 / Shaoying Liu, Zhigang Jiang
审定人 Reviewer(s)	马勇、鲍毅新 / Yong Ma, Yixin Bao
其他贡献人 Other Contributor(s)	李立立、丁晨晨 / Lili Li, Chenchen Ding

理由 Justification: 小狕鼠分布区狭窄，种群数量少，其占有区面积因人类活动而减少。因此，将其列为易危等级 / Lesser Marmoset Rat is narrowly distributed in China, and its populations are small. Its areas of occupancy are declining due to human activities. Thus, Lesser Marmoset Rat is listed as Vulnerable

地理分布 Geographical Distribution

国内分布 Domestic Distribution
海南、广西、云南 / Hainan, Guangxi, Yunnan
世界分布 World Distribution
中国；东南亚 / China; Southeast Asia
分布标注 Distribution Note
非特有种 / Non-endemic

国内分布图
Map of Domestic Distribution

🦌 种群 Population

种群数量 Population Size	未知 / Unknown
种群趋势 Population Trend	下降 / Decreasing

🦌 生境与生态系统 Habitat (s) and Ecosystem (s)

生　　境 Habitat(s)	竹林 / Bamboo Grove
生态系统 Ecosystem(s)	竹林生态系统 / Bamboo Forest Ecosystem

🦌 威胁 Threat (s)

主要威胁 Major Threat(s)	无 / None

🦌 保护级别与保护行动 Protection Category and Conservation Action (s)

国家重点保护野生动物等级 (2021) Category of National Key Protected Wild Animals (2021)	无 / NA
IUCN红色名录(2020-2) IUCN Red List (2020-2)	易危 / VU
CITES 附录 (2019) CITES Appendix (2019)	无 / NA
保护行动 Conservation Action(s)	无 / None

🦌 相关文献 Relevant References

Burgin *et al*., 2018; Jiang *et al*. (蒋志刚等), 2017; Smith *et al*. (史密斯等), 2009; Wang and Xie (汪松和解焱), 2004; Wang (王应祥), 2003

小狨鼠 *Hapalomys delacouri*

褐尾鼠
Niviventer cremoriventer

易危 VU B1; C2a(i)

分类地位 Taxonomic Status

动物界 Animalia	脊索动物门 Chordata	哺乳纲 Mammalia	啮齿目 Rodentia	鼠科 Muridae

学名 Scientific Name	*Niviventer cremoriventer*
命名人 Species Authority	Miller, 1900
英文名 English Name(s)	Sundaic Arboreal Niviventer
同物异名 Synonym(s)	Dark-tailed Tree Rat
种下单元评估 Infra-specific Taxa Assessed	无 / None

评估信息 Assessment Information

评估年份 Year Assessed	2020
评定人 Assessor(s)	刘少英、蒋志刚 / Shaoying Liu, Zhigang Jiang
审定人 Reviewer(s)	马勇、鲍毅新 / Yong Ma, Yixin Bao
其他贡献人 Other Contributor(s)	李立立、丁晨晨 / Lili Li, Chenchen Ding

理由 Justification: 褐尾鼠分布区极为狭窄，其种群数量很少，且栖息地的面积因人类活动而减少。因此，将其列为易危等级 / Sundaic Arboreal Niviventer is narrowly distributed in China, and its populations are very small. The areas of its habitats are declining due to human activities. Thus, it is listed as Vulnerable

地理分布 Geographical Distribution

国内分布 Domestic Distribution
云南 / Yunnan
世界分布 World Distribution
中国、印度尼西亚、马来西亚、新加坡、泰国 / China, Indonesia, Malaysia, Singapore, Thailand
分布标注 Distribution Note
非特有种 / Non-endemic

国内分布图
Map of Domestic Distribution

种群 Population

种群数量 Population Size	未知 / Unknown
种群趋势 Population Trend	下降 / Decreasing

生境与生态系统 Habitat(s) and Ecosystem(s)

生　　境 Habitat(s)	热带和亚热带森林、灌丛、耕地 / Tropical and Subtropical Forest, Shrubland, Arable Land
生态系统 Ecosystem(s)	森林生态系统、灌丛生态系统、农田生态系统 / Forest Ecosystem, Shrubland Ecosystem, Cropland Ecosystem

威胁 Threat(s)

主要威胁 Major Threat(s)	伐木、火灾、耕种 / Logging, Fire, Plantation

保护级别与保护行动 Protection Category and Conservation Action(s)

国家重点保护野生动物等级 (2021) Category of National Key Protected Wild Animals (2021)	无 / NA
IUCN 红色名录 (2020-2) IUCN Red List (2020-2)	无危 / LC
CITES 附录 (2019) CITES Appendix (2019)	无 / NA
保护行动 Conservation Action(s)	无 / None

相关文献 Relevant References

Burgin *et al.*, 2018; Jiang *et al.*（蒋志刚等）, 2017; Wilson *et al.*, 2017; Jing *et al.*, 2007; Pan *et al.*（潘清华等）, 2007; Wang（王应祥）, 2003

褐尾鼠 *Niviventer cremoriventer*

中国生物多样性红色名录

阿尔泰鼢鼠
Myospalax myospalax

易危 VU B2b(i, ii, iii)

| DD 数据缺乏 | LC 无危 | NT 近危 | **VU 易危** | EN 濒危 | CR 极危 | RE 区域灭绝 | EW 野外灭绝 | EX 灭绝 |

分类地位 Taxonomic Status

动物界 Animalia	脊索动物门 Chordata	哺乳纲 Mammalia	啮齿目 Rodentia	鼹型鼠科 Spalacidae

学名 Scientific Name	*Myospalax myospalax*
命名人 Species Authority	Laxmann, 1773
英文名 English Name(s)	Siberian Zokor
同物异名 Synonym(s)	*Myospalax incertus* (Ognev, 1936); *komurai* (Mori, 1927); *laxmanni* (Sherskey, 1873); *tarbagataicus* (Ognev, 1936)
种下单元评估 Infra-specific Taxa Assessed	无 / None

评估信息 Assessment Information

评估年份 Year Assessed	2020
评定人 Assessor(s)	蒋志刚、刘少英 / Zhigang Jiang, Shaoying Liu
审定人 Reviewer(s)	李保国 / Baoguo Li
其他贡献人 Other Contributor(s)	李立立、丁晨晨 / Lili Li, Chenchen Ding

理由 Justification: 阿尔泰鼢鼠分布于阿勒泰地区，占有区小，种群数量少。因此，将其列为易危等级 / Siberian Zokor is found in Altay region but its areas of occupancy are small and its population densities are low. Thus, it is listed as Vulnerable

地理分布 Geographical Distribution

国内分布 Domestic Distribution	新疆 / Xinjiang
世界分布 World Distribution	中国、哈萨克斯坦、俄罗斯 / China, Kazakhstan, Russia
分布标注 Distribution Note	非特有种 / Non-endemic

国内分布图 Map of Domestic Distribution

种群 Population

种群数量 Population Size	未知 / Unknown
种群趋势 Population Trend	未知 / Unknown

生境与生态系统 Habitat (s) and Ecosystem (s)

生　　境 Habitat(s)	灌丛、草地 / Shrubland, Grassland
生态系统 Ecosystem(s)	灌丛生态系统、草地生态系统 / Shrubland Ecosystem, Grassland Ecosystem

威胁 Threat (s)

主要威胁 Major Threat(s)	无 / None

保护级别与保护行动 Protection Category and Conservation Action (s)

国家重点保护野生动物等级 (2021) Category of National Key Protected Wild Animals (2021)	无 / NA
IUCN 红色名录 (2020-2) IUCN Red List (2020-2)	无危 / LC
CITES 附录 (2019) CITES Appendix (2019)	无 / NA
保护行动 Conservation Action(s)	无 / None

相关文献 Relevant References

Burgin *et al.*, 2018; Jiang *et al.*（蒋志刚等），2017; Wilson *et al.*, 2017; Smith *et al.*（史密斯等），2009; Pan *et al.*（潘清华等），2007

阿尔泰鼢鼠 *Myospalax myospalax*

长白山鼠兔
Ochotona coreana

易危 VU B1+2a(i); D2

| 数据缺乏 DD | 无危 LC | 近危 NT | **易危 VU** | 濒危 EN | 极危 CR | 区域灭绝 RE | 野外灭绝 EW | 灭绝 EX |

分类地位 Taxonomic Status

动物界 Animalia	脊索动物门 Chordata	哺乳纲 Mammalia	兔形目 Lagomorpha	鼠兔科 Ochotonidae

学 名 Scientific Name	*Ochotona coreana*
命 名 人 Species Authority	Liu *et al.*, 2017
英 文 名 English Name(s)	Changbaishan Pika
同物异名 Synonym(s)	Northern pika, *Ochotona hyperborea coreana*
种下单元评估 Infra-specific Taxa Assessed	无 / None

评估信息 Assessment Information

评 估 年 份 Year Assessed	2020
评 定 人 Assessor(s)	刘少英、蒋志刚 / Shaoying Liu, Zhigang Jiang
审 定 人 Reviewer(s)	苏建平、冯祚建、宗浩、廖继承 / Jianping Su, Zuojian Feng, Hao Zong, Jicheng Liao
其他贡献人 Other Contributor(s)	李立立、丁晨晨 / Lili Li, Chenchen Ding

理由 Justification: 长白山鼠兔分布区小，其占有区更小。因此，将其列为易危等级 / Changbaishan Pika has a narrow distribution and its area of occupancy is even smaller. Thus, it is listed as Vulnerable

地理分布 Geographical Distribution

国内分布 Domestic Distribution
吉林 / Jilin
世界分布 World Distribution
中国、朝鲜 / China, Korea (the Democratic People's Republic of)
分布标注 Distribution Note
非特有种 / Non-endemic

国内分布图
Map of Domestic Distribution

种群 Population

种群数量 Population Size	未知 / Unknown
种群趋势 Population Trend	未知 / Unknown

生境与生态系统 Habitat (s) and Ecosystem (s)

生　　境 Habitat(s)	草甸、灌丛 / Meadow, Shrubland
生态系统 Ecosystem(s)	灌丛生态系统、草地生态系统 / Shrubland Ecosystem, Grassland Ecosystem

威胁 Threat (s)

主要威胁 Major Threat(s)	无 / None

保护级别与保护行动 Protection Category and Conservation Action (s)

国家重点保护野生动物等级 (2021) Category of National Key Protected Wild Animals (2021)	无 / NA
IUCN 红色名录(2020-2) IUCN Red List (2020-2)	无危 / LC
CITES 附录 (2019) CITES Appendix (2019)	无 / NA
保护行动 Conservation Action(s)	无 / None

相关文献 Relevant References

Burgin *et al.*, 2018; Jiang *et al.* (蒋志刚等), 2017; Liu *et al.* (刘少英等), 2017; Wilson *et al.*, 2016

长白山鼠兔 *Ochotona coreana*

大巴山鼠兔
Ochotona dabashanensis

易危 VU B1+2a(i); D2

分类地位 Taxonomic Status

动物界 Animalia	脊索动物门 Chordata	哺乳纲 Mammalia	兔形目 Lagomorpha	鼠兔科 Ochotonidae
学名 Scientific Name		*Ochotona dabashanensis*		
命名人 Species Authority		Liu *et al.*, 2017		
英文名 English Name(s)		Dabashan Pika		
同物异名 Synonym(s)		无 / None		
种下单元评估 Infra-specific Taxa Assessed		无 / None		

评估信息 Assessment Information

评估年份 Year Assessed	2020
评定人 Assessor(s)	刘少英、蒋志刚 / Shaoying Liu, Zhigang Jiang
审定人 Reviewer(s)	苏建平、冯祚建、宗浩、廖继承 / Jianping Su, Zuojian Feng, Hao Zong, Jicheng Liao
其他贡献人 Other Contributor(s)	李立立、丁晨晨 / Lili Li, Chenchen Ding

理由 Justification: 大巴山鼠兔为新种，其分布区小，数量稀少。因此，将其定为易危 / Dabashan Pika is a new species, its area of occupancy and population size are both small. Thus, it is listed as Vulnerable

地理分布 Geographical Distribution

国内分布 Domestic Distribution
四川 / Sichuan
世界分布 World Distribution
中国 / China
分布标注 Distribution Note
特有种 / Endemic

国内分布图
Map of Domestic Distribution

种群 Population

种群数量 Population Size	未知 / Unknown
种群趋势 Population Trend	未知 / Unknown

生境与生态系统 Habitat (s) and Ecosystem (s)

生 境 Habitat(s)	草甸、灌丛 / Meadow, Shrubland
生态系统 Ecosystem(s)	草地生态系统、灌丛生态系统 / Grassland Ecosystem, Shrubland Ecosystem

威胁 Threat (s)

主要威胁 Major Threat(s)	无 / None

保护级别与保护行动 Protection Category and Conservation Action (s)

国家重点保护野生动物等级 (2021) Category of National Key Protected Wild Animals (2021)	无 / NA
IUCN 红色名录 (2020-2) IUCN Red List (2020-2)	无危 / LC
CITES 附录 (2019) CITES Appendix (2019)	无 / NA
保护行动 Conservation Action(s)	无 / None

相关文献 Relevant References

Burgin *et al*., 2018; Jiang *et al*. (蒋志刚等), 2017; Liu *et al*. (刘少英等), 2017

大巴山鼠兔 *Ochotona dabashanensis*   蒋志刚 绘   Drawn by Zhigang Jiang

黑鼠兔
Ochotona nigritia
易危 VU B1; C2a(i)

| 数据缺乏 DD | 无危 LC | 近危 NT | 易危 VU | 濒危 EN | 极危 CR | 区域灭绝 RE | 野外灭绝 EW | 灭绝 EX |

分类地位 Taxonomic Status

动物界 Animalia	脊索动物门 Chordata	哺乳纲 Mammalia	兔形目 Lagomorpha	鼠兔科 Ochotonidae
学名 Scientific Name		*Ochotona nigritia*		
命名人 Species Authority		Gong, Wang, Li and Li, 2000		
英文名 English Name(s)		Black Pika		
同物异名 Synonym(s)		无 / None		
种下单元评估 Infra-specific Taxa Assessed		无 / None		

评估信息 Assessment Information

评估年份 Year Assessed	2020
评定人 Assessor(s)	刘少英、蒋志刚 / Shaoying Liu, Zhigang Jiang
审定人 Reviewer(s)	苏建平、冯祚建、宗浩、廖继承 / Jianping Su, Zuojian Feng, Hao Zong, Jicheng Liao
其他贡献人 Other Contributor(s)	李立立、丁晨晨 / Lili Li, Chenchen Ding

理由 Justification: 黑鼠兔分布区小，其占有区更小。因此，将其列为易危等级 / Black Pika has a narrow range and its area of occupancy is even smaller. Thus, it is listed as Vulnerable

地理分布 Geographical Distribution

国内分布 Domestic Distribution
云南 / Yunnan
世界分布 World Distribution
中国 / China
分布标注 Distribution Note
特有种 / Endemic

国内分布图
Map of Domestic Distribution

种群 Population

种群数量 Population Size	未知 / Unknown
种群趋势 Population Trend	未知 / Unknown

生境与生态系统 Habitat (s) and Ecosystem (s)

生　　境 Habitat(s)	灌丛、草甸 / Shrubland, Meadow
生态系统 Ecosystem(s)	灌丛生态系统、草地生态系统 / Shrubland Ecosystem, Grassland Ecosystem

威胁 Threat (s)

主要威胁 Major Threat(s)	无 / None

保护级别与保护行动 Protection Category and Conservation Action (s)

国家重点保护野生动物等级 (2021) Category of National Key Protected Wild Animals (2021)	无 / NA
IUCN 红色名录 (2020-2) IUCN Red List (2020-2)	无危 / LC
CITES 附录 (2019) CITES Appendix (2019)	无 / NA
保护行动 Conservation Action(s)	无 / None

相关文献 Relevant References

Burgin *et al*., 2018; Jiang *et al*. (蒋志刚等), 2017; Liu *et al*. (刘少英等), 2017

黑鼠兔 *Ochotona nigritia*　　方红霞 绘　Drawn by Hongxia Fang

峨眉鼠兔
Ochotona sacraria

易危 VU B1+2a(i); D2

| 数据缺乏 DD | 无危 LC | 近危 NT | 易危 VU | 濒危 EN | 极危 CR | 区域灭绝 RE | 野外灭绝 EW | 灭绝 EX |

分类地位 Taxonomic Status

动物界 Animalia	脊索动物门 Chordata	哺乳纲 Mammalia	兔形目 Lagomorpha	鼠兔科 Ochotonidae
学名 Scientific Name		*Ochotona sacraria*		
命名人 Species Authority		Liu *et al.*, 2017		
英文名 English Name(s)		Emei Pika		
同物异名 Synonym(s)		Moupin Pika, *Ochotona thibetana sacraria*		
种下单元评估 Infra-specific Taxa Assessed		无 / None		

评估信息 Assessment Information

评估年份 Year Assessed	2020
评定人 Assessor(s)	刘少英、蒋志刚 / Shaoying Liu, Zhigang Jiang
审定人 Reviewer(s)	苏建平、冯祚建、宗浩、廖继承 / Jianping Su, Zuojian Feng, Hao Zong, Jicheng Liao
其他贡献人 Other Contributor(s)	李立立、丁晨晨 / Lili Li, Chenchen Ding

理由 Justification: 峨眉鼠兔是一个新种，分布区小，数量稀少。因此，将其定为易危 / Emei Pika is a new species, its distribution area is small and its population size also is small. Thus, it is listed as Vulnerable

地理分布 Geographical Distribution

国内分布 Domestic Distribution
四川 / Sichuan
世界分布 World Distribution
中国 / China
分布标注 Distribution Note
特有种 / Endemic

国内分布图
Map of Domestic Distribution

种群 Population

种群数量 Population Size	未知 / Unknown
种群趋势 Population Trend	未知 / Unknown

生境与生态系统 Habitat (s) and Ecosystem (s)

生　　境 Habitat(s)	内陆岩石区域 / Inland Rocky Area
生态系统 Ecosystem(s)	森林生态系统 / Forest Ecosystem

威胁 Threat (s)

主要威胁 Major Threat(s)	无 / None

保护级别与保护行动 Protection Category and Conservation Action (s)

国家重点保护野生动物等级 (2021) Category of National Key Protected Wild Animals (2021)	无 / NA
IUCN 红色名录 (2020-2) IUCN Red List (2020-2)	无危 / LC
CITES 附录 (2019) CITES Appendix (2019)	无 / NA
保护行动 Conservation Action(s)	无 / None

相关文献 Relevant References

Burgin *et al.*, 2018; Jiang *et al.* (蒋志刚等), 2017; Liu *et al.* (刘少英等), 2017; Wilson *et al.*, 2016

峨眉鼠兔 *Ochotona sacraria*

中国生物多样性红色名录：脊椎动物
China's Red List of Biodiversity: Vertebrates

主编 蒋志刚
Chief Editor: Zhigang Jiang

第一卷 哺乳动物（中册）
Volume Ⅰ, Mammals (Ⅱ)

主编 蒋志刚
Chief Editor: Zhigang Jiang

副主编 吴 毅 刘少英 蒋学龙
　　　 周开亚 胡慧建
Vice-Chief Editors: Yi Wu　Shaoying Liu　Xuelong Jiang
　　　　　　　　　Kaiya Zhou　Huijian Hu

科学出版社
北 京

内 容 简 介

本书是"中国生物多样性红色名录：脊椎动物"的"第一卷 哺乳动物"，全书分为总论和各论两部分。总论介绍了哺乳动物的演化与现状、中国哺乳动物多样性与保护现状、本红色名录评估对象的分类系统、中国哺乳动物编目（2021）、中国哺乳动物分布格局和受保护状况；介绍了红色名录评估过程、依据的评估等级和标准，还介绍了咨询专家、评估队伍，以及建立数据库和开展初评、通讯评审、形成评估报告的过程；总结分析了评估结果，介绍了中国哺乳动物濒危状况，分析了野外灭绝的和区域灭绝的物种，分析了受威胁物种比例、哺乳动物的分布和濒危原因等，并将评估结果与《IUCN 受威胁物种红色名录》（2020-2）进行了比较分析；最后，分析了中国哺乳动物保护成效与远景。各论是图书的主体，对评估的 700 种中国哺乳动物物种的相关信息，即分类地位、评估信息、地理分布、种群状况、生境与生态系统、威胁因子、保护级别与保护行动及相关文献进行了详细叙述。

本书适合从事野生动物研究与保护的科研人员参考，适合自然保护区、环境保护、进出口对外贸易、检验检疫等相关各级行政管理部门作为行使管理职能的参照资料，适合作为高年级研究生教学参考书，适合国内大中型图书馆馆藏。

审图号：GS (2020) 2858号

图书在版编目（CIP）数据

中国生物多样性红色名录. 脊椎动物. 第一卷，哺乳动物. 中册 = China's Red List of Biodiversity: Vertebrates, Volume I, Mammals (II)：汉英对照 / 蒋志刚主编. —北京：科学出版社，2021.3
国家出版基金项目
ISBN 978-7-03-065664-3

Ⅰ. ①中… Ⅱ. ①蒋… Ⅲ. ①珍稀动物–中国–名录–汉、英 ②珍稀植物–中国–名录–汉、英 ③哺乳动物纲–中国–名录–汉、英 Ⅳ. ①Q958.52-62 ②Q948.52-62 ③Q959.808

中国版本图书馆CIP数据核字（2020）第131621号

责任编辑：马 俊 孙 青 郝晨扬 / 责任校对：严 娜
责任印制：肖 兴 / 排版设计：北京鑫诚文化传播有限公司

科学出版社 出版
北京东黄城根北街16号
邮政编码：100717
http://www.sciencep.com

中国科学院印刷厂 印刷
科学出版社发行 各地新华书店经销

*

2021年3月第 一 版　开本：889×1194　1/16
2021年3月第一次印刷　印张：108 1/4
字数：3 117 000

定价：1428.00元（全三册）

（如有印装质量问题，我社负责调换）

China's Red List of Biodiversity: Vertebrates, Volume I, Mammals (II)

Chief Editor: Zhigang Jiang
Vice-Chief Editors: Yi Wu　Shaoying Liu　Xuelong Jiang　Kaiya Zhou　Huijian Hu

Abstract

This book is the first volume of the "*China's Red List of Biodiversity: Vertebrates*", i.e. "*Volume I, Mammals*", and is divided into two parts of "General Introduction" and "Species Monograph". General Introduction contains evolution and status of mammals, diversity and conservation status of mammals in China, taxonomic system of this red list, inventory of China's mammals (2021), the distribution patterns and conservation status of China's mammals; the evaluation process, categories and criteria that evaluation refers to, building the database, evaluation teams and how the evaluation teams to establish database, to carry out preliminary evaluation, to review the preliminary results by correspondence and to formulize the evaluation report; analyzes and summarizes the evaluation results, introduces the status of endangered mammal, analyzes the species extinct in the wild and regionally extinct, analyzes the proportion of threatened mammal species in different groups of mammals, in different habitats and its provincial distribution and the threats to endangered mammals, and compares the evaluation results with the *IUCN Red List of Endangered Species* (2020-2). Finally, the implications and prospects of mammal conservation in China are discussed. Species Monograph is the main part of this book, this part elaborates the detailed evaluating information of 700 Chinese mammal species, including taxonomic status, assessment information, geographical distribution, population situation, habitat and ecosystem, threats factor, protection category and conservation action, citations and references.

This book can be used as a reference for wild animal protect staffs and researchers, and can be used as an important data for decisions of government management department, *e.g.* nature reserve, environment protect, overseas trade, inspection and quarantine. This book is appropriate for senior grade graduated students in school, and appropriate to be collected by large- or medium-sized library.

ISBN 978-7-03-065664-3

Copyright© 2021, Science Press (Beijing)
All rights reserved. No part of this publication may be reproduced, stored in a retrieval system, or transmitted in any form or by any means, mechanical, photocopying, recording or otherwise, without the prior written permission of the copyright owner.

Acknowledgement
Illustrations (with mark of "Lynx Edicions") by Toni Liobet from: Wilson, D. E., Lacher, T. E. Jr & Mittermeier, R. A. eds. (2009-2018). *Handbook of the Mammals of the World*. Volumes 1 to 8. Lynx Edicions, Barcelona.

《中国生物多样性红色名录：脊椎动物》编委会

顾 问
陈宜瑜　郑光美　张亚平　金鉴明　马建章　曹文宣

主 编
蒋志刚

副主编
江建平　王跃招　张　鹗　张雁云

编委会成员
蔡　波　曹　亮　车　静　陈小勇　陈晓虹　丁　平
董　路　胡慧建　胡军华　计　翔　江建平　蒋学龙
蒋志刚　李　成　李春旺　李家堂　李丕鹏　梁　伟
刘　阳　刘少英　卢　欣　马　鸣　马　勇　马志军
饶定齐　史海涛　王　斌　王剑伟　王英永　王跃招
吴　华　吴　毅　吴孝兵　谢　锋　杨道德　杨晓君
曾晓茂　张　鹗　张　洁　张保卫　张雁云　张正旺
赵亚辉　周　放　周开亚

责任编辑
马　俊　李　迪　郝晨扬　孙　青

Editorial Committee of China's Red List of Biodiversity: Vertebrates

Consultants of the Editorial Committee

Yiyu Chen   Guangmei Zheng   Yaping Zhang   Jianming Jin   Jianzhang Ma

Wenxuan Cao

Chief Editor of the Editorial Committee

Zhigang Jiang

Vice-Chief Editors of the Editorial Committee

Jianping Jiang   Yuezhao Wang   E Zhang   Yanyun Zhang

Members of the Editorial Committee

Bo Cai   Liang Cao   Jing Che   Xiaoyong Chen   Xiaohong Chen   Ping Ding

Lu Dong   Huijian Hu   Junhua Hu   Xiang Ji   Jianping Jiang   Xuelong Jiang

Zhigang Jiang   Cheng Li   Chunwang Li   Jiatang Li   Pipeng Li   Wei Liang

Yang Liu   Shaoying Liu   Xin Lu   Ming Ma   Yong Ma   Zhijun Ma

Dingqi Rao   Haitao Shi   Bin Wang   Jianwei Wang   Yingyong Wang

Yuezhao Wang   Hua Wu   Yi Wu   Xiaobing Wu   Feng Xie   Daode Yang

Xiaojun Yang   Xiaomao Zeng   E Zhang   Jie Zhang   Baowei Zhang

Yanyun Zhang   Zhengwang Zhang   Yahui Zhao   Fang Zhou   Kaiya Zhou

Responsible Editors

Jun Ma   Di Li   Chenyang Hao   Qing Sun

序 一

地球进入了一个崭新的地质纪元——人类世（Anthropocene），而地球上的人口仍呈指数增长。人类社会进入了全球化、信息化时代。人类的生态足迹日益扩大，人类对自然资源的消耗、对环境的污染达到了一个前所未有的水平，人类的影响已经遍及地球各个角落，导致了全球变化，影响了地球生物圈的结构与功能，危及了许多野生动植物的生存，造成全球范围的生物多样性危机，影响了人类社会的可持续发展。

中国是一个生物多样性大国，是地球上生物多样性最丰富的国家之一。根据《中国生物物种名录》（2019），中国已经记载生物达 106,509 个物种及种下单元，其中物种 94,260 个，种下单元 12,249 个。中国南北纬度跨度大，海拔跨度也大。中国还有多种气候类型、多样生境类型，栖息着丰富的高等生物。这些生物物种是国家重要的战略资源，是社会经济可持续发展中不可替代的物质基础。

濒危物种红色名录已经成为重要的生物多样性保护研究工具。目前，《世界自然保护联盟受威胁物种红色名录》（《IUCN 受威胁物种红色名录》）评估了 98,500 多个物种，发现其中 32,000 多个面临灭绝威胁，包括 41% 的两栖动物、34% 的针叶树、33% 的造礁珊瑚、26% 的哺乳动物和 14% 的鸟类。然而，《IUCN 受威胁物种红色名录》对物种的生存状况的评估是基于全球资料所做的，并不代表物种在各个分布国的生存状况。国家是生物多样性保护的主体，各国须开展自己的濒危物种红色名录研究，对其生物物种的生存状况进行评估。在某种程度上，可以说濒危物种红色名录研究反映了一个国家生物多样性综合研究的能力。

早在 20 世纪 90 年代，中国即引入 IUCN 受威胁物种红色名录等级标准开展了物种濒危状况评估工作，如 1991 年，我国学者发表了《中国植物红皮书》。1998 年，国家环境保护总局联合国家濒危物种科学委员会发表了《中国濒危动物红皮书·鱼类》《中国濒危动物红皮书·两栖类和爬行类》《中国濒危动物红皮书·鸟类》《中国濒危动物红皮书·兽类》等著作。另外，2004 年和 2009 年，相关领域专家开展了不同类群物种的濒危状况评估工作，先后发表了《中国物种红色名录

（第一卷）红色名录》和《中国物种红色名录（第二卷）脊椎动物》。

物种生存状况是变化的，于是，IUCN 每年定期更新 IUCN 红色名录。IUCN 红色名录并不反映一个跨越国家分布的物种在一个分布国家的生存状况。国家是濒危物种的管理主体，各国需要应用国际标准进行红色名录评估。鉴于此，为全面评估中国野生脊椎动物濒危状况，中国研究人员于 2013 年启动了"中国生物多样性红色名录——脊椎动物卷"的物种评估和报告编制工作。这次评估组织全国鱼类、两栖类、爬行类、鸟类与哺乳类专家收集数据，采用综合分析和专家评估相结合的方法，依据中国鱼类、两栖类、爬行类、鸟类和哺乳类野生种群与生境现状，利用 IUCN 红色名录标准第 3.1 版，编制了"中国生物多样性红色名录——脊椎动物卷"。该卷红色名录于 2015 年 5 月 6 日通过环境保护部和中国科学院的联合验收，并于 5 月 22 日以环境保护部、中国科学院 2015 年第 32 号公告形式正式发布。

2016 年以来，中国脊椎动物红色名录工作组组织中国研究人员再次厘定了中国脊椎动物多样性，重新评估了中国脊椎动物的生存状况。完成了《中国生物多样性红色名录：脊椎动物》(2021)。本次脊椎动物红色名录评估发现中国脊椎动物生存状况严峻，中国脊椎动物的灭绝风险高于世界平均水平。中国脊椎动物哺乳类、鸟类、两栖类、爬行类和鱼类等各个类群都发现了野外灭绝或区域灭绝的物种，有许多物种处于极危、濒危和易危的受威胁状态。

中国正处于发展期，人口众多，地貌复杂，区域发展程度差异大。如何拯救这些濒危物种是中国生物多样性保护面临的一项艰巨任务。中国政府十分重视生物多样性保护，缔结了《生物多样性公约》《濒危野生动植物种国际贸易公约》《关于特别是作为水禽栖息地的国际重要湿地公约》（简称《湿地公约》）等国际公约，并积极主动履约。中国大力开展了以自然保护区为主体、以国家公园为龙头的保护地建设。目前，我国建立的各类保护地已达 1.18 万处，面积占国土面积的 18% 以上。其中有 474 个国家级自然保护区。自然保护区保护了 90.5% 的陆地生态系统类型、85% 的野生动植物种类、65% 的高等植物群落。中国的森林覆盖率逐年增加，为濒危物种的种群与栖息地恢复奠定了基础。

生物多样性研究既是一项综合性研究，也是一项组合型研究。生物物种编目与受威胁状态评估需要不同学科的联合研究。为了生物多样性科学研究与保护，来自不同学科的学者走到一起，完成中国生物多样性红色名录研究。这项研究是中国整体生物多样性研究的重要组成部分。《中国生物多样性红色名录：脊椎动物》各卷在《生物多样性公约》第 15 次缔约方大会即将在中国召开之前出版发行，为中国生物多样性保护提供了基础数据，为监测中国生物多样性现状、开展阶段性 IUCN 红色名录指数研究积累了参数，也是中国保护生物学研究成果的展示。

中国科学院院士
国家自然科学基金委员会前主任
国家濒危物种科学委员会主任
2020 年 6 月 26 日

Foreword I

The earth has entered a new geological epoch, the Anthropocene, while the human population on the earth is still increasing exponentially. Human society has entered the era of globalization and information. The human ecological footprint is enlarging while the human consumption of natural resources increases; environmental pollution created by human being has reached an unprecedented level. Impact of human reaches throughout all corners of the earth, causes the global change and affects the structure and function of the earth's biosphere, threatens the survival of many wild animals and plants, causing a global biodiversity crisis, consequently, influences the sustainable development of human society.

China is a country with great biodiversity and one of the countries with the richest biodiversity on the earth. According to the *Species Catalogue of China* (2019), China has already recorded 106,509 species and subspecies taxa, including 94,260 species and 12,249 subspecies. The territory of China spans a large latitude and has huge elevation differences. China also has a variety of climate types, diverse habitat types, rich niches of higher organisms. These biological species are important strategic resources of the country and irreplaceable material basis for sustainable social and economic development.

The red list of endangered species has become an important tool for biodiversity conservation research. Currently, the *IUCN Red List of Threatened Species* has assessed more than 98,500 species and found that more than 32,000 of them are threatened with extinction, including 41% of amphibians, 34% of conifers, 33% of reef-building corals, 26% of mammals and 14% of birds. However, the *IUCN Red List of Threatened Species* is based on global data and does necessarily not represent the status of species in each range country. A country is the main sovereign body of biodiversity conservation, and each country needs to carry out the red list study of its endangered species to assess the survival status of its biological species. To some extent, red list study reflects the comprehensive research capacity of biodiversity in a country.

The red list of endangered species has become an important tool for biodiversity conservation research. As early as in the 1990s, China introduced the IUCN red list criteria for threatened species to carry out the assessment of status of endangered species. For example, in 1991, Chinese scholars published the *Red Data Book of Chinese Plants*. In 1998, Environmental Protection Administration, together with the National Scientific Committee on Endangered Species, published such books as *China Red Data Book of Endangered Animals*: *Pisces*; *China Red Data Book of Endangered Animals*: *Amphibia & Reptilia*; *China Red Data Book of Endangered Animals*: *Aves*; *China Red Data Book of Endangered Animals*: *Mammalia*. In addition, in 2004 and 2009, experts in related fields carried out the assessment of the endangered status of different taxa of species, and published the *China Species Red List Vol I Red List* and the *China Species Red List Vol II Vertebrates*.

Status of species is changing, therefore the IUCN updates its red list every year. However, the IUCN red list does not reflect the status of particular species in a range country if a species distributes in multi-countries. Countries are the main management bodies of endangered species; thus, each country needs to apply international standards for red list assessment. Therefore, in order to comprehensively assess the endangered status of wild vertebrates in China, Chinese researchers launched the compilation of the "China's Red List of Biodiversity: Volume of Vertebrates" in 2013. During that evaluation, by adopting the combination of comprehensive analysis and expert evaluation method, the experts were coordinated to collect data for assessing wild population and habitat status of China's fishes, amphibians, reptiles, birds and mammals and compiled the *China's Red List of Biodiversity*: *Volume of Vertebrates*. The red list was approved by Ministry of Environmental Protection and the Chinese Academy of Sciences on May 6, 2015, and was officially released on May 22 in the form of Announcement No. 32 of 2015 by the Ministry of Environmental Protection and Chinese Academy of Sciences.

Since 2016, the China's vertebrate red list working group has organized Chinese researchers to reassess the diversity and the status of Chinese vertebrates. The working group completed the *China's Red List of Biodiversity: Vertebrates* (2021). The assessment found that China's vertebrate survival situation is still grim, and the risk of extinction of Chinese vertebrates is higher than the world average. Various groups of vertebrates, including mammals, birds, amphibians, reptiles and fishes in China have found species of Extinct in the Wild or Regionally Extinct, and many species are in a state of Critically Endangered, Endangered and Vulnerable.

China is in the process of rapid development. China has the largest human population, complex landforms and huge differences in regional development levels. How to save these endangered species is a difficult task for China's biodiversity conservation. The Chinese government attaches great importance to the protection of biological diversity, and has signed and actively implemented international conventions such as the *Convention on Biological Diversity*, the *Convention on International Trade in Endangered Species of Wild Fauna and Flora*, and the *Convention on Wetlands of International Importance, Especially as Waterfowl Habitats* (*Convention on Wetlands* for short). China has vigorously established protected areas with the nature reserves as the main body and national parks as the leading part. At present, China has set up 11,800 protected areas of various types, accounting for more than 18% of the country's land area. There are 474 national nature reserves. The nature reserves protect 90.5% of terrestrial ecosystem types, 85% of wildlife species, and 65% of higher plant communities. On the other hand, China's forest coverage is increasing year by year, laying a sound foundation for the population and habitat restoration of endangered species.

Biodiversity research is not only a comprehensive research, but also a combined study. The inventory of biological species and the assessment of threatened status require joint efforts of different disciplines. For the scientific research and conservation of biodiversity, scholars from different disciplines came together to complete the red list of China's biodiversity. The study is an important part of China's overall biodiversity research. All volumes of *China's Red List of Biodiversity: Vertebrates* are published before the fifteenth meeting of the Conference of the Parties to the *Convention on Biological Diversity* which will be held in Kunming, China. The set of books will provide basic data for the biodiversity conservation in China, for monitoring the current situation of biological diversity, for accumulating parameters to conduct periodic IUCN red list index research, and is also an outcome of the Chinese conservation biology research.

Yiyu Chen

Member of the Chinese Academy of Sciences

Former Director of National Natural Science Foundation of China

Director of Endangered Species Scientific Commission, P. R. China

June 26, 2020

序 二

1948年，在法国枫丹白露举行的一次由23个政府、126个国家组织和8个国际组织参与的国际会议上，世界自然保护联盟（International Union for the Protection of Nature，IUPN）成立了。当时，这个组织没有财政来源、没有长期预算，甚至没有永久雇员，但是IUPN成为世界政府与非政府组织（Governmental and Nongovernmental Organization，GONGO）的发端。世界自然保护联盟成立后的第一个重大举措是在1950年建立了"生存服务（Survival Service）机构"。"生存服务机构"利用当时筹集到的2,500美元，召集全球的科学家、志愿者为全球濒危物种撰写评估报告，要求各国政府保护其境内的濒危物种。1964年，世界自然保护联盟正式发布《濒危物种红皮书》（Endangered Species Red Book）。今天，世界自然保护联盟已经完全改变了它自己，包括其名称也改变为International Union for Conservation of Nature（IUCN）。IUCN已经成为联合国的观察员、世界范围内主要保护组织。IUCN"生存服务机构"已经演化为物种存续委员会（Species Survival Commission，SSC），IUCN《濒危物种红皮书》也演化为《IUCN受威胁物种红色名录》。现在，IUCN物种存续委员会每年发布《IUCN受威胁物种红色名录》。《IUCN受威胁物种红色名录》已发展成为世界上关于动物、植物和真菌物种全球保护状况最全面的信息源。

世界各国是生物多样性保护的主体。一个物种的IUCN红色名录等级并不一定等同于其在一个国家红色名录中的等级，除非这一物种是该国特有的物种。于是，世界各国也在制定各自的濒危物种红色名录。通过濒危物种红色名录的研究，各国对其境内分布的植物与动物物种的分布、生存状况和保护状况进行调查，然后，对物种的生存状况进行全面评估。因此，濒危物种红色名录是一份物种及其分布的清单，是对物种生存状况、保护状况的客观评估，是生物多样性健康状况的一个重要指标。

濒危物种红色名录被各国政府、自然保护地与野生动植物主管部门、与保护有关的非政府组织、研究人员、自然资源规划人员、教育机构使用。红色名录为生物多样性保护和政策变化提供了信息和促进行动的有力工具，对保护我们赖以生存的自然资源至关重要。通过网络应用，濒危物种红色名录也成为濒危物种信息库，成为保护工作者与研究人员的工具。

中国是 IUCN 的成员，中国也是联合国《生物多样性公约》的最早缔约方之一。中国一直走在生物多样性保护的前沿。中国还是世界上生物多样性最丰富的国家之一，有 7,300 余种脊椎动物，约占全球脊椎动物总数的 11%。中国动物区系组成复杂，空间分布格局差异显著，起源古老，拥有生物演化系统中的各种类群，如有"活化石"之称的大熊猫（*Ailuropoda melanoleuca*）、白鱀豚（*Lipotes vexillifer*）和扬子鳄（*Alligator sinensis*）等。此外，中国还是许多家养动物的起源中心。中国也是生物多样性受威胁最严重的国家之一。人类活动造成的资源过度利用、生境丧失与退化、环境污染及气候变化等因素导致脊椎动物多样性受到严重的威胁。

近年来，党中央和国务院高度重视生物多样性保护工作，将生物多样性保护上升为国家战略，发布了《中国生物多样性保护战略与行动计划（2011－2030 年）》，建立了生物物种资源保护部际联席会议制度，成立了中国生物多样性保护国家委员会，制定和实施了一系列生物多样性保护规划和计划，取得了积极进展。然而，中国生物多样性下降的总体趋势尚未得到有效遏制，保护形势依然严峻，特别是由于目前对中国物种受威胁状况缺乏全面的了解，影响了生物多样性的有效保护。因此，评估物种的受威胁状况，制定红色名录，从而提出针对性的保护策略，对于推动实施《中国生物多样性保护战略与行动计划（2011－2030 年）》和生态文明建设具有重要意义。

到目前，在中国科学家的努力下，中国哺乳动物、鸟类、两栖动物、爬行动物、淡水鱼类都得到了全面的评估。除了评估新发现的物种，中国濒危脊椎动物红色名录还重新评估了一些现存物种的状况，如大熊猫、藏羚（*Pantholops hodgsonii*）等物种，由于中国的保护努力，这些物种的中国濒危脊椎动物红色名录濒危等级下降。然而，中国生物多样性濒危局面仍然严峻。

尽管中国受威胁物种的比例很高，但中国政府正在加强生态环境保护，加强自然保护区、国家公园、世界遗产地及其他类型的保护地建设，加强荒漠化治理、湿地恢复、植树造林，努力扭转或至少制止生物多样性的下降。《生物多样性公约》第 15 次缔约方大会即将在中国昆明召开之际，中国科学家发表最新版《中国生物多样性红色名录：脊椎动物 第一卷 哺乳动物》、《中国生物多样性红色名录：脊椎动物 第二卷 鸟类》、《中国生物多样性红色名录：脊椎动物 第三卷 爬行动物》、《中国生物多样性红色名录：脊椎动物 第四卷 两栖动物》和《中国生物多样性红色名录：脊椎动物 第五卷 淡水鱼类》，全面更新了中国脊椎动物生存状况与种群和栖息地保护状况，从而为确定哪些物种须有针对性地努力恢复，为确定须保护的关键种群和栖息地提供了依据，有助于鉴别未来的濒危脊椎动物保护重点。这套图书的出版是中国自然保护史上的一件大事。

IUCN 主席

2020 年 6 月 26 日

Foreword II

The International Union for the Protection of Nature (IUPN) was established in Fontainebleau of France in 1948 at an international conference that 23 governments, 126 national organizations and 8 international organizations participated in. At that time, the organization had no financial resources, no long-term budget or not even a permanent employee, but it marked the born of the first world Governmental and Nongovernmental Organization (GONGO). Its first move was to establish the "Survival Service" in 1950. Using the $2,500 it raised at the time, the Survival Services called on scientists and volunteers from all around the world to write assessments of the world's endangered species, asking governments to protect those species within their borders. In 1964, the IUCN officially released the *Endangered Species Red Book*. Today, the International Union for Conservation of Nature (IUCN) has completely changed itself, including its name. The IUCN has become an observer of the United Nations and a leading conservation organization worldwide. IUCN Survival Service has evolved into the Species Survival Commission (SSC), the IUCN *Endangered Species Red Book* has been expanded into the website of *IUCN Red List of Threatened Species*, which is now renewed annually by the IUCN Species Survival Committee. The *IUCN Red List of Threatened Species* is the world's most comprehensive source of information on the status of global conservation of animal, plant and fungal species.

Sovereignty countries in the world are the main body of biodiversity protection. The status of a species in IUCN red list is not the affirmatively same in a country's red list except that the species is an endemic species in that country; countries around the world are also developing their own red lists of endangered species. Through the study of the red list of endangered species, countries conduct surveys on the distribution, survival and conservation status of plant and animal species in their territory and then conduct a comprehensive assessment of the survival status of species. Therefore, the red list of endangered species is a list of species and their distribution, an objective assessment of the survival and conservation status of species, and an important indicator of the health status of biodiversity. The red list is used by governments, natural protected areas and wildlife authorities, conservation NGOs, researchers, natural resource planners and educational institutions. It provides information and powerful tools for promoting action on biodiversity conservation and policy formation and is critical to safely guarding the natural resources on which we depend. Through the internet, the red list of endangered species has also become an information base of endangered species and a tool for conservation workers and researchers.

China is a member of the IUCN and one of the earliest parties to the UN *Convention on Biological Diversity*. China has been at the forefront of biodiversity conservation. China is also one of the most biodiverse countries in the world, with more than 7,300 vertebrate species, accounting for about 11% of the total number of vertebrates in the world. China has complex fauna composition, significant differences in spatial distribution pattern, ancient origins and various groups in the biological evolution system, such as Giant Panda (*Ailuropoda melanoleuca*), Baiji (*Lipotes vexillifer*) and Yangtze Alligator (*Alligator sinensis*). In addition, China is the origin center of many domestic animals and plants. China is also one of the countries where biodiversity is most threatened. Due to the overuse of resources, habitat loss and degradation caused by human activities, environmental pollution, climate change and other factors, vertebrate diversity is seriously threatened.

During recent years, the CPC Central Committee and the State Council attach great importance to the protection of biodiversity. Biodiversity conservation is announced as the national strategy, *China's Biodiversity Conservation Strategy and Action Plan (2011-2030)* is issued, the Joint Inter-Ministerial Meeting for Biological Species Resources Protection is regularly held, the China National Committee for Biodiversity Conservation has been set up, a series of biological diversity protection programs and plans have been formulated and implemented, and positive progress in the field has been made. However, the overall trend of biodiversity decline in China has not been effectively stopped, and the conservation situation is still pressing, especially, lack of comprehensive understanding of threatened species in China has hindered the effective conservation of biodiversity. Therefore, it is of great significance to assess the threatened status of species and to formulate the red list of endangered species, and to propose targeted conservation strategies for promoting the implementation of *China's Biodiversity Conservation Strategy and Action Plan (2011-2030)* and the construction of ecological civilization.

Now, the status of Chinese mammals, birds, reptiles, amphibians and freshwater fishes have all been comprehensively assessed by Chinese scientists. In addition to assessing newly discovered species, China's red list of endangered vertebrates has also reassessed the status of some species, including the Giant Panda and Tibetan Antelope (*Pantholops hodgsonii*), whose status on the red list of endangered vertebrates in China has been downgraded due to conservation efforts. However, the overall situation of endangered biodiversity in China is still serious.

Despite the high proportion of threatened species in China, the Chinese government is strengthening ecological protection, stepping up the construction of nature reserves, national parks, World Heritage Sites and other protected areas, working on desertification control, wetland restoration and afforestation, and trying to reverse or at least stop the trend of biodiversity decline. On the occasion that the fifteenth meeting of the Conference of the Parties to the *Convention on Biological Diversity*, which will be held in Kunming, China, Chinese scientists published the latest edition of the "*China's Red List of Biodiversity: Vertebrates, Volume I, Mammals*", "*China's Red List of Biodiversity: Vertebrates, Volume II, Birds*", "*China's Red List of Biodiversity: Vertebrates, Volume III, Reptiles*", "*China's Red List of Biodiversity: Vertebrates, Volume IV, Amphibians*" and "*China's Red List of Biodiversity: Vertebrates, Volume V, Freshwater Fishes*". This set of books comprehensively update the survival status, population and habitat protection of vertebrate, determine the recovery efforts needed for the targeted species, identify key populations and habitats that need to be protected, thus provide a basis for identifying future priorities for endangered vertebrates conservation. Publication of these books is an important event in the history of Chinese nature conservation.

Xinsheng Zhang

President of IUCN,

the International Union for Conservation of Nature

June 26, 2020

总前言

　　物种的濒危现状和濒危机制是保护生物学的核心研究内容,其研究目标是评估人类对生物多样性的影响,提出防止物种灭绝及保护的策略,通过保护生物物种的种群和栖息地,避免物种受到灭绝的威胁。保护生物学研究既关注全球性问题,又具有鲜明的地域特色。中国具有世界上最多的人口,国土面积为世界第三,监测和评估其生物多样性、保护濒危物种,将为中国实现可持续发展提供科学支撑。中国研究人员通过濒危物种红色名录研究,量化了物种灭绝风险,预警了潜在的生态危机,为中国履行《生物多样性公约》等提供科技支撑。

　　世界自然保护联盟(International Union for Conservation of Nature,IUCN)成立后的第一个重大举措是在1950年建立了"生存服务(Survival Service)机构"。"生存服务机构"利用当时募集的2,500美元,召集科学家、志愿者评估全球濒危物种灭绝风险,发表有灭绝风险的物种研究报告,呼吁各国政府保护其境内的濒危物种,这是《IUCN受威胁物种红色名录》的发端。直到1964年,世界自然保护联盟才正式发布《濒危物种红皮书》。今天,世界自然保护联盟完全改变了它自己,"生存服务机构"已经演化成为物种存续委员会(Species Survival Commission,SSC)。IUCN《濒危物种红皮书》已经演变为网络版的《IUCN受威胁物种红色名录》。物种存续委员会从不定期发布IUCN濒危物种红皮书发展到现在每年发布更新的《IUCN受威胁物种红色名录》。

　　《IUCN受威胁物种红色名录》是世界生物多样性健康状况的重要指标。它是世界上最全面的一份动物、植物和真菌物种濒危状况清单。濒危物种红色名录基于物种种群数量、种群数量下降速率、生境破碎程度、生境面积及下降速率、预测灭绝概率等指标估测物种灭绝概率。《IUCN受威胁物种红色名录》(2020-2)发现地球上的32,000多个物种面临着灭绝的风险,占所有被评估物种的27%。其中,41%的两栖类、26%的哺乳类、34%的针叶树、14%的鸟类、30%的鲨鱼、33%的造礁珊瑚，以及28%的特定甲壳类物种面临灭绝风险。《IUCN受威胁物种红色名录》为保护我们赖以生存的自然资源提供了至关重要的信息。

项目背景

自1980年以来，中国经济步入高速发展期。目前，中国已经成为世界第二大经济体。在人口增长、经济发展、全球变化的背景下，中国的生物多样性正面临着前所未有的城镇化、乡村和社会基础设施建设及全球变化的压力，野生生物的生存受到威胁。许多证据显示地球上的生物正面临生物进化中的第六次大灭绝。保护濒危物种是生物多样性保护的核心问题。评估物种濒危等级是生物多样性监测与保护的迫切需要。

虽然《IUCN受威胁物种红色名录》没有国际法和国家法律的效力，但它是专家对全部物种生存状况的评估，它不仅限于评估濒危物种和明星物种，而是最大限度地涵盖了已知的物种，它不仅仅指导世界范围的濒危物种保护，也是指导生物多样性研究的有用工具。《IUCN受威胁物种红色名录》对于政府间组织和非政府组织的保护决策及各国自然与自然保护法律法规的制定都产生了重要影响。

综上所述，濒危物种红色名录是物种灭绝风险的测度，IUCN定期更新其濒危物种红色名录，预警全球物种的生存危机。同时，各国也开展了本国濒危物种红色名录研究。那么，既然已经有《IUCN受威胁物种红色名录》，为什么还要开展国家濒危物种红色名录研究？

《IUCN受威胁物种红色名录》与国家濒危物种红色名录都是物种灭绝风险的测度，前者是全球性评估，后者则是依国别的研究，两者的研究空间尺度不同。《IUCN受威胁物种红色名录》预警了全球物种的濒危状况，为全球生物多样性研究提供了大数据；各国红色名录则确定了各国物种受威胁状况，填补了前者的知识空缺，两份红色名录互为补充。

基于如下原因，应当重视依国别的濒危物种红色名录。①国家是濒危物种保护的行为主体，物种在一个国家的生存状况是确定其保护级别、开展濒危物种保育的依据。②对于仅分布于一个国家的特有物种来说，其按国别的濒危物种红色名录等级即是其全球濒危等级。③《IUCN受威胁物种红色名录》只提供了全球范围的物种濒危信息，并没有评估每一个国家所有物种的生存状况，特别是一些特有物种和跨越国境分布的物种。一些物种跨越国界分布，全球的生存状况并不反映其在个别国家的生存状况，一些全球无危的物种在其边缘分布区的国家里却是极度濒危的或受威胁的物种。世界各国的物种濒危状况有待各国科学家的研究。对于跨国境分布的物种来说，依国别的濒危物种红色名录等级则确定了该物种在本国的生存状况。④结合《IUCN受威胁物种红色名录》，依国别的濒危物种红色名录为建立跨国保护地、保护迁徙物种的栖息地与跨国迁徙洄游通道提供依据。⑤依国别的濒危物种红色名录所特有的"区域灭绝"等级，反映了一个物种边缘种群在该国的区域灭绝，对于一个国家来说，事关重大；恢复"区域灭绝"物种是该物种原分布国家重新引入的相关保育工作的重点。⑥物种濒危状况是不断变化的。近年来，新种、新记录不断被发现。随着人们对生命世界认识的深入，脊椎动物分类系统也发生了变化。依国别的濒危物种红色名录提供了该国物种编目、分类、分布和生存状况的最新信息（蒋志刚等，2020）。

国家红色名录的重要性在许多情况下被忽视了。在研究报告和科普作品中，对国家濒危物种红色名录重视不够。论及物种濒危属性时，作者通常言必《IUCN 受威胁物种红色名录》濒危等级而不提其国家级的红色名录濒危等级。目前正值全球新型冠状病毒肺炎大流行，人们正在重新审视人与野生动物的关系。我国将修订有关野生动物保护与防疫法规和法律、重点保护野生物种名录，防控新的人与野生动物共患疾病再次暴发。对于确定《国家重点保护野生动物名录》而言，物种受威胁程度是物种列为国家重点保护野生物种的特征之一。重视依国别的红色名录有特别的意义。于是，生态环境部（原环境保护部）与中国科学院联合开展了中国生物多样性红色名录研究。

中国动物学家掌握了中国动物分布和生存状况的第一手资料，有必要组织全国淡水鱼类、两栖类、爬行类、鸟类与哺乳类专家及时更新中国脊椎动物分类系统，提供中国脊椎动物多样性的全面、完整的信息；有必要应用统一的国际物种濒危等级标准评估物种生存状况。在国家层面，定期组织全国淡水鱼类、两栖类、爬行类、鸟类与哺乳类专家应用 IUCN 受威胁物种红色名录等级标准和 IUCN 区域受威胁物种红色名录标准，全面评估更新的中国脊椎动物生物多样性红色名录，提供与国际红色名录研究可对比的结果，为红色名录指数的研究积累数据。

经过系统评审制定的中国生物多样性红色名录，由国家权威机构发布。中国生物多样性红色名录淡水鱼类、两栖类、爬行类、鸟类与哺乳类各卷将为监测中国生物多样性现状、为开展阶段性 IUCN 红色名录指数研究和履行《生物多样性公约》提供数据。

中国在 1998 年首次出版了《中国濒危动物红皮书》，2004 年，又出版了《中国物种红色名录》，2009 年，环境保护部组织开展了"中国陆栖脊椎动物物种濒危等级评估"。时隔多年，有必要重新全面评估中国生物多样性的濒危状况。于是，环境保护部委托中国科学院组织有关专家开展了"中国生物多样性红色名录——脊椎动物卷"的研究。在环境保护部和中国科学院的领导下，我们依据 IUCN 受威胁物种红色名录等级标准和 IUCN 区域受威胁物种红色名录标准，全面评估了中国哺乳动物生存状况。

2015 年，环境保护部与中国科学院联合发布了"中国生物多样性红色名录——脊椎动物卷"。现在，历时 6 年，我们全面编研、更新、丰富了此名录，形成了此 2021 版的《中国生物多样性红色名录：脊椎动物》。

项目目标

通过脊椎动物各类群的研究，收集整理中国脊椎动物现有物种种群、生境研究数据、资源监测数据，充实数据库；组织专家，采用综合分析和专家评估相结合的方法，依据中国脊椎动物野生种群与生境现状，利用 IUCN 受威胁物种红色名录等级标准第 3.1 版和 IUCN 区域受威胁物种红色名录标准第 4.0 版综合评价中国脊椎动物濒危状况，编制 2021 版《中国生物多样性红色名录：脊椎动物》。全面评价中国脊椎动物的灭绝风险，对中国濒危物种保护及时提供基础信息。

编研过程

2013 年 5 月 16 日，在中国科学院动物研究所召开了研究启动会。项目聘请陈宜瑜院士、郑光美院士、张亚平院士、金鉴明院士、马建章院士、曹文宣院士为咨询专家，并成立了哺乳类、鸟类、爬行类、两栖类、淡水鱼类课题组。各课题组就评估程序和规范展开了研讨，对典型物种进行了评估并听取了专家委员会的意见。会后总结了专家意见，完善了中国脊椎动物红色名录评估程序和规范。

针对哺乳类、鸟类、爬行类、两栖类和淡水鱼类分别建立了工作组、核心专家组和咨询专家组。工作组负责按照预定的红色名录判定规程开展工作，工作包括资料收集与整理、红色名录初步评定、与通讯评审专家联络及通讯评估结果汇总。核心专家组对红色名录评估的方法、标准使用、数据来源等重要科学问题进行界定，讨论审核有关物种的受威胁等级。工作组在全国范围遴选咨询专家，建立咨询专家库。咨询专家参加了红色名录的通讯评审和会议评审。评审结束后，工作组按照统一格式，整理每个物种包含的信息，形成最终的物种评估说明书。物种评估说明书的内容包括物种的学名、中文名、评估受威胁等级及 IUCN 红色名录等级。"中国生物多样性红色名录——脊椎动物卷"于2015年5月6日通过环境保护部和中国科学院的联合验收，并于5月22日以环境保护部、中国科学院2015年第32号公告形式发布。

红色名录评估的信息来源主要有研究积累、标本数据、文献数据和专家咨询。项目各课题组相关研究团队是工作在中国淡水鱼类、两栖类、爬行类、鸟类和哺乳类研究一线的研究团队，在数十年的研究中积累了大量的科学数据，各分卷主持人还是国家濒危物种科学机构，以及淡水鱼类、两栖类、爬行类、鸟类和哺乳类学术团体的骨干，所在单位是有关动物物种分类、标本收藏、研究的信息交换所，并各自建立了数据库。各分卷主持人还主持或参与了国家有关物种资源本底调查、科学评估、自然保护区生物多样性考察及相关的保护政策制定。

实践意义

《中国生物多样性红色名录：脊椎动物》的出版是一项重大的系统工程。这次生物多样性红色名录评估是迄今评估对象最广、涉及信息最全、参与专家人数最多的一次评估。通过2015版红色名录研究，我们更新了中国脊椎动物编目。中国有2,854种陆生脊椎动物，其中，有407种两栖类，402种爬行类，1,372种鸟类，673种哺乳类。在2021版《中国生物多样性红色名录：脊椎动物》的编研中，我们再次更新了中国脊椎动物分类系统和编目。中国有3,147种陆生脊椎动物，其中，有475种两栖类，527种爬行类，1,445种鸟类，700种哺乳类，比2015年的统计数据增加了293种。我们发现，中国是全球哺乳动物物种数最多的国家。中国陆生脊椎动物中，特有种超过20%。我们还分析了中国脊椎动物的分布格局和特有类群，探讨了其濒危种类的空间分布规律。

中国濒危脊椎物种濒危模式与分布格局在我国生物多样性和生态系统保护中具有指导意义，这将为我国重点保护物种确定、国土空间开发和生态功能区的划分及各类保护地规划设计提供重要参考依据。也是确定中国物种多样性保护热点的依据之一。我们发现，中国脊椎动物生存危机依然严重，中国濒危脊椎物种的分布格局不均衡。物种的空间分布是一种立体格局，除了水平纬度上的物种分布格局，我们也需要物种多样性和濒危种类的垂直分布格局，这些格局对物种多样性保护具有重要参考价值。我们发现，高海拔地区受威胁哺乳动物的比例比低海拔地区高，高海拔地区的濒危物种应受到更多的关注。

展望与致谢

《中国生物多样性红色名录》的编制和发布为生物多样性保护政策和规划的制定提供了科学依据，发挥了中国科学家作为中国《生物多样性公约》履约"智库"的功能，同时，为开展生物多样性科学基础研究积累了基础数据，更新了脊椎动物分类系统与编目，为公众参与生物多样性保护创造了必要条件。《中国生物多样性红色名录》的编制是贯彻实施《中国生物多样性保护战略与行动计划（2011—2030年）》和积极履行《生物多样性公约》的具体行动。通过《中国生物多样性红色名录》的编制，中国在生物多样性评价方面已经在全球先行一步，使我国在履行《生物多样性公约》方面走在世界的前列。

本项目得到了生态环境部（原环境保护部）、中国科学院、国家林业与草原局（原国家林业局）、中国科学院大学、科学出版社的关怀、指导和大力支持；得到了国家出版基金的大力支持。课题组还得到了如下项目的资助：中国科学院战略性先导科技专项（A类）"地球大数据科学工程"（项目编号：XDA19050204）、国家重点研发计划项目（项目编号：2016YFC0503303）、国家科技基础性工作专项（项目编号：2013FY110300）的资助。在此谨致感谢！

蒋志刚

中国科学院动物研究所研究员

中国科学院大学岗位教授

国家濒危物种科学委员会前常务副主任

2020年6月6日

Series' Preface

The current status and threats to species are the key issues in conservation biology. The primary goal of putting forward an endangered species red list is to evaluate human impact on biodiversity, identifying key threats and preventing species extinction by protecting the populations and habitats of threatened species. Conservation research not only pays attention to global issues, but also must focus attention on regional and national problems. China has the largest human population and the third largest terrestrial area in the world. Monitoring and evaluating the country's biodiversity and protecting endangered species will provide scientific support for China's sustainable development. Therefore, Chinese researchers are working to quantify the risk of extinctions through studies related to the red list of threatened species, providing early warning about potential ecological hazards, thus offering scientific support for implementation of the *Convention on Biological Diversity* in China.

The first major move for the International Union for Conservation of Nature (IUCN) after its establishment was to launch the Survival Service in 1950. Using the $2,500 raised at that time, the Survival Service called on scientists and volunteers to dedicate their expertise and time to assess the extinction risk of globally threatened species. The Survival Service then publicized their research reports on species at risk of extinction and called on governments to protect endangered species within their borders. Such an act marked the beginning of the *IUCN Red List of Threatened Species*. However, it was not until 1964 that IUCN officially published its first *Endangered Species Red Book*. Today, IUCN has completely changed itself, the Survival Service has been renamed as the Species Survival Commission (SSC). The IUCN

Endangered Species Red Book has evolved into the online version of *IUCN Red List of Threatened Species*. The Species Survival Commission refreshes and revises the *IUCN Red List of Threatened Species* periodically, and updates the *IUCN Red List of Threatened Species* annually.

The *IUCN Red List of Threatened Species* is an important indicator of the health of the world's biodiversity. It is the world's most comprehensive list of rare and threatened animal, plant and fungal species. The *IUCN Red List of Threatened Species* estimates the extinction probability of species based on population size, population decline rate, degree of habitat fragmentation, rate of decline of habitat area and other indicators. The *IUCN Red List of Threatened Species* (2020-2) estimated that more than 32,000 species on the earth are at risk of extinction, accounting for 27% of all assessed species globally. 41% of amphibians, 26% of mammals, 34% of conifers, 14% of birds, 30% of sharks, 33% of corals, and 28% of certain crustaceans are presently at risk of extinction. The *IUCN Red List of Threatened Species* provides vital information for protecting the biodiversity and natural resources on which we all collectively depend.

The Background

Since the 1980s, China has embarked on a fast track of socioeconomic development. China has become the world's second largest economy. Against a backdrop of population growth, rapid economic development and many global changes, China's biodiversity is under unprecedented pressure from urbanization, infrastructure development and a wide range of other factors, and the survival of wildlife is under threat. Ample evidence shows that life on the earth is facing its Sixth Mass Extinction in its long evolutionary history. Protecting endangered species is the core issue for biodiversity conservation. Thus, assessing the endangerment level of species is the primary and most urgent need that biodiversity monitoring and protection measures seek to address.

Though the *IUCN Red List of Threatened Species* does not possess the power of international or national laws, it is an expert assessment of the survival status of all species, not only endangered or charismatic species, but all known species to the greatest extent possible. It thus serves as a most useful tool not only to guide worldwide protection of endangered species but also for the study of biodiversity. The *IUCN Red List of Threatened Species* has significant impact on the conservation decisions of intergovernmental and non-governmental organizations as well as for the formulation of national laws and regulations regarding wildlife and nature conservation.

As stated above, the red list of threatened species provides a measure of the risk of extinction of species. IUCN regularly updates its global red list of endangered species in order to raise public awareness of the global status of wildlife and the species survival crisis. At the same time, countries also conduct national-level studies on the status of endangered species. However, since there is already an *IUCN Red List of Threatened Species*, the question may arise, why bother to conduct research at country level to produce national red lists of endangered species?

Both the *IUCN Red List of Threatened Species* and country red lists of threatened species assess species' risk of extinction, with the former being global in scope while the latter are regional assessments. The

Series' Preface

IUCN Red List of Threatened Species alerts the world to the status of endangered species, and also serves as a database of global biodiversity. Country red lists, on the other hand, ascertain the status of species in particular countries, filling knowledge gaps in the former. The two lists are thus complementary to each other.

Country-level red lists should be given greater attention for at least the following reasons: (i) A sovereign country is the main authority for taking conservation action in regard to wildlife species within its boundaries, based on the level of endangerment (conservation status) of the species; (ii) For endemic species in a country, the country red list status constitutes its global status; (iii) The *IUCN Red List of Threatened Species* provides only the information on species at risk worldwide and does not assess the status of all species in each country, especially endemic species and those species with transboundary distribution. Some species are distributed across national boundaries and the global conservation status does not entirely reflect the survival status of the species in any particular country. Some global non-threatened species are critically endangered or threatened in the countries where they have peripheral ranges. The endangered status of species in different countries of the world thus remains to be studied by scientists in relation to specific countries. For species whose ranges cross national borders, the country's red list status reflects the survival status of the species in the country; (iv) Combined with the global *IUCN Red List of Threatened Species*, country red lists provide a basis from which to consider the establishment of transnational protected areas, the protection of important habitats for migratory species, and the protection of international migration corridors; (v) The category "Regionally Extinct" is unique to country (regional) red lists of endangered species as it refers only to a subset of the broader geographic distribution of the species, yet the national status is still indicative of the species' overall risk of extinction, this matters a lot for a country, and the restoration of "regionally extinct" species is the focus of conservation efforts for reintroduction in countries where the species originated; (vi) Country red lists provide updated information about endangered species with national inventories as well as with national reviews of classification, geographic distribution, and status of species at national level, which are also relevant for global species descriptions and assessments (Jiang *et al.*, 2020).

Despite these benefits, the significance of country-level red lists is often overlooked. Following onset of the global COVID-19 pandemic, however, people's outlook has been changing in regard to the relationship between people and wildlife. Consequently, China is amending its national laws on wildlife protection, epidemic prevention, and the list of state key protected wild species, in order to better prevent and control emerging zoonoses. The status of wildlife species included in China's red list of threatened species should be one of the defining elements for identifying and updating species on the *List of State Key Protected Wild Animal Species* in China. It is therefore critical to duly recognize the significance of the country red list at this special moment in time. For this purpose, the Ministry of Ecology and Environment (former Ministry of Environmental Protection) and the Chinese Academy of Sciences have jointly launched China's biodiversity red list.

Chinese zoologists have obtained first-hand information on the distribution and living status of animals in China. It is necessary to organize national experts on freshwater fishes, amphibians, reptiles, birds and

mammals to update the taxonomy of vertebrates in China in a timely manner and to provide systematic and comprehensive information on the diversity of vertebrates in China. It is necessary to apply standard international criteria for threatened species to assess the status of species. At the national level, it is necessary to coordinate national experts on freshwater fishes, amphibians, reptiles, birds and mammals to apply the IUCN red list criteria for threatened species and the IUCN regional red list criteria for threatened species, and through this process also to comprehensively update the *China's Red List of Biodiversity: Volume of Vertebrates* and thus to provide a country red list that is comparable to international red lists, and to enable index studies of red lists.

The red list of China's biodiversity, which has been systematically reviewed and formulated, shall be issued by the state authorities. The volumes of freshwater fishes, amphibians, reptiles, birds and mammals of the red list of china's biodiversity will provide data for the implementation of the *Convention on Biological Diversity*, for monitoring the state of biodiversity in China, as well as for conducting periodic IUCN red list index studies in the country.

China firstly published its *China Red Data Book of Endangered Animals* in 1998, followed by the *China Species Red List* in 2004 and in 2009, the Ministry of Environmental Protection coordinated the assessment and publishing of the "Assessment of the Red List of Endangered Species of Terrestrial Vertebrates in China", which is a multi-year project for the comprehensive re-assessment of the threatened status of China's biodiversity. Therefore, the Ministry of Environmental Protection entrusts the Chinese Academy of Sciences to organize experts to carry out research on the *China's Red List of Biodiversity: Volume of Vertebrates*. Under the leadership of the Ministry of Environmental Protection and the Chinese Academy of Sciences, we have conducted a comprehensive assessment of the living status of the vertebrates in China based on the IUCN red list criteria for threatened species and the IUCN regional red list criteria for endangered species.

In 2015, the Ministry of Environmental Protection and the Chinese Academy of Sciences jointly released the *China's Red List of Biodiversity: Volume of Vertebrates*. Now, six years on, we have thoroughly updated and compiled the series of books of *China's Red List of Biodiversity: Vertebrates* (2021).

The Goal

Through the study and preparation for each volume of vertebrates in China's biodiversity red list, we collected and sorted existing information on the population and habitat status of vertebrates in China and completed the database. We systematically and comprehensively evaluated the status of vertebrates in China, using the *IUCN Red List Categories and Criteria* (version 3.1) and the *Guidelines for Application of IUCN Red List Criteria at Regional and National Levels* (version 4.0) based on the status of the species' wild population and habitat. Combined with the empirical analysis and expert evaluation, we compiled the *China's Red List of Biodiversity: Vertebrates* (2021), which is a comprehensive assessment of extinction risk of China's vertebrates, providing the basic information pertinent for the protection of endangered species over the coming years.

The Assessment

The project launch meeting was held at the Institute of Zoology, Chinese Academy of Sciences on May 16, 2013. Academicians Yiyu Chen, Guangmei Zheng, Yaping Zhang, Jianming Jin, Jianzhang Ma and Wenxuan Cao were invited to participate in the meeting as consulting experts. Mammals, birds, reptiles, amphibians and freshwater fishes research groups were formed at the meeting. Each research group held a discussion on evaluation procedures and norms, assessed the typical species of their own taxonomic group, and consulted the opinion of the expert committee. After the meeting, all experts' opinions were summarized and the evaluation procedures and norms of the red list of vertebrates in China were finalized.

Working groups, core expert groups and communication expert groups were formed for each research group, focused respectively on mammals, birds, reptiles, amphibians and fresh water fishes. The working groups were responsible for carrying out work in accordance with the red list category assessment procedures, including data collection and classification, preliminary red list category evaluation, liaison with experts by correspondence, and providing summaries and communicating evaluation results. The core expert group defined the methods, standards, data sources and other important scientific issues of the red list assessment, discussed and reviewed the status of species. The working group selected consulting experts nationwide and established a database of consulting experts. Each consulting expert participated in the red list evaluation and conference review. After the review, the information about every species was sorted and summarized in a unified format to provide the final species evaluation specifications. The species description and assessment includes scientific name, Chinese name, threat level assessment and IUCN red list category criteria. The *"China's Red List of Biodiversity: Volume of Vertebrates"* was jointly approved by the Ministry of Environmental Protection and the Chinese Academy of Sciences on May 6, 2015, and officially released on May 22, 2015, through the Announcement No. 32 of the Ministry of Environmental Protection and the Chinese Academy of Sciences, on International Biodiversity Day of 2015.

The information sources for the evaluation of vertebrates in the red list of China's biodiversity included published and unpublished literature, specimen data, and expert consultation. Experts from the red list working groups for freshwater fishes, amphibians, reptiles, birds and mammals are experts who have accumulated a large amount of scientific data over decades. The coordinators of working groups are people from state endangered species scientific authorities and established academics from scientific communities focused on freshwater fishes, amphibians, reptiles, birds and mammals from across the country. The research institutions are the centers of taxonomy, specimen collections, and databases for animal species. The principal scientists of freshwater fishes, amphibians, reptiles, birds and mammals also often coordinated or participated in background investigations on national species resources, scientific assessments, biodiversity investigations of nature reserves, and the formulation of relevant government conservation policies.

The significance

The publishing of the *China's Red List of Biodiversity*: *Vertebrates* (2021) is a major systematic project. The

biodiversity red list assessment covered the widest range of subjects, providing the most complete information and involving the largest number of experts so far in the country. During the process of developing the 2015 edition of the red list, we updated the Chinese vertebrate inventory. On this basis, it was found that China has 2,854 terrestrial vertebrates, of which 407 are amphibians, 402 are reptiles, 1,372 are birds and 673 are mammals. In the preparation and research of the 2021 edition, we have once again updated the classification system and produced an updated inventory of vertebrates in China. Altogether there are 3,147 terrestrial vertebrates in China, among which there are 475 species of amphibians, 527 reptiles, 1,445 birds and 700 mammals. A further 293 vertebrate species were assessed for preparing the 2021 edition. China is now found to have the largest number of mammal species in the world. Among land vertebrates in China, more than 20% are endemic. We also analyzed the distribution pattern and endemic groups of vertebrates in China and discussed the spatial distribution pattern of endangered species.

The conservation status and distribution patterns of threatened vertebrates in China that are shared in this book provide an important reference for identification of key protected vertebrate species, planning and development of national strategic land use blueprints, the design of ecological functional zones and various protected sites, which are of great significance for biodiversity and ecosystem conservation in China. This information is also one of the criteria for determining hotspot locations for species diversity in China. We have found that the survival crisis of vertebrates is still present in China and the distribution pattern of endangered vertebrates remains uneven. The spatial distribution of species presents a three-dimensional pattern, including their geographic distribution (two dimensions) as well as vertical or elevational dimension where species including threatened species are generally situated. In particular, we found that the proportions of threatened mammals in different families and orders were greater at higher altitudes than those found at lower altitudes, and additionally we found that endangered species that live at higher altitudes often should receive more attention from the public as well as from the government.

The Outlook and Appreciation

The compiling and publishing of *China's Red List of Biodiversity* provides a scientific basis for biodiversity conservation planning and policy in China, based on the long-standing work of Chinese scientists, who constitute a *de facto* "think-tank" for research and implementation of the *Convention on Biological Diversity* in China. At the same time, the red list study has updated the vertebrate taxonomy and inventory in China, accumulated data for basic zoological research in biodiversity science both nationally and globally, and created the necessary conditions for public participation in biodiversity conservation. Producing the *China's Red List of Biodiversity* has been a concrete action in the implementation of *China's Biodiversity Conservation Strategy and Action Plan (2011-2030)* and has also been a key step in implementing the *Convention on Biological Diversity* in its territory. Through the compilation of the *China's Red List of Biodiversity*, China has demonstrated its leading role in the global assessment of the current status of biodiversity.

The project that enabled development and publication of this red list book received guidance and support from the Ministry of Ecology and Environment (former Ministry of Environmental Protection), the Chinese

Academy of Sciences, the National Forestry and Grassland Administration (former National Forestry Administration), the University of Chinese Academy of Sciences, and the Science Press (Beijing). The project received funding from the National Publication Foundation, and each individual research group also received support through the following projects: "Earth Big-Data Scientific Project" (XDA19050204) of the Strategic Leading Science and Technology Project, Chinese Academy of Sciences (Category A); National Key Research and Development Project (2016YFC0503303); Basic Science Special Project of the Ministry of Science and Technology of China (2013FY110300). We express our most sincere gratitude to all of these governmental bodies, institutions and funding agencies for their many different forms of support.

Zhigang Jiang, Ph.D.

Professor of Institute of Zoology, Chinese Academy of Sciences

Professor of University of Chinese Academy of Sciences

Former Executive Director of the Endangered Species Scientific Commission, P. R. China

前言

中国哺乳动物区系有鲜明的特色：中国是世界上哺乳动物种类最多的国家之一，也是特有哺乳动物丰富的国家。进入21世纪后，中国哺乳动物研究得到了长足的发展。由于种种原因，中国的哺乳动物编目工作一直到21世纪初才由王应祥先生完成。王应祥先生在《中国哺乳动物种和亚种分类名录与分布大全》一书报道中国有哺乳动物603种。从2008年起，我们开始评估中国濒危哺乳动物生存状况。我们首先补充了新种与新记录种，删去了无效种，采用最新的哺乳动物分类系统，开展了中国哺乳动物编目研究。我们2015年报道了中国有哺乳动物12目55科245属673种（蒋志刚等，2015a）。2017年，我们再次更新了分类系统，补充了新种与新记录种，删去了无效种，更新了这一数据，记录哺乳动物13目56科248属693种（蒋志刚等，2017）。在本次《中国生物多样性红色名录：脊椎动物 第一卷 哺乳动物》（2021）的评估中，经过查遗补缺，再次更新中国哺乳动物为13目56科248属700种。我们共评估了除智人外的700种哺乳动物。

本书为中英文双语，面向全球读者。为了方便读者提纲挈领，掌握本红色名录研究方法和中国哺乳动物的生存与保护状况。本书在前面部分有介绍中国哺乳动物生存状况评估的"总论"。在总论中，介绍了哺乳动物的演化与现状，中国哺乳动物多样性与保护现状，本次红色名录评估对象的分类系统、分布格局、受保护状况；介绍了中国哺乳动物红色名录的评估过程、依据的评估等级标准，还介绍了咨询专家顾问、评估队伍，以及评估团队建立数据库和开展初步评定、通讯评审、形成评估报告的过程。最后，总结分析了本次红色名录评估结果，比较了《中国生物多样性红色名录：脊椎动物 第一卷 哺乳动物》2021版与2015版的异同，介绍了中国哺乳动物濒危状况，分析了野外灭绝的和区域灭绝的物种，分析了不同哺乳动物类群、不同生境的受威胁哺乳动物物种比例及其省区哺乳动物的分布和濒危原因，并将本次评估结果与《IUCN受威胁物种红色名录》（2020-2）的评估结果进行了比较分析。最后，分析展望了中国哺乳动物保护成效与远景。

本书"各论"中的物种编排顺序基本按照IUCN受威胁物种红色名录濒危等级："极危""濒危""易危""近危""无危""野外灭绝""区域灭绝"（本书特有）"数据缺乏"排列。"各论"编排列出了每一物种中文名Chinese Name和其他物种的分类信息如目Order、科Family、学名Scientific Name、命名人Species Authority、英文名English Name(s)、同物异名Synonym(s)、种下单元评估Infra-specific Taxa Assessed，并配有彩绘插图或照片。各论还列出了评估信息Assessment Information、红色名录等级Red List Category（评估标准版本，已列在物种标题上）、评估年份Year Assessed、评定人Assessor(s)、审定人Reviewer(s)和其他贡献人Other Contributor(s)。并对评估对象进行了评估理由Justification、地理分布Geographical Distribution（国内分布Domestic Distribution和世界分布World Distribution），以及是否特有种的分布标注Distribution Note描述；还给出了国内分布图Map of Domestic Distribution和种群数量Population Size、种群趋势Population Trend、所在生境与生态系统Habitat(s) and Ecosystem(s)，以及威胁Threat(s)、保护级别与保护行动Protection Category and Conservation Action(s)［国家重点保护野生动物等级（2021）Category of National Key Protected Wild Animals（2021）、IUCN红色名录（2020-2）IUCN Red List（2020-2）、CITES附录（2019）CITES Appendix（2019），是否开展了"保护行动Conservation Action(s)"］。最后，列出了相关文献Relevant References。书末附有检索表和参考文献。

在本书的编研过程中，有关工作得到了中国科学院、生态环境部（原环境保护部）、国家林业和草原局（原国家林业局）、中国科学院大学、国家濒危物种科学委员会、中国科学院动物研究所的精心指导、大力支持与帮助。在本书的编辑出版过程中，得到了科学出版社的细心帮助，得到国家出版基金的资助。在本书的编研过程中，我们执行了原环境保护部生物多样性专项：中国脊椎动物红色名录研究（2012—2018年），国家科技基础性工作专项（项目编号：2013FY110300）、国家重点研发计划项目（项目编号：2016YFC0503303）、中国科学院战略性先导科技专项（A类；项目编号：XDA19050204）、"美丽中国"生态文明建设科技工程项目（项目编号：XDA23100203）、全国第二次陆生野生动物资源调查项目、国家自然科学基金项目等项目。

本书的编研历时数年，由于研究团队知识与时间有限，我们在编研过程中深深领会到"吾生也有涯，而知也无涯。以有涯随无涯，殆已"。保护生物学是一门处理危机的学科，濒危物种红色名录是生物多样性预警报告。为了及时拯救濒危物种，尽管缺点、疏漏在所难免，我们仍将这本图书呈示于世，以期有关方面及时采取行动，保护人类赖以生存的基础。希望有关专家、读者对本书存在的问题不吝指正。

蒋志刚

2020年6月6日

Preface

China has distinct characteristics in regard to its mammalian fauna: China has amongst the richest mammal diversity in the world, and it also has one of the greatest levels of endemism of mammal species globally. Since the beginning of the new millennium, mammal research in China has made significant progress. For various reasons, the inventory of mammals in China was not completed until the early 21st century. In his book *Taxonomy and Distribution of Mammal Species and Subspecies in China*, Professor Yingxiang Wang reported that there were 603 mammal species in China. Following this, since 2008, we have been systematically assessing the survival status of threatened mammals in China. First, we have added new species and new records of species, deleted invalid species, and adopted the latest mammal taxonomic system to carry out the inventory research of mammals in China. By 2015, we reported 673 species from 245 genera, 55 families and 12 orders of mammals in China (Jiang *et al.*, 2015a). In 2017, we renewed the classification system, added several more species and new records of species, deleted additional invalid species, and through this process updated the data and inventory, recording in total 693 species from 248 genera, 56 families, 13 orders of mammals (Jiang *et al.*, 2017). In this assessment, in the Mammals Volume of *China's Red List of Biodiversity: Vertebrates* (2021), we now recognize a total of 700 species of mammals (in 248 genera, 56 families, 13 orders) in the national inventory and evaluated 700 species except the *Homo sapiens*.

This book is bilingual in both Chinese and English, and is intended for a global audience, aiming to introduce readers to the basic concepts and grasp research methods concerning the red list of threatened species, and more specifically, to main findings from long-term research about the living conditions and conservation status of mammals in China. This book is preceded by a General Introduction regarding the assessment of the survival or conservation status of mammals in China. In the General Introduction, the evolution and current state of mammals are introduced including their diversity, their taxonomy, and their distribution patterns, along with their conservation or red list status. The General Introduction also describes the evaluation process that was adopted for the mammals of China's red list of biodiversity, including the evaluation criteria and consulting experts, and how the evaluation team established the mammal database, carried out preliminary evaluation, reviewed preliminary results by correspondence, and formalized the evaluation report. Finally, the evaluation results are analyzed and summarized for the readers. The authors compared the Mammal Volume of *China's Red List of Biodiversity: Vertebrates* 2021 edition with the 2015 edition, analyzed the status of threatened mammals, including special note of species that are Extinct in the Wild and Regionally Extinct, as well as the proportion of threatened species in different mammal groups, in different habitats and range provinces, and the threats that are faced by endangered mammals. A comparison of the results of mammals in *China's Red List of Biodiversity: Vertebrates* (2021) with the *IUCN Red List of Threatened Species* (2020-2) also is provided. Finally, the implications and prospects for mammal conservation in China are discussed.

In the book, the species are arranged in "Species Monograph" according to their status in the *IUCN Red List of Threatened Species*: "Critically Endangered", "Endangered", "Vulnerable", "Near Threatened", "Not Threatened", "Extinct in the Wild", "Regionally Extinct" (unique only in this book), and "Data Deficient". In each "Species Monograph", with colored illustrations or photographs, the Chinese Name and taxonomy of the species such as: Order, Family, Scientific Name, Species Authority, English Name(s), Synonym(s), Infra-specific Taxa Assessed of the species are given. In each "Species Monograph", Assessment Information, Red List Category (evaluation criteria version, listed in the title), Year Assessed, Assessor(s), Reviewer(s) and Other Contributor(s) are listed, plus the assessment Justification, Domestic Distribution in China and World Distribution, as well as the Distribution Note of whether the Species is Endemic or not. Map of Domestic Distribution, Population Size and Trend, Habitat(s) and Ecosystem(s) of species are also given. And more, "Species Monograph" presents the Threats, Protection Category and Conservation Action taken such as Category of National Key Protected Wild Animals (2021), IUCN Red List (2020-2), CITES Appendix (2019), and Whether there is a "Conservation Action(s)" for species. Finally, Relevant References are listed. At the end of the book, Index and Reference are listed.

During the compilation and research of the book, the work has been carefully guided, supported and helped by the Chinese Academy of Sciences, the Ministry of Ecology and Environment (formerly the Ministry of Environmental Protection), the State Forestry and Grassland Administration (formerly the State Forestry Administration), the University of Chinese Academy of Sciences, the National Endangered Species Scientific Commission, and the Institute of Zoology of the Chinese Academy of Sciences. During the process of editing and publishing this book, we have received help from Science Press in Beijing, and financial support from the National Publication Foundation. During the process of research for the book, we performed the biodiversity special project: China's Vertebrate Red List of 2012–2018 of the former Ministry of Environmental Protection,

Basic Science Special Project of the Ministry of Science and Technology of China (2013FY110300), National Key Research and Development Project (2016YFC0503303), Strategic Leading Science and Technology Project, Chinese Academy of Sciences (Category A, XDA19050204), Science and Technology Project of Ecological Civilization Construction of "The Beautiful China" (XDA23100203), as well as the special projects of the Second National Survey on Terrestrial Wild Animals and projects of the Natural Science Foundation, *etc*.

Although the compilation of this book took several years, due to the limited knowledge and time of the research team, we deeply understood an ancient proverb "My life is limited, and the knowledge is limitless. To use the limited time to explore the boundlessness, it will be exhausted", during the compilation and research process. Conservation biology is a crisis management discipline, and the red list of endangered species is an early-warning report. In order to save endangered species in a timely manner, despite the inevitable shortcomings, we present this book to the world so that action can be taken to protect the very foundation on which humanity depends. I hope that experts and readers will not hesitate to point out the problems in this book.

Zhigang Jiang

June 6, 2020

目录 Contents

序 一 ································· i	Foreword I ································· iii
序 二 ································· vii	Foreword II ································· ix
总前言 ································· xiii	Series' Preface ································· xix
前 言 ································· xxvii	Preface ································· xxix

各论（中册） Species Monograph (II)

近危 / NT

侯氏猬 ································· 470	*Mesechinus hughi* ································· 470
等齿鼩鼹 ································· 472	*Uropsilus aequodonenia* ································· 472
贡山鼩鼹 ································· 474	*Uropsilus investigator* ································· 474
甘肃鼹 ································· 476	*Scapanulus oweni* ································· 476
小缺齿鼹 ································· 478	*Mogera wogura* ································· 478
白尾鼹 ································· 480	*Parascaptor leucura* ································· 480
麝鼹 ································· 482	*Scaptochirus moschatus* ································· 482
天山鼩鼱 ································· 484	*Sorex asper* ································· 484
中鼩鼱 ································· 486	*Sorex caecutiens* ································· 486
甘肃鼩鼱 ································· 488	*Sorex cansulus* ································· 488
纹背鼩鼱 ································· 490	*Sorex cylindricauda* ································· 490
栗齿鼩鼱 ································· 492	*Sorex daphaenodon* ································· 492
远东鼩鼱 ································· 494	*Sorex isodon* ································· 494
姬鼩鼱 ································· 496	*Sorex minutissimus* ································· 496
小鼩鼱 ································· 498	*Sorex minutus* ································· 498
克什米尔鼩鼱 ································· 500	*Sorex planiceps* ································· 500
陕西鼩鼱 ································· 502	*Sorex sinalis* ································· 502

中文名	页码	学名	页码
藏鼩鼱	504	*Sorex thibetanus*	504
苔原鼩鼱	506	*Sorex tundrensis*	506
长爪鼩鼱	508	*Sorex unguiculatus*	508
狭颅黑齿鼩鼱	510	*Blarinella wardi*	510
大爪长尾鼩鼱	512	*Soriculus nigrescens*	512
滇北长尾鼩	514	*Chodsigoa parva*	514
喜马拉雅水麝鼩	516	*Chimarrogale himalayica*	516
格氏小麝鼩	518	*Crocidura gmelini*	518
印支小麝鼩	520	*Crocidura indochinensis*	520
大麝鼩	522	*Crocidura lasiura*	522
华南中麝鼩	524	*Crocidura rapax*	524
西伯利亚麝鼩	526	*Crocidura sibirica*	526
西南中麝鼩	528	*Crocidura vorax*	528
五指山小麝鼩	530	*Crocidura wuchihensis*	530
棕果蝠	532	*Rousettus leschenaultii*	532
犬蝠	534	*Cynopterus sphinx*	534
台湾菊头蝠	536	*Rhinolophus formosae*	536
大菊头蝠	538	*Rhinolophus luctus*	538
马氏菊头蝠	540	*Rhinolophus marshalli*	540
高鞍菊头蝠	542	*Rhinolophus paradoxolophus*	542
贵州菊头蝠	544	*Rhinolophus rex*	544
清迈菊头蝠	546	*Rhinolophus siamensis*	546
小褐菊头蝠	548	*Rhinolophus stheno*	548
托氏菊头蝠	550	*Rhinolophus thomasi*	550
灰小蹄蝠	552	*Hipposideros cineraceus*	552
普氏蹄蝠	554	*Hipposideros pratti*	554
三叶蹄蝠	556	*Aselliscus stoliczkanus*	556
宽耳犬吻蝠	558	*Tadarida insignis*	558
华北犬吻蝠	560	*Tadarida latouchei*	560
西南鼠耳蝠	562	*Myotis altarium*	562
缺齿鼠耳蝠	564	*Myotis annectans*	564
尖耳鼠耳蝠	566	*Myotis blythii*	566
布氏鼠耳蝠	568	*Myotis brandtii*	568
中华鼠耳蝠	570	*Myotis chinensis*	570
毛腿鼠耳蝠	572	*Myotis fimbriatus*	572
宽吻鼠耳蝠	574	*Myotis latirostris*	574
喜山鼠耳蝠	576	*Myotis muricola*	576

目 录 Contents

高颅鼠耳蝠	578	*Myotis siligorensis*	578
台湾鼠耳蝠	580	*Myotis taiwanensis*	580
爪哇伏翼	582	*Pipistrellus javanicus*	582
小伏翼	584	*Pipistrellus tenuis*	584
灰伏翼	586	*Hypsugo pulveratus*	586
萨氏伏翼	588	*Hypsugo savii*	588
南蝠	590	*Ia io*	590
大山蝠	592	*Nyctalus aviator*	592
褐山蝠	594	*Nyctalus noctula*	594
褐扁颅蝠	596	*Tylonycteris robustula*	596
灰大耳蝠	598	*Plecotus austriacus*	598
台湾大耳蝠	600	*Plecotus taivanus*	600
亚洲长翼蝠	602	*Miniopterus fuliginosus*	602
几内亚长翼蝠	604	*Miniopterus magnater*	604
南长翼蝠	606	*Miniopterus pusillus*	606
金管鼻蝠	608	*Murina aurata*	608
圆耳管鼻蝠	610	*Murina cyclotis*	610
艾氏管鼻蝠	612	*Murina eleryi*	612
台湾管鼻蝠	614	*Murina puta*	614
毛翼管鼻蝠	616	*Harpiocephalus harpia*	616
川金丝猴	618	*Rhinopithecus roxellana*	618
狼	620	*Canis lupus*	620
沙狐	622	*Vulpes corsac*	622
藏狐	624	*Vulpes ferrilata*	624
赤狐	626	*Vulpes vulpes*	626
貉	628	*Nyctereutes procyonoides*	628
北海狮	630	*Eumetopias jubatus*	630
香鼬	632	*Mustela altaica*	632
黄腹鼬	634	*Mustela kathiah*	634
鼬獾	636	*Melogale moschata*	636
亚洲狗獾	638	*Meles leucurus*	638
猪獾	640	*Arctonyx collaris*	640
小灵猫	642	*Viverricula indica*	642
果子狸	644	*Paguma larvata*	644
藏野驴	646	*Equus kiang*	646
毛冠鹿	648	*Elaphodus cephalophus*	648
小麂	650	*Muntiacus reevesi*	650

中文名	页码	学名	页码
赤麂	652	*Muntiacus vaginalis*	652
水鹿	654	*Rusa unicolor*	654
狍	656	*Capreolus pygargus*	656
藏原羚	658	*Procapra picticaudata*	658
藏羚	660	*Pantholops hodgsonii*	660
北山羊	662	*Capra sibirica*	662
西藏盘羊	664	*Ovis hodgsoni*	664
台湾鬣羚	666	*Capricornis swinhoei*	666
松鼠	668	*Sciurus vulgaris*	668
五纹松鼠	670	*Callosciurus quinquestriatus*	670
橙喉长吻松鼠	672	*Dremomys gularis*	672
橙腹长吻松鼠	674	*Dremomys lokriah*	674
红腿长吻松鼠	676	*Dremomys pyrrhomerus*	676
栗褐鼯鼠	678	*Petaurista magnificus*	678
沟牙鼯鼠	680	*Aeretes melanopterus*	680
黑白飞鼠	682	*Hylopetes alboniger*	682
藏仓鼠	684	*Cricetulus kamensis*	684
林旅鼠	686	*Myopus schisticolor*	686
克钦绒鼠	688	*Eothenomys cachinus*	688
中华绒鼠	690	*Eothenomys chinensis*	690
康定绒鼠	692	*Eothenomys hintoni*	692
昭通绒鼠	694	*Eothenomys olitor*	694
玉龙绒鼠	696	*Eothenomys proditor*	696
川西绒鼠	698	*Eothenomys tarquinius*	698
德钦绒鼠	700	*Eothenomys wardi*	700
斯氏高山䶄	702	*Alticola stoliczkanus*	702
聂拉木松田鼠	704	*Neodon nyalamensis*	704
台湾田鼠	706	*Alexandromys kikuchii*	706
黑田鼠	708	*Microtus agrestis*	708
四川田鼠	710	*Volemys millicens*	710
川西田鼠	712	*Volemys musseri*	712
黄兔尾鼠	714	*Eolagurus luteus*	714
蒙古兔尾鼠	716	*Eolagurus przewalskii*	716
凉山沟牙田鼠	718	*Proedromys liangshanensis*	718
长尾攀鼠	720	*Vandeleuria oleracea*	720
云南壮鼠	722	*Hadromys yunnanensis*	722
大齿鼠	724	*Dacnomys millardi*	724

梵鼠	726	*Niviventer brahma*	726
台湾社鼠	728	*Niviventer culturatus*	728
斯氏鼢鼠	730	*Eospalax smithii*	730
林睡鼠	732	*Dryomys nitedula*	732
灰颈鼠兔	734	*Ochotona forresti*	734
灰鼠兔	736	*Ochotona roylei*	736
红鼠兔	738	*Ochotona rutila*	738
狭颅鼠兔	740	*Ochotona thomasi*	740
云南兔	742	*Lepus comus*	742
塔里木兔	744	*Lepus yarkandensis*	744

无危 / LC

中国毛猬	746	*Hylomys suillus*	746
中国鼩猬	748	*Neotetracus sinensis*	748
东北刺猬	750	*Erinaceus amurensis*	750
大耳猬	752	*Hemiechinus auritus*	752
达乌尔猬	754	*Mesechinus dauuricus*	754
林猬	756	*Mesechinus sylvaticus*	756
长吻鼩鼹	758	*Uropsilus gracilis*	758
鼩鼹	760	*Uropsilus soricipes*	760
长尾鼩鼹	762	*Scaptonyx fusicaudus*	762
长吻鼹	764	*Euroscaptor longirostris*	764
华南缺齿鼹	766	*Mogera insularis*	766
缺齿鼹	768	*Mogera robusta*	768
小纹背鼩鼱	770	*Sorex bedfordiae*	770
云南鼩鼱	772	*Sorex excelsus*	772
细鼩鼱	774	*Sorex gracillimus*	774
淡灰黑齿鼩鼱	776	*Blarinella griselda*	776
川鼩	778	*Blarinella quadraticauda*	778
长尾鼩	780	*Episoriculus caudatus*	780
台湾长尾鼩	782	*Episoriculus fumidus*	782
大长尾鼩鼱	784	*Episoriculus leucops*	784
缅甸长尾鼩	786	*Episoriculus macrurus*	786
霍氏缺齿鼩	788	*Chodsigoa hoffmanni*	788
川西缺齿鼩鼱	790	*Chodsigoa hypsibia*	790
云南缺齿鼩鼱	792	*Chodsigoa parca*	792
斯氏缺齿鼩鼱	794	*Chodsigoa smithii*	794

中文名	页码	学名	页码
微尾鼩	796	Anourosorex squamipes	796
蹼足鼩	798	Nectogale elegans	798
臭鼩	800	Suncus murinus	800
灰麝鼩	802	Crocidura attenuata	802
白尾梢麝鼩	804	Crocidura fuliginosa	804
白齿麝鼩	806	Crocidura leucodon	806
山东小麝鼩	808	Crocidura shantungensis	808
台湾长尾麝鼩	810	Crocidura tanakae	810
北树鼩	812	Tupaia belangeri	812
黑髯墓蝠	814	Taphozous melanopogon	814
中菊头蝠	816	Rhinolophus affinis	816
马铁菊头蝠	818	Rhinolophus ferrumequinum	818
大耳菊头蝠	820	Rhinolophus macrotis	820
皮氏菊头蝠	822	Rhinolophus pearsoni	822
小菊头蝠	824	Rhinolophus pusillus	824
中华菊头蝠	826	Rhinolophus sinicus	826
大蹄蝠	828	Hipposideros armiger	828
中蹄蝠	830	Hipposideros larvatus	830
小蹄蝠	832	Hipposideros pomona	832
小犬吻蝠	834	Chaerephon plicatus	834
沼泽鼠耳蝠	836	Myotis dasycneme	836
大卫鼠耳蝠	838	Myotis davidii	838
霍氏鼠耳蝠	840	Myotis horsfieldii	840
伊氏鼠耳蝠	842	Myotis ikonnikovi	842
华南水鼠耳蝠	844	Myotis laniger	844
长指鼠耳蝠	846	Myotis longipes	846
大趾鼠耳蝠	848	Myotis macrodactylus	848
山地鼠耳蝠	850	Myotis montivagus	850
大足鼠耳蝠	852	Myotis pilosus	852
东亚伏翼	854	Pipistrellus abramus	854
锡兰伏翼	856	Pipistrellus ceylonicus	856
印度伏翼	858	Pipistrellus coromandra	858
棒茎伏翼	860	Pipistrellus paterculus	860
普通伏翼	862	Pipistrellus pipistrellus	862
茶褐伏翼	864	Hypsugo affinis	864
双色蝙蝠	866	Vespertilio murinus	866
东方蝙蝠	868	Vespertilio sinensis	868

北棕蝠	870	Eptesicus nilssonii	870
肥耳棕蝠	872	Eptesicus pachyotis	872
大棕蝠	874	Eptesicus serotinus	874
中华山蝠	876	Nyctalus plancyi	876
华南扁颅蝠	878	Tylonycteris fulvidus	878
大黄蝠	880	Scotophilus heathii	880
小黄蝠	882	Scotophilus kuhlii	882
斑蝠	884	Scotomanes ornatus	884
大耳蝠	886	Plecotus auritus	886
黄胸管鼻蝠	888	Murina bicolor	888
东北管鼻蝠	890	Murina hilgendorfi	890
中管鼻蝠	892	Murina huttoni	892
白腹管鼻蝠	894	Murina leucogaster	894
泰坦尼亚彩蝠	896	Kerivoula titania	896
猕猴	898	Macaca mulatta	898
黄鼬	900	Mustela sibirica	900
灰獴	902	Herpestes edwardsii	902
野猪	904	Sus scrofa	904
印度麂	906	Muntiacus muntjak	906
岩羊	908	Pseudois nayaur	908
灰鲸	910	Eschrichtius robustus	910
小须鲸	912	Balaenoptera acutorostrata	912
布氏鲸	914	Balaenoptera edeni	914
大翅鲸	916	Megaptera novaeangliae	916
鹅喙鲸	918	Ziphius cavirostris	918
糙齿海豚	920	Steno bredanensis	920
热带点斑原海豚	922	Stenella attenuata	922
条纹原海豚	924	Stenella coeruleoalba	924
真海豚	926	Delphinus delphis	926
瓶鼻海豚	928	Tursiops truncatus	928
弗氏海豚	930	Lagenodelphis hosei	930
里氏海豚	932	Grampus griseus	932
太平洋斑纹海豚	934	Lagenorhynchus obliquidens	934
瓜头鲸	936	Peponocephala electra	936
金背松鼠	938	Callosciurus caniceps	938
赤腹松鼠	940	Callosciurus erythraeus	940
印支松鼠	942	Callosciurus inornatus	942

中文名	页码	学名	页码
黄足松鼠	944	*Callosciurus phayrei*	944
蓝腹松鼠	946	*Callosciurus pygerythrus*	946
明纹花松鼠	948	*Tamiops macclellandii*	948
倭花鼠	950	*Tamiops maritimus*	950
隐纹花松鼠	952	*Tamiops swinhoei*	952
珀氏长吻松鼠	954	*Dremomys pernyi*	954
红颊长吻松鼠	956	*Dremomys rufigenis*	956
条纹松鼠	958	*Menetes berdmorei*	958
岩松鼠	960	*Sciurotamias davidianus*	960
侧纹岩松鼠	962	*Rupestes forresti*	962
北花松鼠	964	*Tamias sibiricus*	964
阿拉善黄鼠	966	*Spermophilus alashanicus*	966
短尾黄鼠	968	*Spermophilus brevicauda*	968
达乌尔黄鼠	970	*Spermophilus dauricus*	970
淡尾黄鼠	972	*Spermophilus pallidicauda*	972
长尾黄鼠	974	*Spermophilus parryii*	974
天山黄鼠	976	*Spermophilus relictus*	976
灰旱獭	978	*Marmota baibacina*	978
长尾旱獭	980	*Marmota caudata*	980
喜马拉雅旱獭	982	*Marmota himalayana*	982
西伯利亚旱獭	984	*Marmota sibirica*	984
毛耳飞鼠	986	*Belomys pearsonii*	986
红白鼯鼠	988	*Petaurista alborufus*	988
灰头小鼯鼠	990	*Petaurista caniceps*	990
霜背大鼯鼠	992	*Petaurista philippensis*	992
白斑小鼯鼠	994	*Petaurista punctatus*	994
橙色小鼯鼠	996	*Petaurista sybilla*	996
灰鼯鼠	998	*Petaurista xanthotis*	998
海南小飞鼠	1000	*Hylopetes phayrei*	1000
黑线仓鼠	1002	*Cricetulus barabensis*	1002
长尾仓鼠	1004	*Cricetulus longicaudatus*	1004
灰仓鼠	1006	*Cricetulus migratorius*	1006
大仓鼠	1008	*Tscherskia triton*	1008
无斑短尾仓鼠	1010	*Allocricetulus curtatus*	1010
短尾仓鼠	1012	*Allocricetulus eversmanni*	1012
小毛足鼠	1014	*Phodopus roborovskii*	1014
鼹形田鼠	1016	*Ellobius talpinus*	1016

灰棕背䶄	1018	*Myodes centralis*	1018
天山林䶄	1020	*Myodes frater*	1020
棕背䶄	1022	*Myodes rufocanus*	1022
红背䶄	1024	*Myodes rutilus*	1024
西南绒鼠	1026	*Eothenomys custos*	1026
滇绒鼠	1028	*Eothenomys eleusis*	1028
黑腹绒鼠	1030	*Eothenomys melanogaster*	1030
大绒鼠	1032	*Eothenomys miletus*	1032

各 论（中册）
Species Monograph (II)

侯氏猬
Mesechinus hughi

近危 NT

| 数据缺乏 DD | 无危 LC | **近危 NT** | 易危 VU | 濒危 EN | 极危 CR | 区域灭绝 RE | 野外灭绝 EW | 灭绝 EX |

分类地位 Taxonomic Status

动物界 Animalia	脊索动物门 Chordata	哺乳纲 Mammalia	劳亚食虫目 Eulipotyphla	猬科 Erinaceidae
学 名 Scientific Name		*Mesechinus hughi*		
命 名 人 Species Authority		Thomas, 1908		
英 文 名 English Name(s)		Hugh's Hedgehog		
同物异名 Synonym(s)		Central Chinese Hedgehog		
种下单元评估 Infra-specific Taxa Assessed		无 / None		

评估信息 Assessment Information

评估年份 Year Assessed	2020
评 定 人 Assessor(s)	蒋志刚 / Zhigang Jiang
审 定 人 Reviewer(s)	蒋学龙、冯祚建 / Xuelong Jiang, Zuojian Feng
其他贡献人 Other Contributor(s)	李立立、丁晨晨 / Lili Li, Chenchen Ding

理由 Justification: 侯氏猬分布虽广，但其数量较少，且由于人类活动与开发利用，其占有区面积缩小。因此，列为近危等级 / Hugh's Hedgehog is widely distributed, but its population size is small and the areas of its occupancy are shrinking due to human activities and overuse. Thus, it is listed as Near Threatened

地理分布 Geographical Distribution

国内分布 Domestic Distribution
陕西、山西、甘肃、湖北、四川、重庆 / Shaanxi, Shanxi, Gansu, Hubei, Sichuan, Chongqing
世界分布 World Distribution
中国 / China
分布标注 Distribution Note
特有种 / Endemic

国内分布图
Map of Domestic Distribution

种群 Population

种群数量 Population Size	未知 / Unknown
种群趋势 Population Trend	下降 / Decreasing

生境与生态系统 Habitat (s) and Ecosystem (s)

生　　境 Habitat(s)	落叶阔叶林 / Deciduous Broad-leaved Forest
生态系统 Ecosystem(s)	森林生态系统 / Forest Ecosystem

威胁 Threat (s)

主要威胁 Major Threat(s)	狩猎及采集陆生植物 / Hunting and Gathering Terrestrial Plant

保护级别与保护行动 Protection Category and Conservation Action (s)

国家重点保护野生动物等级 (2021) Category of National Key Protected Wild Animals (2021)	无 / NA
IUCN 红色名录 (2020-2) IUCN Red List (2020-2)	无危 / LC
CITES 附录 (2019) CITES Appendix (2019)	无 / NA
保护行动 Conservation Action(s)	宣传教育、生境及栖息地保护 / Communication and education, habitat protection

相关文献 Relevant References

Burgin *et al*., 2018; Jiang *et al*. (蒋志刚等), 2017; Qin *et al*. (秦岭等), 2007; Wilson and Reeder, 2005, 1993; Corbet, 1988

侯氏猬 *Mesechinus hughi*　　　　刘少英 摄　By Shaoying Liu

等齿鼩鼹
Uropsilus aequodonenia

近危 NT

| 数据缺乏 DD | 无危 LC | **近危 NT** | 易危 VU | 濒危 EN | 极危 CR | 区域灭绝 RE | 野外灭绝 EW | 灭绝 EX |

分类地位 Taxonomic Status

动物界 Animalia	脊索动物门 Chordata	哺乳纲 Mammalia	劳亚食虫目 Eulipotyphla	鼹科 Talpidae
学名 Scientific Name		*Uropsilus aequodonenia*		
命名人 Species Authority		Liu, 2013		
英文名 English Name(s)		Equivalent Teeth Shrew Mole		
同物异名 Synonym(s)		无 / None		
种下单元评估 Infra-specific Taxa Assessed		无 / None		

评估信息 Assessment Information

评估年份 Year Assessed	2020
评定人 Assessor(s)	蒋志刚 / Zhigang Jiang
审定人 Reviewer(s)	蒋学龙、冯祚建 / Xuelong Jiang, Zuojian Feng
其他贡献人 Other Contributor(s)	李立立、丁晨晨 / Lili Li, Chenchen Ding

理由 Justification: 等齿鼩鼹是2013年在四川发现的新种，采得11具标本，虽然该种大部分布在自然保护区内，种群较稳定，但其分布狭窄，数量极少（刘洋等，2013a）。因此，列为近危等级 / Equivalent Teeth Shrew Mole is a new species discovered in Sichuan Province in 2013, with 11 specimens have been collected. Although most of its populations are located in nature reserves and are relatively stable, it is found only in a narrow range and its population is small (Liu *et al.*, 2013a). Thus, it is listed as Near Threatened

地理分布 Geographical Distribution

国内分布 Domestic Distribution
四川 / Sichuan
世界分布 World Distribution
中国 / China
分布标注 Distribution Note
特有种 / Endemic

国内分布图
Map of Domestic Distribution

🦌 种群 Population

种群数量 Population Size	未知 / Unknown
种群趋势 Population Trend	未知 / Unknown

🦌 生境与生态系统 Habitat (s) and Ecosystem (s)

生　　境 Habitat(s)	针叶林、灌丛 / Coniferous Forest, Shrubland
生态系统 Ecosystem(s)	森林生态系统、灌丛生态系统 / Forest Ecosystem, Shrubland Ecosystem

🦌 威胁 Threat (s)

主要威胁 Major Threat(s)	未知 / Unknown

🦌 保护级别与保护行动 Protection Category and Conservation Action (s)

国家重点保护野生动物等级 (2021) Category of National Key Protected Wild Animals (2021)	无 / NA
IUCN 红色名录 (2020-2) IUCN Red List (2020-2)	未列入 / NA
CITES 附录 (2019) CITES Appendix (2019)	无 / NA
保护行动 Conservation Action(s)	位于自然保护区内的种群与生境得到保护 / Populations and habitats in nature reserves are protected

🦌 相关文献 Relevant References

Burgin *et al*., 2018; Wilson and Mittermeier, 2018; Jiang *et al*.（蒋志刚等），2017; Liu *et al*.（刘洋等），2013a

等齿鼩鼹 *Uropsilus aequodonenia*

贡山鼩鼹
Uropsilus investigator
近危 NT

| 数据缺乏 DD | 无危 LC | **近危 NT** | 易危 VU | 濒危 EN | 极危 CR | 区域灭绝 RE | 野外灭绝 EW | 灭绝 EX |

分类地位 Taxonomic Status

动物界 Animalia	脊索动物门 Chordata	哺乳纲 Mammalia	劳亚食虫目 Eulipotyphla	鼹科 Talpidae

学名 Scientific Name	*Uropsilus investigator*
命名人 Species Authority	Thomas, 1922
英文名 English Name(s)	Inquisitive Shrew Mole
同物异名 Synonym(s)	无 / None
种下单元评估 Infra-specific Taxa Assessed	无 / None

评估信息 Assessment Information

评估年份 Year Assessed	2020
评定人 Assessor(s)	蒋志刚 / Zhigang Jiang
审定人 Reviewer(s)	蒋学龙、冯祚建 / Xuelong Jiang, Zuojian Feng
其他贡献人 Other Contributor(s)	李立立、丁晨晨 / Lili Li, Chenchen Ding

理由 Justification: 贡山鼩鼹为中国特有种，仅分布在中国云南高黎贡山。虽然贡山鼩鼹有较大的种群数量，但其分布非常狭窄。因此，列为近危等级 / Inquisitive Shrew Mole is an endemic species in China which distributes only in Gaoligong Shan, in Yunnan Province. Although its populations are relatively large, its area of occupancy is quite small. Thus, it is listed as Near Threatened

地理分布 Geographical Distribution

国内分布 Domestic Distribution
云南 / Yunnan
世界分布 World Distribution
中国 / China
分布标注 Distribution Note
特有种 / Endemic

国内分布图
Map of Domestic Distribution

种群 Population

种群数量 Population Size	未知 / Unknown
种群趋势 Population Trend	稳定 / Stable

生境与生态系统 Habitat (s) and Ecosystem (s)

生　　境 Habitat(s)	亚热带高海拔森林、灌丛 / Subtropical High Altitude Forest, Shrubland
生态系统 Ecosystem(s)	森林生态系统、灌丛生态系统 / Forest Ecosystem, Shrubland Ecosystem

威胁 Threat (s)

主要威胁 Major Threat(s)	未知 / Unknown

保护级别与保护行动 Protection Category and Conservation Action (s)

国家重点保护野生动物等级 (2021) Category of National Key Protected Wild Animals (2021)	无 / NA
IUCN 红色名录 (2020-2) IUCN Red List (2020-2)	数据缺乏 / DD
CITES 附录 (2019) CITES Appendix (2019)	无 / NA
保护行动 Conservation Action(s)	位于自然保护区内的种群与生境得到保护 / Populations and habitats in nature reserves are protected

相关文献 Relevant References

Burgin *et al.*, 2018; Wilson and Mittermeier, 2018; Jiang *et al.* (蒋志刚等), 2017; Smith *et al.* (史密斯等), 2009; Wang (王应祥), 2003; Hoffmann, 1984

贡山鼩鼱 *Uropsilus investigator*

甘肃鼹
Scapanulus oweni

近危 NT

| 数据缺乏 DD | 无危 LC | 近危 NT | 易危 VU | 濒危 EN | 极危 CR | 区域灭绝 RE | 野外灭绝 EW | 灭绝 EX |

分类地位 Taxonomic Status

动物界 Animalia	脊索动物门 Chordata	哺乳纲 Mammalia	劳亚食虫目 Eulipotyphla	鼹科 Talpidae
学　　名 Scientific Name		*Scapanulus oweni*		
命 名 人 Species Authority		Thomas, 1912		
英 文 名 English Name(s)		Gansu Mole		
同物异名 Synonym(s)		无 / None		
种下单元评估 Infra-specific Taxa Assessed		无 / None		

评估信息 Assessment Information

评 估 年 份 Year Assessed	2020
评 定 人 Assessor(s)	蒋志刚 / Zhigang Jiang
审 定 人 Reviewer(s)	蒋学龙、冯祚建 / Xuelong Jiang, Zuojian Feng
其他贡献人 Other Contributor(s)	李立立、丁晨晨 / Lili Li, Chenchen Ding

理由 Justification: 甘肃鼹分布广，但数量稀少，目前为止所采到的甘肃鼹标本少于 30 个（刘少英，个人通信）。甘肃鼹部分种群分布在自然保护区内，得到一定保护。因此，列为近危等级 / Gansu Mole is a widespread species, but its population size is small. So far, only 30 specimens have been collected (Shaoying Liu, Personal communications). Some populations are located in nature reserves and are protected. Thus, it is listed as Near Threatened

地理分布 Geographical Distribution

国内分布 Domestic Distribution
陕西、宁夏、甘肃、湖北、青海、重庆、四川 / Shaanxi, Ningxia, Gansu, Hubei, Qinghai, Chongqing, Sichuan
世界分布 World Distribution
中国 / China
分布标注 Distribution Note
特有种 / Endemic

国内分布图
Map of Domestic Distribution

种群 Population

种群数量 Population Size	未知 / Unknown
种群趋势 Population Trend	未知 / Unknown

生境与生态系统 Habitat(s) and Ecosystem(s)

生境 Habitat(s)	亚热带湿润山地森林 / Subtropical Moist Montane Forest
生态系统 Ecosystem(s)	森林生态系统 / Forest Ecosystem

威胁 Threat(s)

主要威胁 Major Threat(s)	未知 / Unknown

保护级别与保护行动 Protection Category and Conservation Action(s)

国家重点保护野生动物等级 (2021) Category of National Key Protected Wild Animals (2021)	无 / NA
IUCN 红色名录 (2020-2) IUCN Red List (2020-2)	无危 / LC
CITES 附录 (2019) CITES Appendix (2019)	无 / NA
保护行动 Conservation Action(s)	位于自然保护区内的种群与生境得到保护 / Populations and habitats in nature reserves are protected

相关文献 Relevant References

Burgin *et al*., 2018; Wilson and Mittermeier, 2018; Jiang *et al*. (蒋志刚等), 2017; Wang and Hu (王酉之和胡锦矗), 1999; Wang (汪松), 1998; Stone, 1995; Northwest Institute of Plateau Biology, Chinese Academy of Sciences (中国科学院西北高原生物研究所), 1989

甘肃鼹 *Scapanulus oweni*

小缺齿鼹
Mogera wogura

近危 NT

| 数据缺乏 DD | 无危 LC | 近危 NT | 易危 VU | 濒危 EN | 极危 CR | 区域灭绝 RE | 野外灭绝 EW | 灭绝 EX |

分类地位 Taxonomic Status

动物界 Animalia	脊索动物门 Chordata	哺乳纲 Mammalia	劳亚食虫目 Eulipotyphla	鼹科 Talpidae
学 名 Scientific Name		*Mogera wogura*		
命 名 人 Species Authority		Temminck, 1842		
英 文 名 English Name(s)		Large Japanese Mole		
同物异名 Synonym(s)		Japanese Mole; Temminck's Mole; *Mogera kobeae* (Thomas, 1905)		
种下单元评估 Infra-specific Taxa Assessed		无 / None		

评估信息 Assessment Information

评 估 年 份 Year Assessed	2020
评 定 人 Assessor(s)	蒋志刚 / Zhigang Jiang
审 定 人 Reviewer(s)	蒋学龙、冯祚建 / Xuelong Jiang, Zuojian Feng
其他贡献人 Other Contributor(s)	李立立、丁晨晨 / Lili Li, Chenchen Ding

理由 Justification: 小缺齿鼹分布狭窄，虽然部分种群分布在自然保护区内，但种群数量稀少。因此，列为近危等级 / Large Japanese Mole is narrowly distributed. Although some populations are located in nature reserves, its population size is small. Thus, it is listed as Near Threatened

地理分布 Geographical Distribution

国内分布 Domestic Distribution
辽宁、河南、安徽 / Liaoning, Henan, Anhui
世界分布 World Distribution
中国、日本 / China, Japan
分布标注 Distribution Note
非特有种 / Non-endemic

国内分布图
Map of Domestic Distribution

种群 Population

种群数量 Population Size	未知 / Unknown
种群趋势 Population Trend	稳定 / Stable

生境与生态系统 Habitat (s) and Ecosystem (s)

生境 Habitat(s)	草地、耕地、森林、冲积平原 / Grassland, Arable Land, Forest, Alluvial Plain
生态系统 Ecosystem(s)	森林生态系统、草地生态系统、农田生态系统、湿地生态系统 / Forest Ecosystem, Grassland Ecosystem, Cropland Ecosystem, Wetland Ecosystem

威胁 Threat (s)

主要威胁 Major Threat(s)	未知 / Unknown

保护级别与保护行动 Protection Category and Conservation Action (s)

国家重点保护野生动物等级 (2021) Category of National Key Protected Wild Animals (2021)	无 / NA
IUCN 红色名录 (2020-2) IUCN Red List (2020-2)	无危 / LC
CITES 附录 (2019) CITES Appendix (2019)	无 / NA
保护行动 Conservation Action(s)	位于自然保护区内的种群与生境得到保护 / Populations and habitats in nature reserves are protected

相关文献 Relevant References

Burgin *et al.*, 2018; Wilson and Mittermeier, 2018; Jiang *et al.*（蒋志刚等）, 2017; Duan *et al.*（段海生等）, 2011; Wang（王应祥）, 2003; Zhang（张荣祖）, 1997

小缺齿鼹 *Mogera wogura*

白尾鼹
Parascaptor leucura

近危 NT

| 数据缺乏 DD | 无危 LC | 近危 NT | 易危 VU | 濒危 EN | 极危 CR | 区域灭绝 RE | 野外灭绝 EW | 灭绝 EX |

分类地位 Taxonomic Status

动物界 Animalia	脊索动物门 Chordata	哺乳纲 Mammalia	劳亚食虫目 Eulipotyphla	鼹科 Talpidae
学名 Scientific Name		*Parascaptor leucura*		
命名人 Species Authority		Blyth, 1850		
英文名 English Name(s)		White-tailed Mole		
同物异名 Synonym(s)		无 / None		
种下单元评估 Infra-specific Taxa Assessed		无 / None		

评估信息 Assessment Information

评估年份 Year Assessed	2020
评定人 Assessor(s)	蒋志刚 / Zhigang Jiang
审定人 Reviewer(s)	蒋学龙、冯祚建 / Xuelong Jiang, Zuojian Feng
其他贡献人 Other Contributor(s)	李立立、丁晨晨 / Lili Li, Chenchen Ding

理由 Justification: 白尾鼹分布较广，但其种群数量较少。总的采集到的标本数少于 20 具（刘少英，个人通信）。其生存环境受到人为开发的影响。因此，列为近危等级 / Although White-tailed Mole is widely distributed, its population sizes are small. The total number of its specimens collected are fewer than 20 (Shaoying Liu, Personal communications). Its habitats have been influenced by human's development. Thus, it is listed as Near Threatened

地理分布 Geographical Distribution

国内分布 Domestic Distribution
四川、云南 / Sichuan, Yunnan
世界分布 World Distribution
孟加拉国、中国、印度、缅甸 / Bangladesh, China, India, Myanmar
分布标注 Distribution Note
非特有种 / Non-endemic

国内分布图
Map of Domestic Distribution

种群 Population

种群数量 Population Size	未知 / Unknown
种群趋势 Population Trend	未知 / Unknown

生境与生态系统 Habitat (s) and Ecosystem (s)

生　　境 Habitat(s)	亚热带湿润山地森林、灌丛、草地 / Subtropical Moist Montane Forest, Shrubland, Grassland
生态系统 Ecosystem(s)	森林生态系统、灌丛生态系统、草地生态系统 / Forest Ecosystem, Shrubland Ecosystem, Grassland Ecosystem

威胁 Threat (s)

主要威胁 Major Threat(s)	伐木、耕种、农业或林业污染 / Logging, Plantation, Agricultural or Forestry Effluent

保护级别与保护行动 Protection Category and Conservation Action (s)

国家重点保护野生动物等级 (2021) Category of National Key Protected Wild Animals (2021)	无 / NA
IUCN 红色名录(2020-2) IUCN Red List (2020-2)	无危 / LC
CITES 附录 (2019) CITES Appendix (2019)	无 / NA
保护行动 Conservation Action(s)	无 / None

相关文献 Relevant References

Burgin *et al.*, 2018; Wilson and Mittermeier, 2018; Jiang *et al.* (蒋志刚等), 2017; Duan *et al.* (段海生等), 2011; Smith *et al.* (史密斯等), 2009; Qin *et al.* (秦岭等), 2007; Hutterer, 2005

白尾鼹 *Parascaptor leucura*

麝鼹
Scaptochirus moschatus

近危 NT

| 数据缺乏 DD | 无危 LC | 近危 NT | 易危 VU | 濒危 EN | 极危 CR | 区域灭绝 RE | 野外灭绝 EW | 灭绝 EX |

分类地位 Taxonomic Status

动物界 Animalia	脊索动物门 Chordata	哺乳纲 Mammalia	劳亚食虫目 Eulipotyphla	鼹科 Talpidae
学名 Scientific Name		*Scaptochirus moschatus*		
命名人 Species Authority		Milne-Edwards, 1867		
英文名 English Name(s)		Short-faced Mole		
同物异名 Synonym(s)		无 / None		
种下单元评估 Infra-specific Taxa Assessed		无 / None		

评估信息 Assessment Information

评估年份 Year Assessed	2020
评定人 Assessor(s)	蒋志刚 / Zhigang Jiang
审定人 Reviewer(s)	蒋学龙、冯祚建 / Xuelong Jiang, Zuojian Feng
其他贡献人 Other Contributor(s)	李立立、丁晨晨 / Lili Li, Chenchen Ding

理由 Justification: 麝鼹分布广，但种群数量少，且栖息地遭到破坏。因此，列为近危等级 / Short-faced Mole is widely distributed, but its population size is small. Additionally, the species is threatened by habitat loss. Thus, it is listed as Near Threatened

地理分布 Geographical Distribution

国内分布 Domestic Distribution

黑龙江、辽宁、内蒙古、北京、甘肃、河北、河南、江苏、宁夏、山西、陕西、山东 / Heilongjiang, Liaoning, Inner Mongolia (Nei Mongol), Beijing, Gansu, Hebei, Henan, Jiangsu, Ningxia, Shanxi, Shaanxi, Shandong

世界分布 World Distribution

中国 / China

分布标注 Distribution Note

特有种 / Endemic

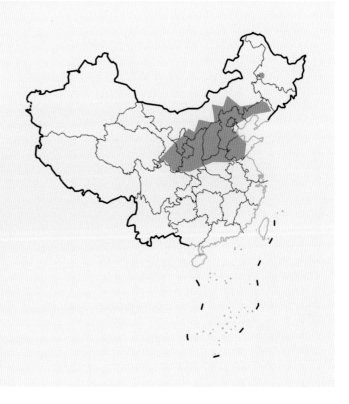

国内分布图 Map of Domestic Distribution

种群 Population

种群数量 Population Size	未知 / Unknown
种群趋势 Population Trend	未知 / Unknown

生境与生态系统 Habitat (s) and Ecosystem (s)

生　　境 Habitat(s)	亚热带干旱低地草地 / Subtropical Dry Lowland Grassland
生态系统 Ecosystem(s)	草地生态系统 / Grassland Ecosystem

威胁 Threat (s)

主要威胁 Major Threat(s)	未知 / Unknown

保护级别与保护行动 Protection Category and Conservation Action (s)

国家重点保护野生动物等级 (2021) Category of National Key Protected Wild Animals (2021)	无 / NA
IUCN 红色名录 (2020-2) IUCN Red List (2020-2)	无危 / LC
CITES 附录 (2019) CITES Appendix (2019)	无 / NA
保护行动 Conservation Action(s)	位于自然保护区内的种群与生境得到保护 / Populations and habitats in nature reserves are protected

相关文献 Relevant References

Burgin *et al*., 2018; Wilson and Mittermeier, 2018; Jiang *et al*. (蒋志刚等), 2017; Smith *et al*. (史密斯等), 2009; Motokawa, 2004; Wang and Xie (汪松和解焱), 2004; Kawada *et al*., 2002

麝鼹 *Scaptochirus moschatus*

天山鼩鼱
Sorex asper

近危 NT

| 数据缺乏 DD | 无危 LC | 近危 NT | 易危 VU | 濒危 EN | 极危 CR | 区域灭绝 RE | 野外灭绝 EW | 灭绝 EX |

分类地位 Taxonomic Status

动物界 Animalia	脊索动物门 Chordata	哺乳纲 Mammalia	劳亚食虫目 Eulipotyphla	鼩鼱科 Soricidae
学名 Scientific Name		*Sorex asper*		
命名人 Species Authority		Thomas, 1914		
英文名 English Name(s)		Tien Shan Shrew		
同物异名 Synonym(s)		无 / None		
种下单元评估 Infra-specific Taxa Assessed		无 / None		

评估信息 Assessment Information

评估年份 Year Assessed	2020
评定人 Assessor(s)	蒋志刚 / Zhigang Jiang
审定人 Reviewer(s)	蒋学龙、冯祚建 / Xuelong Jiang, Zuojian Feng
其他贡献人 Other Contributor(s)	李立立、丁晨晨 / Lili Li, Chenchen Ding

理由 Justification: 天山鼩鼱仅分布在新疆,分布狭窄,数量较少。因此,列为近危等级 / Tien Shan Shrew has a small population and is narrowly distributed only in Xinjiang. Thus, it is listed as Near Threatened

地理分布 Geographical Distribution

国内分布 Domestic Distribution
新疆 / Xinjiang
世界分布 World Distribution
中国、哈萨克斯坦、吉尔吉斯斯坦 / China, Kazakhstan, Kyrgyzstan
分布标注 Distribution Note
非特有种 / Non-endemic

国内分布图
Map of Domestic Distribution

🦌 种群 Population

种群数量 Population Size	未知 / Unknown
种群趋势 Population Trend	未知 / Unknown

🦌 生境与生态系统 Habitat (s) and Ecosystem (s)

生　　境 Habitat(s)	高山草甸、灌丛、泰加林 / Alpine Meadow, Shrubland, Taiga Forest
生态系统 Ecosystem(s)	森林生态系统、灌丛生态系统、草地生态系统 / Forest Ecosystem, Shrubland Ecosystem, Grassland Ecosystem

🦌 威胁 Threat (s)

主要威胁 Major Threat(s)	未知 / Unknown

🦌 保护级别与保护行动 Protection Category and Conservation Action (s)

国家重点保护野生动物等级 (2021) Category of National Key Protected Wild Animals (2021)	无 / NA
IUCN 红色名录 (2020-2) IUCN Red List (2020-2)	无危 / LC
CITES 附录 (2019) CITES Appendix (2019)	无 / NA
保护行动 Conservation Action(s)	无 / None

🦌 相关文献 Relevant References

Burgin *et al.*, 2018; Wilson and Mittermeier, 2018; Jiang *et al.* (蒋志刚等), 2017; Smith *et al.* (史密斯等), 2009; Fumagalli *et al.*, 1999

天山䶄鼩 *Sorex asper*

中鼩鼱
Sorex caecutiens

近危 NT

| 数据缺乏 DD | 无危 LC | 近危 NT | 易危 VU | 濒危 EN | 极危 CR | 区域灭绝 RE | 野外灭绝 EW | 灭绝 EX |

分类地位 Taxonomic Status

动物界 Animalia	脊索动物门 Chordata	哺乳纲 Mammalia	劳亚食虫目 Eulipotyphla	鼩鼱科 Soricidae

学 名 Scientific Name	*Sorex caecutiens*
命 名 人 Species Authority	Laxmann, 1788
英 文 名 English Name(s)	Laxmann's Shrew
同物异名 Synonym(s)	Masked Shrew
种下单元评估 Infra-specific Taxa Assessed	无 / None

评估信息 Assessment Information

评估年份 Year Assessed	2020
评 定 人 Assessor(s)	蒋志刚 / Zhigang Jiang
审 定 人 Reviewer(s)	蒋学龙、冯祚建 / Xuelong Jiang, Zuojian Feng
其他贡献人 Other Contributor(s)	李立立、丁晨晨 / Lili Li, Chenchen Ding

理由 Justification: 中鼩鼱分布广，种群数量较少，且其所在栖息地面积减少。因此，列为近危等级 / Laxmann's Shrew is a widespread species with small populations. Its habitats are shrinking. Thus, it is listed as Near Threatened

地理分布 Geographical Distribution

国内分布 Domestic Distribution
黑龙江、吉林、内蒙古 / Heilongjiang, Jilin, Inner Mongolia (Nei Mongol)
世界分布 World Distribution
亚洲、欧洲 / Asia, Europe
分布标注 Distribution Note
非特有种 / Non-endemic

国内分布图
Map of Domestic Distribution

种群 Population

种群数量 Population Size	未知 / Unknown
种群趋势 Population Trend	稳定 / Stable

生境与生态系统 Habitat (s) and Ecosystem (s)

生境 Habitat(s)	泰加林、苔原、沼泽 / Taiga Forest, Tundra, Swamp
生态系统 Ecosystem(s)	森林生态系统、湿地生态系统、荒漠生态系统 /Forest Ecosystem, Wetland Ecosystem, Desert Ecosystem

威胁 Threat (s)

主要威胁 Major Threat(s)	伐木、耕种 / Logging, Plantation

保护级别与保护行动 Protection Category and Conservation Action (s)

国家重点保护野生动物等级 (2021) Category of National Key Protected Wild Animals (2021)	无 / NA
IUCN 红色名录(2020-2) IUCN Red List (2020-2)	无危 / LC
CITES 附录 (2019) CITES Appendix (2019)	无 / NA
保护行动 Conservation Action(s)	位于自然保护区内的种群与生境得到保护 / Populations and habitats in nature reserves are protected

相关文献 Relevant References

Burgin *et al*., 2018; Wilson and Mittermeier, 2018; Jiang *et al*.（蒋志刚等），2017; Abulimiti Abudukader（阿布力米提·阿布都卡迪尔），2002; Zhang（张荣祖），1997

中鼩鼱 *Sorex caecutiens*

甘肃鼩鼱
Sorex cansulus
近危 NT

| 数据缺乏 DD | 无危 LC | 近危 NT | 易危 VU | 濒危 EN | 极危 CR | 区域灭绝 RE | 野外灭绝 EW | 灭绝 EX |

分类地位 Taxonomic Status

动物界 Animalia	脊索动物门 Chordata	哺乳纲 Mammalia	劳亚食虫目 Eulipotyphla	鼩鼱科 Soricidae

学 名 Scientific Name	*Sorex cansulus*
命 名 人 Species Authority	Thomas, 1912
英 文 名 English Name(s)	Gansu Shrew
同物异名 Synonym(s)	无 / None
种下单元评估 Infra-specific Taxa Assessed	无 / None

评估信息 Assessment Information

评 估 年 份 Year Assessed	2020
评 定 人 Assessor(s)	蒋志刚 / Zhigang Jiang
审 定 人 Reviewer(s)	蒋学龙、冯祚建 / Xuelong Jiang, Zuojian Feng
其他贡献人 Other Contributor(s)	李立立、丁晨晨 / Lili Li, Chenchen Ding

理由 Justification: 甘肃鼩鼱为中国特有种，仅分布在甘肃南部，占有区狭窄且种群数量较少，由于伐木和耕作其栖息地遭到破坏。因此，列为近危等级 / Gansu Shrew is an endemic species distributed only in southern Gansu, China. Its range of occupancy is narrow and population is small. It is threatened by habitat loss due to logging and farming. Thus, it is listed as Near Threatened

地理分布 Geographical Distribution

国内分布 Domestic Distribution
甘肃 / Gansu
世界分布 World Distribution
中国 / China
分布标注 Distribution Note
特有种 / Endemic

国内分布图
Map of Domestic Distribution

种群 Population

种群数量 Population Size	未知 / Unknown
种群趋势 Population Trend	未知 / Unknown

生境与生态系统 Habitat (s) and Ecosystem (s)

生　　境 Habitat(s)	森林、灌丛、沼泽 / Forest, Shrubland, Swamp
生态系统 Ecosystem(s)	森林生态系统、灌丛生态系统、湿地生态系统 / Forest Ecosystem, Shrubland Ecosystem, Wetland Ecosystem

威胁 Threat (s)

主要威胁 Major Threat(s)	未知 / Unknown

保护级别与保护行动 Protection Category and Conservation Action (s)

国家重点保护野生动物等级 (2021) Category of National Key Protected Wild Animals (2021)	无 / NA
IUCN红色名录(2020-2) IUCN Red List (2020-2)	数据缺乏 / DD
CITES 附录 (2019) CITES Appendix (2019)	无 / NA
保护行动 Conservation Action(s)	位于自然保护区内的种群与生境得到保护 / Populations and habitats in nature reserves are protected

相关文献 Relevant References

Burgin *et al.*, 2018; Wilson and Mittermeier, 2018; Jiang *et al.* (蒋志刚等), 2017; Smith *et al.* (史密斯等), 2009; Wang (王香亭), 1991; Hoffmann, 1987; Thomas, 1912b

甘肃鼩鼱 *Sorex cansulus*

纹背鼩鼱
Sorex cylindricauda

近危 NT

| 数据缺乏 DD | 无危 LC | 近危 NT | 易危 VU | 濒危 EN | 极危 CR | 区域灭绝 RE | 野外灭绝 EW | 灭绝 EX |

分类地位 Taxonomic Status

动物界 Animalia	脊索动物门 Chordata	哺乳纲 Mammalia	劳亚食虫目 Eulipotyphla	鼩鼱科 Soricidae
学名 Scientific Name		*Sorex cylindricauda*		
命名人 Species Authority		Milne-Edwards, 1872		
英文名 English Name(s)		Stripe-backed Shrew		
同物异名 Synonym(s)		Greater Stripe-backed Shrew		
种下单元评估 Infra-specific Taxa Assessed		无 / None		

评估信息 Assessment Information

评估年份 Year Assessed	2020
评定人 Assessor(s)	蒋志刚 / Zhigang Jiang
审定人 Reviewer(s)	蒋学龙、冯祚建 / Xuelong Jiang, Zuojian Feng
其他贡献人 Other Contributor(s)	李立立、丁晨晨 / Lili Li, Chenchen Ding

理由 Justification: 纹背鼩鼱的分布区较广，但其种群数量较少，原因不明。因此，列为近危等级 / Stripe-backed Shrew has relatively large distribution range, but its population density is low for unknown reasons. Thus, it is listed as Nearly Threatened

地理分布 Geographical Distribution

国内分布 Domestic Distribution
陕西、甘肃、宁夏、四川、云南 / Shaanxi, Gansu, Ningxia, Sichuan, Yunnan
世界分布 World Distribution
中国 / China
分布标注 Distribution Note
特有种 / Endemic

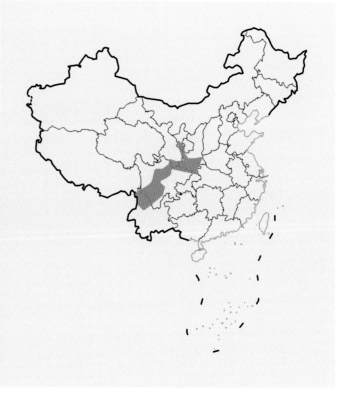

国内分布图
Map of Domestic Distribution

种群 Population

种群数量 Population Size	未知 / Unknown
种群趋势 Population Trend	未知 / Unknown

生境与生态系统 Habitat (s) and Ecosystem (s)

生　　境 Habitat(s)	针阔混交林 / Coniferous and Broad-leaved Mixed Forest
生态系统 Ecosystem(s)	森林生态系统 / Forest Ecosystem

威胁 Threat (s)

主要威胁 Major Threat(s)	未知 / Unknown

保护级别与保护行动 Protection Category and Conservation Action (s)

国家重点保护野生动物等级 (2021) Category of National Key Protected Wild Animals (2021)	无 / NA
IUCN红色名录(2020-2) IUCN Red List (2020-2)	无危 / LC
CITES 附录 (2019) CITES Appendix (2019)	无 / NA
保护行动 Conservation Action(s)	无 / None

相关文献 Relevant References

Burgin *et al*., 2018; Wilson and Mittermeier, 2018; Jiang *et al*. (蒋志刚等), 2017; Tu *et al*. (涂飞云等), 2014; Liu *et al*. (刘少英等), 2005; Zhang (张荣祖), 1997

纹背鼩鼱 *Sorex cylindricauda*

栗齿鼩鼱
Sorex daphaenodon
近危 NT

| 数据缺乏 DD | 无危 LC | 近危 NT | 易危 VU | 濒危 EN | 极危 CR | 区域灭绝 RE | 野外灭绝 EW | 灭绝 EX |

分类地位 Taxonomic Status

动物界 Animalia	脊索动物门 Chordata	哺乳纲 Mammalia	劳亚食虫目 Eulipotyphla	鼩鼱科 Soricidae

学名 Scientific Name	*Sorex daphaenodon*
命名人 Species Authority	Thomas, 1907
英文名 English Name(s)	Large-toothed Siberian Shrew
同物异名 Synonym(s)	Siberian Large-toothed Shrew; *Sorex sanguinidens* (G. Allen, 1914); *scaloni* (Ognev, 1933)
种下单元评估 Infra-specific Taxa Assessed	无 / None

评估信息 Assessment Information

评估年份 Year Assessed	2020
评定人 Assessor(s)	蒋志刚 / Zhigang Jiang
审定人 Reviewer(s)	蒋学龙 / Xuelong Jiang
其他贡献人 Other Contributor(s)	李立立、丁晨晨 / Lili Li, Chenchen Ding

理由 Justification: 栗齿鼩鼱分布区较广，但其种群数量较少。因此，列为近危等级 / Large-toothed Siberian Shrew is widely distributed but its population size is relatively small. Thus, it is listed as Near Threatened

地理分布 Geographical Distribution

国内分布 Domestic Distribution
黑龙江、吉林、内蒙古 / Heilongjiang, Jilin, Inner Mongolia (Nei Mongol)
世界分布 World Distribution
中国、哈萨克斯坦、蒙古、俄罗斯 / China, Kazakhstan, Mongolia, Russia
分布标注 Distribution Note
非特有种 / Non-endemic

国内分布图
Map of Domestic Distribution

🦌 种群 Population

种群数量 Population Size	未知 / Unknown
种群趋势 Population Trend	稳定 / Stable

🦌 生境与生态系统 Habitat (s) and Ecosystem (s)

生 境 Habitat(s)	针阔混交林 / Coniferous and Broad-leaved Mixed Forest
生态系统 Ecosystem(s)	森林生态系统 / Forest Ecosystem

🦌 威胁 Threat (s)

主要威胁 Major Threat(s)	伐木、火灾 / Logging, Fire

🦌 保护级别与保护行动 Protection Category and Conservation Action (s)

国家重点保护野生动物等级 (2021) Category of National Key Protected Wild Animals (2021)	无 / NA
IUCN 红色名录 (2020-2) IUCN Red List (2020-2)	无危 / LC
CITES 附录 (2019) CITES Appendix (2019)	无 / NA
保护行动 Conservation Action(s)	无 / None

🦌 相关文献 Relevant References

Burgin *et al.*, 2018; Wilson and Mittermeier, 2018; Jiang *et al.* (蒋志刚等), 2017; Wilson and Reeder, 1993; Wang (汪松), 1959

栗齿鼩鼱 *Sorex daphaenodon*

远东鼩鼱
Sorex isodon

近危 NT

| 数据缺乏 DD | 无危 LC | 近危 NT | 易危 VU | 濒危 EN | 极危 CR | 区域灭绝 RE | 野外灭绝 EW | 灭绝 EX |

分类地位 Taxonomic Status

动物界 Animalia	脊索动物门 Chordata	哺乳纲 Mammalia	劳亚食虫目 Eulipotyphla	鼩鼱科 Soricidae
学名 Scientific Name		*Sorex isodon*		
命名人 Species Authority		Turov, 1924		
英文名 English Name(s)		Even-toothed Shrew		
同物异名 Synonym(s)		Taiga Shrew		
种下单元评估 Infra-specific Taxa Assessed		无 / None		

评估信息 Assessment Information

评估年份 Year Assessed	2020
评定人 Assessor(s)	蒋志刚 / Zhigang Jiang
审定人 Reviewer(s)	蒋学龙、冯祚建 / Xuelong Jiang, Zuojian Feng
其他贡献人 Other Contributor(s)	李立立、丁晨晨 / Lili Li, Chenchen Ding

理由 Justification: 远东鼩鼱的分布区较狭窄，且种群数量较少。因此，列为近危等级 / Even-toothed Shrew is narrowly distributed with small populations. Thus, it is listed as Near Threatened

地理分布 Geographical Distribution

国内分布 Domestic Distribution

内蒙古、黑龙江 / Inner Mongolia (Nei Mongol), Heilongjiang

世界分布 World Distribution

白俄罗斯、中国、芬兰、哈萨克斯坦、朝鲜、蒙古、挪威、俄罗斯、瑞典 / Belarus, China, Finland, Kazakhstan, Korea (the Democratic People's Republic of), Mongolia, Norway, Russia, Sweden

分布标注 Distribution Note

非特有种 / Non-endemic

国内分布图
Map of Domestic Distribution

种群 Population

种群数量 Population Size	未知 / Unknown
种群趋势 Population Trend	未知 / Unknown

生境与生态系统 Habitat(s) and Ecosystem(s)

生境 Habitat(s)	温带森林 / Temperate Forest
生态系统 Ecosystem(s)	森林生态系统 / Forest Ecosystem

威胁 Threat(s)

主要威胁 Major Threat(s)	农业或林业污染 / Agricultural or Forestry Effluent

保护级别与保护行动 Protection Category and Conservation Action(s)

国家重点保护野生动物等级 (2021) Category of National Key Protected Wild Animals (2021)	无 / NA
IUCN 红色名录 (2020-2) IUCN Red List (2020-2)	无危 / LC
CITES 附录 (2019) CITES Appendix (2019)	无 / NA
保护行动 Conservation Action(s)	无 / None

相关文献 Relevant References

Burgin *et al.*, 2018; Wilson and Mittermeier, 2018; Jiang *et al.* (蒋志刚等), 2017; Zhang and Yu (张杰和于洪伟), 2006; Wilson and Reeder, 1993; Corbet, 1978

远东鼩鼱 *Sorex isodon*

姬鼩鼱
Sorex minutissimus

近危 NT

| 数据缺乏 DD | 无危 LC | 近危 NT | 易危 VU | 濒危 EN | 极危 CR | 区域灭绝 RE | 野外灭绝 EW | 灭绝 EX |

分类地位 Taxonomic Status

动物界 Animalia	脊索动物门 Chordata	哺乳纲 Mammalia	劳亚食虫目 Eulipotyphla	鼩鼱科 Soricidae
学名 Scientific Name		*Sorex minutissimus*		
命名人 Species Authority		Zimmermann, 1780		
英文名 English Name(s)		Eurasian Least Shrew		
同物异名 Synonym(s)		Lesser Pygmy Shrew		
种下单元评估 Infra-specific Taxa Assessed		无 / None		

评估信息 Assessment Information

评估年份 Year Assessed	2020
评定人 Assessor(s)	蒋志刚 / Zhigang Jiang
审定人 Reviewer(s)	蒋学龙、冯祚建 / Xuelong Jiang, Zuojian Feng
其他贡献人 Other Contributor(s)	李立立、丁晨晨 / Lili Li, Chenchen Ding

理由 Justification: 姬鼩鼱的分布范围狭小，且种群数量较少。因此，列为近危等级 / Eurasian Least Shrew is narrowly distributed with small populations. Thus, it is listed as Near Threatened

地理分布 Geographical Distribution

国内分布 Domestic Distribution
黑龙江、四川、云南 / Heilongjiang, Sichuan, Yunnan
世界分布 World Distribution
亚洲、欧洲 / Asia, Europe
分布标注 Distribution Note
非特有种 / Non-endemic

国内分布图
Map of Domestic Distribution

种群 Population

种群数量 Population Size	未知 / Unknown
种群趋势 Population Trend	未知 / Unknown

生境与生态系统 Habitat (s) and Ecosystem (s)

生境 Habitat(s)	森林、灌丛 / Forest, Shrubland
生态系统 Ecosystem(s)	森林生态系统、灌丛生态系统 / Forest Ecosystem, Shrubland Ecosystem

威胁 Threat (s)

主要威胁 Major Threat(s)	未知 / Unknown

保护级别与保护行动 Protection Category and Conservation Action (s)

国家重点保护野生动物等级 (2021) Category of National Key Protected Wild Animals (2021)	无 / NA
IUCN 红色名录 (2020-2) IUCN Red List (2020-2)	无危 / LC
CITES 附录 (2019) CITES Appendix (2019)	无 / NA
保护行动 Conservation Action(s)	无 / None

相关文献 Relevant References

Burgin *et al.*, 2018; Wilson and Mittermeier, 2018; Jiang *et al.*（蒋志刚等）, 2017; Pan *et al.*（潘清华等）, 2007; Wilson and Reeder, 2005, 1993; Wang（王应祥）, 2003

姬鼩鼱 *Sorex minutissimus*

小鼩鼱
Sorex minutus

近危 NT

| 数据缺乏 DD | 无危 LC | 近危 NT | 易危 VU | 濒危 EN | 极危 CR | 区域灭绝 RE | 野外灭绝 EW | 灭绝 EX |

分类地位 Taxonomic Status

动物界 Animalia	脊索动物门 Chordata	哺乳纲 Mammalia	劳亚食虫目 Eulipotyphla	鼩鼱科 Soricidae

学名 Scientific Name	*Sorex minutus*
命名人 Species Authority	Linnaeus, 1766
英文名 English Name(s)	Eurasian Pygmy Shrew
同物异名 Synonym(s)	Pygmy Shrew
种下单元评估 Infra-specific Taxa Assessed	无 / None

评估信息 Assessment Information

评估年份 Year Assessed	2020
评定人 Assessor(s)	蒋志刚 / Zhigang Jiang
审定人 Reviewer(s)	蒋学龙、冯祚建 / Xuelong Jiang, Zuojian Feng
其他贡献人 Other Contributor(s)	李立立、丁晨晨 / Lili Li, Chenchen Ding

理由 Justification: 小鼩鼱分布区较狭窄，且种群数量较少。因此，列为近危等级 / Eurasian Pygmy Shrew is narrowly distributed with small populations. Thus, it is listed as Near Threatened

地理分布 Geographical Distribution

国内分布 Domestic Distribution
新疆 / Xinjiang
世界分布 World Distribution
欧亚大陆 / Eurasian Continent
分布标注 Distribution Note
非特有种 / Non-endemic

国内分布图
Map of Domestic Distribution

种群 Population

种群数量 Population Size	未知 / Unknown
种群趋势 Population Trend	稳定 / Stable

生境与生态系统 Habitat(s) and Ecosystem(s)

生境 Habitat(s)	生活在潮湿的、植被密度高的地区。在沼泽、草地、沙丘、林地边缘、岩石区、灌丛和山地森林都有发现 / Eurasian Pygmy Shrew is found in Damp Area with Dense Vegetation. It occurs in a wide variety of habitat, including Swamp, Grassland, Sand Dune, Woodland Edge, Rocky Area, Shrubland and Montane Forest
生态系统 Ecosystem(s)	森林生态系统、湿地生态系统、灌丛生态系统、草地生态系统、河谷生态系统 / Forest Ecosystem, Wetland Ecosystem, Shrubland Ecosystem, Grassland Ecosystem, Valley Ecosystem

威胁 Threat(s)

主要威胁 Major Threat(s)	农业或林业污染 / Agricultural or Forestry Pollution

保护级别与保护行动 Protection Category and Conservation Action(s)

国家重点保护野生动物等级 (2021) Category of National Key Protected Wild Animals (2021)	无 / NA
IUCN 红色名录 (2020-2) IUCN Red List (2020-2)	无危 / LC
CITES 附录 (2019) CITES Appendix (2019)	无 / NA
保护行动 Conservation Action(s)	无 / None

相关文献 Relevant References

Burgin *et al*., 2018; Wilson and Mittermeier, 2018; Jiang *et al*.（蒋志刚等）, 2017; Pan *et al*.（潘清华等）, 2007; Wang（王应祥）, 2003; Hoffmann, 1987

小鼩鼱 *Sorex minutus*

克什米尔鼩鼱
Sorex planiceps

中国生物多样性红色名录

近危 NT

| 数据缺乏 DD | 无危 LC | 近危 NT | 易危 VU | 濒危 EN | 极危 CR | 区域灭绝 RE | 野外灭绝 EW | 灭绝 EX |

分类地位 Taxonomic Status

动物界 Animalia	脊索动物门 Chordata	哺乳纲 Mammalia	劳亚食虫目 Eulipotyphla	鼩鼱科 Soricidae
学 名 Scientific Name		*Sorex planiceps*		
命 名 人 Species Authority		Miller, 1911		
英 文 名 English Name(s)		Kashmir Shrew		
同物异名 Synonym(s)		Kashmir Pygmy Shrew		
种下单元评估 Infra-specific Taxa Assessed		无 / None		

评估信息 Assessment Information

评估年份 Year Assessed	2020
评 定 人 Assessor(s)	蒋志刚 / Zhigang Jiang
审 定 人 Reviewer(s)	蒋学龙、冯祚建 / Xuelong Jiang, Zuojian Feng
其他贡献人 Other Contributor(s)	李立立、丁晨晨 / Lili Li, Chenchen Ding

理由 Justification: 克什米尔鼩鼱的分布区狭窄，且种群数量较少。因此，列为近危等级 / Kashmir Shrew is narrowly distributed with small populations. Thus, it is listed as Near Threatened

地理分布 Geographical Distribution

国内分布 Domestic Distribution
西藏、新疆 / Tibet (Xizang), Xinjiang
世界分布 World Distribution
中国、印度、巴基斯坦 / China, India, Pakistan
分布标注 Distribution Note
非特有种 / Non-endemic

国内分布图
Map of Domestic Distribution

种群 Population

种群数量 Population Size	未知 / Unknown
种群趋势 Population Trend	未知 / Unknown

生境与生态系统 Habitat(s) and Ecosystem(s)

生　　境 Habitat(s)	泰加林、内陆岩石区域，以无脊椎动物为食 / Taiga Forest, Inland Rocky Area, it feeds on invertebrate
生态系统 Ecosystem(s)	森林生态系统、岩石稀疏植被生态系统 / Forest Ecosystem, Inland Rocky Sparse Vegetation Ecosystem

威胁 Threat(s)

主要威胁 Major Threat(s)	未知 / Unknown

保护级别与保护行动 Protection Category and Conservation Action(s)

国家重点保护野生动物等级 (2021) Category of National Key Protected Wild Animals (2021)	无 / NA
IUCN 红色名录 (2020-2) IUCN Red List (2020-2)	无危 / LC
CITES 附录 (2019) CITES Appendix (2019)	无 / NA
保护行动 Conservation Action(s)	无 / None

相关文献 Relevant References

Burgin *et al.*, 2018; Wilson and Mittermeier, 2018; Jiang *et al.*（蒋志刚等）, 2017; Smith *et al.*（史密斯等）, 2009; Pan *et al.*（潘清华等）, 2007; Hoffmann, 1987

克什米尔鼩鼱 *Sorex planiceps*

陕西鼩鼱
Sorex sinalis
近危 NT

| 数据缺乏 DD | 无危 LC | 近危 NT | 易危 VU | 濒危 EN | 极危 CR | 区域灭绝 RE | 野外灭绝 EW | 灭绝 EX |

分类地位 Taxonomic Status

| 动物界 Animalia | 脊索动物门 Chordata | 哺乳纲 Mammalia | 劳亚食虫目 Eulipotyphla | 鼩鼱科 Soricidae |

学名 Scientific Name	*Sorex sinalis*
命名人 Species Authority	Thomas, 1912
英文名 English Name(s)	Chinese Shrew
同物异名 Synonym(s)	Dusky Shrew
种下单元评估 Infra-specific Taxa Assessed	无 / None

评估信息 Assessment Information

评估年份 Year Assessed	2020
评定人 Assessor(s)	蒋志刚 / Zhigang Jiang
审定人 Reviewer(s)	蒋学龙、冯祚建 / Xuelong Jiang, Zuojian Feng
其他贡献人 Other Contributor(s)	李立立、丁晨晨 / Lili Li, Chenchen Ding

理由 Justification: 陕西鼩鼱分布区较狭窄，且种群数量较少。因此，列为近危等级 / Chinese Shrew is narrowly distributed with small populations. Thus, it is listed as Near Threatened

地理分布 Geographical Distribution

国内分布 Domestic Distribution
陕西、四川、甘肃 / Shaanxi, Sichuan, Gansu
世界分布 World Distribution
中国 / China
分布标注 Distribution Note
特有种 / Endemic

国内分布图
Map of Domestic Distribution

种群 Population

种群数量 Population Size	未知 / Unknown
种群趋势 Population Trend	未知 / Unknown

生境与生态系统 Habitat (s) and Ecosystem (s)

生　　境 Habitat(s)	针阔混交林、温带森林地带的岩石山坡 / Rocky Slope in Coniferous and Broad-leaved Mixed Forest and Temperate Forest
生态系统 Ecosystem(s)	森林生态系统 / Forest Ecosystem

威胁 Threat (s)

主要威胁 Major Threat(s)	未知 / Unknown

保护级别与保护行动 Protection Category and Conservation Action (s)

国家重点保护野生动物等级 (2021) Category of National Key Protected Wild Animals (2021)	无 / NA
IUCN 红色名录 (2020-2) IUCN Red List (2020-2)	数据缺乏 / DD
CITES 附录 (2019) CITES Appendix (2019)	无 / NA
保护行动 Conservation Action(s)	无 / None

相关文献 Relevant References

Burgin *et al*., 2018; Wilson and Mittermeier, 2018; Jiang *et al*. (蒋志刚等), 2017; Jiang *et al*. (姜雪松等), 2013; Liu *et al*. (刘洋等), 2013a; Smith *et al*. (史密斯等), 2009; Qin *et al*. (秦岭等), 2007

陕西鼩鼱 *Sorex sinalis*

藏鼩鼱
Sorex thibetanus
近危 NT

DD	LC	NT	VU	EN	CR	RE	EW	EX
数据缺乏	无危	近危	易危	濒危	极危	区域灭绝	野外灭绝	灭绝

分类地位 Taxonomic Status

动物界 Animalia	脊索动物门 Chordata	哺乳纲 Mammalia	劳亚食虫目 Eulipotyphla	鼩鼱科 Soricidae

学名 Scientific Name	*Sorex thibetanus*
命名人 Species Authority	Kastschenko, 1905
英文名 English Name(s)	Tibetan Shrew
同物异名 Synonym(s)	无 / None
种下单元评估 Infra-specific Taxa Assessed	无 / None

评估信息 Assessment Information

评估年份 Year Assessed	2020
评定人 Assessor(s)	蒋志刚 / Zhigang Jiang
审定人 Reviewer(s)	蒋学龙、冯祚建 / Xuelong Jiang, Zuojian Feng
其他贡献人 Other Contributor(s)	李立立、丁晨晨 / Lili Li, Chenchen Ding

理由 Justification: 藏鼩鼱的分布区小，且种群数量较少。因此，列为近危等级 / Tibetan Shrew is narrowly distributed with small populations in China. Thus, it is listed as Near Threatened

地理分布 Geographical Distribution

国内分布 Domestic Distribution
四川、甘肃、青海、西藏 / Sichuan, Gansu, Qinghai, Tibet (Xizang)
世界分布 World Distribution
中国 / China
分布标注 Distribution Note
特有种 / Endemic

国内分布图
Map of Domestic Distribution

种群 Population

种群数量 Population Size	未知 / Unknown
种群趋势 Population Trend	未知 / Unknown

生境与生态系统 Habitat(s) and Ecosystem(s)

生　　境 Habitat(s)	草地、灌丛、沼泽 / Grassland, Shrubland, Swamp
生态系统 Ecosystem(s)	森林生态系统、湿地生态系统 / Forest Ecosystem, Wetland Ecosystem

威胁 Threat(s)

主要威胁 Major Threat(s)	未知 / Unknown

保护级别与保护行动 Protection Category and Conservation Action(s)

国家重点保护野生动物等级 (2021) Category of National Key Protected Wild Animals (2021)	无 / NA
IUCN 红色名录 (2020-2) IUCN Red List (2020-2)	数据缺乏 / DD
CITES 附录 (2019) CITES Appendix (2019)	无 / NA
保护行动 Conservation Action(s)	无 / None

相关文献 Relevant References

Burgin *et al*., 2018; Wilson and Mittermeier, 2018; Jiang *et al*. (蒋志刚等), 2017; Liao *et al*. (廖锐等), 2015; Xie *et al*. (谢文华等), 2014; Peng *et al*. (彭基泰等), 2007; Wang (王应祥), 2003

藏鼩鼱 *Sorex thibetanus*

苔原鼩鼱
Sorex tundrensis

近危 NT

分类地位 Taxonomic Status

动物界 Animalia	脊索动物门 Chordata	哺乳纲 Mammalia	劳亚食虫目 Eulipotyphla	鼩鼱科 Soricidae
学名 Scientific Name		*Sorex tundrensis*		
命名人 Species Authority		Merriam, 1900		
英文名 English Name(s)		Tundra Shrew		
同物异名 Synonym(s)		无 / None		
种下单元评估 Infra-specific Taxa Assessed		无 / None		

评估信息 Assessment Information

评估年份 Year Assessed	2020
评定人 Assessor(s)	蒋志刚 / Zhigang Jiang
审定人 Reviewer(s)	蒋学龙、冯祚建 / Xuelong Jiang, Zuojian Feng
其他贡献人 Other Contributor(s)	李立立、丁晨晨 / Lili Li, Chenchen Ding

理由 Justification: 苔原鼩鼱的分布区较狭窄，且种群数量较少。因此，列为近危等级 / Tundra Shrew is narrowly distributed with small populations. Thus, it is listed as Near Threatened

地理分布 Geographical Distribution

国内分布 Domestic Distribution

黑龙江、吉林、辽宁、内蒙古、新疆 / Heilongjiang, Jilin, Liaoning, Inner Mongolia (Nei Mongol), Xinjiang

世界分布 World Distribution

加拿大、中国、蒙古、俄罗斯、美国 / Canada, China, Mongolia, Russia, United States

分布标注 Distribution Note

非特有种 / Non-endemic

国内分布图
Map of Domestic Distribution

种群 Population

种群数量 Population Size	未知 / Unknown
种群趋势 Population Trend	稳定 / Stable

生境与生态系统 Habitat(s) and Ecosystem(s)

生　　境 Habitat(s)	灌木丛、草丛或沼泽附近的高地。以昆虫、蠕虫和草本植物为食 / Shrub or Grassy Vegetation or Highland Near Marsh. It eats insect, worm and grass
生态系统 Ecosystem(s)	灌丛生态系统、草地生态系统 / Shrubland Ecosystem, Grassland Ecosystem

威胁 Threat(s)

主要威胁 Major Threat(s)	未知 / Unknown

保护级别与保护行动 Protection Category and Conservation Action(s)

国家重点保护野生动物等级 (2021) Category of National Key Protected Wild Animals (2021)	无 / NA
IUCN 红色名录 (2020-2) IUCN Red List (2020-2)	无危 / LC
CITES 附录 (2019) CITES Appendix (2019)	无 / NA
保护行动 Conservation Action(s)	无 / None

相关文献 Relevant References

Liu *et al.*（刘铸等），2019; Burgin *et al.*, 2018; Wilson and Mittermeier, 2018; Jiang *et al.*（蒋志刚等），2017; Liu *et al.*（刘洋等），2010; Wilson and Reeder, 2005

苔原駒鼩 *Sorex tundrensis*

长爪鼩鼱 *Sorex unguiculatus*

近危 NT

| 数据缺乏 DD | 无危 LC | 近危 NT | 易危 VU | 濒危 EN | 极危 CR | 区域灭绝 RE | 野外灭绝 EW | 灭绝 EX |

分类地位 Taxonomic Status

动物界 Animalia	脊索动物门 Chordata	哺乳纲 Mammalia	劳亚食虫目 Eulipotyphla	鼩鼱科 Soricidae
学名 Scientific Name		*Sorex unguiculatus*		
命名人 Species Authority		Dobson, 1890		
英文名 English Name(s)		Long-clawed Shrew		
同物异名 Synonym(s)		*Sorex shinto* (Thomas, 1907) subsp. *saevus*		
种下单元评估 Infra-specific Taxa Assessed		无 / None		

评估信息 Assessment Information

评估年份 Year Assessed	2020
评定人 Assessor(s)	蒋志刚 / Zhigang Jiang
审定人 Reviewer(s)	蒋学龙、冯祚建 / Xuelong Jiang, Zuojian Feng
其他贡献人 Other Contributor(s)	李立立、丁晨晨 / Lili Li, Chenchen Ding

理由 Justification: 长爪鼩鼱的分布区狭窄，种群数量较少。人为活动对其种群的影响较小，但其分布点少。因此，列为近危等级 / Long-clawed Shrew is narrowly distributed with small populations. The impact of human activities on this species is negligeable, however, it has a few distributed locations. Thus, it is listed as Near Threatened

地理分布 Geographical Distribution

国内分布 Domestic Distribution
黑龙江、内蒙古 / Heilongjiang, Inner Mongolia (Nei Mongol)
世界分布 World Distribution
中国、日本、朝鲜、俄罗斯 / China, Japan, Korea (the Democratic People's Republic of), Russia
分布标注 Distribution Note
非特有种 / Non-endemic

国内分布图
Map of Domestic Distribution

种群 Population

种群数量 Population Size	未知 / Unknown
种群趋势 Population Trend	稳定 / Stable

生境与生态系统 Habitat(s) and Ecosystem(s)

生境 Habitat(s)	草地、泰加林、针阔混交林 / Grassland, Taiga Forest, Coniferous and Broad-leaved Mixed Forest
生态系统 Ecosystem(s)	森林生态系统、草地生态系统 / Forest Ecosystem, Grassland Ecosystem

威胁 Threat(s)

主要威胁 Major Threat(s)	未知 / Unknown

保护级别与保护行动 Protection Category and Conservation Action(s)

国家重点保护野生动物等级 (2021) Category of National Key Protected Wild Animals (2021)	无 / NA
IUCN 红色名录 (2020-2) IUCN Red List (2020-2)	无危 / LC
CITES 附录 (2019) CITES Appendix (2019)	无 / NA
保护行动 Conservation Action(s)	无 / None

相关文献 Relevant References

Liu *et al.*（刘铸等）, 2019; Burgin *et al.*, 2018; Wilson and Mittermeier, 2018; Jiang *et al.*（蒋志刚等）, 2017; Liu *et al.*（刘洋等）, 2013a; Hutterer, 2005

长爪鼩鼱 *Sorex unguiculatus*

狭颅黑齿鼩鼱
Blarinella wardi

近危 NT

| 数据缺乏 DD | 无危 LC | **近危 NT** | 易危 VU | 濒危 EN | 极危 CR | 区域灭绝 RE | 野外灭绝 EW | 灭绝 EX |

分类地位 Taxonomic Status

动物界 Animalia	脊索动物门 Chordata	哺乳纲 Mammalia	劳亚食虫目 Eulipotyphla	鼩鼱科 Soricidae
学名 Scientific Name		*Blarinella wardi*		
命名人 Species Authority		Thomas, 1915		
英文名 English Name(s)		Burmese Short-tailed Shrew		
同物异名 Synonym(s)		无 / None		
种下单元评估 Infra-specific Taxa Assessed		无 / None		

评估信息 Assessment Information

评估年份 Year Assessed	2020
评定人 Assessor(s)	蒋志刚 / Zhigang Jiang
审定人 Reviewer(s)	蒋学龙、冯祚建 / Xuelong Jiang, Zuojian Feng
其他贡献人 Other Contributor(s)	李立立、丁晨晨 / Lili Li, Chenchen Ding

理由 Justification: 狭颅黑齿鼩鼱分布区狭窄，且种群数量较少，分布地农业和林业的发展使其栖息地遭到破坏。因此，列为近危等级 / Burmese Short-tailed Shrew is narrowly distributed with small populations. The development of agriculture and forestry is destroying its habitats. Thus, it is listed as Near Threatened

地理分布 Geographical Distribution

国内分布 Domestic Distribution
云南、四川 / Yunnan, Sichuan
世界分布 World Distribution
中国、缅甸 / China, Myanmar
分布标注 Distribution Note
非特有种 / Non-endemic

国内分布图
Map of Domestic Distribution

种群 Population

种群数量 Population Size	未知 / Unknown
种群趋势 Population Trend	未知 / Unknown

生境与生态系统 Habitat (s) and Ecosystem (s)

生　　境 Habitat(s)	温带森林 / Temperate Forest
生态系统 Ecosystem(s)	森林生态系统 / Forest Ecosystem

威胁 Threat (s)

主要威胁 Major Threat(s)	未知 / Unknown

保护级别与保护行动 Protection Category and Conservation Action (s)

国家重点保护野生动物等级 (2021) Category of National Key Protected Wild Animals (2021)	无 / NA
IUCN 红色名录(2020-2) IUCN Red List (2020-2)	无危 / LC
CITES 附录 (2019) CITES Appendix (2019)	无 / NA
保护行动 Conservation Action(s)	位于自然保护区内的种群与生境得到保护 / Populations and habitats in nature reserves are protected

相关文献 Relevant References

Burgin *et al*., 2018; Wilson and Mittermeier, 2018; Jiang *et al*. (蒋志刚等), 2017; Tu *et al*. (涂飞云等), 2014; Deng *et al*. (邓可等), 2013; Sun *et al*. (孙治宇等), 2013; Hoffmann, 1987

狭颅黑齿鼩鼱 *Blarinella wardi*

大爪长尾鼩鼱
Soriculus nigrescens

近危 NT

| 数据缺乏 DD | 无危 LC | 近危 NT | 易危 VU | 濒危 EN | 极危 CR | 区域灭绝 RE | 野外灭绝 EW | 灭绝 EX |

分类地位 Taxonomic Status

动物界 Animalia	脊索动物门 Chordata	哺乳纲 Mammalia	劳亚食虫目 Eulipotyphla	鼩鼱科 Soricidae
学名 Scientific Name		*Soriculus nigrescens*		
命名人 Species Authority		Gray, 1842		
英文名 English Name(s)		Himalayan Shrew		
同物异名 Synonym(s)		无 / None		
种下单元评估 Infra-specific Taxa Assessed		无 / None		

评估信息 Assessment Information

评估年份 Year Assessed	2020
评定人 Assessor(s)	蒋志刚 / Zhigang Jiang
审定人 Reviewer(s)	蒋学龙、冯祚建 / Xuelong Jiang, Zuojian Feng
其他贡献人 Other Contributor(s)	李立立、丁晨晨 / Lili Li, Chenchen Ding

理由 Justification: 大爪长尾鼩鼱分布区狭窄，且种群数量较少。因此，列为近危等级 / Himalayan Shrew is narrowly distributed with small populations. Thus, it is listed as Near Threatened

地理分布 Geographical Distribution

国内分布 Domestic Distribution
西藏、云南 / Tibet (Xizang), Yunnan
世界分布 World Distribution
不丹、中国、印度、尼泊尔 / Bhutan, China, India, Nepal
分布标注 Distribution Note
非特有种 / Non-endemic

国内分布图
Map of Domestic Distribution

种群 Population

种群数量 Population Size	未知 / Unknown
种群趋势 Population Trend	未知 / Unknown

生境与生态系统 Habitat (s) and Ecosystem (s)

生　　境 Habitat(s)	落叶针叶混交林、针叶杜鹃林、树线以上的高山区 / Mixed Deciduous-Coniferous Forest, Conifer-Rhododendron Forest, Alpine Zone (above timberline)
生态系统 Ecosystem(s)	森林生态系统 / Forest Ecosystem

威胁 Threat (s)

主要威胁 Major Threat(s)	耕种 / Plantation

保护级别与保护行动 Protection Category and Conservation Action (s)

国家重点保护野生动物等级 (2021) Category of National Key Protected Wild Animals (2021)	无 / NA
IUCN 红色名录 (2020-2) IUCN Red List (2020-2)	无危 / LC
CITES 附录 (2019) CITES Appendix (2019)	无 / NA
保护行动 Conservation Action(s)	无 / None

相关文献 Relevant References

Burgin *et al*., 2018; Wilson and Mittermeier, 2018; Jiang *et al*. (蒋志刚等), 2017; Liao *et al*. (廖锐等), 2015; Hu *et al*. (胡一鸣等), 2014

大爪长尾鼩鼱 *Soriculus nigrescens*

滇北长尾鼩 *Chodsigoa parva*

近危 NT

| 数据缺乏 DD | 无危 LC | 近危 NT | 易危 VU | 濒危 EN | 极危 CR | 区域灭绝 RE | 野外灭绝 EW | 灭绝 EX |

分类地位 Taxonomic Status

动物界 Animalia	脊索动物门 Chordata	哺乳纲 Mammalia	劳亚食虫目 Eulipotyphla	鼩鼱科 Soricidae

学名 Scientific Name	*Chodsigoa parva*
命名人 Species Authority	G. M. Allen, 1923
英文名 English Name(s)	Pygmy Brown-toothed Shrew
同物异名 Synonym(s)	无 / None
种下单元评估 Infra-specific Taxa Assessed	无 / None

评估信息 Assessment Information

评估年份 Year Assessed	2020
评定人 Assessor(s)	蒋志刚 / Zhigang Jiang
审定人 Reviewer(s)	蒋学龙 / Xuelong Jiang
其他贡献人 Other Contributor(s)	李立立、丁晨晨 / Lili Li, Chenchen Ding

理由 Justification: 滇北长尾鼩为中国特有种，分布区极为狭窄，种群数量少。因此，列为近危等级 / Pygmy Brown-toothed Shrew is endemic to China, it is narrowly distributed with small populations. Thus, it is listed as Near Threatened

地理分布 Geographical Distribution

国内分布 Domestic Distribution
云南、贵州 / Yunnan, Guizhou
世界分布 World Distribution
中国 / China
分布标注 Distribution Note
特有种 / Endemic

国内分布图
Map of Domestic Distribution

🦌 种群 Population

种群数量 Population Size	未知 / Unknown
种群趋势 Population Trend	未知 / Unknown

🦌 生境与生态系统 Habitat (s) and Ecosystem (s)

生　　境 Habitat(s)	温带森林 / Temperate Forest
生态系统 Ecosystem(s)	森林生态系统 / Forest Ecosystem

🦌 威胁 Threat (s)

主要威胁 Major Threat(s)	未知 / Unknown

🦌 保护级别与保护行动 Protection Category and Conservation Action (s)

国家重点保护野生动物等级 (2021) Category of National Key Protected Wild Animals (2021)	无 / NA
IUCN 红色名录 (2020-2) IUCN Red List (2020-2)	数据缺乏 / DD
CITES 附录 (2019) CITES Appendix (2019)	无 / NA
保护行动 Conservation Action(s)	无 / None

🦌 相关文献 Relevant References

Burgin *et al.*, 2018; Wilson and Mittermeier, 2018; Jiang *et al.* (蒋志刚等), 2017; Shi *et al.* (师蕾等), 2013; Smith *et al.* (史密斯等), 2009; Wang and Xie (汪松和解焱), 2004

滇北长尾鼩 *Chodsigoa parva*

喜马拉雅水麝鼩
Chimarrogale himalayica

近危 NT

| 数据缺乏 DD | 无危 LC | 近危 NT | 易危 VU | 濒危 EN | 极危 CR | 区域灭绝 RE | 野外灭绝 EW | 灭绝 EX |

分类地位 Taxonomic Status

动物界 Animalia	脊索动物门 Chordata	哺乳纲 Mammalia	劳亚食虫目 Eulipotyphla	鼩鼱科 Soricidae

学名 Scientific Name	*Chimarrogale himalayica*
命名人 Species Authority	Gray, 1842
英文名 English Name(s)	Himalayan Water Shrew
同物异名 Synonym(s)	无 / None
种下单元评估 Infra-specific Taxa Assessed	无 / None

评估信息 Assessment Information

评估年份 Year Assessed	2020
评定人 Assessor(s)	蒋志刚 / Zhigang Jiang
审定人 Reviewer(s)	蒋学龙 / Xuelong Jiang
其他贡献人 Other Contributor(s)	李立立、丁晨晨 / Lili Li, Chenchen Ding

理由 Justification: 喜马拉雅水麝鼩生境特殊，栖息于山地森林的洁净溪流生态系统，易受环境污染与破坏影响。其分布区广，但数量少。目前，采集的标本数量少。因此，列为近危等级 / Himalayan Water Shrew requires special habitats, inhabiting only montane forest ecosystem with clean streams; it is vulnerable to environmental pollution and habitat destruction. It is widely distributed, but has small populations. So far, only a few specimens have been collected. Thus, it is listed as Near Threatened

地理分布 Geographical Distribution

国内分布 Domestic Distribution

北京、河北、河南、山西、陕西、宁夏、甘肃、湖北、贵州、四川、重庆、云南、西藏、青海 / Beijing, Hebei, Henan, Shanxi, Shaanxi, Ningxia, Gansu, Hubei, Guizhou, Sichuan, Chongqing, Yunnan, Tibet (Xizang), Qinghai

世界分布 World Distribution

中国；南亚、东南亚 / China; South Asia, Southeast Asia

分布标注 Distribution Note

非特有种 / Non-endemic

国内分布图
Map of Domestic Distribution

种群 Population

种群数量 Population Size	未知 / Unknown
种群趋势 Population Trend	未知 / Unknown

生境与生态系统 Habitat (s) and Ecosystem (s)

生　　境 Habitat(s)	亚热带湿润低地森林、溪流边 / Subtropical Moist Lowland Forest, Near Stream
生态系统 Ecosystem(s)	森林生态系统、湖泊河流生态系统 / Forest Ecosystem, Lake and River Ecosystem

威胁 Threat (s)

主要威胁 Major Threat(s)	未知 / Unknown

保护级别与保护行动 Protection Category and Conservation Action (s)

国家重点保护野生动物等级 (2021) Category of National Key Protected Wild Animals (2021)	无 / NA
IUCN 红色名录 (2020-2) IUCN Red List (2020-2)	无危 / LC
CITES 附录 (2019) CITES Appendix (2019)	无 / NA
保护行动 Conservation Action(s)	位于自然保护区内的种群与生境得到保护 / Populations and habitats in nature reserves are protected

相关文献 Relevant References

Burgin *et al*., 2018; Wilson and Mittermeier, 2018; Jiang *et al*. (蒋志刚等), 2017; Deng *et al*. (邓可等), 2013; He *et al*. (何锴等), 2012; Qin *et al*. (秦岭等), 2007

喜马拉雅水麝鼩 *Chimarrogale himalayica*

格氏小麝鼩
Crocidura gmelini

近危 NT

分类地位 Taxonomic Status

动物界 Animalia	脊索动物门 Chordata	哺乳纲 Mammalia	劳亚食虫目 Eulipotyphla	鼩鼱科 Soricidae
学 名 Scientific Name		*Crocidura gmelini*		
命 名 人 Species Authority		Pallas, 1811		
英 文 名 English Name(s)		Gmelin's Shrew		
同物异名 Synonym(s)		Gmelin's White-toothed Shrew		
种下单元评估 Infra-specific Taxa Assessed		无 / None		

评估信息 Assessment Information

评估年份 Year Assessed	2020
评 定 人 Assessor(s)	蒋志刚 / Zhigang Jiang
审 定 人 Reviewer(s)	蒋学龙 / Xuelong Jiang
其他贡献人 Other Contributor(s)	李立立、丁晨晨 / Lili Li, Chenchen Ding

理由 Justification: 格氏小麝鼩在中国仅分布在新疆，数量较少，且其占有区面积不断减小。因此，列为近危等级 / Gmelin's Shrew is narrowly distributed in Xinjiang of China, and it has a relatively small population. The areas of its occupancy are shrinking. Thus, it is listed as Near Threatened

地理分布 Geographical Distribution

国内分布 Domestic Distribution
新疆 / Xinjiang

世界分布 World Distribution
阿富汗、中国、伊朗、哈萨克斯坦、蒙古、巴基斯坦、土库曼斯坦、乌兹别克斯坦 / Afghanistan, China, Iran, Kazakhstan, Mongolia, Pakistan, Turkmenistan, Uzbekistan

分布标注 Distribution Note
非特有种 / Non-endemic

国内分布图
Map of Domestic Distribution

种群 Population

种群数量 Population Size	未知 / Unknown
种群趋势 Population Trend	未知 / Unknown

生境与生态系统 Habitat (s) and Ecosystem (s)

生境 Habitat(s)	温带灌丛、温带草地 / Temperate Shrubland, Temperate Grassland
生态系统 Ecosystem(s)	灌丛生态系统、草地生态系统 / Shrubland Ecosystem, Grassland Ecosystem

威胁 Threat (s)

主要威胁 Major Threat(s)	未知 / Unknown

保护级别与保护行动 Protection Category and Conservation Action (s)

国家重点保护野生动物等级 (2021) Category of National Key Protected Wild Animals (2021)	无 / NA
IUCN 红色名录 (2020-2) IUCN Red List (2020-2)	无危 / LC
CITES 附录 (2019) CITES Appendix (2019)	无 / NA
保护行动 Conservation Action(s)	无 / None

相关文献 Relevant References

Burgin *et al*., 2018; Wilson and Mittermeier, 2018; Jiang *et al*. (蒋志刚等), 2017; Smith *et al*. (史密斯等), 2009; Hou (侯兰新), 2000

格氏小麝鼩 *Crocidura gmelini*   蒋志刚 绘   Drawn by Zhigang Jiang

印支小麝鼩
Crocidura indochinensis

近危 NT

| 数据缺乏 DD | 无危 LC | 近危 NT | 易危 VU | 濒危 EN | 极危 CR | 区域灭绝 RE | 野外灭绝 EW | 灭绝 EX |

分类地位 Taxonomic Status

动物界 Animalia	脊索动物门 Chordata	哺乳纲 Mammalia	劳亚食虫目 Eulipotyphla	鼩鼱科 Soricidae
学名 Scientific Name		*Crocidura indochinensis*		
命名人 Species Authority		Robinson and Kloss, 1922		
英文名 English Name(s)		Indochinese Shrew		
同物异名 Synonym(s)		无 / None		
种下单元评估 Infra-specific Taxa Assessed		无 / None		

评估信息 Assessment Information

评估年份 Year Assessed	2020
评定人 Assessor(s)	蒋志刚 / Zhigang Jiang
审定人 Reviewer(s)	蒋学龙、冯祚建 / Xuelong Jiang, Zuojian Feng
其他贡献人 Other Contributor(s)	李立立、丁晨晨 / Lili Li, Chenchen Ding

理由 Justification: 印支小麝鼩在中国的分布区较狭窄，种群数量较少，且其占有区面积不断减小。因此，列为近危等级 / Indochinese Shrew is narrowly distributed in China, with relatively small populations. The area of its occupancy is shrinking. Thus, it is listed as Near Threatened

地理分布 Geographical Distribution

国内分布 Domestic Distribution
云南、贵州、四川 / Yunnan, Guizhou, Sichuan
世界分布 World Distribution
中国、老挝、缅甸、泰国、越南 / China, Laos, Myanmar, Thailand, Viet Nam
分布标注 Distribution Note
非特有种 / Non-endemic

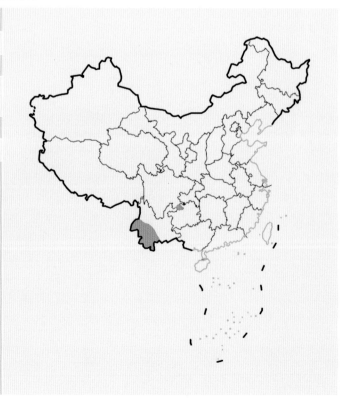

国内分布图
Map of Domestic Distribution

种群 Population

种群数量 Population Size	未知 / Unknown
种群趋势 Population Trend	未知 / Unknown

生境与生态系统 Habitat (s) and Ecosystem (s)

生境 Habitat(s)	亚热带湿润山地森林 / Subtropical Moist Montane Forest
生态系统 Ecosystem(s)	森林生态系统 / Forest Ecosystem

威胁 Threat (s)

主要威胁 Major Threat(s)	未知 / Unknown

保护级别与保护行动 Protection Category and Conservation Action (s)

国家重点保护野生动物等级 (2021) Category of National Key Protected Wild Animals (2021)	无 / NA
IUCN 红色名录 (2020-2) IUCN Red List (2020-2)	无危 / LC
CITES 附录 (2019) CITES Appendix (2019)	无 / NA
保护行动 Conservation Action(s)	位于自然保护区内的种群与生境得到保护 / Populations and habitats in nature reserves are protected

相关文献 Relevant References

Burgin *et al*., 2018; Wilson and Mittermeier, 2018; Jiang *et al*. (蒋志刚等), 2017; Deng *et al*. (邓可等), 2013; Wang (王应祥), 2003; Gong and Wu (龚正达和吴厚永), 2001a, 2001b

印支小麝鼩 *Crocidura indochinensis*

大麝鼩
Crocidura lasiura

近危 NT

| 数据缺乏 DD | 无危 LC | 近危 NT | 易危 VU | 濒危 EN | 极危 CR | 区域灭绝 RE | 野外灭绝 EW | 灭绝 EX |

分类地位 Taxonomic Status

| 动物界 Animalia | 脊索动物门 Chordata | 哺乳纲 Mammalia | 劳亚食虫目 Eulipotyphla | 鼩鼱科 Soricidae |

学名 Scientific Name	*Crocidura lasiura*
命名人 Species Authority	Dobson, 1890
英文名 English Name(s)	Ussuri Shrew
同物异名 Synonym(s)	Ussuri White-toothed Shrew
种下单元评估 Infra-specific Taxa Assessed	无 / None

评估信息 Assessment Information

评估年份 Year Assessed	2020
评定人 Assessor(s)	蒋志刚 / Zhigang Jiang
审定人 Reviewer(s)	蒋学龙、冯祚建 / Xuelong Jiang, Zuojian Feng
其他贡献人 Other Contributor(s)	李立立、丁晨晨 / Lili Li, Chenchen Ding

理由 Justification: 大麝鼩在中国的分布区较狭窄，种群数量较少，且其占有区面积不断减小。因此，列为近危等级 / Ussuri Shrew is narrowly distributed in China, with relatively small populations, and its area of occupancy is shrinking. Thus, it is listed as Near Threatened

地理分布 Geographical Distribution

国内分布 Domestic Distribution	黑龙江、吉林、内蒙古、上海、江苏、四川 / Heilongjiang, Jilin, Inner Mongolia (Nei Mongol), Shanghai, Jiangsu, Sichuan
世界分布 World Distribution	中国；东北亚 / China; Northeast Asia
分布标注 Distribution Note	非特有种 / Non-endemic

国内分布图
Map of Domestic Distribution

种群 Population

种群数量 Population Size	未知 / Unknown
种群趋势 Population Trend	未知 / Unknown

生境与生态系统 Habitat (s) and Ecosystem (s)

生　　　境 Habitat(s)	森林、沼泽、草地、灌丛、耕地 / Forest, Swamp, Grassland, Shrubland, Arable Land
生态系统 Ecosystem(s)	森林生态系统、灌丛生态系统、草地生态系统、农田生态系统、湿地生态系统 / Forest Ecosystem, Shrubland Ecosystem, Grassland Ecosystem, Cropland Ecosystem, Wetland Ecosystem

威胁 Threat (s)

主要威胁 Major Threat(s)	未知 / Unknown

保护级别与保护行动 Protection Category and Conservation Action (s)

国家重点保护野生动物等级 (2021) Category of National Key Protected Wild Animals (2021)	无 / NA
IUCN红色名录(2020-2) IUCN Red List (2020-2)	无危 / LC
CITES 附录 (2019) CITES Appendix (2019)	无 / NA
保护行动 Conservation Action(s)	位于自然保护区内的种群与生境得到保护 / Populations and habitats in nature reserves are protected

相关文献 Relevant References

Burgin *et al.*, 2018; Jiang *et al.* (蒋志刚等), 2017; Ohdachi *et al.*, 2004; Motokawa *et al.*, 2000; Zhang (张荣祖), 1997

大麝鼩 *Crocidura lasiura*

华南中麝鼩 *Crocidura rapax*

近危 NT

| 数据缺乏 DD | 无危 LC | **近危 NT** | 易危 VU | 濒危 EN | 极危 CR | 区域灭绝 RE | 野外灭绝 EW | 灭绝 EX |

分类地位 Taxonomic Status

动物界 Animalia	脊索动物门 Chordata	哺乳纲 Mammalia	劳亚食虫目 Eulipotyphla	鼩鼱科 Soricidae
学名 Scientific Name		*Crocidura rapax*		
命名人 Species Authority		Allen, 1923		
英文名 English Name(s)		Chinese White-toothed Shrew		
同物异名 Synonym(s)		无 / None		
种下单元评估 Infra-specific Taxa Assessed		无 / None		

评估信息 Assessment Information

评估年份 Year Assessed	2020
评定人 Assessor(s)	蒋志刚 / Zhigang Jiang
审定人 Reviewer(s)	蒋学龙、冯祚建 / Xuelong Jiang, Zuojian Feng
其他贡献人 Other Contributor(s)	李立立、丁晨晨 / Lili Li, Chenchen Ding

理由 Justification: 华南中麝鼩在中国南部间断分布，种群数量相对较少，目前采集的标本少。因此，列为近危等级 / Chinese White-toothed Shrew is distributed in southern China in disconnected populations with small populations. Only a few specimens have been collected. Thus, it is listed as Near Threatened

地理分布 Geographical Distribution

国内分布 Domestic Distribution
云南、湖南、广西、海南、四川、贵州、台湾 / Yunnan, Hunan, Guangxi, Hainan, Sichuan, Guizhou, Taiwan
世界分布 World Distribution
中国、越南、缅甸 / China, Viet Nam, Myanmar
分布标注 Distribution Note
非特有种 / Non-endemic

国内分布图
Map of Domestic Distribution

种群 Population

种群数量 Population Size	未知 / Unknown
种群趋势 Population Trend	未知 / Unknown

生境与生态系统 Habitat (s) and Ecosystem (s)

生　　境 Habitat(s)	温带森林 / Temperate Forest
生态系统 Ecosystem(s)	森林生态系统 / Forest Ecosystem

威胁 Threat (s)

主要威胁 Major Threat(s)	未知 / Unknown

保护级别与保护行动 Protection Category and Conservation Action (s)

国家重点保护野生动物等级 (2021) Category of National Key Protected Wild Animals (2021)	无 / NA
IUCN 红色名录 (2020-2) IUCN Red List (2020-2)	数据缺乏 / DD
CITES 附录 (2019) CITES Appendix (2019)	无 / NA
保护行动 Conservation Action(s)	无 / None

相关文献 Relevant References

Burgin *et al*., 2018; Wilson and Mittermeier, 2018; Jiang *et al*. (蒋志刚等), 2017; Liu *et al*. (刘应雄等), 2014; Yu *et al*. (余国睿等), 2014; Wang (王应祥), 2003

华南中麝鼩 *Crocidura rapax*

西伯利亚麝鼩
Crocidura sibirica
近危 NT

| 数据缺乏 DD | 无危 LC | 近危 NT | 易危 VU | 濒危 EN | 极危 CR | 区域灭绝 RE | 野外灭绝 EW | 灭绝 EX |

分类地位 Taxonomic Status

动物界 Animalia	脊索动物门 Chordata	哺乳纲 Mammalia	劳亚食虫目 Eulipotyphla	鼩鼱科 Soricidae

学名 Scientific Name	*Crocidura sibirica*
命名人 Species Authority	Dukelsky, 1930
英文名 English Name(s)	Siberian Shrew
同物异名 Synonym(s)	*Crocidura leucodon* (Stroganov, 1956) subsp. *ognevi*; *C. leucodon* (Dukelsky, 1930) subsp. *sibirica*
种下单元评估 Infra-specific Taxa Assessed	无 / None

评估信息 Assessment Information

评估年份 Year Assessed	2020
评定人 Assessor(s)	蒋志刚 / Zhigang Jiang
审定人 Reviewer(s)	蒋学龙、冯祚建 / Xuelong Jiang, Zuojian Feng
其他贡献人 Other Contributor(s)	李立立、丁晨晨 / Lili Li, Chenchen Ding

理由 Justification: 西伯利亚麝鼩分布较广，但种群数量较少，栖息地水体遭到破坏和污染，且草地过度放牧导致其栖息地质量下降。因此，列为近危等级 / Siberian Shrew is widely distributed, but its population size is relatively small. The water bodies in its habitat are being destroyed and polluted. Furthermore, overgrazing on the grassland is also leading to its habitat degradation. Thus, it is listed as Near Threatened

地理分布 Geographical Distribution

国内分布 Domestic Distribution	新疆、内蒙古 / Xinjiang, Inner Mongolia (Nei Mongol)
世界分布 World Distribution	中国；中亚 / China; Central Asia
分布标注 Distribution Note	非特有种 / Non-endemic

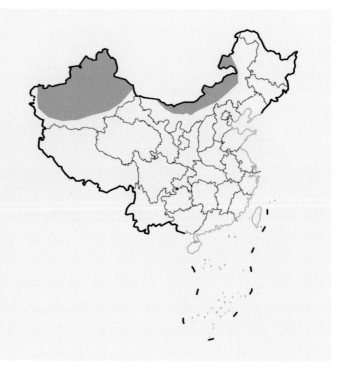

国内分布图
Map of Domestic Distribution

种群 Population

种群数量 Population Size	未知 / Unknown
种群趋势 Population Trend	未知 / Unknown

生境与生态系统 Habitat (s) and Ecosystem (s)

生　　　境 Habitat(s)	森林、沼泽、溪流边 / Forest, Swamp, Near Stream
生态系统 Ecosystem(s)	森林生态系统、湿地生态系统、湖泊河流生态系统 / Forest Ecosystem, Wetland Ecosystem, Lake and River Ecosystem

威胁 Threat (s)

主要威胁 Major Threat(s)	过度放牧、污染 / Over-grazing, Pollution

保护级别与保护行动 Protection Category and Conservation Action (s)

国家重点保护野生动物等级 (2021) Category of National Key Protected Wild Animals (2021)	无 / NA
IUCN 红色名录 (2020-2) IUCN Red List (2020-2)	无危 / LC
CITES 附录 (2019) CITES Appendix (2019)	无 / NA
保护行动 Conservation Action(s)	无 / None

相关文献 Relevant References

Burgin *et al*., 2018; Wilson and Mittermeier, 2018; Jiang *et al*. (蒋志刚等), 2017; Duan *et al*. (段海生等), 2011; Smith *et al*. (史密斯等), 2009; Ohdachi *et al*., 2004; Wang (王应祥), 2003

西伯利亚麝鼩 *Crocidura sibirica*

西南中麝鼩
Crocidura vorax
近危 NT

| 数据缺乏 DD | 无危 LC | **近危 NT** | 易危 VU | 濒危 EN | 极危 CR | 区域灭绝 RE | 野外灭绝 EW | 灭绝 EX |

分类地位 Taxonomic Status

动物界 Animalia	脊索动物门 Chordata	哺乳纲 Mammalia	劳亚食虫目 Eulipotyphla	鼩鼱科 Soricidae
学 名 Scientific Name		*Crocidura vorax*		
命 名 人 Species Authority		Allen, 1923		
英 文 名 English Name(s)		Voracious Shrew		
同物异名 Synonym(s)		无 / None		
种下单元评估 Infra-specific Taxa Assessed		无 / None		

评估信息 Assessment Information

评估年份 Year Assessed	2020
评 定 人 Assessor(s)	蒋志刚 / Zhigang Jiang
审 定 人 Reviewer(s)	蒋学龙、冯祚建 / Xuelong Jiang, Zuojian Feng
其他贡献人 Other Contributor(s)	李立立、丁晨晨 / Lili Li, Chenchen Ding

理由 Justification: 西南中麝鼩分布较广，但种群数量相对较小，目前采集的标本较少。因此，列为近危等级 / Voracious Shrew is widely distributed but has small populations. Only few specimens have been collected. Thus, it is listed as Near Threatened

地理分布 Geographical Distribution

国内分布 Domestic Distribution
四川、贵州、云南、湖南 / Sichuan, Guizhou, Yunnan, Hunan
世界分布 World Distribution
中国；东南亚 / China; Southeast Asia
分布标注 Distribution Note
非特有种 / Non-endemic

国内分布图
Map of Domestic Distribution

种群 Population

种群数量 Population Size	未知 / Unknown
种群趋势 Population Trend	未知 / Unknown

生境与生态系统 Habitat (s) and Ecosystem (s)

生　　境 Habitat(s)	温带森林、亚热带湿润低地森林 / Temperate Forest, Subtropical Moist Lowland Forest
生态系统 Ecosystem(s)	森林生态系统 / Forest Ecosystem

威胁 Threat (s)

主要威胁 Major Threat(s)	未知 / Unknown

保护级别与保护行动 Protection Category and Conservation Action (s)

国家重点保护野生动物等级 (2021) Category of National Key Protected Wild Animals (2021)	无 / NA
IUCN 红色名录 (2020-2) IUCN Red List (2020-2)	无危 / LC
CITES 附录 (2019) CITES Appendix (2019)	无 / NA
保护行动 Conservation Action(s)	无 / None

相关文献 Relevant References

Burgin *et al.*, 2018; Wilson and Mittermeier, 2018; Jiang *et al.*（蒋志刚等）, 2017; Tu *et al.*（涂飞云等）, 2014; Qin *et al.*（秦岭等）, 2007; Wang（王应祥）, 2003

西南中麝鼩 *Crocidura vorax*

五指山小麝鼩
Crocidura wuchihensis

近危 NT

| 数据缺乏 DD | 无危 LC | 近危 NT | 易危 VU | 濒危 EN | 极危 CR | 区域灭绝 RE | 野外灭绝 EW | 灭绝 EX |

分类地位 Taxonomic Status

动物界 Animalia	脊索动物门 Chordata	哺乳纲 Mammalia	劳亚食虫目 Eulipotyphla	鼩鼱科 Soricidae

学名 Scientific Name	*Crocidura wuchihensis*
命名人 Species Authority	Wang, 1966
英文名 English Name(s)	Hainan Island Shrew
同物异名 Synonym(s)	无 / None
种下单元评估 Infra-specific Taxa Assessed	无 / None

评估信息 Assessment Information

评估年份 Year Assessed	2020
评定人 Assessor(s)	蒋志刚 / Zhigang Jiang
审定人 Reviewer(s)	陈辈乐、李飞 / Bosco P. L. Chan, Fei Li
其他贡献人 Other Contributor(s)	李立立、丁晨晨 / Lili Li, Chenchen Ding

理由 Justification: 五指山小麝鼩在中国分布于海南，然而在云南也有发现，但种群数量较少。因此，列为近危等级 / Hainan Island Shrew was originally thought to distributed only in Hainan Province, however, it has also been found in the Yunnan Province with small populations. Thus, it is listed as Near Threatened

地理分布 Geographical Distribution

国内分布 Domestic Distribution
海南、云南 / Hainan, Yunnan
世界分布 World Distribution
中国、越南 / China, Viet Nam
分布标注 Distribution Note
非特有种 / Non-endemic

国内分布图
Map of Domestic Distribution

种群 Population

种群数量 Population Size	未知 / Unknown
种群趋势 Population Trend	未知 / Unknown

生境与生态系统 Habitat (s) and Ecosystem (s)

生　　境 Habitat(s)	热带雨林 / Tropical Rainforest
生态系统 Ecosystem(s)	森林生态系统 / Forest Ecosystem

威胁 Threat (s)

主要威胁 Major Threat(s)	未知 / Unknown

保护级别与保护行动 Protection Category and Conservation Action (s)

国家重点保护野生动物等级 (2021) Category of National Key Protected Wild Animals (2021)	无 / NA
IUCN 红色名录(2020-2) IUCN Red List (2020-2)	数据缺乏 / DD
CITES 附录 (2019) CITES Appendix (2019)	无 / NA
保护行动 Conservation Action(s)	无 / None

相关文献 Relevant References

Burgin *et al*., 2018; Wilson and Mittermeier, 2018; Jiang *et al*.（蒋志刚等）, 2017; Duan *et al*.（段海生等）, 2011; Smith *et al*.（史密斯等）, 2009; Tan（谭邦杰）, 1992

五指山小麝鼩 *Crocidura wuchihensis*

棕果蝠
Rousettus leschenaultii

近危 NT

| 数据缺乏 DD | 无危 LC | 近危 NT | 易危 VU | 濒危 EN | 极危 CR | 区域灭绝 RE | 野外灭绝 EW | 灭绝 EX |

分类地位 Taxonomic Status

动物界 Animalia	脊索动物门 Chordata	哺乳纲 Mammalia	翼手目 Chiroptera	狐蝠科 Pteropodidae
学名 Scientific Name		*Rousettus leschenaultii*		
命名人 Species Authority		Desmarest, 1820		
英文名 English Name(s)		Leschenault's Rousette		
同物异名 Synonym(s)		*Eleutherura fusca* (Gray, 1870); *Pteropus leschenaultii* (Desmarest, 1820); *Rousettus affinis* (Gray, 1843); *R. fuliginosa* (Gray, 1871); *R. fusca* (Gray, 1871): (转下页)		
种下单元评估 Infra-specific Taxa Assessed		无 / None		

评估信息 Assessment Information

评估年份 Year Assessed	2020
评定人 Assessor(s)	吴毅、蒋志刚 / Yi Wu, Zhigang Jiang
审定人 Reviewer(s)	张礼标、毛秀光、张树义 / Libiao Zhang, Xiuguang Mao, Shuyi Zhang
其他贡献人 Other Contributor(s)	李立立、丁晨晨 / Lili Li, Chenchen Ding

理由 Justification: 棕果蝠分布较广，但种群数量较少。因此，列为近危等级 / Leschenault's Rousette is a widespread species with a relatively low population size. Thus, it is listed as Near Threatened

地理分布 Geographical Distribution

国内分布 Domestic Distribution
江西、贵州、广西、广东、海南、福建、云南、西藏、四川、香港、澳门、重庆 / Jiangxi, Guizhou, Guangxi, Guangdong, Hainan, Fujian, Yunnan, Tibet (Xizang), Sichuan, Hong Kong, Macao, Chongqing
世界分布 World Distribution
中国；南亚、东南亚 / China; South Asia, Southeast Asia
分布标注 Distribution Note
非特有种 / Non-endemic

国内分布图
Map of Domestic Distribution

种群 Population

种群数量 Population Size	常见 / Common
种群趋势 Population Trend	稳定 / Stable

生境与生态系统 Habitat(s) and Ecosystem(s)

生 境 Habitat(s)	森林、人造建筑、洞穴 / Forest, Building, Cave
生态系统 Ecosystem(s)	森林生态系统、人类聚落生态系统 / Forest Ecosystem, Human Settlement Ecosystem

威胁 Threat(s)

主要威胁 Major Threat(s)	旅游、采石场、狩猎及采集野生植物 / Tourism, Quarrying Field, Hunting and Collection of Wild Plant

保护级别与保护行动 Protection Category and Conservation Action(s)

国家重点保护野生动物等级 (2021) Category of National Key Protected Wild Animals (2021)	无 / NA
IUCN 红色名录 (2020-2) IUCN Red List (2020-2)	无危 / LC
CITES 附录 (2019) CITES Appendix (2019)	无 / NA
保护行动 Conservation Action(s)	无 / None

相关文献 Relevant References

Wilson and Mittermeier, 2019, 2018; Burgin *et al*., 2018; Jiang *et al*. (蒋志刚等), 2017; Huang *et al*. (黄继展等), 2013; Chen *et al*. (陈忠等), 2007; Shao *et al*. (邵伟伟等), 2007; Zhu (朱光剑), 2007; Tang *et al*. (唐占辉等), 2005; Wang (王应祥), 2003

(接上页)
R. infuscata (Peters, 1873); *R. leschenaulti* (Desmarest, 1820); *R. marginatus* (Gray, 1843); *R. pirivarus* (Hodgson, 1841); *R. pyrivorus* (Hodgson, 1835); *R. seminudus* (Kelaart, 1850); *R. shortridgei* (Thomas and Wroughton, 1909); *Xantharpyia seminuda* (Gray, 1870)

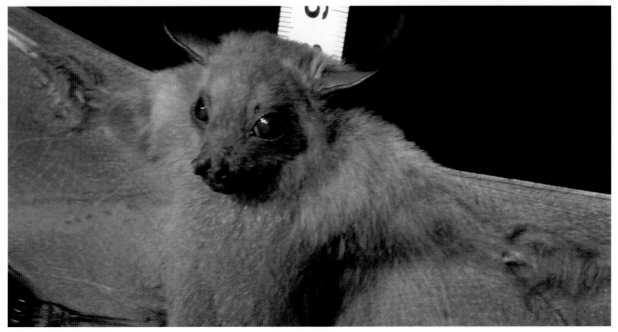

棕果蝠 *Rousettus leschenaultii*

By Alice Hughes

犬蝠
Cynopterus sphinx
近危 NT

| 数据缺乏 DD | 无危 LC | 近危 NT | 易危 VU | 濒危 EN | 极危 CR | 区域灭绝 RE | 野外灭绝 EW | 灭绝 EX |

分类地位 Taxonomic Status

动物界 Animalia	脊索动物门 Chordata	哺乳纲 Mammalia	翼手目 Chiroptera	狐蝠科 Pteropodidae

学名 Scientific Name	*Cynopterus sphinx*
命名人 Species Authority	Vahl, 1797
英文名 English Name(s)	Greater Short-nosed Fruit Bat
同物异名 Synonym(s)	Short-nosed Indian Fruit Bat; *Cynopterus angulatus* (Miller, 1898); *C. brachyotis* (Miller, 1898) subsp. *angulatus*; *C. brachyotis* (Zelebor, 1869) var. *scherzeri*; *C. marginatus* (转下页)
种下单元评估 Infra-specific Taxa Assessed	无 / None

评估信息 Assessment Information

评估年份 Year Assessed	2020
评定人 Assessor(s)	吴毅、蒋志刚 / Yi Wu, Zhigang Jiang
审定人 Reviewer(s)	张礼标、毛秀光、张树义 / Libiao Zhang, Xiuguang Mao, Shuyi Zhang
其他贡献人 Other Contributor(s)	李立立、丁晨晨 / Lili Li, Chenchen Ding

理由 Justification: 犬蝠在中国的分布区较广，但数量较少，且其所在的适宜栖息地面积减小。因此，列为近危等级 / Greater Short-nosed Fruit Bat is widely distributed in China, but its population size is small. Furthermore, the areas of its suitable habitats are declining. Thus, it is listed as Near Threatened

地理分布 Geographical Distribution

国内分布 Domestic Distribution
福建、广西、云南、海南、西藏、广东、香港、澳门 / Fujian, Guangxi, Yunnan, Hainan, Tibet (Xizang), Guangdong, Hong Kong, Macao
世界分布 World Distribution
中国；南亚、东南亚 / China; South Asia, Southeast Asia
分布标注 Distribution Note
非特有种 / Non-endemic

国内分布图
Map of Domestic Distribution

种群 Population

种群数量 Population Size	常见 / Common
种群趋势 Population Trend	稳定 / Stable

生境与生态系统 Habitat (s) and Ecosystem (s)

生　　境 Habitat(s)	森林、城市 / Forest, Urban Area
生态系统 Ecosystem(s)	森林生态系统、人类聚落生态系统 / Forest Ecosystem, Human Settlement Ecosystem

威胁 Threat (s)

主要威胁 Major Threat(s)	耕种、狩猎及采集陆生植物 / Plantation, Hunting and Gathering Terrestrial Plant

保护级别与保护行动 Protection Category and Conservation Action (s)

国家重点保护野生动物等级 (2021) Category of National Key Protected Wild Animals (2021)	无 / NA
IUCN 红色名录 (2020-2) IUCN Red List (2020-2)	无危 / LC
CITES 附录 (2019) CITES Appendix (2019)	无 / NA
保护行动 Conservation Action(s)	无 / None

相关文献 Relevant References

Wilson and Mittermeier, 2019; Burgin *et al.*, 2018; Jiang *et al.* (蒋志刚等), 2017; Huang *et al.* (黄继展等), 2013; Smith *et al.* (史密斯等), 2009; Zhang (张伟), 2008; Zhu (朱光剑), 2007; Feng *et al.* (冯祚建等), 1984; Animal Laboratory of Guangdong Institute of Entomology; Department of Biology, Sun Yat-Sen University (广东省昆虫研究所动物室和中山大学生物系), 1983

(接上页)

(Gray, 1870) var. *ellitoi*; *C. sphnx* (Andersen, 1910) subsp. *gangeticus*; *Pachysoma brevicaudatum* (Temminck, 1837); *Pteropus marginatus* (É. Geoffroy, 1810); *P. pusillus* (É. Geoffroy, 1803); *Vespertilio fibulatus* (Vahl, 1797); *V. sphinx* (Vahl, 1797)

犬蝠 *Cynopterus sphinx* 　　　　吴毅 摄　By Yi Wu

台湾菊头蝠
Rhinolophus formosae

近危 NT

| 数据缺乏 DD | 无危 LC | 近危 NT | 易危 VU | 濒危 EN | 极危 CR | 区域灭绝 RE | 野外灭绝 EW | 灭绝 EX |

分类地位 Taxonomic Status

动物界 Animalia	脊索动物门 Chordata	哺乳纲 Mammalia	翼手目 Chiroptera	菊头蝠科 Rhinolophidae
学名 Scientific Name		*Rhinolophus formosae*		
命名人 Species Authority		Sanborn, 1939		
英文名 English Name(s)		Formosan Woolly Horseshoe Bat		
同物异名 Synonym(s)		无 / None		
种下单元评估 Infra-specific Taxa Assessed		无 / None		

评估信息 Assessment Information

评估年份 Year Assessed	2020
评定人 Assessor(s)	吴毅、蒋志刚 / Yi Wu, Zhigang Jiang
审定人 Reviewer(s)	张礼标、毛秀光、张树义 / Libiao Zhang, Xiuguang Mao, Shuyi Zhang
其他贡献人 Other Contributor(s)	李立立、丁晨晨 / Lili Li, Chenchen Ding

理由 Justification: 台湾菊头蝠为中国特有种，仅分布在台湾，由于该种受到人类活动，如捕猎、旅游和采石等的干扰，其占有区面积缩小。因此，列为近危等级 / Formosan Woolly Horseshoe Bat is an endemic species in China, and is distributed only in Taiwan. Due to disturbance from human encroachment such as hunting, tourism and quarrying, its areas of occupancy are shrinking. Thus, it is listed as Near Threatened

地理分布 Geographical Distribution

国内分布 Domestic Distribution
台湾 / Taiwan
世界分布 World Distribution
中国 / China
分布标注 Distribution Note
特有种 / Endemic

国内分布图
Map of Domestic Distribution

种群 Population

种群数量 Population Size	较常见 / Relatively common
种群趋势 Population Trend	下降 / Decreasing

生境与生态系统 Habitat(s) and Ecosystem(s)

生境 Habitat(s)	洞穴、人造建筑、森林 / Cave, Building, Forest
生态系统 Ecosystem(s)	喀斯特生态系统、人类聚落生态系统 / Karst Ecosystem, Human Settlement Ecosystem

威胁 Threat(s)

主要威胁 Major Threat(s)	住宅区及商业发展、公路和铁路建设、耕种 / Residential and Commercial Development, Construction of Road and Railroad, Plantation

保护级别与保护行动 Protection Category and Conservation Action(s)

国家重点保护野生动物等级 (2021) Category of National Key Protected Wild Animals (2021)	无 / NA
IUCN 红色名录 (2020-2) IUCN Red List (2020-2)	近危 / NT
CITES 附录 (2019) CITES Appendix (2019)	无 / NA
保护行动 Conservation Action(s)	无 / None

相关文献 Relevant References

Wilson and Mittermeier, 2019; Burgin *et al.*, 2018; Jiang *et al.* (蒋志刚等), 2017; Zheng (郑锡奇), 2010; Lin *et al.* (林良恭等), 1997; Chen (陈兼善), 1969; Sanborn, 1939

台湾菊头蝠 *Rhinolophus formosae*

大菊头蝠
Rhinolophus luctus

近危 NT

| 数据缺乏 DD | 无危 LC | 近危 NT | 易危 VU | 濒危 EN | 极危 CR | 区域灭绝 RE | 野外灭绝 EW | 灭绝 EX |

分类地位 Taxonomic Status

动物界 Animalia	脊索动物门 Chordata	哺乳纲 Mammalia	翼手目 Chiroptera	菊头蝠科 Rhinolophidae
学名 Scientific Name		*Rhinolophus luctus*		
命名人 Species Authority		Temminck, 1834		
英文名 English Name(s)		Great Woolly Horsehoe Bat		
同物异名 Synonym(s)		Woolly Horseshoe Bat; *Rhinolophus perniger* (Hodgson, 1843)		
种下单元评估 Infra-specific Taxa Assessed		无 / None		

评估信息 Assessment Information

评估年份 Year Assessed	2020
评定人 Assessor(s)	吴毅、蒋志刚 / Yi Wu, Zhigang Jiang
审定人 Reviewer(s)	张礼标、毛秀光、张树义 / Libiao Zhang, Xiuguang Mao, Shuyi Zhang
其他贡献人 Other Contributor(s)	李立立、丁晨晨 / Lili Li, Chenchen Ding

理由 Justification: 大菊头蝠在中国的分布区较广，但数量较少，且由于人类干扰使其占有区面积不断减小。因此，列为近危等级 / Great Woolly Horseshoe Bat has large distribution range in China, but its population size is small. Furthermore, the areas of its occupancy are declining due to human disturbance. Thus, it is listed as Near Threatened

地理分布 Geographical Distribution

国内分布 Domestic Distribution
浙江、江西、四川、贵州、广东、海南、广西、福建、台湾、重庆 / Zhejiang, Jiangxi, Sichuan, Guizhou, Guangdong, Hainan, Guangxi, Fujian, Taiwan, Chongqing
世界分布 World Distribution
中国；南亚、东南亚 / China; South Asia, Southeast Asia
分布标注 Distribution Note
非特有种 / Non-endemic

国内分布图
Map of Domestic Distribution

种群 Population

种群数量 Population Size	稀少 / Rare
种群趋势 Population Trend	未知 / Unknown

生境与生态系统 Habitat(s) and Ecosystem(s)

生境 Habitat(s)	森林、洞穴、人造建筑 / Forest, Cave, Building
生态系统 Ecosystem(s)	森林生态系统、人类聚落生态系统 / Forest Ecosystem, Human Settlement Ecosystem

威胁 Threat(s)

主要威胁 Major Threat(s)	伐木、火灾、耕种、狩猎及采集 / Logging, Fire, Plantation, Hunting or Collection

保护级别与保护行动 Protection Category and Conservation Action(s)

国家重点保护野生动物等级 (2021) Category of National Key Protected Wild Animals (2021)	无 / NA
IUCN 红色名录 (2020-2) IUCN Red List (2020-2)	无危 / LC
CITES 附录 (2019) CITES Appendix (2019)	无 / NA
保护行动 Conservation Action(s)	位于自然保护区内的种群与生境得到保护 / Populations and habitats in nature reserves are protected

相关文献 Relevant References

Burgin *et al.*, 2018; Zhang *et al.*（张婵等）, 2013; Xu（徐海龙）, 2012; Pei（裴俊峰）, 2011; Yang *et al.*（杨锐等）, 2010; Zhang *et al.*, 2009a, 2009b, 2009c; Zhang *et al.*（张佑祥等）, 2008; Liang and Dong, 1984

大菊头蝠 *Rhinolophus luctus*

By Alice Hughes

马氏菊头蝠
Rhinolophus marshalli
近危 NT

| 数据缺乏 DD | 无危 LC | 近危 NT | 易危 VU | 濒危 EN | 极危 CR | 区域灭绝 RE | 野外灭绝 EW | 灭绝 EX |

分类地位 Taxonomic Status

动物界 Animalia	脊索动物门 Chordata	哺乳纲 Mammalia	翼手目 Chiroptera	菊头蝠科 Rhinolophidae
学　名 Scientific Name		*Rhinolophus marshalli*		
命　名　人 Species Authority		Thonglongya, 1973		
英　文　名 English Name(s)		Marshall's Horseshoe Bat		
同物异名 Synonym(s)		无 / None		
种下单元评估 Infra-specific Taxa Assessed		无 / None		

评估信息 Assessment Information

评 估 年 份 Year Assessed	2020
评　定　人 Assessor(s)	吴毅、蒋志刚 / Yi Wu, Zhigang Jiang
审　定　人 Reviewer(s)	张礼标、毛秀光、张树义 / Libiao Zhang, Xiuguang Mao, Shuyi Zhang
其他贡献人 Other Contributor(s)	李立立、丁晨晨 / Lili Li, Chenchen Ding

理由 Justification: 马氏菊头蝠在中国的分布点狭窄且分散，种群密度低，且其占有区面积减小。因此，列为近危等级 / Marshall's Horseshoe Bat is present only in a few small and scattered localities in China and its population densities are low. Furthermore, the areas of its occupancy are declining. Thus, it is listed as Near Threatened

地理分布 Geographical Distribution

国内分布 Domestic Distribution
广西、云南 / Guangxi, Yunnan
世界分布 World Distribution
中国；东南亚 / China; Southeast Asia
分布标注 Distribution Note
非特有种 / Non-endemic

国内分布图
Map of Domestic Distribution

种群 Population

种群数量 Population Size	稀少 / Rare
种群趋势 Population Trend	未知 / Unknown

生境与生态系统 Habitat(s) and Ecosystem(s)

生境 Habitat(s)	石灰岩洞穴、岩石裂缝 / Limestone Cave, Rock Crevice
生态系统 Ecosystem(s)	喀斯特生态系统、森林生态系统、人类聚落生态系统 / Karst Ecosystem, Forest Ecosystem, Human Settlement Ecosystem

威胁 Threat(s)

主要威胁 Major Threat(s)	伐木、采矿、采石场、旅游 / Logging, Mining, Quarrying Field, Tourism

保护级别与保护行动 Protection Category and Conservation Action(s)

国家重点保护野生动物等级 (2021) Category of National Key Protected Wild Animals (2021)	无 / NA
IUCN 红色名录 (2020-2) IUCN Red List (2020-2)	无危 / LC
CITES 附录 (2019) CITES Appendix (2019)	无 / NA
保护行动 Conservation Action(s)	无 / None

相关文献 Relevant References

Burgin *et al.*, 2018; Jiang *et al.* (蒋志刚等), 2017; Simmons, 2005; Zhang *et al.* (张礼标等), 2005; Wu *et al.* (吴毅等), 2004a

马氏菊头蝠 *Rhinolophus marshalli*　　　　　　　　　　　　　吴毅 摄　By Yi Wu

高鞍菊头蝠
Rhinolophus paradoxolophus

近危 NT

| 数据缺乏 DD | 无危 LC | 近危 NT | 易危 VU | 濒危 EN | 极危 CR | 区域灭绝 RE | 野外灭绝 EW | 灭绝 EX |

分类地位 Taxonomic Status

动物界 Animalia	脊索动物门 Chordata	哺乳纲 Mammalia	翼手目 Chiroptera	菊头蝠科 Rhinolophidae

学 名 Scientific Name	*Rhinolophus paradoxolophus*
命 名 人 Species Authority	Bourret, 1951
英 文 名 English Name(s)	Bourret's Horseshoe Bat
同物异名 Synonym(s)	*Rhinolophus rex*
种下单元评估 Infra-specific Taxa Assessed	无 / None

评估信息 Assessment Information

评估年份 Year Assessed	2020
评 定 人 Assessor(s)	吴毅、蒋志刚 / Yi Wu, Zhigang Jiang
审 定 人 Reviewer(s)	张礼标、毛秀光、张树义 / Libiao Zhang, Xiuguang Mao, Shuyi Zhang
其他贡献人 Other Contributor(s)	李立立、丁晨晨 / Lili Li, Chenchen Ding

理由 Justification: 高鞍菊头蝠在中国的分布区狭窄，且种群数量较少，其占有区面积减小。因此，列为近危等级 / Bourret's Horseshoe Bat is narrowly distributed in China, and its population size is small. Furthermore, the areas of its occupancy are declining. Thus, it is listed as Near Threatened

地理分布 Geographical Distribution

国内分布 Domestic Distribution
广西 / Guangxi
世界分布 World Distribution
中国；东南亚 / China; Southeast Asia
分布标注 Distribution Note
非特有种 / Non-endemic

国内分布图
Map of Domestic Distribution

种群 Population

种群数量 Population Size	稀少 / Rare
种群趋势 Population Trend	未知 / Unknown

生境与生态系统 Habitat(s) and Ecosystem(s)

生　　境 Habitat(s)	洞穴、喀斯特地貌 / Cave, Karst Landscape
生态系统 Ecosystem(s)	喀斯特生态系统、森林生态系统 / Karst Ecosystem, Forest Ecosystem

威胁 Threat(s)

主要威胁 Major Threat(s)	未知 / Unknown

保护级别与保护行动 Protection Category and Conservation Action(s)

国家重点保护野生动物等级 (2021) Category of National Key Protected Wild Animals (2021)	无 / NA
IUCN 红色名录 (2020-2) IUCN Red List (2020-2)	无危 / LC
CITES 附录 (2019) CITES Appendix (2019)	无 / NA
保护行动 Conservation Action(s)	无 / None

相关文献 Relevant References

Wilson and Mittermeier, 2019; Burgin *et al*., 2018; Jiang *et al*.（蒋志刚等）, 2017; Song *et al*.（宋先华等）, 2014; Song（宋华）, 2009; Zhao *et al*.（赵辉华等）, 2002

高鞍菊头蝠 *Rhinolophus paradoxolophus*

贵州菊头蝠
Rhinolophus rex

近危 NT

| 数据缺乏 DD | 无危 LC | 近危 NT | 易危 VU | 濒危 EN | 极危 CR | 区域灭绝 RE | 野外灭绝 EW | 灭绝 EX |

分类地位 Taxonomic Status

动物界 Animalia	脊索动物门 Chordata	哺乳纲 Mammalia	翼手目 Chiroptera	菊头蝠科 Rhinolophidae
学　　名 Scientific Name		*Rhinolophus rex*		
命　名　人 Species Authority		G. M. Allen, 1923		
英　文　名 English Name(s)		King Horseshoe Bat		
同物异名 Synonym(s)		无 / None		
种下单元评估 Infra-specific Taxa Assessed		无 / None		

评估信息 Assessment Information

评估年份 Year Assessed	2020
评　定　人 Assessor(s)	吴毅、蒋志刚 / Yi Wu, Zhigang Jiang
审　定　人 Reviewer(s)	张礼标、毛秀光、张树义 / Libiao Zhang, Xiuguang Mao, Shuyi Zhang
其他贡献人 Other Contributor(s)	李立立、丁晨晨 / Lili Li, Chenchen Ding

理由 Justification: 贵州菊头蝠是中国特有种，种群数量较少，由于该种受到人类活动的影响，且占有区面积缩小。因此，列为近危等级 / King Horseshoe Bat is an endemic species in China with small populations. Its areas of occupancy are declining due to human disturbance. Thus, it is listed as Near Threatened

地理分布 Geographical Distribution

国内分布 Domestic Distribution
广东、重庆、四川、贵州、广西 / Guangdong, Chongqing, Sichuan, Guizhou, Guangxi
世界分布 World Distribution
中国 / China
分布标注 Distribution Note
特有种 / Endemic

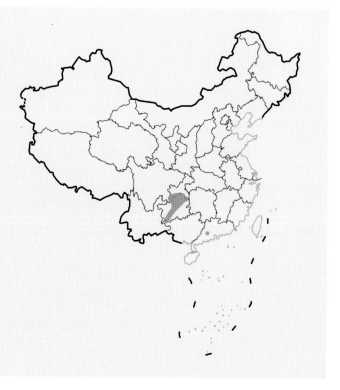

国内分布图
Map of Domestic Distribution

种群 Population

种群数量 Population Size	稀少 / Rare
种群趋势 Population Trend	下降 / Decreasing

生境与生态系统 Habitat (s) and Ecosystem (s)

生境 Habitat(s)	洞穴、人造建筑、森林 / Cave, Building, Forest
生态系统 Ecosystem(s)	喀斯特生态系统、人类聚落生态系统、森林生态系统 / Karst Ecosystem, Human Settlement Ecosystem, Forest Ecosystem

威胁 Threat (s)

主要威胁 Major Threat(s)	伐木、耕种、火灾、采矿、采石场、旅游 / Logging, Plantation, Fire, Mining, Quarrying Field and Tourism

保护级别与保护行动 Protection Category and Conservation Action (s)

国家重点保护野生动物等级 (2021) Category of National Key Protected Wild Animals (2021)	无 / NA
IUCN 红色名录 (2020-2) IUCN Red List (2020-2)	无危 / LC
CITES 附录 (2019) CITES Appendix (2019)	无 / NA
保护行动 Conservation Action(s)	无 / None

相关文献 Relevant References

Burgin *et al.*, 2018; Jiang *et al.* (蒋志刚等), 2017; Song (宋华), 2009; Deng *et al.* (邓庆伟等), 2008; Gu *et al.* (谷晓明等), 2003b

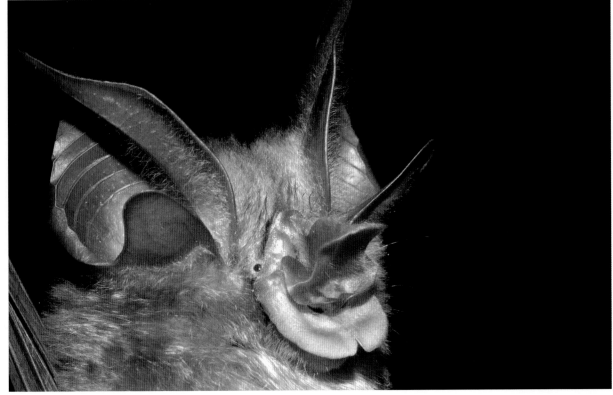

贵州菊头蝠 *Rhinolophus rex*　　　　余文华 摄　By Wenhua Yu

清迈菊头蝠
Rhinolophus siamensis

近危 NT

| 数据缺乏 DD | 无危 LC | 近危 NT | 易危 VU | 濒危 EN | 极危 CR | 区域灭绝 RE | 野外灭绝 EW | 灭绝 EX |

分类地位 Taxonomic Status

动物界 Animalia	脊索动物门 Chordata	哺乳纲 Mammalia	翼手目 Chiroptera	菊头蝠科 Rhinolophidae

学名 Scientific Name	*Rhinolophus siamensis*
命名人 Species Authority	Glydenstolpe, 1917
英文名 English Name(s)	Thai Horseshoe Bat
同物异名 Synonym(s)	Thai Leaf-nosed Bat, Siamese Horseshoe Bat
种下单元评估 Infra-specific Taxa Assessed	*Rhinolophus huananus* Wu *et al.*, 2008

评估信息 Assessment Information

评估年份 Year Assessed	2020
评定人 Assessor(s)	吴毅、蒋志刚 / Yi Wu, Zhigang Jiang
审定人 Reviewer(s)	张礼标、毛秀光、张树义 / Libiao Zhang, Xiuguang Mao, Shuyi Zhang
其他贡献人 Other Contributor(s)	李立立、丁晨晨 / Lili Li, Chenchen Ding

理由 Justification: 清迈菊头蝠 2008 年在中国广东发现，曾命名为华南菊头蝠 (*Rhinolophus huananus*)，后确定为清迈菊头蝠。由于猎杀、采石、旅游等人类活动，其分布区遭到破坏，栖息地面积正在减小。因此，列为近危等级 / In 2008, the population of this species was thought to be a new species, and named the South China Horseshoe Bat (*Rhinolophus huananus*), however it was later ascertained to be the Thai Horseshoe Bat. Due to disturbances from human encroachment such as hunting, tourism and quarrying, the areas of its occupancy are shrinking. Thus, it is listed as Near Threatened

地理分布 Geographical Distribution

国内分布 Domestic Distribution	广东、广西、江西 / Guangdong, Guangxi, Jiangxi
世界分布 World Distribution	中国、老挝、越南、泰国 / China, Laos, Viet Nam, Thailand
分布标注 Distribution Note	非特有种 / Non-endemic

国内分布图
Map of Domestic Distribution

种群 Population

种群数量 Population Size	稀少 / Rare
种群趋势 Population Trend	未知 / Unknown

生境与生态系统 Habitat(s) and Ecosystem(s)

生　　境 Habitat(s)	洞穴 / Cave
生态系统 Ecosystem(s)	森林生态系统、喀斯特生态系统 / Forest Ecosystem, Karst Ecosystem

威胁 Threat(s)

主要威胁 Major Threat(s)	伐木、耕种、火灾、采矿、采石场、旅游休闲区建设 / Logging, Plantation, Fire, Mining, Quarrying Field, Construction of Tourism and Recreation Area

保护级别与保护行动 Protection Category and Conservation Action(s)

国家重点保护野生动物等级 (2021) Category of National Key Protected Wild Animals (2021)	无 / NA
IUCN 红色名录 (2020-2) IUCN Red List (2020-2)	未列入 / NA
CITES 附录 (2019) CITES Appendix (2019)	无 / NA
保护行动 Conservation Action(s)	无 / None

相关文献 Relevant References

Liu and Wu (刘少英和吴毅), 2019; Wilson and Mittermeier, 2019; Burgin *et al.*, 2018; Jiang *et al.* (蒋志刚等), 2017; Zhang *et al.*, 2009a, 2009b, 2009c; Wu *et al.*, 2008

清迈菊头蝠 *Rhinolophus siamensis* 　　　　　吴毅 摄　By Yi Wu

小褐菊头蝠
Rhinolophus stheno

近危 NT

| 数据缺乏 DD | 无危 LC | 近危 NT | 易危 VU | 濒危 EN | 极危 CR | 区域灭绝 RE | 野外灭绝 EW | 灭绝 EX |

分类地位 Taxonomic Status

动物界 Animalia	脊索动物门 Chordata	哺乳纲 Mammalia	翼手目 Chiroptera	菊头蝠科 Rhinolophidae
学名 Scientific Name		*Rhinolophus stheno*		
命名人 Species Authority		K. Andersen, 1905		
英文名 English Name(s)		Lesser Brown Horseshoe Bat		
同物异名 Synonym(s)		无 / None		
种下单元评估 Infra-specific Taxa Assessed		无 / None		

评估信息 Assessment Information

评估年份 Year Assessed	2020
评定人 Assessor(s)	吴毅、蒋志刚 / Yi Wu, Zhigang Jiang
审定人 Reviewer(s)	张礼标、毛秀光、张树义 / Libiao Zhang, Xiuguang Mao, Shuyi Zhang
其他贡献人 Other Contributor(s)	李立立、丁晨晨 / Lili Li, Chenchen Ding

理由 Justification: 小褐菊头蝠在中国云南呈点状分布，种群数量较少，且其占有区面积不断减小。因此，列为近危等级 / Lesser Brown Horseshoe Bat is found in habitat patches in Yunnan, China, and its population size is small. The areas of its occupancy are declining. Thus, it is listed as Near Threatened

地理分布 Geographical Distribution

国内分布 Domestic Distribution
云南 / Yunnan
世界分布 World Distribution
中国；东南亚 / China; Southeast Asia
分布标注 Distribution Note
非特有种 / Non-endemic

国内分布图
Map of Domestic Distribution

种群 Population

种群数量 Population Size	未知 / Unknown
种群趋势 Population Trend	未知 / Unknown

生境与生态系统 Habitat (s) and Ecosystem (s)

生境 Habitat(s)	森林、喀斯特地貌 / Forest, Karst Landscape
生态系统 Ecosystem(s)	森林生态系统 / Forest Ecosystem

威胁 Threat (s)

主要威胁 Major Threat(s)	未知 / Unknown

保护级别与保护行动 Protection Category and Conservation Action (s)

国家重点保护野生动物等级 (2021) Category of National Key Protected Wild Animals (2021)	无 / NA
IUCN 红色名录 (2020-2) IUCN Red List (2020-2)	无危 / LC
CITES 附录 (2019) CITES Appendix (2019)	无 / NA
保护行动 Conservation Action(s)	无 / None

相关文献 Relevant References

Wilson and Mittermeier, 2019; Burgin *et al.*, 2018; Jiang *et al.* (蒋志刚等), 2017; Zhang *et al.*, 2009a, 2009b, 2009c; Zhang *et al.* (张劲硕等), 2005

小褐菊头蝠 *Rhinolophus stheno*

By Alice Hughes

托氏菊头蝠
Rhinolophus thomasi

近危 NT

| 数据缺乏 DD | 无危 LC | 近危 NT | 易危 VU | 濒危 EN | 极危 CR | 区域灭绝 RE | 野外灭绝 EW | 灭绝 EX |

分类地位 Taxonomic Status

| 动物界 Animalia | 脊索动物门 Chordata | 哺乳纲 Mammalia | 翼手目 Chiroptera | 菊头蝠科 Rhinolophidae |

学名 Scientific Name	*Rhinolophus thomasi*
命名人 Species Authority	K. Andersen, 1905
英文名 English Name(s)	Thomas's Horseshoe Bat
同物异名 Synonym(s)	无 / None
种下单元评估 Infra-specific Taxa Assessed	无 / None

评估信息 Assessment Information

评估年份 Year Assessed	2020
评定人 Assessor(s)	吴毅、蒋志刚 / Yi Wu, Zhigang Jiang
审定人 Reviewer(s)	张礼标、毛秀光、张树义 / Libiao Zhang, Xiuguang Mao, Shuyi Zhang
其他贡献人 Other Contributor(s)	李立立、丁晨晨 / Lili Li, Chenchen Ding

理由 Justification: 托氏菊头蝠种群数量较少，且其适宜栖息地面积受到人类活动干扰。因此，列为近危等级 / Thomas's Horseshoe Bat has small populations. The areas of its suitable habitats are disturbed by human activities. Thus, it is listed as Near Threatened

地理分布 Geographical Distribution

国内分布 Domestic Distribution	云南、贵州、广西、四川 / Yunnan, Guizhou, Guangxi, Sichuan
世界分布 World Distribution	中国；东南亚 / China; Southeast Asia
分布标注 Distribution Note	非特有种 / Non-endemic

国内分布图
Map of Domestic Distribution

种群 Population

种群数量 Population Size	未知 / Unknown
种群趋势 Population Trend	未知 / Unknown

生境与生态系统 Habitat(s) and Ecosystem(s)

生境 Habitat(s)	喀斯特地貌、洞穴 / Karst Landscape, Cave
生态系统 Ecosystem(s)	洞穴生态系统 / Cave Ecosystem

威胁 Threat(s)

主要威胁 Major Threat(s)	未知 / Unknown

保护级别与保护行动 Protection Category and Conservation Action(s)

国家重点保护野生动物等级 (2021) Category of National Key Protected Wild Animals (2021)	无 / NA
IUCN 红色名录 (2020-2) IUCN Red List (2020-2)	无危 / LC
CITES 附录 (2019) CITES Appendix (2019)	无 / NA
保护行动 Conservation Action(s)	无 / None

相关文献 Relevant References

Wilson and Mittermeier, 2019; Burgin *et al.*, 2018; Jiang *et al.*（蒋志刚等）, 2017; Smith *et al.*（史密斯等）, 2009; Csorba *et al.*, 2003; Luo *et al.*（罗蓉等）, 1993

托氏菊头蝠 *Rhinolophus thomasi*　　　胡慧建 摄　By Huijian Hu

灰小蹄蝠
Hipposideros cineraceus

近危 NT

| 数据缺乏 DD | 无危 LC | 近危 NT | 易危 VU | 濒危 EN | 极危 CR | 区域灭绝 RE | 野外灭绝 EW | 灭绝 EX |

分类地位 Taxonomic Status

动物界 Animalia	脊索动物门 Chordata	哺乳纲 Mammalia	翼手目 Chiroptera	蹄蝠科 Hipposideridae

学名 Scientific Name	*Hipposideros cineraceus*
命名人 Species Authority	Blyth, 1853
英文名 English Name(s)	Least Leaf-nosed Bat
同物异名 Synonym(s)	Ashy Roundleaf Bat; *Hipposideros cineraceus* (Peters, 1872) subsp. *micropus*; *Phyllorhina micropus* (Peters, 1872)
种下单元评估 Infra-specific Taxa Assessed	无 / None

评估信息 Assessment Information

评估年份 Year Assessed	2020
评定人 Assessor(s)	吴毅、蒋志刚 / Yi Wu, Zhigang Jiang
审定人 Reviewer(s)	张礼标、毛秀光、张树义 / Libiao Zhang, Xiuguang Mao, Shuyi Zhang
其他贡献人 Other Contributor(s)	李立立、丁晨晨 / Lili Li, Chenchen Ding

理由 Justification: 灰小蹄蝠 2009 年在中国首次被记录到，其分布区狭窄，种群数量少，且其占有区面积减小。因此，列为近危等级 / Least Leaf-nosed Bat was first discovered in 2009 in China. Its distribution range in China is narrow, and its population size is small. Furthermore, the areas of its occupancy are declining. Thus, it is listed as Near Threatened

地理分布 Geographical Distribution

国内分布 Domestic Distribution	云南、广西 / Yunnan, Guangxi
世界分布 World Distribution	中国；东南亚 / China; Southeast Asia
分布标注 Distribution Note	非特有种 / Non-endemic

国内分布图
Map of Domestic Distribution

种群 Population

种群数量 Population Size	稀少 / Rare
种群趋势 Population Trend	未知 / Unknown

生境与生态系统 Habitat(s) and Ecosystem(s)

生境 Habitat(s)	洞穴、森林 / Cave, Forest
生态系统 Ecosystem(s)	喀斯特生态系统、森林生态系统 / Karst Ecosystem, Forest Ecosystem

威胁 Threat(s)

主要威胁 Major Threat(s)	未知 / Unknown

保护级别与保护行动 Protection Category and Conservation Action(s)

国家重点保护野生动物等级 (2021) Category of National Key Protected Wild Animals (2021)	无 / NA
IUCN 红色名录 (2020-2) IUCN Red List (2020-2)	无危 / LC
CITES 附录 (2019) CITES Appendix (2019)	无 / NA
保护行动 Conservation Action(s)	无 / None

相关文献 Relevant References

Wilson and Mittermeier, 2019; Burgin *et al*., 2018; Jiang *et al*.（蒋志刚等）, 2017; Tan *et al*.（谭敏等）, 2009; Koopman, 1993

灰小蹄蝠 *Hipposideros cineraceus*　　　　　　　　　　　余文华 摄　By Wenhua Yu

普氏蹄蝠
Hipposideros pratti

近危 NT

| 数据缺乏 DD | 无危 LC | 近危 NT | 易危 VU | 濒危 EN | 极危 CR | 区域灭绝 RE | 野外灭绝 EW | 灭绝 EX |

分类地位 Taxonomic Status

动物界 Animalia	脊索动物门 Chordata	哺乳纲 Mammalia	翼手目 Chiroptera	蹄蝠科 Hipposideridae
学名 Scientific Name		*Hipposideros pratti*		
命名人 Species Authority		Thomas, 1891		
英文名 English Name(s)		Pratt's Leaf-nosed Bat		
同物异名 Synonym(s)		Pratt's Roundleaf Bat		
种下单元评估 Infra-specific Taxa Assessed		无 / None		

评估信息 Assessment Information

评估年份 Year Assessed	2020
评定人 Assessor(s)	吴毅、蒋志刚 / Yi Wu, Zhigang Jiang
审定人 Reviewer(s)	张礼标、毛秀光、张树义 / Libiao Zhang, Xiuguang Mao, Shuyi Zhang
其他贡献人 Other Contributor(s)	李立立、丁晨晨 / Lili Li, Chenchen Ding

理由 Justification: 普氏蹄蝠分布广，但种群数量较少，且其生境受到人类活动干扰。因此，列为近危等级 / Pratt's Leaf-nosed Bat in China is widely distributed, but its population size is small. Furthermore, its habitats are disturbed by human activities. Thus, it is listed as Near Threatened

地理分布 Geographical Distribution

国内分布 Domestic Distribution
陕西、江苏、浙江、安徽、湖南、江西、四川、贵州、云南、广西、福建、重庆、广东、海南 / Shaanxi, Jiangsu, Zhejiang, Anhui, Hunan, Jiangxi, Sichuan, Guizhou, Yunnan, Guangxi, Fujian, Chongqing, Guangdong, Hainan
世界分布 World Distribution
中国；东南亚 / China; Southeast Asia
分布标注 Distribution Note
非特有种 / Non-endemic

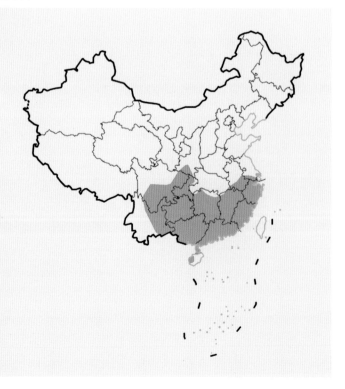

国内分布图
Map of Domestic Distribution

种群 Population

种群数量 Population Size	稀少 / Rare
种群趋势 Population Trend	稳定 / Stable

生境与生态系统 Habitat (s) and Ecosystem (s)

生　　境 Habitat(s)	洞穴 / Cave
生态系统 Ecosystem(s)	喀斯特生态系统 / Karst Ecosystem

威胁 Threat (s)

主要威胁 Major Threat(s)	采矿、采石场 / Mining, Quarrying Field

保护级别与保护行动 Protection Category and Conservation Action (s)

国家重点保护野生动物等级 (2021) Category of National Key Protected Wild Animals (2021)	无 / NA
IUCN 红色名录 (2020-2) IUCN Red List (2020-2)	无危 / LC
CITES 附录 (2019) CITES Appendix (2019)	无 / NA
保护行动 Conservation Action(s)	位于自然保护区内的种群与生境得到保护 / Populations and habitats in nature reserves are protected

相关文献 Relevant References

Burgin *et al.*, 2018; Jiang *et al.*（蒋志刚等），2017; Zeng（曾峰），2012; Liu（刘文超），2009; Wang（王婉莹），2007; Chen *et al.*（陈敏等），2002

普氏蹄蝠 *Hipposideros pratti*　　　　　吴毅 摄　By Yi Wu

三叶蹄蝠
Aselliscus stoliczkanus

近危 NT

| 数据缺乏 DD | 无危 LC | **近危 NT** | 易危 VU | 濒危 EN | 极危 CR | 区域灭绝 RE | 野外灭绝 EW | 灭绝 EX |

分类地位 Taxonomic Status

动物界 Animalia	脊索动物门 Chordata	哺乳纲 Mammalia	翼手目 Chiroptera	蹄蝠科 Hipposideridae
学名 Scientific Name		*Aselliscus stoliczkanus*		
命名人 Species Authority		Dobson, 1871		
英文名 English Name(s)		Stoliczka's Asian Trident Bat		
同物异名 Synonym(s)		Stoliczka's Trident Bat		
种下单元评估 Infra-specific Taxa Assessed		无 / None		

评估信息 Assessment Information

评估年份 Year Assessed	2020
评定人 Assessor(s)	吴毅、蒋志刚 / Yi Wu, Zhigang Jiang
审定人 Reviewer(s)	张礼标、毛秀光、张树义 / Libiao Zhang, Xiuguang Mao, Shuyi Zhang
其他贡献人 Other Contributor(s)	李立立、丁晨晨 / Lili Li, Chenchen Ding

理由 Justification: 三叶蹄蝠分布于中国西南，种群数量较少，且其占有区面积因人类活动有所减小。因此，列为近危等级 / Stoliczka's Asian Trident Bat is found in southwest China with small populations. The areas of its occupancy are declining due to human encroachment. Thus, it is listed as Near Threatened

地理分布 Geographical Distribution

国内分布 Domestic Distribution
云南、广东、广西、贵州 / Yunnan, Guangdong, Guangxi, Guizhou
世界分布 World Distribution
中国；东南亚 / China; Southeast Asia
分布标注 Distribution Note
非特有种 / Non-endemic

国内分布图
Map of Domestic Distribution

种群 Population

种群数量 Population Size	较常见 / Relatively common
种群趋势 Population Trend	稳定 / Stable

生境与生态系统 Habitat (s) and Ecosystem (s)

生境 Habitat(s)	洞穴、耕地、热带和亚热带严重退化的森林 / Cave, Arable Land, Tropical and Subtropical Degraded Forest
生态系统 Ecosystem(s)	喀斯特生态系统、森林生态系统、农田生态系统 / Karst Ecosystem, Forest Ecosystem, Cropland Ecosystem

威胁 Threat (s)

主要威胁 Major Threat(s)	采矿、采石场 / Mining, Quarrying Field

保护级别与保护行动 Protection Category and Conservation Action (s)

国家重点保护野生动物等级 (2021) Category of National Key Protected Wild Animals (2021)	无 / NA
IUCN 红色名录 (2020-2) IUCN Red List (2020-2)	无危 / LC
CITES 附录 (2019) CITES Appendix (2019)	无 / NA
保护行动 Conservation Action(s)	位于自然保护区内的种群与生境得到保护 / Populations and habitats in nature reserves are protected

相关文献 Relevant References

Wilson and Mittermeier, 2019; Burgin *et al.*, 2018; Jiang *et al.*（蒋志刚等），2017; Wang *et al.*, 2013; Wang（王延校），2012; Mao *et al.*, 2010a; Zhang *et al.*（张劲硕等），2009; Wu and Peng（吴毅和彭洪源），2005; Feng *et al.*（冯江等），2002; Zhou（周江），2001

三叶蹄蝠 *Aselliscus stoliczkanus*

By Alice Hughes

宽耳犬吻蝠
Tadarida insignis
近危 NT

| 数据缺乏 DD | 无危 LC | 近危 NT | 易危 VU | 濒危 EN | 极危 CR | 区域灭绝 RE | 野外灭绝 EW | 灭绝 EX |

分类地位 Taxonomic Status

动物界 Animalia	脊索动物门 Chordata	哺乳纲 Mammalia	翼手目 Chiroptera	犬吻蝠科 Molossidae
学名 Scientific Name		*Tadarida insignis*		
命名人 Species Authority		Blyth, 1862		
英文名 English Name(s)		East Asian Free-tailed Bat		
同物异名 Synonym(s)		无 / None		
种下单元评估 Infra-specific Taxa Assessed		无 / None		

评估信息 Assessment Information

评估年份 Year Assessed	2020
评定人 Assessor(s)	吴毅、蒋志刚 / Yi Wu, Zhigang Jiang
审定人 Reviewer(s)	石红艳、江廷磊、余文华 / Hongyan Shi, Tinglei Jiang, Wenhua Yu
其他贡献人 Other Contributor(s)	李立立、丁晨晨 / Lili Li, Chenchen Ding

理由 Justification: 宽耳犬吻蝠分布较狭窄，种群数量较少，且其占有区面积因人类活动减小。因此，列为近危等级 / East Asian Free-tailed Bat is narrowly distributed in China, and its population size is small. Furthermore, the areas of its habitats are constantly declining due to human encroachment. Thus, it is listed as Near Threatened

地理分布 Geographical Distribution

国内分布 Domestic Distribution
安徽、湖北、云南、广西、福建、四川、贵州、台湾 / Anhui, Hubei, Yunnan, Guangxi, Fujian, Sichuan, Guizhou, Taiwan
世界分布 World Distribution
中国；东北亚 / China; Northeast Asia
分布标注 Distribution Note
非特有种 / Non-endemic

国内分布图
Map of Domestic Distribution

种群 Population

种群数量 Population Size	未知 / Unknown
种群趋势 Population Trend	未知 / Unknown

生境与生态系统 Habitat (s) and Ecosystem (s)

生　　境 Habitat(s)	洞穴 / Cave
生态系统 Ecosystem(s)	喀斯特生态系统 / Karst Ecosystem

威胁 Threat (s)

主要威胁 Major Threat(s)	采矿、采石场 / Mining, Quarrying Field

保护级别与保护行动 Protection Category and Conservation Action (s)

国家重点保护野生动物等级 (2021) Category of National Key Protected Wild Animals (2021)	无 / NA
IUCN 红色名录 (2020-2) IUCN Red List (2020-2)	数据缺乏 / DD
CITES 附录 (2019) CITES Appendix (2019)	无 / NA
保护行动 Conservation Action(s)	无 / None

相关文献 Relevant References

Wilson and Mittermeier, 2019; Burgin *et al*., 2018; Jiang *et al*. (蒋志刚等), 2017; Zhou (周现召), 2012; Zheng (郑锡奇), 2010; You *et al*. (由玉岩等), 2009; Gao *et al*. (高武等), 1996; Zhang (张维道), 1985; Xu *et al*. (徐亚君等), 1982

宽耳犬吻蝠 *Tadarida insignis*

华北犬吻蝠
Tadarida latouchei

近危 NT

| 数据缺乏 DD | 无危 LC | 近危 NT | 易危 VU | 濒危 EN | 极危 CR | 区域灭绝 RE | 野外灭绝 EW | 灭绝 EX |

分类地位 Taxonomic Status

动物界 Animalia	脊索动物门 Chordata	哺乳纲 Mammalia	翼手目 Chiroptera	犬吻蝠科 Molossidae
学名 Scientific Name		*Tadarida latouchei*		
命名人 Species Authority		Thomas, 1920		
英文名 English Name(s)		La Touche's Free-tailed Bat		
同物异名 Synonym(s)		无 / None		
种下单元评估 Infra-specific Taxa Assessed		无 / None		

评估信息 Assessment Information

评估年份 Year Assessed	2020
评定人 Assessor(s)	吴毅、蒋志刚 / Yi Wu, Zhigang Jiang
审定人 Reviewer(s)	石红艳、江廷磊、余文华 / Hongyan Shi, Tinglei Jiang, Wenhua Yu
其他贡献人 Other Contributor(s)	李立立、丁晨晨 / Lili Li, Chenchen Ding

理由 Justification: 华北犬吻蝠分布区狭窄，种群数量较少，且其生境受人类活动的干扰。因此，列为近危等级 / La Touche's Free-tailed Bat is narrowly distributed in China, and its population size is small. Its habitats are disturbed by human activities. Thus, it is listed as Near Threatened

地理分布 Geographical Distribution

国内分布 Domestic Distribution
河北、北京、内蒙古、黑龙江、辽宁 / Hebei, Beijing, Inner Mongolia (Nei Mongol), Heilongjiang, Liaoning
世界分布 World Distribution
中国；东南亚 / China; Southeast Asia
分布标注 Distribution Note
非特有种 / Non-endemic

国内分布图
Map of Domestic Distribution

种群 Population

种群数量 Population Size	未知 / Unknown
种群趋势 Population Trend	下降 / Decreasing

生境与生态系统 Habitat(s) and Ecosystem(s)

生境 Habitat(s)	森林、洞穴 / Forest, Cave
生态系统 Ecosystem(s)	喀斯特生态系统、森林生态系统 / Karst Ecosystem, Forest Ecosystem

威胁 Threat(s)

主要威胁 Major Threat(s)	旅游、采矿、采石场 / Tourism, Mining, Quarrying Field

保护级别与保护行动 Protection Category and Conservation Action(s)

国家重点保护野生动物等级 (2021) Category of National Key Protected Wild Animals (2021)	无 / NA
IUCN 红色名录 (2020-2) IUCN Red List (2020-2)	数据缺乏 / DD
CITES 附录 (2019) CITES Appendix (2019)	无 / NA
保护行动 Conservation Action(s)	无 / None

相关文献 Relevant References

Wilson and Mittermeier, 2019; Burgin *et al.*, 2018; Jiang *et al.*（蒋志刚等）, 2017; Smith *et al.*（史密斯等）, 2009; Simmons, 2005; Thomas, 1920

华北犬吻蝠 *Tadarida latouchei*

西南鼠耳蝠
Myotis altarium
近危 NT

| 数据缺乏 DD | 无危 LC | 近危 NT | 易危 VU | 濒危 EN | 极危 CR | 区域灭绝 RE | 野外灭绝 EW | 灭绝 EX |

分类地位 Taxonomic Status

动物界 Animalia	脊索动物门 Chordata	哺乳纲 Mammalia	翼手目 Chiroptera	蝙蝠科 Vespertilionidae
学名 Scientific Name		*Myotis altarium*		
命名人 Species Authority		Thomas, 1911		
英文名 English Name(s)		Szechwan Mouse-eared Bat		
同物异名 Synonym(s)		Szechwan Myotis		
种下单元评估 Infra-specific Taxa Assessed		无 / None		

评估信息 Assessment Information

评估年份 Year Assessed	2020
评定人 Assessor(s)	吴毅、蒋志刚 / Yi Wu, Zhigang Jiang
审定人 Reviewer(s)	江廷磊、余文华、石红艳 / Tinglei Jiang, Wenhua Yu, Hongyan Shi
其他贡献人 Other Contributor(s)	李立立、丁晨晨 / Lili Li, Chenchen Ding

理由 Justification: 西南鼠耳蝠分布广，但种群数量较少，且其占有区面积因人类活动而减小。因此，列为近危等级 / Szechwan Mouse-eared Bat is widely distributed in China, but its population size is small. Furthermore, the areas of its occupancy are declining due to human encroachment. Thus, it is listed as Near Threatened

地理分布 Geographical Distribution

国内分布 Domestic Distribution
安徽、浙江、江西、四川、贵州、重庆、湖南、云南、广西、广东、福建 / Anhui, Zhejiang, Jiangxi, Sichuan, Guizhou, Chongqing, Hunan, Yunnan, Guangxi, Guangdong, Fujian
世界分布 World Distribution
中国；东南亚 / China; Southeast Asia
分布标注 Distribution Note
非特有种 / Non-endemic

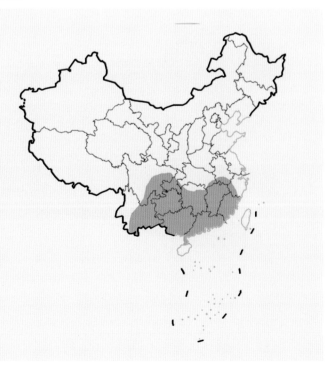

国内分布图
Map of Domestic Distribution

种群 Population

种群数量 Population Size	较常见 / Relatively common
种群趋势 Population Trend	稳定 / Stable

生境与生态系统 Habitat(s) and Ecosystem(s)

生境 Habitat(s)	喀斯特洞穴 / Karst Cave
生态系统 Ecosystem(s)	喀斯特生态系统、人类聚落生态系统 / Karst Ecosystem, Human Settlement Ecosystem

威胁 Threat(s)

主要威胁 Major Threat(s)	采矿、采石场、旅游 / Mining, Quarrying Field and Tourism

保护级别与保护行动 Protection Category and Conservation Action(s)

国家重点保护野生动物等级 (2021) Category of National Key Protected Wild Animals (2021)	无 / NA
IUCN 红色名录 (2020-2) IUCN Red List (2020-2)	无危 / LC
CITES 附录 (2019) CITES Appendix (2019)	无 / NA
保护行动 Conservation Action(s)	无 / None

相关文献 Relevant References

Burgin *et al*., 2018; Jiang *et al*.（蒋志刚等）, 2017; Pei（裴俊峰）, 2012; Fu *et al*.（符丹凤等）, 2010; Zhang *et al*.（张燕均等）, 2010; Wang and Hu（王酉之和胡锦矗）, 1999; Chen *et al*.（陈延熹等）, 1987; Liang and Dong（梁仁济和董永文）, 1984; Xu *et al*.（徐亚君等）, 1984

西南鼠耳蝠 *Myotis altarium*　　　　吴毅 摄　By Yi Wu

缺齿鼠耳蝠
Myotis annectans
近危 NT

| 数据缺乏 DD | 无危 LC | 近危 NT | 易危 VU | 濒危 EN | 极危 CR | 区域灭绝 RE | 野外灭绝 EW | 灭绝 EX |

分类地位 Taxonomic Status

动物界 Animalia	脊索动物门 Chordata	哺乳纲 Mammalia	翼手目 Chiroptera	蝙蝠科 Vespertilionidae
学 名 Scientific Name		*Myotis annectans*		
命 名 人 Species Authority		Dobson, 1871		
英 文 名 English Name(s)		Hairy-faced Mouse-eared Bat		
同物异名 Synonym(s)		Hairy-faced Bat; Interediate Bat; *Myotis annectans* (Thomas, 1920) subsp. *primula*; *primula* (Thomas, 1920); *Vesperugo anectens* (Dobson, 1876)		
种下单元评估 Infra-specific Taxa Assessed		无 / None		

评估信息 Assessment Information

评估年份 Year Assessed	2020
评 定 人 Assessor(s)	吴毅、蒋志刚 / Yi Wu, Zhigang Jiang
审 定 人 Reviewer(s)	江廷磊、余文华、石红艳 / Tinglei Jiang, Wenhua Yu, Hongyan Shi
其他贡献人 Other Contributor(s)	李立立、丁晨晨 / Lili Li, Chenchen Ding

理由 Justification: 缺齿鼠耳蝠分布区小，种群数量少，所采集到的标本不多，且其占有区面积因人类活动而减小。因此，列为近危等级 / Hairy-faced Mouse-eared Bat is narrowly distributed with small populations, only a handful of specimens have been collected so far. Furthermore, the areas of its habitats are declining due to human encroachment. Thus, it is listed as Near Threatened

地理分布 Geographical Distribution

国内分布 Domestic Distribution
云南 / Yunnan
世界分布 World Distribution
中国；南亚、东南亚 / China; South Asia, Southeast Asia
分布标注 Distribution Note
非特有种 / Non-endemic

国内分布图
Map of Domestic Distribution

种群 Population

种群数量 Population Size	未知 / Unknown
种群趋势 Population Trend	未知 / Unknown

生境与生态系统 Habitat (s) and Ecosystem (s)

生 境 Habitat(s)	喀斯特洞穴 / Karst Cave
生态系统 Ecosystem(s)	喀斯特生态系统、人类聚落生态系统 / Karst Ecosystem, Human Settlement Ecosystem

威胁 Threat (s)

主要威胁 Major Threat(s)	旅游、采矿、采石场 / Tourism, Mining, Quarrying Field

保护级别与保护行动 Protection Category and Conservation Action (s)

国家重点保护野生动物等级 (2021) Category of National Key Protected Wild Animals (2021)	无 / NA
IUCN 红色名录(2020-2) IUCN Red List (2020-2)	无危 / LC
CITES 附录 (2019) CITES Appendix (2019)	无 / NA
保护行动 Conservation Action(s)	无 / None

相关文献 Relevant References

Wilson and Mittermeier, 2019; Burgin *et al*., 2018; Jiang *et al*.（蒋志刚等）, 2017; Smith *et al*.（史密斯等）, 2009; Wang（王应祥）, 2003; Luo（罗一宁）, 1987

缺齿鼠耳蝠 *Myotis annectans*

尖耳鼠耳蝠
Myotis blythii

近危 NT

| 数据缺乏 DD | 无危 LC | 近危 NT | 易危 VU | 濒危 EN | 极危 CR | 区域灭绝 RE | 野外灭绝 EW | 灭绝 EX |

分类地位 Taxonomic Status

动物界 Animalia	脊索动物门 Chordata	哺乳纲 Mammalia	翼手目 Chiroptera	蝙蝠科 Vespertilionidae
学 名 Scientific Name		*Myotis blythii*		
命 名 人 Species Authority		Tomes, 1857		
英 文 名 English Name(s)		Lesser Mouse-eared Bat		
同物异名 Synonym(s)		Lesser Mouse-eared Myotis; *Myotis oxygnathus* (Monticelli, 1885)		
种下单元评估 Infra-specific Taxa Assessed		无 / None		

评估信息 Assessment Information

评 估 年 份 Year Assessed	2020
评 定 人 Assessor(s)	吴毅、蒋志刚 / Yi Wu, Zhigang Jiang
审 定 人 Reviewer(s)	江廷磊、余文华、石红艳 / Tinglei Jiang, Wenhua Yu, Hongyan Shi
其他贡献人 Other Contributor(s)	李立立、丁晨晨 / Lili Li, Chenchen Ding

理由 Justification: 尖耳鼠耳蝠在中国的分布区狭窄，且种群数量较少，其占有区面积因人类活动而减小。因此，列为近危等级 / Lesser Mouse-eared Bat is narrowly distributed in China, and its population size is small. Furthermore, the areas of its occupancy are constantly declining due to human encroachment. Thus, it is listed as Near Threatened

地理分布 Geographical Distribution

国内分布 Domestic Distribution
内蒙古、新疆、陕西、广东、广西、山西 / Inner Mongolia (Nei Mongol), Xinjiang, Shaanxi, Guangdong, Guangxi, Shanxi
世界分布 World Distribution
亚洲、欧洲 / Asia, Europe
分布标注 Distribution Note
非特有种 / Non-endemic

国内分布图
Map of Domestic Distribution

种群 Population

种群数量 Population Size	稀少 / Rare
种群趋势 Population Trend	下降 / Decreasing

生境与生态系统 Habitat(s) and Ecosystem(s)

生境 Habitat(s)	洞穴、人造建筑、耕地、乡村花园、灌丛、草地 / Cave, Building, Arable Land, Rural Garden, Shrubland, Grassland
生态系统 Ecosystem(s)	喀斯特生态系统、人类聚落生态系统 / Karst Ecosystem, Human Settlement Ecosystem

威胁 Threat(s)

主要威胁 Major Threat(s)	住房及城市建设、采矿、采石场、耕种 / Construction of House and Urban Area, Mining, Quarrying Field, Plantation

保护级别与保护行动 Protection Category and Conservation Action(s)

国家重点保护野生动物等级 (2021) Category of National Key Protected Wild Animals (2021)	无 / NA
IUCN 红色名录 (2020-2) IUCN Red List (2020-2)	无危 / LC
CITES 附录 (2019) CITES Appendix (2019)	无 / NA
保护行动 Conservation Action(s)	无 / None

相关文献 Relevant References

Wilson and Mittermeier, 2019; Burgin *et al.*, 2018; Jiang *et al.* (蒋志刚等), 2017; Zhou and Yang (周江和杨天友), 2012b; Liu *et al.* (刘少英等), 2001

尖耳鼠耳蝠 *Myotis blythii*

By Amirekulk, CC BY-SA 4.0

布氏鼠耳蝠
Myotis brandtii

近危 NT

| 数据缺乏 DD | 无危 LC | 近危 NT | 易危 VU | 濒危 EN | 极危 CR | 区域灭绝 RE | 野外灭绝 EW | 灭绝 EX |

分类地位 Taxonomic Status

动物界 Animalia	脊索动物门 Chordata	哺乳纲 Mammalia	翼手目 Chiroptera	蝙蝠科 Vespertilionidae
学名 Scientific Name		*Myotis brandtii*		
命名人 Species Authority		Eversmann, 1845		
英文名 English Name(s)		Brandt's Mouse-eared Bat		
同物异名 Synonym(s)		Brandt's Bat; *Myotis brandti* (Eversmann, 1845); *gracilis* (Ognev, 1927)		
种下单元评估 Infra-specific Taxa Assessed		无 / None		

评估信息 Assessment Information

评估年份 Year Assessed	2020
评定人 Assessor(s)	吴毅、蒋志刚 / Yi Wu, Zhigang Jiang
审定人 Reviewer(s)	江廷磊、余文华、石红艳 / Tinglei Jiang, Wenhua Yu, Hongyan Shi
其他贡献人 Other Contributor(s)	李立立、丁晨晨 / Lili Li, Chenchen Ding

理由 Justification: 布氏鼠耳蝠在中国东北呈间断分布，种群数量少，且其占有区面积减小。因此，列为近危等级 / Brandt's Mouse-eared Bat is distributed widely in scattered populations across northeast China, but in relatively small numbers. Furthermore, its areas of occupancy are declining. Thus, it is listed as Near Threatened

地理分布 Geographical Distribution

国内分布 Domestic Distribution
内蒙古、辽宁、吉林、黑龙江、西藏 / Inner Mongolia (Nei Mongol), Liaoning, Jilin, Heilongjiang, Tibet (Xizang)
世界分布 World Distribution
亚洲、欧洲 / Asia, Europe
分布标注 Distribution Note
非特有种 / Non-endemic

国内分布图
Map of Domestic Distribution

种群 Population

种群数量 Population Size	未知 / Unknown
种群趋势 Population Trend	稳定 / Stable

生境与生态系统 Habitat (s) and Ecosystem (s)

生境 Habitat(s)	森林、人造建筑、洞穴 / Forest, Building, Cave
生态系统 Ecosystem(s)	森林生态系统、喀斯特生态系统、人类聚落生态系统 / Forest Ecosystem, Karst Ecosystem, Human Settlement Ecosystem

威胁 Threat (s)

主要威胁 Major Threat(s)	未知 / Unknown

保护级别与保护行动 Protection Category and Conservation Action (s)

国家重点保护野生动物等级 (2021) Category of National Key Protected Wild Animals (2021)	无 / NA
IUCN 红色名录 (2020-2) IUCN Red List (2020-2)	无危 / LC
CITES 附录 (2019) CITES Appendix (2019)	无 / NA
保护行动 Conservation Action(s)	无 / None

相关文献 Relevant References

Wilson and Mittermeier, 2019; Burgin *et al*., 2018; Jiang *et al*.（蒋志刚等）, 2017; Seim *et al*., 2013; Smith *et al*.（史密斯等）, 2009; Wang（王应祥）, 2003; Wilson and Reeder, 1993

布氏鼠耳蝠 *Myotis brandtii*

中华鼠耳蝠
Myotis chinensis

近危 NT

| 数据缺乏 DD | 无危 LC | 近危 NT | 易危 VU | 濒危 EN | 极危 CR | 区域灭绝 RE | 野外灭绝 EW | 灭绝 EX |

分类地位 Taxonomic Status

动物界 Animalia	脊索动物门 Chordata	哺乳纲 Mammalia	翼手目 Chiroptera	蝙蝠科 Vespertilionidae
学名 Scientific Name		*Myotis chinensis*		
命名人 Species Authority		Tomes, 1857		
英文名 English Name(s)		Large Mouse-eared Bat		
同物异名 Synonym(s)		Large Myotis		
种下单元评估 Infra-specific Taxa Assessed		无 / None		

评估信息 Assessment Information

评估年份 Year Assessed	2020
评定人 Assessor(s)	吴毅、蒋志刚 / Yi Wu, Zhigang Jiang
审定人 Reviewer(s)	江廷磊、余文华、石红艳 / Tinglei Jiang, Wenhua Yu, Hongyan Shi
其他贡献人 Other Contributor(s)	李立立、丁晨晨 / Lili Li, Chenchen Ding

理由 Justification: 中华鼠耳蝠分布较广，但种群数量少，且其占有区面积因人类活动而减小。因此，列为近危等级 / Large Mouse-eared Bat is widely distributed in China, but its population size is small. Furthermore, the areas of its occupation are declining due to human encroachment. Thus, it is listed as Near Threatened

地理分布 Geographical Distribution

国内分布 Domestic Distribution

浙江、安徽、江西、广西、四川、香港、贵州、重庆、江苏、福建、广东、海南、湖南、云南 / Zhejiang, Anhui, Jiangxi, Guangxi, Sichuan, Hong Kong, Guizhou, Chongqing, Jiangsu, Fujian, Guangdong, Hainan, Hunan, Yunnan

世界分布 World Distribution

中国；东南亚 / China; Southeast Asia

分布标注 Distribution Note

非特有种 / Non-endemic

国内分布图
Map of Domestic Distribution

种群 Population

种群数量 Population Size	较常见 / Relatively common
种群趋势 Population Trend	未知 / Unknown

生境与生态系统 Habitat (s) and Ecosystem (s)

生　　境 Habitat(s)	洞穴、喀斯特地貌 / Cave, Karst Landscape
生态系统 Ecosystem(s)	喀斯特生态系统 / Karst Ecosystem

威胁 Threat (s)

主要威胁 Major Threat(s)	未知 / Unknown

保护级别与保护行动 Protection Category and Conservation Action (s)

国家重点保护野生动物等级 (2021) Category of National Key Protected Wild Animals (2021)	无 / NA
IUCN 红色名录 (2020-2) IUCN Red List (2020-2)	无危 / LC
CITES 附录 (2019) CITES Appendix (2019)	无 / NA
保护行动 Conservation Action(s)	无 / None

相关文献 Relevant References

Wilson and Mittermeier, 2019; Burgin *et al*., 2018; Jiang *et al*. (蒋志刚等), 2017; Liu *et al*. (刘昊等), 2010; Tian and Jin, 2012; Ma *et al*., 2008; Zhang (张维道), 1984

中华鼠耳蝠 *Myotis chinensis*　　　　　　　　　　　　　　　　　　　　　　　吴毅 摄　By Yi Wu

毛腿鼠耳蝠
Myotis fimbriatus

近危 NT

| 数据缺乏 DD | 无危 LC | 近危 NT | 易危 VU | 濒危 EN | 极危 CR | 区域灭绝 RE | 野外灭绝 EW | 灭绝 EX |

分类地位 Taxonomic Status

动物界 Animalia	脊索动物门 Chordata	哺乳纲 Mammalia	翼手目 Chiroptera	蝙蝠科 Vespertilionidae
学名 Scientific Name		*Myotis fimbriatus*		
命名人 Species Authority		Peters, 1871		
英文名 English Name(s)		Fringed Long-footed Mouse-eared Bat		
同物异名 Synonym(s)		Fringed Long-footed Myotis		
种下单元评估 Infra-specific Taxa Assessed		无 / None		

评估信息 Assessment Information

评估年份 Year Assessed	2020
评定人 Assessor(s)	吴毅、蒋志刚 / Yi Wu, Zhigang Jiang
审定人 Reviewer(s)	江廷磊、余文华、石红艳 / Tinglei Jiang, Wenhua Yu, Hongyan Shi
其他贡献人 Other Contributor(s)	李立立、丁晨晨 / Lili Li, Chenchen Ding

理由 Justification: 毛腿鼠耳蝠分布区有限，种群数量少，且受人类活动，如采石及旅游的影响，其占有区面积减小。因此，列为近危等级 / Fringed Long-footed Mouse-eared Bat has restricted distribution in China, and its population size is small. Furthermore, its areas of occupancy are declining due to human activities such as quarrying and traveling. Thus, it is listed as Near Threatened

地理分布 Geographical Distribution

国内分布 Domestic Distribution
江苏、浙江、安徽、福建、江西、四川、贵州、云南、香港 / Jiangsu, Zhejiang, Anhui, Fujian, Jiangxi, Sichuan, Guizhou, Yunnan, Hong Kong
世界分布 World Distribution
中国 / China
分布标注 Distribution Note
特有种 / Endemic

国内分布图
Map of Domestic Distribution

种群 Population

种群数量 Population Size	未知 / Unknown
种群趋势 Population Trend	未知 / Unknown

生境与生态系统 Habitat(s) and Ecosystem(s)

生境 Habitat(s)	洞穴、喀斯特地貌 / Cave, Karst Landscape
生态系统 Ecosystem(s)	森林生态系统、喀斯特生态系统 / Forest Ecosystem, Karst Ecosystem

威胁 Threat(s)

主要威胁 Major Threat(s)	采石、旅游 / Quarrying, Tourism

保护级别与保护行动 Protection Category and Conservation Action(s)

国家重点保护野生动物等级 (2021) Category of National Key Protected Wild Animals (2021)	无 / NA
IUCN红色名录 (2020-2) IUCN Red List (2020-2)	无危 / LC
CITES 附录 (2019) CITES Appendix (2019)	无 / NA
保护行动 Conservation Action(s)	位于自然保护区内的种群与生境得到保护 / Populations and habitats in nature reserves are protected

相关文献 Relevant References

Burgin *et al*., 2018; Jiang *et al*.（蒋志刚等）, 2017; Wang *et al*.（王会等）, 2009; Liu *et al*.（刘颖等）, 2003; Wang and Hu（王酉之和胡锦矗）, 1999; Dong *et al*.（董聿茂等）, 1989; Chen *et al*.（陈延熹等）, 1987; Xu *et al*.（徐亚君等）, 1984; Zhang *et al*.（张维道等）, 1983

毛腿鼠耳蝠 *Myotis fimbriatus* 　　　　张礼标 摄　By Libiao Zhang

宽吻鼠耳蝠
Myotis latirostris

近危 NT

数据缺乏 DD	无危 LC	近危 NT	易危 VU	濒危 EN	极危 CR	区域灭绝 RE	野外灭绝 EW	灭绝 EX

分类地位 Taxonomic Status

动物界 Animalia	脊索动物门 Chordata	哺乳纲 Mammalia	翼手目 Chiroptera	蝙蝠科 Vespertilionidae

学名 Scientific Name	*Myotis latirostris*
命名人 Species Authority	Kishida, 1932
英文名 English Name(s)	Formosan Broad-muzzled Mouse-eared Bat
同物异名 Synonym(s)	Formosan Broad-muzzled Myotis
种下单元评估 Infra-specific Taxa Assessed	无 / None

评估信息 Assessment Information

评估年份 Year Assessed	2020
评定人 Assessor(s)	吴毅、蒋志刚 / Yi Wu, Zhigang Jiang
审定人 Reviewer(s)	江廷磊、余文华、石红艳 / Tinglei Jiang, Wenhua Yu, Hongyan Shi
其他贡献人 Other Contributor(s)	李立立、丁晨晨 / Lili Li, Chenchen Ding

理由 Justification: 宽吻鼠耳蝠为特有种，仅分布在我国台湾，该种种群数量较少，且其占有区面积因人类活动而减小。因此，列为近危等级 / Formosan Broad-muzzled Mouse-eared Bat is an endemic species distributed only in Taiwan, China. Its population size is small and its areas of occupancy are declining due to human encroachment. Thus, it is listed as Near Threatened

地理分布 Geographical Distribution

国内分布 Domestic Distribution
台湾 / Taiwan
世界分布 World Distribution
中国 / China
分布标注 Distribution Note
特有种 / Endemic

国内分布图
Map of Domestic Distribution

种群 Population

种群数量 Population Size	未知 / Unknown
种群趋势 Population Trend	未知 / Unknown

生境与生态系统 Habitat (s) and Ecosystem (s)

生　　境 Habitat(s)	未知 / Unknown
生态系统 Ecosystem(s)	森林生态系统 / Forest Ecosystem

威胁 Threat (s)

主要威胁 Major Threat(s)	未知 / Unknown

保护级别与保护行动 Protection Category and Conservation Action (s)

国家重点保护野生动物等级 (2021) Category of National Key Protected Wild Animals (2021)	无 / NA
IUCN 红色名录 (2020-2) IUCN Red List (2020-2)	未列入 / NA
CITES 附录 (2019) CITES Appendix (2019)	无 / NA
保护行动 Conservation Action(s)	无 / None

相关文献 Relevant References

Wilson and Mittermeier, 2019; Burgin *et al*., 2018; Jiang *et al*. (蒋志刚等), 2017; Ruedi *et al*., 2013; Wang (王应祥), 2003

宽吻鼠耳蝠 Myotis latirostris　　　　　蒋志刚 绘　Drawn by Zhigang Jiang

喜山鼠耳蝠
Myotis muricola
近危 NT

| 数据缺乏 DD | 无危 LC | 近危 NT | 易危 VU | 濒危 EN | 极危 CR | 区域灭绝 RE | 野外灭绝 EW | 灭绝 EX |

分类地位 Taxonomic Status

动物界 Animalia	脊索动物门 Chordata	哺乳纲 Mammalia	翼手目 Chiroptera	蝙蝠科 Vespertilionidae
学 名 Scientific Name		*Myotis muricola*		
命 名 人 Species Authority		Gray, 1846		
英 文 名 English Name(s)		Nepalese Whiskered Mouse-eared Bat		
同物异名 Synonym(s)		Nepalese Whiskered Myotis; Wall-roosting Mouse-eared Bat; *Myotis mystacinus* (Tomes, 1859) subsp. *caliginosus*; *M. mystacinus* (Gray, 1846) subsp. *muricola*；（转下页）		
种下单元评估 Infra-specific Taxa Assessed		无 / None		

评估信息 Assessment Information

评估年份 Year Assessed	2020
评 定 人 Assessor(s)	吴毅、蒋志刚 / Yi Wu, Zhigang Jiang
审 定 人 Reviewer(s)	江廷磊、余文华、石红艳 / Tinglei Jiang, Wenhua Yu, Hongyan Shi
其他贡献人 Other Contributor(s)	李立立、丁晨晨 / Lili Li, Chenchen Ding

理由 Justification: 喜山鼠耳蝠亚种间生境严重分割，且种群数量不多，受到人类活动的影响，其占有区面积不断减小。因此，列为近危等级 / Nepalese Whiskered Mouse-eared Bat is severely fragmentized among its subspecies, and its population size is small. Furthermore, the areas of its habitats are declining due to human disturbance. Thus, it is listed as Near Threatened

地理分布 Geographical Distribution

国内分布 Domestic Distribution	四川、云南、西藏、台湾 / Sichuan, Yunnan, Tibet (Xizang), Taiwan
世界分布 World Distribution	中国；中亚、南亚、东南亚 / China; Central Asia, South Asia, Southeast Asia
分布标注 Distribution Note	非特有种 / Non-endemic

国内分布图
Map of Domestic Distribution

种群 Population

种群数量 Population Size	未知 / Unknown
种群趋势 Population Trend	稳定 / Stable

生境与生态系统 Habitat (s) and Ecosystem (s)

生　　境 Habitat(s)	热带和亚热带湿润低地森林、次生林、灌丛、乡村花园、洞穴 / Tropical and Subtropical Moist Lowland Forest, Secondary Forest, Shrubland, Rural Garden, Cave
生态系统 Ecosystem(s)	森林生态系统、灌丛生态系统、人类聚落生态系统 / Forest Ecosystem, Shrubland Ecosystem, Human Settlement Ecosystem

威胁 Threat (s)

主要威胁 Major Threat(s)	伐木、火灾、耕种 / Logging, Fire, Plantation

保护级别与保护行动 Protection Category and Conservation Action (s)

国家重点保护野生动物等级 (2021) Category of National Key Protected Wild Animals (2021)	无 / NA
IUCN 红色名录 (2020-2) IUCN Red List (2020-2)	无危 / LC
CITES 附录 (2019) CITES Appendix (2019)	无 / NA
保护行动 Conservation Action(s)	位于自然保护区内的种群与生境得到保护 / Populations and habitats in nature reserves are protected

相关文献 Relevant References

Wilson and Mittermeier, 2019; Burgin *et al.*, 2018; Jiang *et al.* (蒋志刚等), 2017; Hu *et al.* (胡一鸣等), 2014; Zheng (郑锡奇), 2010; Shek and Chan, 2006

(接上页)
Vespertilio blanfordi (Dobson, 1871); *V. caliginosus* (Tomes, 1859); *V. muricola* (Hodgson, 1841); *V. muricola* (Gray, 1846)

喜山鼠耳蝠 *Myotis muricola*　　　　　　　　　　　　　　　　　　　　　　　　　　　　By Alice Hughes

高颅鼠耳蝠
Myotis siligorensis

近危 NT

| 数据缺乏 DD | 无危 LC | 近危 NT | 易危 VU | 濒危 EN | 极危 CR | 区域灭绝 RE | 野外灭绝 EW | 灭绝 EX |

分类地位 Taxonomic Status

动物界 Animalia	脊索动物门 Chordata	哺乳纲 Mammalia	翼手目 Chiroptera	蝙蝠科 Vespertilionidae
学名 Scientific Name		*Myotis siligorensis*		
命名人 Species Authority		Horsfield, 1855		
英文名 English Name(s)		Himalayan Whiskered Mouse-eared Bat		
同物异名 Synonym(s)		Himalayan Whiskered Bat; *Vespertilio darjilingensis* (Horsfield, 1855); *V. siligorensis* (Horsfield, 1855)		
种下单元评估 Infra-specific Taxa Assessed		无 / None		

评估信息 Assessment Information

评估年份 Year Assessed	2020
评定人 Assessor(s)	吴毅、蒋志刚 / Yi Wu, Zhigang Jiang
审定人 Reviewer(s)	江廷磊、余文华、石红艳 / Tinglei Jiang, Wenhua Yu, Hongyan Shi
其他贡献人 Other Contributor(s)	李立立、丁晨晨 / Lili Li, Chenchen Ding

理由 Justification: 高颅鼠耳蝠发生区分散，种群数量少，其占有区面积因人类活动而减小。因此，列为近危等级 / The areas of occurrence of Himalayan Whiskered Mouse-eared Bat are scattered and small, and its population is small. Its areas of occupancy are declining due to human encroachment. Thus, it is listed as Near Threatened

地理分布 Geographical Distribution

国内分布 Domestic Distribution
云南、海南、福建、广东 / Yunnan, Hainan, Fujian, Guangdong
世界分布 World Distribution
中国；南亚、东南亚 / China; South Asia, Southeast Asia
分布标注 Distribution Note
非特有种 / Non-endemic

国内分布图
Map of Domestic Distribution

种群 Population

种群数量 Population Size	未知 / Unknown
种群趋势 Population Trend	未知 / Unknown

生境与生态系统 Habitat (s) and Ecosystem (s)

生　　境 Habitat(s)	次生林、溪流边、洞穴、人造建筑 / Secondary Forest, Near Stream, Cave, Building
生态系统 Ecosystem(s)	森林生态系统、湖泊河流生态系统、人类聚落生态系统 / Forest Ecosystem, Lake and River Ecosystem, Human Settlement Ecosystem

威胁 Threat (s)

主要威胁 Major Threat(s)	未知 / Unknown

保护级别与保护行动 Protection Category and Conservation Action (s)

国家重点保护野生动物等级 (2021) Category of National Key Protected Wild Animals (2021)	无 / NA
IUCN 红色名录(2020-2) IUCN Red List (2020-2)	无危 / LC
CITES 附录 (2019) CITES Appendix (2019)	无 / NA
保护行动 Conservation Action(s)	无 / None

相关文献 Relevant References

Wilson and Mittermeier, 2019; Burgin *et al*., 2018; Jiang *et al*. (蒋志刚等), 2017; Liu *et al*. (刘志霄等), 2013; Lin *et al*. (林洪军等), 2012; Wei *et al*. (韦力等), 2006

高颅鼠耳蝠 *Myotis siligorensis*　　　　　张礼标 摄　By Libiao Zhang

台湾鼠耳蝠
Myotis taiwanensis

近危 NT

| 数据缺乏 DD | 无危 LC | 近危 NT | 易危 VU | 濒危 EN | 极危 CR | 区域灭绝 RE | 野外灭绝 EW | 灭绝 EX |

分类地位 Taxonomic Status

动物界 Animalia	脊索动物门 Chordata	哺乳纲 Mammalia	翼手目 Chiroptera	蝙蝠科 Vespertilionidae

学名 Scientific Name	*Myotis taiwanensis*
命名人 Species Authority	Linde, 1908
英文名 English Name(s)	Taiwanese Mouse-eared Bat
同物异名 Synonym(s)	Taiwanese Myotis
种下单元评估 Infra-specific Taxa Assessed	无 / None

评估信息 Assessment Information

评估年份 Year Assessed	2020
评定人 Assessor(s)	吴毅、蒋志刚 / Yi Wu, Zhigang Jiang
审定人 Reviewer(s)	江廷磊、余文华、石红艳 / Tinglei Jiang, Wenhua Yu, Hongyan Shi
其他贡献人 Other Contributor(s)	李立立、丁晨晨 / Lili Li, Chenchen Ding

理由 Justification：台湾鼠耳蝠为特有种，仅分布在我国台湾，该种种群数量较少，且其生境受到人类活动干扰。因此，列为近危等级 / Taiwanese Mouse-eared Bat is an endemic species, distributed only in Taiwan, China, and its population size is small. Its habitat is disturbed by human activities. Thus, it is listed as Near Threatened

地理分布 Geographical Distribution

国内分布 Domestic Distribution	台湾 / Taiwan
世界分布 World Distribution	中国 / China
分布标注 Distribution Note	特有种 / Endemic

国内分布图
Map of Domestic Distribution

种群 Population

种群数量 Population Size	未知 / Unknown
种群趋势 Population Trend	未知 / Unknown

生境与生态系统 Habitat (s) and Ecosystem (s)

生　　境 Habitat(s)	森林 / Forest
生态系统 Ecosystem(s)	森林生态系统 / Forest Ecosystem

威胁 Threat (s)

主要威胁 Major Threat(s)	伐木、耕种 / Logging, Plantation

保护级别与保护行动 Protection Category and Conservation Action (s)

国家重点保护野生动物等级 (2021) Category of National Key Protected Wild Animals (2021)	无 / NA
IUCN 红色名录 (2020-2) IUCN Red List (2020-2)	未列入 / NA
CITES 附录 (2019) CITES Appendix (2019)	无 / NA
保护行动 Conservation Action(s)	无 / None

相关文献 Relevant References

Wilson and Mittermeier, 2019; Burgin *et al*., 2018; Jiang *et al*.（蒋志刚等）, 2017; Han *et al*., 2010; Zheng（郑锡奇）, 2010; Lee *et al*., 2007; Lin *et al*., 2002a, 2002b

台湾鼠耳蝠 *Myotis taiwanensis*　　　　蒋志刚 绘　Drawn by Zhigang Jiang

爪哇伏翼
Pipistrellus javanicus

近危 NT

| 数据缺乏 DD | 无危 LC | 近危 NT | 易危 VU | 濒危 EN | 极危 CR | 区域灭绝 RE | 野外灭绝 EW | 灭绝 EX |

分类地位 Taxonomic Status

| 动物界 Animalia | 脊索动物门 Chordata | 哺乳纲 Mammalia | 翼手目 Chiroptera | 蝙蝠科 Vespertilionidae |

学名 Scientific Name	*Pipistrellus javanicus*
命名人 Species Authority	Gray, 1838
英文名 English Name(s)	Javan Pipistrelle
同物异名 Synonym(s)	*Pipistrellus babu* (Thomas, 1915); *P. camortae* (Miller, 1902); *P. peguensis* (Sinha, 1969); *Scotophilus javanicus* (Gray, 1838)
种下单元评估 Infra-specific Taxa Assessed	无 / None

评估信息 Assessment Information

评估年份 Year Assessed	2020
评定人 Assessor(s)	吴毅、蒋志刚 / Yi Wu, Zhigang Jiang
审定人 Reviewer(s)	余文华、石红艳、江廷磊 / Wenhua Yu, Hongyan Shi, Tinglei Jiang
其他贡献人 Other Contributor(s)	李立立、丁晨晨 / Lili Li, Chenchen Ding

理由 Justification: 爪哇伏翼在中国的分布区为其边缘分布，种群数量少，且其所在的栖息地受到人类活动的影响。因此，列为近危等级 / The distribution area of Javan Pipistrelle in China is in its peripheral range. Its population sizes are small and its habitats are under affected negatively by human activities. Thus, it is listed as Near Threatened

地理分布 Geographical Distribution

国内分布 Domestic Distribution	西藏、云南 / Tibet (Xizang), Yunnan
世界分布 World Distribution	中国；南亚、东南亚 / China; South Asia, Southeast Asia
分布标注 Distribution Note	非特有种 / Non-endemic

国内分布图
Map of Domestic Distribution

种群 Population

种群数量 Population Size	未知 / Unknown
种群趋势 Population Trend	稳定 / Stable

生境与生态系统 Habitat (s) and Ecosystem (s)

生　　境 Habitat(s)	森林、耕地、种植园、城市、人造建筑 / Forest, Arable Land, Plantation, Urban Area, Building
生态系统 Ecosystem(s)	森林生态系统、农田生态系统、人类聚落生态系统 / Forest Ecosystem, Cropland Ecosystem, Human Settlement Ecosystem

威胁 Threat (s)

主要威胁 Major Threat(s)	伐木、耕种 / Logging, Plantation

保护级别与保护行动 Protection Category and Conservation Action (s)

国家重点保护野生动物等级 (2021) Category of National Key Protected Wild Animals (2021)	无 / NA
IUCN 红色名录(2020-2) IUCN Red List (2020-2)	无危 / LC
CITES 附录 (2019) CITES Appendix (2019)	无 / NA
保护行动 Conservation Action(s)	无 / None

相关文献 Relevant References

Liu and Wu (刘少英和吴毅), 2019; Wilson and Mittermeier, 2019; Burgin *et al.*, 2018; Jiang *et al.* (蒋志刚等), 2017; Zhou and Yang (周江和杨天友), 2009; Hu (胡刚), 1998; Chen *et al.* (陈延熹等), 1987

爪哇伏翼 *Pipistrellus javanicus*

小伏翼
Pipistrellus tenuis
近危 NT

| 数据缺乏 DD | 无危 LC | 近危 NT | 易危 VU | 濒危 EN | 极危 CR | 区域灭绝 RE | 野外灭绝 EW | 灭绝 EX |

分类地位 Taxonomic Status

动物界 Animalia	脊索动物门 Chordata	哺乳纲 Mammalia	翼手目 Chiroptera	蝙蝠科 Vespertilionidae

学名 Scientific Name	*Pipistrellus tenuis*
命名人 Species Authority	Temminck, 1840
英文名 English Name(s)	Least Pipistrelle
同物异名 Synonym(s)	*Pipistrellus mimus* (Wroughton, 1899); *P. mimus* (Wroughton, 1912) subsp. *glaucillus*; *P. mimus* (Wroughton, 1899) subsp. *mimus*; *P. mimus* (Thomas, 1915) subsp.（转下页）
种下单元评估 Infra-specific Taxa Assessed	无 / None

评估信息 Assessment Information

评估年份 Year Assessed	2020
评定人 Assessor(s)	吴毅、蒋志刚 / Yi Wu, Zhigang Jiang
审定人 Reviewer(s)	余文华、石红艳、江廷磊 / Wenhua Yu, Hongyan Shi, Tinglei Jiang
其他贡献人 Other Contributor(s)	李立立、丁晨晨 / Lili Li, Chenchen Ding

理由 Justification: 小伏翼在中国的分布区较广，但数量较少，且其所在的栖息地面积减小。因此，列为近危等级 / Least Pipistrelle is widely distributed in southwest China, but its population size is small. Furthermore, the areas of its habitats are declining. Thus, it is listed as Near Threatened

地理分布 Geographical Distribution

国内分布 Domestic Distribution	浙江、福建、广西、海南、四川、贵州、云南、重庆 / Zhejiang, Fujian, Guangxi, Hainan, Sichuan, Guizhou, Yunnan, Chongqing
世界分布 World Distribution	中国；中亚、东南亚、南亚和太平洋岛屿 / China; Central Asia, Southeast Asia, South Asia and Pacific Ocean Islands
分布标注 Distribution Note	非特有种 / Non-endemic

国内分布图
Map of Domestic Distribution

种群 Population

种群数量 Population Size	未知 / Unknown
种群趋势 Population Trend	稳定 / Stable

生境与生态系统 Habitat(s) and Ecosystem(s)

生　　境 Habitat(s)	人造建筑、森林、城市、次生林 / Building, Forest, Urban Area, Secondary Forest
生态系统 Ecosystem(s)	森林生态系统、人类聚落生态系统 / Forest Ecosystem, Human Settlement Ecosystem

威胁 Threat(s)

主要威胁 Major Threat(s)	人类活动 / Human Activity

保护级别与保护行动 Protection Category and Conservation Action(s)

国家重点保护野生动物等级 (2021) Category of National Key Protected Wild Animals (2021)	无 / NA
IUCN 红色名录 (2020-2) IUCN Red List (2020-2)	无危 / LC
CITES 附录 (2019) CITES Appendix (2019)	无 / NA
保护行动 Conservation Action(s)	无 / None

相关文献 Relevant References

Wilson and Mittermeier, 2019; Burgin *et al*., 2018; Jiang *et al*.（蒋志刚等）, 2017; Zhou *et al*.（周江等）, 2011

（接上页）
principulus; *P. principulus* (Thomas, 1915); *Vespertilio tenuis* (Temminck, 1840)

小伏翼 *Pipistrellus tenuis*

By Dibyendu Ash

灰伏翼
Hypsugo pulveratus

近危 NT

| 数据缺乏 DD | 无危 LC | 近危 NT | 易危 VU | 濒危 EN | 极危 CR | 区域灭绝 RE | 野外灭绝 EW | 灭绝 EX |

分类地位 Taxonomic Status

动物界 Animalia	脊索动物门 Chordata	哺乳纲 Mammalia	翼手目 Chiroptera	蝙蝠科 Vespertilionidae
学名 Scientific Name		*Hypsugo pulveratus*		
命名人 Species Authority		Peters, 1870		
英文名 English Name(s)		Chinese Pipistrelle		
同物异名 Synonym(s)		*Pipistrellus pulveratus* (Peters, 1870)		
种下单元评估 Infra-specific Taxa Assessed		无 / None		

评估信息 Assessment Information

评估年份 Year Assessed	2020
评定人 Assessor(s)	吴毅、蒋志刚 / Yi Wu, Zhigang Jiang
审定人 Reviewer(s)	余文华、石红艳、江廷磊 / Wenhua Yu, Hongyan Shi, Tinglei Jiang
其他贡献人 Other Contributor(s)	李立立、丁晨晨 / Lili Li, Chenchen Ding

理由 Justification： 灰伏翼在中国的分布区较广，但数量较少，且其占有区面积由于人类活动而减小。因此，列为近危等级 / Chinese Pipistrelle is widely distributed in China, but its population size is small. Its areas of occupancy are constantly declining due to human encroachment. Thus, it is listed as Near Threatened

地理分布 Geographical Distribution

国内分布 Domestic Distribution
陕西、上海、安徽、湖南、四川、重庆、贵州、云南、广东、海南、福建、香港 / Shaanxi, Shanghai, Anhui, Hunan, Sichuan, Chongqing, Guizhou, Yunnan, Guangdong, Hainan, Fujian, Hong Kong
世界分布 World Distribution
中国；东南亚 / China; Southeast Asia
分布标注 Distribution Note
非特有种 / Non-endemic

国内分布图
Map of Domestic Distribution

种群 Population

种群数量 Population Size	未知 / Unknown
种群趋势 Population Trend	未知 / Unknown

生境与生态系统 Habitat(s) and Ecosystem(s)

生境 Habitat(s)	森林、人造建筑 / Forest, Building
生态系统 Ecosystem(s)	森林生态系统、人类聚落生态系统 / Forest Ecosystem, Human Settlement Ecosystem

威胁 Threat(s)

主要威胁 Major Threat(s)	人类活动 / Human Activity

保护级别与保护行动 Protection Category and Conservation Action(s)

国家重点保护野生动物等级 (2021) Category of National Key Protected Wild Animals (2021)	无 / NA
IUCN 红色名录 (2020-2) IUCN Red List (2020-2)	无危 / LC
CITES 附录 (2019) CITES Appendix (2019)	无 / NA
保护行动 Conservation Action(s)	无 / None

相关文献 Relevant References

Burgin *et al.*, 2018; Jiang *et al.* (蒋志刚等), 2017; Cen and Peng (岑业文和彭红元), 2010; Wu *et al.* (吴毅等), 2006

灰伏翼 *Hypsugo pulveratus*　　　　　　　　　　　　　　　　张礼标 摄　By Libiao Zhang

萨氏伏翼
Hypsugo savii

近危 NT

| 数据缺乏 DD | 无危 LC | 近危 NT | 易危 VU | 濒危 EN | 极危 CR | 区域灭绝 RE | 野外灭绝 EW | 灭绝 EX |

分类地位 Taxonomic Status

动物界 Animalia	脊索动物门 Chordata	哺乳纲 Mammalia	翼手目 Chiroptera	蝙蝠科 Vespertilionidae
学名 Scientific Name		*Hypsugo savii*		
命名人 Species Authority		Bonaparte, 1837		
英文名 English Name(s)		Savi's Pipistrelle		
同物异名 Synonym(s)		*Pipistrellus savii* (Bonaparte, 1837)		
种下单元评估 Infra-specific Taxa Assessed		无 / None		

评估信息 Assessment Information

评估年份 Year Assessed	2020
评定人 Assessor(s)	吴毅、蒋志刚 / Yi Wu, Zhigang Jiang
审定人 Reviewer(s)	余文华、石红艳、江廷磊 / Wenhua Yu, Hongyan Shi, Tinglei Jiang
其他贡献人 Other Contributor(s)	李立立、丁晨晨 / Lili Li, Chenchen Ding

理由 Justification: 萨氏伏翼在中国的分布区狭窄，且数量少，其所在的栖息地受到人类活动的影响。因此，列为近危等级 / Savi's Pipistrelle is narrowly distributed in China. Its population size is small and its habitats are disturbed by human activities. Thus, it is listed as Near Threatened

地理分布 Geographical Distribution

国内分布 Domestic Distribution
新疆 / Xinjiang
世界分布 World Distribution
亚洲、欧洲 / Asia, Europe
分布标注 Distribution Note
非特有种 / Non-endemic

国内分布图
Map of Domestic Distribution

种群 Population

种群数量 Population Size	未知 / Unknown
种群趋势 Population Trend	稳定 / Stable

生境与生态系统 Habitat(s) and Ecosystem(s)

生境 Habitat(s)	人造建筑、森林、湿地、牧场、内陆岩石区域 / Building, Forest, Wetland, Pasture, Inland Rocky Area
生态系统 Ecosystem(s)	森林生态系统、草地生态系统、湿地生态系统、人类聚落生态系统 / Forest Ecosystem, Grassland Ecosystem, Wetland Ecosystem, Human Settlement Ecosystem

威胁 Threat(s)

主要威胁 Major Threat(s)	未知 / Unknown

保护级别与保护行动 Protection Category and Conservation Action(s)

国家重点保护野生动物等级 (2021) Category of National Key Protected Wild Animals (2021)	无 / NA
IUCN 红色名录 (2020-2) IUCN Red List (2020-2)	无危 / LC
CITES 附录 (2019) CITES Appendix (2019)	无 / NA
保护行动 Conservation Action(s)	无 / None

相关文献 Relevant References

Burgin *et al.*, 2018; Jiang *et al.*（蒋志刚等）, 2017; Smith *et al.*（史密斯等）, 2009; Wang（王应祥）, 2003; Hu and Wu（胡锦矗和吴毅）, 1993

萨氏伏翼 *Hypsugo savii*

By Hypsugo-savi-VE-Trtar

南蝠
Ia io
近危 NT

| 数据缺乏 DD | 无危 LC | 近危 NT | 易危 VU | 濒危 EN | 极危 CR | 区域灭绝 RE | 野外灭绝 EW | 灭绝 EX |

分类地位 Taxonomic Status

动物界 Animalia	脊索动物门 Chordata	哺乳纲 Mammalia	翼手目 Chiroptera	蝙蝠科 Vespertilionidae
学名 Scientific Name		*Ia io*		
命名人 Species Authority		Thomas, 1902		
英文名 English Name(s)		Great Evening Bat		
同物异名 Synonym(s)		*Ia longimana* (Pen, 1962); *Pipistrellus io* (Thomas, 1902)		
种下单元评估 Infra-specific Taxa Assessed		无 / None		

评估信息 Assessment Information

评估年份 Year Assessed	2020
评定人 Assessor(s)	吴毅、蒋志刚 / Yi Wu, Zhigang Jiang
审定人 Reviewer(s)	石红艳、江廷磊、余文华 / Hongyan Shi, Tinglei Jiang, Wenhua Yu
其他贡献人 Other Contributor(s)	李立立、丁晨晨 / Lili Li, Chenchen Ding

理由 Justification: 南蝠在中国的分布区较广，但其占有区面积由于人类活动影响而减小。因此，列为近危等级 / Great Evening Bat is widely distributed in China, but its areas of occupancy are declining due to human activities. Thus, it is listed as Near Threatened

地理分布 Geographical Distribution

国内分布 Domestic Distribution

江苏、安徽、湖南、湖北、江西、四川、云南、贵州、广东、广西、重庆、浙江、陕西 / Jiangsu, Anhui, Hunan, Hubei, Jiangxi, Sichuan, Yunnan, Guizhou, Guangdong, Guangxi, Chongqing, Zhejiang, Shaanxi

世界分布 World Distribution

中国；东南亚 / China; Southeast Asia

分布标注 Distribution Note

非特有种 / Non-endemic

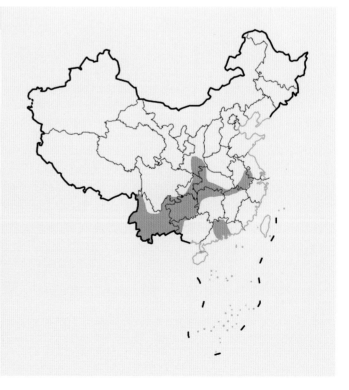

国内分布图
Map of Domestic Distribution

种群 Population

种群数量 Population Size	较常见 / Relatively common
种群趋势 Population Trend	稳定 / Stable

生境与生态系统 Habitat (s) and Ecosystem (s)

生　　境 Habitat(s)	洞穴、森林 / Cave, Forest
生态系统 Ecosystem(s)	喀斯特生态系统、森林生态系统 / Karst Ecosystem, Forest Ecosystem

威胁 Threat (s)

主要威胁 Major Threat(s)	伐木、耕种、火灾、采矿、采石场、旅游 / Logging, Plantation, Fire, Mining, Quarrying Field and Tourism

保护级别与保护行动 Protection Category and Conservation Action (s)

国家重点保护野生动物等级 (2021) Category of National Key Protected Wild Animals (2021)	无 / NA
IUCN红色名录 (2020-2) IUCN Red List (2020-2)	无危 / LC
CITES 附录 (2019) CITES Appendix (2019)	无 / NA
保护行动 Conservation Action(s)	无 / None

相关文献 Relevant References

Wilson and Mittermeier, 2019; Burgin et al., 2018; Jiang et al.（蒋志刚等）, 2017; Chen et al.（陈毅等）, 2013; Han he He（韩宝银和贺红早）, 2012; Smith et al.（史密斯等）, 2009; Zhu et al.（朱光剑等）, 2008b

南蝠 Ia io　　　　　　　　　　　　　　　　　　　吴毅 摄　By Yi Wu

大山蝠
Nyctalus aviator
近危 NT

| 数据缺乏 DD | 无危 LC | 近危 NT | 易危 VU | 濒危 EN | 极危 CR | 区域灭绝 RE | 野外灭绝 EW | 灭绝 EX |

分类地位 Taxonomic Status

动物界 Animalia	脊索动物门 Chordata	哺乳纲 Mammalia	翼手目 Chiroptera	蝙蝠科 Vespertilionidae
学名 Scientific Name		*Nyctalus aviator*		
命名人 Species Authority		Thomas, 1911		
英文名 English Name(s)		Birdlike Noctule		
同物异名 Synonym(s)		*Nyctalus molossus* (Temminck, 1840)		
种下单元评估 Infra-specific Taxa Assessed		无 / None		

评估信息 Assessment Information

评估年份 Year Assessed	2020
评定人 Assessor(s)	吴毅、蒋志刚 / Yi Wu, Zhigang Jiang
审定人 Reviewer(s)	石红艳、江廷磊、余文华 / Hongyan Shi, Tinglei Jiang, Wenhua Yu
其他贡献人 Other Contributor(s)	李立立、丁晨晨 / Lili Li, Chenchen Ding

理由 Justification: 大山蝠在中国的分布区有限，数量较少，且其生境受到人类活动的影响。因此，列为近危等级 / Birdlike Noctule in China has restricted distribution in China, and its population sizes are small. Furthermore, its habitats are disturbed by human activities. Thus, it is listed as Near Threatened

地理分布 Geographical Distribution

国内分布 Domestic Distribution
吉林、黑龙江、浙江、安徽、河南 / Jilin, Heilongjiang, Zhejiang, Anhui, Henan
世界分布 World Distribution
中国；东北亚 / China; Northeast Asia
分布标注 Distribution Note
非特有种 / Non-endemic

国内分布图
Map of Domestic Distribution

种群 Population

种群数量 Population Size	未知 / Unknown
种群趋势 Population Trend	下降 / Decreasing

生境与生态系统 Habitat (s) and Ecosystem (s)

生　　境 Habitat(s)	森林、人造建筑、城市 / Forest, Building, Urban Area
生态系统 Ecosystem(s)	森林生态系统、人类聚落生态系统 / Forest Ecosystem, Human Settlement Ecosystem

威胁 Threat (s)

主要威胁 Major Threat(s)	人类活动干扰、住房及城市建设 / Human Disturbance, Construction of Housing and Urban Area

保护级别与保护行动 Protection Category and Conservation Action (s)

国家重点保护野生动物等级 (2021) Category of National Key Protected Wild Animals (2021)	无 / NA
IUCN红色名录 (2020-2) IUCN Red List (2020-2)	近危 / NT
CITES 附录 (2019) CITES Appendix (2019)	无 / NA
保护行动 Conservation Action(s)	无 / None

相关文献 Relevant References

Wilson and Mittermeier, 2019; Burgin *et al*., 2018; Jiang *et al*. (蒋志刚等), 2017; Zhou and Yang (周江和杨天友), 2012a; Wang *et al*. (王志伟等), 2010; Smith *et al*. (史密斯等), 2009

大山蝠 *Nyctalus aviator*

褐山蝠
Nyctalus noctula
近危 NT

| 数据缺乏 DD | 无危 LC | 近危 NT | 易危 VU | 濒危 EN | 极危 CR | 区域灭绝 RE | 野外灭绝 EW | 灭绝 EX |

分类地位 Taxonomic Status

动物界 Animalia	脊索动物门 Chordata	哺乳纲 Mammalia	翼手目 Chiroptera	蝙蝠科 Vespertilionidae
学名 Scientific Name		*Nyctalus noctula*		
命名人 Species Authority		Schreber, 1774		
英文名 English Name(s)		Noctule		
同物异名 Synonym(s)		Common Noctule		
种下单元评估 Infra-specific Taxa Assessed		无 / None		

评估信息 Assessment Information

评估年份 Year Assessed	2020
评定人 Assessor(s)	吴毅、蒋志刚 / Yi Wu, Zhigang Jiang
审定人 Reviewer(s)	石红艳、江廷磊、余文华 / Hongyan Shi, Tinglei Jiang, Wenhua Yu
其他贡献人 Other Contributor(s)	李立立、丁晨晨 / Lili Li, Chenchen Ding

理由 Justification: 褐山蝠在中国间断分布，数量较少，且其栖息地受到人类活动的扰动。因此，列为近危等级 / Noctule is disconjunctively distributed in China, and its population sizes are small. Furthermore, its habitats are disturbed by human activities. Thus, it is listed as Near Threatened

地理分布 Geographical Distribution

国内分布 Domestic Distribution
北京、新疆、甘肃、山东、陕西、河北 / Beijing, Xinjiang, Gansu, Shandong, Shaanxi, Hebei
世界分布 World Distribution
亚洲、欧洲 / Asia, Europe
分布标注 Distribution Note
非特有种 / Non-endemic

国内分布图
Map of Domestic Distribution

种群 Population

种群数量 Population Size	未知 / Unknown
种群趋势 Population Trend	未知 / Unknown

生境与生态系统 Habitat (s) and Ecosystem (s)

生境 Habitat(s)	森林、人造建筑、城市 / Forest, Building, Urban Area
生态系统 Ecosystem(s)	森林生态系统、人类聚落生态系统 / Forest Ecosystem, Human Settlement Ecosystem

威胁 Threat (s)

主要威胁 Major Threat(s)	人类活动干扰、住房及城市建设 / Human Disturbance, Construction of House and Urban Area

保护级别与保护行动 Protection Category and Conservation Action (s)

国家重点保护野生动物等级 (2021) Category of National Key Protected Wild Animals (2021)	无 / NA
IUCN红色名录 (2020-2) IUCN Red List (2020-2)	无危 / LC
CITES 附录 (2019) CITES Appendix (2019)	无 / NA
保护行动 Conservation Action(s)	无 / None

相关文献 Relevant References

Wilson and Mittermeier, 2019; Burgin *et al.*, 2018; Jiang *et al.* (蒋志刚等), 2017; Wilson and Reeder, 2005; Wang (王应祥), 2003; Zheng and Li (郑生武和李保国), 1999

褐山蝠 *Nyctalus noctula*

褐扁颅蝠
Tylonycteris robustula

近危 NT

| 数据缺乏 DD | 无危 LC | 近危 NT | 易危 VU | 濒危 EN | 极危 CR | 区域灭绝 RE | 野外灭绝 EW | 灭绝 EX |

分类地位 Taxonomic Status

动物界 Animalia	脊索动物门 Chordata	哺乳纲 Mammalia	翼手目 Chiroptera	蝙蝠科 Vespertilionidae
学名 Scientific Name		*Tylonycteris robustula*		
命名人 Species Authority		Thomas, 1915		
英文名 English Name(s)		Greater Bamboo Bat		
同物异名 Synonym(s)		*Tylonycteris malayana* Chasen, 1940		
种下单元评估 Infra-specific Taxa Assessed		无 / None		

评估信息 Assessment Information

评估年份 Year Assessed	2020
评定人 Assessor(s)	吴毅、蒋志刚 / Yi Wu, Zhigang Jiang
审定人 Reviewer(s)	石红艳、江廷磊、余文华 / Hongyan Shi, Tinglei Jiang, Wenhua Yu
其他贡献人 Other Contributor(s)	李立立、丁晨晨 / Lili Li, Chenchen Ding

理由 Justification: 褐扁颅蝠在中国的分布区较广，但其所在的栖息地受到人类活动的影响。因此，列为近危等级 / Greater Bamboo Bat is widely distributed in China, but its habitats are disturbed by human activities. Thus, it is listed as Near Threatened

地理分布 Geographical Distribution

国内分布 Domestic Distribution
云南、广西、四川、海南、贵州、广东、香港、江西 / Yunnan, Guangxi, Sichuan, Hainan, Guizhou, Guangdong, Hong Kong, Jiangxi
世界分布 World Distribution
中国；南亚、东南亚 / China; South Asia, Southeast Asia
分布标注 Distribution Note
非特有种 / Non-endemic

国内分布图
Map of Domestic Distribution

种群 Population

种群数量 Population Size	常见 / Common
种群趋势 Population Trend	未知 / Unknown

生境与生态系统 Habitat(s) and Ecosystem(s)

生境 Habitat(s)	竹林 / Bamboo
生态系统 Ecosystem(s)	竹林生态系统 / Bamboo Ecosystem

威胁 Threat(s)

主要威胁 Major Threat(s)	伐竹、耕种 / Bamboo Logging, Plantation

保护级别与保护行动 Protection Category and Conservation Action(s)

国家重点保护野生动物等级 (2021) Category of National Key Protected Wild Animals (2021)	无 / NA
IUCN 红色名录 (2020-2) IUCN Red List (2020-2)	无危 / LC
CITES 附录 (2019) CITES Appendix (2019)	无 / NA
保护行动 Conservation Action(s)	无 / None

相关文献 Relevant References

Wilson and Mittermeier, 2019; Burgin *et al.*, 2018; Jiang *et al.*（蒋志刚等）, 2017; Wu *et al.*（吴毅等）, 2014; Zhang *et al.*（张秋萍等）, 2014; Yu *et al.*（余文华等）, 2008; Zhang *et al.*（张礼标等）, 2008

褐扁颅蝠 *Tylonycteris robustula*   余文华 摄 By Wenhua Yu

灰大耳蝠
Plecotus austriacus

近危 NT

| 数据缺乏 DD | 无危 LC | 近危 NT | 易危 VU | 濒危 EN | 极危 CR | 区域灭绝 RE | 野外灭绝 EW | 灭绝 EX |

分类地位 Taxonomic Status

动物界 Animalia	脊索动物门 Chordata	哺乳纲 Mammalia	翼手目 Chiroptera	蝙蝠科 Vespertilionidae

学名 Scientific Name	*Plecotus austriacus*
命名人 Species Authority	Fischer, 1829
英文名 English Name(s)	Gray Big-eared Bat
同物异名 Synonym(s)	Gray Long-eared Bat
种下单元评估 Infra-specific Taxa Assessed	无 / None

评估信息 Assessment Information

评估年份 Year Assessed	2020
评定人 Assessor(s)	吴毅、蒋志刚 / Yi Wu, Zhigang Jiang
审定人 Reviewer(s)	石红艳、江廷磊、余文华 / Hongyan Shi, Tinglei Jiang, Wenhua Yu
其他贡献人 Other Contributor(s)	李立立、丁晨晨 / Lili Li, Chenchen Ding

理由 Justification: 灰大耳蝠在中国中部与西部的分布区较广，但数量较少，且其所在的栖息地受到人类活动的影响。因此，列为近危等级 / Gray Big-eared Bat is widely distributed in central and west China, but its population sizes are small. Furthermore, the areas of its habitats are constantly declining due to human activities. Thus, it is listed as Near Threatened

地理分布 Geographical Distribution

国内分布 Domestic Distribution
新疆、四川、青海、西藏、内蒙古、甘肃、宁夏、陕西 / Xinjiang, Sichuan, Qinghai, Tibet (Xizang), Inner Mongolia (Nei Mongol), Gansu, Ningxia, Shaanxi
世界分布 World Distribution
亚洲、欧洲、北非 / Asia, Europe, North Africa
分布标注 Distribution Note
非特有种 / Non-endemic

国内分布图
Map of Domestic Distribution

种群 Population

种群数量 Population Size	未知 / Unknown
种群趋势 Population Trend	未知 / Unknown

生境与生态系统 Habitat (s) and Ecosystem (s)

生境 Habitat(s)	耕地、人造建筑、洞穴 / Arable Land, Building, Cave
生态系统 Ecosystem(s)	森林生态系统、农田生态系统、人类聚落生态系统 / Forests Ecosystem, Cropland Ecosystem, Human Settlement Ecosystem

威胁 Threat (s)

主要威胁 Major Threat(s)	伐木、耕种 / Logging, Plantation

保护级别与保护行动 Protection Category and Conservation Action (s)

国家重点保护野生动物等级 (2021) Category of National Key Protected Wild Animals (2021)	无 / NA
IUCN 红色名录 (2020-2) IUCN Red List (2020-2)	无危 / LC
CITES 附录 (2019) CITES Appendix (2019)	无 / NA
保护行动 Conservation Action(s)	无 / None

相关文献 Relevant References

Wilson and Mittermeier, 2019; Burgin *et al*., 2018; Jiang *et al*. (蒋志刚等), 2017; Wang (王应祥), 2003; Wilson and Reeder, 1993

灰大耳蝠 *Plecotus austriacus*

By Олексій Титов та Ігор Загороднюк

台湾大耳蝠
Plecotus taivanus

近危 NT

| 数据缺乏 DD | 无危 LC | 近危 NT | 易危 VU | 濒危 EN | 极危 CR | 区域灭绝 RE | 野外灭绝 EW | 灭绝 EX |

分类地位 Taxonomic Status

动物界 Animalia	脊索动物门 Chordata	哺乳纲 Mammalia	翼手目 Chiroptera	蝙蝠科 Vespertilionidae
学名 Scientific Name		*Plecotus taivanus*		
命名人 Species Authority		Yoshiyuki, 1991		
英文名 English Name(s)		Taiwan Big-eared Bat		
同物异名 Synonym(s)		Taiwan Long-eared Bat		
种下单元评估 Infra-specific Taxa Assessed		无 / None		

评估信息 Assessment Information

评估年份 Year Assessed	2020
评定人 Assessor(s)	吴毅、蒋志刚 / Yi Wu, Zhigang Jiang
审定人 Reviewer(s)	石红艳、江廷磊、余文华 / Hongyan Shi, Tinglei Jiang, Wenhua Yu
其他贡献人 Other Contributor(s)	李立立、丁晨晨 / Lili Li, Chenchen Ding

理由 Justification: 台湾大耳蝠为中国特有种，仅分布在台湾，其种群数量较少。因此，列为近危等级 / Taiwan Big-eared Bat is an endemic species in China, and distributed only in Taiwan Province. Its population sizes are small. Thus, it is listed as Near Threatened

地理分布 Geographical Distribution

国内分布 Domestic Distribution
台湾 / Taiwan
世界分布 World Distribution
中国 / China
分布标注 Distribution Note
特有种 / Endemic

国内分布图
Map of Domestic Distribution

种群 Population

种群数量 Population Size	未知 / Unknown
种群趋势 Population Trend	下降 / Decreasing

生境与生态系统 Habitat (s) and Ecosystem (s)

生境 Habitat(s)	森林、洞穴、人造建筑 / Forest, Cave, Building
生态系统 Ecosystem(s)	森林生态系统、人类聚落生态系统 / Forest Ecosystem, Human Settlement Ecosystem

威胁 Threat (s)

主要威胁 Major Threat(s)	未知 / Unknown

保护级别与保护行动 Protection Category and Conservation Action (s)

国家重点保护野生动物等级 (2021) Category of National Key Protected Wild Animals (2021)	无 / NA
IUCN 红色名录 (2020-2) IUCN Red List (2020-2)	近危 / NT
CITES 附录 (2019) CITES Appendix (2019)	无 / NA
保护行动 Conservation Action(s)	无 / None

相关文献 Relevant References

Wilson and Mittermeier, 2019; Burgin *et al*., 2018; Jiang *et al*. (蒋志刚等), 2017; Zheng (郑锡奇), 2010; Wang (王应祥), 2003; Yoshiyuki, 1995; Wilson and Reeder, 1993

台湾大耳蝠 *Plecotus taivanus*

亚洲长翼蝠
Miniopterus fuliginosus

近危 NT

| 数据缺乏 DD | 无危 LC | 近危 NT | 易危 VU | 濒危 EN | 极危 CR | 区域灭绝 RE | 野外灭绝 EW | 灭绝 EX |

分类地位 Taxonomic Status

动物界 Animalia	脊索动物门 Chordata	哺乳纲 Mammalia	翼手目 Chiroptera	蝙蝠科 Vespertilionidae
学名 Scientific Name		*Miniopterus fuliginosus*		
命名人 Species Authority		Hodgson, 1835		
英文名 English Name(s)		Eastern Long-fingered Bat		
同物异名 Synonym(s)		Common Bent-wing Bat, Schreibers's Bat, Schreibers's Long-fingered Bat, *Miniopterus schreibersii*		
种下单元评估 Infra-specific Taxa Assessed		无 / None		

评估信息 Assessment Information

评估年份 Year Assessed	2020
评定人 Assessor(s)	吴毅、蒋志刚 / Yi Wu, Zhigang Jiang
审定人 Reviewer(s)	石红艳、江廷磊、余文华 / Hongyan Shi, Tinglei Jiang, Wenhua Yu
其他贡献人 Other Contributor(s)	李立立、丁晨晨 / Lili Li, Chenchen Ding

理由 Justification: 亚洲长翼蝠在中国的发生区较广，但数量较少，且其占有区面积减小。因此，列为近危等级 / Eastern Long-fingered Bat is widely distributed in China, but its population sizes are small. Furthermore, its areas of occurence are constantly declining. Thus, it is listed as Near Threatened

地理分布 Geographical Distribution

国内分布 Domestic Distribution	云南、四川、重庆、贵州、湖南、湖北、广西、海南、广东、江西、福建、台湾、浙江、安徽、山西、陕西、河南、河北、江苏 / Yunnan, Sichuan, Chongqing, Guizhou, Hunan, Hubei, Guangxi, Hainan, Guangdong, Jiangxi, Fujian, Taiwan, Zhejiang, Anhui, Shanxi, Shaanxi, Henan, Hebei, Jiangsu
世界分布 World Distribution	亚洲 / Asia
分布标注 Distribution Note	非特有种 / Non-endemic

国内分布图
Map of Domestic Distribution

种群 Population

种群数量 Population Size	常见 / Common
种群趋势 Population Trend	稳定 / Stable

生境与生态系统 Habitat (s) and Ecosystem (s)

生境 Habitat(s)	洞穴、耕地、热带和亚热带严重退化森林 / Cave, Arable Land, Tropical and Subtropical Severely Degraded Forest
生态系统 Ecosystem(s)	森林生态系统、农田生态系统 / Forest Ecosystem, Cropland Ecosystem

威胁 Threat (s)

主要威胁 Major Threat(s)	伐木、耕种、火灾、采矿、采石场、旅游 / Logging, Plantation, Fire, Mining, Quarrying Field and Tourism

保护级别与保护行动 Protection Category and Conservation Action (s)

国家重点保护野生动物等级 (2021) Category of National Key Protected Wild Animals (2021)	无 / NA
IUCN 红色名录(2020-2) IUCN Red List (2020-2)	无危 / LC
CITES 附录 (2019) CITES Appendix (2019)	无 / NA
保护行动 Conservation Action(s)	无 / None

相关文献 Relevant References

Wilson and Mittermeier, 2019; Burgin *et al.*, 2018; Jiang *et al.*（蒋志刚等）, 2017; Li *et al.*, 2015b; Han *et al.*, 2008a, 2008b; Wang（王应祥）, 2003

亚洲长翼蝠 *Miniopterus fuliginosus*　　　　　　吴毅 摄　By Yi Wu

几内亚长翼蝠
Miniopterus magnater

近危 NT

| 数据缺乏 DD | 无危 LC | 近危 NT | 易危 VU | 濒危 EN | 极危 CR | 区域灭绝 RE | 野外灭绝 EW | 灭绝 EX |

分类地位 Taxonomic Status

动物界 Animalia	脊索动物门 Chordata	哺乳纲 Mammalia	翼手目 Chiroptera	蝙蝠科 Vespertilionidae

学名 Scientific Name	*Miniopterus magnater*
命名人 Species Authority	Sanborn, 1931
英文名 English Name(s)	Western Long-fingered Bat
同物异名 Synonym(s)	Western Bent-winged Bat; *Miniopterus macrodens* Maeda, 1982
种下单元评估 Infra-specific Taxa Assessed	无 / None

评估信息 Assessment Information

评估年份 Year Assessed	2020
评定人 Assessor(s)	吴毅、蒋志刚 / Yi Wu, Zhigang Jiang
审定人 Reviewer(s)	石红艳、江廷磊、余文华 / Hongyan Shi, Tinglei Jiang, Wenhua Yu
其他贡献人 Other Contributor(s)	李立立、丁晨晨 / Lili Li, Chenchen Ding

理由 Justification: 几内亚长翼蝠在中国分布狭窄，主要分布在海南岛。其种群数量极少，占有区面积因人类活动而减少。因此，列为近危等级 / Western Long-fingered Bat is narrowly distributed in China, mainly found on Hainan Island. Its population sizes are very small. Furthermore, its area of occupancy is declining due to human activities. Thus, it is listed as Near Threatened

地理分布 Geographical Distribution

国内分布 Domestic Distribution
香港、海南 / Hong Kong, Hainan
世界分布 World Distribution
中国、印度尼西亚、巴布亚新几内亚 / China, Indonesia, Papua New Guinea
分布标注 Distribution Note
非特有种 / Non-endemic

国内分布图
Map of Domestic Distribution

种群 Population

种群数量 Population Size	未知 / Unknown
种群趋势 Population Trend	未知 / Unknown

生境与生态系统 Habitat(s) and Ecosystem(s)

生　　境 Habitat(s)	洞穴、森林、人工环境 / Cave, Forest, Anthropogenic Environment
生态系统 Ecosystem(s)	森林生态系统、人类聚落生态系统 / Forest Ecosystem, Human Settlement Ecosystem

威胁 Threat(s)

主要威胁 Major Threat(s)	伐木、耕种、火灾、采矿、采石场、旅游 / Logging, Plantation, Fire, Mining, Quarrying Field and Tourism

保护级别与保护行动 Protection Category and Conservation Action(s)

国家重点保护野生动物等级 (2021) Category of National Key Protected Wild Animals (2021)	无 / NA
IUCN 红色名录(2020-2) IUCN Red List (2020-2)	无危 / LC
CITES 附录 (2019) CITES Appendix (2019)	无 / NA
保护行动 Conservation Action(s)	无 / None

相关文献 Relevant References

Wilson and Mittermeier, 2019; Burgin *et al*., 2018; Jiang *et al*. (蒋志刚等), 2017; Li *et al*., 2015b; Smith *et al*. (史密斯等), 2009; Han *et al*., 2008a, 2008b

几内亚长翼蝠 *Miniopterus magnater*

南长翼蝠
Miniopterus pusillus

近危 NT

| 数据缺乏 DD | 无危 LC | **近危 NT** | 易危 VU | 濒危 EN | 极危 CR | 区域灭绝 RE | 野外灭绝 EW | 灭绝 EX |

分类地位 Taxonomic Status

动物界 Animalia	脊索动物门 Chordata	哺乳纲 Mammalia	翼手目 Chiroptera	蝙蝠科 Vespertilionidae
学名 Scientific Name		*Miniopterus pusillus*		
命名人 Species Authority		Dobson, 1876		
英文名 English Name(s)		Small Long-fingered Bat		
同物异名 Synonym(s)		Small Bent-winged Bat; *Miniopterus australis* (Dobson, 1876) subsp. *pusillus*		
种下单元评估 Infra-specific Taxa Assessed		无 / None		

评估信息 Assessment Information

评估年份 Year Assessed	2020
评定人 Assessor(s)	吴毅、蒋志刚 / Yi Wu, Zhigang Jiang
审定人 Reviewer(s)	石红艳、江廷磊、余文华 / Hongyan Shi, Tinglei Jiang, Wenhua Yu
其他贡献人 Other Contributor(s)	李立立、丁晨晨 / Lili Li, Chenchen Ding

理由 Justification: 南长翼蝠在中国的分布区呈斑块状，且占有区小。因此，列为近危等级 / Small Long-fingered Bat's distribution range is scattered across China, and its area of occupancy is very small. Thus, it is listed as Near Threatened

地理分布 Geographical Distribution

国内分布 Domestic Distribution
福建、广东、云南、海南、香港、澳门 / Fujian, Guangdong, Yunnan, Hainan, Hong Kong, Macao
世界分布 World Distribution
中国；南亚、东南亚 / China; South Asia, Southeast Asia
分布标注 Distribution Note
非特有种 / Non-endemic

国内分布图
Map of Domestic Distribution

种群 Population

种群数量 Population Size	常见 / Common
种群趋势 Population Trend	稳定 / Stable

生境与生态系统 Habitat (s) and Ecosystem (s)

生　　境 Habitat(s)	未知 / Unknown
生态系统 Ecosystem(s)	森林生态系统、农田生态系统 / Forest Ecosystem, Cropland Ecosystem

威胁 Threat (s)

主要威胁 Major Threat(s)	未知 / Unknown

保护级别与保护行动 Protection Category and Conservation Action (s)

国家重点保护野生动物等级 (2021) Category of National Key Protected Wild Animals (2021)	无 / NA
IUCN 红色名录 (2020-2) IUCN Red List (2020-2)	无危 / LC
CITES 附录 (2019) CITES Appendix (2019)	无 / NA
保护行动 Conservation Action(s)	无 / None

相关文献 Relevant References

Wilson and Mittermeier, 2019; Burgin *et al.*, 2018; Jiang *et al.* (蒋志刚等), 2017; Huang *et al.* (黄继展等), 2013; Smith *et al.* (史密斯等), 2009; Wang (王应祥), 2003; Koopman, 1993

南长翼蝠 *Miniopterus pusillus*　　余文华 摄　By Wenhua Yu

金管鼻蝠
Murina aurata

近危 NT

| 数据缺乏 DD | 无危 LC | 近危 NT | 易危 VU | 濒危 EN | 极危 CR | 区域灭绝 RE | 野外灭绝 EW | 灭绝 EX |

分类地位 Taxonomic Status

动物界 Animalia	脊索动物门 Chordata	哺乳纲 Mammalia	翼手目 Chiroptera	蝙蝠科 Vespertilionidae
学名 Scientific Name		*Murina aurata*		
命名人 Species Authority		Milne-Edwards, 1872		
英文名 English Name(s)		Little Tube-nosed Bat		
同物异名 Synonym(s)		*Murina aurita* (Miller, 1907)		
种下单元评估 Infra-specific Taxa Assessed		无 / None		

评估信息 Assessment Information

评估年份 Year Assessed	2020
评定人 Assessor(s)	吴毅、蒋志刚 / Yi Wu, Zhigang Jiang
审定人 Reviewer(s)	余文华、石红艳、江廷磊 / Wenhua Yu, Hongyan Shi, Tinglei Jiang
其他贡献人 Other Contributor(s)	李立立、丁晨晨 / Lili Li, Chenchen Ding

理由 Justification: 金管鼻蝠在中国呈间断分布，且其占有区面积小。因此，列为近危等级 / Little Tube-nosed Bat is distributed widely across China but in disconnected regions, with total areas of occupancy relatively small. Thus, it is listed as Near Threatened

地理分布 Geographical Distribution

国内分布 Domestic Distribution
黑龙江、吉林、甘肃、西藏、四川、云南 / Heilongjiang, Jilin, Gansu, Tibet (Xizang), Sichuan, Yunnan
世界分布 World Distribution
中国；南亚、东南亚 / China; South Asia, Southeast Asia
分布标注 Distribution Note
非特有种 / Non-endemic

国内分布图
Map of Domestic Distribution

种群 Population

种群数量 Population Size	未知 / Unknown
种群趋势 Population Trend	未知 / Unknown

生境与生态系统 Habitat(s) and Ecosystem(s)

生　　境 Habitat(s)	森林 / Forest
生态系统 Ecosystem(s)	森林生态系统 / Forest Ecosystem

威胁 Threat(s)

主要威胁 Major Threat(s)	未知 / Unknown

保护级别与保护行动 Protection Category and Conservation Action(s)

国家重点保护野生动物等级 (2021) Category of National Key Protected Wild Animals (2021)	无 / NA
IUCN 红色名录 (2020-2) IUCN Red List (2020-2)	无危 / LC
CITES 附录 (2019) CITES Appendix (2019)	无 / NA
保护行动 Conservation Action(s)	无 / None

相关文献 Relevant References

Liu and Wu (刘少英和吴毅), 2019; Wilson and Mittermeier, 2019; Burgin *et al.*, 2018; Jiang *et al.* (蒋志刚等), 2017; Smith *et al.* (史密斯等), 2009; Wang (王应祥), 2003

金管鼻蝠 *Murina aurata*

圆耳管鼻蝠
Murina cyclotis

近危 NT

| 数据缺乏 DD | 无危 LC | 近危 NT | 易危 VU | 濒危 EN | 极危 CR | 区域灭绝 RE | 野外灭绝 EW | 灭绝 EX |

分类地位 Taxonomic Status

动物界 Animalia	脊索动物门 Chordata	哺乳纲 Mammalia	翼手目 Chiroptera	蝙蝠科 Vespertilionidae

学名 Scientific Name	*Murina cyclotis*
命名人 Species Authority	Dobson, 1872
英文名 English Name(s)	Round-eared Tube-nosed Bat
同物异名 Synonym(s)	*Murina eileenae* (Phillips, 1932)
种下单元评估 Infra-specific Taxa Assessed	无 / None

评估信息 Assessment Information

评估年份 Year Assessed	2020
评定人 Assessor(s)	吴毅、蒋志刚 / Yi Wu, Zhigang Jiang
审定人 Reviewer(s)	余文华、石红艳、江廷磊 / Wenhua Yu, Hongyan Shi, Tinglei Jiang
其他贡献人 Other Contributor(s)	李立立、丁晨晨 / Lili Li, Chenchen Ding

理由 Justification: 圆耳管鼻蝠在中国的分布区范围极为狭窄，种群数量少。因此，列为近危等级 / Round-eared Tube-nosed Bat's distribution range in China is extremely small with small populations. Thus, it is listed as Near Threatened

地理分布 Geographical Distribution

国内分布 Domestic Distribution
福建、海南、广东 / Fujian, Hainan, Guangdong
世界分布 World Distribution
中国；南亚、东南亚 / China; South Asia, Southeast Asia
分布标注 Distribution Note
非特有种 / Non-endemic

国内分布图
Map of Domestic Distribution

种群 Population

种群数量 Population Size	稀少 / Rare
种群趋势 Population Trend	未知 / Unknown

生境与生态系统 Habitat (s) and Ecosystem (s)

生境 Habitat(s)	洞穴、内陆岩石区域、森林 / Cave, Inland Rocky Area, Forest
生态系统 Ecosystem(s)	森林生态系统、喀斯特生态系统 / Forest Ecosystem, Karst Ecosystem

威胁 Threat (s)

主要威胁 Major Threat(s)	伐木、耕种、洞穴开发 / Logging, Plantation, Cave Exploitation

保护级别与保护行动 Protection Category and Conservation Action (s)

国家重点保护野生动物等级 (2021) Category of National Key Protected Wild Animals (2021)	无 / NA
IUCN 红色名录(2020-2) IUCN Red List (2020-2)	无危 / LC
CITES 附录 (2019) CITES Appendix (2019)	无 / NA
保护行动 Conservation Action(s)	无 / None

相关文献 Relevant References

Liu and Wu (刘少英和吴毅), 2019; Wilson and Mittermeier, 2019; Burgin *et al.*, 2018; Jiang *et al.* (蒋志刚等), 2017; Smith *et al.* (史密斯等), 2009; Wang (王应祥), 2003; Xu *et al.* (徐剑等), 2002

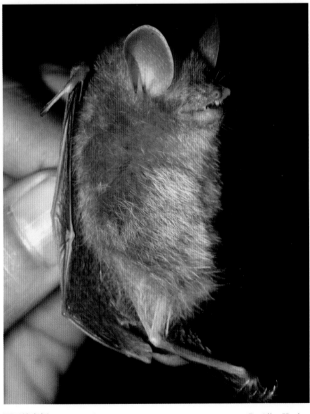

圆耳管鼻蝠 *Murina cyclotis*　　　　　　　　　　　　　　　　By Alice Hughes

艾氏管鼻蝠 *Murina eleryi*

近危 NT

| 数据缺乏 DD | 无危 LC | 近危 NT | 易危 VU | 濒危 EN | 极危 CR | 区域灭绝 RE | 野外灭绝 EW | 灭绝 EX |

分类地位 Taxonomic Status

动物界 Animalia	脊索动物门 Chordata	哺乳纲 Mammalia	翼手目 Chiroptera	蝙蝠科 Vespertilionidae

学名 Scientific Name	*Murina eleryi*
命名人 Species Authority	Furey, 2009
英文名 English Name(s)	Elery's Tube-nosed Bat
同物异名 Synonym(s)	Mekong Bat
种下单元评估 Infra-specific Taxa Assessed	无 / None

评估信息 Assessment Information

评估年份 Year Assessed	2020
评定人 Assessor(s)	吴毅、蒋志刚 / Yi Wu, Zhigang Jiang
审定人 Reviewer(s)	余文华、石红艳、江廷磊 / Wenhua Yu, Hongyan Shi, Tinglei Jiang
其他贡献人 Other Contributor(s)	李立立、丁晨晨 / Lili Li, Chenchen Ding

理由 Justification: 艾氏管鼻蝠在中国的分布区极小，其种群数量亦少。因此，列为近危等级 / Elery's Tube-nosed Bat's distribution range in China is extremely small, and its population size also is small. Thus, it is listed as Near Threatened

地理分布 Geographical Distribution

国内分布 Domestic Distribution
湖南、广东 / Hunan, Guangdong
世界分布 World Distribution
中国、越南 / China, Viet Nam
分布标注 Distribution Note
非特有种 / Non-endemic

国内分布图
Map of Domestic Distribution

种群 Population

种群数量 Population Size	较常见 / Relatively common
种群趋势 Population Trend	未知 / Unknown

生境与生态系统 Habitat (s) and Ecosystem (s)

生境 Habitat(s)	森林、喀斯特地貌 / Forest, Karst Landscape
生态系统 Ecosystem(s)	森林生态系统、喀斯特生态系统 / Forest Ecosystem, Karst Ecosystem

威胁 Threat (s)

主要威胁 Major Threat(s)	伐木、耕种、洞穴开发 / Logging, Plantation, Cave Exploitation

保护级别与保护行动 Protection Category and Conservation Action (s)

国家重点保护野生动物等级 (2021) Category of National Key Protected Wild Animals (2021)	无 / NA
IUCN 红色名录 (2020-2) IUCN Red List (2020-2)	未列入 / NA
CITES 附录 (2019) CITES Appendix (2019)	无 / NA
保护行动 Conservation Action(s)	无 / None

相关文献 Relevant References

Liu and Wu (刘少英和吴毅), 2019; Wilson and Mittermeier, 2019; Burgin *et al*., 2018; Jiang *et al*. (蒋志刚等), 2017; Xu *et al*. (徐忠鲜等), 2014; Liu *et al*. (刘志霄等), 2013; Furey *et al*., 2009

艾氏管鼻蝠 *Murina eleryi* 　　吴毅 摄　By Yi Wu

台湾管鼻蝠
Murina puta

近危 NT

| 数据缺乏 DD | 无危 LC | 近危 NT | 易危 VU | 濒危 EN | 极危 CR | 区域灭绝 RE | 野外灭绝 EW | 灭绝 EX |

分类地位 Taxonomic Status

动物界 Animalia	脊索动物门 Chordata	哺乳纲 Mammalia	翼手目 Chiroptera	蝙蝠科 Vespertilionidae
学 名 Scientific Name		*Murina puta*		
命 名 人 Species Authority		Kishida, 1924		
英 文 名 English Name(s)		Taiwanese Tube-nosed Bat		
同物异名 Synonym(s)		Taiwan Tube-nosed Bat		
种下单元评估 Infra-specific Taxa Assessed		无 / None		

评估信息 Assessment Information

评估年份 Year Assessed	2020
评 定 人 Assessor(s)	吴毅、蒋志刚 / Yi Wu, Zhigang Jiang
审 定 人 Reviewer(s)	余文华、石红艳、江廷磊 / Wenhua Yu, Hongyan Shi, Tinglei Jiang
其他贡献人 Other Contributor(s)	李立立、丁晨晨 / Lili Li, Chenchen Ding

理由 Justification: 台湾管鼻蝠为中国特有种，仅分布在台湾，其数量少。因此，列为近危等级 / Taiwanese Tube-nosed Bat is endemic to China, distributed only in Taiwan. Its population sizes are small. Thus, it is listed as Near Threatened

地理分布 Geographical Distribution

国内分布 Domestic Distribution
台湾 / Taiwan
世界分布 World Distribution
中国 / China
分布标注 Distribution Note
特有种 / Endemic

国内分布图
Map of Domestic Distribution

🦌 种群 Population

种群数量 Population Size	未知 / Unknown
种群趋势 Population Trend	未知 / Unknown

🦌 生境与生态系统 Habitat(s) and Ecosystem(s)

生　　　境 Habitat(s)	温带森林 / Temperate Forest
生态系统 Ecosystem(s)	森林生态系统 / Forest Ecosystem

🦌 威胁 Threat(s)

主要威胁 Major Threat(s)	砍伐 / Logging

🦌 保护级别与保护行动 Protection Category and Conservation Action(s)

国家重点保护野生动物等级 (2021) Category of National Key Protected Wild Animals (2021)	无 / NA
IUCN红色名录 (2020-2) IUCN Red List (2020-2)	数据缺乏 / DD
CITES 附录 (2019) CITES Appendix (2019)	无 / NA
保护行动 Conservation Action(s)	无 / None

🦌 相关文献 Relevant References

Wilson and Mittermeier, 2019; Burgin *et al*., 2018; Jiang *et al*. (蒋志刚等), 2017; Cheng *et al*., 2015; Zheng (郑锡奇), 2010; Simmons, 2005; Lin *et al*., 2002a, 2002b; Lin *et al*. (林良恭等), 1997

台湾管鼻蝠 *Murina puta*

毛翼管鼻蝠
Harpiocephalus harpia
近危 NT

| 数据缺乏 DD | 无危 LC | 近危 NT | 易危 VU | 濒危 EN | 极危 CR | 区域灭绝 RE | 野外灭绝 EW | 灭绝 EX |

分类地位 Taxonomic Status

动物界 Animalia	脊索动物门 Chordata	哺乳纲 Mammalia	翼手目 Chiroptera	蝙蝠科 Vespertilionidae
学名 Scientific Name		*Harpiocephalus harpia*		
命名人 Species Authority		Temminck, 1840		
英文名 English Name(s)		Lesser Hairy-winged Bat		
同物异名 Synonym(s)		无 / None		
种下单元评估 Infra-specific Taxa Assessed		无 / None		

评估信息 Assessment Information

评估年份 Year Assessed	2020
评定人 Assessor(s)	吴毅、蒋志刚 / Yi Wu, Zhigang Jiang
审定人 Reviewer(s)	余文华、石红艳、江廷磊 / Wenhua Yu, Hongyan Shi, Tinglei Jiang
其他贡献人 Other Contributor(s)	李立立、丁晨晨 / Lili Li, Chenchen Ding

理由 Justification: 毛翼管鼻蝠在中国的分布区较广，但数量较少，且其占有区面积减小。因此，列为近危等级 / Lesser Hairy-winged Bat's distribution range in China is large, but its population sizes are small. Furthermore, its areas of occupancy are declining. Thus, it is listed as Near Threatened

地理分布 Geographical Distribution

国内分布 Domestic Distribution
云南、广东、福建、台湾 / Yunnan, Guangdong, Fujian, Taiwan
世界分布 World Distribution
中国；东南亚、大洋洲 / China; Southeast Asia, Oceania
分布标注 Distribution Note
非特有种 / Non-endemic

国内分布图
Map of Domestic Distribution

种群 Population

种群数量 Population Size	稀少 / Rare
种群趋势 Population Trend	下降 / Decreasing

生境与生态系统 Habitat (s) and Ecosystem (s)

生　　境 Habitat(s)	森林、次生林 / Forest, Secondary Forest
生态系统 Ecosystem(s)	森林生态系统 / Forest Ecosystem

威胁 Threat (s)

主要威胁 Major Threat(s)	伐木 / Logging

保护级别与保护行动 Protection Category and Conservation Action (s)

国家重点保护野生动物等级 (2021) Category of National Key Protected Wild Animals (2021)	无 / NA
IUCN红色名录(2020-2) IUCN Red List (2020-2)	数据缺乏 / DD
CITES 附录 (2019) CITES Appendix (2019)	无 / NA
保护行动 Conservation Action(s)	无 / None

相关文献 Relevant References

Liu and Wu (刘少英和吴毅), 2019; Wilson and Mittermeier, 2019; Burgin *et al*., 2018; Jiang *et al*. (蒋志刚等), 2017; Chen *et al*. (陈柏承等), 2015; Zhou *et al*. (周全等), 2014; Lin *et al*., 2006; Wang (王应祥), 2003

毛翼管鼻蝠 *Harpiocephalus harpia*　　余文华 摄　By Wenhua Yu

川金丝猴
Rhinopithecus roxellana

近危 NT

| 数据缺乏 DD | 无危 LC | 近危 NT | 易危 VU | 濒危 EN | 极危 CR | 区域灭绝 RE | 野外灭绝 EW | 灭绝 EX |

分类地位 Taxonomic Status

动物界 Animalia	脊索动物门 Chordata	哺乳纲 Mammalia	灵长目 Primates	猴科 Cercopithecidae
学名 Scientific Name		*Rhinopithecus roxellana*		
命名人 Species Authority		Milne-Edwards, 1870		
英文名 English Name(s)		Golden Snub-nosed Monkey		
同物异名 Synonym(s)		无 / None		
种下单元评估 Infra-specific Taxa Assessed		无 / None		

评估信息 Assessment Information

评估年份 Year Assessed	2020
评定人 Assessor(s)	蒋志刚、蒋学龙 / Zhigang Jiang, Xuelong Jiang
审定人 Reviewer(s)	李明、龙勇诚、范朋飞、蒋学龙、李保国、李义明、王应祥、李言阔 / Ming Li, Yongcheng Long, Pengfei Fan, Xuelong Jiang, Baoguo Li, Yiming Li, Yingxiang Wang, Yankuo Li
其他贡献人 Other Contributor(s)	李立立、丁晨晨 / Lili Li, Chenchen Ding

理由 Justification: 川金丝猴分布在陕西、四川和湖北，种群数量较大且在增长。因此，列为近危等级 / The Golden Snub-nosed Monkey is found in Shaanxi, Sichuan and Hubei with large populations and its populations are increasing. Thus, it is listed as Near Threatened

地理分布 Geographical Distribution

国内分布 Domestic Distribution
陕西、四川、甘肃、湖北、重庆 / Shaanxi, Sichuan, Gansu, Hubei, Chongqing
世界分布 World Distribution
中国 / China
分布标注 Distribution Note
特有种 / Endemic

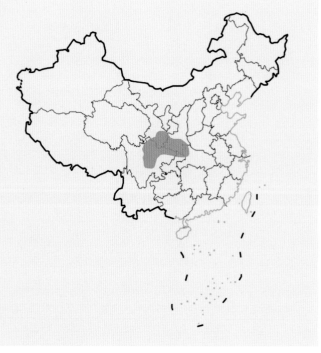

国内分布图
Map of Domestic Distribution

种群 Population

种群数量 Population Size	12,000 只 / 12,000 individuals
种群趋势 Population Trend	稳定 / Stable

生境与生态系统 Habitat(s) and Ecosystem(s)

生　　境 Habitat(s)	泰加林、针阔混交林 / Taiga Forest, Coniferous and Broad-leaved Mixed Forest
生态系统 Ecosystem(s)	森林生态系统 / Forest Ecosystem

威胁 Threat(s)

主要威胁 Major Threat(s)	旅游 / Tourism

保护级别与保护行动 Protection Category and Conservation Action(s)

国家重点保护野生动物等级 (2021) Category of National Key Protected Wild Animals (2021)	一级 / Category I
IUCN 红色名录 (2020-2) IUCN Red List (2020-2)	濒危 / EN
CITES 附录 (2019) CITES Appendix (2019)	I
保护行动 Conservation Action(s)	大部分栖息地已被自然保护区保护 / Most of its habitats are protected by nature reserves

相关文献 Relevant References

Burgin *et al.*, 2018; Jiang *et al.* (蒋志刚等), 2017; Wu and Liu (吴逸群和刘科科), 2010; Jin *et al.* (金崑等), 2005; Li *et al.* (李义明等), 2005

川金丝猴 *Rhinopithecus roxellana*　　　　蒋志刚 摄　By Zhigang Jiang

狼 Canis lupus

近危 NT

| 数据缺乏 DD | 无危 LC | 近危 NT | 易危 VU | 濒危 EN | 极危 CR | 区域灭绝 RE | 野外灭绝 EW | 灭绝 EX |

分类地位 Taxonomic Status

动物界 Animalia	脊索动物门 Chordata	哺乳纲 Mammalia	食肉目 Carnivora	犬科 Canidae
学名 Scientific Name		*Canis lupus*		
命名人 Species Authority		Linnaeus, 1758		
英文名 English Name(s)		Gray Wolf		
同物异名 Synonym(s)		无 / None		
种下单元评估 Infra-specific Taxa Assessed		无 / None		

评估信息 Assessment Information

评估年份 Year Assessed	2020
评定人 Assessor(s)	蒋志刚 / Zhigang Jiang
审定人 Reviewer(s)	马建章、鲍伟东、姜广顺、胡慧建、徐爱春 / Jianzhang Ma, Weidong Bao, Guangshun Jiang, Huijian Hu, Aichun Xu
其他贡献人 Other Contributor(s)	李立立、丁晨晨 / Lili Li, Chenchen Ding

理由 Justification: 狼在中国的分布区较广，但种群密度较低。由于狼捕食人类放牧的家畜，在一些地区造成人兽冲突，受到人类报复性猎杀。因此，列为近危等级 / Gray Wolf's distribution range in China is large, but its population densities are low. Because Gray Wolf preys on livestock in some areas of the country, it is threatened by human retaliative killing. Thus, it is listed as Near Threatened

地理分布 Geographical Distribution

国内分布 Domestic Distribution

山东、山西、河南、新疆、青海、河北、内蒙古、辽宁、吉林、黑龙江、江苏、浙江、湖北、广东、广西、四川、贵州、云南、西藏、陕西、甘肃、宁夏、福建、湖南、江西、天津、北京、重庆 / Shandong, Shanxi, Henan, Xinjiang, Qinghai, Hebei, Inner Mongolia (Nei Mongol), Liaoning, Jilin, Heilongjiang, Jiangsu, Zhejiang, Hubei, Guangdong, Guangxi, Sichuan, Guizhou, Yunnan, Tibet (Xizang), Shaanxi, Gansu, Ningxia, Fujian, Hunan, Jiangxi, Tianjin, Beijing, Chongqing

世界分布 World Distribution

亚洲、非洲 / Asia, Europe

分布标注 Distribution Note

非特有种 / Non-endemic

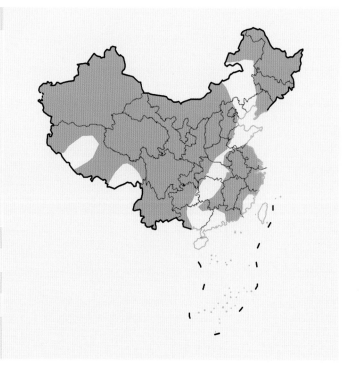

国内分布图
Map of Domestic Distribution

种群 Population

种群数量 Population Size	35,000 只 / 35,000 individuals
种群趋势 Population Trend	稳定 / Stable

生境与生态系统 Habitat(s) and Ecosystem(s)

生境 Habitat(s)	苔原、森林、草甸、沙漠、耕地 / Tundra, Forest, Meadow, Desert, Arable Land
生态系统 Ecosystem(s)	森林生态系统、草地生态系统、农田生态系统、荒漠生态系统 / Forest Ecosystem, Grassland Ecosystem, Cropland Ecosystem, Desert Ecosystem

威胁 Threat(s)

主要威胁 Major Threat(s)	投毒、狩猎 / Poisoning, Hunting

保护级别与保护行动 Protection Category and Conservation Action(s)

国家重点保护野生动物等级 (2021) Category of National Key Protected Wild Animals (2021)	二级 / Category II
IUCN 红色名录 (2020-2) IUCN Red List (2020-2)	无危 / LC
CITES 附录 (2019) CITES Appendix (2019)	II
保护行动 Conservation Action(s)	无 / None

相关文献 Relevant References

Burgin *et al.*, 2018; Jiang *et al.* (蒋志刚等), 2017; Zhang *et al.*, 2015; Liu *et al.* (刘姝等), 2013; Wilson and Mittermeier, 2009; Meng *et al.* (孟超等), 2008; Gao (高中信), 2006; Vilà *et al.*, 1999; Gao *et al.* (高耀亭等), 1987

狼 *Canis lupus*

沙狐

Vulpes corsac

近危 NT

| 数据缺乏 DD | 无危 LC | 近危 NT | 易危 VU | 濒危 EN | 极危 CR | 区域灭绝 RE | 野外灭绝 EW | 灭绝 EX |

分类地位 Taxonomic Status

动物界 Animalia	脊索动物门 Chordata	哺乳纲 Mammalia	食肉目 Carnivora	犬科 Canidae
学名 Scientific Name		*Vulpes corsac*		
命名人 Species Authority		Linnaeus, 1768		
英文名 English Name(s)		Corsac Fox		
同物异名 Synonym(s)		无 / None		
种下单元评估 Infra-specific Taxa Assessed		无 / None		

评估信息 Assessment Information

评估年份 Year Assessed	2020
评定人 Assessor(s)	蒋志刚 / Zhigang Jiang
审定人 Reviewer(s)	马建章、鲍伟东、姜广顺、胡慧建、徐爱春 / Jianzhang Ma, Weidong Bao, Guangshun Jiang, Huijian Hu, Aichun Xu
其他贡献人 Other Contributor(s)	李立立、丁晨晨 / Lili Li, Chenchen Ding

理由 Justification: 沙狐是中国北部干旱区常见的一个物种。该种分布区广，数量较多，但其生存受到人类干扰。因此，列为近危等级 / Corsac Fox is a common species in arid areas in northern China. Its distribution range and its population sizes are both large, but its survival is threatened by human activities. Thus, it is listed as Near Threatened

地理分布 Geographical Distribution

国内分布 Domestic Distribution
宁夏、新疆、内蒙古、甘肃、青海 / Ningxia, Xinjiang, Inner Mongolia (Nei Mongol), Gansu, Qinghai
世界分布 World Distribution
中国；西亚、中亚 / China; West Asia, Central Asia
分布标注 Distribution Note
非特有种 / Non-endemic

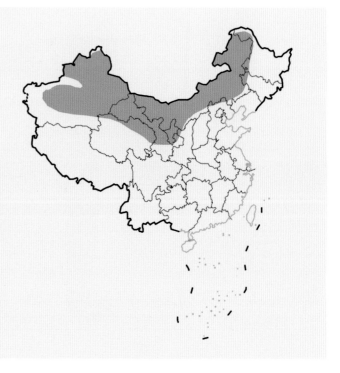

国内分布图
Map of Domestic Distribution

种群 Population

种群数量 Population Size	160,000 只 / 160,000 individuals
种群趋势 Population Trend	未知 / Unknown

生境与生态系统 Habitat (s) and Ecosystem (s)

生　　境 Habitat(s)	草甸、半荒漠地区 / Meadow, Semi-desert Area
生态系统 Ecosystem(s)	草地生态系统、荒漠生态系统 / Grassland Ecosystem, Desert Ecosystem

威胁 Threat (s)

主要威胁 Major Threat(s)	狩猎 / Hunting

保护级别与保护行动 Protection Category and Conservation Action (s)

国家重点保护野生动物等级 (2021) Category of National Key Protected Wild Animals (2021)	二级 / Category II
IUCN 红色名录 (2020-2) IUCN Red List (2020-2)	无危 / LC
CITES 附录 (2019) CITES Appendix (2019)	无 / NA
保护行动 Conservation Action(s)	无 / None

相关文献 Relevant References

Burgin *et al.*, 2018; Jiang *et al.* (蒋志刚等), 2017; Zhang (张进), 2014; Clark *et al.*, 2009; Wilson and Mittermeier, 2009; Zhang *et al.* (张洪海等), 2006; Gao *et al.* (高耀亭等), 1987

沙狐 *Vulpes corsac*

藏狐
Vulpes ferrilata

近危 NT

| 数据缺乏 DD | 无危 LC | 近危 NT | 易危 VU | 濒危 EN | 极危 CR | 区域灭绝 RE | 野外灭绝 EW | 灭绝 EX |

分类地位 Taxonomic Status

动物界 Animalia	脊索动物门 Chordata	哺乳纲 Mammalia	食肉目 Carnivora	犬科 Canidae
学名 Scientific Name		*Vulpes ferrilata*		
命名人 Species Authority		Hodgson, 1842		
英文名 English Name(s)		Tibetan Fox		
同物异名 Synonym(s)		Tibetan Sand Fox		
种下单元评估 Infra-specific Taxa Assessed		无 / None		

评估信息 Assessment Information

评估年份 Year Assessed	2020
评定人 Assessor(s)	蒋志刚 / Zhigang Jiang
审定人 Reviewer(s)	胡慧建、徐爱春 / Huijian Hu, Aichun Xu
其他贡献人 Other Contributor(s)	李立立、丁晨晨 / Lili Li, Chenchen Ding

理由 Justification: 藏狐是青藏高原腹地常见的物种，分布区较广，但数量较少。因此，列为近危等级 / Tibetan Fox is a common species in the heartland of Qinghai-Tibet (Xizang) Plateau. Its distribution range on the plateau is large, but its population size is small. Thus, it is listed as Near Threatened

地理分布 Geographical Distribution

国内分布 Domestic Distribution
新疆、青海、甘肃、四川、云南、西藏 / Xinjiang, Qinghai, Gansu, Sichuan, Yunnan, Tibet (Xizang)
世界分布 World Distribution
中国、印度、尼泊尔 / China, India, Nepal
分布标注 Distribution Note
非特有种 / Non-endemic

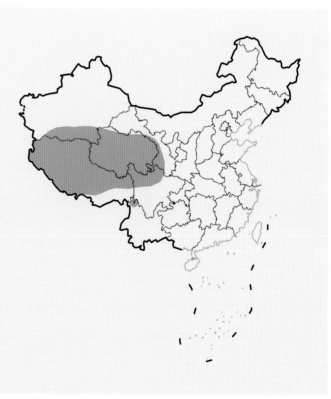

国内分布图
Map of Domestic Distribution

种群 Population

种群数量 Population Size	37,000 只 / 37,000 individuals
种群趋势 Population Trend	未知 / Unknown

生境与生态系统 Habitat(s) and Ecosystem(s)

生　　境 Habitat(s)	草甸、荒漠 / Meadow, Desert
生态系统 Ecosystem(s)	草地生态系统、荒漠生态系统 / Grassland Ecosystem, Desert Ecosystem

威胁 Threat(s)

主要威胁 Major Threat(s)	狩猎、投毒、食物缺乏 / Hunting, Poisoning, Food Scarcity

保护级别与保护行动 Protection Category and Conservation Action(s)

国家重点保护野生动物等级 (2021) Category of National Key Protected Wild Animals (2021)	二级 / Category II
IUCN 红色名录 (2020-2) IUCN Red List (2020-2)	无危 / LC
CITES 附录 (2019) CITES Appendix (2019)	无 / NA
保护行动 Conservation Action(s)	部分种群位于自然保护区内 / Part of populations are covered by nature reserve

相关文献 Relevant References

Burgin *et al*., 2018; Jiang *et al*. (蒋志刚等), 2017; Wilson and Mittermeier, 2009; Clark *et al*., 2008; Wang *et al*., 2008; Wang *et al*. (王正寰等), 2004; Gao *et al*. (高耀亭等), 1987

藏狐 *Vulpes ferrilata*

赤狐
Vulpes vulpes
近危 NT

| 数据缺乏 DD | 无危 LC | 近危 NT | 易危 VU | 濒危 EN | 极危 CR | 区域灭绝 RE | 野外灭绝 EW | 灭绝 EX |

分类地位 Taxonomic Status

动物界 Animalia	脊索动物门 Chordata	哺乳纲 Mammalia	食肉目 Carnivora	犬科 Canidae
学名 Scientific Name		*Vulpes vulpes*		
命名人 Species Authority		Linnaeus, 1758		
英文名 English Name(s)		Red Fox		
同物异名 Synonym(s)		无 / None		
种下单元评估 Infra-specific Taxa Assessed		无 / None		

评估信息 Assessment Information

评估年份 Year Assessed	2020
评定人 Assessor(s)	蒋志刚 / Zhigang Jiang
审定人 Reviewer(s)	胡慧建、徐爱春 / Huijian Hu, Aichun Xu
其他贡献人 Other Contributor(s)	李立立、丁晨晨 / Lili Li, Chenchen Ding

理由 Justification: 赤狐在中国的分布区广，但数量较少。因此，列为近危等级 / Red Fox is widely distributed in China, but its population size is small. Thus, it is listed as Near Threatened

地理分布 Geographical Distribution

国内分布 Domestic Distribution

吉林、山西、河南、黑龙江、内蒙古、湖南、北京、河北、辽宁、江苏、浙江、安徽、福建、江西、山东、湖北、广东、广西、四川、贵州、云南、西藏、陕西、甘肃、青海、宁夏、新疆、香港、重庆 / Jilin, Shanxi, Henan, Heilongjiang, Inner Mongolia (Nei Mongol), Hunan, Beijing, Hebei, Liaoning, Jiangsu, Zhejiang, Anhui, Fujian, Jiangxi, Shandong, Hubei, Guangdong, Guangxi, Sichuan, Guizhou, Yunnan, Tibet (Xizang), Shaanxi, Gansu, Qinghai, Ningxia, Xinjiang, Hong Kong, Chongqing

世界分布 World Distribution

亚洲、欧洲 / Asia, Europe

分布标注 Distribution Note

非特有种 / Non-endemic

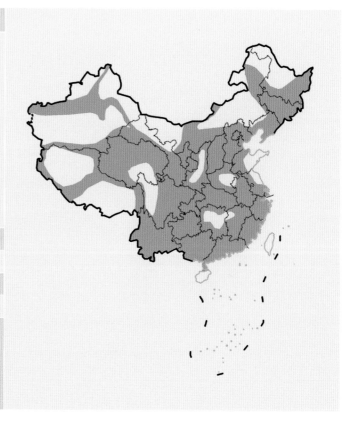

国内分布图
Map of Domestic Distribution

种群 Population

种群数量 Population Size	150,000 只 / 150,000 individuals
种群趋势 Population Trend	稳定 / Stable

生境与生态系统 Habitat (s) and Ecosystem (s)

生　　境 Habitat(s)	苔原、沙漠、森林、城市、灌丛、耕地 / Tundra, Desert, Forest, Urban Area, Shrubland, Arable Land
生态系统 Ecosystem(s)	森林生态系统、灌丛生态系统、农田生态系统、荒漠生态系统、人类聚落生态系统 / Forest Ecosystem, Shrubland Ecosystem, Cropland Ecosystem, Desert Ecosystem, Human Settlement Ecosystem

威胁 Threat (s)

主要威胁 Major Threat(s)	狩猎、人类活动干扰、家畜养殖或放牧 / Hunting, Human Disturbance, Livestock Farming or Ranching

保护级别与保护行动 Protection Category and Conservation Action (s)

国家重点保护野生动物等级 (2021) Category of National Key Protected Wild Animals (2021)	二级 / Category II
IUCN 红色名录(2020-2) IUCN Red List (2020-2)	无危 / LC
CITES 附录 (2019) CITES Appendix (2019)	III
保护行动 Conservation Action(s)	无 / None

相关文献 Relevant References

Burgin *et al.*, 2018; Jiang *et al.*（蒋志刚等）, 2017; Li（李成涛）, 2011; Smith *et al.*（史密斯等）, 2009; Wilson and Mittermeier, 2009; Pan *et al.*（潘清华等）, 2007; Wang（王应祥）, 2003; Gao *et al.*（高耀亭等）, 1987

赤狐 *Vulpes vulpes*

貂
Nyctereutes procyonoides

近危 NT

| 数据缺乏 DD | 无危 LC | **近危 NT** | 易危 VU | 濒危 EN | 极危 CR | 区域灭绝 RE | 野外灭绝 EW | 灭绝 EX |

分类地位 Taxonomic Status

动物界 Animalia	脊索动物门 Chordata	哺乳纲 Mammalia	食肉目 Carnivora	犬科 Canidae
学名 Scientific Name		*Nyctereutes procyonoides*		
命名人 Species Authority		Gray, 1834		
英文名 English Name(s)		Raccoon Dog		
同物异名 Synonym(s)		无 / None		
种下单元评估 Infra-specific Taxa Assessed		无 / None		

评估信息 Assessment Information

评估年份 Year Assessed	2020
评定人 Assessor(s)	蒋志刚 / Zhigang Jiang
审定人 Reviewer(s)	胡慧建、徐爱春 / Huijian Hu, Aichun Xu
其他贡献人 Other Contributor(s)	李立立、丁晨晨 / Lili Li, Chenchen Ding

理由 Justification: 貂原在中国东部广泛分布，由于栖息地破坏、过度捕杀和非法贸易，其野生种群数量下降。因此，列为近危等级 / Raccoon Dog was historically widely distributed in eastern China. However, due to habitat destruction, over-hunting and illegal trades, its populations are declining in the wild. Thus, it is listed as Near Threatened

地理分布 Geographical Distribution

国内分布 Domestic Distribution

河南、湖南、陕西、北京、河北、山西、内蒙古、辽宁、吉林、黑龙江、江苏、浙江、安徽、福建、江西、湖北、广东、广西、四川、贵州、云南、甘肃、重庆 / Henan, Hunan, Shaanxi, Beijing, Hebei, Shanxi, Inner Mongolia (Nei Mongol), Liaoning, Jilin, Heilongjiang, Jiangsu, Zhejiang, Anhui, Fujian, Jiangxi, Hubei, Guangdong, Guangxi, Sichuan, Guizhou, Yunnan, Gansu, Chongqing

世界分布 World Distribution

中国；东南亚、东北亚 / China; Southeast Asia, Northeast Asia

分布标注 Distribution Note

非特有种 / Non-endemic

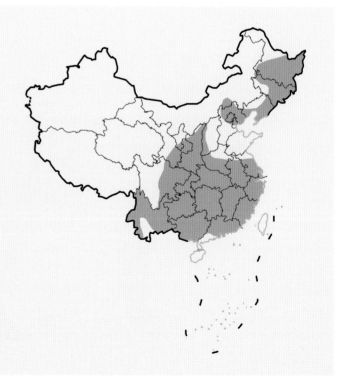

国内分布图
Map of Domestic Distribution

 种群 Population

种群数量 Population Size	未知 / Unknown
种群趋势 Population Trend	稳定 / Stable

生境与生态系统 Habitat (s) and Ecosystem (s)

生　　境 Habitat(s)	森林、溪流边、灌丛、草甸 / Forest, Near Stream, Shrubland, Meadow
生态系统 Ecosystem(s)	森林生态系统、灌丛生态系统、草地生态系统、湖泊河流生态系统 / Forest Ecosystem, Shrubland Ecosystem, Grassland Ecosystem, Lake and River Ecosystem

威胁 Threat (s)

主要威胁 Major Threat(s)	猎杀、投毒、公路碾压事件、疾病 / Hunting, Poisoning, Road Kill, Disease

保护级别与保护行动 Protection Category and Conservation Action (s)

国家重点保护野生动物等级 (2021) Category of National Key Protected Wild Animals (2021)	二级 (仅限野外种群) / Category II (Only Wild Population)
IUCN 红色名录 (2020-2) IUCN Red List (2020-2)	无危 / LC
CITES 附录 (2019) CITES Appendix (2019)	无 / NA
保护行动 Conservation Action(s)	无 / None

相关文献 Relevant References

Burgin *et al*., 2018; Jiang *et al*. (蒋志刚等), 2017; Smith *et al*. (史密斯等), 2009; Wilson and Mittermeier, 2009; Pieńkowska *et al*., 2002; Zhuang (庄炜), 1991; Gao *et al*. (高耀亭等), 1987

貉 *Nyctereutes procyonoides*

北海狮
Eumetopias jubatus

近危 NT

| 数据缺乏 DD | 无危 LC | 近危 NT | 易危 VU | 濒危 EN | 极危 CR | 区域灭绝 RE | 野外灭绝 EW | 灭绝 EX |

分类地位 Taxonomic Status

动物界 Animalia	脊索动物门 Chordata	哺乳纲 Mammalia	食肉目 Carnivora	海狮科 Otariidae
学名 Scientific Name		*Eumetopias jubatus*		
命名人 Species Authority		Schreber, 1776		
英文名 English Name(s)		Steller Sea Lion		
同物异名 Synonym(s)		Steller's Sea Lion		
种下单元评估 Infra-specific Taxa Assessed		无 / None		

评估信息 Assessment Information

评估年份 Year Assessed	2020
评定人 Assessor(s)	周开亚 / Kaiya Zhou
审定人 Reviewer(s)	张先锋、王克雄、王丁、祝茜、蒋志刚 / Xianfeng Zhang, Kexiong Wang, Ding Wang, Qian Zhu, Zhigang Jiang
其他贡献人 Other Contributor(s)	李立立、丁晨晨 / Lili Li, Chenchen Ding

理由 Justification: 北海狮分布在从日本到南加利福尼亚的北太平洋沿岸。仅漫游个体偶然到达中国的渤海、黄海。因此，列为近危等级 / Steller Sea Lion occurs throughout the North Pacific Ocean rim from Japan to southern California. Some roaming individuals accidentally reached China's Bohai Sea, Yellow Sea. Thus, it is listed as Near Threatened

地理分布 Geographical Distribution

国内分布 Domestic Distribution
渤海、黄海 / Bohai Sea, Yellow Sea
世界分布 World Distribution
北太平洋 / North Pacific Ocean
分布标注 Distribution Note
非特有种 / Non-endemic

国内分布图
Map of Domestic Distribution

种群 Population

种群数量 Population Size	未知 / Unknown
种群趋势 Population Trend	上升 / Increasing

生境与生态系统 Habitat (s) and Ecosystem (s)

生　境 Habitat(s)	近海岸 / Near Coast
生态系统 Ecosystem(s)	海岸生态系统 / Coast Ecosystem

威胁 Threat (s)

主要威胁 Major Threat(s)	渔业、狩猎 / Fishery, Hunting

保护级别与保护行动 Protection Category and Conservation Action (s)

国家重点保护野生动物等级 (2021) Category of National Key Protected Wild Animals (2021)	二级 / Category II
IUCN 红色名录 (2020-2) IUCN Red List (2020-2)	近危 / NT
CITES 附录 (2019) CITES Appendix (2019)	无 / NA
保护行动 Conservation Action(s)	部分种群位于自然保护区内 / Part of populations are covered by nature reserves

相关文献 Relevant References

Burgin *et al.*, 2018; Jiang *et al.* (蒋志刚等), 2017, 2015a; Mittermeier and Wilson, 2014; Smith *et al.* (史密斯等), 2009; Zhou (周开亚), 2008, 2004; Wilson and Reeder, 2005

北海狮 *Eumetopias jubatus*

香鼬
Mustela altaica

近危 NT

| 数据缺乏 DD | 无危 LC | 近危 NT | 易危 VU | 濒危 EN | 极危 CR | 区域灭绝 RE | 野外灭绝 EW | 灭绝 EX |

分类地位 Taxonomic Status

| 动物界 Animalia | 脊索动物门 Chordata | 哺乳纲 Mammalia | 食肉目 Carnivora | 鼬科 Mustelidae |

学 名 Scientific Name	*Mustela altaica*
命 名 人 Species Authority	Pallas, 1811
英 文 名 English Name(s)	Altai Weasel
同物异名 Synonym(s)	Mountain Weasel, Pale Weasel
种下单元评估 Infra-specific Taxa Assessed	无 / None

评估信息 Assessment Information

评 估 年 份 Year Assessed	2020
评 定 人 Assessor(s)	蒋志刚 / Zhigang Jiang
审 定 人 Reviewer(s)	李晟、王昊、周友兵 / Sheng Li, Hao Wang, Youbing Zhou
其他贡献人 Other Contributor(s)	李立立、丁晨晨 / Lili Li, Chenchen Ding

理由 Justification: 香鼬曾是传统狩猎对象。香鼬在中国的分布区较广，但种群增长受到人类活动的影响。因此，列为近危等级 / Altai Weasel used to be a common game hunting species. Its distribution range in China is large, but its rate of population growth has been affected by human activities. Thus, it is listed as Near Threatened

地理分布 Geographical Distribution

国内分布 Domestic Distribution

山西、青海、新疆、内蒙古、辽宁、吉林、黑龙江、四川、西藏、甘肃、宁夏、湖北、重庆 / Shanxi, Qinghai, Xinjiang, Inner Mongolia (Nei Mongol), Liaoning, Jilin, Heilongjiang, Sichuan, Tibet (Xizang), Gansu, Ningxia, Hubei, Chongqing

世界分布 World Distribution

中国；南亚、中亚 / China; South Asia, Central Asia

分布标注 Distribution Note

非特有种 / Non-endemic

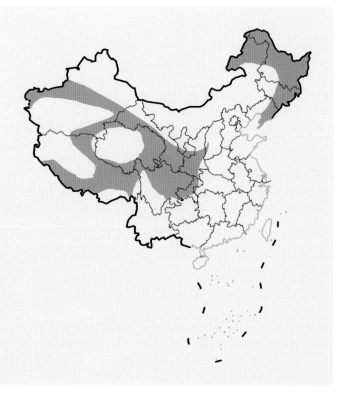

国内分布图
Map of Domestic Distribution

种群 Population

种群数量 Population Size	未知 / Unknown
种群趋势 Population Trend	下降 / Decreasing

生境与生态系统 Habitat (s) and Ecosystem (s)

生　　境 Habitat(s)	草甸、内陆岩石区域 / Meadow, Inland Rocky Area
生态系统 Ecosystem(s)	草地生态系统 / Grassland Ecosystem

威胁 Threat (s)

主要威胁 Major Threat(s)	狩猎、家畜放牧、二次中毒 / Hunting, Livestock Ranching, Secondary Poisoning

保护级别与保护行动 Protection Category and Conservation Action (s)

国家重点保护野生动物等级 (2021) Category of National Key Protected Wild Animals (2021)	无 / NA
IUCN红色名录(2020-2) IUCN Red List (2020-2)	近危 / NT
CITES 附录 (2019) CITES Appendix (2019)	III
保护行动 Conservation Action(s)	无 / None

相关文献 Relevant References

Burgin *et al.*, 2018; Jiang *et al.* (蒋志刚等), 2017; Hu *et al.* (胡一鸣等), 2014; Liu *et al.* (刘洋等), 2013b; Wilson and Mittermeier, 2009; Gao *et al.* (高耀亭等), 1987

香鼬 *Mustela altaica*

黄腹鼬 *Mustela kathiah*

近危 NT

| 数据缺乏 DD | 无危 LC | 近危 NT | 易危 VU | 濒危 EN | 极危 CR | 区域灭绝 RE | 野外灭绝 EW | 灭绝 EX |

分类地位 Taxonomic Status

动物界 Animalia	脊索动物门 Chordata	哺乳纲 Mammalia	食肉目 Carnivora	鼬科 Mustelidae

学名 Scientific Name	*Mustela kathiah*
命名人 Species Authority	Hodgson, 1835
英文名 English Name(s)	Yellow-bellied Weasel
同物异名 Synonym(s)	无 / None
种下单元评估 Infra-specific Taxa Assessed	无 / None

评估信息 Assessment Information

评估年份 Year Assessed	2020
评定人 Assessor(s)	蒋志刚 / Zhigang Jiang
审定人 Reviewer(s)	李晟、胡慧建、王昊、周友兵 / Sheng Li, Huijian Hu, Hao Wang, Youbing Zhou
其他贡献人 Other Contributor(s)	李立立、丁晨晨 / Lili Li, Chenchen Ding

理由 Justification: 黄腹鼬广泛分布于中国南方，但由于种种原因，其种群数量较少。因此，列为近危等级 / Yellow-bellied Weasel is widely distributed in south China, but its population sizes are small due to various reasons. Thus, it is listed as Near Threatened

地理分布 Geographical Distribution

国内分布 Domestic Distribution
海南、云南、浙江、安徽、福建、江西、湖北、广东、广西、四川、贵州、陕西、湖南、重庆 / Hainan, Yunnan, Zhejiang, Anhui, Fujian, Jiangxi, Hubei, Guangdong, Guangxi, Sichuan, Guizhou, Shaanxi, Hunan, Chongqing
世界分布 World Distribution
中国；南亚、东南亚 / China; South Asia, Southeast Asia
分布标注 Distribution Note
非特有种 / Non-endemic

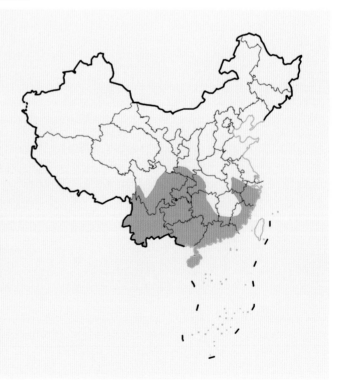

国内分布图
Map of Domestic Distribution

种群 Population

种群数量 Population Size	未知 / Unknown
种群趋势 Population Trend	未知 / Unknown

生境与生态系统 Habitat (s) and Ecosystem (s)

生　　境 Habitat(s)	森林、次生林、热带和亚热带退化森林 / Forest, Secondary Forest, Tropical and Subtropical Degraded Forest
生态系统 Ecosystem(s)	森林生态系统 / Forest Ecosystem

威胁 Threat (s)

主要威胁 Major Threat(s)	未知 / Unknown

保护级别与保护行动 Protection Category and Conservation Action (s)

国家重点保护野生动物等级 (2021) Category of National Key Protected Wild Animals (2021)	无 / NA
IUCN 红色名录 (2020-2) IUCN Red List (2020-2)	无危 / LC
CITES 附录 (2019) CITES Appendix (2019)	III
保护行动 Conservation Action(s)	无 / None

相关文献 Relevant References

Burgin *et al.*, 2018; Jiang *et al.*（蒋志刚等）, 2017; Lau *et al.*, 2010; Smith *et al.*（史密斯等）, 2009; Wilson and Mittermeier, 2009; Wang（王应祥）, 2003; Gao *et al.*（高耀亭等）, 1987

黄腹鼬 *Mustela kathiah*

鼬獾
Melogale moschata

近危 NT

| 数据缺乏 DD | 无危 LC | 近危 NT | 易危 VU | 濒危 EN | 极危 CR | 区域灭绝 RE | 野外灭绝 EW | 灭绝 EX |

分类地位 Taxonomic Status

动物界 Animalia	脊索动物门 Chordata	哺乳纲 Mammalia	食肉目 Carnivora	鼬科 Mustelidae
学名 Scientific Name		*Melogale moschata*		
命名人 Species Authority		Gray, 1831		
英文名 English Name(s)		Small-toothed Ferret-badger		
同物异名 Synonym(s)		Chinese Ferret-badger		
种下单元评估 Infra-specific Taxa Assessed		无 / None		

评估信息 Assessment Information

评估年份 Year Assessed	2020
评定人 Assessor(s)	蒋志刚 / Zhigang Jiang
审定人 Reviewer(s)	李晟、胡慧建、王昊、张明海、周友兵 / Sheng Li, Huijian Hu, Hao Wang, Minghai Zhang, Youbing Zhou
其他贡献人 Other Contributor(s)	李立立、丁晨晨 / Lili Li, Chenchen Ding

理由 Justification: 鼬獾在中国南方的分布区较广，但数量稀少，且其占有区面积减小。因此，列为近危等级 / Small-toothed Ferret-badger is widely distributed in south China, but its population sizes are small. Furthermore, its areas of occupancy are declining. Thus, it is listed as Near Threatened

地理分布 Geographical Distribution

国内分布 Domestic Distribution

山西、河南、上海、江苏、浙江、安徽、福建、江西、湖北、湖南、广东、广西、海南、四川、贵州、云南、陕西、台湾、香港、重庆 / Shanxi, Henan, Shanghai, Jiangsu, Zhejiang, Anhui, Fujian, Jiangxi, Hubei, Hunan, Guangdong, Guangxi, Hainan, Sichuan, Guizhou, Yunnan, Shaanxi, Taiwan, Hong Kong, Chongqing

世界分布 World Distribution

中国；南亚、东南亚 / China; South Asia, Southeast Asia

分布标注 Distribution Note

非特有种 / Non-endemic

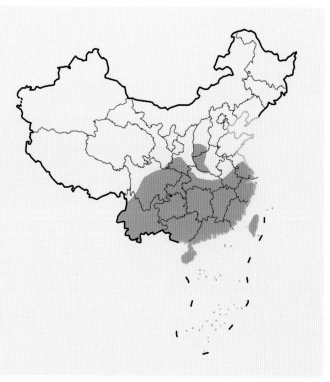

国内分布图
Map of Domestic Distribution

种群 Population

种群数量 Population Size	未知 / Unknown
种群趋势 Population Trend	未知 / Unknown

生境与生态系统 Habitat (s) and Ecosystem (s)

生　　境 Habitat(s)	亚热带湿润低地森林、草地、耕地 / Subtropical Moist Lowland Forest, Grassland, Arable Land
生态系统 Ecosystem(s)	森林生态系统、草地生态系统、农田生态系统 / Forest Ecosystem, Grassland Ecosystem, Cropland Ecosystem

威胁 Threat (s)

主要威胁 Major Threat(s)	栖息地变化、猎捕 / Habitat Change, Trapping

保护级别与保护行动 Protection Category and Conservation Action (s)

国家重点保护野生动物等级 (2021) Category of National Key Protected Wild Animals (2021)	无 / NA
IUCN 红色名录 (2020-2) IUCN Red List (2020-2)	无危 / LC
CITES 附录 (2019) CITES Appendix (2019)	无 / NA
保护行动 Conservation Action(s)	无 / None

相关文献 Relevant References

Burgin *et al*., 2018; Jiang *et al*. (蒋志刚等), 2017; Zhang *et al*., 2010; Wilson and Mittermeier, 2009; Zhou (周开亚), 2008; Wang (王应祥), 2003; Gao *et al*. (高耀亭等), 1987

鼬獾 *Melogale moschata*

中国生物多样性红色名录
亚洲狗獾
Meles leucurus

近危 NT

| 数据缺乏 DD | 无危 LC | **近危 NT** | 易危 VU | 濒危 EN | 极危 CR | 区域灭绝 RE | 野外灭绝 EW | 灭绝 EX |

分类地位 Taxonomic Status

| 动物界 Animalia | 脊索动物门 Chordata | 哺乳纲 Mammalia | 食肉目 Carnivora | 鼬科 Mustelidae |

学名 Scientific Name	*Meles leucurus*
命名人 Species Authority	Linnaeus, 1758
英文名 English Name(s)	Asian Badger
同物异名 Synonym(s)	Sand Badger
种下单元评估 Infra-specific Taxa Assessed	无 / None

评估信息 Assessment Information

评估年份 Year Assessed	2020
评定人 Assessor(s)	蒋志刚 / Zhigang Jiang
审定人 Reviewer(s)	徐宏发、李峰 / Hongfa Xu, Feng Li
其他贡献人 Other Contributor(s)	李立立、丁晨晨 / Lili Li, Chenchen Ding

理由 Justification: 亚洲狗獾在中国的分布区较广，但数量较少，人类活动影响其种群增长，亚洲狗獾已经在原分布区的许多地点消失。因此，列为近危等级 / Asian Badger's distribution range in China is extensive, but its population densities are low. Furthermore, human encroachment negatively affects its population growth; it has been extirpated in many localities. Thus, it is listed as Near Threatened

地理分布 Geographical Distribution

国内分布 Domestic Distribution

黑龙江、吉林、辽宁、内蒙古、新疆、安徽、北京、福建、甘肃、广东、广西、贵州、河北、河南、湖北、湖南、江苏、江西、青海、陕西、山东、山西、四川、云南、浙江、重庆、宁夏、西藏 / Heilongjiang, Jilin, Liaoning, Inner Mongolia (Nei Mongol), Xinjiang, Anhui, Beijing, Fujian, Gansu, Guangdong, Guangxi, Guizhou, Hebei, Henan, Hubei, Hunan, Jiangsu, Jiangxi, Qinghai, Shaanxi, Shandong, Shanxi, Sichuan, Yunnan, Zhejiang, Chongqing, Ningxia, Tibet (Xizang)

世界分布 World Distribution

中国；中亚、东北亚 / China; Central Asia, Northeast Asia

分布标注 Distribution Note

非特有种 / Non-endemic

国内分布图
Map of Domestic Distribution

种群 Population

种群数量 Population Size	未知 / Unknown
种群趋势 Population Trend	未知 / Unknown

生境与生态系统 Habitat(s) and Ecosystem(s)

生 境 Habitat(s)	森林、灌丛、耕地、草地、半荒漠 / Forest, Shrubland, Arable Land, Grassland, Semi-desert
生态系统 Ecosystem(s)	森林生态系统、灌丛生态系统、草地生态系统、农田生态系统、荒漠生态系统 / Forest Ecosystem, Shrubland Ecosystem, Grassland Ecosystem, Cropland Ecosystem, Desert Ecosystem

威胁 Threat(s)

主要威胁 Major Threat(s)	猎捕、生境改变 / Trapping, Habitat Change

保护级别与保护行动 Protection Category and Conservation Action(s)

国家重点保护野生动物等级 (2021) Category of National Key Protected Wild Animals (2021)	无 / NA
IUCN 红色名录 (2020-2) IUCN Red List (2020-2)	无危 / LC
CITES 附录 (2019) CITES Appendix (2019)	无 / NA
保护行动 Conservation Action(s)	无 / None

相关文献 Relevant References

Burgin *et al.*, 2018; Jiang *et al.*（蒋志刚等）, 2017; Luo *et al.*（罗晓等）, 2016 ; Jiang *et al.*（姜雪松等）, 2013; Yao *et al.*, 2013; Smith *et al.*（史密斯等）, 2009; Wilson and Mittermeier, 2009; Gao *et al.*（高耀亭等）, 1987

亚洲狗獾 *Meles leucurus*

中国生物多样性红色名录 China's Red List of Biodiversity

猪獾 Arctonyx collaris

近危 NT

| 数据缺乏 DD | 无危 LC | 近危 NT | 易危 VU | 濒危 EN | 极危 CR | 区域灭绝 RE | 野外灭绝 EW | 灭绝 EX |

分类地位 Taxonomic Status

动物界 Animalia	脊索动物门 Chordata	哺乳纲 Mammalia	食肉目 Carnivora	鼬科 Mustelidae

学名 Scientific Name	*Arctonyx collaris*
命名人 Species Authority	F. G. Cuvier, 1825
英文名 English Name(s)	Hog Badger
同物异名 Synonym(s)	无 / None
种下单元评估 Infra-specific Taxa Assessed	无 / None

评估信息 Assessment Information

评估年份 Year Assessed	2020
评定人 Assessor(s)	蒋志刚 / Zhigang Jiang
审定人 Reviewer(s)	徐宏发、李峰 / Hongfa Xu, Feng Li
其他贡献人 Other Contributor(s)	李立立、丁晨晨 / Lili Li, Chenchen Ding

理由 Justification: 猪獾在中国南部与中部的分布区较广，但数量较少，且其所在的栖息地受到人类活动的扰动。因此，列为近危等级 / Hog Badger's area of occurrence in south and central China is large, but its population density is low. Furthermore, its habitats are disturbed by human activities. Thus, it is listed as Near Threatened

地理分布 Geographical Distribution

国内分布 Domestic Distribution

山西、河南、湖南、河北、辽宁、山东、江苏、浙江、安徽、福建、江西、湖北、广东、广西、四川、贵州、云南、西藏、陕西、甘肃、青海、宁夏、内蒙古、北京、重庆 / Shanxi, Henan, Hunan, Hebei, Liaoning, Shandong, Jiangsu, Zhejiang, Anhui, Fujian, Jiangxi, Hubei, Guangdong, Guangxi, Sichuan, Guizhou, Yunnan, Tibet (Xizang), Shaanxi, Gansu, Qinghai, Ningxia, Inner Mongolia (Nei Mongol), Beijing, Chongqing

世界分布 World Distribution

中国；南亚、中亚、东南亚 / China; South Asia, Central Asia, Southeast Asia

分布标注 Distribution Note

非特有种 / Non-endemic

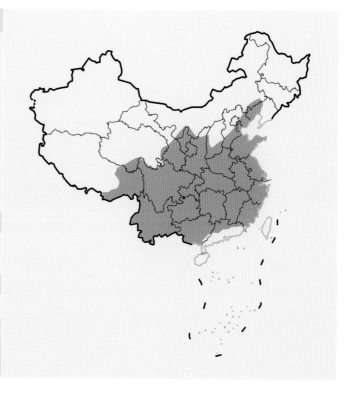

国内分布图
Map of Domestic Distribution

种群 Population

种群数量 Population Size	未知 / Unknown
种群趋势 Population Trend	下降 / Decreasing

生境与生态系统 Habitat (s) and Ecosystem (s)

生　　境 Habitat(s)	森林 / Forest
生态系统 Ecosystem(s)	森林生态系统 / Forest Ecosystem

威胁 Threat (s)

主要威胁 Major Threat(s)	狩猎 / Hunting

保护级别与保护行动 Protection Category and Conservation Action (s)

国家重点保护野生动物等级 (2021) Category of National Key Protected Wild Animals (2021)	无 / NA
IUCN红色名录(2020-2) IUCN Red List (2020-2)	近危 / NT
CITES 附录 (2019) CITES Appendix (2019)	无 / NA
保护行动 Conservation Action(s)	无 / None

相关文献 Relevant References

Burgin *et al.*, 2018; Jiang *et al.* (蒋志刚等), 2017; Smith *et al.* (史密斯等), 2009; Wilson and Mittermeier, 2009; Cheng and Liu (程泽信和刘武), 2000; Zhang (张荣祖), 1997; Gao *et al.* (高耀亭等), 1987

猪獾 *Arctonyx collaris*

小灵猫
Viverricula indica

近危 NT

| 数据缺乏 DD | 无危 LC | **近危 NT** | 易危 VU | 濒危 EN | 极危 CR | 区域灭绝 RE | 野外灭绝 EW | 灭绝 EX |

分类地位 Taxonomic Status

动物界 Animalia	脊索动物门 Chordata	哺乳纲 Mammalia	食肉目 Carnivora	灵猫科 Viverridae
学 名 Scientific Name		*Viverricula indica*		
命 名 人 Species Authority		É. Geoffroy Saint-Hilaire, 1803		
英 文 名 English Name(s)		Small Indian Civet		
同物异名 Synonym(s)		无 / None		
种下单元评估 Infra-specific Taxa Assessed		无 / None		

评估信息 Assessment Information

评估年份 Year Assessed	2020
评定人 Assessor(s)	蒋志刚 / Zhigang Jiang
审定人 Reviewer(s)	陈辈乐、李飞、李晟、胡慧建、王大军、王昊、周友兵 / Bosco P. L. Chan, Fei Li, Sheng Li, Huijian Hu, Dajun Wang, Hao Wang, Youbing Zhou
其他贡献人 Other Contributor(s)	李立立、丁晨晨 / Lili Li, Chenchen Ding

理由 Justification: 小灵猫曾广泛分布于中国黄河以南地区，曾被大规模捕杀，且栖息地受到人类活动干扰，其种群数量少。然而，近年来在多个保护区的红外相机调查中记录到小灵猫，表明其种群正在恢复。因此，列为近危等级 / Small Indian Civet was historically widely distributed in south of the Huanghe River in China. However, it was long hunted as a game species and its habitat has been disturbed by human activities, therefore, its present population size is small. In recent years, Small Indian Civet has been recorded by infrared camera traps in many nature reserves, indicating that its populations are recovering. Thus, it is listed as Near Threatened

地理分布 Geographical Distribution

国内分布 Domestic Distribution

河南、湖南、上海、江苏、浙江、安徽、福建、江西、湖北、广东、广西、四川、贵州、云南、西藏、陕西、甘肃、香港、重庆 / Henan, Hunan, Shanghai, Jiangsu, Zhejiang, Anhui, Fujian, Jiangxi, Hubei, Guangdong, Guangxi, Sichuan, Guizhou, Yunnan, Tibet (Xizang), Shaanxi, Gansu, Hong Kong, Chongqing

世界分布 World Distribution

中国；南亚、东南亚 / China; South Asia, Southeast Asia

分布标注 Distribution Note

非特有种 / Non-endemic

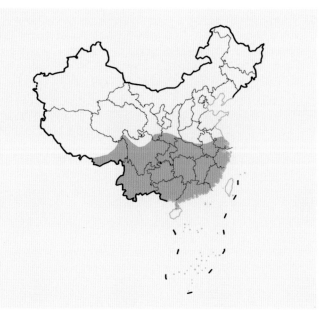

国内分布图
Map of Domestic Distribution

种群 Population

种群数量 Population Size	未知 / Unknown
种群趋势 Population Trend	稳定 / Stable

生境与生态系统 Habitat (s) and Ecosystem (s)

生境 Habitat(s)	草地、灌丛、农田 / Grassland, Shrubland, Cropland
生态系统 Ecosystem(s)	森林生态系统、灌丛生态系统、农田生态系统 / Forest Ecosystem, Shrubland Ecosystem, Cropland Ecosystem

威胁 Threat (s)

主要威胁 Major Threat(s)	狩猎、伐木、耕种 / Hunting, Logging, Plantation

保护级别与保护行动 Protection Category and Conservation Action (s)

国家重点保护野生动物等级 (2021) Category of National Key Protected Wild Animals (2021)	一级 / Category I
IUCN红色名录(2020-2) IUCN Red List (2020-2)	近危 / NT
CITES 附录 (2019) CITES Appendix (2019)	III
保护行动 Conservation Action(s)	部分种群位于自然保护区内 / Part of populations are covered by nature reserve

相关文献 Relevant References

Burgin *et al*., 2018; Jiang *et al*. (蒋志刚等), 2017; Hu *et al*. (胡一鸣等), 2014; Lau *et al*., 2010; Wilson and Mittermeier, 2009; Gao *et al*. (高耀亭等), 1987

小灵猫 *Viverricula indica*

果子狸
Paguma larvata

近危 NT

| 数据缺乏 DD | 无危 LC | 近危 NT | 易危 VU | 濒危 EN | 极危 CR | 区域灭绝 RE | 野外灭绝 EW | 灭绝 EX |

分类地位 Taxonomic Status

动物界 Animalia	脊索动物门 Chordata	哺乳纲 Mammalia	食肉目 Carnivora	灵猫科 Viverridae
学名 Scientific Name		*Paguma larvata*		
命名人 Species Authority		C. E. H. Smith, 1827		
英文名 English Name(s)		Masked Palm Civet		
同物异名 Synonym(s)		花面狸, Gem-faced Civet		
种下单元评估 Infra-specific Taxa Assessed		无 / None		

评估信息 Assessment Information

评估年份 Year Assessed	2020
评定人 Assessor(s)	蒋志刚 / Zhigang Jiang
审定人 Reviewer(s)	李晟、胡慧建、王大军、王昊、周友兵 / Sheng Li, Huijian Hu, Dajun Wang, Hao Wang, Youbing Zhou
其他贡献人 Other Contributor(s)	李立立、丁晨晨 / Lili Li, Chenchen Ding

理由 Justification: 果子狸在中国的分布区较广，野外常见且能耐受被干扰的生境，但果子狸被猎杀作为肉用、捕捉作为宠物，或被果农捕杀。因此，列为近危等级 / Masked Palm Civet is widely distributed in China. The species is common in wild and it could tolerates disturbed habitat, but is hunted for meat, captured as pet, or killed by fruit farmers. Thus, it is listed as Near Threatened

地理分布 Geographical Distribution

国内分布 Domestic Distribution

山西、河南、湖南、河北、海南、陕西、北京、江苏、浙江、安徽、福建、江西、湖北、广东、广西、四川、贵州、云南、西藏、甘肃、香港、重庆、上海、台湾 / Shanxi, Henan, Hunan, Hebei, Hainan, Shaanxi, Beijing, Jiangsu, Zhejiang, Anhui, Fujian, Jiangxi, Hubei, Guangdong, Guangxi, Sichuan, Guizhou, Yunnan, Tibet (Xizang), Gansu, Hong Kong, Chongqing, Shanghai, Taiwan

世界分布 World Distribution

中国；南亚、东南亚 / China; South Asia, Southeast Asia

分布标注 Distribution Note

非特有种 / Non-endemic

国内分布图
Map of Domestic Distribution

种群 Population

种群数量 Population Size	未知 / Unknown
种群趋势 Population Trend	稳定 / Stable

生境与生态系统 Habitat (s) and Ecosystem (s)

生　　境 Habitat(s)	森林、农田 / Forest, Cropland
生态系统 Ecosystem(s)	森林生态系统、农田生态系统 / Forest Ecosystem, Cropland Ecosystem

威胁 Threat (s)

主要威胁 Major Threat(s)	狩猎 / Hunting

保护级别与保护行动 Protection Category and Conservation Action (s)

国家重点保护野生动物等级 (2021) Category of National Key Protected Wild Animals (2021)	无 / NA
IUCN 红色名录 (2020-2) IUCN Red List (2020-2)	无危 / LC
CITES 附录 (2019) CITES Appendix (2019)	III
保护行动 Conservation Action(s)	无 / None

相关文献 Relevant References

Burgin *et al*., 2018; Jiang *et al*. (蒋志刚等), 2017, 2003; Lau *et al*., 2010; Zheng *et al*. (曾国仕等), 2010; Zhu *et al*. (朱红艳等), 2010; Wang *et al*. (王健等), 2009; Wilson and Mittermeier, 2009; Gao *et al*. (高耀亭等), 1987

果子狸 *Paguma larvata*

藏野驴
Equus kiang

近危 NT

| 数据缺乏 DD | 无危 LC | 近危 NT | 易危 VU | 濒危 EN | 极危 CR | 区域灭绝 RE | 野外灭绝 EW | 灭绝 EX |

分类地位 Taxonomic Status

动物界 Animalia	脊索动物门 Chordata	哺乳纲 Mammalia	奇蹄目 Perissodactyla	马科 Equidae

学名 Scientific Name	*Equus kiang*
命名人 Species Authority	Moorcroft, 1841
英文名 English Name(s)	Kiang
同物异名 Synonym(s)	西藏野驴；*Equus equioides* (Hodgson, 1842); *holdereri* (Matschie, 1911); *kyang* (Kinloch, 1869); *nepalensis* (Trumler, 1959); *polyodon* (Hodgson, 1847); *tafeli* (Matschie, 1924)
种下单元评估 Infra-specific Taxa Assessed	无 / None

评估信息 Assessment Information

评估年份 Year Assessed	2020
评定人 Assessor(s)	蒋志刚 / Zhigang Jiang
审定人 Reviewer(s)	苏建平、徐爱春、胡慧建 / Jianping Su, Aichun Xu, Huijian Hu
其他贡献人 Other Contributor(s)	李立立、丁晨晨 / Lili Li, Chenchen Ding

理由 Justification: 藏野驴是青藏高原的特有种，种群数量大且增长快，但与当地牧民的牲畜竞争草地。很多区域建立围栏作为牧场，此外，道路建设也对藏野驴有一定影响。因此，基于其占有区面积，藏野驴列为近危等级 / Kiang is endemic to Qinghai-Tibet (Xizang) Plateau, its populations sizes are large and increasing rapidly, it competes with livestocks for forage. The construction of fences paddocks on the plateau and construction of roads influence the Kiang. Thus, Kiang is listed as Near Threatened

地理分布 Geographical Distribution

国内分布 Domestic Distribution
新疆、青海、甘肃、四川、西藏 / Xinjiang, Qinghai, Gansu, Sichuan, Tibet (Xizang)
世界分布 World Distribution
中国、印度、尼泊尔、巴基斯坦 / China, India, Nepal, Pakistan
分布标注 Distribution Note
非特有种 / Non-endemic

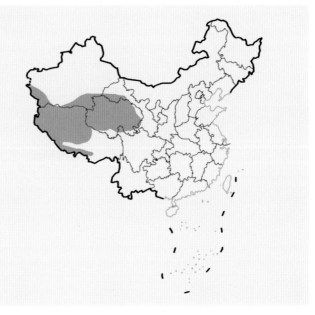

国内分布图
Map of Domestic Distribution

种群 Population

种群数量 Population Size	56,500 ~ 68,500 只 / 56,500 ~ 68,500 individuals
种群趋势 Population Trend	稳定 / Stable

生境与生态系统 Habitat(s) and Ecosystem(s)

生境 Habitat(s)	草甸、荒漠 / Meadow, Desert
生态系统 Ecosystem(s)	草地生态系统、荒漠生态系统 / Grassland Ecosystem, Desert Ecosystem

威胁 Threat(s)

主要威胁 Major Threat(s)	人兽冲突、家畜放牧、狩猎、道路建设 / Human-Animal Conflict, Livestock Ranching, Hunting, Road Construction

保护级别与保护行动 Protection Category and Conservation Action(s)

国家重点保护野生动物等级 (2021) Category of National Key Protected Wild Animals (2021)	一级 / Category I
IUCN 红色名录 (2020-2) IUCN Red List (2020-2)	无危 / LC
CITES 附录 (2019) CITES Appendix (2019)	II
保护行动 Conservation Action(s)	自然保护区内种群得到保护 / Populations in nature reserves are protected

相关文献 Relevant References

Burgin *et al.*, 2018; Jiang *et al.* (蒋志刚等), 2017; Su *et al.* (苏旭坤等), 2014; Luo *et al.*, 2011; Smith *et al.* (史密斯等), 2009

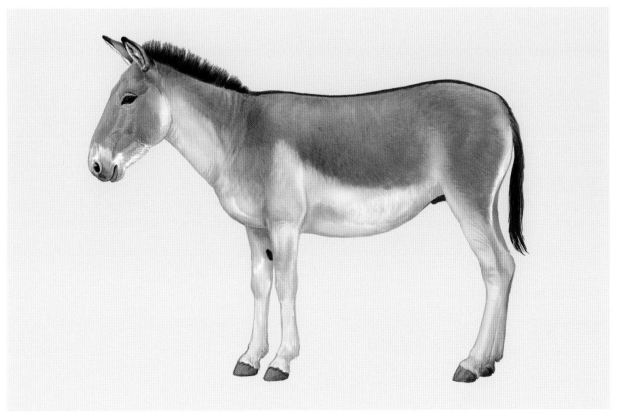

藏野驴 *Equus kiang*

毛冠鹿
Elaphodus cephalophus
近危 NT

| 数据缺乏 DD | 无危 LC | 近危 NT | 易危 VU | 濒危 EN | 极危 CR | 区域灭绝 RE | 野外灭绝 EW | 灭绝 EX |

分类地位 Taxonomic Status

动物界 Animalia	脊索动物门 Chordata	哺乳纲 Mammalia	偶蹄目 Artiodactyla	鹿科 Cervidae

学　名 Scientific Name	*Elaphodus cephalophus*
命名人 Species Authority	Milne-Edwards, 1872
英文名 English Name(s)	Tufted Deer
同物异名 Synonym(s)	无 / None
种下单元评估 Infra-specific Taxa Assessed	无 / None

评估信息 Assessment Information

评估年份 Year Assessed	2020
评定人 Assessor(s)	蒋志刚 / Zhigang Jiang
审定人 Reviewer(s)	杨道德 / Daode Yang
其他贡献人 Other Contributor(s)	李立立、丁晨晨 / Lili Li, Chenchen Ding

理由 Justification: 毛冠鹿分布广泛，且有一定的种群数量，但毛冠鹿栖息地受人为干扰严重，受到的盗猎压力大。因此，列为近危等级 / Tufted Deer has an extensive distribution in the country and its wild population is large, but its areas of occupancy are declining due to human encroachment and the species is threatened by hunting. Thus, it is listed as Near Threatened

地理分布 Geographical Distribution

国内分布 Domestic Distribution
湖南、浙江、安徽、福建、江西、湖北、广东、广西、四川、贵州、云南、西藏、陕西、甘肃、青海、重庆 / Hunan, Zhejiang, Anhui, Fujian, Jiangxi, Hubei, Guangdong, Guangxi, Sichuan, Guizhou, Yunnan, Tibet (Xizang), Shaanxi, Gansu, Qinghai, Chongqing
世界分布 World Distribution
中国、缅甸 / China, Myanmar
分布标注 Distribution Note
非特有种 / Non-endemic

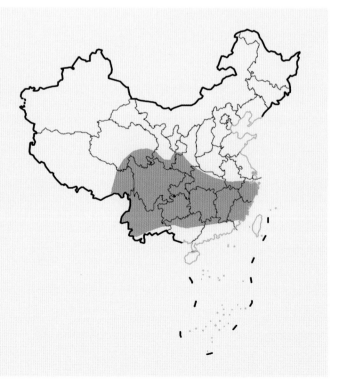

国内分布图
Map of Domestic Distribution

种群 Population

种群数量 Population Size	300,000～500,000 只 / 300,000～500,000 individuals
种群趋势 Population Trend	下降 / Decreasing

生境与生态系统 Habitat(s) and Ecosystem(s)

生　　境 Habitat(s)	森林、草甸 / Forest, Meadow
生态系统 Ecosystem(s)	森林生态系统、草地生态系统 / Forest Ecosystem, Grassland Ecosystem

威胁 Threat(s)

主要威胁 Major Threat(s)	狩猎 / Hunting

保护级别与保护行动 Protection Category and Conservation Action(s)

国家重点保护野生动物等级 (2021) Category of National Key Protected Wild Animals (2021)	二级 / Category II
IUCN 红色名录 (2020-2) IUCN Red List (2020-2)	近危 / NT
CITES 附录 (2019) CITES Appendix (2019)	无 / NA
保护行动 Conservation Action(s)	未知 / Unknown

相关文献 Relevant References

Burgin *et al.*, 2018; Jiang *et al.* (蒋志刚等), 2017; Leslie *et al.*, 2013; Wilson and Mittermeier, 2012; Groves and Grubb, 2011; Smith *et al.* (史密斯等), 2009; Wang (王应祥), 2003; Sheng and Lu (盛和林和陆厚基), 1982

毛冠鹿 *Elaphodus cephalophus*

小麂
Muntiacus reevesi

近危 NT

| 数据缺乏 DD | 无危 LC | 近危 NT | 易危 VU | 濒危 EN | 极危 CR | 区域灭绝 RE | 野外灭绝 EW | 灭绝 EX |

分类地位 Taxonomic Status

动物界 Animalia	脊索动物门 Chordata	哺乳纲 Mammalia	偶蹄目 Artiodactyla	鹿科 Cervidae

学 名 Scientific Name	*Muntiacus reevesi*
命 名 人 Species Authority	Ogilby, 1839
英 文 名 English Name(s)	Reeves' Muntjac
同物异名 Synonym(s)	Chinese Muntjac; *Cervulus bridgemani* (Lydekker, 1910); *C. micrurus* (Sclater, 1875); *C. reevesi* (Hilzheimer, 1906) subsp. *pingshiangicus*; *C. sclateri* (Swinhoe, 1872);（转下页）
种下单元评估 Infra-specific Taxa Assessed	无 / None

评估信息 Assessment Information

评估年份 Year Assessed	2020
评定人 Assessor(s)	蒋志刚 / Zhigang Jiang
审定人 Reviewer(s)	蒋学龙、陈辈乐、李飞 / Xuelong Jiang, Bosco P. L. Chan, Fei Li
其他贡献人 Other Contributor(s)	李立立、丁晨晨 / Lili Li, Chenchen Ding

理由 Justification: 小麂在中国南方广泛分布，种群数量大。其种群数量在浙江等省份的自然保护区内有增长。但作为狩猎对象而存在偷猎压力，其位于自然保护区外的种群在下降。因此，列为近危等级 / Reeves' Muntjac's population sizes are increasing in some nature reserves of Zhejiang and other provinces in recent years. In southern China, the distribution range and populations of Reeves' Muntjac are large, but Reeves' Muntjac confronts poaching pressure, its populations outside of nature reserves are declining. Thus, it is listed as Near Threatened

地理分布 Geographical Distribution

国内分布 Domestic Distribution
河南、贵州、江苏、浙江、安徽、福建、江西、湖北、湖南、广东、广西、四川、云南、陕西、甘肃、台湾、香港、重庆 / Henan, Guizhou, Jiangsu, Zhejiang, Anhui, Fujian, Jiangxi, Hubei, Hunan, Guangdong, Guangxi, Sichuan, Yunnan, Shaanxi, Gansu, Taiwan, Hong Kong, Chongqing
世界分布 World Distribution
中国 / China
分布标注 Distribution Note
特有种 / Endemic

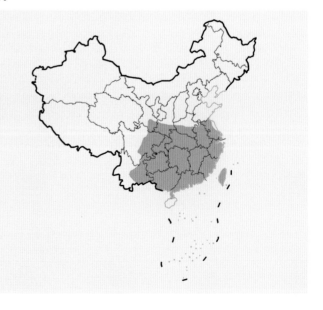

国内分布图
Map of Domestic Distribution

种群 Population

种群数量 Population Size	未知 / Unknown
种群趋势 Population Trend	下降 / Decreasing

生境与生态系统 Habitat (s) and Ecosystem (s)

生　　境 Habitat(s)	灌丛、内陆岩石区域、森林 / Shrubland, Inland Rocky Area, Forest
生态系统 Ecosystem(s)	森林生态系统、灌丛生态系统 / Forest Ecosystem, Shrubland Ecosystem

威胁 Threat (s)

主要威胁 Major Threat(s)	狩猎、耕种、伐木、住宅区及商业发展 / Hunting, Plantation, Logging, Residential and Commercial Development

保护级别与保护行动 Protection Category and Conservation Action (s)

国家重点保护野生动物等级 (2021) Category of National Key Protected Wild Animals (2021)	无 / NA
IUCN 红色名录 (2020-2) IUCN Red List (2020-2)	无危 / LC
CITES 附录 (2019) CITES Appendix (2019)	无 / NA
保护行动 Conservation Action(s)	未知 / Unknown

相关文献 Relevant References

Burgin *et al.*, 2018; Jiang *et al.* (蒋志刚等), 2017; Wilson and Mittermeier, 2012; Groves and Grubb, 2011; Shi *et al.* (史文博等), 2010; Zhang *et al.*, 2010; Smith *et al.* (史密斯等), 2009; Wang (王应祥), 2003

（接上页）
C. sinensis (Hilzheimer, 1905); *C. lachrymans* (Milne-Edwards, 1871); *C. reevesi* (Ogilby, 1839); *Muntiacus lachrymans* (Wroughton, 1914) subsp. *teesdalei*

小麂 *Muntiacus reevesi*

中国生物多样性红色名录

赤麂
Muntiacus vaginalis

近危 NT

| 数据缺乏 DD | 无危 LC | 近危 NT | 易危 VU | 濒危 EN | 极危 CR | 区域灭绝 RE | 野外灭绝 EW | 灭绝 EX |

分类地位 Taxonomic Status

动物界 Animalia	脊索动物门 Chordata	哺乳纲 Mammalia	偶蹄目 Artiodactyla	鹿科 Cervidae

学名 Scientific Name	*Muntiacus vaginalis*
命名人 Species Authority	Boddaert, 1785
英文名 English Name(s)	Northern Red Muntjac
同物异名 Synonym(s)	*Cervus melas* (Ogilby, 1839); *C. moschatus* (Smith, 1827); *C. ratwa* (Hodgson, 1833); *C. stylocerus* (Schinz, 1844); *C. vaginalis* (Daert, 1785); *Prox ratva* (Sundevall, 1846); (转下页)
种下单元评估 Infra-specific Taxa Assessed	无 / None

评估信息 Assessment Information

评估年份 Year Assessed	2020
评定人 Assessor(s)	蒋志刚 / Zhigang Jiang
审定人 Reviewer(s)	蒋学龙、陈辈乐、李飞 / Xuelong Jiang, Bosco P. L. Chan, Fei Li
其他贡献人 Other Contributor(s)	李立立、丁晨晨 / Lili Li, Chenchen Ding

理由 Justification: 赤麂分布较广泛，自然保护区内有一定的种群数量，但仍面临偷猎的压力，自然保护区外的种群数量下降。因此，列为近危等级 / Northern Red Muntjac is widely distributed, and its populations are increasing in nature reserves. But Northern Red Muntjac still faces poaching pressure, and its population outside of nature reserves are declining. Thus, it is listed as Near Threatened

地理分布 Geographical Distribution

国内分布 Domestic Distribution	湖南、云南、海南、福建、江西、广东、广西、四川、贵州、西藏、香港 / Hunan, Yunnan, Hainan, Fujian, Jiangxi, Guangdong, Guangxi, Sichuan, Guizhou, Tibet (Xizang), Hong Kong
世界分布 World Distribution	中国；南亚、东南亚 / China; South Asia, Southeast Asia
分布标注 Distribution Note	非特有种 / Non-endemic

国内分布图
Map of Domestic Distribution

种群 Population

种群数量 Population Size	220,000 只 / 220,000 individuals
种群趋势 Population Trend	上升 / Increasing

生境与生态系统 Habitat(s) and Ecosystem(s)

生境 Habitat(s)	森林、灌丛 / Forest, Shrubland
生态系统 Ecosystem(s)	森林生态系统、灌丛生态系统 / Forest Ecosystem, Shrubland Ecosystem

威胁 Threat(s)

主要威胁 Major Threat(s)	狩猎 / Hunting

保护级别与保护行动 Protection Category and Conservation Action(s)

国家重点保护野生动物等级 (2021) Category of National Key Protected Wild Animals (2021)	无 / NA
IUCN 红色名录 (2020-2) IUCN Red List (2020-2)	无危 / LC
CITES 附录 (2019) CITES Appendix (2019)	无 / NA
保护行动 Conservation Action(s)	未知 / Unknown

相关文献 Relevant References

Burgin *et al*., 2018; 蒋志刚等 (Jiang *et al*.), 2017; Deng *et al*. (邓可等), 2013; Wilson and Mittermeier, 2012; Groves and Grubb, 2011; Smith *et al*. (史密斯等), 2009; Teng *et al*. (滕丽微等), 2005

(接上页)
Stylocerus muntjac (Cantor, 1846); *S. muntjacus* (Kelaart, 1852)

赤麂 *Muntiacus vaginalis* 嘉道理农场暨植物园 提供 By Kadoorie Farm and Botanic Garden

水鹿
Rusa unicolor

近危 NT

| 数据缺乏 DD | 无危 LC | 近危 NT | 易危 VU | 濒危 EN | 极危 CR | 区域灭绝 RE | 野外灭绝 EW | 灭绝 EX |

分类地位 Taxonomic Status

动物界 Animalia	脊索动物门 Chordata	哺乳纲 Mammalia	偶蹄目 Artiodactyla	鹿科 Cervidae
学名 Scientific Name		*Rusa unicolor*		
命名人 Species Authority		G. Cuvier, 1823		
英文名 English Name(s)		Southeast Asian Sambar		
同物异名 Synonym(s)		*Cervus equinus*		
种下单元评估 Infra-specific Taxa Assessed		无 / None		

评估信息 Assessment Information

评估年份 Year Assessed	2020
评定人 Assessor(s)	蒋志刚 / Zhigang Jiang
审定人 Reviewer(s)	王小明、蒋学龙、陈辈乐、李飞 / Xiaoming Wang, Xuelong Jiang, Bosco P. L. Chan, Fei Li
其他贡献人 Other Contributor(s)	李立立、丁晨晨 / Lili Li, Chenchen Ding

理由 Justification: 水鹿曾遍布于中国南部和西南部。近年的红外相机调查显示四川亚种及台湾亚种生存状况略好,而海南亚种和华南亚种仅在一些保护状况良好的保护区中才有发现。在很多地区,尤其是热带,水鹿依然受到盗猎及栖息地丧失的威胁。因此,列为近危等级 / Southeast Asian Sambar used to be found throughout south and southwest China. In recent years, infrared camera surveys have shown that the population of the Sichuan and Taiwan subspecies of Southeast Asian Sambar are in a better status, while the Hainan and south China subspecies are found only in some well-protected areas. In many areas, especially in the tropics, Southeast Asian Sambar is threatened by poaching and habitat loss. Thus, it is listed as Near Threatened

地理分布 Geographical Distribution

国内分布 Domestic Distribution
西藏、青海、云南、四川、重庆、贵州、广西、海南、广东、湖南、江西、福建、台湾 / Tibet (Xizang), Qinghai, Yunnan, Sichuan, Chongqing, Guizhou, Guangxi, Hainan, Guangdong, Hunan, Jiangxi, Fujian, Taiwan
世界分布 World Distribution
中国;南亚、东南亚 / China; South Asia, Southeast Asia
分布标注 Distribution Note
非特有种 / Non-endemic

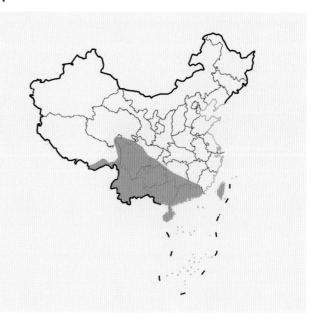

国内分布图
Map of Domestic Distribution

种群 Population

种群数量 Population Size	未知 / Unknown
种群趋势 Population Trend	未知 / Unknown

生境与生态系统 Habitat (s) and Ecosystem (s)

生境 Habitat(s)	森林、灌丛、沼泽、农田 / Forest, Shrubland, Swamp, Cropland
生态系统 Ecosystem(s)	森林生态系统、灌丛生态系统、农田生态系统、湿地生态系统 / Forest Ecosystem, Shrubland Ecosystem, Cropland Ecosystem, Wetland Ecosystem

威胁 Threat (s)

主要威胁 Major Threat(s)	狩猎、耕种、伐木、住宅区及商业发展 / Hunting, Plantation, Logging, Residential and Commercial Development

保护级别与保护行动 Protection Category and Conservation Action (s)

国家重点保护野生动物等级 (2021) Category of National Key Protected Wild Animals (2021)	二级 / Category II
IUCN 红色名录(2020-2) IUCN Red List (2020-2)	未列入 / NA
CITES 附录 (2019) CITES Appendix (2019)	无 / NA
保护行动 Conservation Action(s)	自然保护区内种群得到保护 / Populations in nature reserves are protected

相关文献 Relevant References

Burgin *et al*., 2018; Jiang *et al*. (蒋志刚等), 2017; Groves and Grubb, 2011; Wilson and Mittermeier, 2011

水鹿 *Rusa unicolor* 　　　　　　　　　　　　　　　　　　李晟 摄　By Sheng Li

狍
Capreolus pygargus

近危 NT

| 数据缺乏 DD | 无危 LC | 近危 NT | 易危 VU | 濒危 EN | 极危 CR | 区域灭绝 RE | 野外灭绝 EW | 灭绝 EX |

分类地位 Taxonomic Status

动物界 Animalia	脊索动物门 Chordata	哺乳纲 Mammalia	偶蹄目 Artiodactyla	鹿科 Cervidae

学名 Scientific Name	*Capreolus pygargus*
命名人 Species Authority	Pallas, 1771
英文名 English Name(s)	Roe Deer
同物异名 Synonym(s)	Siberian Roe Deer; Eastern Roe Deer; *Capreolus bedfordi* (Thomas, 1908); *C. capreolus* (Barclay, 1935); subsp. *ochracea*; *C. melanotis* (Miller, 1911); *C. pygargus* (转下页)
种下单元评估 Infra-specific Taxa Assessed	无 / None

评估信息 Assessment Information

评估年份 Year Assessed	2020
评定人 Assessor(s)	蒋志刚 / Zhigang Jiang
审定人 Reviewer(s)	张明海 / Minghai Zhang
其他贡献人 Other Contributor(s)	李立立、丁晨晨 / Lili Li, Chenchen Ding

理由 Justification: 狍在中国的分布区较广，但数量较少，且其种群片段化，并受到不同威胁因子的威胁。因此，列为近危等级 / Roe Deer's distribution range in China is large, but its population sizes are small. Furthermore, populations of Roe Deer are fragmented, and their survival is threatened by various threats. Thus, it is listed as Near Threatened

地理分布 Geographical Distribution

国内分布 Domestic Distribution
吉林、山西、内蒙古、黑龙江、宁夏、新疆、陕西、青海、北京、河北、辽宁、河南、湖北、四川、西藏、甘肃、重庆 / Jilin, Shanxi, Inner Mongolia (Nei Mongol), Heilongjiang, Ningxia, Xinjiang, Shaanxi, Qinghai, Beijing, Hebei, Liaoning, Henan, Hubei, Sichuan, Tibet (Xizang), Gansu, Chongqing
世界分布 World Distribution
中国；中亚、东亚 / China; Central Asia, East Asia
分布标注 Distribution Note
非特有种 / Non-endemic

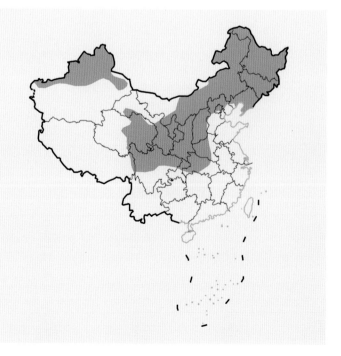

国内分布图
Map of Domestic Distribution

种群 Population

种群数量 Population Size	440,000 只 / 440,000 individuals
种群趋势 Population Trend	下降 / Decreasing

生境与生态系统 Habitat(s) and Ecosystem(s)

生境 Habitat(s)	森林、草原 / Forest, Steppe Forest
生态系统 Ecosystem(s)	森林生态系统、草地生态系统 / Forest Ecosystem, Grassland Ecosystem

威胁 Threat(s)

主要威胁 Major Threat(s)	狩猎、生境丧失 / Hunting, Habitat Loss

保护级别与保护行动 Protection Category and Conservation Action(s)

国家重点保护野生动物等级 (2021) Category of National Key Protected Wild Animals (2021)	无 / NA
IUCN 红色名录 (2020-2) IUCN Red List (2020-2)	无危 / LC
CITES 附录 (2019) CITES Appendix (2019)	无 / NA
保护行动 Conservation Action(s)	未知 / Unknown

相关文献 Relevant References

Burgin *et al.*, 2018; Jiang *et al.* (蒋志刚等), 2017; Wilson and Mittermeier, 2012; Groves and Grubb, 2011; Liu and Zhang (刘艳华和张海明), 2009; Smith *et al.* (史密斯等), 2009; Wang (王应祥), 2003

(接上页)
(Dinnik, 1910) var. *caucasica*; *C. pygargus* (Rasewig, 1909) var. *ferganicus*; *C. tianschanicus* (Saturnin, 1906); *Cervus pygargus* (Pallas, 1771)

狍 *Capreolus pygargus*

藏原羚
Procapra picticaudata

近危 NT

| 数据缺乏 DD | 无危 LC | 近危 NT | 易危 VU | 濒危 EN | 极危 CR | 区域灭绝 RE | 野外灭绝 EW | 灭绝 EX |

分类地位 Taxonomic Status

动物界 Animalia	脊索动物门 Chordata	哺乳纲 Mammalia	偶蹄目 Artiodactyla	牛科 Bovidae
学 名 Scientific Name		*Procapra picticaudata*		
命 名 人 Species Authority		Hodgson, 1846		
英 文 名 English Name(s)		Tibetan Gazelle		
同物异名 Synonym(s)		Goa		
种下单元评估 Infra-specific Taxa Assessed		无 / None		

评估信息 Assessment Information

评估年份 Year Assessed	2020
评 定 人 Assessor(s)	蒋志刚 / Zhigang Jiang
审 定 人 Reviewer(s)	胡慧建、李忠秋、李春旺 / Huijian Hu, Zhongqiu Li, Chunwang Li
其他贡献人 Other Contributor(s)	李立立、丁晨晨 / Lili Li, Chenchen Ding

理由 Justification: 藏原羚在中国的分布区较广，但数量较少，且其所在的栖息地面积基本稳定。因此，列为近危等级 / Tibetan Gazelle's distribution range in China is large, but its population sizes are small. Furthermore, the areas of its habitats are stable. Thus, it is listed as Near Threatened

地理分布 Geographical Distribution

国内分布 Domestic Distribution
新疆、青海、甘肃、四川、西藏 / Xinjiang, Qinghai, Gansu, Sichuan, Tibet (Xizang)
世界分布 World Distribution
中国 / China
分布标注 Distribution Note
特有种 / Endemic

国内分布图
Map of Domestic Distribution

🦌 种群 Population

种群数量 Population Size	100,000 只 / 100,000 individuals
种群趋势 Population Trend	下降 / Decreasing

🦌 生境与生态系统 Habitat (s) and Ecosystem (s)

生　　境 Habitat(s)	荒漠、半荒漠、草地、灌丛 / Desert, Semi-desert, Grassland, Shrubland
生态系统 Ecosystem(s)	灌丛生态系统、草地生态系统、荒漠生态系统 / Shrubland Ecosystem, Grassland Ecosystem, Desert Ecosystem

🦌 威胁 Threat (s)

主要威胁 Major Threat(s)	未知 / Unknown

🦌 保护级别与保护行动 Protection Category and Conservation Action (s)

国家重点保护野生动物等级 (2021) Category of National Key Protected Wild Animals (2021)	二级 / Category II
IUCN 红色名录 (2020-2) IUCN Red List (2020-2)	近危 / NT
CITES 附录 (2019) CITES Appendix (2019)	无 / NA
保护行动 Conservation Action(s)	自然保护区内种群得到保护 / Populations in nature reserves are protected

🦌 相关文献 Relevant References

Burgin *et al*., 2018; Jiang *et al*.（蒋志刚等）, 2017; Castelló, 2016; Groves and Grubb, 2011; Wilson and Mittermeier, 2011; Smith *et al*.（史密斯等）, 2009; Yin *et al*.（殷宝法等）, 2007

藏原羚 *Procapra picticaudata*

藏羚 *Pantholops hodgsonii*

近危 NT

| 数据缺乏 DD | 无危 LC | 近危 NT | 易危 VU | 濒危 EN | 极危 CR | 区域灭绝 RE | 野外灭绝 EW | 灭绝 EX |

分类地位 Taxonomic Status

动物界 Animalia	脊索动物门 Chordata	哺乳纲 Mammalia	偶蹄目 Artiodactyla	牛科 Bovidae
学名 Scientific Name		*Pantholops hodgsonii*		
命名人 Species Authority		Abel, 1826		
英文名 English Name(s)		Chiru		
同物异名 Synonym(s)		Tibetan Antelope		
种下单元评估 Infra-specific Taxa Assessed		无 / None		

评估信息 Assessment Information

评估年份 Year Assessed	2020
评定人 Assessor(s)	蒋志刚 / Zhigang Jiang
审定人 Reviewer(s)	苏建平、胡慧建 / Jianping Su, Huijian Hu
其他贡献人 Other Contributor(s)	李立立、丁晨晨 / Lili Li, Chenchen Ding

理由 Justification: 藏羚主要分布在青藏高原腹地，20 世纪末，偷猎藏羚的势头被遏制，近年藏羚种群数量呈恢复趋势，种群恢复到 20 万只。但还存在一定的盗猎压力。因此，列为近危等级 / Chiru inhabits the heartland of the Qinghai-Tibet (Xizang) Plateau. Although poaching of Chiru was curbed at the end of 20th century, and the populations of this species have been recovering to 200,000 individuals in recent years, pressure of poaching still exists. Thus, it is listed as Near Threatened

地理分布 Geographical Distribution

国内分布 Domestic Distribution
新疆、西藏、青海、四川 / Xinjiang, Tibet (Xizang), Qinghai, Sichuan
世界分布 World Distribution
中国 / China
分布标注 Distribution Note
特有种 / Endemic

国内分布图
Map of Domestic Distribution

种群 Population

种群数量 Population Size	200,000 只 / 200,000 individuals
种群趋势 Population Trend	稳定 / Stable

生境与生态系统 Habitat (s) and Ecosystem (s)

生　　境 Habitat(s)	荒漠、高寒草地 / Desert, Alpine Steppe
生态系统 Ecosystem(s)	高寒草地生态系统、荒漠生态系统 / Alpine Steppe Ecosystem, Desert Ecosystem

威胁 Threat (s)

主要威胁 Major Threat(s)	狩猎、家畜养殖或放牧、公路和铁路、极端天气 / Hunting, Livestock Farming or Ranching, Road and Railroad, Extreme Weather Event

保护级别与保护行动 Protection Category and Conservation Action (s)

国家重点保护野生动物等级 (2021) Category of National Key Protected Wild Animals (2021)	一级 / Category I
IUCN 红色名录 (2020-2) IUCN Red List (2020-2)	濒危 / EN
CITES 附录 (2019) CITES Appendix (2019)	I
保护行动 Conservation Action(s)	自然保护区内种群得到保护 / Populations in nature reserves are protected

相关文献 Relevant References

Burgin *et al.*, 2018; Jiang *et al.*（蒋志刚等）, 2017; Castelló, 2016; Groves and Grubb, 2011; Wilson and Mittermeier, 2011; Bleisch *et al.*, 2009; Fox *et al.*, 2009; Qiu and Feng（裘丽和冯祚建）, 2004

藏羚 *Pantholops hodgsonii*

北山羊
Capra sibirica
近危 NT

| 数据缺乏 DD | 无危 LC | 近危 NT | 易危 VU | 濒危 EN | 极危 CR | 区域灭绝 RE | 野外灭绝 EW | 灭绝 EX |

分类地位 Taxonomic Status

动物界 Animalia	脊索动物门 Chordata	哺乳纲 Mammalia	偶蹄目 Artiodactyla	牛科 Bovidae

学 名 Scientific Name	*Capra sibirica*
命 名 人 Species Authority	Pallas, 1776
英 文 名 English Name(s)	Siberian Ibex
同物异名 Synonym(s)	Altai Ibex; Gobi Ibex; *Capra sibrica* (Pallas, 1776)
种下单元评估 Infra-specific Taxa Assessed	无 / None

评估信息 Assessment Information

评估年份 Year Assessed	2020
评 定 人 Assessor(s)	蒋志刚 / Zhigang Jiang
审 定 人 Reviewer(s)	初红军、杨维康 / Hongjun Chu, Weikang Yang
其他贡献人 Other Contributor(s)	李立立、丁晨晨 / Lili Li, Chenchen Ding

理由 Justification: 北山羊种群在新疆恢复较快。2013～2014年冬季，杨维康等在乌鲁木齐天山后峡不足15km地段见到50～60群，500余只，最大群有100只以上。2010年8月，在中蒙边境北塔山桑塔斯边防公路不足10km路段两旁见到30群，60～70只北山羊。然而，尽管种群数量稳定，目前矿山开发对北山羊构成威胁。因此，列为近危等级 / In recent years, the populations of Siberian Ibex have recovered quickly in Xinjiang. Weikang Yang *et al.* counted over 500 Siberian Ibex in 50~60 groups in a gorge under 15km long in the Tian Shan area near Ürümqi in the winter of 2013~2014, with the largest group including over 100 individuals. In August 2010, 60~70 individuals in around 30 groups were sighted in a survey under 10km along the Santas National Border Highway in the Baytik Shan near the China-Mongolia border. Although the populations of this species are stable, mining poses a serious threat to it. Thus, it is listed as Near Threatened

地理分布 Geographical Distribution

国内分布 Domestic Distribution
新疆、西藏、甘肃、内蒙古 / Xinjiang, Tibet (Xizang), Gansu, Inner Mongolia (Nei Mongol)
世界分布 World Distribution
中国；中亚 / China; Central Asia
分布标注 Distribution Note
非特有种 / Non-endemic

国内分布图
Map of Domestic Distribution

种群 Population

种群数量 Population Size	未知 / Unknown
种群趋势 Population Trend	上升 / Increasing

生境与生态系统 Habitat (s) and Ecosystem (s)

生 境 Habitat(s)	草甸、内陆岩石区域 / Meadow, Inland Rocky Area
生态系统 Ecosystem(s)	草地生态系统 / Grassland Ecosystem

威胁 Threat (s)

主要威胁 Major Threat(s)	家畜放牧压力增加、狩猎 / Increasing of Livestock Grazing Pressure, Hunting

保护级别与保护行动 Protection Category and Conservation Action (s)

国家重点保护野生动物等级 (2021) Category of National Key Protected Wild Animals (2021)	二级 / Category II
IUCN 红色名录(2020-2) IUCN Red List (2020-2)	无危 / LC
CITES 附录 (2019) CITES Appendix (2019)	III
保护行动 Conservation Action(s)	自然保护区内种群得到保护 / Populations in nature reserves are protected

相关文献 Relevant References

Burgin *et al.*, 2018; Jiang *et al.* (蒋志刚等), 2017; Castelló, 2016; Wang *et al.* (王君等), 2012; Groves and Grubb, 2011; Wilson and Mittermeier, 2011; Abulimiti Abudukader *et al.* (阿布力米提·阿布都卡迪尔等), 2010; Smith *et al.* (史密斯等), 2009

北山羊 *Capra sibirica*　　蒋志刚 摄　By Zhigang Jiang

西藏盘羊
Ovis hodgsoni

近危 NT

| 数据缺乏 DD | 无危 LC | 近危 NT | 易危 VU | 濒危 EN | 极危 CR | 区域灭绝 RE | 野外灭绝 EW | 灭绝 EX |

分类地位 Taxonomic Status

动物界 Animalia	脊索动物门 Chordata	哺乳纲 Mammalia	偶蹄目 Artiodactyla	牛科 Bovidae

学名 Scientific Name	*Ovis hodgsoni*
命名人 Species Authority	Blyth, 1841
英文名 English Name(s)	Tibetan Argali
同物异名 Synonym(s)	无 / None
种下单元评估 Infra-specific Taxa Assessed	无 / None

评估信息 Assessment Information

评估年份 Year Assessed	2020
评定人 Assessor(s)	蒋志刚 / Zhigang Jiang
审定人 Reviewer(s)	初红军、杨维康 / Hongjun Chu, Weikang Yang
其他贡献人 Other Contributor(s)	李立立、丁晨晨 / Lili Li, Chenchen Ding

理由 Justification: 西藏盘羊在中国青藏高原的分布区较广，种群数量大，但面临着全球变暖和放牧家畜的威胁。因此，列为近危等级 / Tibetan Argali is widely distributed with large populations on the Qinghai-Tibet (Xizang) Plateau in China. However, it faces various threats from global warming, poaching and human encroachment. Thus, it is listed as Near Threatened

地理分布 Geographical Distribution

国内分布 Domestic Distribution
西藏、青海、新疆、甘肃、四川 / Tibet (Xizang), Qinghai, Xinjiang, Gansu, Sichuan
世界分布 World Distribution
中国 / China
分布标注 Distribution Note
特有种 / Endemic

国内分布图
Map of Domestic Distribution

种群 Population

种群数量 Population Size	1998 年曾估计野外有 7,000 只，现在仍有一定规模 / It was estimated that there were 7,000 individuals in the field in 1998. Now it still has fairly large populations
种群趋势 Population Trend	稳定 / Stable

生境与生态系统 Habitat (s) and Ecosystem (s)

生　　境 Habitat(s)	高寒草原、疏林地 / Alpine Steppe, Sparse Shrubland
生态系统 Ecosystem(s)	草地生态系统 / Grassland Ecosystem

威胁 Threat (s)

主要威胁 Major Threat(s)	家畜放牧压力增加，全球变化导致的温度上升 / Increasing Livestock Grazing Pressure, Warming Caused by Global Change

保护级别与保护行动 Protection Category and Conservation Action (s)

国家重点保护野生动物等级 (2021) Category of National Key Protected Wild Animals (2021)	一级 / Category I
IUCN 红色名录 (2020-2) IUCN Red List (2020-2)	未列入 / NA
CITES 附录 (2019) CITES Appendix (2019)	I
保护行动 Conservation Action(s)	自然保护区内种群得到保护 / Populations in nature reserves are protected

相关文献 Relevant References

Jiang *et al.*（蒋志刚等），2017; Wilson and Mittermeier, 2012; Groves and Grubb, 2011; Chu and Jiang（初红军和蒋志刚），2009

西藏盘羊 *Ovis hodgsoni* 　　　　　　　　　　　　　　　　　　　　　蒋志刚 摄　By Zhigang Jiang

台湾鬣羚 *Capricornis swinhoei*

近危 NT

| 数据缺乏 DD | 无危 LC | 近危 NT | 易危 VU | 濒危 EN | 极危 CR | 区域灭绝 RE | 野外灭绝 EW | 灭绝 EX |

分类地位 Taxonomic Status

动物界 Animalia	脊索动物门 Chordata	哺乳纲 Mammalia	偶蹄目 Artiodactyla	牛科 Bovidae

学名 Scientific Name	*Capricornis swinhoei*
命名人 Species Authority	Gray, 1862
英文名 English Name(s)	Formosan Serow
同物异名 Synonym(s)	Taiwan Serow, *Naemorhedus swinhoei*
种下单元评估 Infra-specific Taxa Assessed	无 / None

评估信息 Assessment Information

评估年份 Year Assessed	2019
评定人 Assessor(s)	蒋志刚 / Zhigang Jiang
审定人 Reviewer(s)	陈辈乐、李飞 / Bosco P. L. Chan, Fei Li
其他贡献人 Other Contributor(s)	李立立、丁晨晨 / Lili Li, Chenchen Ding

理由 Justification: 台湾鬣羚常见于台湾中部山脉中高海拔地区，低海拔山区偶尔也有发现。种群数量较为稳定。因其分布区仅限台湾岛局部地区。因此，列为近危等级 / Formosan Serow is commonly found in the central mountains of Taiwan at middle and high elevations, and it is occasionally found in the lower mountains as well. The population is relatively stable. its distribution is limited to only part of Taiwan Island. Thus, it is listed as Near Threatened

地理分布 Geographical Distribution

国内分布 Domestic Distribution	
台湾 / Taiwan	
世界分布 World Distribution	
中国 / China	
分布标注 Distribution Note	
特有种 / Endemic	

国内分布图
Map of Domestic Distribution

种群 Population

种群数量 Population Size	未知 / Unknown
种群趋势 Population Trend	未知 / Unknown

生境与生态系统 Habitat (s) and Ecosystem (s)

生　　境 Habitat(s)	森林 / Forest
生态系统 Ecosystem(s)	森林生态系统 / Forest Ecosystem

威胁 Threat (s)

主要威胁 Major Threat(s)	未知 / Unknown

保护级别与保护行动 Protection Category and Conservation Action (s)

国家重点保护野生动物等级 (2021) Category of National Key Protected Wild Animals (2021)	一级 / Category I
IUCN 红色名录 (2020-2) IUCN Red List (2020-2)	无危 / LC
CITES 附录 (2019) CITES Appendix (2019)	无 / NA
保护行动 Conservation Action(s)	无 / None

相关文献 Relevant References

Burgin *et al.*, 2018; Jiang *et al.* (蒋志刚等), 2017; Castelló, 2016; Groves and Grubb, 2011; Wilson and Mittermeier, 2011; Lu *et al.* (陆雪等), 2007

台湾鬣羚 *Capricornis swinhoei*

松鼠
Sciurus vulgaris

近危 NT

数据缺乏 DD	无危 LC	近危 NT	易危 VU	濒危 EN	极危 CR	区域灭绝 RE	野外灭绝 EW	灭绝 EX

分类地位 Taxonomic Status

动物界 Animalia	脊索动物门 Chordata	哺乳纲 Mammalia	啮齿目 Rodentia	松鼠科 Sciuridae

学 名 Scientific Name	*Sciurus vulgaris*
命名人 Species Authority	Linnaeus, 1758
英文名 English Name(s)	Eurasian Red Squirrel
同物异名 Synonym(s)	*Sciurus fuscorubens* (Dwigubski, 1804); *nadymensis* (Serebrennikov, 1928); *subalpinus* (Burg, 1920)
种下单元评估 Infra-specific Taxa Assessed	无 / None

评估信息 Assessment Information

评估年份 Year Assessed	2020
评定人 Assessor(s)	蒋志刚、刘少英 / Zhigang Jiang, Shaoying Liu
审定人 Reviewer(s)	马勇、鲁长虎、鲍毅新 / Yong Ma, Changhu Lu, Yixin Bao
其他贡献人 Other Contributor(s)	李立立、丁晨晨 / Lili Li, Chenchen Ding

理由 Justification: 松鼠在中国的分布区较广，但数量较少，且其占有区面积不断减小。因此，列为近危等级 / Eurasian Red Squirrel's distribution range in China is large, but its population sizes are small. Furthermore, the areas of its occupancy are declining. Thus, it is listed as Near Threatened

地理分布 Geographical Distribution

国内分布 Domestic Distribution	黑龙江、吉林、辽宁、内蒙古、河北、河南、新疆 / Heilongjiang, Jilin, Liaoning, Inner Mongolia (Nei Mongol), Hebei, Henan, Xinjiang
世界分布 World Distribution	亚洲、欧洲 / Asia, Europe
分布标注 Distribution Note	非特有种 / Non-endemic

国内分布图
Map of Domestic Distribution

种群 Population

种群数量 Population Size	未知 / Unknown
种群趋势 Population Trend	下降 / Decreasing

生境与生态系统 Habitat(s) and Ecosystem(s)

生　　境 Habitat(s)	温带森林 / Temperate Forest
生态系统 Ecosystem(s)	森林生态系统 / Forest Ecosystem

威胁 Threat(s)

主要威胁 Major Threat(s)	松子采集、狩猎 / Harvesting Pine Cone, Hunting

保护级别与保护行动 Protection Category and Conservation Action(s)

国家重点保护野生动物等级 (2021) Category of National Key Protected Wild Animals (2021)	无 / NA
IUCN 红色名录(2020-2) IUCN Red List (2020-2)	无危 / LC
CITES 附录 (2019) CITES Appendix (2019)	无 / NA
保护行动 Conservation Action(s)	无 / None

相关文献 Relevant References

Burgin *et al*., 2018; Jiang *et al*.（蒋志刚等）, 2017; Wilson *et al*., 2016; Zhang *et al*.（张立志等）, 2011; Smith *et al*.（史密斯等）, 2009; Ma *et al*.（马建章等）, 2008; Wang（王应祥）, 2003

松鼠 *Sciurus vulgaris*

© Lynx Edicions

五纹松鼠
Callosciurus quinquestriatus

近危 NT

| 数据缺乏 DD | 无危 LC | 近危 NT | 易危 VU | 濒危 EN | 极危 CR | 区域灭绝 RE | 野外灭绝 EW | 灭绝 EX |

分类地位 Taxonomic Status

动物界 Animalia	脊索动物门 Chordata	哺乳纲 Mammalia	啮齿目 Rodentia	松鼠科 Sciuridae
学 名 Scientific Name		*Callosciurus quinquestriatus*		
命 名 人 Species Authority		Anderson, 1871		
英 文 名 English Name(s)		Anderson's Squirrel		
同物异名 Synonym(s)		无 / None		
种下单元评估 Infra-specific Taxa Assessed		无 / None		

评估信息 Assessment Information

评估年份 Year Assessed	2020
评 定 人 Assessor(s)	蒋志刚、刘少英 / Zhigang Jiang, Shaoying Liu
审 定 人 Reviewer(s)	马勇、鲁长虎、鲍毅新 / Yong Ma, Changhu Lu, Yixin Bao
其他贡献人 Other Contributor(s)	李立立、丁晨晨 / Lili Li, Chenchen Ding

理由 Justification: 五纹松鼠发生区小，种群数量较少，且由于人类活动其所在的栖息地面积减小。因此，列为近危等级 / Anderson's Squirrel's area of occurrence and population sizes are small, and the areas of its habitats are declining due to human activities. Thus, it is listed as Near Threatened

地理分布 Geographical Distribution

国内分布 Domestic Distribution
云南 / Yunnan
世界分布 World Distribution
中国、缅甸 / China, Myanmar
分布标注 Distribution Note
非特有种 / Non-endemic

国内分布图
Map of Domestic Distribution

种群 Population

种群数量 Population Size	未知 / Unknown
种群趋势 Population Trend	下降 / Decreasing

生境与生态系统 Habitat (s) and Ecosystem (s)

生　　境 Habitat(s)	森林 / Forest
生态系统 Ecosystem(s)	森林生态系统 / Forest Ecosystem

威胁 Threat (s)

主要威胁 Major Threat(s)	未知 / Unknown

保护级别与保护行动 Protection Category and Conservation Action (s)

国家重点保护野生动物等级 (2021) Category of National Key Protected Wild Animals (2021)	无 / NA
IUCN 红色名录(2020-2) IUCN Red List (2020-2)	近危 / NT
CITES 附录 (2019) CITES Appendix (2019)	无 / NA
保护行动 Conservation Action(s)	无 / None

相关文献 Relevant References

Burgin *et al*., 2018; Jiang *et al*. (蒋志刚等), 2017; Wilson *et al*., 2016; Zheng *et al*. (郑智民等), 2012; Smith *et al*. (史密斯等), 2009

五纹松鼠 *Callosciurus quinquestriatus*

橙喉长吻松鼠
Dremomys gularis

近危 NT

| 数据缺乏 DD | 无危 LC | 近危 NT | 易危 VU | 濒危 EN | 极危 CR | 区域灭绝 RE | 野外灭绝 EW | 灭绝 EX |

分类地位 Taxonomic Status

动物界 Animalia	脊索动物门 Chordata	哺乳纲 Mammalia	啮齿目 Rodentia	松鼠科 Sciuridae

学名 Scientific Name	*Dremomys gularis*
命名人 Species Authority	Osgood, 1932
英文名 English Name(s)	Red-throated Squirrel
同物异名 Synonym(s)	无 / None
种下单元评估 Infra-specific Taxa Assessed	无 / None

评估信息 Assessment Information

评估年份 Year Assessed	2020
评定人 Assessor(s)	蒋志刚、刘少英 / Zhigang Jiang, Shaoying Liu
审定人 Reviewer(s)	马勇、鲁长虎、鲍毅新 / Yong Ma, Changhu Lu, Yixin Bao
其他贡献人 Other Contributor(s)	李立立、丁晨晨 / Lili Li, Chenchen Ding

理由 Justification: 橙喉长吻松鼠的发生区小，种群数量较少，且其生境受到人类活动的干扰，占有区面积缩小。因此，列为近危等级 / Red-throated Squirrel's area of occurrence and population sizes are small. Furthermore, its habitats are declining due to human activities and its areas of occupancy are declining. Thus, it is listed as Near Threatened

地理分布 Geographical Distribution

国内分布 Domestic Distribution
云南 / Yunnan
世界分布 World Distribution
中国、越南 / China, Viet Nam
分布标注 Distribution Note
非特有种 / Non-endemic

国内分布图
Map of Domestic Distribution

种群 Population

种群数量 Population Size	未知 / Unknown
种群趋势 Population Trend	稳定 / Stable

生境与生态系统 Habitat (s) and Ecosystem (s)

生 境 Habitat(s)	森林 / Forest
生态系统 Ecosystem(s)	森林生态系统 / Forest Ecosystem

威胁 Threat (s)

主要威胁 Major Threat(s)	未知 / Unknown

保护级别与保护行动 Protection Category and Conservation Action (s)

国家重点保护野生动物等级 (2021) Category of National Key Protected Wild Animals (2021)	无 / NA
IUCN 红色名录 (2020-2) IUCN Red List (2020-2)	无危 / LC
CITES 附录 (2019) CITES Appendix (2019)	无 / NA
保护行动 Conservation Action(s)	无 / None

相关文献 Relevant References

Burgin *et al.*, 2018; Jiang *et al.*（蒋志刚等）, 2017; Wilson *et al.*, 2016; Zheng *et al.*（郑智民等）, 2012; Smith *et al.*（史密斯等）, 2009; Wang（王应祥）, 2003

橙喉长吻松鼠 *Dremomys gularis*

橙腹长吻松鼠
Dremomys lokriah

近危 NT

| 数据缺乏 DD | 无危 LC | 近危 NT | 易危 VU | 濒危 EN | 极危 CR | 区域灭绝 RE | 野外灭绝 EW | 灭绝 EX |

分类地位 Taxonomic Status

动物界 Animalia	脊索动物门 Chordata	哺乳纲 Mammalia	啮齿目 Rodentia	松鼠科 Sciuridae

学名 Scientific Name	*Dremomys lokriah*
命名人 Species Authority	Hodgson, 1836
英文名 English Name(s)	Orange-bellied Himalayan Squirrel
同物异名 Synonym(s)	无 / None
种下单元评估 Infra-specific Taxa Assessed	无 / None

评估信息 Assessment Information

评估年份 Year Assessed	2020
评定人 Assessor(s)	蒋志刚、刘少英 / Zhigang Jiang, Shaoying Liu
审定人 Reviewer(s)	马勇、鲁长虎、鲍毅新 / Yong Ma, Changhu Lu, Yixin Bao
其他贡献人 Other Contributor(s)	李立立、丁晨晨 / Lili Li, Chenchen Ding

理由 Justification: 橙腹长吻松鼠的分布区小，种群数量较少，且其所在的栖息地面积由于人类活动有所减小。因此，列为近危等级 / Orange-bellied Himalayan Squirrel's area of occurrence and population sizes are small and the areas of its habitats are declining due to human activities. Thus, it is listed as Near Threatened

地理分布 Geographical Distribution

国内分布 Domestic Distribution
西藏、云南 / Tibet (Xizang), Yunnan
世界分布 World Distribution
孟加拉国、中国、印度、缅甸、尼泊尔 / Bangladesh, China, India, Myanmar, Nepal
分布标注 Distribution Note
非特有种 / Non-endemic

国内分布图
Map of Domestic Distribution

种群 Population

种群数量 Population Size	未知 / Unknown
种群趋势 Population Trend	下降 / Decreasing

生境与生态系统 Habitat (s) and Ecosystem (s)

生　　境 Habitat(s)	亚热带湿润山地森林、泰加林 / Subtropical Moist Montane Forest, Taiga Forest
生态系统 Ecosystem(s)	森林生态系统 / Forest Ecosystem

威胁 Threat (s)

主要威胁 Major Threat(s)	狩猎 / Hunting

保护级别与保护行动 Protection Category and Conservation Action (s)

国家重点保护野生动物等级 (2021) Category of National Key Protected Wild Animals (2021)	无 / NA
IUCN 红色名录 (2020-2) IUCN Red List (2020-2)	无危 / LC
CITES 附录 (2019) CITES Appendix (2019)	无 / NA
保护行动 Conservation Action(s)	无 / None

相关文献 Relevant References

Burgin *et al.*, 2018; Jiang *et al.*（蒋志刚等）, 2017; Wilson *et al.*, 2016; Zheng *et al.*（郑智民等）, 2012; Smith *et al.*（史密斯等）, 2009; Wang（王应祥）, 2003; Li and Wang（李健雄和王应祥）, 1992

橙腹长吻松鼠 *Dremomys lokriah*

© Lynx Edicions

红腿长吻松鼠
Dremomys pyrrhomerus

近危 NT

| 数据缺乏 DD | 无危 LC | **近危 NT** | 易危 VU | 濒危 EN | 极危 CR | 区域灭绝 RE | 野外灭绝 EW | 灭绝 EX |

分类地位 Taxonomic Status

动物界 Animalia	脊索动物门 Chordata	哺乳纲 Mammalia	啮齿目 Rodentia	松鼠科 Sciuridae
学名 Scientific Name		*Dremomys pyrrhomerus*		
命名人 Species Authority		Thomas, 1895		
英文名 English Name(s)		Red-hipped Squirrel		
同物异名 Synonym(s)		无 / None		
种下单元评估 Infra-specific Taxa Assessed		无 / None		

评估信息 Assessment Information

评估年份 Year Assessed	2020
评定人 Assessor(s)	蒋志刚、刘少英 / Zhigang Jiang, Shaoying Liu
审定人 Reviewer(s)	马勇、鲁长虎、鲍毅新 / Yong Ma, Changhu Lu, Yixin Bao
其他贡献人 Other Contributor(s)	李立立、丁晨晨 / Lili Li, Chenchen Ding

理由 Justification: 红腿长吻松鼠的分布区与中国西南喀斯特地区重合，但其种群数量较少。因此，列为近危等级 / Red-hipped Squirrel's area of occurrence is overlapped with karst region in southwest China, but its population sizes are small. Thus, it is listed as Near Threatened

地理分布 Geographical Distribution

国内分布 Domestic Distribution

湖北、安徽、四川、重庆、湖南、江西、贵州、云南、广西、广东、海南 / Hubei, Anhui, Sichuan, Chongqing, Hunan, Jiangxi, Guizhou, Yunnan, Guangxi, Guangdong, Hainan

世界分布 World Distribution

中国、越南 / China, Viet Nam

分布标注 Distribution Note

非特有种 / Non-endemic

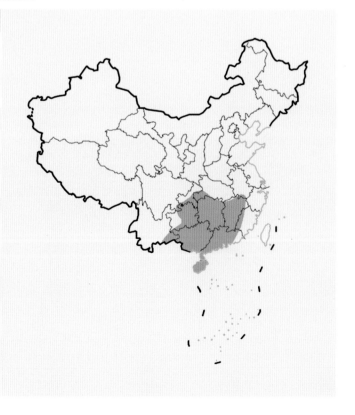

国内分布图
Map of Domestic Distribution

种群 Population

种群数量 Population Size	未知 / Unknown
种群趋势 Population Trend	未知 / Unknown

生境与生态系统 Habitat(s) and Ecosystem(s)

生境 Habitat(s)	森林 / Forest
生态系统 Ecosystem(s)	森林生态系统 / Forest Ecosystem

威胁 Threat(s)

主要威胁 Major Threat(s)	未知 / Unknown

保护级别与保护行动 Protection Category and Conservation Action(s)

国家重点保护野生动物等级 (2021) Category of National Key Protected Wild Animals (2021)	无 / NA
IUCN 红色名录 (2020-2) IUCN Red List (2020-2)	无危 / LC
CITES 附录 (2019) CITES Appendix (2019)	无 / NA
保护行动 Conservation Action(s)	无 / None

相关文献 Relevant References

Burgin *et al*., 2018; Jiang *et al*. (蒋志刚等), 2017; Wilson *et al*., 2016; Smith *et al*. (史密斯等), 2009; Thorington and Hoffmann, 2005; Wang and Xie (汪松和解焱), 2004

红腿长吻松鼠 *Dremomys pyrrhomerus*

栗褐鼯鼠 Petaurista magnificus

近危 NT

分类地位 Taxonomic Status

动物界 Animalia	脊索动物门 Chordata	哺乳纲 Mammalia	啮齿目 Rodentia	松鼠科 Sciuridae
学名 Scientific Name		*Petaurista magnificus*		
命名人 Species Authority		Hodgson, 1836		
英文名 English Name(s)		Hodgson's Giant Flying Squirrel		
同物异名 Synonym(s)		丽鼯鼠; *Petaurista hodgsoni* (Ghose and Saha, 1981)		
种下单元评估 Infra-specific Taxa Assessed		无 / None		

评估信息 Assessment Information

评估年份 Year Assessed	2020
评定人 Assessor(s)	蒋志刚、刘少英 / Zhigang Jiang, Shaoying Liu
审定人 Reviewer(s)	马勇、鲍毅新 / Yong Ma, Yixin Bao
其他贡献人 Other Contributor(s)	李立立、丁晨晨 / Lili Li, Chenchen Ding

理由 Justification: 栗褐鼯鼠在中国的发生区为其边缘分布区，种群数量较少。因此，列为近危等级 / Hodgson's Giant Flying Squirrel's area of occurrence in China is its peripheral range, and its population sizes are small. Thus, it is listed as Near Threatened

地理分布 Geographical Distribution

国内分布 Domestic Distribution
西藏 / Tibet (Xizang)
世界分布 World Distribution
中国；南亚 / China; South Asia
分布标注 Distribution Note
非特有种 / Non-endemic

国内分布图 Map of Domestic Distribution

种群 Population

种群数量 Population Size	未知 / Unknown
种群趋势 Population Trend	下降 / Decreasing

生境与生态系统 Habitat (s) and Ecosystem (s)

生　　境 Habitat(s)	森林 / Forest
生态系统 Ecosystem(s)	森林生态系统 / Forest Ecosystem

威胁 Threat (s)

主要威胁 Major Threat(s)	未知 / Unknown

保护级别与保护行动 Protection Category and Conservation Action (s)

国家重点保护野生动物等级 (2021) Category of National Key Protected Wild Animals (2021)	无 / NA
IUCN 红色名录 (2020-2) IUCN Red List (2020-2)	无危 / LC
CITES 附录 (2019) CITES Appendix (2019)	无 / NA
保护行动 Conservation Action(s)	未知 / Unknown

相关文献 Relevant References

Burgin *et al.*, 2018; Jiang *et al.* (蒋志刚等), 2017; Wilson *et al.*, 2016; Pan *et al.* (潘清华等), 2007; Wang (王应祥), 2003

栗褐鼯鼠 *Petaurista magnificus*

沟牙鼯鼠
Aeretes melanopterus

近危 NT

| 数据缺乏 DD | 无危 LC | 近危 NT | 易危 VU | 濒危 EN | 极危 CR | 区域灭绝 RE | 野外灭绝 EW | 灭绝 EX |

分类地位 Taxonomic Status

| 动物界 Animalia | 脊索动物门 Chordata | 哺乳纲 Mammalia | 啮齿目 Rodentia | 松鼠科 Sciuridae |

学 名 Scientific Name	*Aeretes melanopterus*
命 名 人 Species Authority	Milne-Edwards, 1867
英 文 名 English Name(s)	Northern Chinese Flying Squirrel
同物异名 Synonym(s)	North Chinese Flying Squirrel, Groove-toothed Flying Squirrel
种下单元评估 Infra-specific Taxa Assessed	无 / None

评估信息 Assessment Information

评 估 年 份 Year Assessed	2020
评 定 人 Assessor(s)	蒋志刚、刘少英 / Zhigang Jiang, Shaoying Liu
审 定 人 Reviewer(s)	马勇、鲍毅新 / Yong Ma, Yixin Bao
其他贡献人 Other Contributor(s)	李立立、丁晨晨 / Lili Li, Chenchen Ding

理由 Justification: 沟牙鼯鼠发生区小，种群数量较少。因此，列为近危等级 / Northern Chinese Flying Squirrel's area of occurrence and population sizes are small. Thus, it is listed as Near Threatened

地理分布 Geographical Distribution

国内分布 Domestic Distribution	河北、四川、甘肃 / Hebei, Sichuan, Gansu
世界分布 World Distribution	中国 / China
分布标注 Distribution Note	特有种 / Endemic

国内分布图
Map of Domestic Distribution

种群 Population

种群数量 Population Size	未知 / Unknown
种群趋势 Population Trend	下降 / Decreasing

生境与生态系统 Habitat(s) and Ecosystem(s)

生　　境 Habitat(s)	森林 / Forest
生态系统 Ecosystem(s)	森林生态系统 / Forest Ecosystem

威胁 Threat(s)

主要威胁 Major Threat(s)	无 / None

保护级别与保护行动 Protection Category and Conservation Action(s)

国家重点保护野生动物等级 (2021) Category of National Key Protected Wild Animals (2021)	无 / NA
IUCN 红色名录 (2020-2) IUCN Red List (2020-2)	无危 / LC
CITES 附录 (2019) CITES Appendix (2019)	无 / NA
保护行动 Conservation Action(s)	无 / None

相关文献 Relevant References

Burgin *et al*., 2018; Jiang *et al*.（蒋志刚等）, 2017; Wilson *et al*., 2016; Zheng *et al*.（郑智民等）, 2012; Smith *et al*.（史密斯等）, 2009; Wang（王应祥）, 2003

沟牙鼯鼠 *Aeretes melanopterus*

中国生物多样性红色名录

黑白飞鼠
Hylopetes alboniger

近危 NT

| 数据缺乏 DD | 无危 LC | 近危 NT | 易危 VU | 濒危 EN | 极危 CR | 区域灭绝 RE | 野外灭绝 EW | 灭绝 EX |

分类地位 Taxonomic Status

动物界 Animalia	脊索动物门 Chordata	哺乳纲 Mammalia	啮齿目 Rodentia	松鼠科 Sciuridae
学　名 Scientific Name		*Hylopetes alboniger*		
命名人 Species Authority		Hodgson, 1836		
英文名 English Name(s)		Particolored Flying Squirrel		
同物异名 Synonym(s)		无 / None		
种下单元评估 Infra-specific Taxa Assessed		无 / None		

评估信息 Assessment Information

评估年份 Year Assessed	2020
评定人 Assessor(s)	刘少英、蒋志刚 / Shaoying Liu, Zhigang Jiang
审定人 Reviewer(s)	马勇、鲍毅新 / Yong Ma, Yixin Bao
其他贡献人 Other Contributor(s)	李立立、丁晨晨 / Lili Li, Chenchen Ding

理由 Justification：黑白飞鼠在中国的分布区较广，但数量较少，且其占有区面积小。因此，列为近危等级 / Particolored Flying Squirrel's distribution range in China is relatively large, but its population sizes are small. Furthermore, its areas of occupancy are small. Thus, it is listed as Near Threatened

地理分布 Geographical Distribution

国内分布 Domestic Distribution
四川、云南、贵州、浙江、福建、江西、西藏、海南、重庆 / Sichuan, Yunnan, Guizhou, Zhejiang, Fujian, Jiangxi, Tibet (Xizang), Hainan, Chongqing
世界分布 World Distribution
中国；南亚、东南亚 / China; South Asia, Southeast Asia
分布标注 Distribution Note
非特有种 / Non-endemic

国内分布图
Map of Domestic Distribution

种群 Population

种群数量 Population Size	未知 / Unknown
种群趋势 Population Trend	下降 / Decreasing

生境与生态系统 Habitat (s) and Ecosystem (s)

生 境 Habitat(s)	森林 / Forest
生态系统 Ecosystem(s)	森林生态系统 / Forest Ecosystem

威胁 Threat (s)

主要威胁 Major Threat(s)	未知 / Unknown

保护级别与保护行动 Protection Category and Conservation Action (s)

国家重点保护野生动物等级 (2021) Category of National Key Protected Wild Animals (2021)	无 / NA
IUCN 红色名录 (2020-2) IUCN Red List (2020-2)	无危 / LC
CITES 附录 (2019) CITES Appendix (2019)	无 / NA
保护行动 Conservation Action(s)	无 / None

相关文献 Relevant References

Burgin *et al*., 2018; Jiang *et al*. (蒋志刚等), 2017; Wilson *et al*., 2016; Zheng *et al*. (郑智民等), 2012; Smith *et al*. (史密斯等), 2009; Wang (王应祥), 2003

黑白飞鼠 *Hylopetes alboniger*

藏仓鼠 *Cricetulus kamensis*

近危 NT

| 数据缺乏 DD | 无危 LC | 近危 NT | 易危 VU | 濒危 EN | 极危 CR | 区域灭绝 RE | 野外灭绝 EW | 灭绝 EX |

分类地位 Taxonomic Status

动物界 Animalia	脊索动物门 Chordata	哺乳纲 Mammalia	啮齿目 Rodentia	仓鼠科 Cricetidae

学名 Scientific Name	*Cricetulus kamensis*
命名人 Species Authority	Satunin, 1903
英文名 English Name(s)	Kam Dwarf Hamster
同物异名 Synonym(s)	*Cricetulus kozlovi* (Satunin, 1903); *lama* (Bonhote, 1905); *xizangus* (Thomas, 1922)
种下单元评估 Infra-specific Taxa Assessed	无 / None

评估信息 Assessment Information

评估年份 Year Assessed	2020
评定人 Assessor(s)	刘少英、蒋志刚 / Shaoying Liu, Zhigang Jiang
审定人 Reviewer(s)	马勇、路纪琪、鲍毅新 / Yong Ma, Jiqi Lu, Yixin Bao
其他贡献人 Other Contributor(s)	李立立、丁晨晨 / Lili Li, Chenchen Ding

理由 Justification: 藏仓鼠呈斑块状分布于青藏高原，发生区彼此隔离，种群数量较少。因此，列为近危等级 / Kam Dwarf Hamster is patchily distributed on the Qinghai-Tibet (Xizang) Plateau. Its areas of occurrence are isolated from each other, and its population sizes are small. Thus, it is listed as Near Threatened

地理分布 Geographical Distribution

国内分布 Domestic Distribution
甘肃、青海、西藏、新疆 / Gansu, Qinghai, Tibet (Xizang), Xinjiang
世界分布 World Distribution
中国 / China
分布标注 Distribution Note
特有种 / Endemic

国内分布图
Map of Domestic Distribution

种群 Population

种群数量 Population Size	未知 / Unknown
种群趋势 Population Trend	未知 / Unknown

生境与生态系统 Habitat (s) and Ecosystem (s)

生　　境 Habitat(s)	草地、灌丛、沼泽 / Grassland, Shrubland, Swamp
生态系统 Ecosystem(s)	灌丛生态系统、草地生态系统 / Shrubland Ecosystem, Grassland Ecosystem

威胁 Threat (s)

主要威胁 Major Threat(s)	未知 / Unknown

保护级别与保护行动 Protection Category and Conservation Action (s)

国家重点保护野生动物等级 (2021) Category of National Key Protected Wild Animals (2021)	无 / NA
IUCN红色名录 (2020-2) IUCN Red List (2020-2)	无危 / LC
CITES 附录 (2019) CITES Appendix (2019)	无 / NA
保护行动 Conservation Action(s)	无 / None

相关文献 Relevant References

Burgin *et al.*, 2018; Jiang *et al.* (蒋志刚等), 2017; Wilson *et al.*, 2017; Zheng (郑智民等), 2012; Smith *et al.* (史密斯等), 2009; Wang (王应祥), 2003; Luo (罗泽珣), 2000

藏仓鼠 *Cricetulus kamensis*

林旅鼠
Myopus schisticolor

近危 NT

| 数据缺乏 DD | 无危 LC | 近危 NT | 易危 VU | 濒危 EN | 极危 CR | 区域灭绝 RE | 野外灭绝 EW | 灭绝 EX |

分类地位 Taxonomic Status

动物界 Animalia	脊索动物门 Chordata	哺乳纲 Mammalia	啮齿目 Rodentia	仓鼠科 Cricetidae
学名 Scientific Name		*Myopus schisticolor*		
命名人 Species Authority		Lilljeborg, 1844		
英文名 English Name(s)		Wood Lemming		
同物异名 Synonym(s)		*Myopus morulus* (Hollister, 1912); *saianicus* (Hinton, 1914); *thayeri* (G. M. Allen, 1914); *vinogradovi* (Skalon and Raevski, 1940)		
种下单元评估 Infra-specific Taxa Assessed		无 / None		

评估信息 Assessment Information

评估年份 Year Assessed	2020
评定人 Assessor(s)	刘少英、蒋志刚 / Shaoying Liu, Zhigang Jiang
审定人 Reviewer(s)	马勇、路纪琪、鲍毅新 / Yong Ma, Jiqi Lu, Yixin Bao
其他贡献人 Other Contributor(s)	李立立、丁晨晨 / Lili Li, Chenchen Ding

理由 Justification: 林旅鼠分布在大小兴安岭，其发生区面积小，且种群数量较少。因此，列为近危等级 / Wood Lemming is found in Da Hinggan Ling and Xiao Hinggan Ling. Its area of occurrence is small and its population sizes are small. Thus, it is listed as Near Threatened

地理分布 Geographical Distribution

国内分布 Domestic Distribution	黑龙江、内蒙古 / Heilongjiang, Inner Mongolia (Nei Mongol)
世界分布 World Distribution	中国；环北极圈 / China; Circum-Arctic
分布标注 Distribution Note	非特有种 / Non-endemic

国内分布图
Map of Domestic Distribution

种群 Population

种群数量 Population Size	未知 / Unknown
种群趋势 Population Trend	未知 / Unknown

生境与生态系统 Habitat (s) and Ecosystem (s)

生　　境 Habitat(s)	泰加林 / Taiga Forest
生态系统 Ecosystem(s)	森林生态系统 / Forest Ecosystem

威胁 Threat (s)

主要威胁 Major Threat(s)	无 / None

保护级别与保护行动 Protection Category and Conservation Action (s)

国家重点保护野生动物等级 (2021) Category of National Key Protected Wild Animals (2021)	无 / NA
IUCN 红色名录 (2020-2) IUCN Red List (2020-2)	无危 / LC
CITES 附录 (2019) CITES Appendix (2019)	无 / NA
保护行动 Conservation Action(s)	无 / None

相关文献 Relevant References

Burgin *et al*., 2018; Jiang *et al*.（蒋志刚等）, 2017; Wilson *et al*., 2017; Zheng *et al*.（郑智民等）, 2012; Smith *et al*.（史密斯等）, 2009; Luo（罗泽珣）, 2000; Sha and Guo（莎莉和郭凤清）, 1999

林旅鼠 *Myopus schisticolor*

克钦绒鼠 Eothenomys cachinus

近危 NT

| 数据缺乏 DD | 无危 LC | 近危 NT | 易危 VU | 濒危 EN | 极危 CR | 区域灭绝 RE | 野外灭绝 EW | 灭绝 EX |

分类地位 Taxonomic Status

动物界 Animalia	脊索动物门 Chordata	哺乳纲 Mammalia	啮齿目 Rodentia	仓鼠科 Cricetidae
学名 Scientific Name		*Eothenomys cachinus*		
命名人 Species Authority		Thomas, 1921		
英文名 English Name(s)		Kachin Red-backed Vole		
同物异名 Synonym(s)		*Eothenomys confinii* (Hinton, 1923)		
种下单元评估 Infra-specific Taxa Assessed		无 / None		

评估信息 Assessment Information

评估年份 Year Assessed	2020
评定人 Assessor(s)	刘少英、蒋志刚 / Shaoying Liu, Zhigang Jiang
审定人 Reviewer(s)	马勇、路纪琪、鲍毅新 / Yong Ma, Jiqi Lu, Yixin Bao
其他贡献人 Other Contributor(s)	李立立、丁晨晨 / Lili Li, Chenchen Ding

理由 Justification: 克钦绒鼠在中国的发生区为其边缘分布区，仅分布于高黎贡山，种群数量较少。因此，列为近危等级 / Kachin Red-backed Vole's area of occurrence in China is its peripheral range, it occurs only in Gaoligong Shan, and its population sizes are small. Thus, it is listed as Near Threatened

地理分布 Geographical Distribution

国内分布 Domestic Distribution
云南 / Yunnan
世界分布 World Distribution
中国、缅甸 / China, Myanmar
分布标注 Distribution Note
非特有种 / Non-endemic

国内分布图 Map of Domestic Distribution

种群 Population

种群数量 Population Size	未知 / Unknown
种群趋势 Population Trend	未知 / Unknown

生境与生态系统 Habitat (s) and Ecosystem (s)

生 境 Habitat(s)	亚热带湿润山地森林、溪流边 / Subtropical Moist Montane Forest, Near Stream
生态系统 Ecosystem(s)	森林生态系统、湖泊河流生态系统 / Forest Ecosystem, Lake and River Ecosystem

威胁 Threat (s)

主要威胁 Major Threat(s)	无 / None

保护级别与保护行动 Protection Category and Conservation Action (s)

国家重点保护野生动物等级 (2021) Category of National Key Protected Wild Animals (2021)	无 / NA
IUCN红色名录 (2020-2) IUCN Red List (2020-2)	无危 / LC
CITES 附录 (2019) CITES Appendix (2019)	无 / NA
保护行动 Conservation Action(s)	无 / None

相关文献 Relevant References

Burgin *et al*., 2018; Jiang *et al*. (蒋志刚等), 2017; Fu *et al*. (符建荣等), 2012; Zheng *et al*. (郑智民等), 2012; Smith *et al*. (史密斯等), 2009; Luo (罗泽珣), 2000

克钦绒鼠 *Eothenomys cachinus*   蒋志刚 绘   Drawn by Zhigang Jiang

中华绒鼠
Eothenomys chinensis

近危 NT

| 数据缺乏 DD | 无危 LC | 近危 NT | 易危 VU | 濒危 EN | 极危 CR | 区域灭绝 RE | 野外灭绝 EW | 灭绝 EX |

分类地位 Taxonomic Status

动物界 Animalia	脊索动物门 Chordata	哺乳纲 Mammalia	啮齿目 Rodentia	仓鼠科 Cricetidae

学名 Scientific Name	*Eothenomys chinensis*
命名人 Species Authority	Thomas, 1891
英文名 English Name(s)	Sichuan Red-backed Vole
同物异名 Synonym(s)	Pratt's Vole; *Eothenomys tarquinus* (Thomas, 1912)
种下单元评估 Infra-specific Taxa Assessed	无 / None

评估信息 Assessment Information

评估年份 Year Assessed	2020
评定人 Assessor(s)	刘少英、蒋志刚 / Shaoying Liu, Zhigang Jiang
审定人 Reviewer(s)	马勇、路纪琪、鲍毅新 / Yong Ma, Jiqi Lu, Yixin Bao
其他贡献人 Other Contributor(s)	李立立、丁晨晨 / Lili Li, Chenchen Ding

理由 Justification: 中华绒鼠的发生区面积小，种群数量较少。因此，列为近危等级 / Sichuan Red-backed Vole's area of occurrence is small and its population size also is small. Thus, it is listed as Near Threatened

地理分布 Geographical Distribution

国内分布 Domestic Distribution
四川 / Sichuan
世界分布 World Distribution
中国 / China
分布标注 Distribution Note
特有种 / Endemic

国内分布图
Map of Domestic Distribution

种群 Population

种群数量 Population Size	未知 / Unknown
种群趋势 Population Trend	未知 / Unknown

生境与生态系统 Habitat (s) and Ecosystem (s)

生　　境 Habitat(s)	亚热带湿润山地森林、溪流边 / Subtropical Moist Montane Forest, Near Stream
生态系统 Ecosystem(s)	森林生态系统、湖泊河流生态系统 / Forest Ecosystem, Lake and River Ecosystem

威胁 Threat (s)

主要威胁 Major Threat(s)	无 / None

保护级别与保护行动 Protection Category and Conservation Action (s)

国家重点保护野生动物等级 (2021) Category of National Key Protected Wild Animals (2021)	无 / NA
IUCN 红色名录 (2020-2) IUCN Red List (2020-2)	无危 / LC
CITES 附录 (2019) CITES Appendix (2019)	无 / NA
保护行动 Conservation Action(s)	无 / None

相关文献 Relevant References

Burgin *et al*., 2018; Jiang *et al*. (蒋志刚等), 2017; Wilson *et al*., 2017; Zheng *et al*. (郑智民等), 2012; Wang (王应祥), 2003; Luo (罗泽珣), 2000; Kaneko, 1996

中华绒鼠 *Eothenomys chinensis*

康定绒鼠
Eothenomys hintoni

近危 NT

| 数据缺乏 DD | 无危 LC | 近危 NT | 易危 VU | 濒危 EN | 极危 CR | 区域灭绝 RE | 野外灭绝 EW | 灭绝 EX |

分类地位 Taxonomic Status

动物界 Animalia	脊索动物门 Chordata	哺乳纲 Mammalia	啮齿目 Rodentia	仓鼠科 Cricetidae

学名 Scientific Name	*Eothenomys hintoni*
命名人 Species Authority	Osgood, 1932
英文名 English Name(s)	Kangting Red-backed Vole
同物异名 Synonym(s)	无 / None
种下单元评估 Infra-specific Taxa Assessed	无 / None

评估信息 Assessment Information

评估年份 Year Assessed	2020
评定人 Assessor(s)	刘少英、蒋志刚 / Shaoying Liu, Zhigang Jiang
审定人 Reviewer(s)	马勇、路纪琪、鲍毅新 / Yong Ma, Jiqi Lu, Yixin Bao
其他贡献人 Other Contributor(s)	李立立、丁晨晨 / Lili Li, Chenchen Ding

理由 Justification: 康定绒鼠的发生区面积小，种群数量较少。因此，列为近危等级 / Kangting Red-backed Vole's area of occurrence is small and its population size is small. Thus, it is listed as Near Threatened

地理分布 Geographical Distribution

国内分布 Domestic Distribution
四川 / Sichuan
世界分布 World Distribution
中国 / China
分布标注 Distribution Note
特有种 / Endemic

国内分布图
Map of Domestic Distribution

种群 Population

种群数量 Population Size	未知 / Unknown
种群趋势 Population Trend	未知 / Unknown

生境与生态系统 Habitat(s) and Ecosystem(s)

生　　境 Habitat(s)	山地森林 / Montane Forest
生态系统 Ecosystem(s)	森林生态系统 / Forest Ecosystem

威胁 Threat(s)

主要威胁 Major Threat(s)	耕种、伐木 / Plantation, Logging

保护级别与保护行动 Protection Category and Conservation Action(s)

国家重点保护野生动物等级 (2021) Category of National Key Protected Wild Animals (2021)	无 / NA
IUCN 红色名录 (2020-2) IUCN Red List (2020-2)	无危 / LC
CITES 附录 (2019) CITES Appendix (2019)	无 / NA
保护行动 Conservation Action(s)	无 / None

相关文献 Relevant References

Burgin *et al*., 2018; Jiang *et al*.（蒋志刚等）, 2017; Liu *et al*., 2012a; Luo（罗泽珣）, 2000

康定绒鼠 *Eothenomys hintoni*

昭通绒鼠
Eothenomys olitor

近危 NT

| 数据缺乏 DD | 无危 LC | 近危 NT | 易危 VU | 濒危 EN | 极危 CR | 区域灭绝 RE | 野外灭绝 EW | 灭绝 EX |

分类地位 Taxonomic Status

动物界 Animalia	脊索动物门 Chordata	哺乳纲 Mammalia	啮齿目 Rodentia	仓鼠科 Cricetidae
学 名 Scientific Name		*Eothenomys olitor*		
命名人 Species Authority		Thomas, 1911		
英文名 English Name(s)		Black-eared Red-backed Vole		
同物异名 Synonym(s)		Chaotung Vole; *Eothenomys hypolitor* (Wang and Li, 2000)		
种下单元评估 Infra-specific Taxa Assessed		无 / None		

评估信息 Assessment Information

评估年份 Year Assessed	2020
评定人 Assessor(s)	刘少英、蒋志刚 / Shaoying Liu, Zhigang Jiang
审定人 Reviewer(s)	马勇、路纪琪、鲍毅新 / Yong Ma, Jiqi Lu, Yixin Bao
其他贡献人 Other Contributor(s)	李立立、丁晨晨 / Lili Li, Chenchen Ding

理由 Justification: 昭通绒鼠发生区小，种群数量较少。因此，列为近危等级 / Black-eared Red-backed Vole's area of occurrence and population sizes are small. Thus, it is listed as Near Threatened

地理分布 Geographical Distribution

国内分布 Domestic Distribution
云南、贵州 / Yunnan, Guizhou
世界分布 World Distribution
中国 / China
分布标注 Distribution Note
特有种 / Endemic

国内分布图
Map of Domestic Distribution

种群 Population

种群数量 Population Size	未知 / Unknown
种群趋势 Population Trend	未知 / Unknown

生境与生态系统 Habitat (s) and Ecosystem (s)

生　　境 Habitat(s)	热带和亚热带山地森林 / Tropical and Subtropical Montane Forest
生态系统 Ecosystem(s)	森林生态系统 / Forest Ecosystem

威胁 Threat (s)

主要威胁 Major Threat(s)	无 / None

保护级别与保护行动 Protection Category and Conservation Action (s)

国家重点保护野生动物等级 (2021) Category of National Key Protected Wild Animals (2021)	无 / NA
IUCN 红色名录 (2020-2) IUCN Red List (2020-2)	无危 / LC
CITES 附录 (2019) CITES Appendix (2019)	无 / NA
保护行动 Conservation Action(s)	无 / None

相关文献 Relevant References

Burgin *et al.*, 2018; Jiang *et al.* (蒋志刚等), 2017; Wilson *et al.*, 2017; Zheng *et al.* (郑智民等), 2012; Wang (王应祥), 2003; Zhang *et al.* (张云智等), 2002; Luo (罗泽珣), 2000; Kaneko, 1996; Lu *et al.* (陆长坤等), 1965

昭通绒鼠 *Eothenomys olitor*

玉龙绒鼠
Eothenomys proditor

近危 NT

| 数据缺乏 DD | 无危 LC | 近危 NT | 易危 VU | 濒危 EN | 极危 CR | 区域灭绝 RE | 野外灭绝 EW | 灭绝 EX |

分类地位 Taxonomic Status

动物界 Animalia	脊索动物门 Chordata	哺乳纲 Mammalia	啮齿目 Rodentia	仓鼠科 Cricetidae
学　名 Scientific Name		*Eothenomys proditor*		
命名人 Species Authority		Hinton, 1923		
英文名 English Name(s)		Yulongxuen Red-backed Vole		
同物异名 Synonym(s)		Yulungxuen Chinese Vole, Yulungshan Vole, Yulong Chinese Vole		
种下单元评估 Infra-specific Taxa Assessed		无 / None		

评估信息 Assessment Information

评估年份 Year Assessed	2020
评定人 Assessor(s)	刘少英、蒋志刚 / Shaoying Liu, Zhigang Jiang
审定人 Reviewer(s)	马勇、路纪琪、鲍毅新 / Yong Ma, Jiqi Lu, Yixin Bao
其他贡献人 Other Contributor(s)	李立立、丁晨晨 / Lili Li, Chenchen Ding

理由 Justification: 玉龙绒鼠的发生区小，种群数量较少。因此，列为近危等级 / Yulongxuen Red-backed Vole's area of occurrence and population sizes are small. Thus, it is listed as Near Threatened

地理分布 Geographical Distribution

国内分布 Domestic Distribution
四川、云南 / Sichuan, Yunnan
世界分布 World Distribution
中国 / China
分布标注 Distribution Note
特有种 / Endemic

国内分布图
Map of Domestic Distribution

种群 Population

种群数量 Population Size	未知 / Unknown
种群趋势 Population Trend	未知 / Unknown

生境与生态系统 Habitat (s) and Ecosystem (s)

生　　境 Habitat(s)	草地 / Grassland
生态系统 Ecosystem(s)	草地生态系统 / Grassland Ecosystem

威胁 Threat (s)

主要威胁 Major Threat(s)	无 / None

保护级别与保护行动 Protection Category and Conservation Action (s)

国家重点保护野生动物等级 (2021) Category of National Key Protected Wild Animals (2021)	无 / NA
IUCN 红色名录 (2020-2) IUCN Red List (2020-2)	无危 / LC
CITES 附录 (2019) CITES Appendix (2019)	无 / NA
保护行动 Conservation Action(s)	无 / None

相关文献 Relevant References

Burgin *et al*., 2018; Jiang *et al*. (蒋志刚等), 2017; Wilson *et al*., 2017; Zheng *et al*. (郑智民等), 2012; Smith *et al*. (史密斯等), 2009; Wang and Xie (汪松和解焱), 2004; Luo (罗泽珣), 2000; Kaneko, 1996

玉龙绒鼠 *Eothenomys proditor*

川西绒鼠
Eothenomys tarquinius

近危 NT

| 数据缺乏 DD | 无危 LC | 近危 NT | 易危 VU | 濒危 EN | 极危 CR | 区域灭绝 RE | 野外灭绝 EW | 灭绝 EX |

分类地位 Taxonomic Status

动物界 Animalia	脊索动物门 Chordata	哺乳纲 Mammalia	啮齿目 Rodentia	仓鼠科 Cricetidae
学名 Scientific Name		*Eothenomys tarquinius*		
命名人 Species Authority		Thomas, 1912		
英文名 English Name(s)		Western Sichuan Red-backed Vole		
同物异名 Synonym(s)		无 / None		
种下单元评估 Infra-specific Taxa Assessed		无 / None		

评估信息 Assessment Information

评估年份 Year Assessed	2020
评定人 Assessor(s)	刘少英、蒋志刚 / Shaoying Liu, Zhigang Jiang
审定人 Reviewer(s)	马勇、路纪琪、鲍毅新 / Yong Ma, Jiqi Lu, Yixin Bao
其他贡献人 Other Contributor(s)	李立立、丁晨晨 / Lili Li, Chenchen Ding

理由 Justification: 川西绒鼠种群数量较少，且其栖息地面积减小。因此，列为近危等级 / Western Sichuan Red-backed Vole's population sizes are small and the areas of its habitats are declining. Thus, it is listed as Near Threatened

地理分布 Geographical Distribution

国内分布 Domestic Distribution
四川 / Sichuan
世界分布 World Distribution
中国 / China
分布标注 Distribution Note
特有种 / Endemic

国内分布图
Map of Domestic Distribution

种群 Population

种群数量 Population Size	未知 / Unknown
种群趋势 Population Trend	未知 / Unknown

生境与生态系统 Habitat (s) and Ecosystem (s)

生　　境 Habitat(s)	山地森林 / Montane Forest
生态系统 Ecosystem(s)	森林生态系统 / Forest Ecosystem

威胁 Threat (s)

主要威胁 Major Threat(s)	无 / None

保护级别与保护行动 Protection Category and Conservation Action (s)

国家重点保护野生动物等级 (2021) Category of National Key Protected Wild Animals (2021)	无 / NA
IUCN 红色名录 (2020-2) IUCN Red List (2020-2)	无危 / LC
CITES 附录 (2019) CITES Appendix (2019)	无 / NA
保护行动 Conservation Action(s)	无 / None

相关文献 Relevant References

Burgin *et al*., 2018; Jiang *et al*. (蒋志刚等), 2017; Wilson *et al*., 2017; Liu *et al*., 2012a

川西绒鼠 *Eothenomys tarquinius*

中国生物多样性红色名录

德钦绒鼠
Eothenomys wardi

近危 NT

| 数据缺乏 DD | 无危 LC | 近危 NT | 易危 VU | 濒危 EN | 极危 CR | 区域灭绝 RE | 野外灭绝 EW | 灭绝 EX |

分类地位 Taxonomic Status

动物界 Animalia	脊索动物门 Chordata	哺乳纲 Mammalia	啮齿目 Rodentia	仓鼠科 Cricetidae
学名 Scientific Name		*Eothenomys wardi*		
命名人 Species Authority		Thomas, 1912		
英文名 English Name(s)		Ward's Red-backed Vole		
同物异名 Synonym(s)		无 / None		
种下单元评估 Infra-specific Taxa Assessed		无 / None		

评估信息 Assessment Information

评估年份 Year Assessed	2020
评定人 Assessor(s)	刘少英、蒋志刚 / Shaoying Liu, Zhigang Jiang
审定人 Reviewer(s)	马勇、路纪琪、鲍毅新 / Yong Ma, Jiqi Lu, Yixin Bao
其他贡献人 Other Contributor(s)	李立立、丁晨晨 / Lili Li, Chenchen Ding

理由 Justification: 德钦绒鼠的发生区面积小，种群数量较少。因此，列为近危等级 / Ward's Red-backed Vole's area of occurrence and population sizes are small. Thus, it is listed as Near Threatened

地理分布 Geographical Distribution

国内分布 Domestic Distribution
云南 / Yunnan
世界分布 World Distribution
中国 / China
分布标注 Distribution Note
特有种 / Endemic

国内分布图
Map of Domestic Distribution

种群 Population

种群数量 Population Size	未知 / Unknown
种群趋势 Population Trend	未知 / Unknown

生境与生态系统 Habitat(s) and Ecosystem(s)

生境 Habitat(s)	森林、溪流边、草甸 / Forest, Near Stream, Meadow
生态系统 Ecosystem(s)	森林生态系统、草地生态系统、湖泊河流生态系统 / Forest Ecosystem, Grassland Ecosystem, Lake and River Ecosystem

威胁 Threat(s)

主要威胁 Major Threat(s)	无 / None

保护级别与保护行动 Protection Category and Conservation Action(s)

国家重点保护野生动物等级 (2021) Category of National Key Protected Wild Animals (2021)	无 / NA
IUCN 红色名录 (2020-2) IUCN Red List (2020-2)	无危 / LC
CITES 附录 (2019) CITES Appendix (2019)	无 / NA
保护行动 Conservation Action(s)	无 / None

相关文献 Relevant References

Burgin *et al.*, 2018; Jiang *et al.* (蒋志刚等), 2017; Wilson *et al.*, 2017; Zheng *et al.* (郑智民等), 2012; Smith *et al.* (史密斯等), 2009; Wang and Xie (汪松和解焱), 2004; Luo (罗泽珣), 2000; Kaneko, 1996

德钦绒鼠 *Eothenomys wardi*

斯氏高山䶄
Alticola stoliczkanus
近危 NT

| 数据缺乏 DD | 无危 LC | 近危 NT | 易危 VU | 濒危 EN | 极危 CR | 区域灭绝 RE | 野外灭绝 EW | 灭绝 EX |

分类地位 Taxonomic Status

动物界 Animalia	脊索动物门 Chordata	哺乳纲 Mammalia	啮齿目 Rodentia	仓鼠科 Cricetidae

学名 Scientific Name	*Alticola stoliczkanus*
命名人 Species Authority	Blanford, 1875
英文名 English Name(s)	Stoliczka's Mountain Vole
同物异名 Synonym(s)	Stolička's Mountain Vole; *Alticola acrophilus* (Miller, 1899); *bhatnagari* (Biswas and Khajuria, 1955); *cricetulus* (Miller, 1899); *kaznakovi* (Satunin, 1903); *lama* (转下页)
种下单元评估 Infra-specific Taxa Assessed	无 / None

评估信息 Assessment Information

评估年份 Year Assessed	2020
评定人 Assessor(s)	刘少英、蒋志刚 / Shaoying Liu, Zhigang Jiang
审定人 Reviewer(s)	马勇、路纪琪、鲍毅新 / Yong Ma, Jiqi Lu, Yixin Bao
其他贡献人 Other Contributor(s)	李立立、丁晨晨 / Lili Li, Chenchen Ding

理由 Justification: 斯氏高山䶄在中国的分布区较广，但其占有区面积减小，其数量较少。因此，列为近危等级 / Stoliczka's Mountain Vole's distribution area in China is relatively large, however, its areas of occupancy are declining. Its population sizes are small. Thus, it is listed as Near Threatened

地理分布 Geographical Distribution

国内分布 Domestic Distribution	西藏、青海、新疆、甘肃 / Tibet (Xizang), Qinghai, Xinjiang, Gansu
世界分布 World Distribution	中国、印度、尼泊尔 / China, India, Nepal
分布标注 Distribution Note	非特有种 / Non-endemic

国内分布图
Map of Domestic Distribution

种群 Population

种群数量 Population Size	未知 / Unknown
种群趋势 Population Trend	未知 / Unknown

生境与生态系统 Habitat (s) and Ecosystem (s)

生　　境 Habitat(s)	泰加林、草地、灌丛 / Taiga Forest, Grassland, Shrubland
生态系统 Ecosystem(s)	森林生态系统、灌丛生态系统、草地生态系统 / Forest Ecosystem, Shrubland Ecosystem, Grassland Ecosystem

威胁 Threat (s)

主要威胁 Major Threat(s)	无 / None

保护级别与保护行动 Protection Category and Conservation Action (s)

国家重点保护野生动物等级 (2021) Category of National Key Protected Wild Animals (2021)	无 / NA
IUCN 红色名录 (2020-2) IUCN Red List (2020-2)	无危 / LC
CITES 附录 (2019) CITES Appendix (2019)	无 / NA
保护行动 Conservation Action(s)	无 / None

相关文献 Relevant References

Burgin *et al.*, 2018; Jiang *et al.* (蒋志刚等), 2017; Wilson *et al.*, 2017; Zheng *et al.* (郑智民等), 2012; Smith *et al.* (史密斯等), 2009; Luo (罗泽珣), 2000; Rossolimo *et al.*, 1994

(接上页)
(Barret-Hamilton, 1900); *nanschanicus* (Satunin, 1903); *stracheyi* (Thomas, 1880)

斯氏高山䶄 *Alticola stoliczkanus*

聂拉木松田鼠
Neodon nyalamensis

近危 NT

| 数据缺乏 DD | 无危 LC | **近危 NT** | 易危 VU | 濒危 EN | 极危 CR | 区域灭绝 RE | 野外灭绝 EW | 灭绝 EX |

分类地位 Taxonomic Status

动物界 Animalia	脊索动物门 Chordata	哺乳纲 Mammalia	啮齿目 Rodentia	仓鼠科 Cricetidae

学名 Scientific Name	*Neodon nyalamensis*
命名人 Species Authority	Liu *et al.*, 2017
英文名 English Name(s)	Nyalam Mountain Vole
同物异名 Synonym(s)	无 / None
种下单元评估 Infra-specific Taxa Assessed	无 / None

评估信息 Assessment Information

评估年份 Year Assessed	2020
评定人 Assessor(s)	刘少英、蒋志刚 / Shaoying Liu, Zhigang Jiang
审定人 Reviewer(s)	马勇、鲍毅新 / Yong Ma, Yixin Bao
其他贡献人 Other Contributor(s)	李立立、丁晨晨 / Lili Li, Chenchen Ding

理由 Justification: 聂拉木松田鼠的发生区极小，种群数量亦少。因此，列为近危等级 / Nyalam Mountain Vole's area of occurrence and population sizes are small. Thus, it is listed as Near Threatened

地理分布 Geographical Distribution

国内分布 Domestic Distribution
西藏 / Tibet (Xizang)
世界分布 World Distribution
中国；不详 / China; Unknown
分布标注 Distribution Note
非特有种 / Non-endemic

国内分布图
Map of Domestic Distribution

种群 Population

种群数量 Population Size	未知 / Unknown
种群趋势 Population Trend	未知 / Unknown

生境与生态系统 Habitat (s) and Ecosystem (s)

生　　境 Habitat(s)	森林、灌丛 / Forest, Shrub
生态系统 Ecosystem(s)	森林生态系统、灌丛生态系统 / Forest Ecosystem, Shrubland Ecosystem

威胁 Threat (s)

主要威胁 Major Threat(s)	无 / None

保护级别与保护行动 Protection Category and Conservation Action (s)

国家重点保护野生动物等级 (2021) Category of National Key Protected Wild Animals (2021)	无 / NA
IUCN 红色名录 (2020-2) IUCN Red List (2020-2)	无危 / LC
CITES 附录 (2019) CITES Appendix (2019)	无 / NA
保护行动 Conservation Action(s)	无 / None

相关文献 Relevant References

Burgin *et al*., 2018; Jiang *et al*. (蒋志刚等), 2017; Liu *et al*. (刘少英等), 2017; Wilson *et al*., 2017

聂拉木松田鼠 *Neodon nyalamensis*

台湾田鼠
Alexandromys kikuchii
近危 NT

| 数据缺乏 DD | 无危 LC | 近危 NT | 易危 VU | 濒危 EN | 极危 CR | 区域灭绝 RE | 野外灭绝 EW | 灭绝 EX |

分类地位 Taxonomic Status

动物界 Animalia	脊索动物门 Chordata	哺乳纲 Mammalia	啮齿目 Rodentia	仓鼠科 Cricetidae
学名 Scientific Name		*Alexandromys kikuchii*		
命名人 Species Authority		Kuroda, 1920		
英文名 English Name(s)		Taiwan Vole		
同物异名 Synonym(s)		无 / None		
种下单元评估 Infra-specific Taxa Assessed		无 / None		

评估信息 Assessment Information

评估年份 Year Assessed	2020
评定人 Assessor(s)	刘少英、蒋志刚 / Shaoying Liu, Zhigang Jiang
审定人 Reviewer(s)	马勇、鲍毅新 / Yong Ma, Yixin Bao
其他贡献人 Other Contributor(s)	李立立、丁晨晨 / Lili Li, Chenchen Ding

理由 Justification: 台湾田鼠的发生区极小，种群数量少。因此，列为近危等级 / Taiwan Vole's area of occurrence and population sizes are small. Thus, it is listed as Near Threatened

地理分布 Geographical Distribution

国内分布 Domestic Distribution
台湾 / Taiwan
世界分布 World Distribution
中国 / China
分布标注 Distribution Note
特有种 / Endemic

国内分布图
Map of Domestic Distribution

种群 Population

种群数量 Population Size	未知 / Unknown
种群趋势 Population Trend	未知 / Unknown

生境与生态系统 Habitat(s) and Ecosystem(s)

生境 Habitat(s)	亚热带湿润山地森林 / Subtropical Moist Montane Forest
生态系统 Ecosystem(s)	森林生态系统 / Forest Ecosystem

威胁 Threat(s)

主要威胁 Major Threat(s)	无 / None

保护级别与保护行动 Protection Category and Conservation Action(s)

国家重点保护野生动物等级 (2021) Category of National Key Protected Wild Animals (2021)	无 / NA
IUCN 红色名录 (2020-2) IUCN Red List (2020-2)	无危 / LC
CITES 附录 (2019) CITES Appendix (2019)	无 / NA
保护行动 Conservation Action(s)	无 / None

相关文献 Relevant References

Burgin *et al.*, 2018; Jiang *et al.* (蒋志刚等), 2017; Wilson *et al.*, 2017; Zheng *et al.* (郑智民等), 2012; Luo (罗泽珣), 2000; Kaneko, 1987; Yu, 1995, 1994, 1993

台湾田鼠 *Alexandromys kikuchii*

黑田鼠
Microtus agrestis
近危 NT

| 数据缺乏 DD | 无危 LC | 近危 NT | 易危 VU | 濒危 EN | 极危 CR | 区域灭绝 RE | 野外灭绝 EW | 灭绝 EX |

分类地位 Taxonomic Status

动物界 Animalia	脊索动物门 Chordata	哺乳纲 Mammalia	啮齿目 Rodentia	仓鼠科 Cricetidae

学名 Scientific Name	*Microtus agrestis*
命名人 Species Authority	Linnaeus, 1761
英文名 English Name(s)	Field Vole
同物异名 Synonym(s)	Short-tailed Vole; *Microtus agrestoides* (Hinton, 1910); *angustifrons* (Fatio, 1905); *arcturus* (Thomas, 1912); *argyropoli* (Ognev, 1944); *argyropuli* (Ognev, 1950); (转下页)
种下单元评估 Infra-specific Taxa Assessed	无 / None

评估信息 Assessment Information

评估年份 Year Assessed	2020
评定人 Assessor(s)	刘少英、蒋志刚 / Shaoying Liu, Zhigang Jiang
审定人 Reviewer(s)	马勇、鲍毅新 / Yong Ma, Yixin Bao
其他贡献人 Other Contributor(s)	李立立、丁晨晨 / Lili Li, Chenchen Ding

理由 Justification: 黑田鼠的发生区小，种群数量少。因此，列为近危等级 / Field Vole's area of occurrence and population sizes are small. Thus, it is listed as Near Threatened

地理分布 Geographical Distribution

国内分布 Domestic Distribution
新疆 / Xinjiang
世界分布 World Distribution
亚洲、欧洲 / Asia, Europe
分布标注 Distribution Note
非特有种 / Non-endemic

国内分布图
Map of Domestic Distribution

种群 Population

种群数量 Population Size	未知 / Unknown
种群趋势 Population Trend	稳定 / Stable

生境与生态系统 Habitat (s) and Ecosystem (s)

生境 Habitat(s)	草地、淡水湖、江河 / Grassland, Freshwater Lake, River
生态系统 Ecosystem(s)	草地生态系统、湖泊河流生态系统 / Grassland Ecosystem, Lake and River Ecosystem

威胁 Threat (s)

主要威胁 Major Threat(s)	无 / None

保护级别与保护行动 Protection Category and Conservation Action (s)

国家重点保护野生动物等级 (2021) Category of National Key Protected Wild Animals (2021)	无 / NA
IUCN 红色名录(2020-2) IUCN Red List (2020-2)	无危 / LC
CITES 附录 (2019) CITES Appendix (2019)	无 / NA
保护行动 Conservation Action(s)	无 / None

相关文献 Relevant References

Burgin *et al*., 2018; Jiang *et al*. (蒋志刚等), 2017; Wilson *et al*., 2017; Zheng *et al*. (郑智民等), 2012; Smith *et al*. (史密斯等), 2009; Pan *et al*. (潘清华等), 2007; Luo (罗泽珣), 2000

(接上页)

argyropuloi (Ognev, 1952); *armoricanus* (Heim de Balsac and Beaufort, 1966); *bailloni* (de Sélys Longchamps, 1841); *britannicus* (de Sélys Longchamps, 1847); *bucklandii* (Giebel, 1847); *carinthiacus* (Kretzoi, 1958); *enezgroezi* (Heim de Balsac and Beaufort, 1966); *estiae* (Reinwaldt, 1927); *exsul* (Miller, 1908); *fiona* (Montagu, 1922); *gregarius* (Linneaus, 1766); *hirta* (Bellamy, 1839); *insul* (Lydekker, 1909); *insularis* (Nilsson, 1844); *intermedia* (Bonaparte, 1845); *latifrons* (Fatio, 1905); *levernedii* (Crespon, 1844); *luch* (Barrett-Hamilton and Hinton, 1913); *macgillivrayi* (Barrett-Hamilton and Hinton, 1913); *mial* (Barrett-Hamilton and Hinton, 1913); *mongol* (Thomas, 1911); *neglectus* (Jenyns, 1841); *nigra* (Fatio, 1869); *nigricans* (Kerr, 1792); *ognevi* (Scalon, 1935); *orioecus* (Cabrera, 1924); *pannonicus* (Ehik, 1924); *pallida* (Melander, 1938); *punctus* (Montagu, 1923); *rozianus* (Bocage, 1865); *rufa* (Fatio, 1900); *scaloni* (Heptner, 1948); *tridentinus* (Dal Piaz, 1924); *wettsteini* (Ehik, 1928)

黑田鼠 *Microtus agrestis*

四川田鼠
Volemys millicens

近危 NT

| 数据缺乏 DD | 无危 LC | 近危 NT | 易危 VU | 濒危 EN | 极危 CR | 区域灭绝 RE | 野外灭绝 EW | 灭绝 EX |

分类地位 Taxonomic Status

动物界 Animalia	脊索动物门 Chordata	哺乳纲 Mammalia	啮齿目 Rodentia	仓鼠科 Cricetidae
学名 Scientific Name		*Volemys millicens*		
命名人 Species Authority		Thomas, 1911		
英文名 English Name(s)		Sichuan Vole		
同物异名 Synonym(s)		Szechuan Vole; *Microtus millicens* (Thomas, 1991)		
种下单元评估 Infra-specific Taxa Assessed		无 / None		

评估信息 Assessment Information

评估年份 Year Assessed	2020
评定人 Assessor(s)	刘少英、蒋志刚 / Shaoying Liu, Zhigang Jiang
审定人 Reviewer(s)	马勇、鲍毅新 / Yong Ma, Yixin Bao
其他贡献人 Other Contributor(s)	李立立、丁晨晨 / Lili Li, Chenchen Ding

理由 Justification: 四川田鼠发生区呈斑块状，种群数量较少。因此，列为近危等级 / Sichuan Vole are patchily distributed across its range and its population sizes are small. Thus, it is listed as Near Threatened

地理分布 Geographical Distribution

国内分布 Domestic Distribution
四川、西藏、云南 / Sichuan, Tibet (Xizang), Yunnan
世界分布 World Distribution
中国 / China
分布标注 Distribution Note
特有种 / Endemic

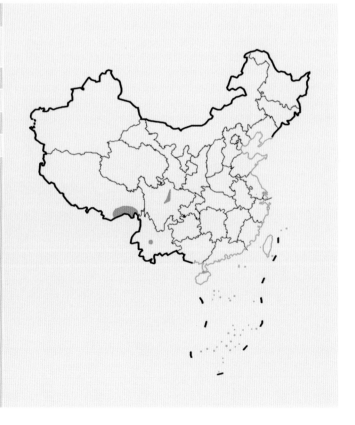

国内分布图
Map of Domestic Distribution

🦌 种群 Population

种群数量 Population Size	未知 / Unknown
种群趋势 Population Trend	未知 / Unknown

🦌 生境与生态系统 Habitat (s) and Ecosystem (s)

生　　境 Habitat(s)	森林 / Forest
生态系统 Ecosystem(s)	森林生态系统 / Forest Ecosystem

🦌 威胁 Threat (s)

主要威胁 Major Threat(s)	无 / None

🦌 保护级别与保护行动 Protection Category and Conservation Action (s)

国家重点保护野生动物等级 (2021) Category of National Key Protected Wild Animals (2021)	无 / NA
IUCN 红色名录(2020-2) IUCN Red List (2020-2)	无危 / LC
CITES 附录 (2019) CITES Appendix (2019)	无 / NA
保护行动 Conservation Action(s)	无 / None

🦌 相关文献 Relevant References

Burgin *et al.*, 2018; Jiang *et al.* (蒋志刚等), 2017; Liu *et al.* (刘少英等), 2017; Wilson *et al.*, 2017; Zheng *et al.* (郑智民等), 2012; Wang and Xie (汪松和解焱), 2004; Wang (王应祥), 2003; Luo (罗泽珣), 2000; Honacki *et al.*, 1982

四川田鼠 *Volemys millicens*

川西田鼠
Volemys musseri

近危 NT

数据缺乏 DD	无危 LC	近危 NT	易危 VU	濒危 EN	极危 CR	区域灭绝 RE	野外灭绝 EW	灭绝 EX

分类地位 Taxonomic Status

动物界 Animalia	脊索动物门 Chordata	哺乳纲 Mammalia	啮齿目 Rodentia	仓鼠科 Cricetidae
学名 Scientific Name		*Volemys musseri*		
命名人 Species Authority		Lawrence, 1982		
英文名 English Name(s)		Marie's Vole		
同物异名 Synonym(s)		无 / None		
种下单元评估 Infra-specific Taxa Assessed		无 / None		

评估信息 Assessment Information

评估年份 Year Assessed	2020
评定人 Assessor(s)	刘少英、蒋志刚 / Shaoying Liu, Zhigang Jiang
审定人 Reviewer(s)	马勇、鲍毅新 / Yong Ma, Yixin Bao
其他贡献人 Other Contributor(s)	李立立、丁晨晨 / Lili Li, Chenchen Ding

理由 Justification: 川西田鼠的发生区小，种群数量少。因此，列为近危等级 / Marie's Vole's area of occurrence and population sizes are small. Thus, it is listed as Near Threatened

地理分布 Geographical Distribution

国内分布 Domestic Distribution
四川 / Sichuan
世界分布 World Distribution
中国 / China
分布标注 Distribution Note
特有种 / Endemic

国内分布图
Map of Domestic Distribution

种群 Population

种群数量 Population Size	未知 / Unknown
种群趋势 Population Trend	稳定 / Stable

生境与生态系统 Habitat (s) and Ecosystem (s)

生　　境 Habitat(s)	高海拔草地、灌丛 / High Altitude Grassland, Shrubland
生态系统 Ecosystem(s)	灌丛生态系统、草地生态系统 / Shrubland Ecosystem, Grassland Ecosystem

威胁 Threat (s)

主要威胁 Major Threat(s)	无 / None

保护级别与保护行动 Protection Category and Conservation Action (s)

国家重点保护野生动物等级 (2021) Category of National Key Protected Wild Animals (2021)	无 / NA
IUCN红色名录 (2020-2) IUCN Red List (2020-2)	无危 / LC
CITES 附录 (2019) CITES Appendix (2019)	无 / NA
保护行动 Conservation Action(s)	无 / None

相关文献 Relevant References

Burgin *et al.*, 2018; Jiang *et al.* (蒋志刚等), 2017; Liu *et al.*, 2017; Wilson *et al.*, 2017; Zheng *et al.* (郑智民等), 2012; Smith *et al.* (史密斯等), 2009; Wang (王应祥), 2003; Luo (罗泽珣), 2000

川西田鼠 *Volemys musseri*

黄兔尾鼠 *Eolagurus luteus*

近危 NT

| 数据缺乏 DD | 无危 LC | 近危 NT | 易危 VU | 濒危 EN | 极危 CR | 区域灭绝 RE | 野外灭绝 EW | 灭绝 EX |

分类地位 Taxonomic Status

动物界 Animalia	脊索动物门 Chordata	哺乳纲 Mammalia	啮齿目 Rodentia	仓鼠科 Cricetidae
学名 Scientific Name		*Eolagurus luteus*		
命名人 Species Authority		Eversmann, 1840		
英文名 English Name(s)		Yellow Steppe Lemming		
同物异名 Synonym(s)		*Eolagurus praeluteus* (Schevtschenko, 1965); *volgensis* (Alexandrova, 1976); *gromovi* (Topatchevski, 1963)		
种下单元评估 Infra-specific Taxa Assessed		无 / None		

评估信息 Assessment Information

评估年份 Year Assessed	2020
评定人 Assessor(s)	刘少英、蒋志刚 / Shaoying Liu, Zhigang Jiang
审定人 Reviewer(s)	马勇、鲍毅新 / Yong Ma, Yixin Bao
其他贡献人 Other Contributor(s)	李立立、丁晨晨 / Lili Li, Chenchen Ding

理由 Justification: 黄兔尾鼠在新疆北部有一定面积发生区，但其种群数量较少。因此，列为近危等级 / Yellow Steppe Lemming has a certain area of occurrence in northern Xinjiang, but its population sizes are small. Thus, it is listed as Near Threatened

地理分布 Geographical Distribution

国内分布 Domestic Distribution
新疆 / Xinjiang
世界分布 World Distribution
中国、哈萨克斯坦、蒙古、俄罗斯 / China, Kazakhstan, Mongolia, Russia
分布标注 Distribution Note
非特有种 / Non-endemic

国内分布图
Map of Domestic Distribution

种群 Population

种群数量 Population Size	未知 / Unknown
种群趋势 Population Trend	未知 / Unknown

生境与生态系统 Habitat (s) and Ecosystem (s)

生 境 Habitat(s)	干旱草地、半荒漠 / Dry Grassland, Semi-desert
生态系统 Ecosystem(s)	草地生态系统、荒漠生态系统 / Grassland Ecosystem, Desert Ecosystem

威胁 Threat (s)

主要威胁 Major Threat(s)	无 / None

保护级别与保护行动 Protection Category and Conservation Action (s)

国家重点保护野生动物等级 (2021) Category of National Key Protected Wild Animals (2021)	无 / NA
IUCN 红色名录 (2020-2) IUCN Red List (2020-2)	无危 / LC
CITES 附录 (2019) CITES Appendix (2019)	无 / NA
保护行动 Conservation Action(s)	无 / None

相关文献 Relevant References

Burgin *et al.*, 2018; Jiang *et al.* (蒋志刚等), 2017; Wilson *et al.*, 2017; Zheng *et al.* (郑智民等), 2012; Mi *et al.* (米景川等), 2003; Luo (罗泽珣), 2000; Ma *et al.* (马勇等), 1982

黄兔尾鼠 *Eolagurus luteus*

蒙古兔尾鼠
Eolagurus przewalskii

近危 NT

| 数据缺乏 DD | 无危 LC | 近危 NT | 易危 VU | 濒危 EN | 极危 CR | 区域灭绝 RE | 野外灭绝 EW | 灭绝 EX |

分类地位 Taxonomic Status

动物界 Animalia	脊索动物门 Chordata	哺乳纲 Mammalia	啮齿目 Rodentia	仓鼠科 Cricetidae

学 名 Scientific Name	*Eolagurus przewalskii*
命名人 Species Authority	Büchner, 1889
英文名 English Name(s)	Przewalski's Steppe Lemming
同物异名 Synonym(s)	无 / None
种下单元评估 Infra-specific Taxa Assessed	无 / None

评估信息 Assessment Information

评估年份 Year Assessed	2020
评定人 Assessor(s)	刘少英、蒋志刚 / Shaoying Liu, Zhigang Jiang
审定人 Reviewer(s)	马勇、鲍毅新 / Yong Ma, Yixin Bao
其他贡献人 Other Contributor(s)	李立立、丁晨晨 / Lili Li, Chenchen Ding

理由 Justification: 蒙古兔尾鼠在中国新疆、青海、内蒙古、甘肃有一定分布面积，但其数量较少。因此，列为近危等级 / Przewalski's Steppe Lemming is found in several localities in Xinjiang, Qinghai, Inner Mongolia (Nei Mongol) and Gansu in western China, but its total population size remains small. Thus, it is listed as Near Threatened

地理分布 Geographical Distribution

国内分布 Domestic Distribution	新疆、青海、内蒙古、甘肃 / Xinjiang, Qinghai, Inner Mongolia (Nei Mongol), Gansu
世界分布 World Distribution	中国、蒙古 / China, Mongolia
分布标注 Distribution Note	非特有种 / Non-endemic

国内分布图
Map of Domestic Distribution

种群 Population

种群数量 Population Size	未知 / Unknown
种群趋势 Population Trend	未知 / Unknown

生境与生态系统 Habitat(s) and Ecosystem(s)

生　　境 Habitat(s)	草地、江河 / Grassland, River
生态系统 Ecosystem(s)	草地生态系统、湖泊河流生态系统 / Grassland Ecosystem, Lake and River Ecosystem

威胁 Threat(s)

主要威胁 Major Threat(s)	无 / None

保护级别与保护行动 Protection Category and Conservation Action(s)

国家重点保护野生动物等级 (2021) Category of National Key Protected Wild Animals (2021)	无 / NA
IUCN 红色名录 (2020-2) IUCN Red List (2020-2)	无危 / LC
CITES 附录 (2019) CITES Appendix (2019)	无 / NA
保护行动 Conservation Action(s)	无 / None

相关文献 Relevant References

Burgin *et al.*, 2018; Jiang *et al.*（蒋志刚等）, 2017; Wilson *et al.*, 2017; Zheng *et al.*（郑智民等）, 2012; Smith *et al.*（史密斯等）, 2009; Wang and Xie（汪松和解焱）, 2004; Luo（罗泽珣）, 2000; Zhao（赵肯堂）, 1984

蒙古兔尾鼠 *Eolagurus przewalskii*

凉山沟牙田鼠
Proedromys liangshanensis

中国生物多样性红色名录

近危 NT

| 数据缺乏 DD | 无危 LC | 近危 NT | 易危 VU | 濒危 EN | 极危 CR | 区域灭绝 RE | 野外灭绝 EW | 灭绝 EX |

分类地位 Taxonomic Status

动物界 Animalia	脊索动物门 Chordata	哺乳纲 Mammalia	啮齿目 Rodentia	仓鼠科 Cricetidae
学名 Scientific Name		*Proedromys liangshanensis*		
命名人 Species Authority		Liu, Sun, Zeng and Zhao, 2007		
英文名 English Name(s)		Liangshan Vole		
同物异名 Synonym(s)		无 / None		
种下单元评估 Infra-specific Taxa Assessed		无 / None		

评估信息 Assessment Information

评估年份 Year Assessed	2020
评定人 Assessor(s)	刘少英、蒋志刚 / Shaoying Liu, Zhigang Jiang
审定人 Reviewer(s)	马勇、鲍毅新 / Yong Ma, Yixin Bao
其他贡献人 Other Contributor(s)	李立立、丁晨晨 / Lili Li, Chenchen Ding

理由 Justification: 凉山沟牙田鼠的发生区小，种群数量少。因此，列为近危等级 / Liangshan Vole's area of occurrence and population size are small. Thus, it is listed as Near Threatened

地理分布 Geographical Distribution

国内分布 Domestic Distribution
四川 / Sichuan
世界分布 World Distribution
中国 / China
分布标注 Distribution Note
特有种 / Endemic

国内分布图
Map of Domestic Distribution

种群 Population

种群数量 Population Size	未知 / Unknown
种群趋势 Population Trend	稳定 / Stable

生境与生态系统 Habitat (s) and Ecosystem (s)

生　　境 Habitat(s)	亚热带湿润山地森林 / Subtropical Moist Montane Forest
生态系统 Ecosystem(s)	森林生态系统 / Forest Ecosystem

威胁 Threat (s)

主要威胁 Major Threat(s)	无 / None

保护级别与保护行动 Protection Category and Conservation Action (s)

国家重点保护野生动物等级 (2021) Category of National Key Protected Wild Animals (2021)	无 / NA
IUCN 红色名录 (2020-2) IUCN Red List (2020-2)	无危 / LC
CITES 附录 (2019) CITES Appendix (2019)	无 / NA
保护行动 Conservation Action(s)	无 / None

相关文献 Relevant References

Burgin *et al.*, 2018; Jiang *et al.* (蒋志刚等), 2017; Wilson *et al.*, 2017; Hao *et al.* (郝海邦等), 2011; Liu *et al.*, 2007

凉山沟牙田鼠 *Proedromys liangshanensis*

长尾攀鼠 *Vandeleuria oleracea*

近危 NT

| 数据缺乏 DD | 无危 LC | 近危 NT | 易危 VU | 濒危 EN | 极危 CR | 区域灭绝 RE | 野外灭绝 EW | 灭绝 EX |

分类地位 Taxonomic Status

动物界 Animalia	脊索动物门 Chordata	哺乳纲 Mammalia	啮齿目 Rodentia	鼠科 Muridae

学名 Scientific Name	*Vandeleuria oleracea*
命名人 Species Authority	Bennett, 1832
英文名 English Name(s)	Asiatic Long-tailed Climbing Mouse
同物异名 Synonym(s)	Indomalayan Vandeleuria; *Vandeleuria badius* (Blyth, 1859); *domecolus* (Hodgson, 1841); *dumeticola* (Hodgson, 1845); *marica* (Thomas, 1915); *modesta* (Thomas, 1914); (转下页)
种下单元评估 Infra-specific Taxa Assessed	无 / None

评估信息 Assessment Information

评估年份 Year Assessed	2020
评定人 Assessor(s)	刘少英、蒋志刚 / Shaoying Liu, Zhigang Jiang
审定人 Reviewer(s)	马勇、鲍毅新 / Yong Ma, Yixin Bao
其他贡献人 Other Contributor(s)	李立立、丁晨晨 / Lili Li, Chenchen Ding

理由 Justification: 长尾攀鼠种群数量较少，且其占有区面积小。因此，列为近危等级 / Asiatic Long-tailed Climbing Mouse's population sizes and the areas of its occupancy are small. Thus, it is listed as Near Threatened

地理分布 Geographical Distribution

国内分布 Domestic Distribution
云南、西藏 / Yunnan, Tibet (Xizang)
世界分布 World Distribution
孟加拉国、不丹、柬埔寨、中国、印度、缅甸、尼泊尔、斯里兰卡、泰国、越南 / Bangladesh, Bhutan, Cambodia, China, India, Myanmar, Nepal, Sri Lanka, Thailand, Viet Nam
分布标注 Distribution Note
非特有种 / Non-endemic

国内分布图
Map of Domestic Distribution

种群 Population

种群数量 Population Size	未知 / Unknown
种群趋势 Population Trend	下降 / Decreasing

生境与生态系统 Habitat(s) and Ecosystem(s)

生境 Habitat(s)	森林、灌丛 / Forest, Shrubland
生态系统 Ecosystem(s)	森林生态系统、灌丛生态系统 / Forest Ecosystem, Shrubland Ecosystem

威胁 Threat(s)

主要威胁 Major Threat(s)	无 / None

保护级别与保护行动 Protection Category and Conservation Action(s)

国家重点保护野生动物等级 (2021) Category of National Key Protected Wild Animals (2021)	无 / NA
IUCN 红色名录 (2020-2) IUCN Red List (2020-2)	无危 / LC
CITES 附录 (2019) CITES Appendix (2019)	无 / NA
保护行动 Conservation Action(s)	无 / None

相关文献 Relevant References

Burgin *et al.*, 2018; Jiang *et al.* (蒋志刚等), 2017; Wilson *et al.*, 2017; Zheng *et al.* (郑智民等), 2012; Smith *et al.* (史密斯等), 2009; Wang (王应祥), 2003

(接上页)

povensis (Hodgson, 1845); *rubida* (Thomas, 1914); *sibylla* (Thomas, 1914); *scandens* (Osgood, 1932); *spadicea* (Ryley, 1914); *wroughtoni* (Ryley, 1914)

长尾攀鼠 *Vandeleuria oleracea*

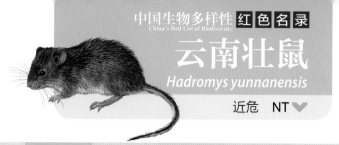

云南壮鼠 *Hadromys yunnanensis*

近危 NT

分类地位 Taxonomic Status

动物界 Animalia	脊索动物门 Chordata	哺乳纲 Mammalia	啮齿目 Rodentia	鼠科 Muridae
学名 Scientific Name		*Hadromys yunnanensis*		
命名人 Species Authority		Yang and Wang, 1987		
英文名 English Name(s)		Yunnan Hadromys		
同物异名 Synonym(s)		Yunnan Bush Rat		
种下单元评估 Infra-specific Taxa Assessed		无 / None		

评估信息 Assessment Information

评估年份 Year Assessed	2020
评定人 Assessor(s)	刘少英、蒋志刚 / Shaoying Liu, Zhigang Jiang
审定人 Reviewer(s)	马勇、鲍毅新 / Yong Ma, Yixin Bao
其他贡献人 Other Contributor(s)	李立立、丁晨晨 / Lili Li, Chenchen Ding

理由 Justification: 云南壮鼠占有区面积小，种群数量少。因此，列为近危等级 / Yunnan Hadromys' area of occupancy and population size are small. Thus, it is listed as Near Threatened

地理分布 Geographical Distribution

国内分布 Domestic Distribution
云南 / Yunnan
世界分布 World Distribution
中国 / China
分布标注 Distribution Note
特有种 / Endemic

国内分布图 Map of Domestic Distribution

种群 Population

种群数量 Population Size	未知 / Unknown
种群趋势 Population Trend	未知 / Unknown

生境与生态系统 Habitat (s) and Ecosystem (s)

生　　境 Habitat(s)	热带季雨林、灌丛 / Tropical Monsoon Forest, Shrubland
生态系统 Ecosystem(s)	森林生态系统、灌丛生态系统 / Forest Ecosystem, Shrubland Ecosystem

威胁 Threat (s)

主要威胁 Major Threat(s)	无 / None

保护级别与保护行动 Protection Category and Conservation Action (s)

国家重点保护野生动物等级 (2021) Category of National Key Protected Wild Animals (2021)	无 / NA
IUCN 红色名录 (2020-2) IUCN Red List (2020-2)	无危 / LC
CITES 附录 (2019) CITES Appendix (2019)	无 / NA
保护行动 Conservation Action(s)	无 / None

相关文献 Relevant References

Burgin *et al*., 2018; Jiang *et al*. (蒋志刚等), 2017; Wilson *et al*., 2017; Zheng *et al*. (郑智民等), 2012; Smith *et al*. (史密斯等), 2009; Wang and Xie (汪松和解焱), 2004; Luo (罗泽珣), 2000; Yang and Wang (杨光荣和王应祥), 1987

云南壮鼠 *Hadromys yunnanensis*

大齿鼠
Dacnomys millardi

近危 NT

| 数据缺乏 DD | 无危 LC | 近危 NT | 易危 VU | 濒危 EN | 极危 CR | 区域灭绝 RE | 野外灭绝 EW | 灭绝 EX |

分类地位 Taxonomic Status

动物界 Animalia	脊索动物门 Chordata	哺乳纲 Mammalia	啮齿目 Rodentia	鼠科 Muridae

学名 Scientific Name	*Dacnomys millardi*
命名人 Species Authority	Thomas, 1916
英文名 English Name(s)	Millard's Rat
同物异名 Synonym(s)	*Dacnomys ingens* (Osgood, 1932); *wroughtoni* (Thomas, 1922)
种下单元评估 Infra-specific Taxa Assessed	无 / None

评估信息 Assessment Information

评估年份 Year Assessed	2020
评定人 Assessor(s)	刘少英、蒋志刚 / Shaoying Liu, Zhigang Jiang
审定人 Reviewer(s)	马勇、鲍毅新 / Yong Ma, Yixin Bao
其他贡献人 Other Contributor(s)	李立立、丁晨晨 / Lili Li, Chenchen Ding

理由 Justification: 大齿鼠种群数量较少，且其占有的栖息地面积不断减小。因此，列为近危等级 / Millard's Rat's population sizes are small and the areas of its occupancy are small. Thus, it is listed as Near Threatened

地理分布 Geographical Distribution

国内分布 Domestic Distribution
云南 / Yunnan
世界分布 World Distribution
中国、印度、老挝、尼泊尔、越南 / China, India, Laos, Nepal, Viet Nam
分布标注 Distribution Note
非特有种 / Non-endemic

国内分布图
Map of Domestic Distribution

种群 Population

种群数量 Population Size	未知 / Unknown
种群趋势 Population Trend	未知 / Unknown

生境与生态系统 Habitat (s) and Ecosystem (s)

生　　境 Habitat(s)	森林 / Forest
生态系统 Ecosystem(s)	森林生态系统 / Forest Ecosystem

威胁 Threat (s)

主要威胁 Major Threat(s)	无 / None

保护级别与保护行动 Protection Category and Conservation Action (s)

国家重点保护野生动物等级 (2021) Category of National Key Protected Wild Animals (2021)	无 / NA
IUCN 红色名录 (2020-2) IUCN Red List (2020-2)	无危 / LC
CITES 附录 (2019) CITES Appendix (2019)	无 / NA
保护行动 Conservation Action(s)	无 / None

相关文献 Relevant References

Burgin *et al.*, 2018; Jiang *et al.* (蒋志刚等), 2017; Zheng *et al.* (郑智民等), 2012; Smith *et al.* (史密斯等), 2009; Wang (王应祥), 2003

大齿鼠 *Dacnomys millardi*

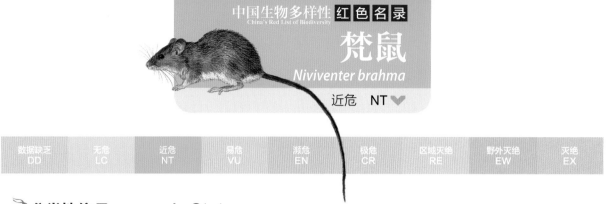

梵鼠 *Niviventer brahma*

近危 NT

| 数据缺乏 DD | 无危 LC | 近危 NT | 易危 VU | 濒危 EN | 极危 CR | 区域灭绝 RE | 野外灭绝 EW | 灭绝 EX |

分类地位 Taxonomic Status

动物界 Animalia	脊索动物门 Chordata	哺乳纲 Mammalia	啮齿目 Rodentia	鼠科 Muridae

学名 Scientific Name	*Niviventer brahma*
命名人 Species Authority	Thomas, 1914
英文名 English Name(s)	Brahma White-bellied Rat
同物异名 Synonym(s)	*Rattus fulvescens* (Thomas, 1914) subsp. *brahma*
种下单元评估 Infra-specific Taxa Assessed	无 / None

评估信息 Assessment Information

评估年份 Year Assessed	2020
评定人 Assessor(s)	刘少英、蒋志刚 / Shaoying Liu, Zhigang Jiang
审定人 Reviewer(s)	马勇、鲍毅新 / Yong Ma, Yixin Bao
其他贡献人 Other Contributor(s)	李立、丁晨晨 / Lili Li, Chenchen Ding

理由 Justification: 梵鼠种群在中国的发生区为其边缘分布区，种群数量少。因此，列为近危等级 / Brahma White-bellied Rat's area of occurrence in China is its peripheral range, its population sizes are small. Thus, it is listed as Near Threatened

地理分布 Geographical Distribution

国内分布 Domestic Distribution
云南 / Yunnan
世界分布 World Distribution
中国、印度、缅甸 / China, India, Myanmar
分布标注 Distribution Note
非特有种 / Non-endemic

国内分布图
Map of Domestic Distribution

种群 Population

种群数量 Population Size	未知 / Unknown
种群趋势 Population Trend	稳定 / Stable

生境与生态系统 Habitat (s) and Ecosystem (s)

生　　　境 Habitat(s)	亚热带湿润山地森林 / Subtropical Moist Montane Forest
生态系统 Ecosystem(s)	森林生态系统 / Forest Ecosystem

威胁 Threat (s)

主要威胁 Major Threat(s)	无 / None

保护级别与保护行动 Protection Category and Conservation Action (s)

国家重点保护野生动物等级 (2021) Category of National Key Protected Wild Animals (2021)	无 / NA
IUCN 红色名录 (2020-2) IUCN Red List (2020-2)	无危 / LC
CITES 附录 (2019) CITES Appendix (2019)	无 / NA
保护行动 Conservation Action(s)	无 / None

相关文献 Relevant References

Burgin *et al.*, 2018; Jiang *et al.* (蒋志刚等), 2017; Wilson *et al.*, 2017; Zheng *et al.* (郑智民等), 2012; Smith *et al.* (史密斯等), 2009; Wang (王应祥), 2003; Gong *et al.* (龚正达等), 2001a, 2001b

梵鼠 *Niviventer brahma*

台湾社鼠 *Niviventer culturatus*

近危 NT

| 数据缺乏 DD | 无危 LC | 近危 NT | 易危 VU | 濒危 EN | 极危 CR | 区域灭绝 RE | 野外灭绝 EW | 灭绝 EX |

分类地位 Taxonomic Status

动物界 Animalia	脊索动物门 Chordata	哺乳纲 Mammalia	啮齿目 Rodentia	鼠科 Muridae

学名 Scientific Name	*Niviventer culturatus*
命名人 Species Authority	Thomas, 1917
英文名 English Name(s)	Soft-furred Taiwan Niviventer
同物异名 Synonym(s)	Oldfield White-bellied Rat
种下单元评估 Infra-specific Taxa Assessed	无 / None

评估信息 Assessment Information

评估年份 Year Assessed	2020
评定人 Assessor(s)	刘少英、蒋志刚 / Shaoying Liu, Zhigang Jiang
审定人 Reviewer(s)	马勇、鲍毅新 / Yong Ma, Yixin Bao
其他贡献人 Other Contributor(s)	李立立、丁晨晨 / Lili Li, Chenchen Ding

理由 Justification: 台湾社鼠占有区面积小，种群数量少。因此，列为近危等级 / Soft-furred Taiwan Niviventer's area of occupancy and population size are small. Thus, it is listed as Near Threatened

地理分布 Geographical Distribution

国内分布 Domestic Distribution
台湾 / Taiwan
世界分布 World Distribution
中国 / China
分布标注 Distribution Note
特有种 / Endemic

国内分布图 Map of Domestic Distribution

种群 Population

种群数量 Population Size	未知 / Unknown
种群趋势 Population Trend	下降 / Decreasing

生境与生态系统 Habitat (s) and Ecosystem (s)

生　　境 Habitat(s)	森林、次生林 / Forest, Secondary Forest
生态系统 Ecosystem(s)	森林生态系统 / Forest Ecosystem

威胁 Threat (s)

主要威胁 Major Threat(s)	无 / None

保护级别与保护行动 Protection Category and Conservation Action (s)

国家重点保护野生动物等级 (2021) Category of National Key Protected Wild Animals (2021)	无 / NA
IUCN 红色名录 (2020-2) IUCN Red List (2020-2)	无危 / LC
CITES 附录 (2019) CITES Appendix (2019)	无 / NA
保护行动 Conservation Action(s)	无 / None

相关文献 Relevant References

Burgin *et al*., 2018; Jiang *et al*.（蒋志刚等）, 2017; Wilson *et al*., 2017; Zheng *et al*.（郑智民等）, 2012; Wang and Xie（汪松和解焱）, 2004; Adler, 1996; Yu, 1995, 1994

台湾社鼠 *Niviventer culturatus*

斯氏鼢鼠
Eospalax smithii

近危 NT

| 数据缺乏 DD | 无危 LC | 近危 NT | 易危 VU | 濒危 EN | 极危 CR | 区域灭绝 RE | 野外灭绝 EW | 灭绝 EX |

分类地位 Taxonomic Status

动物界 Animalia	脊索动物门 Chordata	哺乳纲 Mammalia	啮齿目 Rodentia	鼹型鼠科 Spalacidae

学 名 Scientific Name	*Eospalax smithii*
命 名 人 Species Authority	Thomas, 1911
英 文 名 English Name(s)	Smith's Zokor
同物异名 Synonym(s)	无 / None
种下单元评估 Infra-specific Taxa Assessed	无 / None

评估信息 Assessment Information

评 估 年 份 Year Assessed	2020
评 定 人 Assessor(s)	蒋志刚、刘少英 / Zhigang Jiang, Shaoying Liu
审 定 人 Reviewer(s)	李保国 / Baoguo Li
其他贡献人 Other Contributor(s)	李立立、丁晨晨 / Lili Li, Chenchen Ding

理由 Justification: 斯氏鼢鼠分布在甘肃、宁夏、陕西和四川，种群数量较少。因此，列为近危等级 / Smith's Zokor is found with small populations in Gansu, Ningxia, Shaanxi and Sichuan. Thus, it is listed as Near Threatened

地理分布 Geographical Distribution

国内分布 Domestic Distribution
甘肃、宁夏、陕西、四川 / Gansu, Ningxia, Shaanxi, Sichuan
世界分布 World Distribution
中国 / China
分布标注 Distribution Note
特有种 / Endemic

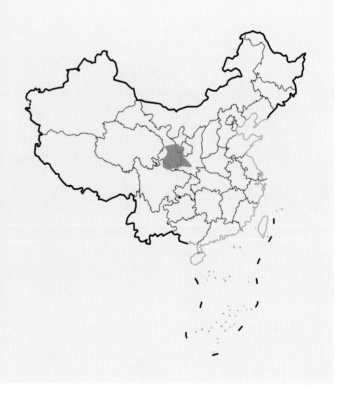

国内分布图 Map of Domestic Distribution

种群 Population

种群数量 Population Size	未知 / Unknown
种群趋势 Population Trend	未知 / Unknown

生境与生态系统 Habitat (s) and Ecosystem (s)

生　　境 Habitat(s)	干旱草地、耕地 / Dry Grassland, Arable Land
生态系统 Ecosystem(s)	草地生态系统、农田生态系统 / Grassland Ecosystem, Cropland Ecosystem

威胁 Threat (s)

主要威胁 Major Threat(s)	无 / None

保护级别与保护行动 Protection Category and Conservation Action (s)

国家重点保护野生动物等级 (2021) Category of National Key Protected Wild Animals (2021)	无 / NA
IUCN红色名录 (2020-2) IUCN Red List (2020-2)	无危 / LC
CITES 附录 (2019) CITES Appendix (2019)	无 / NA
保护行动 Conservation Action(s)	无 / None

相关文献 Relevant References

Burgin *et al*., 2018; Jiang *et al*. (蒋志刚等), 2017; Wilson *et al*., 2017; He *et al*. (何娅等), 2012; Zheng *et al*. (郑智民等), 2012; Smith *et al*. (史密斯等), 2009; Qin (秦长育), 1991

斯氏鼢鼠 *Eospalax smithii*

林睡鼠
Dryomys nitedula

近危 NT

| 数据缺乏 DD | 无危 LC | 近危 NT | 易危 VU | 濒危 EN | 极危 CR | 区域灭绝 RE | 野外灭绝 EW | 灭绝 EX |

分类地位 Taxonomic Status

动物界 Animalia	脊索动物门 Chordata	哺乳纲 Mammalia	啮齿目 Rodentia	睡鼠科 Gliridae

学名 Scientific Name	*Dryomys nitedula*
命名人 Species Authority	Pallas, 1778
英文名 English Name(s)	Forest Dormouse
同物异名 Synonym(s)	*Dryomys angelus* (Thomas, 1906); *aspromontis* (Lehmann, 1963); *bilkjewiczi* (Ognev and Heptner, 1928); *carpathicus* (Brohmer, 1927); *caucasicus* (Ognev and Turov, 1935); (转下页)
种下单元评估 Infra-specific Taxa Assessed	无 / None

评估信息 Assessment Information

评估年份 Year Assessed	2020
评定人 Assessor(s)	蒋志刚、刘少英 / Zhigang Jiang, Shaoying Liu
审定人 Reviewer(s)	马勇、刘伟 / Yong Ma, Wei Liu
其他贡献人 Other Contributor(s)	李立立、丁晨晨 / Lili Li, Chenchen Ding

理由 Justification: 林睡鼠种群数量较少，且其所在的栖息地面积小。因此，列为近危等级 / The population of Forest Dormouse is small and its area of occupancy is restricted. Thus, it is listed as Near Threatened

地理分布 Geographical Distribution

国内分布 Domestic Distribution
新疆 / Xinjiang
世界分布 World Distribution
亚洲、欧洲 / Asia, Europe
分布标注 Distribution Note
非特有种 / Non-endemic

国内分布图
Map of Domestic Distribution

种群 Population

种群数量 Population Size	未知 / Unknown
种群趋势 Population Trend	未知 / Unknown

生境与生态系统 Habitat (s) and Ecosystem (s)

生　　境 Habitat(s)	针阔混交林 / Coniferous and Broad-leaved Mixed Forest
生态系统 Ecosystem(s)	森林生态系统 / Forest Ecosystem

威胁 Threat (s)

主要威胁 Major Threat(s)	无 / None

保护级别与保护行动 Protection Category and Conservation Action (s)

国家重点保护野生动物等级 (2021) Category of National Key Protected Wild Animals (2021)	无 / NA
IUCN 红色名录 (2020-2) IUCN Red List (2020-2)	无危 / LC
CITES 附录 (2019) CITES Appendix (2019)	无 / NA
保护行动 Conservation Action(s)	无 / None

相关文献 Relevant References

Burgin *et al.*, 2018; Jiang *et al.*（蒋志刚等）, 2017; Wilson *et al.*, 2017; An *et al.*（安冉等）, 2015; Zheng *et al.*（郑智民等）, 2012; Wang（王应祥）, 2003; Ma *et al.*（马勇等）, 1981

（接上页）

daghestanicus (Ognev and Turov, 1935); *diamesus* (Lehmann, 1959); *dryas* (Schreber, 1782); *intermedius* (Nehring, 1902); *kurdistanicus* (Ognev and Turov, 1935); *milleri* (Thomas, 1912); *obolenskii* (Ognev and Worobiev, 1923); *ognevi* (Heptner and Formozov, 1928); *pallidus* (Ognev and Turov, 1935); *phrygius* (Thomas, 1907); *pictus* (Blanford, 1875); *ravijojla* (Paspalev *et al.*, 1952); *robustus* (Miller, 1910); *saxatilis* (Rosanov, 1935); *tanaiticus* (Ognev and Turov, 1935); *tichomirowi* (Satunin, 1920); *wingei* (Nehring, 1902)

林睡鼠 *Dryomys nitedula*

灰颈鼠兔
Ochotona forresti

近危 NT

| 数据缺乏 DD | 无危 LC | 近危 NT | 易危 VU | 濒危 EN | 极危 CR | 区域灭绝 RE | 野外灭绝 EW | 灭绝 EX |

分类地位 Taxonomic Status

动物界 Animalia	脊索动物门 Chordata	哺乳纲 Mammalia	兔形目 Lagomorpha	鼠兔科 Ochotonidae

学名 Scientific Name	*Ochotona forresti*
命名人 Species Authority	Thomas, 1923
英文名 English Name(s)	Forrest's Pika
同物异名 Synonym(s)	无 / None
种下单元评估 Infra-specific Taxa Assessed	无 / None

评估信息 Assessment Information

评估年份 Year Assessed	2020
评定人 Assessor(s)	刘少英、蒋志刚 / Shaoying Liu, Zhigang Jiang
审定人 Reviewer(s)	苏建平、冯祚建、宗浩、廖继承 / Jianping Su, Zuojian Feng, Hao Zong, Jicheng Liao
其他贡献人 Other Contributor(s)	李立立、丁晨晨 / Lili Li, Chenchen Ding

理由 Justification: 灰颈鼠兔仅在云南、西藏发现,种群数量少。因此,列为近危等级 / Forrest's Pika is only found in Yunnan, Tibet (Xizang), with small populations. Thus, it is listed as Near Threatened

地理分布 Geographical Distribution

国内分布 Domestic Distribution
云南、西藏 / Yunnan, Tibet (Xizang)

世界分布 World Distribution
不丹、中国、印度、缅甸 / Bhutan, China, India, Myanmar

分布标注 Distribution Note
非特有种 / Non-endemic

国内分布图
Map of Domestic Distribution

种群 Population

种群数量 Population Size	未知 / Unknown
种群趋势 Population Trend	下降 / Decreasing

生境与生态系统 Habitat (s) and Ecosystem (s)

生境 Habitat(s)	针阔混交林、灌丛、泰加林 / Coniferous and Broad-leaved Mixed Forest, Shrubland, Taiga Forest
生态系统 Ecosystem(s)	森林生态系统、灌丛生态系统 / Forest Ecosystem, Shrubland Ecosystem

威胁 Threat (s)

主要威胁 Major Threat(s)	无 / None

保护级别与保护行动 Protection Category and Conservation Action (s)

国家重点保护野生动物等级 (2021) Category of National Key Protected Wild Animals (2021)	无 / NA
IUCN 红色名录 (2020-2) IUCN Red List (2020-2)	无危 / LC
CITES 附录 (2019) CITES Appendix (2019)	无 / NA
保护行动 Conservation Action(s)	无 / None

相关文献 Relevant References

Burgin *et al.*, 2018; Jiang *et al.* (蒋志刚等), 2017; Wilson *et al.*, 2017; Ge *et al.*, 2012; Zheng *et al.* (郑智民等), 2012; Chen and Li (陈晓澄和李文靖), 2009; Huang *et al.* (黄薇等), 2008; Wang (王应祥), 2003

灰颈鼠兔 *Ochotona forresti*

灰鼠兔
Ochotona roylei

近危 NT

| 数据缺乏 DD | 无危 LC | 近危 NT | 易危 VU | 濒危 EN | 极危 CR | 区域灭绝 RE | 野外灭绝 EW | 灭绝 EX |

分类地位 Taxonomic Status

动物界 Animalia	脊索动物门 Chordata	哺乳纲 Mammalia	兔形目 Lagomorpha	鼠兔科 Ochotonidae

学名 Scientific Name	*Ochotona roylei*
命名人 Species Authority	Ogilby, 1839
英文名 English Name(s)	Royle's Pika
同物异名 Synonym(s)	无 / None
种下单元评估 Infra-specific Taxa Assessed	无 / None

评估信息 Assessment Information

评估年份 Year Assessed	2020
评定人 Assessor(s)	刘少英、蒋志刚 / Shaoying Liu, Zhigang Jiang
审定人 Reviewer(s)	苏建平、冯祚建、宗浩、廖继承 / Jianping Su, Zuojian Feng, Hao Zong, Jicheng Liao
其他贡献人 Other Contributor(s)	李立立、丁晨晨 / Lili Li, Chenchen Ding

理由 Justification: 灰鼠兔沿喜马拉雅山脉分布，占有区面积小，种群数量少。因此，列为近危等级 / Royle's Pika occurs only in a narrow range in the Himalayas, its area of occupancy and population sizes are small. Thus, it is listed as Near Threatened

地理分布 Geographical Distribution

国内分布 Domestic Distribution
西藏 / Tibet (Xizang)
世界分布 World Distribution
中国、尼泊尔、巴基斯坦 / China, Nepal, Pakistan
分布标注 Distribution Note
非特有种 / Non-endemic

国内分布图
Map of Domestic Distribution

种群 Population

种群数量 Population Size	未知 / Unknown
种群趋势 Population Trend	稳定 / Stable

生境与生态系统 Habitat (s) and Ecosystem (s)

生　　境 Habitat(s)	内陆岩石区域、人造建筑 / Inland Rocky Area, Building
生态系统 Ecosystem(s)	高原生态系统 / Plateau Ecosystem

威胁 Threat (s)

主要威胁 Major Threat(s)	家畜放牧 / Livestock Ranching

保护级别与保护行动 Protection Category and Conservation Action (s)

国家重点保护野生动物等级 (2021) Category of National Key Protected Wild Animals (2021)	无 / NA
IUCN 红色名录 (2020-2) IUCN Red List (2020-2)	无危 / LC
CITES 附录 (2019) CITES Appendix (2019)	无 / NA
保护行动 Conservation Action(s)	无 / None

相关文献 Relevant References

Burgin *et al*., 2018; Jiang *et al*. (蒋志刚等), 2017; Wilson *et al*., 2017; Zheng *et al*. (郑智民等), 2012; Huang *et al*. (黄薇等), 2008; Zhou *et al*. (周立志等), 2002; Yu *et al*., 2000

灰鼠兔 *Ochotona roylei*

By Tungnath

红鼠兔 *Ochotona rutila*

近危 NT

数据缺乏 DD	无危 LC	近危 NT	易危 VU	濒危 EN	极危 CR	区域灭绝 RE	野外灭绝 EW	灭绝 EX

分类地位 Taxonomic Status

动物界 Animalia	脊索动物门 Chordata	哺乳纲 Mammalia	兔形目 Lagomorpha	鼠兔科 Ochotonidae
学 名 Scientific Name		*Ochotona rutila*		
命 名 人 Species Authority		Severtzov, 1873		
英 文 名 English Name(s)		Turkestan Red Pika		
同物异名 Synonym(s)		无 / None		
种下单元评估 Infra-specific Taxa Assessed		无 / None		

评估信息 Assessment Information

评估年份 Year Assessed	2020
评 定 人 Assessor(s)	刘少英、蒋志刚 / Shaoying Liu, Zhigang Jiang
审 定 人 Reviewer(s)	苏建平、冯祚建、宗浩、廖继承 / Jianping Su, Zuojian Feng, Hao Zong, Jicheng Liao
其他贡献人 Other Contributor(s)	李立立、丁晨晨 / Lili Li, Chenchen Ding

理由 Justification: 红鼠兔属于边缘性分布，种群数量少。因此，列为近危等级 / The distribution area of Turkestan Red Pika in China is its peripheral range and its population is very small. Thus, it is listed as Near Threatened

地理分布 Geographical Distribution

国内分布 Domestic Distribution
新疆 / Xinjiang
世界分布 World Distribution
中国；中亚 / China; Central Asia
分布标注 Distribution Note
非特有种 / Non-endemic

国内分布图
Map of Domestic Distribution

种群 Population

种群数量 Population Size	未知 / Unknown
种群趋势 Population Trend	稳定 / Stable

生境与生态系统 Habitat (s) and Ecosystem (s)

生 境 Habitat(s)	内陆岩石区域 / Inland Rocky Area
生态系统 Ecosystem(s)	荒漠草地生态系统 / Desert Grassland Ecosystem

威胁 Threat (s)

主要威胁 Major Threat(s)	无 / None

保护级别与保护行动 Protection Category and Conservation Action (s)

国家重点保护野生动物等级 (2021) Category of National Key Protected Wild Animals (2021)	无 / NA
IUCN 红色名录 (2020-2) IUCN Red List (2020-2)	无危 / LC
CITES 附录 (2019) CITES Appendix (2019)	无 / NA
保护行动 Conservation Action(s)	无 / None

相关文献 Relevant References

Burgin *et al*., 2018; Jiang *et al*. (蒋志刚等), 2017; Smith *et al*. (史密斯等), 2009; Huang *et al*. (黄薇等), 2007; Niu *et al*., 2004; Wang (王应祥), 2003

红鼠兔 *Ochotona rutila*

狭颅鼠兔
Ochotona thomasi
近危 NT

| 数据缺乏 DD | 无危 LC | 近危 NT | 易危 VU | 濒危 EN | 极危 CR | 区域灭绝 RE | 野外灭绝 EW | 灭绝 EX |

分类地位 Taxonomic Status

动物界 Animalia	脊索动物门 Chordata	哺乳纲 Mammalia	兔形目 Lagomorpha	鼠兔科 Ochotonidae
学名 Scientific Name		*Ochotona thomasi*		
命名人 Species Authority		Argyropulo, 1948		
英文名 English Name(s)		Thomas' Pika		
同物异名 Synonym(s)		无 / None		
种下单元评估 Infra-specific Taxa Assessed		无 / None		

评估信息 Assessment Information

评估年份 Year Assessed	2020
评定人 Assessor(s)	刘少英、蒋志刚 / Shaoying Liu, Zhigang Jiang
审定人 Reviewer(s)	苏建平、冯祚建、宗浩、廖继承 / Jianping Su, Zuojian Feng, Hao Zong, Jicheng Liao
其他贡献人 Other Contributor(s)	李立立、丁晨晨 / Lili Li, Chenchen Ding

理由 Justification: 狭颅鼠兔在中国的分布区较广，但数量较少，且其栖息地面积小。因此，列为近危等级 / The geographical range of Thomas' Pika in China is large, but its actual areas of occupancy and its population are small. Thus, it is listed as Near Threatened

地理分布 Geographical Distribution

国内分布 Domestic Distribution
甘肃、青海、四川 / Gansu, Qinghai, Sichuan
世界分布 World Distribution
中国 / China
分布标注 Distribution Note
特有种 / Endemic

国内分布图
Map of Domestic Distribution

种群 Population

种群数量 Population Size	未知 / Unknown
种群趋势 Population Trend	未知 / Unknown

生境与生态系统 Habitat (s) and Ecosystem (s)

生　　境 Habitat(s)	灌丛 / Shrubland
生态系统 Ecosystem(s)	灌丛生态系统 / Shrubland Ecosystem

威胁 Threat (s)

主要威胁 Major Threat(s)	无 / None

保护级别与保护行动 Protection Category and Conservation Action (s)

国家重点保护野生动物等级 (2021) Category of National Key Protected Wild Animals (2021)	无 / NA
IUCN红色名录(2020-2) IUCN Red List (2020-2)	无危 / LC
CITES 附录 (2019) CITES Appendix (2019)	无 / NA
保护行动 Conservation Action(s)	无 / None

相关文献 Relevant References

Burgin *et al*., 2018; Jiang *et al*.（蒋志刚等）, 2017; Liu *et al*.（刘少英等）, 2017; Zheng *et al*.（郑智民等）, 2012; Smith *et al*.（史密斯等）, 2009; Xing *et al*.（邢雅俊等）, 2008; Zhou *et al*.（周立志等）, 2002

狭颅鼠兔 *Ochotona thomasi*　　方红霞 绘　Drawn by Hongxia Fang

云南兔 *Lepus comus*

近危 NT

| 数据缺乏 DD | 无危 LC | 近危 NT | 易危 VU | 濒危 EN | 极危 CR | 区域灭绝 RE | 野外灭绝 EW | 灭绝 EX |

分类地位 Taxonomic Status

动物界 Animalia	脊索动物门 Chordata	哺乳纲 Mammalia	兔形目 Lagomorpha	兔科 Leporidae
学名 Scientific Name		*Lepus comus*		
命名人 Species Authority		Allen, 1927		
英文名 English Name(s)		Yunnan Hare		
同物异名 Synonym(s)		无 / None		
种下单元评估 Infra-specific Taxa Assessed		无 / None		

评估信息 Assessment Information

评估年份 Year Assessed	2020
评定人 Assessor(s)	蒋志刚 / Zhigang Jiang
审定人 Reviewer(s)	夏霖、杨奇森 / Lin Xia, Qisen Yang
其他贡献人 Other Contributor(s)	李立立、丁晨晨 / Lili Li, Chenchen Ding

理由 Justification: 云南兔在中国云贵高原的分布区较广，但其占有区面积小，种群数量较小。因此，列为近危等级 / Yunnan Hare is widely distributed in the Yunnan-Guizhou Plateau, but its areas of occupancy and population are small. Thus, it is listed as Near Threatened

地理分布 Geographical Distribution

国内分布 Domestic Distribution
云南、四川、贵州 / Yunnan, Sichuan, Guizhou
世界分布 World Distribution
中国、缅甸 / China, Myanmar
分布标注 Distribution Note
非特有种 / Non-endemic

国内分布图
Map of Domestic Distribution

种群 Population

种群数量 Population Size	未知 / Unknown
种群趋势 Population Trend	未知 / Unknown

生境与生态系统 Habitat (s) and Ecosystem (s)

生　　境 Habitat(s)	草甸、灌丛 / Meadow, Shrubland
生态系统 Ecosystem(s)	灌丛生态系统、草地生态系统 / Shrubland Ecosystem, Grassland Ecosystem

威胁 Threat (s)

主要威胁 Major Threat(s)	无 / None

保护级别与保护行动 Protection Category and Conservation Action (s)

国家重点保护野生动物等级 (2021) Category of National Key Protected Wild Animals (2021)	无 / NA
IUCN 红色名录 (2020-2) IUCN Red List (2020-2)	无危 / LC
CITES 附录 (2019) CITES Appendix (2019)	无 / NA
保护行动 Conservation Action(s)	无 / None

相关文献 Relevant References

Burgin *et al*., 2018; Jiang *et al*. (蒋志刚等), 2017; Wilson *et al*., 2016; Zheng *et al*. (郑智民等), 2012; Wu *et al*., 2000; Chen *et al*. (陈志平等), 1993; Wang *et al*. (王应祥等), 1985

云南兔 *Lepus comus*

塔里木兔 Lepus yarkandensis
近危 NT

| 数据缺乏 DD | 无危 LC | 近危 NT | 易危 VU | 濒危 EN | 极危 CR | 区域灭绝 RE | 野外灭绝 EW | 灭绝 EX |

分类地位 Taxonomic Status

动物界 Animalia	脊索动物门 Chordata	哺乳纲 Mammalia	兔形目 Lagomorpha	兔科 Leporidae
学名 Scientific Name		*Lepus yarkandensis*		
命名人 Species Authority		Günther, 1875		
英文名 English Name(s)		Yarkand Hare		
同物异名 Synonym(s)		无 / None		
种下单元评估 Infra-specific Taxa Assessed		无 / None		

评估信息 Assessment Information

评估年份 Year Assessed	2020
评定人 Assessor(s)	蒋志刚 / Zhigang Jiang
审定人 Reviewer(s)	苏建平、冯祚建、宗浩、廖继承 / Jianping Su, Zuojian Feng, Hao Zong, Jicheng Liao
其他贡献人 Other Contributor(s)	李立立、丁晨晨 / Lili Li, Chenchen Ding

理由 Justification: 塔里木兔是中国特有种，种群数量较少，且其占有面积减小。因此，列为近危等级 / Yarkand Hare is endemic to China, its population size is small and its areas of occupancy are declining. Thus, it is listed as Near Threatened

地理分布 Geographical Distribution

国内分布 Domestic Distribution
新疆 / Xinjiang
世界分布 World Distribution
中国 / China
分布标注 Distribution Note
特有种 / Endemic

国内分布图
Map of Domestic Distribution

种群 Population

种群数量 Population Size	未知 / Unknown
种群趋势 Population Trend	未知 / Unknown

生境与生态系统 Habitat (s) and Ecosystem (s)

生　　境 Habitat(s)	荒漠草原 / Desert Steppe
生态系统 Ecosystem(s)	荒漠草原生态系统 / Desert Steppe Ecosystem

威胁 Threat (s)

主要威胁 Major Threat(s)	无 / None

保护级别与保护行动 Protection Category and Conservation Action (s)

国家重点保护野生动物等级 (2021) Category of National Key Protected Wild Animals (2021)	二级 / Category II
IUCN 红色名录(2020-2) IUCN Red List (2020-2)	无危 / LC
CITES 附录 (2019) CITES Appendix (2019)	无 / NA
保护行动 Conservation Action(s)	无 / None

相关文献 Relevant References

Burgin *et al*., 2018; Jiang *et al*. (蒋志刚等), 2017; Wilson *et al*., 2016; Shan and Halik (单文娟和马合木提•哈力克), 2013; Zheng *et al*. (郑智民等), 2012; Wu *et al*., 2010; Smith *et al*. (史密斯等), 2009

塔里木兔 *Lepus yarkandensis*

中国毛猬 *Hylomys suillus*

无危 LC

| 数据缺乏 DD | 无危 LC | 近危 NT | 易危 VU | 濒危 EN | 极危 CR | 区域灭绝 RE | 野外灭绝 EW | 灭绝 EX |

分类地位 Taxonomic Status

动物界 Animalia	脊索动物门 Chordata	哺乳纲 Mammalia	劳亚食虫目 Eulipotyphla	猬科 Erinaceidae
学 名 Scientific Name		*Hylomys suillus*		
命 名 人 Species Authority		Müller, 1840		
英 文 名 English Name(s)		Short-tailed Gymnure		
同物异名 Synonym(s)		无 / None		
种下单元评估 Infra-specific Taxa Assessed		无 / None		

评估信息 Assessment Information

评 估 年 份 Year Assessed	2020
评 定 人 Assessor(s)	蒋志刚 / Zhigang Jiang
审 定 人 Reviewer(s)	蒋学龙 / Xuelong Jiang
其他贡献人 Other Contributor(s)	李立立、丁晨晨 / Lili Li, Chenchen Ding

理由 Justification: 中国毛猬广泛分布于东南亚地区，在国内仅分布于云南，但其栖息于海拔较低、温度适宜的区域，在合适栖息地中易于采集到。因此，列为无危等级 / Short-tailed Gymnure is widely distributed in Yunnan Province and across Southeast Asia, especially in low altitude and warm habitats. It is easily collected in its habitat. Thus, it is listed as Least Concern

地理分布 Geographical Distribution

国内分布 Domestic Distribution
云南 / Yunnan
世界分布 World Distribution
中国；东南亚 / China; Southeast Asia
分布标注 Distribution Note
非特有种 / Non-endemic

国内分布图
Map of Domestic Distribution

种群 Population

种群数量 Population Size	未知 / Unknown
种群趋势 Population Trend	稳定 / Stable

生境与生态系统 Habitat (s) and Ecosystem (s)

生　　境 Habitat(s)	热带和亚热带湿润低地森林，热带和亚热带湿润山地森林、灌丛 / Tropical and Subtropical Moist Lowland Forest; Tropical and Subtropical Moist Montane Forest, Shrubland
生态系统 Ecosystem(s)	森林生态系统、灌丛生态系统 / Forest Ecosystem, Shrubland Ecosystem

威胁 Threat (s)

主要威胁 Major Threat(s)	未知 / Unknown

保护级别与保护行动 Protection Category and Conservation Action (s)

国家重点保护野生动物等级 (2021) Category of National Key Protected Wild Animals (2021)	无 / NA
IUCN 红色名录 (2020-2) IUCN Red List (2020-2)	无危 / LC
CITES 附录 (2019) CITES Appendix (2019)	无 / NA
保护行动 Conservation Action(s)	无 / None

相关文献 Relevant References

Burgin *et al.*, 2018; Wilson and Mittermeier, 2018; Jiang *et al.* (蒋志刚等), 2017; Tong and Lu (仝磊和路纪琪), 2010b; Smith *et al.* (史密斯等), 2009

中国毛猬 *Hylomys suillus*

中国鼩猬 *Neotetracus sinensis*

无危 LC

| 数据缺乏 DD | 无危 LC | 近危 NT | 易危 VU | 濒危 EN | 极危 CR | 区域灭绝 RE | 野外灭绝 EW | 灭绝 EX |

分类地位 Taxonomic Status

动物界 Animalia	脊索动物门 Chordata	哺乳纲 Mammalia	劳亚食虫目 Eulipotyphla	猬科 Erinaceidae
学名 Scientific Name		*Neotetracus sinensis*		
命名人 Species Authority		Trouessart, 1909		
英文名 English Name(s)		Shrew Gymnure		
同物异名 Synonym(s)		Shrew Hedgehog; *Hylomys sinensis* (Trouessart, 1909)		
种下单元评估 Infra-specific Taxa Assessed		无 / None		

评估信息 Assessment Information

评估年份 Year Assessed	2020
评定人 Assessor(s)	蒋志刚 / Zhigang Jiang
审定人 Reviewer(s)	蒋学龙 / Xuelong Jiang
其他贡献人 Other Contributor(s)	李立立、丁晨晨 / Lili Li, Chenchen Ding

理由 Justification: 中国鼩猬分布较广，为常见物种。中国种群数量大且稳定，部分种群分布在保护区范围内。因此，列为无危等级 / Shrew Gymnure is a common and widespread species with stable and large populations in China, and some of its populations are located in nature reserves. Thus, it is listed as Least Concern

地理分布 Geographical Distribution

国内分布 Domestic Distribution	四川、贵州、云南、广东 / Sichuan, Guizhou, Yunnan, Guangdong
世界分布 World Distribution	中国、缅甸、越南 / China, Myanmar, Viet Nam
分布标注 Distribution Note	非特有种 / Non-endemic

国内分布图 Map of Domestic Distribution

种群 Population

种群数量 Population Size	未知 / Unknown
种群趋势 Population Trend	稳定 / Stable

生境与生态系统 Habitat (s) and Ecosystem (s)

生　　　境 Habitat(s)	亚热带湿润山地森林 / Subtropical Moist Montane Forest
生态系统 Ecosystem(s)	森林生态系统 / Forest Ecosystem

威胁 Threat (s)

主要威胁 Major Threat(s)	未知 / Unknown

保护级别与保护行动 Protection Category and Conservation Action (s)

国家重点保护野生动物等级 (2021) Category of National Key Protected Wild Animals (2021)	无 / NA
IUCN 红色名录 (2020-2) IUCN Red List (2020-2)	无危 / LC
CITES 附录 (2019) CITES Appendix (2019)	无 / NA
保护行动 Conservation Action(s)	位于自然保护区内的种群与生境得到保护 / Populations and habitats in nature reserves are protected

相关文献 Relevant References

Burgin *et al*., 2018; Wilson and Mittermeier, 2018; Jiang *et al*. (蒋志刚等), 2017; Deng *et al*. (邓可等), 2013; Gong *et al*. (龚晓俊等), 2013; Sun *et al*. (孙治宇等), 2013; Tu *et al*., 2012; Wu *et al*. (吴毅等), 2011

中国鼩猬 *Neotetracus sinensis*

东北刺猬 *Erinaceus amurensis*

无危 LC

| 数据缺乏 DD | 无危 LC | 近危 NT | 易危 VU | 濒危 EN | 极危 CR | 区域灭绝 RE | 野外灭绝 EW | 灭绝 EX |

分类地位 Taxonomic Status

动物界 Animalia	脊索动物门 Chordata	哺乳纲 Mammalia	劳亚食虫目 Eulipotyphla	猬科 Erinaceidae
学名 Scientific Name		*Erinaceus amurensis*		
命名人 Species Authority		Schrenk, 1859		
英文名 English Name(s)		Amur Hedgehog		
同物异名 Synonym(s)		Manchurian Hedgehog; *Erinaceus chinensis* (Satunin, 1907); *dealbatus* (Swinhoe, 1870); *hanensis* (Matschie, 1907); *koreanus* (Lönnberg, 1922); *koreensis* (转下页)		
种下单元评估 Infra-specific Taxa Assessed		无 / None		

评估信息 Assessment Information

评估年份 Year Assessed	2020
评定人 Assessor(s)	蒋志刚 / Zhigang Jiang
审定人 Reviewer(s)	蒋学龙、冯祚建 / Xuelong Jiang, Zuojian Feng
其他贡献人 Other Contributor(s)	李立立、丁晨晨 / Lili Li, Chenchen Ding

理由 Justification: 东北刺猬分布较广，种群数量大且稳定，部分种群分布在保护区范围内。因此，列为无危等级 / Amur Hedgehog is a widespread species with a large and stable population. Parts of its populations are included in nature reserves. Thus, it is listed as Least Concern

地理分布 Geographical Distribution

国内分布 Domestic Distribution

黑龙江、吉林、辽宁、内蒙古、北京、宁夏、河北、河南、陕西、甘肃、山西、山东、上海、江苏、浙江、安徽、江西、湖北、湖南、福建、广东、贵州、重庆 / Heilongjiang, Jilin, Liaoning, Inner Mongolia (Nei Mongol), Beijing, Ningxia, Hebei, Henan, Shaanxi, Gansu, Shanxi, Shandong, Shanghai, Jiangsu, Zhejiang, Anhui, Jiangxi, Hubei, Hunan, Fujian, Guangdong, Guizhou, Chongqing

世界分布 World Distribution

中国；东北亚 / China; Northeast Asia

分布标注 Distribution Note

非特有种 / Non-endemic

国内分布图
Map of Domestic Distribution

种群 Population

种群数量 Population Size	未知 / Unknown
种群趋势 Population Trend	稳定 / Stable

生境与生态系统 Habitat (s) and Ecosystem (s)

生　　境 Habitat(s)	亚热带湿润低地森林，亚热带湿润高地森林、灌丛、草地、泰加林、耕地、城市 / Subtropical Moist Lowland Forest; Subtropical Moist Highland Forest, Shrubland, Grassland, Taiga Forest, Arable Land, Urban Area
生态系统 Ecosystem(s)	森林生态系统、灌丛生态系统、草地生态系统、农田生态系统、人类聚落生态系统 / Forest Ecosystem, Shrubland Ecosystem, Grassland Ecosystem, Cropland Ecosystem, Human Settlement Ecosystem

威胁 Threat (s)

主要威胁 Major Threat(s)	农药、环境污染 / Pesticide, Environmental Pollution

保护级别与保护行动 Protection Category and Conservation Action (s)

国家重点保护野生动物等级 (2021) Category of National Key Protected Wild Animals (2021)	无 / NA
IUCN 红色名录 (2020-2) IUCN Red List (2020-2)	无危 / LC
CITES 附录 (2019) CITES Appendix (2019)	无 / NA
保护行动 Conservation Action(s)	无 / None

相关文献 Relevant References

Burgin *et al.*, 2018; Wilson and Mittermeier, 2018; Jiang *et al.* (蒋志刚等), 2017; Smith *et al.* (史密斯等), 2009; Wang (王应祥), 2003; Corbet, 1988

(接上页)
(Mori, 1922); *kreyenbergi* (Matschie, 1907); *orientalis* (Allen, 1903); *tschifuensis* (Matschie, 1907); *ussuriensis* (Satunin, 1907)

东北刺猬 *Erinaceus amurensis*

大耳猬 *Hemiechinus auritus*

无危 LC

| 数据缺乏 DD | 无危 LC | 近危 NT | 易危 VU | 濒危 EN | 极危 CR | 区域灭绝 RE | 野外灭绝 EW | 灭绝 EX |

分类地位 Taxonomic Status

动物界 Animalia	脊索动物门 Chordata	哺乳纲 Mammalia	劳亚食虫目 Eulipotyphla	猬科 Erinaceidae

学名 Scientific Name	*Hemiechinus auritus*
命名人 Species Authority	Gmelin, 1770
英文名 English Name(s)	Long-eared Hedgehog
同物异名 Synonym(s)	无 / None
种下单元评估 Infra-specific Taxa Assessed	无 / None

评估信息 Assessment Information

评估年份 Year Assessed	2020
评定人 Assessor(s)	蒋志刚 / Zhigang Jiang
审定人 Reviewer(s)	蒋学龙、冯祚建 / Xuelong Jiang, Zuojian Feng
其他贡献人 Other Contributor(s)	李立立、丁晨晨 / Lili Li, Chenchen Ding

理由 Justification: 大耳猬广泛分布于中国西北部，种群数量大且稳定，没有受到来自人类活动和其他方面的威胁。因此，列为无危等级 / Long-eared Hedgehog is widely distributed with large and stable populations. Furthermore, the species is not threatened by anthropogenic activities and other factors. Thus, it is listed as Least Concern

地理分布 Geographical Distribution

国内分布 Domestic Distribution	内蒙古、新疆、陕西、宁夏、甘肃、青海、四川 / Inner Mongolia (Nei Mongol), Xinjiang, Shaanxi, Ningxia, Gansu, Qinghai, Sichuan
世界分布 World Distribution	中国；中亚、西亚 / China; Central Asia, Western Asia
分布标注 Distribution Note	非特有种 / Non-endemic

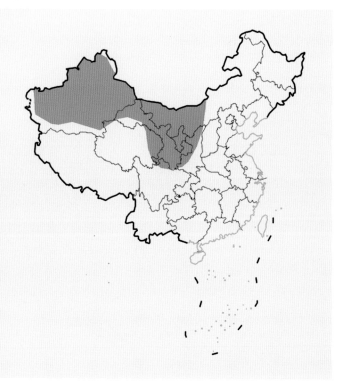

国内分布图
Map of Domestic Distribution

种群 Population

种群数量 Population Size	未知 / Unknown
种群趋势 Population Trend	未知 / Unknown

生境与生态系统 Habitat (s) and Ecosystem (s)

生　　境 Habitat(s)	草地、灌丛、耕地、温带沙漠 / Grassland, Shrubland, Arable Land, Temperate Desert
生态系统 Ecosystem(s)	灌丛生态系统、草地生态系统、农田生态系统、荒漠生态系统 / Shrubland Ecosystem, Grassland Ecosystem, Cropland Ecosystem, Desert Ecosystem

威胁 Threat (s)

主要威胁 Major Threat(s)	无 / None

保护级别与保护行动 Protection Category and Conservation Action (s)

国家重点保护野生动物等级 (2021) Category of National Key Protected Wild Animals (2021)	无 / NA
IUCN红色名录(2020-2) IUCN Red List (2020-2)	无危 / LC
CITES 附录 (2019) CITES Appendix (2019)	无 / NA
保护行动 Conservation Action(s)	政策性措施和立法保护、宣传教育、栖息地保护 / Policy-based action and legislation, communication and education, habitat protection

相关文献 Relevant References

Burgin *et al*., 2018; Wilson and Mittermeier, 2018; Jiang *et al*.（蒋志刚等），2017; Smith *et al*.（史密斯等），2009; Abulimiti Abudukader（阿布力米提·阿都卡迪尔），2002; Hou（侯兰新），2000; Wang（王香亭），1991; Corbet, 1988

大耳猬 *Hemiechinus auritus*

达乌尔猬
Mesechinus dauuricus

无危 LC

| 数据缺乏 DD | 无危 LC | 近危 NT | 易危 VU | 濒危 EN | 极危 CR | 区域灭绝 RE | 野外灭绝 EW | 灭绝 EX |

分类地位 Taxonomic Status

动物界 Animalia	脊索动物门 Chordata	哺乳纲 Mammalia	劳亚食虫目 Eulipotyphla	猬科 Erinaceidae

学名 Scientific Name	*Mesechinus dauuricus*
命名人 Species Authority	Sundevall, 1842
英文名 English Name(s)	Daurian Hedgehog
同物异名 Synonym(s)	*Erinaceus manchuricus* (Mori, 1926); *E. przewalskii* (Satunin, 1907); *E. sibiricus* (Erxleben, 1777)
种下单元评估 Infra-specific Taxa Assessed	无 / None

评估信息 Assessment Information

评估年份 Year Assessed	2020
评定人 Assessor(s)	蒋志刚 / Zhigang Jiang
审定人 Reviewer(s)	蒋学龙、冯祚建 / Xuelong Jiang, Zuojian Feng
其他贡献人 Other Contributor(s)	李立立、丁晨晨 / Lili Li, Chenchen Ding

理由 Justification: 达乌尔猬分布较广，种群数量大且稳定。因此，列为无危等级 / Daurian Hedgehog is a widespread species with stable and large populations. Thus, it is listed as Least Concern

地理分布 Geographical Distribution

国内分布 Domestic Distribution

黑龙江、吉林、辽宁、内蒙古、北京、河北、河南、山西、陕西、宁夏、甘肃、湖北 / Heilongjiang, Jilin, Liaoning, Inner Mongolia (Nei Mongol), Beijing, Hebei, Henan, Shanxi, Shaanxi, Ningxia, Gansu, Hubei

世界分布 World Distribution

中国、蒙古、俄罗斯 / China, Mongolia, Russia

分布标注 Distribution Note

非特有种 / Non-endemic

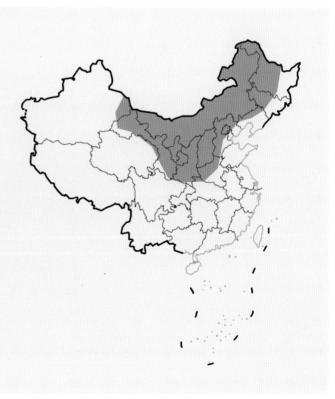

国内分布图
Map of Domestic Distribution

种群 Population

种群数量 Population Size	未知 / Unknown
种群趋势 Population Trend	稳定 / Stable

生境与生态系统 Habitat(s) and Ecosystem(s)

生境 Habitat(s)	草地 / Grassland
生态系统 Ecosystem(s)	草地生态系统 / Grassland Ecosystem

威胁 Threat(s)

主要威胁 Major Threat(s)	采矿、家畜养殖或放牧、公路碾压事件 / Mining, Livestock Farming or Ranching, Road Kill

保护级别与保护行动 Protection Category and Conservation Action(s)

国家重点保护野生动物等级 (2021) Category of National Key Protected Wild Animals (2021)	无 / NA
IUCN 红色名录 (2020-2) IUCN Red List (2020-2)	无危 / LC
CITES 附录 (2019) CITES Appendix (2019)	无 / NA
保护行动 Conservation Action(s)	政策性措施和立法保护、宣传教育、栖息地保护 / Policy-based action and legislation, communication and education, habitat protection

相关文献 Relevant References

Burgin *et al*., 2018; Wilson and Mittermeier, 2018; Jiang *et al*. (蒋志刚等), 2017; Zhang and Yu (张杰和于洪伟), 2005; Corbet, 1988

达乌尔猬 *Mesechinus dauuricus*

林猬
Mesechinus sylvaticus

无危 LC

| 数据缺乏 DD | 无危 LC | 近危 NT | 易危 VU | 濒危 EN | 极危 CR | 区域灭绝 RE | 野外灭绝 EW | 灭绝 EX |

分类地位 Taxonomic Status

动物界 Animalia	脊索动物门 Chordata	哺乳纲 Mammalia	劳亚食虫目 Eulipotyphla	猬科 Erinaceidae
学 名 Scientific Name		*Mesechinus sylvaticus*		
命 名 人 Species Authority		Ma, 1964		
英 文 名 English Name(s)		Forest Hedgehog		
同物异名 Synonym(s)		无 / None		
种下单元评估 Infra-specific Taxa Assessed		无 / None		

评估信息 Assessment Information

评 估 年 份 Year Assessed	2020
评 定 人 Assessor(s)	蒋志刚 / Zhigang Jiang
审 定 人 Reviewer(s)	蒋学龙、冯祚建 / Xuelong Jiang, Zuojian Feng
其他贡献人 Other Contributor(s)	李立立、丁晨晨 / Lili Li, Chenchen Ding

理由 Justification: 林猬为中国特有种，分布于山西中条山，种群相对较大且所受威胁较小。因此，列为无危等级 / Forest Hedgehog is an endemic species in China, distributed in the vicinity of Zhongtiao Shan in Shanxi, with a relatively large populations and bears little threat to its survival. Thus, it is listed as Least Concern

地理分布 Geographical Distribution

国内分布 Domestic Distribution
山西 / Shanxi
世界分布 World Distribution
中国 / China
分布标注 Distribution Note
特有种 / Endemic

国内分布图
Map of Domestic Distribution

🦌 种群 Population

种群数量 Population Size	未知 / Unknown
种群趋势 Population Trend	未知 / Unknown

🦌 生境与生态系统 Habitat (s) and Ecosystem (s)

生　　境 Habitat(s)	落叶阔叶林 / Deciduous Broad-leaved Forest
生态系统 Ecosystem(s)	森林生态系统 / Forest Ecosystem

🦌 威胁 Threat (s)

主要威胁 Major Threat(s)	无 / None

🦌 保护级别与保护行动 Protection Category and Conservation Action (s)

国家重点保护野生动物等级 (2021) Category of National Key Protected Wild Animals (2021)	无 / NA
IUCN 红色名录 (2020-2) IUCN Red List (2020-2)	未列入 / NA
CITES 附录 (2019) CITES Appendix (2019)	无 / NA
保护行动 Conservation Action(s)	无 / None

🦌 相关文献 Relevant References

Burgin *et al*., 2018; Wilson and Mittermeier, 2018; Jiang *et al*. (蒋志刚等), 2017; Ma (马勇), 1964

林猬 *Mesechinus sylvaticus*

© Lynx Edicions

长吻鼩鼹 *Uropsilus gracilis*

无危 LC

| 数据缺乏 DD | 无危 LC | 近危 NT | 易危 VU | 濒危 EN | 极危 CR | 区域灭绝 RE | 野外灭绝 EW | 灭绝 EX |

分类地位 Taxonomic Status

动物界 Animalia	脊索动物门 Chordata	哺乳纲 Mammalia	劳亚食虫目 Eulipotyphla	鼹科 Talpidae
学名 Scientific Name		*Uropsilus gracilis*		
命名人 Species Authority		Thomas, 1911		
英文名 English Name(s)		Gracile Shrew Mole		
同物异名 Synonym(s)		*Uropsilus atronates* (Allen, 1923); *nivatus* (Allen, 1923)		
种下单元评估 Infra-specific Taxa Assessed		无 / None		

评估信息 Assessment Information

评估年份 Year Assessed	2020
评定人 Assessor(s)	蒋志刚 / Zhigang Jiang
审定人 Reviewer(s)	蒋学龙、冯祚建 / Xuelong Jiang, Zuojian Feng
其他贡献人 Other Contributor(s)	李立立、丁晨晨 / Lili Li, Chenchen Ding

理由 Justification: 长吻鼩鼹分布较广，种群数量大。因此，列为无危等级 / Gracile Shrew Mole is widely distributed with large populations. Thus, it is listed as Least Concern

地理分布 Geographical Distribution

国内分布 Domestic Distribution	湖北、湖南、陕西、四川、重庆、贵州、云南 / Hubei, Hunan, Shaanxi, Sichuan, Chongqing, Guizhou, Yunnan
世界分布 World Distribution	中国 / China
分布标注 Distribution Note	特有种 / Endemic

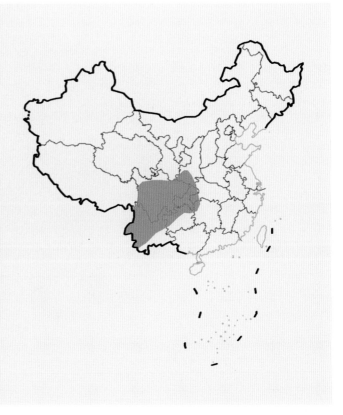

国内分布图
Map of Domestic Distribution

种群 Population

种群数量 Population Size	未知 / Unknown
种群趋势 Population Trend	未知 / Unknown

生境与生态系统 Habitat (s) and Ecosystem (s)

生　　境 Habitat(s)	亚热带高海拔森林、灌丛 / Subtropical High Altitude Forest and Shrubland
生态系统 Ecosystem(s)	森林生态系统、灌丛生态系统 / Forest Ecosystem, Shrubland Ecosystem

威胁 Threat (s)

主要威胁 Major Threat(s)	未知 / Unknown

保护级别与保护行动 Protection Category and Conservation Action (s)

国家重点保护野生动物等级 (2021) Category of National Key Protected Wild Animals (2021)	无 / NA
IUCN 红色名录 (2020-2) IUCN Red List (2020-2)	无危 / LC
CITES 附录 (2019) CITES Appendix (2019)	无 / NA
保护行动 Conservation Action(s)	无 / None

相关文献 Relevant References

Burgin *et al.*, 2018; Wilson and Mittermeier, 2018; Jiang *et al.* (蒋志刚等), 2017; Smith *et al.* (史密斯等), 2009; Wang (王应祥), 2003; Hoffmann, 1984

长吻鼩鼹 *Uropsilus gracilis*

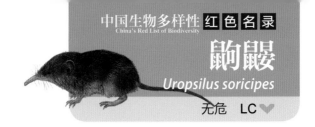

鼩鼱
Uropsilus soricipes

无危 LC

| 数据缺乏 DD | 无危 LC | 近危 NT | 易危 VU | 濒危 EN | 极危 CR | 区域灭绝 RE | 野外灭绝 EW | 灭绝 EX |

分类地位 Taxonomic Status

动物界 Animalia	脊索动物门 Chordata	哺乳纲 Mammalia	劳亚食虫目 Eulipotyphla	鼹科 Talpidae

学名 Scientific Name	*Uropsilus soricipes*
命名人 Species Authority	Milne-Edwards, 1871
英文名 English Name(s)	Chinese Shrew Mole
同物异名 Synonym(s)	无 / None
种下单元评估 Infra-specific Taxa Assessed	无 / None

评估信息 Assessment Information

评估年份 Year Assessed	2020
评定人 Assessor(s)	蒋志刚 / Zhigang Jiang
审定人 Reviewer(s)	蒋学龙、冯祚建 / Xuelong Jiang, Zuojian Feng
其他贡献人 Other Contributor(s)	李立立、丁晨晨 / Lili Li, Chenchen Ding

理由 Justification: 鼩鼱为中国特有种，其分布较广，种群数量大。因此，列为无危等级 / Chinese Shrew Mole is an endemic species in China. It is widely distributed and has large populations. Thus, it is listed as Least Concern

地理分布 Geographical Distribution

国内分布 Domestic Distribution
陕西、甘肃、四川 / Shaanxi, Gansu, Sichuan
世界分布 World Distribution
中国 / China
分布标注 Distribution Note
特有种 / Endemic

国内分布图 Map of Domestic Distribution

种群 Population

种群数量 Population Size	未知 / Unknown
种群趋势 Population Trend	稳定 / Stable

生境与生态系统 Habitat (s) and Ecosystem (s)

生　　境 Habitat(s)	森林 / Forest
生态系统 Ecosystem(s)	森林生态系统 / Forest Ecosystem

威胁 Threat (s)

主要威胁 Major Threat(s)	无 / None

保护级别与保护行动 Protection Category and Conservation Action (s)

国家重点保护野生动物等级 (2021) Category of National Key Protected Wild Animals (2021)	无 / NA
IUCN红色名录(2020-2) IUCN Red List (2020-2)	无危 / LC
CITES 附录 (2019) CITES Appendix (2019)	无 / NA
保护行动 Conservation Action(s)	位于自然保护区内的种群与生境得到保护 / Populations and habitats in nature reserves are protected

相关文献 Relevant References

Burgin *et al*., 2018; Wilson and Mittermeier, 2018; Jiang *et al*. (蒋志刚等), 2017; Wang (王应祥), 2003; Zhang (张荣祖), 1997; Hoffmann, 1984; Shi and Zhao (施白南和赵尔宓), 1980

鼩鼱 *Uropsilus soricipes*

中国生物多样性红色名录 China's Red List of Biodiversity

长尾鼩鼹
Scaptonyx fusicaudus

无危 LC

| 数据缺乏 DD | 无危 LC | 近危 NT | 易危 VU | 濒危 EN | 极危 CR | 区域灭绝 RE | 野外灭绝 EW | 灭绝 EX |

分类地位 Taxonomic Status

动物界 Animalia	脊索动物门 Chordata	哺乳纲 Mammalia	劳亚食虫目 Eulipotyphla	鼹科 Talpidae

学 名 Scientific Name	*Scaptonyx fusicaudus*
命名人 Species Authority	Milne-Edwards, 1872
英文名 English Name(s)	Long-tailed Mole
同物异名 Synonym(s)	*Scaptonyx affinis* (Thomas, 1912); *fusicaudatus* (Milne-Edwards, 1872)
种下单元评估 Infra-specific Taxa Assessed	无 / None

评估信息 Assessment Information

评估年份 Year Assessed	2020
评定人 Assessor(s)	蒋志刚 / Zhigang Jiang
审定人 Reviewer(s)	蒋学龙、冯祚建 / Xuelong Jiang, Zuojian Feng
其他贡献人 Other Contributor(s)	李立立、丁晨晨 / Lili Li, Chenchen Ding

理由 Justification: 长尾鼩鼹分布较广，有一定种群数量。因此，列为无危等级 / Long-tailed Mole is widely distributed and has certain populations. Thus, it is listed as Least Concern

地理分布 Geographical Distribution

国内分布 Domestic Distribution	陕西、四川、湖北、重庆、贵州、云南 / Shaanxi, Sichuan, Hubei, Chongqing, Guizhou, Yunnan
世界分布 World Distribution	中国、越南、缅甸 / China, Viet Nam, Myanmar
分布标注 Distribution Note	
非特有种 / Non-endemic	

国内分布图
Map of Domestic Distribution

种群 Population

种群数量 Population Size	未知 / Unknown
种群趋势 Population Trend	稳定 / Stable

生境与生态系统 Habitat (s) and Ecosystem (s)

生　　境 Habitat(s)	阔叶林、针阔混交林 / Broad-leaved Forest, Coniferous and Broad-leaved Mixed Forest
生态系统 Ecosystem(s)	森林生态系统 / Forest Ecosystem

威胁 Threat (s)

主要威胁 Major Threat(s)	未知 / Unknown

保护级别与保护行动 Protection Category and Conservation Action (s)

国家重点保护野生动物等级 (2021) Category of National Key Protected Wild Animals (2021)	无 / NA
IUCN红色名录(2020-2) IUCN Red List (2020-2)	无危 / LC
CITES 附录 (2019) CITES Appendix (2019)	无 / NA
保护行动 Conservation Action(s)	位于自然保护区内的种群与生境得到保护 / Populations and habitats in nature reserves are protected

相关文献 Relevant References

Burgin *et al.*, 2018; Wilson and Mittermeier, 2018; Jiang *et al.*（蒋志刚等）, 2017; Wang（王应祥）, 2003; Thomas, 1912a

长尾鼩鼹 *Scaptonyx fusicaudus*

长吻鼹
Euroscaptor longirostris

无危 LC

| 数据缺乏 DD | 无危 LC | 近危 NT | 易危 VU | 濒危 EN | 极危 CR | 区域灭绝 RE | 野外灭绝 EW | 灭绝 EX |

分类地位 Taxonomic Status

动物界 Animalia	脊索动物门 Chordata	哺乳纲 Mammalia	劳亚食虫目 Eulipotyphla	鼹科 Talpidae
学名 Scientific Name		*Euroscaptor longirostris*		
命名人 Species Authority		Milne-Edwards, 1870		
英文名 English Name(s)		Long-nosed Mole		
同物异名 Synonym(s)		无 / None		
种下单元评估 Infra-specific Taxa Assessed		无 / None		

评估信息 Assessment Information

评估年份 Year Assessed	2020
评定人 Assessor(s)	蒋志刚 / Zhigang Jiang
审定人 Reviewer(s)	蒋学龙、冯祚建 / Xuelong Jiang, Zuojian Feng
其他贡献人 Other Contributor(s)	李立立、丁晨晨 / Lili Li, Chenchen Ding

理由 Justification: 长吻鼹呈斑块状分布，分布区较广，种群数量大。因此，列为无危等级 / Long-nosed Mole is a widespread species with large populations inhabiting patchily across its range. Thus, it is listed as Least Concern

地理分布 Geographical Distribution

国内分布 Domestic Distribution
湖南、陕西、四川、重庆、福建、广西、贵州、云南 / Hunan, Shaanxi, Sichuan, Chongqing, Fujian, Guangxi, Guizhou, Yunnan
世界分布 World Distribution
中国、越南 / China, Viet Nam
分布标注 Distribution Note
非特有种 / Non-endemic

国内分布图
Map of Domestic Distribution

种群 Population

种群数量 Population Size	未知 / Unknown
种群趋势 Population Trend	未知 / Unknown

生境与生态系统 Habitat (s) and Ecosystem (s)

生　　境 Habitat(s)	亚热带湿润山地森林、草地 / Subtropical Moist Montane Forest, Grassland
生态系统 Ecosystem(s)	森林生态系统 / Forest Ecosystem

威胁 Threat (s)

主要威胁 Major Threat(s)	无 / None

保护级别与保护行动 Protection Category and Conservation Action (s)

国家重点保护野生动物等级 (2021) Category of National Key Protected Wild Animals (2021)	无 / NA
IUCN 红色名录 (2020-2) IUCN Red List (2020-2)	无危 / LC
CITES 附录 (2019) CITES Appendix (2019)	无 / NA
保护行动 Conservation Action(s)	无 / None

相关文献 Relevant References

Burgin *et al.*, 2018; Wilson and Mittermeier, 2018; Jiang *et al.* (蒋志刚等), 2017; Qin *et al.* (秦岭等), 2007; Tu *et al.*, 2012; Wang (王应祥), 2003

长吻鼹 *Euroscaptor longirostris*

华南缺齿鼹
Mogera insularis

无危 LC

| 数据缺乏 DD | **无危 LC** | 近危 NT | 易危 VU | 濒危 EN | 极危 CR | 区域灭绝 RE | 野外灭绝 EW | 灭绝 EX |

分类地位 Taxonomic Status

动物界 Animalia	脊索动物门 Chordata	哺乳纲 Mammalia	劳亚食虫目 Eulipotyphla	鼹科 Talpidae

学名 Scientific Name	*Mogera insularis*
命名人 Species Authority	Swinhoe, 1863
英文名 English Name(s)	Insular Mole
同物异名 Synonym(s)	Formosan Blind Mole
种下单元评估 Infra-specific Taxa Assessed	无 / None

评估信息 Assessment Information

评估年份 Year Assessed	2020
评定人 Assessor(s)	蒋志刚 / Zhigang Jiang
审定人 Reviewer(s)	蒋学龙、冯祚建 / Xuelong Jiang, Zuojian Feng
其他贡献人 Other Contributor(s)	李立立、丁晨晨 / Lili Li, Chenchen Ding

理由 Justification: 华南缺齿鼹在长江以南地区分布较广，种群数量大。因此，列为无危等级 / Insular Mole has large populations and is widely distributed in south of the Changjiang River. Thus, it is listed as Least Concern

地理分布 Geographical Distribution

国内分布 Domestic Distribution
江苏、浙江、安徽、江西、湖南、四川、贵州、广西、广东、福建、海南、台湾 / Jiangsu, Zhejiang, Anhui, Jiangxi, Hunan, Sichuan, Guizhou, Guangxi, Guangdong, Fujian, Hainan, Taiwan
世界分布 World Distribution
中国、越南 / China, Viet Nam
分布标注 Distribution Note
非特有种 / Non-endemic

国内分布图
Map of Domestic Distribution

种群 Population

种群数量 Population Size	未知 / Unknown
种群趋势 Population Trend	未知 / Unknown

生境与生态系统 Habitat (s) and Ecosystem (s)

生　　境 Habitat(s)	热带和亚热带湿润山地森林 / Tropical and Subtropical Moist Montane Forest
生态系统 Ecosystem(s)	森林生态系统 / Forest Ecosystem

威胁 Threat (s)

主要威胁 Major Threat(s)	无 / None

保护级别与保护行动 Protection Category and Conservation Action (s)

国家重点保护野生动物等级 (2021) Category of National Key Protected Wild Animals (2021)	无 / NA
IUCN 红色名录 (2020-2) IUCN Red List (2020-2)	无危 / LC
CITES 附录 (2019) CITES Appendix (2019)	无 / NA
保护行动 Conservation Action(s)	位于自然保护区内的种群与生境得到保护 / Populations and habitats in nature reserves are protected

相关文献 Relevant References

Burgin *et al.*, 2018; Wilson and Mittermeier, 2018; Jiang *et al.* (蒋志刚等), 2017; Tu *et al.* (涂飞云等), 2014; Su *et al.*(粟海军等), 2013; Pan *et al.*(潘清华等), 2007

华南缺齿鼹 *Mogera insularis*

缺齿鼹
Mogera robusta

无危 LC

| 数据缺乏 DD | 无危 LC | 近危 NT | 易危 VU | 濒危 EN | 极危 CR | 区域灭绝 RE | 野外灭绝 EW | 灭绝 EX |

分类地位 Taxonomic Status

动物界 Animalia	脊索动物门 Chordata	哺乳纲 Mammalia	劳亚食虫目 Eulipotyphla	鼹科 Talpidae
学名 Scientific Name		*Mogera robusta*		
命名人 Species Authority		Nehring, 1891		
英文名 English Name(s)		Large Mole		
同物异名 Synonym(s)		Ussuri Mole		
种下单元评估 Infra-specific Taxa Assessed		无 / None		

评估信息 Assessment Information

评估年份 Year Assessed	2020
评定人 Assessor(s)	蒋志刚 / Zhigang Jiang
审定人 Reviewer(s)	蒋学龙、冯祚建 / Xuelong Jiang, Zuojian Feng
其他贡献人 Other Contributor(s)	李立立、丁晨晨 / Lili Li, Chenchen Ding

理由 Justification: 缺齿鼹分布在中国东北，种群数量大。因此，列为无危等级 / Large Mole is found in northeast China with large populations. Thus, it is listed as Least Concern

地理分布 Geographical Distribution

国内分布 Domestic Distribution
黑龙江、吉林、辽宁 / Heilongjiang, Jilin, Liaoning
世界分布 World Distribution
中国；东北亚 / China; Northeast Asia
分布标注 Distribution Note
非特有种 / Non-endemic

国内分布图
Map of Domestic Distribution

种群 Population

种群数量 Population Size	未知 / Unknown
种群趋势 Population Trend	未知 / Unknown

生境与生态系统 Habitat (s) and Ecosystem (s)

生境 Habitat(s)	森林、草地、耕地 / Forest, Grassland, Arable Land
生态系统 Ecosystem(s)	森林生态系统、草地生态系统、农田生态系统 / Forest Ecosystem, Grassland Ecosystem, Cropland Ecosystem

威胁 Threat (s)

主要威胁 Major Threat(s)	无 / None

保护级别与保护行动 Protection Category and Conservation Action (s)

国家重点保护野生动物等级 (2021) Category of National Key Protected Wild Animals (2021)	无 / NA
IUCN 红色名录 (2020-2) IUCN Red List (2020-2)	无危 / LC
CITES 附录 (2019) CITES Appendix (2019)	无 / NA
保护行动 Conservation Action(s)	位于自然保护区内的种群与生境得到保护 / Populations and habitats in nature reserves are protected

相关文献 Relevant References

Burgin *et al.*, 2018; Wilson and Mittermeier, 2018; Jiang *et al.* (蒋志刚等), 2017; Smith *et al.* (史密斯等), 2009; Abe, 1995; Shou (寿振黄), 1962; Mammal Research Group, Institute of Zoology, Chinese Academy of Sciences (中国科学院动物研究所兽类学研究组), 1958

缺齿鼹 *Mogera robusta*

小纹背鼩鼱
Sorex bedfordiae

无危 LC

| 数据缺乏 DD | 无危 LC | 近危 NT | 易危 VU | 濒危 EN | 极危 CR | 区域灭绝 RE | 野外灭绝 EW | 灭绝 EX |

分类地位 Taxonomic Status

动物界 Animalia	脊索动物门 Chordata	哺乳纲 Mammalia	劳亚食虫目 Eulipotyphla	鼩鼱科 Soricidae
学名 Scientific Name		*Sorex bedfordiae*		
命名人 Species Authority		Thomas, 1911		
英文名 English Name(s)		Lesser Stripe-backed Shrew		
同物异名 Synonym(s)		Lesser Striped Shrew		
种下单元评估 Infra-specific Taxa Assessed		无 / None		

评估信息 Assessment Information

评估年份 Year Assessed	2020
评定人 Assessor(s)	蒋志刚 / Zhigang Jiang
审定人 Reviewer(s)	蒋学龙、冯祚建 / Xuelong Jiang, Zuojian Feng
其他贡献人 Other Contributor(s)	李立立、丁晨晨 / Lili Li, Chenchen Ding

理由 Justification: 小纹背鼩鼱分布较广，种群数量大。因此，列为无危等级 / Lesser Stripe-backed Shrew is a widespread species with large populations. Thus, it is listed as Least Concern

地理分布 Geographical Distribution

国内分布 Domestic Distribution
陕西、甘肃、青海、四川、云南 / Shaanxi, Gansu, Qinghai, Sichuan, Yunnan
世界分布 World Distribution
中国；南亚 / China; South Asia
分布标注 Distribution Note
非特有种 / Non-endemic

国内分布图
Map of Domestic Distribution

种群 Population

种群数量 Population Size	未知 / Unknown
种群趋势 Population Trend	未知 / Unknown

生境与生态系统 Habitat(s) and Ecosystem(s)

生　　境 Habitat(s)	泰加林 / Taiga Forest
生态系统 Ecosystem(s)	森林生态系统 / Forest Ecosystem

威胁 Threat(s)

主要威胁 Major Threat(s)	无 / None

保护级别与保护行动 Protection Category and Conservation Action(s)

国家重点保护野生动物等级 (2021) Category of National Key Protected Wild Animals (2021)	无 / NA
IUCN 红色名录 (2020-2) IUCN Red List (2020-2)	无危 / LC
CITES 附录 (2019) CITES Appendix (2019)	无 / NA
保护行动 Conservation Action(s)	位于自然保护区内的种群与生境得到保护 / Populations and habitats in nature reserves are protected

相关文献 Relevant References

Burgin *et al*., 2018; Wilson and Mittermeier, 2018; Jiang *et al*.（蒋志刚等）, 2017; Smith *et al*.（史密斯等）, 2009; Zhang（张荣祖）, 1997; Wu *et al*.（吴毅等）, 1990

小纹背鼩鼱 *Sorex bedfordiae*

云南鼩鼱
Sorex excelsus

无危 LC

| 数据缺乏 DD | 无危 LC | 近危 NT | 易危 VU | 濒危 EN | 极危 CR | 区域灭绝 RE | 野外灭绝 EW | 灭绝 EX |

分类地位 Taxonomic Status

动物界 Animalia	脊索动物门 Chordata	哺乳纲 Mammalia	劳亚食虫目 Eulipotyphla	鼩鼱科 Soricidae
学名 Scientific Name		*Sorex excelsus*		
命名人 Species Authority		Allen, 1923		
英文名 English Name(s)		Highland Shrew		
同物异名 Synonym(s)		Chinese Highland Shrew		
种下单元评估 Infra-specific Taxa Assessed		无 / None		

评估信息 Assessment Information

评估年份 Year Assessed	2020
评定人 Assessor(s)	蒋志刚 / Zhigang Jiang
审定人 Reviewer(s)	蒋学龙 / Xuelong Jiang
其他贡献人 Other Contributor(s)	李立立、丁晨晨 / Lili Li, Chenchen Ding

理由 Justification: 云南鼩鼱分布在横断山区中高海拔地区，种群数量较大，受人为影响较弱。因此，列为无危等级 / Highland Shrew is distributed in the middle and high altitude areas of the Hengduan Shan, its population sizes are large. The impact of human activities on this species is relatively weak. Thus, it is listed as Least Concern

地理分布 Geographical Distribution

国内分布 Domestic Distribution
青海、四川、云南、西藏 / Qinghai, Sichuan, Yunnan, Tibet (Xizang)
世界分布 World Distribution
中国 / China
分布标注 Distribution Note
特有种 / Endemic

国内分布图
Map of Domestic Distribution

种群 Population

种群数量 Population Size	未知 / Unknown
种群趋势 Population Trend	未知 / Unknown

生境与生态系统 Habitat (s) and Ecosystem (s)

生 境 Habitat(s)	亚热带湿润山地森林 / Subtropical Moist Montane Forest
生态系统 Ecosystem(s)	森林生态系统 / Forest Ecosystem

威胁 Threat (s)

主要威胁 Major Threat(s)	未知 / Unknown

保护级别与保护行动 Protection Category and Conservation Action (s)

国家重点保护野生动物等级 (2021) Category of National Key Protected Wild Animals (2021)	无 / NA
IUCN 红色名录 (2020-2) IUCN Red List (2020-2)	无危 / LC
CITES 附录 (2019) CITES Appendix (2019)	无 / NA
保护行动 Conservation Action(s)	位于自然保护区内的种群与生境得到保护 / Populations and habitats in nature reserves are protected

相关文献 Relevant References

Burgin *et al*., 2018; Wilson and Mittermeier, 2018; Jiang *et al*. (蒋志刚等), 2015a; Deng *et al*. (邓可等), 2013; Zhang (张荣祖), 1997; Corbet, 1988

云南鼩鼱 *Sorex excelsus*

细鼩鼱
Sorex gracillimus
无危 LC

| 数据缺乏 DD | 无危 LC | 近危 NT | 易危 VU | 濒危 EN | 极危 CR | 区域灭绝 RE | 野外灭绝 EW | 灭绝 EX |

分类地位 Taxonomic Status

动物界 Animalia	脊索动物门 Chordata	哺乳纲 Mammalia	劳亚食虫目 Eulipotyphla	鼩鼱科 Soricidae
学 名 Scientific Name		*Sorex gracillimus*		
命 名 人 Species Authority		Thomas, 1907		
英 文 名 English Name(s)		Slender Shrew		
同物异名 Synonym(s)		*Sorex gracillimus* (Okhotina, 1993) subsp. *granti*; *S. gracillimus* (Okhotina, 1993) subsp. *minor*; *S. gracillimus* (Okhotina, 1993) subsp. *nataliae*; *S. minutus* (转下页)		
种下单元评估 Infra-specific Taxa Assessed		无 / None		

评估信息 Assessment Information

评估年份 Year Assessed	2020
评定人 Assessor(s)	蒋志刚 / Zhigang Jiang
审定人 Reviewer(s)	蒋学龙、冯祚建 / Xuelong Jiang, Zuojian Feng
其他贡献人 Other Contributor(s)	李立立、丁晨晨 / Lili Li, Chenchen Ding

理由 Justification: 细鼩鼱分布区小，但在其分布区常见。因此，列为无危等级 / Slender Shrew is common in its range but is narrowly distributed. Thus, it is listed as Near Threatened

地理分布 Geographical Distribution

国内分布 Domestic Distribution
内蒙古、黑龙江、吉林 / Inner Mongolia (Nei Mongol), Heilongjiang, Jilin
世界分布 World Distribution
中国；东北亚 / China; Northeast Asia
分布标注 Distribution Note
非特有种 / Non-endemic

国内分布图
Map of Domestic Distribution

种群 Population

种群数量 Population Size	未知 / Unknown
种群趋势 Population Trend	稳定 / Stable

生境与生态系统 Habitat (s) and Ecosystem (s)

生　　境 Habitat(s)	泰加林、次生林、针阔混交林 / Taiga Forest, Secondary Forest, Coniferous and Broad-leaved Mixed Forest
生态系统 Ecosystem(s)	森林生态系统 / Forest Ecosystem

威胁 Threat (s)

主要威胁 Major Threat(s)	未知 / Unknown

保护级别与保护行动 Protection Category and Conservation Action (s)

国家重点保护野生动物等级 (2021) Category of National Key Protected Wild Animals (2021)	无 / NA
IUCN 红色名录(2020-2) IUCN Red List (2020-2)	无危 / LC
CITES 附录 (2019) CITES Appendix (2019)	无 / NA
保护行动 Conservation Action(s)	无 / None

相关文献 Relevant References

Liu *et al.* (刘铸等), 2019; Burgin *et al.*, 2018; Wilson and Mittermeier, 2018; Jiang *et al.* (蒋志刚等), 2017; Smith *et al.* (史密斯等), 2009; Hutterer, 2005; Hoffmann, 1987

(接上页)
(Thomas, 1907) subsp. *gracillimus*

细鼩鼱 *Sorex gracillimus*

淡灰黑齿鼩鼱
Blarinella griselda

无危 LC

| 数据缺乏 DD | 无危 LC | 近危 NT | 易危 VU | 濒危 EN | 极危 CR | 区域灭绝 RE | 野外灭绝 EW | 灭绝 EX |

分类地位 Taxonomic Status

动物界 Animalia	脊索动物门 Chordata	哺乳纲 Mammalia	劳亚食虫目 Eulipotyphla	鼩鼱科 Soricidae

学名 Scientific Name	*Blarinella griselda*
命名人 Species Authority	Thomas, 1912
英文名 English Name(s)	Indochinese Short-tailed Shrew
同物异名 Synonym(s)	无 / None
种下单元评估 Infra-specific Taxa Assessed	无 / None

评估信息 Assessment Information

评估年份 Year Assessed	2020
评定人 Assessor(s)	蒋志刚 / Zhigang Jiang
审定人 Reviewer(s)	蒋学龙、冯祚建 / Xuelong Jiang, Zuojian Feng
其他贡献人 Other Contributor(s)	李立立、丁晨晨 / Lili Li, Chenchen Ding

理由 Justification: 淡灰黑齿鼩鼱分布广，种群数量大。因此，列为无危等级 / Indochinese Short-tailed Shrew is widely distributed with large populations. Thus, it is listed as Least Concern

地理分布 Geographical Distribution

国内分布 Domestic Distribution
四川、陕西、湖北、甘肃、贵州、云南 / Sichuan, Shaanxi, Hubei, Gansu, Guizhou, Yunnan
世界分布 World Distribution
中国、越南 / China, Viet Nam
分布标注 Distribution Note
非特有种 / Non-endemic

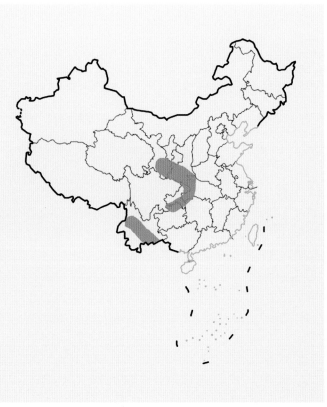

国内分布图
Map of Domestic Distribution

种群 Population

种群数量 Population Size	未知 / Unknown
种群趋势 Population Trend	未知 / Unknown

生境与生态系统 Habitat(s) and Ecosystem(s)

生　境 Habitat(s)	森林 / Forest
生态系统 Ecosystem(s)	森林生态系统 / Forest Ecosystem

威胁 Threat(s)

主要威胁 Major Threat(s)	未知 / Unknown

保护级别与保护行动 Protection Category and Conservation Action(s)

国家重点保护野生动物等级 (2021) Category of National Key Protected Wild Animals (2021)	无 / NA
IUCN 红色名录 (2020-2) IUCN Red List (2020-2)	无危 / LC
CITES 附录 (2019) CITES Appendix (2019)	无 / NA
保护行动 Conservation Action(s)	无 / None

相关文献 Relevant References

Burgin *et al*., 2018; Wilson and Mittermeier, 2018; Jiang *et al*. (蒋志刚等), 2017; Xie *et al*. (谢文华等), 2014; Deng *et al*. (邓可等), 2013; Chen *et al*., 2012; Smith *et al*. (史密斯等), 2009

淡灰黑齿鼩鼱 *Blarinella griselda*

川䶄
Blarinella quadraticauda
无危 LC

| 数据缺乏 DD | 无危 LC | 近危 NT | 易危 VU | 濒危 EN | 极危 CR | 区域灭绝 RE | 野外灭绝 EW | 灭绝 EX |

分类地位 Taxonomic Status

动物界 Animalia	脊索动物门 Chordata	哺乳纲 Mammalia	劳亚食虫目 Eulipotyphla	䶄鼩科 Soricidae

学名 Scientific Name	*Blarinella quadraticauda*
命名人 Species Authority	Milne-Edwards, 1872
英文名 English Name(s)	Asiatic Short-tailed Shrew
同物异名 Synonym(s)	无 / None
种下单元评估 Infra-specific Taxa Assessed	无 / None

评估信息 Assessment Information

评估年份 Year Assessed	2020
评定人 Assessor(s)	蒋志刚 / Zhigang Jiang
审定人 Reviewer(s)	蒋学龙、冯祚建 / Xuelong Jiang, Zuojian Feng
其他贡献人 Other Contributor(s)	李立立、丁晨晨 / Lili Li, Chenchen Ding

理由 Justification: 川䶄分布较广，种群数量大，其主要的栖息地均位于保护区内。因此，列为无危等级 / Asiatic Short-tailed Shrew is widely distributed with large populations. The species' main habitats are situated in nature reserves. Thus, it is listed as Least Concern

地理分布 Geographical Distribution

国内分布 Domestic Distribution	四川、陕西、甘肃、湖北、贵州、重庆、云南 / Sichuan, Shaanxi, Gansu, Hubei, Guizhou, Chongqing, Yunnan
世界分布 World Distribution	中国 / China
分布标注 Distribution Note	特有种 / Endemic

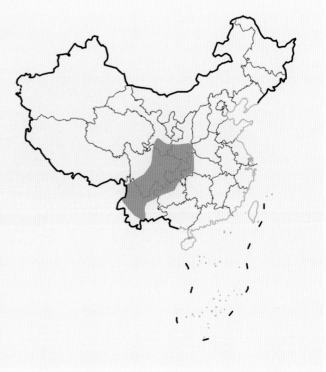

国内分布图
Map of Domestic Distribution

种群 Population

种群数量 Population Size	未知 / Unknown
种群趋势 Population Trend	下降 / Decreasing

生境与生态系统 Habitat (s) and Ecosystem (s)

生境 Habitat(s)	泰加林、次生林 / Taiga Forest, Secondary Forest
生态系统 Ecosystem(s)	森林生态系统 / Forest Ecosystem

威胁 Threat (s)

主要威胁 Major Threat(s)	伐木、耕作、住房及城市建设 / Logging, Farming, House and Urban Area Construction

保护级别与保护行动 Protection Category and Conservation Action (s)

国家重点保护野生动物等级 (2021) Category of National Key Protected Wild Animals (2021)	无 / NA
IUCN红色名录(2020-2) IUCN Red List (2020-2)	近危 / NT
CITES 附录 (2019) CITES Appendix (2019)	无 / NA
保护行动 Conservation Action(s)	位于自然保护区内的种群与生境得到保护 / Populations and habitats in nature reserves are protected

相关文献 Relevant References

Burgin *et al.*, 2018; Wilson and Mittermeier, 2018; Jiang *et al.* (蒋志刚等), 2017; Tu *et al.* (涂飞云等), 2014; Sun *et al.* (孙治宇等), 2013; Smith *et al.* (史密斯等), 2009

川鼩 *Blarinella quadraticauda*

长尾鼩
Episoriculus caudatus
无危 LC

| 数据缺乏 DD | 无危 LC | 近危 NT | 易危 VU | 濒危 EN | 极危 CR | 区域灭绝 RE | 野外灭绝 EW | 灭绝 EX |

分类地位 Taxonomic Status

动物界 Animalia	脊索动物门 Chordata	哺乳纲 Mammalia	劳亚食虫目 Eulipotyphla	鼩鼱科 Soricidae
学名 Scientific Name		*Episoriculus caudatus*		
命名人 Species Authority		Horsfield, 1851		
英文名 English Name(s)		Hodgson's Brown-toothed Shrew		
同物异名 Synonym(s)		*Soriculus caudatus* (Horsfield, 1851)		
种下单元评估 Infra-specific Taxa Assessed		无 / None		

评估信息 Assessment Information

评估年份 Year Assessed	2020
评定人 Assessor(s)	蒋志刚 / Zhigang Jiang
审定人 Reviewer(s)	蒋学龙、冯祚建 / Xuelong Jiang, Zuojian Feng
其他贡献人 Other Contributor(s)	李立立、丁晨晨 / Lili Li, Chenchen Ding

理由 Justification: 长尾鼩分布广、数量大，在四川、云南灌丛和森林广泛分布。因此，列为无危等级 / Hodgson's Brown-toothed Shrew is widely distributed with large populations. In Sichuan and Yunnan, the species is widely distributed in shrublands and forests. Thus, it is listed as Least Concern

地理分布 Geographical Distribution

国内分布 Domestic Distribution
四川、西藏、云南 / Sichuan, Tibet (Xizang), Yunnan
世界分布 World Distribution
中国；南亚 / China; South Asia
分布标注 Distribution Note
非特有种 / Non-endemic

国内分布图
Map of Domestic Distribution

种群 Population

种群数量 Population Size	未知 / Unknown
种群趋势 Population Trend	未知 / Unknown

生境与生态系统 Habitat (s) and Ecosystem (s)

生境 Habitat(s)	耕地、草甸、泰加林 / Arable Land, Meadow, Taiga Forest
生态系统 Ecosystem(s)	森林生态系统、草地生态系统、农田生态系统 / Forest Ecosystem, Grassland Ecosystem, Cropland Ecosystem

威胁 Threat (s)

主要威胁 Major Threat(s)	无 / None

保护级别与保护行动 Protection Category and Conservation Action (s)

国家重点保护野生动物等级 (2021) Category of National Key Protected Wild Animals (2021)	无 / NA
IUCN红色名录(2020-2) IUCN Red List (2020-2)	无危 / LC
CITES 附录 (2019) CITES Appendix (2019)	无 / NA
保护行动 Conservation Action(s)	无 / None

相关文献 Relevant References

Burgin *et al.*, 2018; Wilson and Mittermeier, 2018; Jiang *et al.* (蒋志刚等), 2017; Deng *et al.* (邓可等), 2013; Hutterer, 2005

长尾鼩 *Episoriculus caudatus*

台湾长尾鼩 Episoriculus fumidus

无危 LC

| 数据缺乏 DD | 无危 LC | 近危 NT | 易危 VU | 濒危 EN | 极危 CR | 区域灭绝 RE | 野外灭绝 EW | 灭绝 EX |

分类地位 Taxonomic Status

动物界 Animalia	脊索动物门 Chordata	哺乳纲 Mammalia	劳亚食虫目 Eulipotyphla	鼩鼱科 Soricidae

学名 Scientific Name	*Episoriculus fumidus*
命名人 Species Authority	Thomas, 1913
英文名 English Name(s)	Taiwan Brown-toothed Shrew
同物异名 Synonym(s)	*Soriculus fumidus* (Thomas, 1913)
种下单元评估 Infra-specific Taxa Assessed	无 / None

评估信息 Assessment Information

评估年份 Year Assessed	2020
评定人 Assessor(s)	蒋志刚 / Zhigang Jiang
审定人 Reviewer(s)	蒋学龙、冯祚建 / Xuelong Jiang, Zuojian Feng
其他贡献人 Other Contributor(s)	李立立、丁晨晨 / Lili Li, Chenchen Ding

理由 Justification: 台湾长尾鼩为中国特有种，仅分布在台湾，种群数量较大。因此，列为无危等级 / Taiwan Brown-toothed Shrew is an endemic species in China, and only narrowly distributed in Taiwan. It has fairly large populations in field. Thus, it is listed as Least concern

地理分布 Geographical Distribution

国内分布 Domestic Distribution
台湾 / Taiwan
世界分布 World Distribution
中国 / China
分布标注 Distribution Note
特有种 / Endemic

国内分布图 Map of Domestic Distribution

种群 Population

种群数量 Population Size	未知 / Unknown
种群趋势 Population Trend	未知 / Unknown

生境与生态系统 Habitat(s) and Ecosystem(s)

生境 Habitat(s)	泰加林、灌丛、亚热带湿润山地森林 / Taiga Forest, Shrubland, Subtropical Moist Montane Forest
生态系统 Ecosystem(s)	森林生态系统、灌丛生态系统 / Forest Ecosystem, Shrubland Ecosystem

威胁 Threat(s)

主要威胁 Major Threat(s)	未知 / Unknown

保护级别与保护行动 Protection Category and Conservation Action(s)

国家重点保护野生动物等级 (2021) Category of National Key Protected Wild Animals (2021)	无 / NA
IUCN 红色名录 (2020-2) IUCN Red List (2020-2)	无危 / LC
CITES 附录 (2019) CITES Appendix (2019)	无 / NA
保护行动 Conservation Action(s)	无 / None

相关文献 Relevant References

Burgin *et al.*, 2018; Wilson and Mittermeier, 2018; Jiang *et al.* (蒋志刚等), 2017; Motokawa *et al.*, 1998, 1997; Yu, 1994, 1993; Jameson and Jones, 1977

台湾长尾鼩 *Episoriculus fumidus*

大长尾鼩鼱
Episoriculus leucops
无危 LC

| 数据缺乏 DD | 无危 LC | 近危 NT | 易危 VU | 濒危 EN | 极危 CR | 区域灭绝 RE | 野外灭绝 EW | 灭绝 EX |

分类地位 Taxonomic Status

动物界 Animalia	脊索动物门 Chordata	哺乳纲 Mammalia	劳亚食虫目 Eulipotyphla	鼩鼱科 Soricidae

学 名 Scientific Name	*Episoriculus leucops*
命名人 Species Authority	Horsfield, 1855
英文名 English Name(s)	Long-tailed Brown-toothed Shrew
同物异名 Synonym(s)	*Soriculus leucops* (Horsfield, 1855)
种下单元评估 Infra-specific Taxa Assessed	无 / None

评估信息 Assessment Information

评估年份 Year Assessed	2020
评定人 Assessor(s)	蒋志刚 / Zhigang Jiang
审定人 Reviewer(s)	蒋学龙、冯祚建 / Xuelong Jiang, Zuojian Feng
其他贡献人 Other Contributor(s)	李立立、丁晨晨 / Lili Li, Chenchen Ding

理由 Justification: 大长尾鼩鼱分布在中国西南，种群数量较大。因此，列为无危等级 / Long-tailed Brown-toothed Shrew is distributed in southwest China with a fairly large populations. Thus, it is listed as Least Concern

地理分布 Geographical Distribution

国内分布 Domestic Distribution
四川、西藏、云南 / Sichuan, Tibet (Xizang), Yunnan
世界分布 World Distribution
中国、印度、缅甸、尼泊尔、越南 / China, India, Myanmar, Nepal, Viet Nam
分布标注 Distribution Note
非特有种 / Non-endemic

国内分布图
Map of Domestic Distribution

种群 Population

种群数量 Population Size	未知 / Unknown
种群趋势 Population Trend	未知 / Unknown

生境与生态系统 Habitat (s) and Ecosystem (s)

生　　境 Habitat(s)	泰加林、灌丛、亚热带湿润山地森林、耕地、草地 / Taiga Forest, Shrubland, Subtropical Moist Montane Forest, Arable Land, Grassland
生态系统 Ecosystem(s)	森林生态系统、灌丛生态系统、草地生态系统、农田生态系统 / Forest Ecosystem, Shrubland Ecosystem, Grassland Ecosystem, Cropland Ecosystem

威胁 Threat (s)

主要威胁 Major Threat(s)	未知 / Unknown

保护级别与保护行动 Protection Category and Conservation Action (s)

国家重点保护野生动物等级 (2021) Category of National Key Protected Wild Animals (2021)	无 / NA
IUCN 红色名录 (2020-2) IUCN Red List (2020-2)	无危 / LC
CITES 附录 (2019) CITES Appendix (2019)	无 / NA
保护行动 Conservation Action(s)	无 / None

相关文献 Relevant References

Burgin *et al*., 2018; Wilson and Mittermeier, 2018; Jiang *et al*. (蒋志刚等), 2017; Smith *et al*. (史密斯等), 2009; Hutterer, 2005; Chakraborty *et al*., 2004

大长尾鼩鼱 *Episoriculus leucops*

缅甸长尾鼩
Episoriculus macrurus

无危 LC

| 数据缺乏 DD | 无危 LC | 近危 NT | 易危 VU | 濒危 EN | 极危 CR | 区域灭绝 RE | 野外灭绝 EW | 灭绝 EX |

分类地位 Taxonomic Status

动物界 Animalia	脊索动物门 Chordata	哺乳纲 Mammalia	劳亚食虫目 Eulipotyphla	鼩鼱科 Soricidae

学名 Scientific Name	*Episoriculus macrurus*
命名人 Species Authority	Blanford, 1888
英文名 English Name(s)	Arboreal Brown-toothed Shrew
同物异名 Synonym(s)	Long-tailed Mountain Shrew; *Sorex macrurus* (Hodgson, 1863); *Soriculus macrurus* (Blanford, 1888)
种下单元评估 Infra-specific Taxa Assessed	无 / None

评估信息 Assessment Information

评估年份 Year Assessed	2020
评定人 Assessor(s)	蒋志刚 / Zhigang Jiang
审定人 Reviewer(s)	蒋学龙、冯祚建 / Xuelong Jiang, Zuojian Feng
其他贡献人 Other Contributor(s)	李立立、丁晨晨 / Lili Li, Chenchen Ding

理由 Justification: 缅甸长尾鼩分布区大，种群数量较大。因此，列为无危等级 / Arboreal Brown-toothed Shrew is widely distributed with relatively large populations. Thus, it is listed as Least Concern

地理分布 Geographical Distribution

国内分布 Domestic Distribution
四川、云南 / Sichuan, Yunnan
世界分布 World Distribution
中国；南亚、东南亚 / China; South Asia, Southeast Asia
分布标注 Distribution Note
非特有种 / Non-endemic

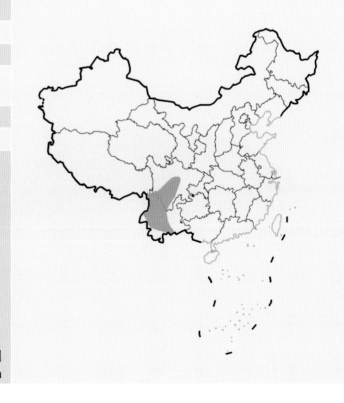

国内分布图
Map of Domestic Distribution

种群 Population

种群数量 Population Size	未知 / Unknown
种群趋势 Population Trend	未知 / Unknown

生境与生态系统 Habitat (s) and Ecosystem (s)

生　　境 Habitat(s)	温带森林、灌丛 / Temperate Forest, Shrubland
生态系统 Ecosystem(s)	森林生态系统、灌丛生态系统 / Forest Ecosystem, Shrubland Ecosystem

威胁 Threat (s)

主要威胁 Major Threat(s)	未知 / Unknown

保护级别与保护行动 Protection Category and Conservation Action (s)

国家重点保护野生动物等级 (2021) Category of National Key Protected Wild Animals (2021)	无 / NA
IUCN 红色名录 (2020-2) IUCN Red List (2020-2)	无危 / LC
CITES 附录 (2019) CITES Appendix (2019)	无 / NA
保护行动 Conservation Action(s)	无 / None

相关文献 Relevant References

Burgin *et al*., 2018; Wilson and Mittermeier, 2018; Jiang *et al*. (蒋志刚等), 2017; Deng *et al*. (邓可等), 2013; Tu *et al*., 2012; Wang (王应祥), 2003

缅甸长尾鼩 *Episoriculus macrurus*

霍氏缺齿鼩
Chodsigoa hoffmanni

无危 LC

分类地位 Taxonomic Status

动物界 Animalia	脊索动物门 Chordata	哺乳纲 Mammalia	劳亚食虫目 Eulipotyphla	鼩鼱科 Soricidae
学名 Scientific Name		*Chodsigoa hoffmanni*		
命名人 Species Authority		Chen *et al.*, 2017		
英文名 English Name(s)		Hoffmann's Long-tailed Shrew		
同物异名 Synonym(s)		无 / None		
种下单元评估 Infra-specific Taxa Assessed		无 / None		

评估信息 Assessment Information

评估年份 Year Assessed	2020
评定人 Assessor(s)	蒋志刚 / Zhigang Jiang
审定人 Reviewer(s)	蒋学龙、冯祚建 / Xuelong Jiang, Zuojian Feng
其他贡献人 Other Contributor(s)	丁晨晨 / Chenchen Ding

理由 Justification: 霍氏缺齿鼩分布于云南中部哀牢山、无量山与越南北部常绿阔叶林，有一定的数量，易被发现。因此，列为无危等级 / Hoffmann's Long-tailed Shrew is easily found in the evergreen broad-leaved forests of Ailao Shan and Wuliang Shan in central Yunnan, China, and in northern Viet Nam. Thus, it is listed as Least Concern

地理分布 Geographical Distribution

国内分布 Domestic Distribution
云南 / Yunnan
世界分布 World Distribution
中国、越南 / China, Viet Nam
分布标注 Distribution Note
非特有种 / Non-endemic

国内分布图
Map of Domestic Distribution

种群 Population

种群数量 Population Size	未知 / Unknown
种群趋势 Population Trend	未知 / Unknown

生境与生态系统 Habitat (s) and Ecosystem (s)

生　　境 Habitat(s)	云南哀牢山和无量山山区海拔 1,500～2,000m 的常绿阔叶林 / Evergreen Broad-leaved Forests from 1,500~2000m of Ailao Shan and Wuliang Shan in Yunnan
生态系统 Ecosystem(s)	森林生态系统 / Forest Ecosystem

威胁 Threat (s)

主要威胁 Major Threat(s)	未知 / Unknown

保护级别与保护行动 Protection Category and Conservation Action (s)

国家重点保护野生动物等级 (2021) Category of National Key Protected Wild Animals (2021)	无 / NA
IUCN 红色名录 (2020-2) IUCN Red List (2020-2)	未列入 / NA
CITES 附录 (2019) CITES Appendix (2019)	无 / NA
保护行动 Conservation Action(s)	无 / None

相关文献 Relevant References

Burgin *et al*., 2018; Wilson and Mittermeier, 2018; Chen *et al*., 2017b; Jiang *et al*. (蒋志刚等), 2017

霍氏缺齿鼩 *Chodsigoa hoffmanni*

川西缺齿鼩鼱
Chodsigoa hypsibia

无危 LC

| 数据缺乏 DD | 无危 LC | 近危 NT | 易危 VU | 濒危 EN | 极危 CR | 区域灭绝 RE | 野外灭绝 EW | 灭绝 EX |

分类地位 Taxonomic Status

动物界 Animalia	脊索动物门 Chordata	哺乳纲 Mammalia	劳亚食虫目 Eulipotyphla	鼩鼱科 Soricidae

学名 Scientific Name	*Chodsigoa hypsibia*
命名人 Species Authority	De Winton, 1899
英文名 English Name(s)	De Winton's Shrew
同物异名 Synonym(s)	*Chodsigoa parca furva* (Hoffmann, 1985)
种下单元评估 Infra-specific Taxa Assessed	无 / None

评估信息 Assessment Information

评估年份 Year Assessed	2020
评定人 Assessor(s)	蒋志刚 / Zhigang Jiang
审定人 Reviewer(s)	蒋学龙、冯祚建 / Xuelong Jiang, Zuojian Feng
其他贡献人 Other Contributor(s)	李立立、丁晨晨 / Lili Li, Chenchen Ding

理由 Justification: 川西缺齿鼩鼱分布广，种群数量大。因此，列为无危等级 / De Winton's Shrew is a widespread species with large populations. Thus, it is listed as Least Concern

地理分布 Geographical Distribution

国内分布 Domestic Distribution
北京、四川、云南、甘肃、西藏、陕西 / Beijing, Sichuan, Yunnan, Gansu, Tibet (Xizang), Shaanxi
世界分布 World Distribution
中国 / China
分布标注 Distribution Note
特有种 / Endemic

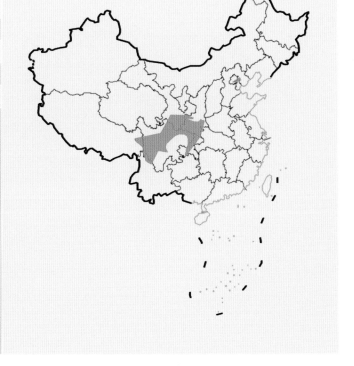

国内分布图
Map of Domestic Distribution

种群 Population

种群数量 Population Size	未知 / Unknown
种群趋势 Population Trend	未知 / Unknown

生境与生态系统 Habitat(s) and Ecosystem(s)

生　　境 Habitat(s)	亚热带湿润山地森林、溪流边 / Subtropical Moist Montane Forest, Near Stream
生态系统 Ecosystem(s)	森林生态系统、湖泊河流生态系统 / Forest Ecosystem, Lake and River Ecosystem

威胁 Threat(s)

主要威胁 Major Threat(s)	未知 / Unknown

保护级别与保护行动 Protection Category and Conservation Action(s)

国家重点保护野生动物等级 (2021) Category of National Key Protected Wild Animals (2021)	无 / NA
IUCN 红色名录 (2020-2) IUCN Red List (2020-2)	无危 / LC
CITES 附录 (2019) CITES Appendix (2019)	无 / NA
保护行动 Conservation Action(s)	位于自然保护区内的种群与生境得到保护 / Populations and habitats in nature reserves are protected

相关文献 Relevant References

Burgin *et al*., 2018; Wilson and Mittermeier, 2018; Jiang *et al*. (蒋志刚等), 2017; Tu *et al*., 2012; Liu *et al*. (刘洋等), 2011; Liu *et al*. (刘少英等), 2005; Lunde *et al*., 2003

川西缺齿鼩鼱 *Chodsigoa hypsibia*

云南缺齿鼩鼱
Chodsigoa parca

无危 LC

| 数据缺乏 DD | 无危 LC | 近危 NT | 易危 VU | 濒危 EN | 极危 CR | 区域灭绝 RE | 野外灭绝 EW | 灭绝 EX |

分类地位 Taxonomic Status

| 动物界 Animalia | 脊索动物门 Chordata | 哺乳纲 Mammalia | 劳亚食虫目 Eulipotyphla | 鼩鼱科 Soricidae |

学名 Scientific Name	*Chodsigoa parca*
命名人 Species Authority	G. M. Allen, 1923
英文名 English Name(s)	Lowe's Shrew
同物异名 Synonym(s)	*Soriculus parca* (G. M. Allen, 1923)
种下单元评估 Infra-specific Taxa Assessed	无 / None

评估信息 Assessment Information

评估年份 Year Assessed	2020
评定人 Assessor(s)	蒋志刚 / Zhigang Jiang
审定人 Reviewer(s)	蒋学龙 / Xuelong Jiang
其他贡献人 Other Contributor(s)	李立立、丁晨晨 / Lili Li, Chenchen Ding

理由 Justification: 云南缺齿鼩鼱分布于云南高黎贡山森林中，在其分布区常见。因此，列为无危等级 / Lowe's Shrew is narrowly distributed in forested areas in the Gaoligong Shan, and is common in its range. Thus, it is listed as Least Concern

地理分布 Geographical Distribution

国内分布 Domestic Distribution
四川、云南 / Sichuan, Yunnan
世界分布 World Distribution
中国、缅甸、泰国、越南 / China, Myanmar, Thailand, Viet Nam
分布标注 Distribution Note
非特有种 / Non-endemic

国内分布图
Map of Domestic Distribution

种群 Population

种群数量 Population Size	未知 / Unknown
种群趋势 Population Trend	未知 / Unknown

生境与生态系统 Habitat (s) and Ecosystem (s)

生　　境 Habitat(s)	亚热带湿润山地森林 / Subtropical Moist Montane Forest
生态系统 Ecosystem(s)	森林生态系统 / Forest Ecosystem

威胁 Threat (s)

主要威胁 Major Threat(s)	未知 / Unknown

保护级别与保护行动 Protection Category and Conservation Action (s)

国家重点保护野生动物等级 (2021) Category of National Key Protected Wild Animals (2021)	无 / NA
IUCN 红色名录 (2020-2) IUCN Red List (2020-2)	无危 / LC
CITES 附录 (2019) CITES Appendix (2019)	无 / NA
保护行动 Conservation Action(s)	无 / None

相关文献 Relevant References

Burgin *et al.*, 2018; Wilson and Mittermeier, 2018; Jiang *et al.* (蒋志刚等), 2017; Smith *et al.* (史密斯等), 2009; Allen, 1923

云南缺齿鼩鼱 *Chodsigoa parca*

斯氏缺齿鼩鼱
Chodsigoa smithii

无危 LC

| 数据缺乏 DD | 无危 LC | 近危 NT | 易危 VU | 濒危 EN | 极危 CR | 区域灭绝 RE | 野外灭绝 EW | 灭绝 EX |

分类地位 Taxonomic Status

动物界 Animalia	脊索动物门 Chordata	哺乳纲 Mammalia	劳亚食虫目 Eulipotyphla	鼩鼱科 Soricidae

学名 Scientific Name	*Chodsigoa smithii*
命名人 Species Authority	Thomas, 1911
英文名 English Name(s)	Smith's Shrew
同物异名 Synonym(s)	*Soriculus smithii* (Thomas, 1911)
种下单元评估 Infra-specific Taxa Assessed	无 / None

评估信息 Assessment Information

评估年份 Year Assessed	2020
评定人 Assessor(s)	蒋志刚 / Zhigang Jiang
审定人 Reviewer(s)	蒋学龙 / Xuelong Jiang
其他贡献人 Other Contributor(s)	李立立、丁晨晨 / Lili Li, Chenchen Ding

理由 Justification: 斯氏缺齿鼩鼱为中国特有种，在其分布区常见。因此，列为无危等级 / Smith's Shrew is an endemic species and is common in its range. Thus, it is listed as Least Concern

地理分布 Geographical Distribution

国内分布 Domestic Distribution
四川、陕西、湖北 / Sichuan, Shaanxi, Hubei
世界分布 World Distribution
中国 / China
分布标注 Distribution Note
特有种 / Endemic

国内分布图
Map of Domestic Distribution

种群 Population

种群数量 Population Size	未知 / Unknown
种群趋势 Population Trend	下降 / Decreasing

生境与生态系统 Habitat (s) and Ecosystem (s)

生　　境 Habitat(s)	温带森林 / Temperate Forest
生态系统 Ecosystem(s)	森林生态系统 / Forest Ecosystem

威胁 Threat (s)

主要威胁 Major Threat(s)	未知 / Unknown

保护级别与保护行动 Protection Category and Conservation Action (s)

国家重点保护野生动物等级 (2021) Category of National Key Protected Wild Animals (2021)	无 / NA
IUCN 红色名录 (2020-2) IUCN Red List (2020-2)	近危 / NT
CITES 附录 (2019) CITES Appendix (2019)	无 / NA
保护行动 Conservation Action(s)	无 / None

相关文献 Relevant References

Burgin *et al.*, 2018; Wilson and Mittermeier, 2018; Chen *et al.*, 2017b; Jiang *et al.* (蒋志刚等), 2017; Tu *et al.*, 2012; Qin *et al.* (秦岭等), 2007

斯氏缺齿鼩鼱 *Chodsigoa smithii*

微尾鼩
Anourosorex squamipes

无危 LC

| 数据缺乏 DD | 无危 LC | 近危 NT | 易危 VU | 濒危 EN | 极危 CR | 区域灭绝 RE | 野外灭绝 EW | 灭绝 EX |

分类地位 Taxonomic Status

动物界 Animalia	脊索动物门 Chordata	哺乳纲 Mammalia	劳亚食虫目 Eulipotyphla	鼩鼱科 Soricidae
学名 Scientific Name		*Anourosorex squamipes*		
命名人 Species Authority		Milne-Edwards, 1872		
英文名 English Name(s)		Mole Shrew		
同物异名 Synonym(s)		Chinese Mole Shrew		
种下单元评估 Infra-specific Taxa Assessed		无 / None		

评估信息 Assessment Information

评估年份 Year Assessed	2020
评定人 Assessor(s)	蒋志刚 / Zhigang Jiang
审定人 Reviewer(s)	蒋学龙 / Xuelong Jiang
其他贡献人 Other Contributor(s)	李立立、丁晨晨 / Lili Li, Chenchen Ding

理由 Justification: 微尾鼩分布广，种群数量大。因此，列为无危等级 / Mole Shrew is a widespread species with large populations. Thus, it is listed as Least Concern

地理分布 Geographical Distribution

国内分布 Domestic Distribution

陕西、甘肃、湖北、重庆、四川、贵州、广东、云南 / Shaanxi, Gansu, Hubei, Chongqing, Sichuan, Guizhou, Guangdong, Yunnan

世界分布 World Distribution

中国、印度、老挝、缅甸、泰国、越南 / China, India, Laos, Myanmar, Thailand, Viet Nam

分布标注 Distribution Note

非特有种 / Non-endemic

国内分布图
Map of Domestic Distribution

种群 Population

种群数量 Population Size	未知 / Unknown
种群趋势 Population Trend	未知 / Unknown

生境与生态系统 Habitat (s) and Ecosystem (s)

生 境 Habitat(s)	亚热带湿润山地森林 / Subtropical Moist Montane Forest
生态系统 Ecosystem(s)	森林生态系统 / Forest Ecosystem

威胁 Threat (s)

主要威胁 Major Threat(s)	无 / None

保护级别与保护行动 Protection Category and Conservation Action (s)

国家重点保护野生动物等级 (2021) Category of National Key Protected Wild Animals (2021)	无 / NA
IUCN 红色名录 (2020-2) IUCN Red List (2020-2)	无危 / LC
CITES 附录 (2019) CITES Appendix (2019)	无 / NA
保护行动 Conservation Action(s)	无 / None

相关文献 Relevant References

Burgin *et al.*, 2018; Wilson and Mittermeier, 2018; Jiang *et al.* (蒋志刚等), 2017; Wang *et al.* (王于玫等), 2014; Xie *et al.* (谢文华等), 2014; Wu (吴毅), 2011; Motokawa and Lin, 2002

微尾鼩 *Anourosorex squamipes*

蹼足鼩
Nectogale elegans

无危 LC

| 数据缺乏 DD | 无危 LC | 近危 NT | 易危 VU | 濒危 EN | 极危 CR | 区域灭绝 RE | 野外灭绝 EW | 灭绝 EX |

分类地位 Taxonomic Status

动物界 Animalia	脊索动物门 Chordata	哺乳纲 Mammalia	劳亚食虫目 Eulipotyphla	鼩鼱科 Soricidae

学名 Scientific Name	*Nectogale elegans*
命名人 Species Authority	Milne-Edwards, 1870
英文名 English Name(s)	Elegant Water Shrew
同物异名 Synonym(s)	无 / None
种下单元评估 Infra-specific Taxa Assessed	无 / None

评估信息 Assessment Information

评估年份 Year Assessed	2020
评定人 Assessor(s)	蒋志刚 / Zhigang Jiang
审定人 Reviewer(s)	蒋学龙 / Xuelong Jiang
其他贡献人 Other Contributor(s)	李立立、丁晨晨 / Lili Li, Chenchen Ding

理由 Justification: 蹼足鼩分布较广，适于水生生活，且种群数量大。因此，列为无危等级 / Elegant Water Shrew is a widespread species with large populations, and is adapted for aquatic living. Thus, it is listed as Least Concern

地理分布 Geographical Distribution

国内分布 Domestic Distribution
四川、云南、西藏、陕西、甘肃、青海 / Sichuan, Yunnan, Tibet (Xizang), Shaanxi, Gansu, Qinghai
世界分布 World Distribution
中国、印度、缅甸、尼泊尔 / China, India, Myanmar, Nepal
分布标注 Distribution Note
非特有种 / Non-endemic

国内分布图
Map of Domestic Distribution

种群 Population

种群数量 Population Size	未知 / Unknown
种群趋势 Population Trend	未知 / Unknown

生境与生态系统 Habitat (s) and Ecosystem (s)

生　　境 Habitat(s)	溪流边 / Near Stream
生态系统 Ecosystem(s)	森林生态系统、湖泊河流生态系统 / Forest Ecosystem, Lake and River Ecosystem

威胁 Threat (s)

主要威胁 Major Threat(s)	未知 / Unknown

保护级别与保护行动 Protection Category and Conservation Action (s)

国家重点保护野生动物等级 (2021) Category of National Key Protected Wild Animals (2021)	无 / NA
IUCN 红色名录(2020-2) IUCN Red List (2020-2)	无危 / LC
CITES 附录 (2019) CITES Appendix (2019)	无 / NA
保护行动 Conservation Action(s)	位于自然保护区内的种群与生境得到保护 / Populations and habitats in nature reserves are protected

相关文献 Relevant References

Burgin *et al*., 2018; Wilson and Mittermeier, 2018; Jiang *et al*. (蒋志刚等), 2017; Deng *et al*. (邓可等), 2013; Jiang *et al*. (姜雪松等), 2013; Gong *et al*. (龚正达等), 2001a, 2001b; Chen and Zhang (陈耕和张林源), 1998; Chu (褚新洛), 1989; Feng *et al*. (冯祚建等), 1986

蹼足鼩 *Nectogale elegans*

中国生物多样性红色名录

臭鼩
Suncus murinus

无危 LC

| 数据缺乏 DD | 无危 LC | 近危 NT | 易危 VU | 濒危 EN | 极危 CR | 区域灭绝 RE | 野外灭绝 EW | 灭绝 EX |

分类地位 Taxonomic Status

| 动物界 Animalia | 脊索动物门 Chordata | 哺乳纲 Mammalia | 劳亚食虫目 Eulipotyphla | 鼩鼱科 Soricidae |

学 名 Scientific Name	*Suncus murinus*
命 名 人 Species Authority	Linnaeus, 1766
英 文 名 English Name(s)	House Shrew
同物异名 Synonym(s)	Asian House Shrew, Gray Musk Shrew, Asian Musk Shrew, Indian Musk Shrew
种下单元评估 Infra-specific Taxa Assessed	无 / None

评估信息 Assessment Information

评估年份 Year Assessed	2020
评定人 Assessor(s)	蒋志刚 / Zhigang Jiang
审定人 Reviewer(s)	蒋学龙、冯祚建 / Xuelong Jiang, Zuojian Feng
其他贡献人 Other Contributor(s)	李立立、丁晨晨 / Lili Li, Chenchen Ding

理由 Justification: 臭鼩分布较广，种群数量大。因此，列为无危等级 / House Shrew is a widespread species with large populations. Thus, it is listed as Least Concern

地理分布 Geographical Distribution

国内分布 Domestic Distribution

广西、福建、浙江、西藏、江西、湖南、广东、海南、四川、贵州、云南、甘肃、台湾、香港、澳门 / Guangxi, Fujian, Zhejiang, Tibet (Xizang), Jiangxi, Hunan, Guangdong, Hainan, Sichuan, Guizhou, Yunnan, Gansu, Taiwan, Hong Kong, Macao

世界分布 World Distribution

中国；中亚、南亚、东南亚 / China; Central Asia, South Asia, Southeast Asia

分布标注 Distribution Note

非特有种 / Non-endemic

国内分布图
Map of Domestic Distribution

种群 Population

种群数量 Population Size	未知 / Unknown
种群趋势 Population Trend	稳定 / Stable

生境与生态系统 Habitat (s) and Ecosystem (s)

生　　境 Habitat(s)	耕地、池塘、灌丛、森林、沼泽 / Arable Land, Pond, Shrubland, Forest, Swamp
生态系统 Ecosystem(s)	森林生态系统、灌丛生态系统、农田生态系统 / Forest Ecosystem, Shrubland Ecosystem, Cropland Ecosystem

威胁 Threat (s)

主要威胁 Major Threat(s)	未知 / Unknown

保护级别与保护行动 Protection Category and Conservation Action (s)

国家重点保护野生动物等级 (2021) Category of National Key Protected Wild Animals (2021)	无 / NA
IUCN 红色名录 (2020-2) IUCN Red List (2020-2)	无危 / LC
CITES 附录 (2019) CITES Appendix (2019)	无 / NA
保护行动 Conservation Action(s)	无 / None

相关文献 Relevant References

Burgin *et al.*, 2018; Wilson and Mittermeier, 2018; Jiang *et al.* (蒋志刚等), 2017; Chen and Zhuge (陈水华和诸葛阳), 1993; Xin and Qiu (辛景禧和邱梦辞), 1990; Yang and Zhuge (杨士剑和诸葛阳), 1989

臭鼩 *Suncus murinus*

灰麝鼩
Crocidura attenuata

无危 LC

| 数据缺乏 DD | 无危 LC | 近危 NT | 易危 VU | 濒危 EN | 极危 CR | 区域灭绝 RE | 野外灭绝 EW | 灭绝 EX |

分类地位 Taxonomic Status

动物界 Animalia	脊索动物门 Chordata	哺乳纲 Mammalia	劳亚食虫目 Eulipotyphla	鼩鼱科 Soricidae
学名 Scientific Name		*Crocidura attenuata*		
命名人 Species Authority		Milne-Edwards, 1872		
英文名 English Name(s)		Grey Shrew		
同物异名 Synonym(s)		Asian Gray Shrew		
种下单元评估 Infra-specific Taxa Assessed		无 / None		

评估信息 Assessment Information

评估年份 Year Assessed	2020
评定人 Assessor(s)	蒋志刚 / Zhigang Jiang
审定人 Reviewer(s)	蒋学龙、冯祚建 / Xuelong Jiang, Zuojian Feng
其他贡献人 Other Contributor(s)	李立立、丁晨晨 / Lili Li, Chenchen Ding

理由 Justification: 灰麝鼩分布较广，种群数量大。因此，列为无危等级 / Grey Shrew is a widespread species with large populations. Thus, it is listed as Least Concern

地理分布 Geographical Distribution

国内分布 Domestic Distribution

广西、云南、湖南、陕西、江苏、浙江、安徽、福建、台湾、江西、湖北、广东、海南、四川、贵州、西藏、甘肃、香港、重庆 / Guangxi, Yunnan, Hunan, Shaanxi, Jiangsu, Zhejiang, Anhui, Fujian, Taiwan, Jiangxi, Hubei, Guangdong, Hainan, Sichuan, Guizhou, Tibet (Xizang), Gansu, Hong Kong, Chongqing

世界分布 World Distribution

中国；南亚、东南亚 / China; South Asia, Southeast Asia

分布标注 Distribution Note

非特有种 / Non-endemic

国内分布图
Map of Domestic Distribution

种群 Population

种群数量 Population Size	未知 / Unknown
种群趋势 Population Trend	未知 / Unknown

生境与生态系统 Habitat (s) and Ecosystem (s)

生　　境 Habitat(s)	热带和亚热带湿润低地山地森林、草地、灌丛 / Tropical and Subtropical Moist Lowland Montane Forest, Grassland, Shrubland
生态系统 Ecosystem(s)	森林生态系统、灌丛生态系统、草地生态系统 / Forest Ecosystem, Shrubland Ecosystem, Grassland Ecosystem

威胁 Threat (s)

主要威胁 Major Threat(s)	未知 / Unknown

保护级别与保护行动 Protection Category and Conservation Action (s)

国家重点保护野生动物等级 (2021) Category of National Key Protected Wild Animals (2021)	无 / NA
IUCN 红色名录 (2020-2) IUCN Red List (2020-2)	无危 / LC
CITES 附录 (2019) CITES Appendix (2019)	无 / NA
保护行动 Conservation Action(s)	无 / None

相关文献 Relevant References

Burgin *et al*., 2018; Wilson and Mittermeier, 2018; Jiang *et al*. (蒋志刚等), 2017; Xie *et al*. (谢文华等), 2014; Liu *et al*., 2008; Wang *et al*. (王湉等), 2006

灰麝鼩 *Crocidura attenuata*

白尾梢麝鼩 Crocidura fuliginosa

无危 LC

| 数据缺乏 DD | 无危 LC | 近危 NT | 易危 VU | 濒危 EN | 极危 CR | 区域灭绝 RE | 野外灭绝 EW | 灭绝 EX |

分类地位 Taxonomic Status

动物界 Animalia	脊索动物门 Chordata	哺乳纲 Mammalia	劳亚食虫目 Eulipotyphla	鼩鼱科 Soricidae

学名 Scientific Name	*Crocidura fuliginosa*
命名人 Species Authority	Blyth, 1855
英文名 English Name(s)	Southeast Asian Shrew
同物异名 Synonym(s)	*Sorex fuliginosus* (Blyth, 1855)
种下单元评估 Infra-specific Taxa Assessed	无 / None

评估信息 Assessment Information

评估年份 Year Assessed	2020
评定人 Assessor(s)	蒋志刚 / Zhigang Jiang
审定人 Reviewer(s)	蒋学龙、冯祚建 / Xuelong Jiang, Zuojian Feng
其他贡献人 Other Contributor(s)	李立立、丁晨晨 / Lili Li, Chenchen Ding

理由 Justification: 白尾梢麝鼩分布较广，种群数量大。因此，列为无危等级 / Southeast Asian Shrew is a widespread species with large populations. Thus, it is listed as Least Concern

地理分布 Geographical Distribution

国内分布 Domestic Distribution

四川、湖北、云南、广西、贵州、西藏、陕西、重庆 / Sichuan, Hubei, Yunnan, Guangxi, Guizhou, Tibet (Xizang), Shaanxi, Chongqing

世界分布 World Distribution

中国；南亚、东南亚 / China; South Asia, Southeast Asia

分布标注 Distribution Note

非特有种 / Non-endemic

国内分布图
Map of Domestic Distribution

种群 Population

种群数量 Population Size	未知 / Unknown
种群趋势 Population Trend	未知 / Unknown

生境与生态系统 Habitat (s) and Ecosystem (s)

生　　境 Habitat(s)	热带湿润低地森林 / Tropical Moist Lowland Forest
生态系统 Ecosystem(s)	森林生态系统 / Forest Ecosystem

威胁 Threat (s)

主要威胁 Major Threat(s)	无 / None

保护级别与保护行动 Protection Category and Conservation Action (s)

国家重点保护野生动物等级 (2021) Category of National Key Protected Wild Animals (2021)	无 / NA
IUCN 红色名录 (2020-2) IUCN Red List (2020-2)	无危 / LC
CITES 附录 (2019) CITES Appendix (2019)	无 / NA
保护行动 Conservation Action(s)	无 / None

相关文献 Relevant References

Burgin *et al*., 2018; Wilson and Mittermeier, 2018; Jiang *et al*. (蒋志刚等), 2017; Cui *et al*. (崔茂欢等), 2014; Zhang *et al*. (张斌等), 2014; Smith *et al*. (史密斯等), 2009; Hutterer, 2005; Wang (王应祥), 2003

白尾梢麝鼩 *Crocidura fuliginosa*

白齿麝鼩

Crocidura leucodon

无危 LC

| 数据缺乏 DD | 无危 LC | 近危 NT | 易危 VU | 濒危 EN | 极危 CR | 区域灭绝 RE | 野外灭绝 EW | 灭绝 EX |

分类地位 Taxonomic Status

动物界 Animalia	脊索动物门 Chordata	哺乳纲 Mammalia	劳亚食虫目 Eulipotyphla	鼩鼱科 Soricidae

学名 Scientific Name	*Crocidura leucodon*
命名人 Species Authority	Hermann, 1780
英文名 English Name(s)	Bicolored Shrew
同物异名 Synonym(s)	Bicoloured White-toothed Shrew
种下单元评估 Infra-specific Taxa Assessed	无 / None

评估信息 Assessment Information

评估年份 Year Assessed	2020
评定人 Assessor(s)	蒋志刚 / Zhigang Jiang
审定人 Reviewer(s)	蒋学龙、冯祚建 / Xuelong Jiang, Zuojian Feng
其他贡献人 Other Contributor(s)	李立立、丁晨晨 / Lili Li, Chenchen Ding

理由 Justification: 白齿麝鼩种群数量大。因此，列为无危等级 / Bicolored Shrew has large populations. Thus, it is listed as Least Concern

地理分布 Geographical Distribution

国内分布 Domestic Distribution
内蒙古 / Inner Mongolia (Nei Mongol)
世界分布 World Distribution
亚洲、欧洲 / Asia, Europe
分布标注 Distribution Note
非特有种 / Non-endemic

国内分布图
Map of Domestic Distribution

种群 Population

种群数量 Population Size	未知 / Unknown
种群趋势 Population Trend	未知 / Unknown

生境与生态系统 Habitat(s) and Ecosystem(s)

生　　境 Habitat(s)	草地、沼泽 / Grassland, Swamp
生态系统 Ecosystem(s)	草地生态系统、湿地生态系统 / Grassland Ecosystem, Wetland Ecosystem

威胁 Threat(s)

主要威胁 Major Threat(s)	无 / None

保护级别与保护行动 Protection Category and Conservation Action(s)

国家重点保护野生动物等级 (2021) Category of National Key Protected Wild Animals (2021)	无 / NA
IUCN红色名录 (2020-2) IUCN Red List (2020-2)	无危 / LC
CITES 附录 (2019) CITES Appendix (2019)	无 / NA
保护行动 Conservation Action(s)	无 / None

相关文献 Relevant References

Burgin *et al*., 2018; Wilson and Mittermeier, 2018; Jiang *et al*. (蒋志刚等), 2017; Zheng and Li (郑生武和李保国), 1999; Zhang (张荣祖), 1997

白齿麝鼩 *Crocidura leucodon*

山东小麝鼩 Crocidura shantungensis

无危 LC

分类地位 Taxonomic Status

动物界 Animalia	脊索动物门 Chordata	哺乳纲 Mammalia	劳亚食虫目 Eulipotyphla	鼩鼱科 Soricidae
学 名 Scientific Name		Crocidura shantungensis		
命 名 人 Species Authority		Miller, 1901		
英 文 名 English Name(s)		Asian Lesser White-toothed Shrew		
同物异名 Synonym(s)		无 / None		
种下单元评估 Infra-specific Taxa Assessed		无 / None		

评估信息 Assessment Information

评估年份 Year Assessed	2020
评 定 人 Assessor(s)	蒋志刚 / Zhigang Jiang
审 定 人 Reviewer(s)	蒋学龙、冯祚建 / Xuelong Jiang, Zuojian Feng
其他贡献人 Other Contributor(s)	李立立、丁晨晨 / Lili Li, Chenchen Ding

理由 Justification: 山东小麝鼩分布广，种群数量大。因此，列为无危等级 / Asian Lesser White-toothed Shrew is a widespread species with large populations. Thus, it is listed as Least Concern

地理分布 Geographical Distribution

国内分布 Domestic Distribution

台湾、黑龙江、辽宁、吉林、内蒙古、四川、重庆、贵州、湖北、陕西、甘肃、青海、河北、北京、天津、山东、山西、安徽、浙江、江苏、宁夏 / Taiwan, Heilongjiang, Liaoning, Jilin, Inner Mongolia (Nei Mongol), Sichuan, Chongqing, Guizhou, Hubei, Shaanxi, Gansu, Qinghai, Hebei, Beijing, Tianjin, Shandong, Shanxi, Anhui, Zhejiang, Jiangsu, Ningxia

世界分布 World Distribution

中国；东北亚 / China; Northeast Asia

分布标注 Distribution Note

非特有种 / Non-endemic

国内分布图
Map of Domestic Distribution

种群 Population

种群数量 Population Size	未知 / Unknown
种群趋势 Population Trend	稳定 / Stable

生境与生态系统 Habitat(s) and Ecosystem(s)

生境 Habitat(s)	草地、森林 / Grassland, Forest
生态系统 Ecosystem(s)	森林生态系统、草地生态系统 / Forest Ecosystem, Grassland Ecosystem

威胁 Threat(s)

主要威胁 Major Threat(s)	未知 / Unknown

保护级别与保护行动 Protection Category and Conservation Action(s)

国家重点保护野生动物等级 (2021) Category of National Key Protected Wild Animals (2021)	无 / NA
IUCN 红色名录 (2020-2) IUCN Red List (2020-2)	无危 / LC
CITES 附录 (2019) CITES Appendix (2019)	无 / NA
保护行动 Conservation Action(s)	无 / None

相关文献 Relevant References

Burgin *et al.*, 2018; Wilson and Mittermeier, 2018; Jiang *et al.* (蒋志刚等), 2017; Sun *et al.* (孙治宇等), 2013; Qin *et al.* (秦岭等), 2007

山东小麝鼩 *Crocidura shantungensis*

台湾长尾麝鼩 Crocidura tanakae

无危 LC

| 数据缺乏 DD | 无危 LC | 近危 NT | 易危 VU | 濒危 EN | 极危 CR | 区域灭绝 RE | 野外灭绝 EW | 灭绝 EX |

分类地位 Taxonomic Status

动物界 Animalia	脊索动物门 Chordata	哺乳纲 Mammalia	劳亚食虫目 Eulipotyphla	鼩鼱科 Soricidae

学名 Scientific Name	*Crocidura tanakae*
命名人 Species Authority	Kuroda, 1938
英文名 English Name(s)	Taiwanese Gray Shrew
同物异名 Synonym(s)	无 / None
种下单元评估 Infra-specific Taxa Assessed	无 / None

评估信息 Assessment Information

评估年份 Year Assessed	2020
评定人 Assessor(s)	蒋志刚 / Zhigang Jiang
审定人 Reviewer(s)	蒋学龙、冯祚建 / Xuelong Jiang, Zuojian Feng
其他贡献人 Other Contributor(s)	李立立、丁晨晨 / Lili Li, Chenchen Ding

理由 Justification: 台湾长尾麝鼩为中国特有种，种群数量大。因此，列为无危等级 / Taiwanese Gray Shrew is an endemic species in China. It has large populations. Thus, it is listed as Least Concern

地理分布 Geographical Distribution

国内分布 Domestic Distribution
台湾 / Taiwan
世界分布 World Distribution
中国 / China
分布标注 Distribution Note
特有种 / Endemic

国内分布图
Map of Domestic Distribution

种群 Population

种群数量 Population Size	未知 / Unknown
种群趋势 Population Trend	未知 / Unknown

生境与生态系统 Habitat (s) and Ecosystem (s)

生　　境 Habitat(s)	草地、次生林、牧场 / Grassland, Secondary Forest, Pastureland
生态系统 Ecosystem(s)	森林生态系统、草地生态系统 / Forest Ecosystem, Grassland Ecosystem

威胁 Threat (s)

主要威胁 Major Threat(s)	无 / None

保护级别与保护行动 Protection Category and Conservation Action (s)

国家重点保护野生动物等级 (2021) Category of National Key Protected Wild Animals (2021)	无 / NA
IUCN 红色名录 (2020-2) IUCN Red List (2020-2)	无危 / LC
CITES 附录 (2019) CITES Appendix (2019)	无 / NA
保护行动 Conservation Action(s)	无 / None

相关文献 Relevant References

Burgin *et al.*, 2018; Wilson and Mittermeier, 2018; Jiang *et al.* (蒋志刚等), 2017; Fang and Lee, 2002; Motokawa *et al.*, 2001

台湾长尾麝鼩 *Crocidura tanakae*

北树鼩 *Tupaia belangeri*

无危 LC

| 数据缺乏 DD | 无危 LC | 近危 NT | 易危 VU | 濒危 EN | 极危 CR | 区域灭绝 RE | 野外灭绝 EW | 灭绝 EX |

分类地位 Taxonomic Status

动物界 Animalia	脊索动物门 Chordata	哺乳纲 Mammalia	攀鼩目 Scandentia	树鼩科 Tupaiidae
学名 Scientific Name		*Tupaia belangeri*		
命名人 Species Authority		Wagner, 1841		
英文名 English Name(s)		Northern Tree Shrew		
同物异名 Synonym(s)		无 / None		
种下单元评估 Infra-specific Taxa Assessed		无 / None		

评估信息 Assessment Information

评估年份 Year Assessed	2020
评定人 Assessor(s)	蒋志刚 / Zhigang Jiang
审定人 Reviewer(s)	蒋学龙 / Xuelong Jiang
其他贡献人 Other Contributor(s)	李立立、丁晨晨 / Lili Li, Chenchen Ding

理由 Justification: 北树鼩分布较广，种群数量大。因此，列为无危等级 / Northern Tree Shrew is a widespread species with large populations. Thus, it is listed as Least Concern

地理分布 Geographical Distribution

国内分布 Domestic Distribution
云南、四川、西藏、海南、广东、广西、贵州 / Yunnan, Sichuan, Tibet (Xizang), Hainan, Guangdong, Guangxi, Guizhou
世界分布 World Distribution
中国；东南亚 / China; Southeast Asia
分布标注 Distribution Note
非特有种 / Non-endemic

国内分布图
Map of Domestic Distribution

种群 Population

种群数量 Population Size	未知 / Unknown
种群趋势 Population Trend	未知 / Unknown

生境与生态系统 Habitat (s) and Ecosystem (s)

生　　境 Habitat(s)	热带和亚热带森林、喀斯特地区 / Tropical and Subtropical Forest, Karst Landscape
生态系统 Ecosystem(s)	森林生态系统 / Forest Ecosystem

威胁 Threat (s)

主要威胁 Major Threat(s)	未知 / Unknown

保护级别与保护行动 Protection Category and Conservation Action (s)

国家重点保护野生动物等级 (2021) Category of National Key Protected Wild Animals (2021)	无 / NA
IUCN 红色名录 (2020-2) IUCN Red List (2020-2)	无危 / LC
CITES 附录 (2019) CITES Appendix (2019)	无 / NA
保护行动 Conservation Action(s)	无 / None

相关文献 Relevant References

Burgin *et al*., 2018; Wilson and Mittermeier, 2018; Jiang *et al*. (蒋志刚等), 2017; Liu and Yao, 2013; Xu *et al*. (许凌等), 2013; Helgen, 2005

北树鼩 *Tupaia belangeri*

黑髯墓蝠 *Taphozous melanopogon*

无危 LC

| 数据缺乏 DD | 无危 LC | 近危 NT | 易危 VU | 濒危 EN | 极危 CR | 区域灭绝 RE | 野外灭绝 EW | 灭绝 EX |

分类地位 Taxonomic Status

| 动物界 Animalia | 脊索动物门 Chordata | 哺乳纲 Mammalia | 翼手目 Chiroptera | 鞘尾蝠科 Emballonuridae |

学 名 Scientific Name	*Taphozous melanopogon*
命名人 Species Authority	Temminck, 1841
英文名 English Name(s)	Black-bearded Tomb Bat
同物异名 Synonym(s)	*Taphozous bicolor* (Temminck, 1841); *phillipenensis* (Waterhouse, 1845); *solifer* (Hollister, 1913)
种下单元评估 Infra-specific Taxa Assessed	无 / None

评估信息 Assessment Information

评估年份 Year Assessed	2020
评定人 Assessor(s)	吴毅、蒋志刚 / Yi Wu, Zhigang Jiang
审定人 Reviewer(s)	石红艳、江廷磊、余文华 / Hongyan Shi, Tinglei Jiang, Wenhua Yu
其他贡献人 Other Contributor(s)	李立立、丁晨晨 / Lili Li, Chenchen Ding

理由 Justification: 黑髯墓蝠分布较广，种群数量大。因此，列为无危等级 / Black-bearded Tomb Bat is a widely distributed species with large populations. Thus, it is listed as Least Concern

地理分布 Geographical Distribution

国内分布 Domestic Distribution	广西、海南、贵州、云南、香港、澳门、北京、广东 / Guangxi, Hainan, Guizhou, Yunnan, Hong Kong, Macao, Beijing, Guangdong
世界分布 World Distribution	中国；南亚、东南亚 / China; South Asia, Southeast Asia
分布标注 Distribution Note	非特有种 / Non-endemic

国内分布图
Map of Domestic Distribution

种群 Population

种群数量 Population Size	常见 / Common
种群趋势 Population Trend	未知 / Unknown

生境与生态系统 Habitat(s) and Ecosystem(s)

生境 Habitat(s)	热带和亚热带湿润低地森林，洞穴 / Tropical and Subtropical Moist Lowland Forest, Cave
生态系统 Ecosystem(s)	森林生态系统、喀斯特生态系统 / Forest Ecosystem, Karst Ecosystem

威胁 Threat(s)

主要威胁 Major Threat(s)	伐木、耕作、狩猎及采集 / Logging, Farming, Hunting and Collection

保护级别与保护行动 Protection Category and Conservation Action(s)

国家重点保护野生动物等级 (2021) Category of National Key Protected Wild Animals (2021)	无 / NA
IUCN 红色名录 (2020-2) IUCN Red List (2020-2)	无危 / LC
CITES 附录 (2019) CITES Appendix (2019)	无 / NA
保护行动 Conservation Action(s)	无 / None

相关文献 Relevant References

Wilson and Mittermeier, 2019; Burgin *et al.*, 2018; Jiang *et al.* (蒋志刚等), 2017; Zhang *et al.* (张礼标等), 2014; Huang *et al.* (黄继展等), 2013; Wei (韦力), 2007; Luo *et al.* (罗蓉等), 1993; Chu (褚新洛), 1989; Animal Laboratory of Guangdong Institute of Entomology; Department of Biology, Sun Yat-Sen University (广东省昆虫研究所动物室和中山大学生物系), 1983; Gao *et al.* (高耀亭等), 1962

黑髯墓蝠 *Taphozous melanopogon* 　　吴毅 摄　By Yi Wu

中菊头蝠
Rhinolophus affinis

无危 LC

分类地位 Taxonomic Status

动物界 Animalia	脊索动物门 Chordata	哺乳纲 Mammalia	翼手目 Chiroptera	菊头蝠科 Rhinolophidae

学名 Scientific Name	*Rhinolophus affinis*
命名人 Species Authority	Horsfield, 1823
英文名 English Name(s)	Intermediate Horseshoe Bat
同物异名 Synonym(s)	*Rhinolophus andamanensis* (Dobson, 1872)
种下单元评估 Infra-specific Taxa Assessed	无 / None

评估信息 Assessment Information

评估年份 Year Assessed	2020
评定人 Assessor(s)	吴毅、蒋志刚 / Yi Wu, Zhigang Jiang
审定人 Reviewer(s)	张礼标、毛秀光、张树义 / Libiao Zhang, Xiuguang Mao, Shuyi Zhang
其他贡献人 Other Contributor(s)	李立立、丁晨晨 / Lili Li, Chenchen Ding

理由 Justification: 中菊头蝠分布较广，种群数量大。因此，列为无危等级 / Intermediate Horseshoe Bat is a widespread species with large populations. Thus, it is listed as Least Concern

地理分布 Geographical Distribution

国内分布 Domestic Distribution	陕西、江苏、浙江、安徽、湖北、湖南、江西、四川、贵州、云南、广西、海南、福建、香港、重庆、广东 / Shaanxi, Jiangsu, Zhejiang, Anhui, Hubei, Hunan, Jiangxi, Sichuan, Guizhou, Yunnan, Guangxi, Hainan, Fujian, Hong Kong, Chongqing, Guangdong
世界分布 World Distribution	中国；南亚、东南亚 / China; South Asia, Southeast Asia
分布标注 Distribution Note	非特有种 / Non-endemic

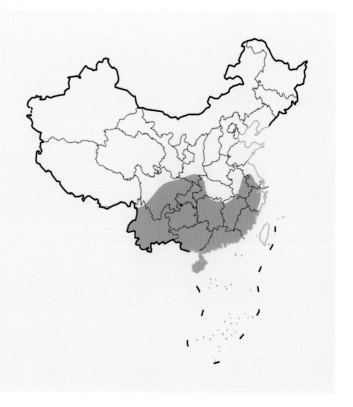

国内分布图
Map of Domestic Distribution

种群 Population

种群数量 Population Size	常见 / Common
种群趋势 Population Trend	稳定 / Stable

生境与生态系统 Habitat (s) and Ecosystem (s)

生　　境 Habitat(s)	洞穴、种植园、耕地、森林 / Cave, Plantation, Arable Land, Forest
生态系统 Ecosystem(s)	森林生态系统、农田生态系统、喀斯特生态系统 / Forest Ecosystem, Cropland Ecosystem, Karst Ecosystem

威胁 Threat (s)

主要威胁 Major Threat(s)	伐木、耕作、火灾、采矿、采石场、旅游 / Logging, Farming, Fire, Mining, Quarrying Field and Tourism

保护级别与保护行动 Protection Category and Conservation Action (s)

国家重点保护野生动物等级 (2021) Category of National Key Protected Wild Animals (2021)	无 / NA
IUCN 红色名录 (2020-2) IUCN Red List (2020-2)	无危 / LC
CITES 附录 (2019) CITES Appendix (2019)	无 / NA
保护行动 Conservation Action(s)	位于自然保护区内的种群与生境得到保护 / Populations and habitats in nature reserves are protected

相关文献 Relevant References

Wilson and Mittermeier, 2019; Burgin *et al*., 2018; Jiang *et al*. (蒋志刚等), 2017; Mao *et al*., 2010b; Zhang *et al*., 2009a; Wu *et al*., 2009a, 2009b; Niu *et al*., 2007; Liu *et al*. (刘延德等), 2006; Zhou *et al*. (周昭敏等), 2005b; Gu *et al*. (谷晓明等), 2003a, 2003b

中菊头蝠 *Rhinolophus affinis*　　　　　　　　　　　　　　　　　　　　余文华 摄　By Wenhua Yu

马铁菊头蝠
Rhinolophus ferrumequinum
无危 LC

| 数据缺乏 DD | 无危 LC | 近危 NT | 易危 VU | 濒危 EN | 极危 CR | 区域灭绝 RE | 野外灭绝 EW | 灭绝 EX |

分类地位 Taxonomic Status

动物界 Animalia	脊索动物门 Chordata	哺乳纲 Mammalia	翼手目 Chiroptera	菊头蝠科 Rhinolophidae
学 名 Scientific Name		*Rhinolophus ferrumequinum*		
命 名 人 Species Authority		Schreber, 1774		
英 文 名 English Name(s)		Greater Horseshoe Bat		
同物异名 Synonym(s)		无 / None		
种下单元评估 Infra-specific Taxa Assessed		无 / None		

评估信息 Assessment Information

评估年份 Year Assessed	2020
评 定 人 Assessor(s)	吴毅、蒋志刚 / Yi Wu, Zhigang Jiang
审 定 人 Reviewer(s)	张礼标、毛秀光、张树义 / Libiao Zhang, Xiuguang Mao, Shuyi Zhang
其他贡献人 Other Contributor(s)	李立立、丁晨晨 / Lili Li, Chenchen Ding

理由 Justification: 马铁菊头蝠分布较广，种群数量大。因此，列为无危等级 / Greater Horseshoe Bat is a widespread species with large populations. Thus, it is listed as Least Concern

地理分布 Geographical Distribution

国内分布 Domestic Distribution

吉林、辽宁、河北、北京、山东、河南、陕西、山西、上海、江苏、浙江、安徽、江西、湖北、湖南、四川、甘肃、贵州、云南、广西、福建、重庆 / Jilin, Liaoning, Hebei, Beijing, Shandong, Henan, Shaanxi, Shanxi, Shanghai, Jiangsu, Zhejiang, Anhui, Jiangxi, Hubei, Hunan, Sichuan, Gansu, Guizhou, Yunnan, Guangxi, Fujian, Chongqing

世界分布 World Distribution

亚洲、欧洲 / Asia, Europe

分布标注 Distribution Note

非特有种 / Non-endemic

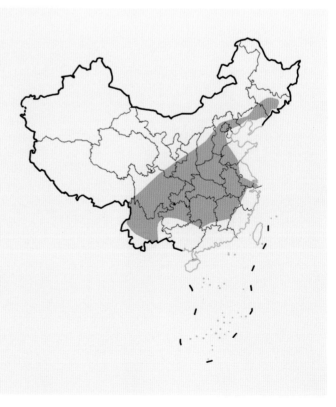

国内分布图
Map of Domestic Distribution

种群 Population

种群数量 Population Size	较常见 / Relatively common
种群趋势 Population Trend	未知 / Unknown

生境与生态系统 Habitat (s) and Ecosystem (s)

生境 Habitat(s)	牧场、地中海灌木植被、温带森林、洞穴、人造建筑 / Pastureland, Mediterranean Shrub Vegetation, Temperate Forest, Cave, Building
生态系统 Ecosystem(s)	森林生态系统、草地生态系统、喀斯特生态系统 / Forest Ecosystem, Grassland Ecosystem, Karst Ecosystem

威胁 Threat (s)

主要威胁 Major Threat(s)	伐木、耕作、火灾、采矿、采石场、旅游 / Logging, Farming, Fire, Mining, Quarrying Field and Tourism

保护级别与保护行动 Protection Category and Conservation Action (s)

国家重点保护野生动物等级 (2021) Category of National Key Protected Wild Animals (2021)	无 / NA
IUCN红色名录(2020-2) IUCN Red List (2020-2)	无危 / LC
CITES 附录 (2019) CITES Appendix (2019)	无 / NA
保护行动 Conservation Action(s)	无 / None

相关文献 Relevant References

Wilson and Mittermeier, 2019; Burgin *et al.*, 2018; Jiang *et al.*（蒋志刚等）, 2017; He（何新焕）, 2011; Luo（罗丽）, 2011; Wang *et al.*（王静等）, 2010; Wang（王静）, 2009; Wang（王新华）, 2008; Gao（高晶）, 2006; Liu *et al.*（刘延德）, 2006

马铁菊头蝠 *Rhinolophus ferrumequinum* 　　　　　　　　　　　　　　吴毅 摄　By Yi Wu

大耳菊头蝠
Rhinolophus macrotis

无危 LC

分类地位 Taxonomic Status

动物界 Animalia	脊索动物门 Chordata	哺乳纲 Mammalia	翼手目 Chiroptera	菊头蝠科 Rhinolophidae

学名 Scientific Name	*Rhinolophus macrotis*
命名人 Species Authority	Blyth, 1844
英文名 English Name(s)	Big-eared Horseshoe Bat
同物异名 Synonym(s)	无 / None
种下单元评估 Infra-specific Taxa Assessed	无 / None

评估信息 Assessment Information

评估年份 Year Assessed	2020
评定人 Assessor(s)	吴毅、蒋志刚 / Yi Wu, Zhigang Jiang
审定人 Reviewer(s)	张礼标、毛秀光、张树义 / Libiao Zhang, Xiuguang Mao, Shuyi Zhang
其他贡献人 Other Contributor(s)	李立立、丁晨晨 / Lili Li, Chenchen Ding

理由 Justification: 大耳菊头蝠分布较广，种群数量大。因此，列为无危等级 / Big-eared Horseshoe Bat is a widespread species with large populations. Thus, it is listed as Least Concern

地理分布 Geographical Distribution

国内分布 Domestic Distribution
陕西、浙江、江西、四川、贵州、广西、云南、福建、重庆 / Shaanxi, Zhejiang, Jiangxi, Sichuan, Guizhou, Guangxi, Yunnan, Fujian, Chongqing
世界分布 World Distribution
中国；南亚、东南亚 / China; South Asia, Southeast Asia
分布标注 Distribution Note
非特有种 / Non-endemic

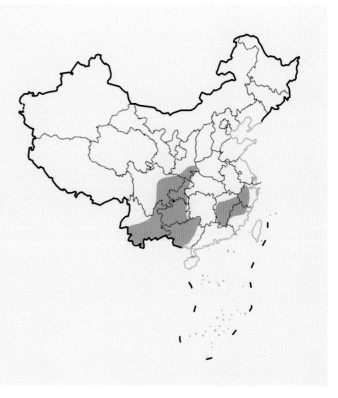

国内分布图
Map of Domestic Distribution

种群 Population

种群数量 Population Size	较常见 / Relatively common
种群趋势 Population Trend	稳定 / Stable

生境与生态系统 Habitat (s) and Ecosystem (s)

生境 Habitat(s)	洞穴 / Cave
生态系统 Ecosystem(s)	森林生态系统、喀斯特生态系统 / Forest Ecosystem, Karst Ecosystem

威胁 Threat (s)

主要威胁 Major Threat(s)	伐木、火灾、耕作 / Logging, Fire, Farming

保护级别与保护行动 Protection Category and Conservation Action (s)

国家重点保护野生动物等级 (2021) Category of National Key Protected Wild Animals (2021)	无 / NA
IUCN 红色名录 (2020-2) IUCN Red List (2020-2)	无危 / LC
CITES 附录 (2019) CITES Appendix (2019)	无 / NA
保护行动 Conservation Action(s)	无 / None

相关文献 Relevant References

Wilson and Mittermeier, 2019; Burgin *et al*., 2018; Jiang *et al*. (蒋志刚等), 2017; Wei (魏学文), 2013; Wang *et al*. (王延校等), 2012; Li *et al*. (李艳丽等), 2012; Liu *et al*., 2011; Zhang *et al*., 2009a

大耳菊头蝠 *Rhinolophus macrotis*   罗冬玮 摄（西南山地供图） By Dongwei Luo (Swild.cn)

皮氏菊头蝠
Rhinolophus pearsoni

无危 LC

| 数据缺乏 DD | 无危 LC | 近危 NT | 易危 VU | 濒危 EN | 极危 CR | 区域灭绝 RE | 野外灭绝 EW | 灭绝 EX |

分类地位 Taxonomic Status

动物界 Animalia	脊索动物门 Chordata	哺乳纲 Mammalia	翼手目 Chiroptera	菊头蝠科 Rhinolophidae
学 名 Scientific Name		*Rhinolophus pearsoni*		
命 名 人 Species Authority		Horsfield, 1851		
英 文 名 English Name(s)		Pearson's Horseshoe Bat		
同物异名 Synonym(s)		无 / None		
种下单元评估 Infra-specific Taxa Assessed		无 / None		

评估信息 Assessment Information

评估年份 Year Assessed	2020
评 定 人 Assessor(s)	吴毅、蒋志刚 / Yi Wu, Zhigang Jiang
审 定 人 Reviewer(s)	张礼标、毛秀光、张树义 / Libiao Zhang, Xiuguang Mao, Shuyi Zhang
其他贡献人 Other Contributor(s)	李立立、丁晨晨 / Lili Li, Chenchen Ding

理由 Justification: 皮氏菊头蝠分布较广，种群数量大。因此，列为无危等级 / Pearson's Horseshoe Bat is a widespread species with large populations. Thus, it is listed as Least Concern

地理分布 Geographical Distribution

国内分布 Domestic Distribution
浙江、江苏、陕西、西藏、湖南、江西、四川、贵州、云南、广西、广东、福建、重庆 / Zhejiang, Jiangsu, Shaanxi, Tibet (Xizang), Hunan, Jiangxi, Sichuan, Guizhou, Yunnan, Guangxi, Guangdong, Fujian, Chongqing
世界分布 World Distribution
中国；南亚、东南亚 / China; South Asia, Southeast Asia
分布标注 Distribution Note
非特有种 / Non-endemic

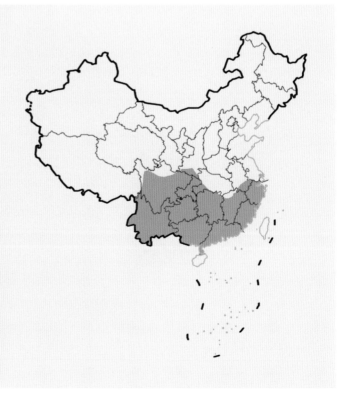

国内分布图
Map of Domestic Distribution

种群 Population

种群数量 Population Size	较常见 / Relatively common
种群趋势 Population Trend	未知 / Unknown

生境与生态系统 Habitat(s) and Ecosystem(s)

生境 Habitat(s)	洞穴、森林、种植园、竹林 / Cave, Forest, Plantation, Bamboo Grove
生态系统 Ecosystem(s)	森林生态系统、农田生态系统、喀斯特生态系统 / Forest Ecosystem, Cropland Ecosystem, Karst Ecosystem

威胁 Threat(s)

主要威胁 Major Threat(s)	旅游、采矿、采石场 / Tourism, Mining, Quarrying Field

保护级别与保护行动 Protection Category and Conservation Action(s)

国家重点保护野生动物等级 (2021) Category of National Key Protected Wild Animals (2021)	无 / NA
IUCN 红色名录 (2020-2) IUCN Red List (2020-2)	无危 / LC
CITES 附录 (2019) CITES Appendix (2019)	无 / NA
保护行动 Conservation Action(s)	无 / None

相关文献 Relevant References

Wilson and Mittermeier, 2019; Burgin *et al.*, 2018; Jiang *et al.*（蒋志刚等）, 2017; Mao（毛秀光）, 2010; Lin *et al.*（林爱青等）, 2009; Niu *et al.*（牛红星等）, 2008; Zhou *et al.*（周江等）, 2002

皮氏菊头蝠 *Rhinolophus pearsoni* 吴毅 摄 By Yi Wu

小菊头蝠
Rhinolophus pusillus

无危 LC

| 数据缺乏 DD | 无危 LC | 近危 NT | 易危 VU | 濒危 EN | 极危 CR | 区域灭绝 RE | 野外灭绝 EW | 灭绝 EX |

分类地位 Taxonomic Status

动物界 Animalia	脊索动物门 Chordata	哺乳纲 Mammalia	翼手目 Chiroptera	菊头蝠科 Rhinolophidae

学名 Scientific Name	*Rhinolophus pusillus*
命名人 Species Authority	Temminck, 1834
英文名 English Name(s)	Least Horseshoe Bat
同物异名 Synonym(s)	*Rhinolophus blythi* (Andersen, 1918); *cornutus* (Temminck, 1835); *R. cornutus* (Andersen, 1918) subsp. *blythi*; *gracilis* (Andersen, 1905); *imaizumii* (Hill and Yoshiyuki, (转下页)
种下单元评估 Infra-specific Taxa Assessed	无 / None

评估信息 Assessment Information

评估年份 Year Assessed	2020
评定人 Assessor(s)	吴毅、蒋志刚 / Yi Wu, Zhigang Jiang
审定人 Reviewer(s)	张礼标、毛秀光、张树义 / Libiao Zhang, Xiuguang Mao, Shuyi Zhang
其他贡献人 Other Contributor(s)	李立立、丁晨晨 / Lili Li, Chenchen Ding

理由 Justification: 小菊头蝠分布较广，种群数量大。因此，列为无危等级 / Least Horseshoe Bat is a widespread species with large populations. Thus, it is listed as Least Concern

地理分布 Geographical Distribution

国内分布 Domestic Distribution	江苏、湖北、湖南、浙江、安徽、江西、贵州、四川、云南、广西、海南、福建、重庆、广东、澳门 / Jiangsu, Hubei, Hunan, Zhejiang, Anhui, Jiangxi, Guizhou, Sichuan, Yunnan, Guangxi, Hainan, Fujian, Chongqing, Guangdong, Macao
世界分布 World Distribution	南亚、东南亚 / South Asia, Southeast Asia
分布标注 Distribution Note	非特有种 / Non-endemic

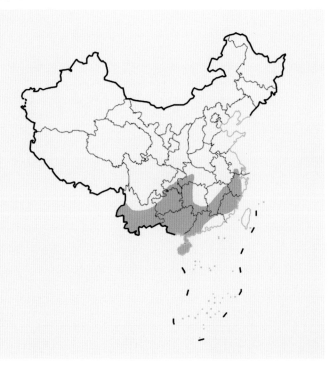

国内分布图
Map of Domestic Distribution

种群 Population

种群数量 Population Size	常见 / Common
种群趋势 Population Trend	未知 / Unknown

生境与生态系统 Habitat(s) and Ecosystem(s)

生境 Habitat(s)	洞穴、森林、人造建筑、竹林、喀斯特地貌 / Cave, Forest, Building, Bamboo Grove, Karst Landscape
生态系统 Ecosystem(s)	喀斯特生态系统、森林生态系统 / Karst Ecosystem, Forest Ecosystem

威胁 Threat(s)

主要威胁 Major Threat(s)	伐木、耕作、火灾、采矿、采石场、旅游 / Logging, Farming, Fire, Mining, Quarrying Field and Tourism

保护级别与保护行动 Protection Category and Conservation Action(s)

国家重点保护野生动物等级 (2021) Category of National Key Protected Wild Animals (2021)	无 / NA
IUCN 红色名录 (2020-2) IUCN Red List (2020-2)	无危 / LC
CITES 附录 (2019) CITES Appendix (2019)	无 / NA
保护行动 Conservation Action(s)	位于自然保护区内的种群与生境得到保护 / Populations and habitats in nature reserves are protected

相关文献 Relevant References

Burgin *et al.*, 2018; Wilson and Mittermeier, 2018; Jiang *et al.*（蒋志刚等），2017; Huang（黄继展等），2013; Wu *et al.*, 2012b; Jiang *et al.*, 2010b; Chen and Peng, 2010; Xu *et al.*（许立杰等），2008; Wu and Harada, 2005; Gu *et al.*（谷晓明等），2003a, 2003b; Zhou（周江），2001

（接上页）
1980); *minor* (Horsfield, 1823); *monoceros* (K. Andersen, 1905); *perditus* (K. Andersen, 1918); *pumilus* (K. Andersen, 1905)

小菊头蝠 *Rhinolophus pusillus* 吴毅 摄 By Yi Wu

中华菊头蝠
Rhinolophus sinicus

无危 LC

| 数据缺乏 DD | 无危 LC | 近危 NT | 易危 VU | 濒危 EN | 极危 CR | 区域灭绝 RE | 野外灭绝 EW | 灭绝 EX |

分类地位 Taxonomic Status

动物界 Animalia	脊索动物门 Chordata	哺乳纲 Mammalia	翼手目 Chiroptera	菊头蝠科 Rhinolophidae

学名 Scientific Name	*Rhinolophus sinicus*
命名人 Species Authority	K. Andersen, 1905
英文名 English Name(s)	Chinese Horseshoe Bat
同物异名 Synonym(s)	Chinese Rufous Horseshoe Bat; *Rhinolophus rouxii* (Andersen, 1905) subsp. *sinicus*
种下单元评估 Infra-specific Taxa Assessed	无 / None

评估信息 Assessment Information

评估年份 Year Assessed	2020
评定人 Assessor(s)	吴毅、蒋志刚 / Yi Wu, Zhigang Jiang
审定人 Reviewer(s)	张礼标、毛秀光、张树义 / Libiao Zhang, Xiuguang Mao, Shuyi Zhang
其他贡献人 Other Contributor(s)	李立立、丁晨晨 / Lili Li, Chenchen Ding

理由 Justification: 中华菊头蝠分布较广，种群数量大。因此，列为无危等级 / Chinese Horseshoe Bat is a widespread species with large populations. Thus, it is listed as Least Concern

地理分布 Geographical Distribution

国内分布 Domestic Distribution

福建、江苏、浙江、安徽、江西、湖北、湖南、广东、广西、四川、贵州、云南、陕西、香港、海南、重庆 / Fujian, Jiangsu, Zhejiang, Anhui, Jiangxi, Hubei, Hunan, Guangdong, Guangxi, Sichuan, Guizhou, Yunnan, Shaanxi, Hong Kong, Hainan, Chongqing

世界分布 World Distribution

中国；南亚、东南亚 / China; South Asia, Southeast Asia

分布标注 Distribution Note

非特有种 / Non-endemic

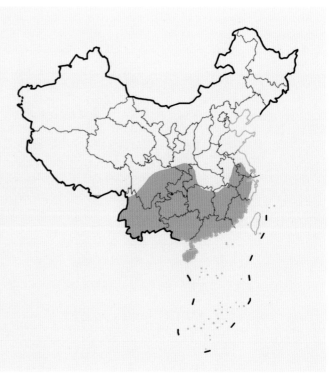

国内分布图
Map of Domestic Distribution

种群 Population

种群数量 Population Size	常见 / Common
种群趋势 Population Trend	下降 / Decreasing

生境与生态系统 Habitat(s) and Ecosystem(s)

生境 Habitat(s)	热带和亚热带湿润山地森林、洞穴，人造建筑 / Tropical and Subtropical Moist Montane Forest, Cave; Building
生态系统 Ecosystem(s)	森林生态系统、人类聚落生态系统、喀斯特生态系统 / Forest Ecosystem, Human Settlement Ecosystem, Karst Ecosystem

威胁 Threat(s)

主要威胁 Major Threat(s)	伐木、耕作、火灾 / Logging, Farming, Fire

保护级别与保护行动 Protection Category and Conservation Action(s)

国家重点保护野生动物等级 (2021) Category of National Key Protected Wild Animals (2021)	无 / NA
IUCN 红色名录(2020-2) IUCN Red List (2020-2)	无危 / LC
CITES 附录 (2019) CITES Appendix (2019)	无 / NA
保护行动 Conservation Action(s)	无 / None

相关文献 Relevant References

Wilson and Mittermeier, 2019; Burgin *et al*., 2018; Jiang *et al*. (蒋志刚等), 2017; Mao *et al*., 2013; Song *et al*., 2009; Zhang *et al*., 2009a; Wu and Harada, 2005; Xu *et al*. (徐伟霞等), 2005

中华菊头蝠 *Rhinolophus sinicus* 　　　　　　吴毅 摄　By Yi Wu

大蹄蝠
Hipposideros armiger
无危 LC

| 数据缺乏 DD | 无危 LC | 近危 NT | 易危 VU | 濒危 EN | 极危 CR | 区域灭绝 RE | 野外灭绝 EW | 灭绝 EX |

分类地位 Taxonomic Status

动物界 Animalia	脊索动物门 Chordata	哺乳纲 Mammalia	翼手目 Chiroptera	蹄蝠科 Hipposideridae
学名 Scientific Name		*Hipposideros armiger*		
命名人 Species Authority		Hodgson, 1835		
英文名 English Name(s)		Great Leaf-nosed Bat		
同物异名 Synonym(s)		Great Roundleaf Bat; Great Himalayan Leaf-nosed Bat; *Rhinolophus armiger* (Hodgson, 1835)		
种下单元评估 Infra-specific Taxa Assessed		无 / None		

评估信息 Assessment Information

评估年份 Year Assessed	2020
评定人 Assessor(s)	吴毅、蒋志刚 / Yi Wu, Zhigang Jiang
审定人 Reviewer(s)	张礼标、毛秀光、张树义 / Libiao Zhang, Xiuguang Mao, Shuyi Zhang
其他贡献人 Other Contributor(s)	李立立、丁晨晨 / Lili Li, Chenchen Ding

理由 Justification: 大蹄蝠分布较广，种群数量大。因此，列为无危等级 / Great Leaf-nosed Bat is a widespread species with large populations. Thus, it is listed as Least Concern

地理分布 Geographical Distribution

国内分布 Domestic Distribution

陕西、江苏、浙江、安徽、湖南、江西、四川、贵州、云南、广西、广东、海南、台湾、香港、重庆、澳门 / Shaanxi, Jiangsu, Zhejiang, Anhui, Hunan, Jiangxi, Sichuan, Guizhou, Yunnan, Guangxi, Guangdong, Hainan, Taiwan, Hong Kong, Chongqing, Macao

世界分布 World Distribution

中国；南亚、东南亚 / China; South Asia, Southeast Asia

分布标注 Distribution Note

非特有种 / Non-endemic

国内分布图
Map of Domestic Distribution

种群 Population

种群数量 Population Size	常见 / Common
种群趋势 Population Trend	下降 / Decreasing

生境与生态系统 Habitat(s) and Ecosystem(s)

生境 Habitat(s)	洞穴、人造建筑 / Cave, Building
生态系统 Ecosystem(s)	喀斯特生态系统、人类聚落生态系统 / Karst Ecosystem, Human Settlement Ecosystem

威胁 Threat(s)

主要威胁 Major Threat(s)	伐木、耕作、采矿、采石场 / Logging, Farming, Mining, Quarrying Field

保护级别与保护行动 Protection Category and Conservation Action(s)

国家重点保护野生动物等级 (2021) Category of National Key Protected Wild Animals (2021)	无 / NA
IUCN 红色名录 (2020-2) IUCN Red List (2020-2)	无危 / LC
CITES 附录 (2019) CITES Appendix (2019)	无 / NA
保护行动 Conservation Action(s)	无 / None

相关文献 Relevant References

Wilson and Mittermeier, 2019; Burgin *et al*., 2018; Jiang *et al*. (蒋志刚等), 2017; Huang *et al*. (黄继展等), 2013; Zheng (郑锡奇), 2010; Wu *et al*. (吴毅等), 2000

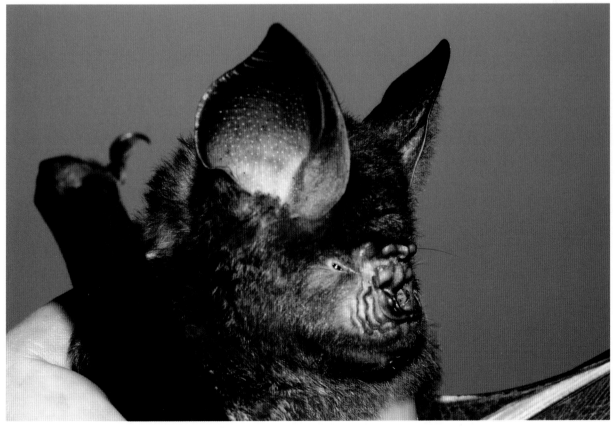

大蹄蝠 *Hipposideros armiger* 吴毅 摄 By Yi Wu

中蹄蝠
Hipposideros larvatus

无危 LC

| 数据缺乏 DD | 无危 LC | 近危 NT | 易危 VU | 濒危 EN | 极危 CR | 区域灭绝 RE | 野外灭绝 EW | 灭绝 EX |

分类地位 Taxonomic Status

动物界 Animalia	脊索动物门 Chordata	哺乳纲 Mammalia	翼手目 Chiroptera	蹄蝠科 Hipposideridae

学名 Scientific Name	*Hipposideros larvatus*
命名人 Species Authority	Horsfield, 1823
英文名 English Name(s)	Horsfield's Leaf-nosed Bat
同物异名 Synonym(s)	Intermediate Roundleaf Bat; *Phyllorhina leptophylla* (Dobson, 1874); *Rhinolophus larvatus* (Horsfield, 1823)
种下单元评估 Infra-specific Taxa Assessed	无 / None

评估信息 Assessment Information

评估年份 Year Assessed	2020
评定人 Assessor(s)	吴毅、蒋志刚 / Yi Wu, Zhigang Jiang
审定人 Reviewer(s)	张礼标、毛秀光、张树义 / Libiao Zhang, Xiuguang Mao, Shuyi Zhang
其他贡献人 Other Contributor(s)	李立立、丁晨晨 / Lili Li, Chenchen Ding

理由 Justification: 中蹄蝠分布较广，种群数量大。因此，列为无危等级 / Horsfield's Leaf-nosed Bat is a widespread species with large populations. Thus, it is listed as Least Concern

地理分布 Geographical Distribution

国内分布 Domestic Distribution
贵州、海南、广西、广东 / Guizhou, Hainan, Guangxi, Guangdong
世界分布 World Distribution
中国；南亚、东南亚 / China; South Asia, Southeast Asia
分布标注 Distribution Note
非特有种 / Non-endemic

国内分布图
Map of Domestic Distribution

种群 Population

种群数量 Population Size	常见 / Common
种群趋势 Population Trend	下降 / Decreasing

生境与生态系统 Habitat(s) and Ecosystem(s)

生境 Habitat(s)	洞穴、人造建筑 / Cave, Building
生态系统 Ecosystem(s)	喀斯特生态系统、人类聚落生态系统 / Karst Ecosystem, Human Settlement Ecosystem

威胁 Threat(s)

主要威胁 Major Threat(s)	旅游、采矿、采石场 / Tourism, Mining, Quarrying Field

保护级别与保护行动 Protection Category and Conservation Action(s)

国家重点保护野生动物等级 (2021) Category of National Key Protected Wild Animals (2021)	无 / NA
IUCN 红色名录 (2020-2) IUCN Red List (2020-2)	无危 / LC
CITES 附录 (2019) CITES Appendix (2019)	无 / NA
保护行动 Conservation Action(s)	无 / None

相关文献 Relevant References

Wilson and Mittermeier, 2019; Burgin *et al.*, 2018; Jiang *et al.* (蒋志刚等), 2017; Yuan *et al.*, 2012; Wei *et al.* (韦力等), 2011; Jiang *et al.*, 2010a; Hong *et al.* (洪体玉等), 2009

中蹄蝠 *Hipposideros larvatus* 吴毅 摄 By Yi Wu

小蹄蝠
Hipposideros pomona

无危 LC

| 数据缺乏 DD | 无危 LC | 近危 NT | 易危 VU | 濒危 EN | 极危 CR | 区域灭绝 RE | 野外灭绝 EW | 灭绝 EX |

分类地位 Taxonomic Status

动物界 Animalia	脊索动物门 Chordata	哺乳纲 Mammalia	翼手目 Chiroptera	蹄蝠科 Hipposideridae

学名 Scientific Name	*Hipposideros pomona*
命名人 Species Authority	K. Andersen, 1918
英文名 English Name(s)	Andersen's Leaf-nosed Bat
同物异名 Synonym(s)	Pomona Roundleaf Bat, Pomona Leaf-nosed Bat
种下单元评估 Infra-specific Taxa Assessed	无 / None

评估信息 Assessment Information

评估年份 Year Assessed	2020
评定人 Assessor(s)	吴毅、蒋志刚 / Yi Wu, Zhigang Jiang
审定人 Reviewer(s)	张礼标、毛秀光、张树义 / Libiao Zhang, Xiuguang Mao, Shuyi Zhang
其他贡献人 Other Contributor(s)	李立立、丁晨晨 / Lili Li, Chenchen Ding

理由 Justification: 小蹄蝠分布较广，种群数量大。因此，列为无危等级 / Andersen's Leaf-nosed Bat is a widespread species with large populations. Thus, it is listed as Least Concern

地理分布 Geographical Distribution

国内分布 Domestic Distribution
福建、湖南、广西、广东、海南、四川、贵州、云南、香港 / Fujian, Hunan, Guangxi, Guangdong, Hainan, Sichuan, Guizhou, Yunnan, Hong Kong
世界分布 World Distribution
中国；南亚、东南亚 / China; South Asia, Southeast Asia
分布标注 Distribution Note
非特有种 / Non-endemic

国内分布图
Map of Domestic Distribution

种群 Population

种群数量 Population Size	常见 / Common
种群趋势 Population Trend	下降 / Decreasing

生境与生态系统 Habitat (s) and Ecosystem (s)

生境 Habitat(s)	栖息在地下洞穴和裂缝中，能耐受高度改变的栖息地，甚至栖息在城区 / It inhabits cave and crevice underground, and it is tolerant of modified habitat and can even inhabit in urban area
生态系统 Ecosystem(s)	喀斯特生态系统、人类聚落生态系统 / Karst Ecosystem, Human Settlement Ecosystem

威胁 Threat (s)

主要威胁 Major Threat(s)	旅游、采矿、采石场 / Tourism, Mining, Quarrying Field

保护级别与保护行动 Protection Category and Conservation Action (s)

国家重点保护野生动物等级 (2021) Category of National Key Protected Wild Animals (2021)	无 / NA
IUCN 红色名录 (2020-2) IUCN Red List (2020-2)	无危 / LC
CITES 附录 (2019) CITES Appendix (2019)	无 / NA
保护行动 Conservation Action(s)	位于自然保护区内的种群与生境得到保护 / Populations and habitats in nature reserves are protected

相关文献 Relevant References

Wilson and Mittermeier, 2019; Burgin *et al*., 2018; Jiang *et al*.（蒋志刚等）, 2017; Yang *et al*., 2012; Yuan *et al*., 2012; Hong *et al*., 2011; Tan *et al*.（谭敏等）, 2009; Zhang *et al*., 2009a; Shek and Lau, 2006

小蹄蝠 *Hipposideros pomona*

By Alice Hughes

小犬吻蝠 *Chaerephon plicatus*

无危 LC

| 数据缺乏 DD | 无危 LC | 近危 NT | 易危 VU | 濒危 EN | 极危 CR | 区域灭绝 RE | 野外灭绝 EW | 灭绝 EX |

分类地位 Taxonomic Status

动物界 Animalia	脊索动物门 Chordata	哺乳纲 Mammalia	翼手目 Chiroptera	犬吻蝠科 Molossidae
学 名 Scientific Name		*Chaerephon plicatus*		
命 名 人 Species Authority		Buchanan, 1800		
英 文 名 English Name(s)		Wrinkle-lipped Bat		
同物异名 Synonym(s)		Wrinkle-lipped Free-tailed Bat; *Chaerephon luzonus* (Hill, 1961); *C. plicata* (Buchanan, 1800); *Dysopes murinus* (Gray, 1830); *Nyctinomus bengaldeshensis* (转下页)		
种下单元评估 Infra-specific Taxa Assessed		无 / None		

评估信息 Assessment Information

评估年份 Year Assessed	2020
评定人 Assessor(s)	吴毅、蒋志刚 / Yi Wu, Zhigang Jiang
审定人 Reviewer(s)	石红艳、江廷磊、余文华 / Hongyan Shi, Tinglei Jiang, Wenhua Yu
其他贡献人 Other Contributor(s)	李立立、丁晨晨 / Lili Li, Chenchen Ding

理由 Justification: 小犬吻蝠分布较广, 种群数量大。因此, 列为无危等级 / Wrinkle-lipped Bat is a widespread species with large populations. Thus, it is listed as Least Concern

地理分布 Geographical Distribution

国内分布 Domestic Distribution
云南、广东、香港、海南、贵州、甘肃、宁夏、广西 / Yunnan, Guangdong, Hong Kong, Hainan, Guizhou, Gansu, Ningxia, Guangxi
世界分布 World Distribution
中国; 南亚、东南亚 / China; South Asia, Southeast Asia
分布标注 Distribution Note
非特有种 / Non-endemic

国内分布图
Map of Domestic Distribution

种群 Population

种群数量 Population Size	未知 / Unknown
种群趋势 Population Trend	未知 / Unknown

生境与生态系统 Habitat (s) and Ecosystem (s)

生　　境 Habitat(s)	洞穴、人造建筑、内陆岩石区域 / Cave, Building, Inland Rocky Area
生态系统 Ecosystem(s)	喀斯特生态系统、人类聚落生态系统 / Karst Ecosystem, Human Settlement Ecosystem

威胁 Threat (s)

主要威胁 Major Threat(s)	采矿、采石场、旅游 / Mining, Quarrying Field and Tourism

保护级别与保护行动 Protection Category and Conservation Action (s)

国家重点保护野生动物等级 (2021) Category of National Key Protected Wild Animals (2021)	无 / NA
IUCN 红色名录 (2020-2) IUCN Red List (2020-2)	无危 / LC
CITES 附录 (2019) CITES Appendix (2019)	无 / NA
保护行动 Conservation Action(s)	无 / None

相关文献 Relevant References

Wilson and Mittermeier, 2019; Burgin *et al.*, 2018; Zhang and Yu（张显理和于有志）, 1994; Luo *et al.*（罗蓉等）, 1993; Wang（王香亭）, 1991; Animal Laboratory of Guangdong Institute of Entomology; Department of Biology, Sun Yat-Sen University（广东省昆虫研究所动物室和中山大学生物系）, 1983

（接上页）
(Desmarest, 1820); *Tadarida plicata* (Phillips, 1932) subsp. *insularis*; *Vespertilio plicatus* (Buchannan, 1800)

小犬吻蝠 *Chaerephon plicatus*　　吴毅 摄　By Yi Wu

沼泽鼠耳蝠
Myotis dasycneme

无危 LC

| 数据缺乏 DD | 无危 LC | | 易危 VU | 濒危 EN | 极危 CR | 区域灭绝 RE | 野外灭绝 EW | 灭绝 EX |

分类地位 Taxonomic Status

动物界 Animalia	脊索动物门 Chordata	哺乳纲 Mammalia	翼手目 Chiroptera	蝙蝠科 Vespertilionidae
学名 Scientific Name		*Myotis dasycneme*		
命名人 Species Authority		Boie, 1825		
英文名 English Name(s)		Pond Mouse-eared Bat		
同物异名 Synonym(s)		Pond Bat		
种下单元评估 Infra-specific Taxa Assessed		无 / None		

评估信息 Assessment Information

评估年份 Year Assessed	2020
评定人 Assessor(s)	吴毅、蒋志刚 / Yi Wu, Zhigang Jiang
审定人 Reviewer(s)	江廷磊、余文华、石红艳 / Tinglei Jiang, Wenhua Yu, Hongyan Shi
其他贡献人 Other Contributor(s)	李立立、丁晨晨 / Lili Li, Chenchen Ding

理由 Justification: 沼泽鼠耳蝠为常见种，种群数量大。因此，列为无危等级 / Pond Mouse-eared Bat is a common species with large populations. Thus, it is listed as Least Concern

地理分布 Geographical Distribution

国内分布 Domestic Distribution
山东 / Shandong
世界分布 World Distribution
亚洲、欧洲 / Asia, Europe
分布标注 Distribution Note
非特有种 / Non-endemic

国内分布图
Map of Domestic Distribution

种群 Population

种群数量 Population Size	未知 / Unknown
种群趋势 Population Trend	下降 / Decreasing

生境与生态系统 Habitat (s) and Ecosystem (s)

生境 Habitat(s)	溪流边、淡水湖、洞穴、人造建筑 / Near Stream, Freshwater Lake, Cave, Building
生态系统 Ecosystem(s)	森林生态系统、湖泊河流生态系统、人类聚落生态系统 / Forest Ecosystem, Lake and River Ecosystem, Human Settlement Ecosystem

威胁 Threat (s)

主要威胁 Major Threat(s)	农业或林业污染、城市污水、工业废水 / Agricultural or Forestry Effluent, Urban Waste Water, Industrial Waste Water

保护级别与保护行动 Protection Category and Conservation Action (s)

国家重点保护野生动物等级 (2021) Category of National Key Protected Wild Animals (2021)	无 / NA
IUCN 红色名录 (2020-2) IUCN Red List (2020-2)	近危 / NT
CITES 附录 (2019) CITES Appendix (2019)	无 / NA
保护行动 Conservation Action(s)	无 / None

相关文献 Relevant References

Wilson and Mittermeier, 2019; Burgin *et al.*, 2018; Jiang *et al.*（蒋志刚等）, 2017; Seim *et al.*, 2013; Smith *et al.*（史密斯等）, 2009; Wang（王应祥）, 2003; Wilson and Reeder, 1993

沼泽鼠耳蝠 *Myotis dasycneme*　　　　By Gilles San Martin

大卫鼠耳蝠 *Myotis davidii*

无危 LC

| 数据缺乏 DD | 无危 LC | 近危 NT | 易危 VU | 濒危 EN | 极危 CR | 区域灭绝 RE | 野外灭绝 EW | 灭绝 EX |

分类地位 Taxonomic Status

动物界 Animalia	脊索动物门 Chordata	哺乳纲 Mammalia	翼手目 Chiroptera	蝙蝠科 Vespertilionidae
学名 Scientific Name		*Myotis davidii*		
命名人 Species Authority		Peters, 1869		
英文名 English Name(s)		David's Mouse-eared Bat		
同物异名 Synonym(s)		David's Myotis; *Myotis hajastanicus* (Argyropulo, 1939)		
种下单元评估 Infra-specific Taxa Assessed		无 / None		

评估信息 Assessment Information

评估年份 Year Assessed	2020
评定人 Assessor(s)	吴毅、蒋志刚 / Yi Wu, Zhigang Jiang
审定人 Reviewer(s)	江廷磊、余文华、石红艳 / Tinglei Jiang, Wenhua Yu, Hongyan Shi
其他贡献人 Other Contributor(s)	李立立、丁晨晨 / Lili Li, Chenchen Ding

理由 Justification: 大卫鼠耳蝠为常见种，种群数量大。因此，列为无危等级 / David's Mouse-eared Bat is a common speices with large populations. Thus, it is listed as Least Concern

地理分布 Geographical Distribution

国内分布 Domestic Distribution

北京、河北、山西、内蒙古、甘肃、江西、福建、贵州、海南、广东、重庆 / Beijing, Hebei, Shanxi, Inner Mongolia (Nei Mongol), Gansu, Jiangxi, Fujian, Guizhou, Hainan, Guangdong, Chongqing

世界分布 World Distribution

中国 / China

分布标注 Distribution Note

特有种 / Endemic

国内分布图
Map of Domestic Distribution

种群 Population

种群数量 Population Size	未知 / Unknown
种群趋势 Population Trend	下降 / Decreasing

生境与生态系统 Habitat (s) and Ecosystem (s)

生境 Habitat(s)	未知 / Unknown
生态系统 Ecosystem(s)	森林生态系统 / Forest Ecosystem

威胁 Threat (s)

主要威胁 Major Threat(s)	未知 / Unknown

保护级别与保护行动 Protection Category and Conservation Action (s)

国家重点保护野生动物等级 (2021) Category of National Key Protected Wild Animals (2021)	无 / NA
IUCN 红色名录 (2020-2) IUCN Red List (2020-2)	无危 / LC
CITES 附录 (2019) CITES Appendix (2019)	无 / NA
保护行动 Conservation Action(s)	位于自然保护区内的种群与生境得到保护 / Populations and habitats in nature reserves are protected

相关文献 Relevant References

Wilson and Mittermeier, 2019; Burgin *et al*., 2018; Jiang *et al*. (蒋志刚等), 2017; You (由玉岩), 2013; Jiang and Feng (江廷磊和冯江), 2011; Yin *et al*. (尹皓等), 2011; You and Du (由玉岩和杜江峰), 2011; You *et al*., 2010; Zhang *et al*., 2009a; Wu and Harada, 2006; Phillips and Wilson, 1968

大卫鼠耳蝠 *Myotis davidii*

霍氏鼠耳蝠
Myotis horsfieldii
无危 LC

| 数据缺乏 DD | 无危 LC | 近危 NT | 易危 VU | 濒危 EN | 极危 CR | 区域灭绝 RE | 野外灭绝 EW | 灭绝 EX |

分类地位 Taxonomic Status

动物界 Animalia	脊索动物门 Chordata	哺乳纲 Mammalia	翼手目 Chiroptera	蝙蝠科 Vespertilionidae

学名 Scientific Name	*Myotis horsfieldii*
命名人 Species Authority	Temminck, 1840
英文名 English Name(s)	Horsfield's Mouse-eared Bat
同物异名 Synonym(s)	Horsfield's Bat; *Leuconoe peshwa* (Thomas, 1915); *Myotis adversus* (Andersen, 1907) subsp. *dryas*; *M. adversus* (Thomas, 1915) subsp. *peshwa*; *M. dryas* (Andersen, 1907); (转下页)
种下单元评估 Infra-specific Taxa Assessed	无 / None

评估信息 Assessment Information

评估年份 Year Assessed	2020
评定人 Assessor(s)	吴毅、蒋志刚 / Yi Wu, Zhigang Jiang
审定人 Reviewer(s)	江廷磊、余文华、石红艳 / Tinglei Jiang, Wenhua Yu, Hongyan Shi
其他贡献人 Other Contributor(s)	李立立、丁晨晨 / Lili Li, Chenchen Ding

理由 Justification: 霍氏鼠耳蝠主要分布在中国岭南地区，种群数量大。因此，列为无危等级 / Horsfield's Mouse-eared Bat is mainly distributed in the Lingnan area, China, it has large populations. Thus, it is listed as Least Concern

地理分布 Geographical Distribution

国内分布 Domestic Distribution	香港、广东、海南 / Hong Kong, Guangdong, Hainan
世界分布 World Distribution	中国；南亚、东南亚 / China; South Asia, Southeast Asia
分布标注 Distribution Note	非特有种 / Non-endemic

国内分布图
Map of Domestic Distribution

种群 Population

种群数量 Population Size	未知 / Unknown
种群趋势 Population Trend	稳定 / Stable

生境与生态系统 Habitat (s) and Ecosystem (s)

生 境 Habitat(s)	洞穴、森林、人造建筑、种植园、溪流边 / Cave, Forest, Building, Plantation, Near Stream
生态系统 Ecosystem(s)	森林生态系统、农田生态系统、人类聚落生态系统、湖泊河流生态系统 / Forest Ecosystem, Cropland Ecosystem, Human Settlement Ecosystem, Lake and River Ecosystem

威胁 Threat (s)

主要威胁 Major Threat(s)	山洞开发、伐木、耕种 / Cave Development, Logging, Plantation

保护级别与保护行动 Protection Category and Conservation Action (s)

国家重点保护野生动物等级 (2021) Category of National Key Protected Wild Animals (2021)	无 / NA
IUCN 红色名录 (2020-2) IUCN Red List (2020-2)	无危 / LC
CITES 附录 (2019) CITES Appendix (2019)	无 / NA
保护行动 Conservation Action(s)	无 / None

相关文献 Relevant References

Burgin *et al.*, 2018; Jiang *et al.*（蒋志刚等），2017; Smith *et al.*（史密斯等），2009; Zhu（朱斌良），2008; Wang（王应祥），2003

（接上页）
M. jeannei (Hill, 1983); *Vespertilio horsfieldi* (Temminck, 1840)

霍氏鼠耳蝠 *Myotis horsfieldii* 张礼标 摄 By Libiao Zhang

伊氏鼠耳蝠
Myotis ikonnikovi

无危 LC

| 数据缺乏 DD | 无危 LC | 近危 NT | 易危 VU | 濒危 EN | 极危 CR | 区域灭绝 RE | 野外灭绝 EW | 灭绝 EX |

分类地位 Taxonomic Status

动物界 Animalia	脊索动物门 Chordata	哺乳纲 Mammalia	翼手目 Chiroptera	蝙蝠科 Vespertilionidae
学 名 Scientific Name		*Myotis ikonnikovi*		
命 名 人 Species Authority		Ognev, 1912		
英 文 名 English Name(s)		Ikonnikov's Mouse-eared Bat		
同物异名 Synonym(s)		Ikonnikov's Bat; *Myotis fujiensis* (Imaizumi, 1954); *hosonoi* (Imaizumi, 1954); *ozensis* (Imaizumi, 1954); *yesoensis* (Yoshiyuki, 1984)		
种下单元评估 Infra-specific Taxa Assessed		无 / None		

评估信息 Assessment Information

评估年份 Year Assessed	2020
评 定 人 Assessor(s)	吴毅、蒋志刚 / Yi Wu, Zhigang Jiang
审 定 人 Reviewer(s)	江廷磊、余文华、石红艳 / Tinglei Jiang, Wenhua Yu, Hongyan Shi
其他贡献人 Other Contributor(s)	李立立、丁晨晨 / Lili Li, Chenchen Ding

理由 Justification: 伊氏鼠耳蝠分布较广，种群数量大。因此，列为无危等级 / Ikonnikov's Mouse-eared Bat is a widespread species with large populations. Thus, it is listed as Least Concern

地理分布 Geographical Distribution

国内分布 Domestic Distribution
内蒙古、黑龙江、吉林、辽宁、甘肃、陕西 / Inner Mongolia (Nei Mongol), Heilongjiang, Jilin, Liaoning, Gansu, Shaanxi
世界分布 World Distribution
中国；东北亚 / China; Northeast Asia
分布标注 Distribution Note
非特有种 / Non-endemic

国内分布图
Map of Domestic Distribution

种群 Population

种群数量 Population Size	未知 / Unknown
种群趋势 Population Trend	未知 / Unknown

生境与生态系统 Habitat(s) and Ecosystem(s)

生境 Habitat(s)	洞穴、热带和亚热带湿润山地森林 / Cave, Tropical and Subtropical Moist Montane Forest
生态系统 Ecosystem(s)	喀斯特生态系统、森林生态系统 / Karst Ecosystem, Forest Ecosystem

威胁 Threat(s)

主要威胁 Major Threat(s)	无 / None

保护级别与保护行动 Protection Category and Conservation Action(s)

国家重点保护野生动物等级 (2021) Category of National Key Protected Wild Animals (2021)	无 / NA
IUCN 红色名录 (2020-2) IUCN Red List (2020-2)	无危 / LC
CITES 附录 (2019) CITES Appendix (2019)	无 / NA
保护行动 Conservation Action(s)	无 / None

相关文献 Relevant References

Wilson and Mittermeier, 2019; Burgin *et al.*, 2018; Jiang *et al.* (蒋志刚等), 2017; Piao *et al.* (朴龙国等), 2013; Wu and Pei (吴家炎和裴俊峰), 2011; Zhang *et al.*, 2009a; Gao (高志英), 2008; Sun *et al.* (孙克萍等), 2006

伊氏鼠耳蝠 *Myotis ikonnikovi*

华南水鼠耳蝠
Myotis laniger

无危 LC

| 数据缺乏 DD | 无危 LC | 近危 NT | 易危 VU | 濒危 EN | 极危 CR | 区域灭绝 RE | 野外灭绝 EW | 灭绝 EX |

分类地位 Taxonomic Status

动物界 Animalia	脊索动物门 Chordata	哺乳纲 Mammalia	翼手目 Chiroptera	蝙蝠科 Vespertilionidae
学名 Scientific Name		*Myotis laniger*		
命名人 Species Authority		Peters, 1871		
英文名 English Name(s)		Chinese Water Mouse-eared Bat		
同物异名 Synonym(s)		Chinese Water Myotis		
种下单元评估 Infra-specific Taxa Assessed		无 / None		

评估信息 Assessment Information

评估年份 Year Assessed	2020
评定人 Assessor(s)	吴毅、蒋志刚 / Yi Wu, Zhigang Jiang
审定人 Reviewer(s)	江廷磊、余文华、石红艳 / Tinglei Jiang, Wenhua Yu, Hongyan Shi
其他贡献人 Other Contributor(s)	李立立、丁晨晨 / Lili Li, Chenchen Ding

理由 Justification: 华南水鼠耳蝠分布较广，种群数量大。因此，列为无危等级 / Chinese Water Mouse-eared Bat is a widespread species with large populations. Thus, it is listed as Least Concern

地理分布 Geographical Distribution

国内分布 Domestic Distribution

山东、江苏、安徽、浙江、福建、江西、广东、广西、香港、海南、贵州、重庆、云南、西藏、四川、陕西 / Shandong, Jiangsu, Anhui, Zhejiang, Fujian, Jiangxi, Guangdong, Guangxi, Hong Kong, Hainan, Guizhou, Chongqing, Yunnan, Tibet (Xizang), Sichuan, Shaanxi

世界分布 World Distribution

中国、印度、越南 / China, India, Viet Nam

分布标注 Distribution Note

非特有种 / Non-endemic

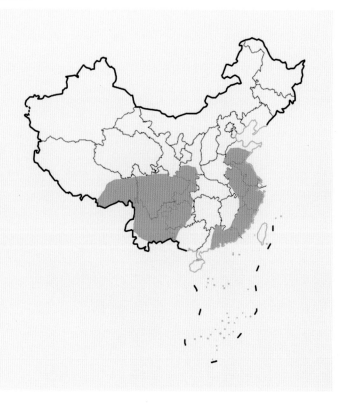

国内分布图
Map of Domestic Distribution

种群 Population

种群数量 Population Size	常见 / Common
种群趋势 Population Trend	稳定 / Stable

生境与生态系统 Habitat(s) and Ecosystem(s)

生境 Habitat(s)	洞穴、森林 / Cave, Forest
生态系统 Ecosystem(s)	喀斯特生态系统、森林生态系统 / Karst Ecosystem, Forest Ecosystem

威胁 Threat(s)

主要威胁 Major Threat(s)	未知 / Unknown

保护级别与保护行动 Protection Category and Conservation Action(s)

国家重点保护野生动物等级 (2021) Category of National Key Protected Wild Animals (2021)	无 / NA
IUCN 红色名录 (2020-2) IUCN Red List (2020-2)	无危 / LC
CITES 附录 (2019) CITES Appendix (2019)	无 / NA
保护行动 Conservation Action(s)	无 / None

相关文献 Relevant References

Wilson and Mittermeier, 2019; Burgin *et al*., 2018; Jiang *et al*. (蒋志刚等), 2017; Tu *et al*. (涂飞云等), 2014; Hu *et al*. (胡开良等), 2012; Wu and Pei (吴家炎和裴俊峰), 2011; Zhang (张维道), 1985

华南水鼠耳蝠 *Myotis laniger* 余文华 摄 By Wenhua Yu

长指鼠耳蝠 *Myotis longipes*

无危 LC

| 数据缺乏 DD | 无危 LC | 近危 NT | 易危 VU | 濒危 EN | 极危 CR | 区域灭绝 RE | 野外灭绝 EW | 灭绝 EX |

分类地位 Taxonomic Status

动物界 Animalia	脊索动物门 Chordata	哺乳纲 Mammalia	翼手目 Chiroptera	蝙蝠科 Vespertilionidae

学名 Scientific Name	*Myotis longipes*
命名人 Species Authority	Dobson, 1873
英文名 English Name(s)	Kashmir Cave Mouse-eared Bat
同物异名 Synonym(s)	Kashmir Cave Bat; *Myotis theobaldi* (Blyth, 1855); *Vespertilio longipes* (Dobson, 1873); *V. macropus* (Dobson, 1872); *V. megalopus* (Dobson, 1875)
种下单元评估 Infra-specific Taxa Assessed	无 / None

评估信息 Assessment Information

评估年份 Year Assessed	2020
评定人 Assessor(s)	吴毅、蒋志刚 / Yi Wu, Zhigang Jiang
审定人 Reviewer(s)	江廷磊、余文华、石红艳 / Tinglei Jiang, Wenhua Yu, Hongyan Shi
其他贡献人 Other Contributor(s)	李立立、丁晨晨 / Lili Li, Chenchen Ding

理由 Justification: 长指鼠耳蝠分布在云南、贵州、湖南、广东，分布范围较小，但其种群数量较多。因此，列为无危等级 / In China, Kashmir Cave Mouse-eared Bat is found in Yunnan, Guizhou, Hunan and Guangdong. Its distribution area is relatively small. However, its population is large. Thus, it is listed as Least Concern

地理分布 Geographical Distribution

国内分布 Domestic Distribution	云南、贵州、湖南、广东 / Yunnan, Guizhou, Hunan, Guangdong
世界分布 World Distribution	中国；南亚、中亚 / China; South Asia, Central Asia
分布标注 Distribution Note	非特有种 / Non-endemic

国内分布图
Map of Domestic Distribution

种群 Population

种群数量 Population Size	常见 / Common
种群趋势 Population Trend	未知 / Unknown

生境与生态系统 Habitat(s) and Ecosystem(s)

生境 Habitat(s)	洞穴、人造建筑 / Cave, Building
生态系统 Ecosystem(s)	喀斯特生态系统、人类聚落生态系统 / Karst Ecosystem, Human Settlement Ecosystem

威胁 Threat(s)

主要威胁 Major Threat(s)	未知 / Unknown

保护级别与保护行动 Protection Category and Conservation Action(s)

国家重点保护野生动物等级 (2021) Category of National Key Protected Wild Animals (2021)	无 / NA
IUCN红色名录 (2020-2) IUCN Red List (2020-2)	数据缺乏 / DD
CITES 附录 (2019) CITES Appendix (2019)	无 / NA
保护行动 Conservation Action(s)	无 / None

相关文献 Relevant References

Burgin *et al.*, 2018; Yu *et al.* (余子寒等), 2018; Jiang *et al.* (蒋志刚等), 2017; Zhang *et al.* (张琴等), 2017; Wang (王应祥), 2003

长指鼠耳蝠 *Myotis longipes* 张礼标 摄 By Libiao Zhang

大趾鼠耳蝠
Myotis macrodactylus
无危 LC

| 数据缺乏 DD | 无危 LC | 近危 NT | 易危 VU | 濒危 EN | 极危 CR | 区域灭绝 RE | 野外灭绝 EW | 灭绝 EX |

分类地位 Taxonomic Status

动物界 Animalia	脊索动物门 Chordata	哺乳纲 Mammalia	翼手目 Chiroptera	蝙蝠科 Vespertilionidae

学 名 Scientific Name	*Myotis macrodactylus*
命名人 Species Authority	Temminck, 1840
英文名 English Name(s)	Big-footed Mouse-eared Bat
同物异名 Synonym(s)	Big-footed Myotis; Eastern Long-fingered Bat; *Myotis macrodactylus* (Tiunov, 1997) subsp. *continentalis*; *M. macrodactylus* (Tiunov, 1997) subsp. *insularis*
种下单元评估 Infra-specific Taxa Assessed	无 / None

评估信息 Assessment Information

评估年份 Year Assessed	2020
评定人 Assessor(s)	吴毅、蒋志刚 / Yi Wu, Zhigang Jiang
审定人 Reviewer(s)	江廷磊、余文华、石红艳 / Tinglei Jiang, Wenhua Yu, Hongyan Shi
其他贡献人 Other Contributor(s)	李立立、丁晨晨 / Lili Li, Chenchen Ding

理由 Justification: 大趾鼠耳蝠仅分布在东北，发生区小，但种群数量达到了约2,000只。因此，列为无危等级 / Big-footed Mouse-eared Bat is only found in northeast China with small area of occurrence, however, its population size reaches about 2,000 individuals. Thus, it is listed as Least Concern

地理分布 Geographical Distribution

国内分布 Domestic Distribution
黑龙江、吉林 / Heilongjiang, Jilin
世界分布 World Distribution
中国；东北亚 / China; Northeast Asia
分布标注 Distribution Note
非特有种 / Non-endemic

国内分布图
Map of Domestic Distribution

种群 Population

种群数量 Population Size	约 2,000 只 / About 2,000 individuals
种群趋势 Population Trend	下降 / Decreasing

生境与生态系统 Habitat (s) and Ecosystem (s)

生境 Habitat(s)	洞穴、人造建筑、溪流边、淡水湖 / Cave, Building, Near Stream, Freshwater Lake
生态系统 Ecosystem(s)	喀斯特生态系统、人类聚落生态系统、湖泊河流生态系统 / Karst Ecosystem, Human Settlement Ecosystem, Lake and River Ecosystem

威胁 Threat (s)

主要威胁 Major Threat(s)	未知 / Unknown

保护级别与保护行动 Protection Category and Conservation Action (s)

国家重点保护野生动物等级 (2021) Category of National Key Protected Wild Animals (2021)	无 / NA
IUCN 红色名录 (2020-2) IUCN Red List (2020-2)	无危 / LC
CITES 附录 (2019) CITES Appendix (2019)	无 / NA
保护行动 Conservation Action(s)	无 / None

相关文献 Relevant References

Wilson and Mittermeier, 2019; Burgin *et al*., 2018; Jiang *et al*. (蒋志刚等), 2017; Wang *et al*., 2014; Wang (王红娜), 2010; Liu *et al*. (刘丰等), 2009; Luo *et al*. (罗金红等), 2009; Jiang *et al*. (江廷磊等), 2008a

大趾鼠耳蝠 *Myotis macrodactylus* 　　郭东革 摄　By Dongge Guo

山地鼠耳蝠
Myotis montivagus

无危 LC

| 数据缺乏 DD | 无危 LC | 近危 NT | 易危 VU | 濒危 EN | 极危 CR | 区域灭绝 RE | 野外灭绝 EW | 灭绝 EX |

分类地位 Taxonomic Status

动物界 Animalia	脊索动物门 Chordata	哺乳纲 Mammalia	翼手目 Chiroptera	蝙蝠科 Vespertilionidae
学　　名 Scientific Name		*Myotis montivagus*		
命 名 人 Species Authority		Dobson, 1874		
英 文 名 English Name(s)		Burmese Whiskered Mouse-eared Bat		
同物异名 Synonym(s)		Burmese Whiskered Bat; *Myotis mystacinus* (Dobson, 1874) subsp. *montivagus*; *M. peytoni* (Wroughton and Ryley, 1913); *Vespertilio montivagus* (Dobson, 1874)		
种下单元评估 Infra-specific Taxa Assessed		无 / None		

评估信息 Assessment Information

评 估 年 份 Year Assessed	2020
评 定 人 Assessor(s)	吴毅、蒋志刚 / Yi Wu, Zhigang Jiang
审 定 人 Reviewer(s)	江廷磊、余文华、石红艳 / Tinglei Jiang, Wenhua Yu, Hongyan Shi
其他贡献人 Other Contributor(s)	李立立、丁晨晨 / Lili Li, Chenchen Ding

理由 Justification: 山地鼠耳蝠分布较广，种群数量大。因此，列为无危等级 / Burmese Whiskered Mouse-eared Bat is a widespread species with large populations. Thus, it is listed as Least Concern

地理分布 Geographical Distribution

国内分布 Domestic Distribution
江苏、上海、福建、云南、浙江 / Jiangsu, Shanghai, Fujian, Yunnan, Zhejiang
世界分布 World Distribution
中国、印度、老挝、马来西亚、缅甸、越南 / China, India, Laos, Malaysia, Myanmar, Viet Nam
分布标注 Distribution Note
非特有种 / Non-endemic

国内分布图
Map of Domestic Distribution

种群 Population

种群数量 Population Size	未知 / Unknown
种群趋势 Population Trend	未知 / Unknown

生境与生态系统 Habitat (s) and Ecosystem (s)

生 境 Habitat(s)	森林 / Forest
生态系统 Ecosystem(s)	森林生态系统 / Forest Ecosystem

威胁 Threat (s)

主要威胁 Major Threat(s)	旅游休闲 / Tourism and Recreation

保护级别与保护行动 Protection Category and Conservation Action (s)

国家重点保护野生动物等级 (2021) Category of National Key Protected Wild Animals (2021)	无 / NA
IUCN 红色名录 (2020-2) IUCN Red List (2020-2)	无危 / LC
CITES 附录 (2019) CITES Appendix (2019)	无 / NA
保护行动 Conservation Action(s)	无 / None

相关文献 Relevant References

Wilson and Mittermeier, 2019; Burgin *et al.*, 2018; Jiang *et al.* (蒋志刚等), 2017; Goerfoel *et al.*, 2013; Smith *et al.* (史密斯等), 2009; Wang (王应祥), 2003

山地鼠耳蝠 *Myotis montivagus* 　　蒋志刚 绘　Drawn by Zhigang Jiang

大足鼠耳蝠
Myotis pilosus

无危 LC

| 数据缺乏 DD | 无危 LC | 近危 NT | 易危 VU | 濒危 EN | 极危 CR | 区域灭绝 RE | 野外灭绝 EW | 灭绝 EX |

分类地位 Taxonomic Status

动物界 Animalia	脊索动物门 Chordata	哺乳纲 Mammalia	翼手目 Chiroptera	蝙蝠科 Vespertilionidae
学 名 Scientific Name		*Myotis pilosus*		
命 名 人 Species Authority		Peters, 1869		
英 文 名 English Name(s)		Rickett's Big-footed Mouse-eared Bat		
同物异名 Synonym(s)		Rickett's Big-footed Bat; *Myotis ricketti* (Thomas, 1894)		
种下单元评估 Infra-specific Taxa Assessed		无 / None		

评估信息 Assessment Information

评估年份 Year Assessed	2020
评 定 人 Assessor(s)	吴毅、蒋志刚 / Yi Wu, Zhigang Jiang
审 定 人 Reviewer(s)	江廷磊、余文华、石红艳 / Tinglei Jiang, Wenhua Yu, Hongyan Shi
其他贡献人 Other Contributor(s)	李立立、丁晨晨 / Lili Li, Chenchen Ding

理由 Justification: 大足鼠耳蝠在中国东部分布较广，受人类活动，如采石和旅游影响其种群数量有所下降，但该种仍然还是较常见。因此，列为无危等级 / Rickett's Big-footed Mouse-eared Bat in east China is widely distributed, its population size is declining due to human encroachment and activities such as quarrying and traveling. However, it is still a common species. Thus, it is listed as Least Concern

地理分布 Geographical Distribution

国内分布 Domestic Distribution
山东、山西、浙江、安徽、江西、云南、广西、福建、香港、澳门、北京、海南、广东 / Shandong, Shanxi, Zhejiang, Anhui, Jiangxi, Yunnan, Guangxi, Fujian, Hong Kong, Macao, Beijing, Hainan, Guangdong
世界分布 World Distribution
东南亚 / Southeast Asia
分布标注 Distribution Note
非特有种 / Non-endemic

国内分布图
Map of Domestic Distribution

种群 Population

种群数量 Population Size	较常见 / Relatively common
种群趋势 Population Trend	下降 / Decreasing

生境与生态系统 Habitat(s) and Ecosystem(s)

生境 Habitat(s)	次生林、喀斯特地貌、淡水湖、洞穴 / Secondary Forest, Karst Landscape, Freshwater Lake, Cave
生态系统 Ecosystem(s)	森林生态系统、喀斯特生态系统、湖泊河流生态系统 / Forest Ecosystem, Karst Ecosystem, Lake and River Ecosystem

威胁 Threat(s)

主要威胁 Major Threat(s)	采石、污染 / Quarrying, Pollution

保护级别与保护行动 Protection Category and Conservation Action(s)

国家重点保护野生动物等级 (2021) Category of National Key Protected Wild Animals (2021)	无 / NA
IUCN 红色名录 (2020-2) IUCN Red List (2020-2)	近危 / NT
CITES 附录 (2019) CITES Appendix (2019)	无 / NA
保护行动 Conservation Action(s)	无 / None

相关文献 Relevant References

Wilson and Mittermeier, 2019; Burgin *et al*., 2018; Jiang *et al*. (蒋志刚等), 2017; Pei and Feng (裴俊峰和冯祁君), 2014; Huang *et al*. (黄继展等), 2013; Lu *et al*., 2013; Luo *et al*. (罗丽等), 2011; Jiang *et al*. (江廷磊等), 2008a; Li *et al*. (李玉春等), 2006; Ma *et al*. (马杰等), 2005; Feng *et al*. (冯江等), 2003

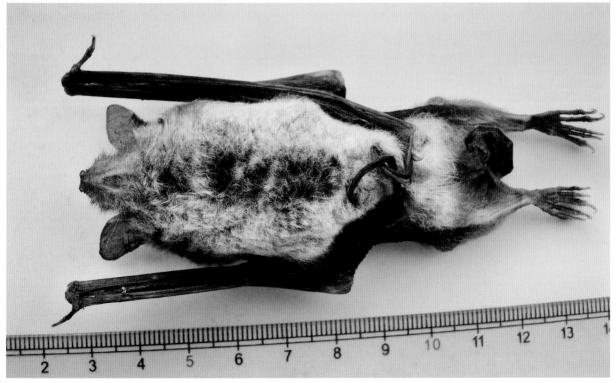

大足鼠耳蝠 *Myotis pilosus* 　　　　吴毅 摄　By Yi Wu

东亚伏翼 Pipistrellus abramus

无危 LC

| DD 数据缺乏 | LC 无危 | NT 近危 | VU 易危 | EN 濒危 | CR 极危 | RE 区域灭绝 | EW 野外灭绝 | EX 灭绝 |

分类地位 Taxonomic Status

动物界 Animalia	脊索动物门 Chordata	哺乳纲 Mammalia	翼手目 Chiroptera	蝙蝠科 Vespertilionidae

学名 Scientific Name	*Pipistrellus abramus*
命名人 Species Authority	Temminck, 1838
英文名 English Name(s)	Japanese Pipistrelle
同物异名 Synonym(s)	Japanese House Bat; *Vespertilio akokomuli* (Temminck, 1840); *V. irretilus* (Cantor, 1842); *V. pumiloides* (Tomes, 1857)
种下单元评估 Infra-specific Taxa Assessed	无 / None

评估信息 Assessment Information

评估年份 Year Assessed	2020
评定人 Assessor(s)	吴毅、蒋志刚 / Yi Wu, Zhigang Jiang
审定人 Reviewer(s)	余文华、石红艳、江廷磊 / Wenhua Yu, Hongyan Shi, Tinglei Jiang
其他贡献人 Other Contributor(s)	李立立、丁晨晨 / Lili Li, Chenchen Ding

理由 Justification: 东亚伏翼分布较广，种群数量稳定。因此，列为无危等级 / Japanese Pipistrelle is a widespread species with stable populations. Thus, it is listed as Least Concern

地理分布 Geographical Distribution

国内分布 Domestic Distribution

黑龙江、吉林、辽宁、河北、天津、山东、山西、陕西、甘肃、西藏、浙江、江苏、湖北、湖南、四川、重庆、贵州、云南、海南、福建、台湾、广东、广西、香港、澳门 / Heilongjiang, Jilin, Liaoning, Hebei, Tianjin, Shandong, Shanxi, Shaanxi, Gansu, Tibet (Xizang), Zhejiang, Jiangsu, Hubei, Hunan, Sichuan, Chongqing, Guizhou, Yunnan, Hainan, Fujian, Taiwan, Guangdong, Guangxi, Hong Kong, Macao

世界分布 World Distribution

中国；东南亚、东北亚 / China; Southeast Asia, Northeast Asia

分布标注 Distribution Note

非特有种 / Non-endemic

国内分布图
Map of Domestic Distribution

种群 Population

种群数量 Population Size	常见 / Common
种群趋势 Population Trend	稳定 / Stable

生境与生态系统 Habitat (s) and Ecosystem (s)

生境 Habitat(s)	人造建筑、森林 / Building, Forest
生态系统 Ecosystem(s)	森林生态系统、人类聚落生态系统 / Forest Ecosystem, Human Settlement Ecosystem

威胁 Threat (s)

主要威胁 Major Threat(s)	人类干扰、住房及城市建设 / Human Disturbance, Construction of House and Urban Area

保护级别与保护行动 Protection Category and Conservation Action (s)

国家重点保护野生动物等级 (2021) Category of National Key Protected Wild Animals (2021)	无 / NA
IUCN 红色名录 (2020-2) IUCN Red List (2020-2)	无危 / LC
CITES 附录 (2019) CITES Appendix (2019)	无 / NA
保护行动 Conservation Action(s)	无 / None

相关文献 Relevant References

Liu and Wu (刘少英和吴毅), 2019; Wilson and Mittermeier, 2019; Burgin *et al.*, 2018; Jiang *et al.* (蒋志刚等), 2017; Huang *et al.* (黄继展等), 2013; Cen and Peng (岑业文和彭红元), 2010; Wei *et al.*, 2010; Wu *et al.*, 2009a, 2009b; Niu (牛红星), 2008; Wu *et al.* (吴毅等), 2007

东亚伏翼 *Pipistrellus abramus* 吴毅 摄 By Yi Wu

锡兰伏翼
Pipistrellus ceylonicus

无危 LC

| 数据缺乏 DD | 无危 LC | 近危 NT | 易危 VU | 濒危 EN | 极危 CR | 区域灭绝 RE | 野外灭绝 EW | 灭绝 EX |

分类地位 Taxonomic Status

动物界 Animalia	脊索动物门 Chordata	哺乳纲 Mammalia	翼手目 Chiroptera	蝙蝠科 Vespertilionidae

学名 Scientific Name	*Pipistrellus ceylonicus*
命名人 Species Authority	Kelaart, 1852
英文名 English Name(s)	Kelaart's Pipistrelle
同物异名 Synonym(s)	*Pipistrellus ceylonicus* (Wroughton, 1899) subsp. *chrysothrix*; *P. ceylonicus* (Thomas, 1915) subsp. *subcanus*; *P. chrysothrix* (Wroughton, 1899); *Scotophilus ceylonicus* (转下页)
种下单元评估 Infra-specific Taxa Assessed	无 / None

评估信息 Assessment Information

评估年份 Year Assessed	2020
评定人 Assessor(s)	吴毅、蒋志刚 / Yi Wu, Zhigang Jiang
审定人 Reviewer(s)	余文华、石红艳、江廷磊 / Wenhua Yu, Hongyan Shi, Tinglei Jiang
其他贡献人 Other Contributor(s)	李立立、丁晨晨 / Lili Li, Chenchen Ding

理由 Justification: 锡兰伏翼种群数量较多。因此，列为无危等级 / Kelaart's Pipistrelle has large populations. Thus, it is listed as Least Concern

地理分布 Geographical Distribution

国内分布 Domestic Distribution
海南、广西、广东 / Hainan, Guangxi, Guangdong
世界分布 World Distribution
中国；南亚、东南亚 / China; South Asia, Southeast Asia
分布标注 Distribution Note
非特有种 / Non-endemic

国内分布图
Map of Domestic Distribution

种群 Population

种群数量 Population Size	未知 / Unknown
种群趋势 Population Trend	稳定 / Stable

生境与生态系统 Habitat (s) and Ecosystem (s)

生　　境 Habitat(s)	热带和亚热带湿润山地森林、洞穴，人造建筑 / Tropical and Subtropical Moist Montane Forest, Cave; Building
生态系统 Ecosystem(s)	森林生态系统、人类聚落生态系统 / Forest Ecosystem, Human Settlement Ecosystem

威胁 Threat (s)

主要威胁 Major Threat(s)	狩猎及采集 / Hunting and Collection

保护级别与保护行动 Protection Category and Conservation Action (s)

国家重点保护野生动物等级 (2021) Category of National Key Protected Wild Animals (2021)	无 / NA
IUCN 红色名录 (2020-2) IUCN Red List (2020-2)	无危 / LC
CITES 附录 (2019) CITES Appendix (2019)	无 / NA
保护行动 Conservation Action(s)	无 / None

相关文献 Relevant References

Liu and Wu（刘少英和吴毅）, 2019; Wilson and Mittermeier, 2019; Burgin *et al*., 2018; Jiang *et al*.（蒋志刚等）, 2017; Chen, 2009; Zhu（朱斌良）, 2008

（接上页）
(Kelaart, 1852); *Vesperugo indicus* (Dobson, 1878)

锡兰伏翼 *Pipistrellus ceylonicus*

印度伏翼 *Pipistrellus coromandra*

无危 LC

| 数据缺乏 DD | 无危 LC | 近危 NT | 易危 VU | 濒危 EN | 极危 CR | 区域灭绝 RE | 野外灭绝 EW | 灭绝 EX |

分类地位 Taxonomic Status

| 动物界 Animalia | 脊索动物门 Chordata | 哺乳纲 Mammalia | 翼手目 Chiroptera | 蝙蝠科 Vespertilionidae |

学 名 Scientific Name	*Pipistrellus coromandra*
命 名 人 Species Authority	Gray, 1838
英 文 名 English Name(s)	Coromandel Pipistrelle
同物异名 Synonym(s)	Indian Pipistrelle; *Myotis parvipes* (Blyth, 1853); *Pipistrellus coromandra* (Gaisler, 1870) subsp. *afghanus*; *Scotophilus coromandelianus* (Blyth, 1863); *S. coromandra* (转下页)
种下单元评估 Infra-specific Taxa Assessed	无 / None

评估信息 Assessment Information

评 估 年 份 Year Assessed	2020
评 定 人 Assessor(s)	吴毅、蒋志刚 / Yi Wu, Zhigang Jiang
审 定 人 Reviewer(s)	余文华、石红艳、江廷磊 / Wenhua Yu, Hongyan Shi, Tinglei Jiang
其他贡献人 Other Contributor(s)	李立立、丁晨晨 / Lili Li, Chenchen Ding

理由 Justification: 印度伏翼种群数量较多。因此，列为无危等级 / Coromandel Pipistrelle has large populations. Thus, it is listed as Least Concern

地理分布 Geographical Distribution

国内分布 Domestic Distribution	西藏、广东 / Tibet (Xizang), Guangdong
世界分布 World Distribution	中国；南亚、东南亚 / China; South Asia, Southeast Asia
分布标注 Distribution Note	非特有种 / Non-endemic

国内分布图
Map of Domestic Distribution

种群 Population

种群数量 Population Size	未知 / Unknown
种群趋势 Population Trend	未知 / Unknown

生境与生态系统 Habitat (s) and Ecosystem (s)

生　　境 Habitat(s)	人造建筑、森林、耕地 / Building, Forest, Arable Land
生态系统 Ecosystem(s)	森林生态系统、人类聚落生态系统 / Forest Ecosystem, Human Settlement Ecosystem

威胁 Threat (s)

主要威胁 Major Threat(s)	无 / None

保护级别与保护行动 Protection Category and Conservation Action (s)

国家重点保护野生动物等级 (2021) Category of National Key Protected Wild Animals (2021)	无 / NA
IUCN 红色名录 (2020-2) IUCN Red List (2020-2)	无危 / LC
CITES 附录 (2019) CITES Appendix (2019)	无 / NA
保护行动 Conservation Action(s)	无 / None

相关文献 Relevant References

Liu and Wu (刘少英和吴毅), 2019; Wilson and Mittermeier, 2019; Burgin *et al.*, 2018; Jiang *et al.* (蒋志刚等), 2017; Gu *et al.* (谷晓明等), 2003a, 2003b; Hu and Wu (胡锦矗和吴毅), 1993

(接上页)
(Gray, 1838); *Vespertilio blythii* (Wagner, 1855); *V. coromandelicus* (Blyth, 1851); *V. micropus* (Peters, 1872); *V. nicobaricus* (Fitzinger, 1861)

印度伏翼 *Pipistrellus coromandra*　　余文华 摄　By Wenhua Yu

棒茎伏翼
Pipistrellus paterculus
无危 LC

| 数据缺乏 DD | 无危 LC | 近危 NT | 易危 VU | 濒危 EN | 极危 CR | 区域灭绝 RE | 野外灭绝 EW | 灭绝 EX |

分类地位 Taxonomic Status

动物界 Animalia	脊索动物门 Chordata	哺乳纲 Mammalia	翼手目 Chiroptera	蝙蝠科 Vespertilionidae

学名 Scientific Name	*Pipistrellus paterculus*
命名人 Species Authority	Thomas, 1915
英文名 English Name(s)	Mount Popa Pipistrelle
同物异名 Synonym(s)	*Pipistrellus abramus paterculus* (Thomas, 1915)
种下单元评估 Infra-specific Taxa Assessed	无 / None

评估信息 Assessment Information

评估年份 Year Assessed	2020
评定人 Assessor(s)	吴毅、蒋志刚 / Yi Wu, Zhigang Jiang
审定人 Reviewer(s)	余文华、石红艳、江廷磊 / Wenhua Yu, Hongyan Shi, Tinglei Jiang
其他贡献人 Other Contributor(s)	李立立、丁晨晨 / Lili Li, Chenchen Ding

理由 Justification: 棒茎伏翼种群数量较多。因此，列为无危等级 / Mount Popa Pipistrelle has large populations. Thus, it is listed as Least Concern

地理分布 Geographical Distribution

国内分布 Domestic Distribution
云南、贵州 / Yunnan, Guizhou
世界分布 World Distribution
中国；南亚、东南亚 / China; South Asia, Southeast Asia
分布标注 Distribution Note
非特有种 / Non-endemic

国内分布图
Map of Domestic Distribution

种群 Population

种群数量 Population Size	未知 / Unknown
种群趋势 Population Trend	未知 / Unknown

生境与生态系统 Habitat (s) and Ecosystem (s)

生　　境 Habitat(s)	森林、种植园 / Forest, Plantation
生态系统 Ecosystem(s)	森林生态系统、农田生态系统 / Forest Ecosystem, Cropland Ecosystem

威胁 Threat (s)

主要威胁 Major Threat(s)	无 / None

保护级别与保护行动 Protection Category and Conservation Action (s)

国家重点保护野生动物等级 (2021) Category of National Key Protected Wild Animals (2021)	无 / NA
IUCN 红色名录 (2020-2) IUCN Red List (2020-2)	无危 / LC
CITES 附录 (2019) CITES Appendix (2019)	无 / NA
保护行动 Conservation Action(s)	无 / None

相关文献 Relevant References

Wilson and Mittermeier, 2019; Burgin *et al.*, 2018; Jiang *et al.* (蒋志刚等), 2017; Zhou and Yang (周江和杨天友), 2010, 2009; Smith *et al.* (史密斯等), 2009; Wang (王应祥), 2003

棒茎伏翼 *Pipistrellus paterculus*

普通伏翼
Pipistrellus pipistrellus
无危 LC

| 数据缺乏 DD | 无危 LC | 近危 NT | 易危 VU | 濒危 EN | 极危 CR | 区域灭绝 RE | 野外灭绝 EW | 灭绝 EX |

分类地位 Taxonomic Status

动物界 Animalia	脊索动物门 Chordata	哺乳纲 Mammalia	翼手目 Chiroptera	蝙蝠科 Vespertilionidae
学名 Scientific Name		*Pipistrellus pipistrellus*		
命名人 Species Authority		Schreber, 1774		
英文名 English Name(s)		Common Pipistrelle		
同物异名 Synonym(s)		无 / None		
种下单元评估 Infra-specific Taxa Assessed		无 / None		

评估信息 Assessment Information

评估年份 Year Assessed	2020
评定人 Assessor(s)	吴毅、蒋志刚 / Yi Wu, Zhigang Jiang
审定人 Reviewer(s)	余文华、石红艳、江廷磊 / Wenhua Yu, Hongyan Shi, Tinglei Jiang
其他贡献人 Other Contributor(s)	李立立、丁晨晨 / Lili Li, Chenchen Ding

理由 Justification: 普通伏翼种群数量较多。因此，列为无危等级 / Common Pipistrelle has large populations. Thus, it is listed as Least Concern

地理分布 Geographical Distribution

国内分布 Domestic Distribution

陕西、新疆、江西、云南、台湾、四川、重庆、山东、广西、广东、福建、浙江、澳门 / Shaanxi, Xinjiang, Jiangxi, Yunnan, Taiwan, Sichuan, Chongqing, Shandong, Guangxi, Guangdong, Fujian, Zhejiang, Macao

世界分布 World Distribution

亚洲、欧洲、北非 / Asia, Europe, North Africa

分布标注 Distribution Note

非特有种 / Non-endemic

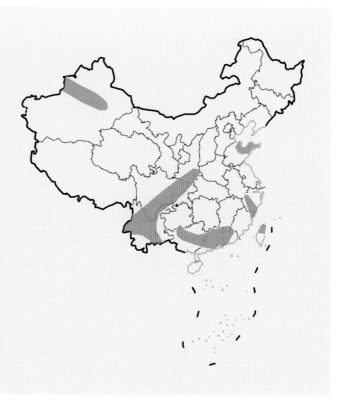

国内分布图
Map of Domestic Distribution

种群 Population

种群数量 Population Size	未知 / Unknown
种群趋势 Population Trend	稳定 / Stable

生境与生态系统 Habitat (s) and Ecosystem (s)

生　　境 Habitat(s)	人造建筑、森林 / Building, Forest
生态系统 Ecosystem(s)	森林生态系统、人类聚落生态系统 / Forest Ecosystem, Human Settlement Ecosystem

威胁 Threat (s)

主要威胁 Major Threat(s)	人类活动干扰、住房及城市建设 / Human Disturbance, Construction of House and Urban Area

保护级别与保护行动 Protection Category and Conservation Action (s)

国家重点保护野生动物等级 (2021) Category of National Key Protected Wild Animals (2021)	无 / NA
IUCN 红色名录(2020-2) IUCN Red List (2020-2)	无危 / LC
CITES 附录 (2019) CITES Appendix (2019)	无 / NA
保护行动 Conservation Action(s)	无 / None

相关文献 Relevant References

Liu and Wu (刘少英和吴毅), 2019; Wilson and Mittermeier, 2019; Burgin *et al*., 2018; Jiang *et al*. (蒋志刚等), 2017; Huang *et al*. (黄继展等), 2013; International Commission on Zoological Nomenclature, 2003

普通伏翼 *Pipistrellus pipistrellus*　　　　　　　　　　　　　　　　　　　　　　袁屏 摄　By Ping Yuan

茶褐伏翼
Hypsugo affinis

无危 LC

| 数据缺乏 DD | 无危 LC | 近危 NT | 易危 VU | 濒危 EN | 极危 CR | 区域灭绝 RE | 野外灭绝 EW | 灭绝 EX |

分类地位 Taxonomic Status

| 动物界 Animalia | 脊索动物门 Chordata | 哺乳纲 Mammalia | 翼手目 Chiroptera | 蝙蝠科 Vespertilionidae |

学名 Scientific Name	*Hypsugo affinis*
命名人 Species Authority	Dobson, 1871
英文名 English Name(s)	Chocolate Pipistrelle
同物异名 Synonym(s)	*Pipistrellus affinis* (Dobson, 1871); *Vesperugo affinis* (Dobson, 1871)
种下单元评估 Infra-specific Taxa Assessed	无 / None

评估信息 Assessment Information

评估年份 Year Assessed	2020
评定人 Assessor(s)	吴毅、蒋志刚 / Yi Wu, Zhigang Jiang
审定人 Reviewer(s)	余文华、石红艳、江廷磊 / Wenhua Yu, Hongyan Shi, Tinglei Jiang
其他贡献人 Other Contributor(s)	李立立、丁晨晨 / Lili Li, Chenchen Ding

理由 Justification: 茶褐伏翼分布广，种群数量较多。因此，列为无危等级 / Chocolate Pipistrelle is a widespread species with large populations. Thus, it is listed as Least Concern

地理分布 Geographical Distribution

国内分布 Domestic Distribution	西藏、云南、广西 / Tibet (Xizang), Yunnan, Guangxi
世界分布 World Distribution	中国；南亚 / China; South Asia
分布标注 Distribution Note	非特有种 / Non-endemic

国内分布图
Map of Domestic Distribution

种群 Population

种群数量 Population Size	未知 / Unknown
种群趋势 Population Trend	未知 / Unknown

生境与生态系统 Habitat (s) and Ecosystem (s)

生 境 Habitat(s)	人造建筑、森林 / Building, Forest
生态系统 Ecosystem(s)	森林生态系统、人类聚落生态系统 / Forest Ecosystem, Human Settlement Ecosystem

威胁 Threat (s)

主要威胁 Major Threat(s)	伐木、耕种 / Logging, Plantation

保护级别与保护行动 Protection Category and Conservation Action (s)

国家重点保护野生动物等级 (2021) Category of National Key Protected Wild Animals (2021)	无 / NA
IUCN红色名录 (2020-2) IUCN Red List (2020-2)	无危 / LC
CITES 附录 (2019) CITES Appendix (2019)	无 / NA
保护行动 Conservation Action(s)	无 / None

相关文献 Relevant References

Wilson and Mittermeier, 2019; Burgin *et al*., 2018; Jiang *et al*. (蒋志刚等), 2017; Huang *et al*. (黄薇等), 2008; Corbet and Hill, 1992

茶褐伏翼 *Hypsugo affinis*

双色蝙蝠 Vespertilio murinus

无危 LC

分类地位 Taxonomic Status

动物界 Animalia	脊索动物门 Chordata	哺乳纲 Mammalia	翼手目 Chiroptera	蝙蝠科 Vespertilionidae
学名 Scientific Name		Vespertilio murinus		
命名人 Species Authority		Linnaeus, 1758		
英文名 English Name(s)		Particoloured Bat		
同物异名 Synonym(s)		Eurasian Particolored Bat, Rearmouse		
种下单元评估 Infra-specific Taxa Assessed		无 / None		

评估信息 Assessment Information

评估年份 Year Assessed	2020
评定人 Assessor(s)	吴毅、蒋志刚 / Yi Wu, Zhigang Jiang
审定人 Reviewer(s)	石红艳、江廷磊、余文华 / Hongyan Shi, Tinglei Jiang, Wenhua Yu
其他贡献人 Other Contributor(s)	李立立、丁晨晨 / Lili Li, Chenchen Ding

理由 Justification: 双色蝙蝠种群数量较多。因此，列为无危等级 / Particoloured Bat has large populations. Thus, it is listed as Least Concern

地理分布 Geographical Distribution

国内分布 Domestic Distribution	黑龙江、内蒙古、新疆、甘肃 / Heilongjiang, Inner Mongolia (Nei Mongol), Xinjiang, Gansu
世界分布 World Distribution	亚洲、欧洲 / Asia, Europe
分布标注 Distribution Note	非特有种 / Non-endemic

国内分布图
Map of Domestic Distribution

种群 Population

种群数量 Population Size	未知 / Unknown
种群趋势 Population Trend	未知 / Unknown

生境与生态系统 Habitat (s) and Ecosystem (s)

生　　境 Habitat(s)	森林、草地、耕地、人造建筑 / Forest, Grassland, Arable Land, Building
生态系统 Ecosystem(s)	森林生态系统、草地生态系统、农田生态系统、人类聚落生态系统 / Forest Ecosystem, Grassland Ecosystem, Cropland Ecosystem, Human Settlement Ecosystem

威胁 Threat (s)

主要威胁 Major Threat(s)	伐木、耕种 / Logging, Plantation

保护级别与保护行动 Protection Category and Conservation Action (s)

国家重点保护野生动物等级 (2021) Category of National Key Protected Wild Animals (2021)	无 / NA
IUCN红色名录(2020-2) IUCN Red List (2020-2)	无危 / LC
CITES 附录 (2019) CITES Appendix (2019)	无 / NA
保护行动 Conservation Action(s)	无 / None

相关文献 Relevant References

Wilson and Mittermeier, 2019; Burgin *et al*., 2018; Jiang *et al*. (蒋志刚等), 2017; Wu *et al*. (武明录等), 2006; Rydell and Baagøe, 1994; Wilson and Reeder, 1993

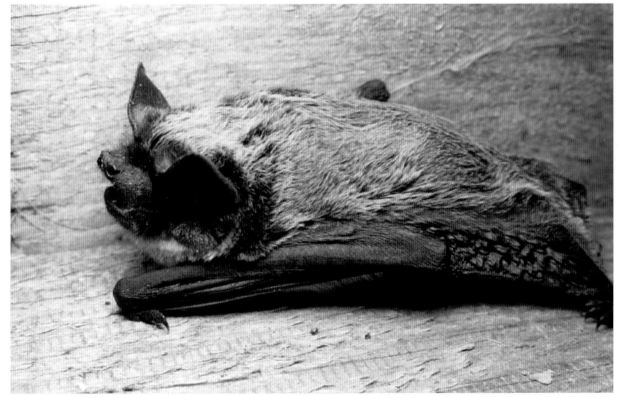

双色蝙蝠 *Vespertilio murinus*　　By Mnolf

东方蝙蝠
Vespertilio sinensis
无危 LC

| 数据缺乏 DD | 无危 LC | 近危 NT | 易危 VU | 濒危 EN | 极危 CR | 区域灭绝 RE | 野外灭绝 EW | 灭绝 EX |

分类地位 Taxonomic Status

动物界 Animalia	脊索动物门 Chordata	哺乳纲 Mammalia	翼手目 Chiroptera	蝙蝠科 Vespertilionidae

学 名 Scientific Name	*Vespertilio sinensis*
命 名 人 Species Authority	Peters, 1880
英 文 名 English Name(s)	Asian Particolored Bat
同物异名 Synonym(s)	*Vespertilio andersoni* (Wallin, 1962); *aurijunctus* (Mori, 1928); *montanus* (Kishida, 1931); *motoyoshii* (Kuroda, 1934); *namyiei* (Kuroda, 1920); *orientalis* (转下页)
种下单元评估 Infra-specific Taxa Assessed	无 / None

评估信息 Assessment Information

评 估 年 份 Year Assessed	2020
评 定 人 Assessor(s)	吴毅、蒋志刚 / Yi Wu, Zhigang Jiang
审 定 人 Reviewer(s)	石红艳、江廷磊、余文华 / Hongyan Shi, Tinglei Jiang, Wenhua Yu
其他贡献人 Other Contributor(s)	李立立、丁晨晨 / Lili Li, Chenchen Ding

理由 Justification: 东方蝙蝠种群数量较多。因此，列为无危等级 / Asian Particolored Bat has large populations. Thus, it is listed as Least Concern

地理分布 Geographical Distribution

国内分布 Domestic Distribution
广西、北京、内蒙古、黑龙江、江西、湖北、湖南、四川、陕西、甘肃、台湾、福建、辽宁、山东、云南、浙江、天津、重庆 / Guangxi, Beijing, Inner Mongolia (Nei Mongol), Heilongjiang, Jiangxi, Hubei, Hunan, Sichuan, Shaanxi, Gansu, Taiwan, Fujian, Liaoning, Shandong, Yunnan, Zhejiang, Tianjin, Chongqing

世界分布 World Distribution
中国；东北亚 / China; Northeast Asia

分布标注 Distribution Note
非特有种 / Non-endemic

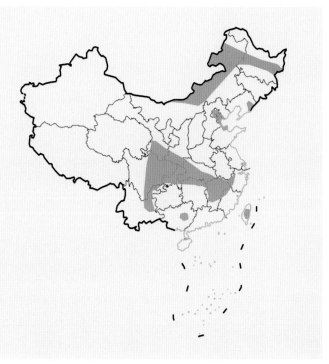

国内分布图
Map of Domestic Distribution

🦌 种群 Population

种群数量 Population Size	未知 / Unknown
种群趋势 Population Trend	未知 / Unknown

🦌 生境与生态系统 Habitat (s) and Ecosystem (s)

生　　　境 Habitat(s)	洞穴、草地、沙漠 / Cave, Grassland, Desert
生态系统 Ecosystem(s)	草地生态系统、荒漠生态系统 / Grassland Ecosystem, Desert Ecosystem

🦌 威胁 Threat (s)

主要威胁 Major Threat(s)	伐木、耕种 / Logging, Plantation

🦌 保护级别与保护行动 Protection Category and Conservation Action (s)

国家重点保护野生动物等级 (2021) Category of National Key Protected Wild Animals (2021)	无 / NA
IUCN 红色名录 (2020-2) IUCN Red List (2020-2)	无危 / LC
CITES 附录 (2019) CITES Appendix (2019)	无 / NA
保护行动 Conservation Action(s)	无 / None

🦌 相关文献 Relevant References

Wilson and Mittermeier, 2019; Burgin *et al.*, 2018; Jiang *et al.* (蒋志刚等), 2017; Li *et al.* (李文靖等), 2009; Smith *et al.* (史密斯等), 2009; Wang *et al.* (王静等), 2009; Wang (王应祥), 2003

(接上页)
(Wallin, 1969); *superans* (Thomas, 1899)

东方蝙蝠 *Vespertilio sinensis*　　　　　　　　　　江廷磊 摄　By Tinglei Jiang

北棕蝠
Eptesicus nilssonii

无危 LC

| 数据缺乏 DD | 无危 LC | 近危 NT | 易危 VU | 濒危 EN | 极危 CR | 区域灭绝 RE | 野外灭绝 EW | 灭绝 EX |

分类地位 Taxonomic Status

动物界 Animalia	脊索动物门 Chordata	哺乳纲 Mammalia	翼手目 Chiroptera	蝙蝠科 Vespertilionidae
学名 Scientific Name		*Eptesicus nilssonii*		
命名人 Species Authority		Keyserling and Blasius, 1839		
英文名 English Name(s)		Northern Bat		
同物异名 Synonym(s)		无 / None		
种下单元评估 Infra-specific Taxa Assessed		无 / None		

评估信息 Assessment Information

评估年份 Year Assessed	2020
评定人 Assessor(s)	吴毅、蒋志刚 / Yi Wu, Zhigang Jiang
审定人 Reviewer(s)	石红艳、江廷磊、余文华 / Hongyan Shi, Tinglei Jiang, Wenhua Yu
其他贡献人 Other Contributor(s)	李立立、丁晨晨 / Lili Li, Chenchen Ding

理由 Justification: 北棕蝠种群数量较多。因此，列为无危等级 / Northern Bat has large populations. Thus, it is listed as Least Concern

地理分布 Geographical Distribution

国内分布 Domestic Distribution

黑龙江、新疆、山东、宁夏、甘肃、青海、西藏、四川、吉林、辽宁、河北、河南、云南、内蒙古、北京、山西、陕西 / Heilongjiang, Xinjiang, Shandong, Ningxia, Gansu, Qinghai, Tibet (Xizang), Sichuan, Jilin, Liaoning, Hebei, Henan, Yunnan, Inner Mongolia (Nei Mongol), Beijing, Shanxi, Shaanxi

世界分布 World Distribution

亚洲、欧洲 / Asia, Europe

分布标注 Distribution Note

非特有种 / Non-endemic

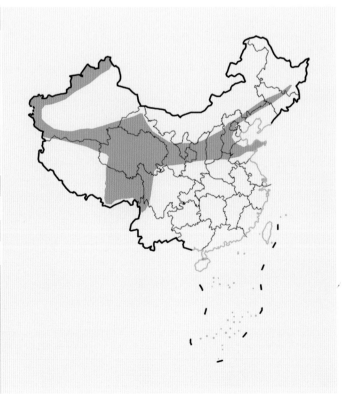

国内分布图
Map of Domestic Distribution

种群 Population

种群数量 Population Size	未知 / Unknown
种群趋势 Population Trend	稳定 / Stable

生境与生态系统 Habitat (s) and Ecosystem (s)

生　　境 Habitat(s)	森林、人造建筑 / Forest, Building
生态系统 Ecosystem(s)	森林生态系统、人类聚落生态系统 / Forest Ecosystem, Human Settlement Ecosystem

威胁 Threat (s)

主要威胁 Major Threat(s)	未知 / Unknown

保护级别与保护行动 Protection Category and Conservation Action (s)

国家重点保护野生动物等级 (2021) Category of National Key Protected Wild Animals (2021)	无 / NA
IUCN 红色名录 (2020-2) IUCN Red List (2020-2)	无危 / LC
CITES 附录 (2019) CITES Appendix (2019)	无 / NA
保护行动 Conservation Action(s)	无 / None

相关文献 Relevant References

Wilson and Mittermeier, 2019; Burgin *et al*., 2018; Jiang *et al*. (蒋志刚等), 2017; Smith *et al*. (史密斯等), 2009; Wang (王应祥), 2003; Rydell, 1993

北棕蝠 *Eptesicus nilssonii*

By Mnolf

肥耳棕蝠
Eptesicus pachyotis

无危 LC

分类地位 Taxonomic Status

| 动物界 Animalia | 脊索动物门 Chordata | 哺乳纲 Mammalia | 翼手目 Chiroptera | 蝙蝠科 Vespertilionidae |

学名 Scientific Name	*Eptesicus pachyotis*
命名人 Species Authority	Dobson, 1871
英文名 English Name(s)	Thick-eared Bat
同物异名 Synonym(s)	*Vesperugo pachyotis* (Dobson, 1871)
种下单元评估 Infra-specific Taxa Assessed	无 / None

评估信息 Assessment Information

评估年份 Year Assessed	2020
评定人 Assessor(s)	吴毅、蒋志刚 / Yi Wu, Zhigang Jiang
审定人 Reviewer(s)	石红艳、江廷磊、余文华 / Hongyan Shi, Tinglei Jiang, Wenhua Yu
其他贡献人 Other Contributor(s)	李立立、丁晨晨 / Lili Li, Chenchen Ding

理由 Justification: 肥耳棕蝠种群数量较多。因此，列为无危等级 / Thick-eared Bat has large populations. Thus, it is listed as Least Concern

地理分布 Geographical Distribution

国内分布 Domestic Distribution	宁夏、四川、西藏、甘肃、青海 / Ningxia, Sichuan, Tibet (Xizang), Gansu, Qinghai
世界分布 World Distribution	中国；南亚、东南亚 / China; South Asia, Southeast Asia
分布标注 Distribution Note	非特有种 / Non-endemic

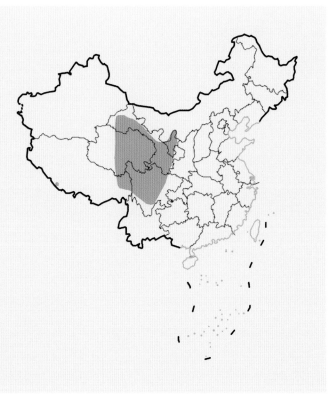

国内分布图
Map of Domestic Distribution

种群 Population

种群数量 Population Size	未知 / Unknown
种群趋势 Population Trend	未知 / Unknown

生境与生态系统 Habitat (s) and Ecosystem (s)

生　　境 Habitat(s)	未知 / Unknown
生态系统 Ecosystem(s)	森林生态系统 / Forest Ecosystem

威胁 Threat (s)

主要威胁 Major Threat(s)	未知 / Unknown

保护级别与保护行动 Protection Category and Conservation Action (s)

国家重点保护野生动物等级 (2021) Category of National Key Protected Wild Animals (2021)	无 / NA
IUCN 红色名录 (2020-2) IUCN Red List (2020-2)	无危 / LC
CITES 附录 (2019) CITES Appendix (2019)	无 / NA
保护行动 Conservation Action(s)	无 / None

相关文献 Relevant References

Wilson and Mittermeier, 2019; Burgin *et al*., 2018; Jiang *et al*. (蒋志刚等), 2017; Smith *et al*. (史密斯等), 2009; Wang (王应祥), 2003

肥耳棕蝠 *Eptesicus pachyotis*

大棕蝠
Eptesicus serotinus

无危 LC

| 数据缺乏 DD | 无危 LC | 近危 NT | 易危 VU | 濒危 EN | 极危 CR | 区域灭绝 RE | 野外灭绝 EW | 灭绝 EX |

分类地位 Taxonomic Status

动物界 Animalia	脊索动物门 Chordata	哺乳纲 Mammalia	翼手目 Chiroptera	蝙蝠科 Vespertilionidae

学 名 Scientific Name	*Eptesicus serotinus*
命名人 Species Authority	Schreber, 1774
英文名 English Name(s)	Common Serotine
同物异名 Synonym(s)	Serotine Bat; *Eptesicus isabellinus* (Temminck, 1840)
种下单元评估 Infra-specific Taxa Assessed	无 / None

评估信息 Assessment Information

评估年份 Year Assessed	2020
评定人 Assessor(s)	吴毅、蒋志刚 / Yi Wu, Zhigang Jiang
审定人 Reviewer(s)	石红艳、江廷磊、余文华 / Hongyan Shi, Tinglei Jiang, Wenhua Yu
其他贡献人 Other Contributor(s)	李立立、丁晨晨 / Lili Li, Chenchen Ding

理由 Justification: 大棕蝠种群数量较多。因此，列为无危等级 / Common Serotine has large populations. Thus, it is listed as Least Concern

地理分布 Geographical Distribution

国内分布 Domestic Distribution

黑龙江、吉林、辽宁、内蒙古、河北、北京、天津、山东、河南、山西、安徽、江苏、浙江、上海、江西、福建、台湾、湖北、湖南、贵州、云南、四川、陕西、甘肃、宁夏、新疆、西藏、重庆 / Heilongjiang, Jilin, Liaoning, Inner Mongolia (Nei Mongol), Hebei, Beijing, Tianjin, Shandong, Henan, Shanxi, Anhui, Jiangsu, Zhejiang, Shanghai, Jiangxi, Fujian, Taiwan, Hubei, Hunan, Guizhou, Yunnan, Sichuan, Shaanxi, Gansu, Ningxia, Xinjiang, Tibet (Xizang), Chongqing

世界分布 World Distribution

亚洲、欧洲、北非 / Asia, Europe, North Africa

分布标注 Distribution Note

非特有种 / Non-endemic

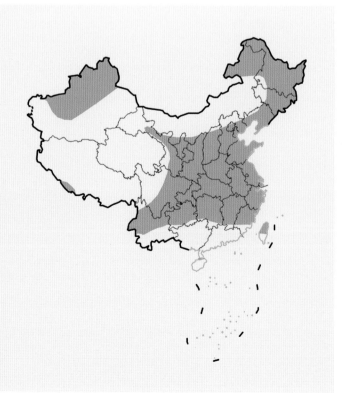

国内分布图
Map of Domestic Distribution

种群 Population

种群数量 Population Size	未知 / Unknown
种群趋势 Population Trend	未知 / Unknown

生境与生态系统 Habitat (s) and Ecosystem (s)

生　　境 Habitat(s)	温带森林、亚热带干旱森林、地中海灌木植被、耕地、种植园、乡村花园、人造建筑 / Temperate Forest, Subtropical Dry Forest, Mediterranean Shrub Vegetation, Arable Land, Plantation, Rural Garden, Building
生态系统 Ecosystem(s)	森林生态系统、农田生态系统、人类聚落生态系统 / Forest Ecosystem, Cropland Ecosystem, Human Settlement Ecosystem

威胁 Threat (s)

主要威胁 Major Threat(s)	食物缺乏 / Food Scarcity

保护级别与保护行动 Protection Category and Conservation Action (s)

国家重点保护野生动物等级 (2021) Category of National Key Protected Wild Animals (2021)	无 / NA
IUCN红色名录(2020-2) IUCN Red List (2020-2)	无危 / LC
CITES 附录 (2019) CITES Appendix (2019)	无 / NA
保护行动 Conservation Action(s)	无 / None

相关文献 Relevant References

Wilson and Mittermeier, 2019; Burgin *et al*., 2018; Jiang *et al*.（蒋志刚等）, 2017; Smith *et al*.（史密斯等）, 2009; Zhu *et al*.（朱旭等）, 2009; Mayer *et al*., 2007; Wang（王应祥）, 2003

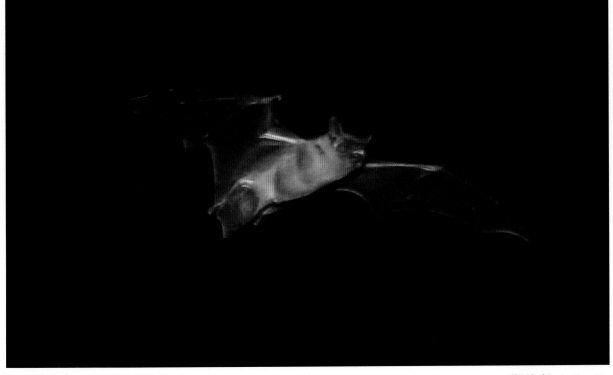

大棕蝠 *Eptesicus serotinus*　　　　　薄顺奇 摄　By Shunqi Bo

中华山蝠 Nyctalus plancyi

无危 LC

分类地位 Taxonomic Status

动物界 Animalia	脊索动物门 Chordata	哺乳纲 Mammalia	翼手目 Chiroptera	蝙蝠科 Vespertilionidae
学名 Scientific Name		Nyctalus plancyi		
命名人 Species Authority		Gerbe, 1880		
英文名 English Name(s)		Chinese Noctule		
同物异名 Synonym(s)		Nyctalus velutinus (G. M. Allen, 1923)		
种下单元评估 Infra-specific Taxa Assessed		无 / None		

评估信息 Assessment Information

评估年份 Year Assessed	2020
评定人 Assessor(s)	吴毅、蒋志刚 / Yi Wu, Zhigang Jiang
审定人 Reviewer(s)	石红艳、江廷磊、余文华 / Hongyan Shi, Tinglei Jiang, Wenhua Yu
其他贡献人 Other Contributor(s)	李立立、丁晨晨 / Lili Li, Chenchen Ding

理由 Justification: 中华山蝠的种群数量较多。因此，列为无危等级 / Chinese Noctule has large populations. Thus, it is listed as Least Concern

地理分布 Geographical Distribution

国内分布 Domestic Distribution

辽宁、吉林、北京、河南、山东、山西、陕西、甘肃、浙江、安徽、湖北、湖南、四川、重庆、贵州、云南、福建、台湾、广东、广西、江西、香港 / Liaoning, Jilin, Beijing, Henan, Shandong, Shanxi, Shaanxi, Gansu, Zhejiang, Anhui, Hubei, Hunan, Sichuan, Chongqing, Guizhou, Yunnan, Fujian, Taiwan, Guangdong, Guangxi, Jiangxi, Hong Kong

世界分布 World Distribution

中国 / China

分布标注 Distribution Note

特有种 / Endemic

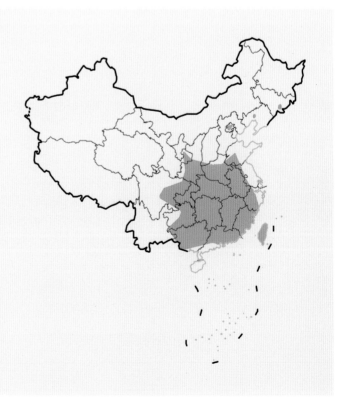

国内分布图
Map of Domestic Distribution

种群 Population

种群数量 Population Size	常见 / Common
种群趋势 Population Trend	稳定 / Stable

生境与生态系统 Habitat(s) and Ecosystem(s)

生境 Habitat(s)	人造建筑、洞穴、森林 / Building, Cave, Forest
生态系统 Ecosystem(s)	森林生态系统、人类聚落生态系统 / Forest Ecosystem, Human Settlement Ecosystem

威胁 Threat(s)

主要威胁 Major Threat(s)	未知 / Unknown

保护级别与保护行动 Protection Category and Conservation Action(s)

国家重点保护野生动物等级 (2021) Category of National Key Protected Wild Animals (2021)	无 / NA
IUCN 红色名录 (2020-2) IUCN Red List (2020-2)	无危 / LC
CITES 附录 (2019) CITES Appendix (2019)	无 / NA
保护行动 Conservation Action(s)	无 / None

相关文献 Relevant References

Burgin *et al*., 2018; Jiang *et al*. (蒋志刚等), 2017; Zheng *et al*. (郑锡奇等), 2010; Wang and Xie (汪松和解焱), 2004; Zhu *et al*. (朱光剑等), 2008a; Shi *et al*. (石红艳等), 2000; Liang and Dong (梁仁济和董永文), 1985

中华山蝠 *Nyctalus plancyi* 吴毅 摄 By Yi Wu

华南扁颅蝠
Tylonycteris fulvidus
无危 LC

| 数据缺乏 DD | 无危 LC | 近危 NT | 易危 VU | 濒危 EN | 极危 CR | 区域灭绝 RE | 野外灭绝 EW | 灭绝 EX |

分类地位 Taxonomic Status

动物界 Animalia	脊索动物门 Chordata	哺乳纲 Mammalia	翼手目 Chiroptera	蝙蝠科 Vespertilionidae

学 名 Scientific Name	*Tylonycteris fulvidus*
命名人 Species Authority	Temminck, 1840
英文名 English Name(s)	Lesser Bamboo Bat
同物异名 Synonym(s)	*Scotophilus fulvidus* (Blyth, 1850); *Tylonycteris aurex* (Thomas, 1915); *T. pachypus* (Blyth, 1850) subsp. *fulvida*; *T. rubidus* (Thomas, 1915); *Vespertilio pachypus* (Temminck, 1840)
种下单元评估 Infra-specific Taxa Assessed	无 / None

评估信息 Assessment Information

评估年份 Year Assessed	2020
评定人 Assessor(s)	吴毅、蒋志刚 / Yi Wu, Zhigang Jiang
审定人 Reviewer(s)	石红艳、江廷磊、余文华 / Hongyan Shi, Tinglei Jiang, Wenhua Yu
其他贡献人 Other Contributor(s)	李立立、丁晨晨 / Lili Li, Chenchen Ding

理由 Justification: 华南扁颅蝠种群数量较多。因此，列为无危等级 / Lesser Bamboo Bat has large populations. Thus, it is listed as Least Concern

地理分布 Geographical Distribution

国内分布 Domestic Distribution
广东、广西、贵州、云南、香港 / Guangdong, Guangxi, Guizhou, Yunnan, Hong Kong
世界分布 World Distribution
中国；南亚、东南亚 / China; South Asia, Southeast Asia
分布标注 Distribution Note
非特有种 / Non-endemic

国内分布图
Map of Domestic Distribution

种群 Population

种群数量 Population Size	常见 / Common
种群趋势 Population Trend	稳定 / Stable

生境与生态系统 Habitat (s) and Ecosystem (s)

生　　境 Habitat(s)	竹林、洞穴 / Bamboo Grove, Cave
生态系统 Ecosystem(s)	竹林生态系统 / Bamboo Ecosystem

威胁 Threat (s)

主要威胁 Major Threat(s)	伐竹、耕种 / Bamboo Logging, Plantation

保护级别与保护行动 Protection Category and Conservation Action (s)

国家重点保护野生动物等级 (2021) Category of National Key Protected Wild Animals (2021)	无 / NA
IUCN 红色名录 (2020-2) IUCN Red List (2020-2)	无危 / LC
CITES 附录 (2019) CITES Appendix (2019)	无 / NA
保护行动 Conservation Action(s)	无 / None

相关文献 Relevant References

Burgin *et al.*, 2018; Huang *et al.*, 2014; Zhang *et al.* (张礼标等), 2011; Smith *et al.* (史密斯等), 2009; Zhou *et al.* (周全等), 2005; Wu *et al.* (吴毅等), 2004a

华南扁颅蝠 *Tylonycteris fulvidus*　　　　余文华 摄　By Wenhua Yu

大黄蝠
Scotophilus heathii
无危 LC

| 数据缺乏 DD | 无危 LC | 近危 NT | 易危 VU | 濒危 EN | 极危 CR | 区域灭绝 RE | 野外灭绝 EW | 灭绝 EX |

分类地位 Taxonomic Status

| 动物界 Animalia | 脊索动物门 Chordata | 哺乳纲 Mammalia | 翼手目 Chiroptera | 蝙蝠科 Vespertilionidae |

学名 Scientific Name	*Scotophilus heathii*
命名人 Species Authority	Horsfield, 1831
英文名 English Name(s)	Greater Asiatic Yellow House Bat
同物异名 Synonym(s)	Greater Asiatic Yellow Bat; Common Yellow Bat; *Nycticejus heathii* (Horsfield, 1831); *N. luteus* (Blyth, 1851); *S. heathii* (Horsfield, 1831); *Scotophilus heathii* (Geoffroy, (转下页)
种下单元评估 Infra-specific Taxa Assessed	无 / None

评估信息 Assessment Information

评估年份 Year Assessed	2020
评定人 Assessor(s)	吴毅、蒋志刚 / Yi Wu, Zhigang Jiang
审定人 Reviewer(s)	石红艳、江廷磊、余文华 / Hongyan Shi, Tinglei Jiang, Wenhua Yu
其他贡献人 Other Contributor(s)	李立立、丁晨晨 / Lili Li, Chenchen Ding

理由 Justification: 大黄蝠的种群数量较多。因此，列为无危等级 / Greater Asiatic Yellow House Bat has large populations. Thus, it is listed as Least Concern

地理分布 Geographical Distribution

国内分布 Domestic Distribution	云南、广西、广东、海南、福建 / Yunnan, Guangxi, Guangdong, Hainan, Fujian
世界分布 World Distribution	中国；中亚、南亚、东南亚 / China; Central Asia, South Asia, Southeast Asia
分布标注 Distribution Note	非特有种 / Non-endemic

国内分布图
Map of Domestic Distribution

种群 Population

种群数量 Population Size	稀少 / Rare
种群趋势 Population Trend	稳定 / Stable

生境与生态系统 Habitat (s) and Ecosystem (s)

生境 Habitat(s)	城市、人造建筑、森林 / Urban Area, Building, Forest
生态系统 Ecosystem(s)	森林生态系统、人类聚落生态系统 / Forest Ecosystem, Human Settlement Ecosystem

威胁 Threat (s)

主要威胁 Major Threat(s)	伐木、耕种 / Logging, Plantation

保护级别与保护行动 Protection Category and Conservation Action (s)

国家重点保护野生动物等级 (2021) Category of National Key Protected Wild Animals (2021)	无 / NA
IUCN 红色名录 (2020-2) IUCN Red List (2020-2)	无危 / LC
CITES 附录 (2019) CITES Appendix (2019)	无 / NA
保护行动 Conservation Action(s)	无 / None

相关文献 Relevant References

Wilson and Mittermeier, 2019; Burgin *et al*., 2018; Jiang *et al*. (蒋志刚等), 2017; Smith *et al*. (史密斯等), 2009; Wu *et al*. (吴毅等), 2007

(接上页)
1834) subsp. *belangeri*; *Vespertilio belangeri* (Geoffroy, 1834)

大黄蝠 *Scotophilus heathii*   吴毅 摄 By Yi Wu

小黄蝠
Scotophilus kuhlii
无危 LC

| 数据缺乏 DD | 无危 LC | 近危 NT | 易危 VU | 濒危 EN | 极危 CR | 区域灭绝 RE | 野外灭绝 EW | 灭绝 EX |

分类地位 Taxonomic Status

动物界 Animalia	脊索动物门 Chordata	哺乳纲 Mammalia	翼手目 Chiroptera	蝙蝠科 Vespertilionidae

学 名 Scientific Name	*Scotophilus kuhlii*
命 名 人 Species Authority	Leach, 1821
英 文 名 English Name(s)	Lesser Asiatic Yellow House Bat
同物异名 Synonym(s)	Lesser Asiatic Yellow Bat; Asiatic Lesser Yellow House Bat; *Scotophilus fulvus* (Gray, 1843); *S. kuhlii* (Thomas, 1897) subsp. *wroughtoni*; *S. temmincki* (Thomas, 1897) (转下页)
种下单元评估 Infra-specific Taxa Assessed	无 / None

评估信息 Assessment Information

评 估 年 份 Year Assessed	2020
评 定 人 Assessor(s)	吴毅、蒋志刚 / Yi Wu, Zhigang Jiang
审 定 人 Reviewer(s)	石红艳、江廷磊、余文华 / Hongyan Shi, Tinglei Jiang, Wenhua Yu
其他贡献人 Other Contributor(s)	李立立、丁晨晨 / Lili Li, Chenchen Ding

理由 Justification: 小黄蝠种群数量较多，列为无危等级 / Lesser Asiatic Yellow House Bat has large populations. Thus it is listed as Least Concern

地理分布 Geographical Distribution

国内分布 Domestic Distribution
广西、广东、海南、云南、福建、台湾 / Guangxi, Guangdong, Hainan, Yunnan, Fujian, Taiwan
世界分布 World Distribution
中国；南亚、东南亚 / China; South Asia, Southeast Asia
分布标注 Distribution Note
非特有种 / Non-endemic

国内分布图
Map of Domestic Distribution

种群 Population

种群数量 Population Size	常见 / Common
种群趋势 Population Trend	稳定 / Stable

生境与生态系统 Habitat (s) and Ecosystem (s)

生　　　境 Habitat(s)	森林、次生林、城市、人造建筑 / Forest, Secondary Forest, Urban Area, Building
生态系统 Ecosystem(s)	森林生态系统、人类聚落生态系统 / Forest Ecosystem, Human Settlement Ecosystem

威胁 Threat (s)

主要威胁 Major Threat(s)	未知 / Unknown

保护级别与保护行动 Protection Category and Conservation Action (s)

国家重点保护野生动物等级 (2021) Category of National Key Protected Wild Animals (2021)	无 / NA
IUCN 红色名录 (2020-2) IUCN Red List (2020-2)	无危 / LC
CITES 附录 (2019) CITES Appendix (2019)	无 / NA
保护行动 Conservation Action(s)	无 / None

相关文献 Relevant References

Wilson and Mittermeier, 2019; Burgin *et al.*, 2018; Jiang *et al.*（蒋志刚等）, 2017; Zheng *et al.*（郑锡奇等）, 2010; Zhu *et al.*（朱光剑等）, 2008; Wu *et al.*（吴毅等）, 2005; Wang（王应祥）, 2003

（接上页）
subsp. *wroughtoni*; *wroughtoni* (Thomas, 1897)

小黄蝠 *Scotophilus kuhlii*　　　吴毅 摄　By Yi Wu

斑蝠
Scotomanes ornatus
无危 LC

分类地位 Taxonomic Status

动物界 Animalia	脊索动物门 Chordata	哺乳纲 Mammalia	翼手目 Chiroptera	蝙蝠科 Vespertilionidae

学名 Scientific Name	*Scotomanes ornatus*
命名人 Species Authority	Blyth, 1851
英文名 English Name(s)	Harlequin Bat
同物异名 Synonym(s)	*Nycticejus emarginatus* (Dobson, 1871); *N. nivicolus* (Hodgson, 1855); *N. ornatus* (Blyth, 1851); *Scotomanes emarginatus* (Dobson, 1871)
种下单元评估 Infra-specific Taxa Assessed	无 / None

评估信息 Assessment Information

评估年份 Year Assessed	2020
评定人 Assessor(s)	吴毅、蒋志刚 / Yi Wu, Zhigang Jiang
审定人 Reviewer(s)	石红艳、江廷磊、余文华 / Hongyan Shi, Tinglei Jiang, Wenhua Yu
其他贡献人 Other Contributor(s)	李立立、丁晨晨 / Lili Li, Chenchen Ding

理由 Justification: 斑蝠分布广，种群数量较多。因此，列为无危等级 / Harlequin Bat is widely distributed and has large populations. Thus, it is listed as Least Concern

地理分布 Geographical Distribution

国内分布 Domestic Distribution
安徽、浙江、湖北、湖南、四川、贵州、云南、广西、广东、海南、福建、重庆 / Anhui, Zhejiang, Hubei, Hunan, Sichuan, Guizhou, Yunnan, Guangxi, Guangdong, Hainan, Fujian, Chongqing
世界分布 World Distribution
中国；南亚、东南亚 / China; South Asia, Southeast Asia
分布标注 Distribution Note
非特有种 / Non-endemic

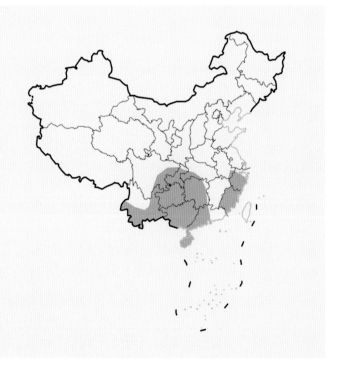

国内分布图
Map of Domestic Distribution

种群 Population

种群数量 Population Size	稀少 / Rare
种群趋势 Population Trend	稳定 / Stable

生境与生态系统 Habitat (s) and Ecosystem (s)

生　　境 Habitat(s)	森林、洞穴 / Forest, Cave
生态系统 Ecosystem(s)	森林生态系统 / Forest Ecosystem

威胁 Threat (s)

主要威胁 Major Threat(s)	伐木、耕种 / Logging, Plantation

保护级别与保护行动 Protection Category and Conservation Action (s)

国家重点保护野生动物等级 (2021) Category of National Key Protected Wild Animals (2021)	无 / NA
IUCN 红色名录 (2020-2) IUCN Red List (2020-2)	无危 / LC
CITES 附录 (2019) CITES Appendix (2019)	无 / NA
保护行动 Conservation Action(s)	无 / None

相关文献 Relevant References

Wilson and Mittermeier, 2019; Burgin *et al.*, 2018; Jiang *et al.* (蒋志刚等), 2017; Smith *et al.* (史密斯等), 2009; Ding and Wang (丁铁明和王作义), 1989; Liang *et al.* (梁仁济等), 1983

斑蝠 *Scotomanes ornatus*　　　　　吴毅 摄　By Yi Wu

大耳蝠
Plecotus auritus

无危 LC

| 数据缺乏 DD | 无危 LC | 近危 NT | 易危 VU | 濒危 EN | 极危 CR | 区域灭绝 RE | 野外灭绝 EW | 灭绝 EX |

分类地位 Taxonomic Status

动物界 Animalia	脊索动物门 Chordata	哺乳纲 Mammalia	翼手目 Chiroptera	蝙蝠科 Vespertilionidae

学名 Scientific Name	*Plecotus auritus*
命名人 Species Authority	Linnaeus, 1758
英文名 English Name(s)	Brown Big-eared Bat
同物异名 Synonym(s)	Brown Long-eared Bat
种下单元评估 Infra-specific Taxa Assessed	无 / None

评估信息 Assessment Information

评估年份 Year Assessed	2020
评定人 Assessor(s)	吴毅、蒋志刚 / Yi Wu, Zhigang Jiang
审定人 Reviewer(s)	石红艳、江廷磊、余文华 / Hongyan Shi, Tinglei Jiang, Wenhua Yu
其他贡献人 Other Contributor(s)	李立立、丁晨晨 / Lili Li, Chenchen Ding

理由 Justification: 大耳蝠种群数量较多。因此，列为无危等级 / Brown Big-eared Bat has large populations. Thus, it is listed as Least Concern

地理分布 Geographical Distribution

国内分布 Domestic Distribution
黑龙江、吉林、内蒙古、甘肃、河北、山西、陕西 / Heilongjiang, Jilin, Inner Mongolia (Nei Mongol), Gansu, Hebei, Shanxi, Shaanxi
世界分布 World Distribution
亚洲、欧洲 / Asia, Europe
分布标注 Distribution Note
非特有种 / Non-endemic

国内分布图
Map of Domestic Distribution

种群 Population

种群数量 Population Size	未知 / Unknown
种群趋势 Population Trend	稳定 / Stable

生境与生态系统 Habitat (s) and Ecosystem (s)

生　　境 Habitat(s)	森林、乡村花园、人造建筑、洞穴 / Forest, Rural Garden, Building, Cave
生态系统 Ecosystem(s)	森林生态系统、人类聚落生态系统 / Forest Ecosystem, Human Settlement Ecosystem

威胁 Threat (s)

主要威胁 Major Threat(s)	伐木、耕种 / Logging, Plantation

保护级别与保护行动 Protection Category and Conservation Action (s)

国家重点保护野生动物等级 (2021) Category of National Key Protected Wild Animals (2021)	无 / NA
IUCN红色名录(2020-2) IUCN Red List (2020-2)	无危 / LC
CITES 附录 (2019) CITES Appendix (2019)	无 / NA
保护行动 Conservation Action(s)	无 / None

相关文献 Relevant References

Wilson and Mittermeier, 2019; Burgin *et al*., 2018; Jiang *et al*. (蒋志刚等), 2017; Wang (王应祥), 2003; Wilson and Reeder, 1993

大耳蝠 *Plecotus auritus*　　　　冯利民 摄　By Limin Feng

黄胸管鼻蝠
Murina bicolor
无危 LC

| DD 数据缺乏 | **LC 无危** | NT 近危 | VU 易危 | EN 濒危 | CR 极危 | RE 区域灭绝 | EW 野外灭绝 | EX 灭绝 |

分类地位 Taxonomic Status

动物界 Animalia	脊索动物门 Chordata	哺乳纲 Mammalia	翼手目 Chiroptera	蝙蝠科 Vespertilionidae
学名 Scientific Name		*Murina bicolor*		
命名人 Species Authority		Kuo, 2009		
英文名 English Name(s)		Yellow-chested Tube-nosed Bat		
同物异名 Synonym(s)		Bicolored Tube-nosed Bat		
种下单元评估 Infra-specific Taxa Assessed		无 / None		

评估信息 Assessment Information

评估年份 Year Assessed	2020
评定人 Assessor(s)	吴毅、蒋志刚 / Yi Wu, Zhigang Jiang
审定人 Reviewer(s)	余文华、石红艳、江廷磊 / Wenhua Yu, Hongyan Shi, Tinglei Jiang
其他贡献人 Other Contributor(s)	李立立、丁晨晨 / Lili Li, Chenchen Ding

理由 Justification: 黄胸管鼻蝠为中国特有种，仅分布于台湾。目前，其种群数量稳定。因此，列为无危等级 / Yellow-chested Tube-nosed Bat is an endemic species and distributed only in Taiwan, China. Its population is stable currently. Thus, it is listed as Least Concern

地理分布 Geographical Distribution

国内分布 Domestic Distribution
台湾 / Taiwan
世界分布 World Distribution
中国 / China
分布标注 Distribution Note
特有种 / Endemic

国内分布图
Map of Domestic Distribution

种群 Population

种群数量 Population Size	未知 / Unknown
种群趋势 Population Trend	未知 / Unknown

生境与生态系统 Habitat(s) and Ecosystem(s)

生　　境 Habitat(s)	森林、人造建筑 / Forest, Building
生态系统 Ecosystem(s)	森林生态系统、人类聚落生态系统 / Forest Ecosystem, Human Settlement Ecosystem

威胁 Threat(s)

主要威胁 Major Threat(s)	未知 / Unknown

保护级别与保护行动 Protection Category and Conservation Action(s)

国家重点保护野生动物等级 (2021) Category of National Key Protected Wild Animals (2021)	无 / NA
IUCN 红色名录 (2020-2) IUCN Red List (2020-2)	无危 / LC
CITES 附录 (2019) CITES Appendix (2019)	无 / NA
保护行动 Conservation Action(s)	无 / None

相关文献 Relevant References

Burgin *et al.*, 2018; Jiang *et al.*（蒋志刚等），2017; Cheng *et al.*, 2015; Kuo *et al.*, 2009

黄胸管鼻蝠 *Murina bicolor*

东北管鼻蝠
Murina hilgendorfi

无危 LC

| 数据缺乏 DD | 无危 LC | 近危 NT | 易危 VU | 濒危 EN | 极危 CR | 区域灭绝 RE | 野外灭绝 EW | 灭绝 EX |

分类地位 Taxonomic Status

| 动物界 Animalia | 脊索动物门 Chordata | 哺乳纲 Mammalia | 翼手目 Chiroptera | 蝙蝠科 Vespertilionidae |

学名 Scientific Name	*Murina hilgendorfi*
命名人 Species Authority	Peters, 1880
英文名 English Name(s)	Greater Tube-nosed Bat
同物异名 Synonym(s)	Hilgendorf's Tube-nosed Bat
种下单元评估 Infra-specific Taxa Assessed	无 / None

评估信息 Assessment Information

评估年份 Year Assessed	2020
评定人 Assessor(s)	吴毅、蒋志刚 / Yi Wu, Zhigang Jiang
审定人 Reviewer(s)	余文华、石红艳、江廷磊 / Wenhua Yu, Hongyan Shi, Tinglei Jiang
其他贡献人 Other Contributor(s)	李立立、丁晨晨 / Lili Li, Chenchen Ding

理由 Justification: 东北管鼻蝠种群数量较多。因此，列为无危等级 / Greater Tube-nosed Bat has large populations. Thus, it is listed as Least Concern

地理分布 Geographical Distribution

国内分布 Domestic Distribution
黑龙江、内蒙古 / Heilongjiang, Inner Mongolia (Nei Mongol)
世界分布 World Distribution
中国；东北亚 / China; Northeast Asia
分布标注 Distribution Note
非特有种 / Non-endemic

国内分布图
Map of Domestic Distribution

种群 Population

种群数量 Population Size	未知 / Unknown
种群趋势 Population Trend	未知 / Unknown

生境与生态系统 Habitat (s) and Ecosystem (s)

生境 Habitat(s)	洞穴、森林、人造建筑 / Cave, Forest, Building
生态系统 Ecosystem(s)	森林生态系统、人类聚落生态系统 / Forest Ecosystem, Human Settlement Ecosystem

威胁 Threat (s)

主要威胁 Major Threat(s)	砍伐 / Logging

保护级别与保护行动 Protection Category and Conservation Action (s)

国家重点保护野生动物等级 (2021) Category of National Key Protected Wild Animals (2021)	无 / NA
IUCN 红色名录 (2020-2) IUCN Red List (2020-2)	数据缺乏 / DD
CITES 附录 (2019) CITES Appendix (2019)	无 / NA
保护行动 Conservation Action(s)	无 / None

相关文献 Relevant References

Liu and Wu (刘少英和吴毅), 2019; Wilson and Mittermeier, 2019; Burgin *et al*., 2018; Jiang *et al*. (蒋志刚等), 2017; Smith *et al*. (史密斯等), 2009; Wang (王应祥), 2003

东北管鼻蝠 *Murina hilgendorfi*

中管鼻蝠
Murina huttoni
无危 LC

数据缺乏 DD | 无危 LC | 近危 NT | 易危 VU | 濒危 EN | 极危 CR | 区域灭绝 RE | 野外灭绝 EW | 灭绝 EX

分类地位 Taxonomic Status

动物界 Animalia	脊索动物门 Chordata	哺乳纲 Mammalia	翼手目 Chiroptera	蝙蝠科 Vespertilionidae

学名 Scientific Name	*Murina huttoni*
命名人 Species Authority	Peters, 1872
英文名 English Name(s)	Hutton's Tube-nosed Bat
同物异名 Synonym(s)	*Harpyiocephalus huttonii* (Peters, 1872)
种下单元评估 Infra-specific Taxa Assessed	无 / None

评估信息 Assessment Information

评估年份 Year Assessed	2020
评定人 Assessor(s)	吴毅、蒋志刚 / Yi Wu, Zhigang Jiang
审定人 Reviewer(s)	余文华、石红艳、江廷磊 / Wenhua Yu, Hongyan Shi, Tinglei Jiang
其他贡献人 Other Contributor(s)	李立立、丁晨晨 / Lili Li, Chenchen Ding

理由 Justification: 中管鼻蝠种群数量较多。因此，列为无危等级 / Hutton's Tube-nosed Bat has large populations. Thus, it is listed as Least Concern

地理分布 Geographical Distribution

国内分布 Domestic Distribution
广西、广东、福建 / Guangxi, Guangdong, Fujian
世界分布 World Distribution
中国；南亚、东南亚 / China; South Asia, Southeast Asia
分布标注 Distribution Note
非特有种 / Non-endemic

国内分布图
Map of Domestic Distribution

种群 Population

种群数量 Population Size	常见 / Common
种群趋势 Population Trend	稳定 / Stable

生境与生态系统 Habitat(s) and Ecosystem(s)

生 境 Habitat(s)	森林、种植园 / Forest, Plantation
生态系统 Ecosystem(s)	森林生态系统、农田生态系统 / Forest Ecosystem, Cropland Ecosystem

威胁 Threat(s)

主要威胁 Major Threat(s)	砍伐 / Logging

保护级别与保护行动 Protection Category and Conservation Action(s)

国家重点保护野生动物等级 (2021) Category of National Key Protected Wild Animals (2021)	无 / NA
IUCN红色名录(2020-2) IUCN Red List (2020-2)	数据缺乏 / DD
CITES 附录 (2019) CITES Appendix (2019)	无 / NA
保护行动 Conservation Action(s)	无 / None

相关文献 Relevant References

Wilson and Mittermeier, 2019; Burgin *et al*., 2018; Huang *et al*. (黄正澜懿等), 2018; Jiang *et al*. (蒋志刚等), 2017; Zhou *et al*. (周全等), 2011; Smith *et al*. (史密斯等), 2009

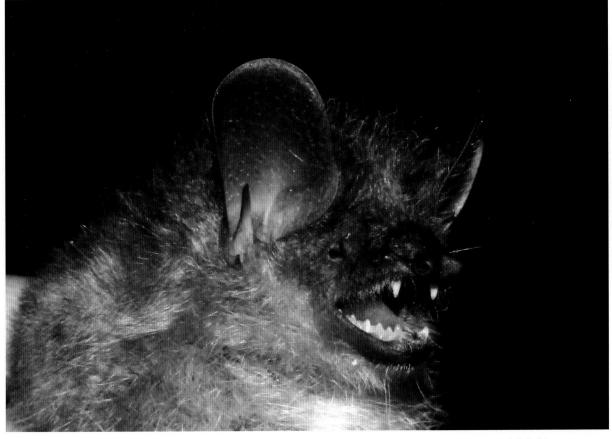

中管鼻蝠 *Murina huttoni* 余文华 摄 By Wenhua Yu

白腹管鼻蝠
Murina leucogaster
无危 LC

| 数据缺乏 DD | 无危 LC | 近危 NT | 易危 VU | 濒危 EN | 极危 CR | 区域灭绝 RE | 野外灭绝 EW | 灭绝 EX |

分类地位 Taxonomic Status

动物界 Animalia	脊索动物门 Chordata	哺乳纲 Mammalia	翼手目 Chiroptera	蝙蝠科 Vespertilionidae

学名 Scientific Name	*Murina leucogaster*
命名人 Species Authority	Milne-Edwards, 1872
英文名 English Name(s)	Rufous Tube-nosed Bat
同物异名 Synonym(s)	Greater Tube-nosed Bat; *Murina leucogastra* (Thomas, 1899); *rubex* (Thomas, 1916)
种下单元评估 Infra-specific Taxa Assessed	无 / None

评估信息 Assessment Information

评估年份 Year Assessed	2020
评定人 Assessor(s)	吴毅、蒋志刚 / Yi Wu, Zhigang Jiang
审定人 Reviewer(s)	余文华、石红艳、江廷磊 / Wenhua Yu, Hongyan Shi, Tinglei Jiang
其他贡献人 Other Contributor(s)	李立立、丁晨晨 / Lili Li, Chenchen Ding

理由 Justification: 白腹管鼻蝠种群数量较多。因此，列为无危等级 / Rufous Tube-nosed Bat has large populations. Thus, it is listed as Least Concern

地理分布 Geographical Distribution

国内分布 Domestic Distribution
北京、河北、山西、辽宁、吉林、黑龙江、福建、广西、四川、贵州、西藏、陕西、云南 / Beijing, Hebei, Shanxi, Liaoning, Jilin, Heilongjiang, Fujian, Guangxi, Sichuan, Guizhou, Tibet (Xizang), Shaanxi, Yunnan

世界分布 World Distribution
中国；南亚、东南亚、东北亚 / China; South Asia, Southeast Asia, Northeast Asia

分布标注 Distribution Note
非特有种 / Non-endemic

国内分布图
Map of Domestic Distribution

种群 Population

种群数量 Population Size	未知 / Unknown
种群趋势 Population Trend	未知 / Unknown

生境与生态系统 Habitat (s) and Ecosystem (s)

生境 Habitat(s)	洞穴、森林、人造建筑 / Cave, Forest, Building
生态系统 Ecosystem(s)	森林生态系统、人类聚落生态系统 / Forest Ecosystem, Human Settlement Ecosystem

威胁 Threat (s)

主要威胁 Major Threat(s)	伐木 / Logging

保护级别与保护行动 Protection Category and Conservation Action (s)

国家重点保护野生动物等级 (2021) Category of National Key Protected Wild Animals (2021)	无 / NA
IUCN 红色名录 (2020-2) IUCN Red List (2020-2)	数据缺乏 / DD
CITES 附录 (2019) CITES Appendix (2019)	无 / NA
保护行动 Conservation Action(s)	无 / None

相关文献 Relevant References

Liu and Wu (刘少英和吴毅), 2019; Wilson and Mittermeier, 2019; Burgin *et al.*, 2018; Jiang *et al.* (蒋志刚等), 2017; Smith *et al.* (史密斯等), 2009; Wang (王应祥), 2003

白腹管鼻蝠 *Murina leucogaster*

泰坦尼亚彩蝠
Kerivoula titania

无危 LC

| 数据缺乏 DD | 无危 LC | 近危 NT | 易危 VU | 濒危 EN | 极危 CR | 区域灭绝 RE | 野外灭绝 EW | 灭绝 EX |

分类地位 Taxonomic Status

| 动物界 Animalia | 脊索动物门 Chordata | 哺乳纲 Mammalia | 翼手目 Chiroptera | 蝙蝠科 Vespertilionidae |

学名 Scientific Name	*Kerivoula titania*
命名人 Species Authority	Bates, 2007
英文名 English Name(s)	Titania's Woolly Bat
同物异名 Synonym(s)	无 / None
种下单元评估 Infra-specific Taxa Assessed	无 / None

评估信息 Assessment Information

评估年份 Year Assessed	2020
评定人 Assessor(s)	吴毅、蒋志刚 / Yi Wu, Zhigang Jiang
审定人 Reviewer(s)	张礼标、毛秀光、张树义 / Libiao Zhang, Xiuguang Mao, Shuyi Zhang
其他贡献人 Other Contributor(s)	李立立、丁晨晨 / Lili Li, Chenchen Ding

理由 Justification: 泰坦尼亚彩蝠在中国主要分布于台湾和海南，其种群数量少但稳定。因此，列为无危等级 / Titania's Woolly Bat is found mainly in Taiwan and Hainan in China, its population sizes are small but stable. Thus, it is listed as Least Concern

地理分布 Geographical Distribution

国内分布 Domestic Distribution	海南、台湾 / Hainan, Taiwan
世界分布 World Distribution	中国；东南亚 / China; Southeast Asia
分布标注 Distribution Note	非特有种 / Non-endemic

国内分布图
Map of Domestic Distribution

种群 Population

种群数量 Population Size	稳定 / Stable
种群趋势 Population Trend	未知 / Unknown

生境与生态系统 Habitat(s) and Ecosystem(s)

生　　境 Habitat(s)	森林、喀斯特地貌 / Forest, Karst Landscape
生态系统 Ecosystem(s)	森林生态系统、喀斯特生态系统 / Forest Ecosystem, Karst Ecosystem

威胁 Threat(s)

主要威胁 Major Threat(s)	砍伐 / Logging

保护级别与保护行动 Protection Category and Conservation Action(s)

国家重点保护野生动物等级 (2021) Category of National Key Protected Wild Animals (2021)	无 / NA
IUCN 红色名录 (2020-2) IUCN Red List (2020-2)	数据缺乏 / DD
CITES 附录 (2019) CITES Appendix (2019)	无 / NA
保护行动 Conservation Action(s)	无 / None

相关文献 Relevant References

Wilson and Mittermeier, 2019; Burgin *et al*., 2018; Jiang *et al*.（蒋志刚等），2017; Wu *et al*., 2012a

泰坦尼亚彩蝠 *Kerivoula titania*　　　　吴毅 摄　By Yi Wu

猕猴
Macaca mulatta

无危 LC

| 数据缺乏 DD | 无危 LC | 近危 NT | 易危 VU | 濒危 EN | 极危 CR | 区域灭绝 RE | 野外灭绝 EW | 灭绝 EX |

分类地位 Taxonomic Status

动物界 Animalia	脊索动物门 Chordata	哺乳纲 Mammalia	灵长目 Primates	猴科 Cercopithecidae
学　　名 Scientific Name		*Macaca mulatta*		
命 名 人 Species Authority		Zimmermann, 1780		
英 文 名 English Name(s)		Rhesus Monkey		
同物异名 Synonym(s)		Rhesus Macaque; *Macaca brachyurus* (Elliot, 1909); *brevicaudatus* (Elliot, 1913); *erythraea* (Shaw, 1800); *fulvus* (Kerr, 1792); *lasiotus* (Gray, 1868); *littoralis*（转下页）		
种下单元评估 Infra-specific Taxa Assessed		无 / None		

评估信息 Assessment Information

评估年份 Year Assessed	2020
评 定 人 Assessor(s)	蒋志刚、蒋学龙 / Zhigang Jiang, Xuelong Jiang
审 定 人 Reviewer(s)	黄乘明、蒋学龙、龙勇诚、范朋飞 / Chengming Huang, Xuelong Jiang, Yongcheng Long, Pengfei Fan
其他贡献人 Other Contributor(s)	李立立、丁晨晨 / Lili Li, Chenchen Ding

理由 Justification: 猕猴在中国分布广，种群数量多。因此，列为无危等级 / Rhesus Monkey is widely distributed with large populations in China. Thus, it is listed as Least Concern

地理分布 Geographical Distribution

国内分布 Domestic Distribution

山西、湖南、四川、云南、广西、浙江、安徽、福建、江西、河南、湖北、广东、海南、贵州、西藏、陕西、甘肃、青海、香港、重庆 / Shanxi, Hunan, Sichuan, Yunnan, Guangxi, Zhejiang, Anhui, Fujian, Jiangxi, Henan, Hubei, Guangdong, Hainan, Guizhou, Tibet (Xizang), Shaanxi, Gansu, Qinghai, Hong Kong, Chongqing

世界分布 World Distribution

中国；南亚、东南亚 / China; South Asia, Southeast Asia

分布标注 Distribution Note

非特有种 / Non-endemic

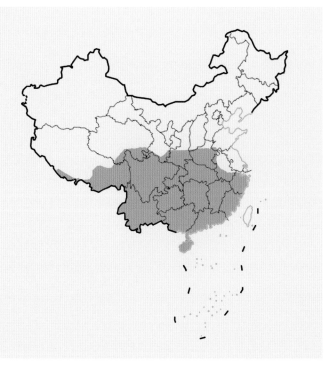

国内分布图
Map of Domestic Distribution

种群 Population

种群数量 Population Size	100,000 只（非精准估计）/ 100,000 (Approximation)
种群趋势 Population Trend	未知 / Unknown

生境与生态系统 Habitat(s) and Ecosystem(s)

生境 Habitat(s)	森林、海岸、灌丛、红树林、人造建筑、种植园 / Forest, Coast, Shrubland, Mangrove, Building, Plantation
生态系统 Ecosystem(s)	森林生态系统、灌丛生态系统、农田生态系统、人类聚落生态系统 / Forest Ecosystem, Shrubland Ecosystem, Cropland Ecosystem, Human Settlement Ecosystem

威胁 Threat(s)

主要威胁 Major Threat(s)	未知 / Unknown

保护级别与保护行动 Protection Category and Conservation Action(s)

国家重点保护野生动物等级 (2021) Category of National Key Protected Wild Animals (2021)	二级 / Category II
IUCN 红色名录 (2020-2) IUCN Red List (2020-2)	无危 / LC
CITES 附录 (2019) CITES Appendix (2019)	II
保护行动 Conservation Action(s)	在自然保护区内的种群及栖息地得到保护 / Populations and habitats in nature reserves are protected

相关文献 Relevant References

Burgin *et al*., 2018; Jiang *et al*.（蒋志刚等），2017; Wilson and Mittermeier, 2012; Li *et al*.（李友邦等），2009; Yang（杨光照），2007; Tan *et al*.（谈建文等），2005

（接上页）
(Elliot, 1909); *mcmahoni* (Pocock, 1932); *nipalensis* (Hodgson, 1840); *oinops* (Hodgson, 1840); *rhesus* (Audebert, 1798); *sancti-johannis* (Swinhoe, 1866); *siamica* (Kloss, 1917); *tcheliensis* (Milne-Edwards, 1872); *vestita* (Milne-Edwards, 1892); *villosa* (True, 1894)

猕猴 *Macaca mulatta*　　　　　　　　　　　　　　蒋志刚 摄　By Zhigang Jiang

中国生物多样性 红色名录

黄鼬
Mustela sibirica

无危 LC

| DD 数据缺乏 | LC 无危 | NT 近危 | VU 易危 | EN 濒危 | CR 极危 | RE 区域灭绝 | EW 野外灭绝 | EX 灭绝 |

分类地位 Taxonomic Status

动物界 Animalia	脊索动物门 Chordata	哺乳纲 Mammalia	食肉目 Carnivora	鼬科 Mustelidae

学名 Scientific Name	*Mustela sibirica*
命名人 Species Authority	Pallas, 1773
英文名 English Name(s)	Siberian Weasel
同物异名 Synonym(s)	无 / None
种下单元评估 Infra-specific Taxa Assessed	无 / None

评估信息 Assessment Information

评估年份 Year Assessed	2020
评定人 Assessor(s)	蒋志刚 / Zhigang Jiang
审定人 Reviewer(s)	马建章、徐爱春、李晟、胡慧建、王昊、张明海、马逸清、周友兵 / Jianzhang Ma, Aichun Xu, Sheng Li, Huijian Hu, Hao Wang, Minghai Zhang, Yiqing Ma, Youbing Zhou
其他贡献人 Other Contributor(s)	李立立、丁晨晨 / Lili Li, Chenchen Ding

理由 Justification: 黄鼬分布较广，种群数量较大。因此，列为无危等级 / Siberian Weasel is widely distributed with large populations. Thus, it is listed as Least Concern

地理分布 Geographical Distribution

国内分布 Domestic Distribution

吉林、山西、河南、云南、湖南、新疆、北京、河北、内蒙古、辽宁、黑龙江、上海、江苏、浙江、安徽、福建、江西、山东、湖北、广东、广西、四川、贵州、西藏、陕西、甘肃、青海、宁夏、重庆、台湾 / Jilin, Shanxi, Henan, Yunnan, Hunan, Xinjiang, Beijing, Hebei, Inner Mongolia (Nei Mongol), Liaoning, Heilongjiang, Shanghai, Jiangsu, Zhejiang, Anhui, Fujian, Jiangxi, Shandong, Hubei, Guangdong, Guangxi, Sichuan, Guizhou, Tibet (Xizang), Shaanxi, Gansu, Qinghai, Ningxia, Chongqing, Taiwan

世界分布 World Distribution

亚洲 / Asia

分布标注 Distribution Note

非特有种 / Non-endemic

国内分布图
Map of Domestic Distribution

种群 Population

种群数量 Population Size	未知 / Unknown
种群趋势 Population Trend	稳定 / Stable

生境与生态系统 Habitat (s) and Ecosystem (s)

生　　境 Habitat(s)	森林、次生林、沼泽、耕地 / Forest, Secondary Forest, Swamp, Arable Land
生态系统 Ecosystem(s)	森林生态系统、农田生态系统、湿地生态系统 / Forest Ecosystem, Cropland Ecosystem, Wetland Ecosystem

威胁 Threat (s)

主要威胁 Major Threat(s)	猎捕 / Trapping

保护级别与保护行动 Protection Category and Conservation Action (s)

国家重点保护野生动物等级 (2021) Category of National Key Protected Wild Animals (2021)	无 / NA
IUCN 红色名录 (2020-2) IUCN Red List (2020-2)	无危 / LC
CITES 附录 (2019) CITES Appendix (2019)	III
保护行动 Conservation Action(s)	无 / None

相关文献 Relevant References

Burgin *et al.*, 2018; Jiang *et al.* (蒋志刚等), 2017; Deng *et al.* (邓可等), 2013; Hunter and Barrett, 2011; Smith *et al.* (史密斯等), 2009; Wilson and Mittermeier, 2009; Wang (王应祥), 2003

黄鼬 *Mustela sibirica*

灰獴
Herpestes edwardsii
无危 LC

| DD 数据缺乏 | **LC 无危** | NT 近危 | VU 易危 | EN 濒危 | CR 极危 | RE 区域灭绝 | EW 野外灭绝 | EX 灭绝 |

分类地位 Taxonomic Status

动物界 Animalia	脊索动物门 Chordata	哺乳纲 Mammalia	食肉目 Carnivora	獴科 Herpestidae

学名 Scientific Name	*Herpestes edwardsii*
命名人 Species Authority	É. Geoffroy Saint-Hilaire, 1818
英文名 English Name(s)	Indian Grey Mongoose
同物异名 Synonym(s)	Common Mongoose; Grey Mongoose; *Urva edwardsii* (É. Geoffroy Saint-Hilaire, 1818)
种下单元评估 Infra-specific Taxa Assessed	无 / None

评估信息 Assessment Information

评估年份 Year Assessed	2020
评定人 Assessor(s)	蒋志刚 / Zhigang Jiang
审定人 Reviewer(s)	李晟、胡慧建、王大军、王昊、周友兵 / Sheng Li, Huijian Hu, Dajun Wang, Hao Wang, Youbing Zhou
其他贡献人 Other Contributor(s)	丁晨晨 / Chenchen Ding

理由 Justification: 灰獴是一个常见的、广布的种。灰獴适应于人居环境，尽管常被捕捉，但灰獴处于无危状态。因此，列为无危等级 / Indian Grey Mongoose is a common and widespread species. It is adapted to human environments, including settlements, and it is often captured. Thus, it is listed as Least Concern

地理分布 Geographical Distribution

国内分布 Domestic Distribution
西藏 / Tibet (Xizang)

世界分布 World Distribution
中国；中亚、西亚、北非 / China; Central Asia, West Asia, North Africa

分布标注 Distribution Note
非特有种 / Non-endemic

国内分布图
Map of Domestic Distribution

种群 Population

种群数量 Population Size	未知 / Unknown
种群趋势 Population Trend	稳定 / Stable

生境与生态系统 Habitat (s) and Ecosystem (s)

生境 Habitat(s)	人居生境、次生林和热带棘林 / Human Living Area, Secondary Forest and Tropical Thorn Forest
生态系统 Ecosystem(s)	人类聚落生态系统、次生林生态系统 / Human Settlement Ecosystem, Secondary Forest Ecosystem

威胁 Threat (s)

主要威胁 Major Threat(s)	未知 / Unknown

保护级别与保护行动 Protection Category and Conservation Action (s)

国家重点保护野生动物等级 (2021) Category of National Key Protected Wild Animals (2021)	未列入 / NA
IUCN 红色名录 (2020-2) IUCN Red List (2020-2)	无危 / LC
CITES 附录 (2019) CITES Appendix (2019)	无 / NA
保护行动 Conservation Action(s)	无 / None

相关文献 Relevant References

Burgin *et al*., 2018; Jiang *et al*. (蒋志刚等), 2017; Hunter and Barrett, 2011; Wilson and Mittermeier, 2009; Choudhury, 2003

灰獴 *Herpestes edwardsii*

中国生物多样性红色名录

野猪
Sus scrofa

无危 LC

| DD 数据缺乏 | LC 无危 | NT 近危 | VU 易危 | EN 濒危 | CR 极危 | RE 区域灭绝 | EW 野外灭绝 | EX 灭绝 |

分类地位 Taxonomic Status

| 动物界 Animalia | 脊索动物门 Chordata | 哺乳纲 Mammalia | 偶蹄目 Artiodactyla | 猪科 Suidae |

学名 Scientific Name	*Sus scrofa*
命名人 Species Authority	Linnaeus, 1758
英文名 English Name(s)	Wild Boar
同物异名 Synonym(s)	*Sus andamanensis* (Blyth, 1858); *aruensis* (Rosenberg, 1878); *babi* (Miller, 1906); *ceramensis* (Rosenberg, 1878); *enganus* (Lyon, 1916); *floresianus* (Jentink, 1905); (转下页)
种下单元评估 Infra-specific Taxa Assessed	无 / None

评估信息 Assessment Information

评估年份 Year Assessed	2020
评定人 Assessor(s)	蒋志刚 / Zhigang Jiang
审定人 Reviewer(s)	胡慧建、刘丙万 / Huijian Hu, Bingwan Liu
其他贡献人 Other Contributor(s)	李立立、丁晨晨 / Lili Li, Chenchen Ding

理由 Justification: 野猪在全国均有分布，且种群呈现增长趋势，在东北地区监测部分区域表明野猪种群密度可约达 0.3 头 /km²。在东北等地区，野猪毁坏庄稼已成为难题。因此，列为无危等级 / Wild Boar is distributed across the entire country and its populations have shown increasing trends. Regional monitoring in northeast China has shown population densities reaching about 0.3 individuals/km². In northeast China regions *etc*., crop destruction by Wild Boar has become a problem. Thus, based on its population and habitat status, Wild Boar is listed as Least Concern

地理分布 Geographical Distribution

国内分布 Domestic Distribution

山西、湖南、海南、新疆、北京、河北、内蒙古、辽宁、吉林、黑龙江、上海、江苏、浙江、安徽、福建、江西、河南、湖北、广东、广西、四川、贵州、云南、西藏、陕西、甘肃、青海、宁夏、台湾、香港、天津、重庆 / Shanxi, Hunan, Hainan, Xinjiang, Beijing, Hebei, Inner Mongolia (Nei Mongol), Liaoning, Jilin, Heilongjiang, Shanghai, Jiangsu, Zhejiang, Anhui, Fujian, Jiangxi, Henan, Hubei, Guangdong, Guangxi, Sichuan, Guizhou, Yunnan, Tibet (Xizang), Shaanxi, Gansu, Qinghai, Ningxia, Taiwan, Hong Kong, Tianjin, Chongqing

世界分布 World Distribution

全球除极端干旱和高寒地区之外的陆地 / Lands of the world except for extremely arid and alpine regions

分布标注 Distribution Note

非特有种 / Non-endemic

国内分布图
Map of Domestic Distribution

种群 Population

种群数量 Population Size	1,000,000 只 / 1,000,000 individuals
种群趋势 Population Trend	稳定 / Stable

生境与生态系统 Habitat(s) and Ecosystem(s)

生　　境 Habitat(s)	森林、灌丛、草地、沼泽、农田 / Forest, Shrubland, Grassland, Swamp, Cropland
生态系统 Ecosystem(s)	森林生态系统、灌丛生态系统、草地生态系统、农田生态系统、湿地生态系统 / Forest Ecosystem, Shrubland Ecosystem, Grassland Ecosystem, Cropland Ecosystem, Wetland Ecosystem

威胁 Threat(s)

主要威胁 Major Threat(s)	无 / None

保护级别与保护行动 Protection Category and Conservation Action(s)

国家重点保护野生动物等级 (2021) Category of National Key Protected Wild Animals (2021)	无 / NA
IUCN 红色名录 (2020-2) IUCN Red List (2020-2)	无危 / LC
CITES 附录 (2019) CITES Appendix (2019)	无 / NA
保护行动 Conservation Action(s)	无管理计划 / No management plan

相关文献 Relevant References

Burgin *et al.*, 2018; Jiang *et al.* (蒋志刚等), 2017; Castelló, 2016; Wilson and Mittermeier, 2011; Liu *et al.* (刘鹤等), 2011; Groves and Grubb, 2011; Smith *et al.* (史密斯等), 2009; Wang (王应祥), 2003

(接上页)
goramensis (De Beaux, 1924); *natunensis* (Miller, 1901); *nicobaricus* (Miller, 1902); *niger* (Finsch, 1886); *papuensis* (Lesson and Garnot, 1826); *ternatensis* (Rolleston, 1877); *tuancus* (Lyon, 1916)

野猪 *Sus scrofa*

印度麂 *Muntiacus muntjak*
无危 LC

| 数据缺乏 DD | 无危 LC | 近危 NT | 易危 VU | 濒危 EN | 极危 CR | 区域灭绝 RE | 野外灭绝 EW | 灭绝 EX |

分类地位 Taxonomic Status

动物界 Animalia	脊索动物门 Chordata	哺乳纲 Mammalia	偶蹄目 Artiodactyla	鹿科 Cervidae

学名 Scientific Name	*Muntiacus muntjak*
命名人 Species Authority	Zimmermann, 1780
英文名 English Name(s)	Southern Red Muntjac
同物异名 Synonym(s)	Indian Muntjac; *Cervus moschatus* (Blainville, 1816); *C. muntjak* (Zimmermann, 1780); *C. pleiharicus* (Kohlbrugge, 1896); *Muntiacus bancanus* (Lyon, 1906); (转下页)
种下单元评估 Infra-specific Taxa Assessed	无 / None

评估信息 Assessment Information

评估年份 Year Assessed	2020
评定人 Assessor(s)	蒋志刚 / Zhigang Jiang
审定人 Reviewer(s)	蒋学龙 / Xuelong Jiang
其他贡献人 Other Contributor(s)	丁晨晨 / Chenchen Ding

理由 Justification: 在中国藏南地区，印度麂在其分布区较常见。因此，列为无危等级 / Southern Red Muntjac is relatively common in Zangnan area, China. Thus, it is listed as Least Concern

地理分布 Geographical Distribution

国内分布 Domestic Distribution
西藏 / Tibet (Xizang)
世界分布 World Distribution
文莱、中国、印度、印度尼西亚、马来西亚、泰国 / Brunei, China, India, Indonesia, Malaysia, Thailand
分布标注 Distribution Note
非特有种 / Non-endemic

国内分布图
Map of Domestic Distribution

种群 Population

种群数量 Population Size	未知 / Unknown
种群趋势 Population Trend	下降 / Decreasing

生境与生态系统 Habitat (s) and Ecosystem (s)

生　　境 Habitat(s)	森林、严重退化森林和森林附近的种植园 / Forest, Heavily Degraded Forest and Plantation Near Forest
生态系统 Ecosystem(s)	森林生态系统与种植园生态系统 / Forest Ecosystem and Plantation Ecosystem

威胁 Threat (s)

主要威胁 Major Threat(s)	狩猎、开垦 / Hunting, Land Reclaimed for Plantation

保护级别与保护行动 Protection Category and Conservation Action (s)

国家重点保护野生动物等级 (2021) Category of National Key Protected Wild Animals (2021)	无 / NA
IUCN 红色名录 (2020-2) IUCN Red List (2020-2)	无危 / LC
CITES 附录 (2019) CITES Appendix (2019)	无 / NA
保护行动 Conservation Action(s)	未知 / Unknown

相关文献 Relevant References

Burgin *et al.*, 2018; Jiang *et al.*（蒋志刚等）, 2017; Castelló, 2016; Groves and Grubb, 2011; Wilson and Mittermeier, 2011; Choudhury, 2003; Wang（王应祥）, 2003

（接上页）
M. rubidus (Lyon, 1911)

印度麂 *Muntiacus muntjak*

中国生物多样性红色名录

岩羊
Pseudois nayaur

无危 LC

| 数据缺乏 DD | 无危 LC | 近危 NT | 易危 VU | 濒危 EN | 极危 CR | 区域灭绝 RE | 野外灭绝 EW | 灭绝 EX |

分类地位 Taxonomic Status

动物界 Animalia	脊索动物门 Chordata	哺乳纲 Mammalia	偶蹄目 Artiodactyla	牛科 Bovidae

学名 Scientific Name	*Pseudois nayaur*
命名人 Species Authority	Hodgson, 1833
英文名 English Name(s)	Bharal
同物异名 Synonym(s)	Blue Sheep, Himalayan Blue Sheep, Naur
种下单元评估 Infra-specific Taxa Assessed	无 / None

评估信息 Assessment Information

评估年份 Year Assessed	2020
评定人 Assessor(s)	蒋志刚 / Zhigang Jiang
审定人 Reviewer(s)	初红军、杨维康 / Hongjun Chu, Weikang Yang
其他贡献人 Other Contributor(s)	李立立、丁晨晨 / Lili Li, Chenchen Ding

理由 Justification: 岩羊在中国分布广泛，种群数量多。因此，列为无危等级 / Bharal is widely distributed in China with large populations. Thus, it is listed as Least Concern

地理分布 Geographical Distribution

国内分布 Domestic Distribution
内蒙古、新疆、青海、甘肃、四川、云南、西藏、陕西、宁夏 / Inner Mongolia (Nei Mongol), Xinjiang, Qinghai, Gansu, Sichuan, Yunnan, Tibet (Xizang), Shaanxi, Ningxia
世界分布 World Distribution
中国；南亚 / China; South Asia
分布标注 Distribution Note
非特有种 / Non-endemic

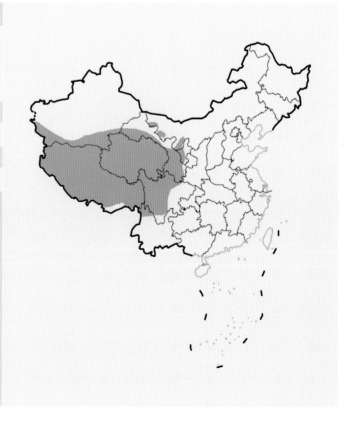

国内分布图
Map of Domestic Distribution

种群 Population

种群数量 Population Size	未知 / Unknown
种群趋势 Population Trend	上升 / Increasing

生境与生态系统 Habitat (s) and Ecosystem (s)

生　　境 Habitat(s)	草甸 / Meadow
生态系统 Ecosystem(s)	草地生态系统 / Grassland Ecosystem

威胁 Threat (s)

主要威胁 Major Threat(s)	无 / None

保护级别与保护行动 Protection Category and Conservation Action (s)

国家重点保护野生动物等级 (2021) Category of National Key Protected Wild Animals (2021)	二级 / Category II
IUCN红色名录(2020-2) IUCN Red List (2020-2)	无危 / LC
CITES 附录 (2019) CITES Appendix (2019)	III
保护行动 Conservation Action(s)	无 / None

相关文献 Relevant References

Burgin *et al*., 2018; Jiang *et al*. (蒋志刚等), 2017; Castelló, 2016; Groves and Grubb, 2011; Wilson and Mittermeier, 2011

岩羊 *Pseudois nayaur*　　　　　　　　　　　　　　　　　吴秀山 摄　By Xiushan Wu

灰鲸
Eschrichtius robustus

无危 LC

| 数据缺乏 DD | 无危 LC | 近危 NT | 易危 VU | 濒危 EN | 极危 CR | 区域灭绝 RE | 野外灭绝 EW | 灭绝 EX |

分类地位 Taxonomic Status

动物界 Animalia	脊索动物门 Chordata	哺乳纲 Mammalia	鲸目 Cetacea	灰鲸科 Eschrichtiidae
学名 Scientific Name		*Eschrichtius robustus*		
命名人 Species Authority		Lilljeborg, 1861		
英文名 English Name(s)		Gray Whale		
同物异名 Synonym(s)		Gray Back Whale, Pacific Gray Whale		
种下单元评估 Infra-specific Taxa Assessed		无 / None		

评估信息 Assessment Information

评估年份 Year Assessed	2020
评定人 Assessor(s)	周开亚 / Kaiya Zhou
审定人 Reviewer(s)	张先锋、王克雄、王丁、祝茜、蒋志刚 / Xianfeng Zhang, Kexiong Wang, Ding Wang, Qian Zhu, Zhigang Jiang
其他贡献人 Other Contributor(s)	李立立、丁晨晨 / Lili Li, Chenchen Ding

理由 Justification: 灰鲸的北太平洋西部种群少于250头，但其北太平洋东部种群数量较多。考虑灰鲸的种群迁移，将灰鲸列为无危等级 / There are fewer than 250 Gray Whales in the western North Pacific Ocean population. However, the population in the eastern North Pacific Ocean is larger. Thus, considering its migratory habits, it is listed as Least Concern

地理分布 Geographical Distribution

国内分布 Domestic Distribution
渤海、黄海、东海、南海 / Bohai Sea, Yellow Sea, East China Sea, South China Sea
世界分布 World Distribution
北太平洋 / North Pacific Ocean
分布标注 Distribution Note
非特有种 / Non-endemic

国内分布图 Map of Domestic Distribution

种群 Population

种群数量 Population Size	北太平洋东部种群约 2 万头，北太平洋西部种群少于 250 头 / The eastern North Pacific Ocean population has around 20,000 individuals, whereas the western North Pacific Ocean population has fewer than 250 individuals
种群趋势 Population Trend	稳定 / Stable

生境与生态系统 Habitat (s) and Ecosystem (s)

生 境 Habitat(s)	海洋 / Ocean
生态系统 Ecosystem(s)	海洋生态系统 / Ocean Ecosystem

威胁 Threat (s)

主要威胁 Major Threat(s)	海洋污染、渔具缠绕 / Ocean Pollution, Entanglement in Fishing Gear

保护级别与保护行动 Protection Category and Conservation Action (s)

国家重点保护野生动物等级 (2021) Category of National Key Protected Wild Animals (2021)	一级 / Category I
IUCN 红色名录 (2020-2) IUCN Red List (2020-2)	无危 / LC
CITES 附录 (2019) CITES Appendix (2019)	I
保护行动 Conservation Action(s)	法律保护物种 / Legally protected species

相关文献 Relevant References

Burgin *et al.*, 2018; Jiang *et al.* (蒋志刚等), 2017, 2015a; Mittermeier and Wilson, 2014; Wang *et al.* (王先艳等), 2013; Wang and Lu (王丕烈和鹿志创), 2009; Zhou (周开亚), 2008, 2004

灰鲸 *Eschrichtius robustus*

小须鲸
Balaenoptera acutorostrata

无危 LC

| 数据缺乏 DD | 无危 LC | 近危 NT | 易危 VU | 濒危 EN | 极危 CR | 区域灭绝 RE | 野外灭绝 EW | 灭绝 EX |

分类地位 Taxonomic Status

动物界 Animalia	脊索动物门 Chordata	哺乳纲 Mammalia	鲸目 Cetacea	须鲸科 Balaenopteridae

学名 Scientific Name	*Balaenoptera acutorostrata*
命名人 Species Authority	Lacépède, 1804
英文名 English Name(s)	Common Minke Whale
同物异名 Synonym(s)	Minke Whale, Lesser Rorqual
种下单元评估 Infra-specific Taxa Assessed	无 / None

评估信息 Assessment Information

评估年份 Year Assessed	2020
评定人 Assessor(s)	周开亚 / Kaiya Zhou
审定人 Reviewer(s)	张先锋、王克雄、王丁、祝茜、蒋志刚 / Xianfeng Zhang, Kexiong Wang, Ding Wang, Qian Zhu, Zhigang Jiang
其他贡献人 Other Contributor(s)	李立立、丁晨晨 / Lili Li, Chenchen Ding

理由 Justification: 小须鲸的种群数量超过200,000头。因此，列为无危等级 / Common Minke Whale has more than 200,000 individuals. Thus, it is listed as Least Concern

地理分布 Geographical Distribution

国内分布 Domestic Distribution
渤海、黄海、东海、台湾海峡、南海 / Bohai Sea, Yellow Sea, East China Sea, Taiwan Strait, South China Sea
世界分布 World Distribution
北半球海域 / Oceans of North Hemisphere
分布标注 Distribution Note
非特有种 / Non-endemic

国内分布图
Map of Domestic Distribution

种群 Population

种群数量 Population Size	超过 200,000 头 / Over 200,000 individuals
种群趋势 Population Trend	稳定 / Stable

生境与生态系统 Habitat (s) and Ecosystem (s)

生　　境 Habitat(s)	海洋 / Ocean
生态系统 Ecosystem(s)	海洋生态系统 / Ocean Ecosystem

威胁 Threat (s)

主要威胁 Major Threat(s)	海洋污染、渔具纠缠 / Ocean Pollution, Fishery Entanglement

保护级别与保护行动 Protection Category and Conservation Action (s)

国家重点保护野生动物等级 (2021) Category of National Key Protected Wild Animals (2021)	一级 / Category I
IUCN 红色名录 (2020-2) IUCN Red List (2020-2)	无危 / LC
CITES 附录 (2019) CITES Appendix (2019)	I
保护行动 Conservation Action(s)	法律保护物种 / Legally protected species

相关文献 Relevant References

Burgin *et al.*, 2018; Jiang *et al.* (蒋志刚等), 2017, 2015a; Mittermeier and Wilson, 2014; Wang *et al.* (王先艳等), 2013; Wang and Lu (王丕烈和鹿志创), 2009; Zhou (周开亚), 2008, 2004

小须鲸 *Balaenoptera acutorostrata*

布氏鲸
Balaenoptera edeni

无危 LC

| 数据缺乏 DD | 无危 LC | 近危 NT | 易危 VU | 濒危 EN | 极危 CR | 区域灭绝 RE | 野外灭绝 EW | 灭绝 EX |

分类地位 Taxonomic Status

动物界 Animalia	脊索动物门 Chordata	哺乳纲 Mammalia	鲸目 Cetacea	须鲸科 Balaenopteridae

学 名 Scientific Name	*Balaenoptera edeni*
命名人 Species Authority	Anderson, 1879
英文名 English Name(s)	Bryde's Whale
同物异名 Synonym(s)	Sittang; Eden's Whale; *Balaenoptera brydei* (Olsen, 1913)
种下单元评估 Infra-specific Taxa Assessed	无 / None

评估信息 Assessment Information

评估年份 Year Assessed	2020
评定人 Assessor(s)	周开亚 / Kaiya Zhou
审定人 Reviewer(s)	张先锋、王克雄、王丁、祝茜、蒋志刚 / Xianfeng Zhang, Kexiong Wang, Ding Wang, Qian Zhu, Zhigang Jiang
其他贡献人 Other Contributor(s)	李立立、丁晨晨 / Lili Li, Chenchen Ding

理由 Justification: 布氏鲸的分类仍不清楚，它可能是 *Balaenoptera brydei* (Olsen, 1913) 和 *B. edeni* (Anderson, 1879) 两个种的复合体。*B. brydei* 体型较大，常见于全球暖温带和热带水域，而 *B. edeni* 体型较小，仅分布于印度洋 - 太平洋地区。基于其分布区和数量，布氏鲸列为无危等级 / Taxonomy of Bryde's Whale remains unclear, it may comprise two species of *Balaenoptera brydei* (Olsen, 1913) and *B. edeni* (Anderson, 1879). The former is a larger form that occurs worldwide in warm-temperate and tropical waters, and the latter is a smaller form that restricted to the Indo-Pacific Oceans. Thus, based on its distribution range and population number, it is listed as Least Concern

地理分布 Geographical Distribution

国内分布 Domestic Distribution
黄海、东海、台湾海峡、南海 / Yellow Sea, East China Sea, Taiwan Strait, South China Sea
世界分布 World Distribution
大西洋、太平洋、印度洋 / Atlantic Ocean, Pacific Ocean, Indian Ocean
分布标注 Distribution Note
非特有种 / Non-endemic

国内分布图
Map of Domestic Distribution

🦌 种群 Population

种群数量 Population Size	全球布氏鲸多达 90,000 ～ 100,000 头 / Up to 90,000~100,000 individuals worldwide
种群趋势 Population Trend	未知 / Unknown

🦌 生境与生态系统 Habitat(s) and Ecosystem(s)

生　　境 Habitat(s)	海洋 / Ocean
生态系统 Ecosystem(s)	海洋生态系统 / Ocean Ecosystem

🦌 威胁 Threat(s)

主要威胁 Major Threat(s)	海洋污染 / Ocean Pollution

🦌 保护级别与保护行动 Protection Category and Conservation Action(s)

国家重点保护野生动物等级 (2021) Category of National Key Protected Wild Animals (2021)	一级 / Category I
IUCN 红色名录 (2020-2) IUCN Red List (2020-2)	数据缺乏 / DD
CITES 附录 (2019) CITES Appendix (2019)	I
保护行动 Conservation Action(s)	法律保护物种 / Legally protected species

🦌 相关文献 Relevant References

Burgin *et al*., 2018; Kato and Perrin, 2018; Jiang *et al*.（蒋志刚等）, 2017; Mittermeier and Wilson, 2014; Zhou（周开亚）, 2008, 2004

布氏鲸 *Balaenoptera edeni*

中国生物多样性红色名录

大翅鲸
Megaptera novaeangliae

无危 LC

| 数据缺乏 DD | 无危 LC | 近危 NT | 易危 VU | 濒危 EN | 极危 CR | 区域灭绝 RE | 野外灭绝 EW | 灭绝 EX |

分类地位 Taxonomic Status

动物界 Animalia	脊索动物门 Chordata	哺乳纲 Mammalia	鲸目 Cetacea	须鲸科 Balaenopteridae

学 名 Scientific Name	*Megaptera novaeangliae*
命 名 人 Species Authority	Borowski, 1781
英 文 名 English Name(s)	Humpback Whale
同物异名 Synonym(s)	*Balaena novaeangliae* (Borowski, 1781); *B. nodosa* (Bonnaterre, 1789); *Megaptera nodosa* (Lahille, 1905)
种下单元评估 Infra-specific Taxa Assessed	无 / None

评估信息 Assessment Information

评估年份 Year Assessed	2020
评 定 人 Assessor(s)	周开亚 / Kaiya Zhou
审 定 人 Reviewer(s)	张先锋、王克雄、王丁、祝茜、蒋志刚 / Xianfeng Zhang, Kexiong Wang, Ding Wang, Qian Zhu, Zhigang Jiang
其他贡献人 Other Contributor(s)	李立立、丁晨晨 / Lili Li, Chenchen Ding

理由 Justification: 大翅鲸种群目前估计共有 60,000 多头，有些大翅鲸种群在持续增长。因此，列为无危等级 / The current population of Humpback Whale is estimated to be more than 60,000 individuals. Some populations of Humpback Whale have been observed to be continually increasing. Thus, it is listed as Least Concern

地理分布 Geographical Distribution

国内分布 Domestic Distribution
黄海、东海、南海、台湾海峡 / Yellow Sea, East China Sea, South China Sea, Taiwan Strait
世界分布 World Distribution
从南极洲冰缘到北纬 81° 范围内的所有海洋 / All oceans from the Antarctica ice edge to 81° of Northern Latitude
分布标注 Distribution Note
非特有种 / Non-endemic

国内分布图
Map of Domestic Distribution

种群 Population

种群数量 Population Size	超过 60,000 头 / More than 60,000 individuals
种群趋势 Population Trend	上升 / Increasing

生境与生态系统 Habitat(s) and Ecosystem(s)

生境 Habitat(s)	海洋 / Ocean
生态系统 Ecosystem(s)	海洋生态系统 / Ocean Ecosystem

威胁 Threat(s)

主要威胁 Major Threat(s)	船舶撞击、渔具纠缠 / Ship Collision, Fishery Entanglement

保护级别与保护行动 Protection Category and Conservation Action(s)

国家重点保护野生动物等级 (2021) Category of National Key Protected Wild Animals (2021)	一级 / Category I
IUCN 红色名录 (2020-2) IUCN Red List (2020-2)	无危 / LC
CITES 附录 (2019) CITES Appendix (2019)	I
保护行动 Conservation Action(s)	法律保护物种 / Legally protected species

相关文献 Relevant References

Burgin *et al.*, 2018; Jiang *et al.*（蒋志刚等）, 2015a; Mittermeier and Wilson, 2014; Zhou（周开亚）, 2008, 2004; Zhou *et al.*（周开亚等）, 2001; Yang（杨鸿嘉）, 1976

大翅鲸 *Megaptera novaeangliae*

鹅喙鲸
Ziphius cavirostris
无危 LC

| 数据缺乏 DD | 无危 LC | 近危 NT | 易危 VU | 濒危 EN | 极危 CR | 区域灭绝 RE | 野外灭绝 EW | 灭绝 EX |

分类地位 Taxonomic Status

动物界 Animalia	脊索动物门 Chordata	哺乳纲 Mammalia	鲸目 Cetacea	喙鲸科 Ziphiidae

学名 Scientific Name	*Ziphius cavirostris*
命名人 Species Authority	G. Cuvier, 1823
英文名 English Name(s)	Cuvier's Beaked Whale
同物异名 Synonym(s)	Goose-beaked Whale
种下单元评估 Infra-specific Taxa Assessed	无 / None

评估信息 Assessment Information

评估年份 Year Assessed	2020
评定人 Assessor(s)	周开亚 / Kaiya Zhou
审定人 Reviewer(s)	张先锋、王克雄、王丁、祝茜、蒋志刚 / Xianfeng Zhang, Kexiong Wang, Ding Wang, Qian Zhu, Zhigang Jiang
其他贡献人 Other Contributor(s)	李立立、丁晨晨 / Lili Li, Chenchen Ding

理由 Justification: 鹅喙鲸是各种喙鲸中数量最多的一个物种。因此，列为无危等级 / Cuvier's Beaked Whale is among the most abundant species of all the beaked whales. Thus, it is listed as Least Concern

地理分布 Geographical Distribution

国内分布 Domestic Distribution	黄海、东海、台湾海峡、南海 / Yellow Sea, East China Sea, Taiwan Strait, South China Sea
世界分布 World Distribution	北太平洋阿留申群岛和北大西洋加拿大沿岸至澳大利亚南部、新西兰和南半球的查塔姆群岛之间的海域 / In oceans from the Aleutian Islands in the North Pacific Ocean and Canadian coast in the North Atlantic, to southern Australia, New Zealand and the Chatham Islands in the Southern Hemisphere
分布标注 Distribution Note	非特有种 / Non-endemic

国内分布图
Map of Domestic Distribution

种群 Population

种群数量 Population Size	全球种群数量可能超过 100,000 头 / Worldwide abundance is likely to be over 100,000 individuals
种群趋势 Population Trend	未知 / Unknown

生境与生态系统 Habitat(s) and Ecosystem(s)

生　　境 Habitat(s)	海洋 / Ocean
生态系统 Ecosystem(s)	海洋生态系统 / Ocean Ecosystem

威胁 Threat(s)

主要威胁 Major Threat(s)	水下噪声，尤其是海军声呐；流刺网渔业误捕 / Under Water Noise (especially Naval Sonar), By-catch in Driftnet Fishery

保护级别与保护行动 Protection Category and Conservation Action(s)

国家重点保护野生动物等级 (2021) Category of National Key Protected Wild Animals (2021)	二级 / Category II
IUCN 红色名录 (2020-2) IUCN Red List (2020-2)	无危 / LC
CITES 附录 (2019) CITES Appendix (2019)	II
保护行动 Conservation Action(s)	法律保护物种 / Legally protected species

相关文献 Relevant References

Burgin *et al.*, 2018; Jiang *et al.*（蒋志刚等）, 2015a; Mittermeier and Wilson, 2014; Wang（王丕烈）, 2011; Zhou（周开亚）, 2008, 2004; Zhou *et al.*（周开亚等）, 2001

鹅喙鲸 *Ziphius cavirostris*

糙齿海豚
Steno bredanensis

无危 LC

| 数据缺乏 DD | 无危 LC | 近危 NT | 易危 VU | 濒危 EN | 极危 CR | 区域灭绝 RE | 野外灭绝 EW | 灭绝 EX |

分类地位 Taxonomic Status

| 动物界 Animalia | 脊索动物门 Chordata | 哺乳纲 Mammalia | 鲸目 Cetacea | 海豚科 Delphinidae |

学名 Scientific Name	*Steno bredanensis*
命名人 Species Authority	G. Cuvier, 1828
英文名 English Name(s)	Rough-toothed Dolphin
同物异名 Synonym(s)	无 / None
种下单元评估 Infra-specific Taxa Assessed	无 / None

评估信息 Assessment Information

评估年份 Year Assessed	2020
评定人 Assessor(s)	周开亚 / Kaiya Zhou
审定人 Reviewer(s)	张先锋、王克雄、王丁、祝茜、蒋志刚 / Xianfeng Zhang, Kexiong Wang, Ding Wang, Qian Zhu, Zhigang Jiang
其他贡献人 Other Contributor(s)	李立立、丁晨晨 / Lili Li, Chenchen Ding

理由 Justification: 糙齿海豚在全球热带和温带海洋广泛分布，在东太平洋热带有146,000头左右。因此，列为无危等级 / Rough-toothed Dolphin is widely distributed in the tropical and temperate oceans, there are about 146,000 individuals in the tropical eastern Pacific. Thus, it is listed as Least Concern

地理分布 Geographical Distribution

国内分布 Domestic Distribution	东海、台湾海峡、南海 / East China Sea, Taiwan Strait, South China Sea
世界分布 World Distribution	太平洋、大西洋、印度洋及地中海 / Pacific Ocean, Atlantic Ocean, Indian Ocean, and the Mediterranean Sea
分布标注 Distribution Note	非特有种 / Non-endemic

国内分布图
Map of Domestic Distribution

种群 Population

种群数量 Population Size	未知 / Unknown
种群趋势 Population Trend	未知 / Unknown

生境与生态系统 Habitat (s) and Ecosystem (s)

生 境 Habitat(s)	海洋 / Ocean
生态系统 Ecosystem(s)	海洋生态系统 / Ocean Ecosystem

威胁 Threat (s)

主要威胁 Major Threat(s)	金枪鱼围网误捕、污染 / Incidental Catch in Tuna Seines, Pollution

保护级别与保护行动 Protection Category and Conservation Action (s)

国家重点保护野生动物等级 (2021) Category of National Key Protected Wild Animals (2021)	二级 / Category II
IUCN 红色名录 (2020-2) IUCN Red List (2020-2)	无危 / LC
CITES 附录 (2019) CITES Appendix (2019)	II
保护行动 Conservation Action(s)	法律保护物种 / Legally protected species

相关文献 Relevant References

Burgin *et al.*, 2018; Jiang *et al.* (蒋志刚等), 2015a; Mittermeier and Wilson, 2014; Culik, 2011; Wang and Han (王丕烈和韩家波), 2007; Zhou (周开亚), 2008, 2004

糙齿海豚 *Steno bredanensis*

中国生物多样性 红色名录

热带点斑原海豚
Stenella attenuata

无危 LC

| 数据缺乏 DD | 无危 LC | 近危 NT | 易危 VU | 濒危 EN | 极危 CR | 区域灭绝 RE | 野外灭绝 EW | 灭绝 EX |

分类地位 Taxonomic Status

| 动物界 Animalia | 脊索动物门 Chordata | 哺乳纲 Mammalia | 鲸目 Cetacea | 海豚科 Delphinidae |

学　名 Scientific Name	*Stenella attenuata*
命名人 Species Authority	Gray, 1846
英文名 English Name(s)	Pantropical Spotted Dolphin
同物异名 Synonym(s)	*Stenella graffmani* (Lönnberg, 1934)
种下单元评估 Infra-specific Taxa Assessed	无 / None

评估信息 Assessment Information

评估年份 Year Assessed	2020
评定人 Assessor(s)	周开亚 / Kaiya Zhou
审定人 Reviewer(s)	张先锋、王克雄、王丁、祝茜、蒋志刚 / Xianfeng Zhang, Kexiong Wang, Ding Wang, Qian Zhu, Zhigang Jiang
其他贡献人 Other Contributor(s)	李立立、丁晨晨 / Lili Li, Chenchen Ding

理由 Justification: 热带点斑原海豚是热带太平洋中数量最多的海豚之一。因此，列为无危等级 / Pantropical Spotted Dolphin is among the most abundant dolphins in the tropical Pacific Ocean. Thus, it is listed as Least Concern

地理分布 Geographical Distribution

国内分布 Domestic Distribution	东海、台湾海峡、南海 / East China Sea, Taiwan Strait, South China Sea
世界分布 World Distribution	北纬40°至南纬40°之间的热带海洋 / Tropical oceans between 40°N and 40°S
分布标注 Distribution Note	非特有种 / Non-endemic

国内分布图
Map of Domestic Distribution

种群 Population

种群数量 Population Size	总计超过 250 万头 / More than 2.5 million individuals totally
种群趋势 Population Trend	未知 / Unknown

生境与生态系统 Habitat(s) and Ecosystem(s)

生　　境 Habitat(s)	海洋 / Ocean
生态系统 Ecosystem(s)	海洋生态系统 / Ocean Ecosystem

威胁 Threat(s)

主要威胁 Major Threat(s)	渔业捕获、渔业误捕、海洋污染 / Fishery Catch, Incidental Catch, Ocean Pollution

保护级别与保护行动 Protection Category and Conservation Action(s)

国家重点保护野生动物等级 (2021) Category of National Key Protected Wild Animals (2021)	二级 / Category II
IUCN 红色名录 (2020-2) IUCN Red List (2020-2)	无危 / LC
CITES 附录 (2019) CITES Appendix (2019)	II
保护行动 Conservation Action(s)	法律保护物种 / Legally protected species

相关文献 Relevant References

Burgin *et al.*, 2018; Jiang *et al.* (蒋志刚等), 2015a; Mittermeier and Wilson, 2014; Culik, 2011; Wang (王丕烈), 2011; Zhou (周开亚), 2008, 2004

热带点斑原海豚 *Stenella attenuata*

条纹原海豚
Stenella coeruleoalba
无危 LC

| 数据缺乏 DD | 无危 LC | 近危 NT | 易危 VU | 濒危 EN | 极危 CR | 区域灭绝 RE | 野外灭绝 EW | 灭绝 EX |

分类地位 Taxonomic Status

动物界 Animalia	脊索动物门 Chordata	哺乳纲 Mammalia	鲸目 Cetacea	海豚科 Delphinidae

学名 Scientific Name	*Stenella coeruleoalba*
命名人 Species Authority	Meyen, 1833
英文名 English Name(s)	Striped Dolphin
同物异名 Synonym(s)	*Stenella euphrosyne* (Gray, 1846); *styx* (Gray, 1846)
种下单元评估 Infra-specific Taxa Assessed	无 / None

评估信息 Assessment Information

评估年份 Year Assessed	2020
评定人 Assessor(s)	周开亚 / Kaiya Zhou
审定人 Reviewer(s)	张先锋、王克雄、王丁、祝茜、蒋志刚 / Xianfeng Zhang, Kexiong Wang, Ding Wang, Qian Zhu, Zhigang Jiang
其他贡献人 Other Contributor(s)	李立立、丁晨晨 / Lili Li, Chenchen Ding

理由 Justification: 条纹原海豚的全球种群数量估计超过 200 万头。因此，列为无危等级 / The population size of Striped Dolphin is estimated over 2 million individuals worldwide. Thus, it is listed as Least Concern

地理分布 Geographical Distribution

国内分布 Domestic Distribution	东海、台湾海峡、南海 / East China Sea, Taiwan Strait, South China Sea
世界分布 World Distribution	从北纬 40° 到南纬 30° 温带或热带海洋离岸水域 / Off-shore waters of temperate or tropical ocean from 40°N to 30°S
分布标注 Distribution Note	非特有种 / Non-endemic

国内分布图
Map of Domestic Distribution

种群 Population

种群数量 Population Size	全球估计超过 200 万头 / Estimated over 2 million individuals globally
种群趋势 Population Trend	未知 / Unknown

生境与生态系统 Habitat (s) and Ecosystem (s)

生　　境 Habitat(s)	海洋 / Ocean
生态系统 Ecosystem(s)	海洋生态系统 / Ocean Ecosystem

威胁 Threat (s)

主要威胁 Major Threat(s)	渔业捕获、渔业误捕、海洋污染、海军声呐噪声 / Fishery Catching, Fishery Incidental Catch, Ocean Pollution, Navy Sonar Noise

保护级别与保护行动 Protection Category and Conservation Action (s)

国家重点保护野生动物等级 (2021) Category of National Key Protected Wild Animals (2021)	二级 / Category II
IUCN 红色名录 (2020-2) IUCN Red List (2020-2)	无危 / LC
CITES 附录 (2019) CITES Appendix (2019)	II
保护行动 Conservation Action(s)	法律保护物种 / Legally protected species

相关文献 Relevant References

Burgin *et al.*, 2018; Jiang *et al.* (蒋志刚等), 2015a; Mittermeier and Wilson, 2014; Culik, 2011; Wang (王丕烈), 2011; Zhou (周开亚), 2008, 2004

条纹原海豚 *Stenella coeruleoalba*

真海豚
Delphinus delphis

无危 LC

| 数据缺乏 DD | 无危 LC | 近危 NT | 易危 VU | 濒危 EN | 极危 CR | 区域灭绝 RE | 野外灭绝 EW | 灭绝 EX |

分类地位 Taxonomic Status

动物界 Animalia	脊索动物门 Chordata	哺乳纲 Mammalia	鲸目 Cetacea	海豚科 Delphinidae

学名 Scientific Name	*Delphinus delphis*
命名人 Species Authority	Linnaeus, 1758
英文名 English Name(s)	Common Dolphin
同物异名 Synonym(s)	Short-beaked Common Dolphin
种下单元评估 Infra-specific Taxa Assessed	无 / None

评估信息 Assessment Information

评估年份 Year Assessed	2020
评定人 Assessor(s)	周开亚 / Kaiya Zhou
审定人 Reviewer(s)	张先锋、王克雄、王丁、祝茜、蒋志刚 / Xianfeng Zhang, Kexiong Wang, Ding Wang, Qian Zhu, Zhigang Jiang
其他贡献人 Other Contributor(s)	李立立、丁晨晨 / Lili Li, Chenchen Ding

理由 Justification: 真海豚的种群数量估计超过300万头。因此，列为无危等级 / The population size of Common Dolphin is estimated over 3 million individuals. Thus, it is listed as Least Concer

地理分布 Geographical Distribution

国内分布 Domestic Distribution
属于以下省份或地区的海域：浙江、台湾、福建、广东、香港、广西、海南 / Oceans of the provinces or regions: Zhejiang, Taiwan, Fujian, Guangdong, Hong Kong, Guangxi, Hainan

世界分布 World Distribution
大西洋和太平洋的暖温带部分，以及加勒比海和地中海 / Warm-temperate regions of the Atlantic and Pacific Oceans, as well as the Caribbean and Mediterranean Seas

分布标注 Distribution Note
非特有种 / Non-endemic

国内分布图
Map of Domestic Distribution

种群 Population

种群数量 Population Size	全球各种群的估计数量合计超过 300 万头 / The estimated number of several populations exceed 3 million individuals globally
种群趋势 Population Trend	未知 / Unknown

生境与生态系统 Habitat(s) and Ecosystem(s)

生　　境 Habitat(s)	海洋 / Ocean
生态系统 Ecosystem(s)	海洋生态系统 / Ocean Ecosystem

威胁 Threat(s)

主要威胁 Major Threat(s)	渔业、兼捕、污染 / Fishery, By-catch, Pollution

保护级别与保护行动 Protection Category and Conservation Action(s)

国家重点保护野生动物等级 (2021) Category of National Key Protected Wild Animals (2021)	二级 / Category II
IUCN 红色名录 (2020-2) IUCN Red List (2020-2)	无危 / LC
CITES 附录 (2019) CITES Appendix (2019)	II
保护行动 Conservation Action(s)	法律保护物种 / Legally protected species

相关文献 Relevant References

Burgin *et al*., 2018; Perrin, 2018; Jiang *et al*.(蒋志刚等), 2017; Mittermeier and Wilson, 2014; Wang and Fan (王火根和范忠勇), 2004; Zhou (周开亚), 2008, 2004

真海豚 *Delphinus delphis*

瓶鼻海豚
Tursiops truncatus

无危 LC

| 数据缺乏 DD | 无危 LC | 近危 NT | 易危 VU | 濒危 EN | 极危 CR | 区域灭绝 RE | 野外灭绝 EW | 灭绝 EX |

分类地位 Taxonomic Status

动物界 Animalia	脊索动物门 Chordata	哺乳纲 Mammalia	鲸目 Cetacea	海豚科 Delphinidae

学名 Scientific Name	*Tursiops truncatus*
命名人 Species Authority	Montagu, 1821
英文名 English Name(s)	Common Bottlenose Dolphin
同物异名 Synonym(s)	Atlantic Bottlenose Dolphin; *Tursiops gephyreus* (Lahille, 1908); *gilli* (Dall, 1873); *nuuanu* (Andrews, 1911)
种下单元评估 Infra-specific Taxa Assessed	无 / None

评估信息 Assessment Information

评估年份 Year Assessed	2020
评定人 Assessor(s)	周开亚 / Kaiya Zhou
审定人 Reviewer(s)	张先锋、王克雄、王丁、祝茜、蒋志刚 / Xianfeng Zhang, Kexiong Wang, Ding Wang, Qian Zhu, Zhigang Jiang
其他贡献人 Other Contributor(s)	李立立、丁晨晨 / Lili Li, Chenchen Ding

理由 Justification: 瓶鼻海豚的全球种群数量最低估计值为 600,000 头。因此，列为无危等级 / A minimum estimate of the global population size of Common Bottlenose Dolphin is 600,000 individuals. Thus, it is listed as Least Concern

地理分布 Geographical Distribution

国内分布 Domestic Distribution

属于以下省份或地区的海域：辽宁、山东、江苏、上海、浙江、福建、台湾、广东、香港、广西 / Oceans of the provinces or regions: Liaoning, Shandong, Jiangsu, Shanghai, Zhejiang, Fujian, Taiwan, Guangdong, Hong Kong, Guangxi

世界分布 World Distribution

全球温带、热带和亚热带海洋 / Temperate, tropical and subtropical oceans worldwide

分布标注 Distribution Note

非特有种 / Non-endemic

国内分布图
Map of Domestic Distribution

种群 Population

种群数量 Population Size	全球最低估计值为 600,000 头 / A minimum estimate of the global population size of Common Bottlenose Dolphin is 600,000 individuals
种群趋势 Population Trend	未知 / Unknown

生境与生态系统 Habitat (s) and Ecosystem (s)

生　　境 Habitat(s)	海洋 / Ocean
生态系统 Ecosystem(s)	海洋生态系统 / Ocean Ecosystem

威胁 Threat (s)

主要威胁 Major Threat(s)	捕获、兼捕、栖息地退化、军事用途 / Catch, By-catch, Habitat Degradation, Military Operation

保护级别与保护行动 Protection Category and Conservation Action (s)

国家重点保护野生动物等级 (2021) Category of National Key Protected Wild Animals (2021)	二级 / Category II
IUCN 红色名录 (2020-2) IUCN Red List (2020-2)	无危 / LC
CITES 附录 (2019) CITES Appendix (2019)	II
保护行动 Conservation Action(s)	法律保护物种 / Legally protected species

相关文献 Relevant References

Burgin *et al*., 2018; Jiang *et al*. (蒋志刚等), 2015a; Culik, 2011; Zhou (周开亚), 2008, 2004

瓶鼻海豚 *Tursiops truncatus*

弗氏海豚
Lagenodelphis hosei

无危 LC

| 数据缺乏 DD | 无危 LC | 近危 NT | 易危 VU | 濒危 EN | 极危 CR | 区域灭绝 RE | 野外灭绝 EW | 灭绝 EX |

分类地位 Taxonomic Status

| 动物界 Animalia | 脊索动物门 Chordata | 哺乳纲 Mammalia | 鲸目 Cetacea | 海豚科 Delphinidae |

学名 Scientific Name	*Lagenodelphis hosei*
命名人 Species Authority	Fraser, 1956
英文名 English Name(s)	Fraser's Dolphin
同物异名 Synonym(s)	Sarawak Dolphin
种下单元评估 Infra-specific Taxa Assessed	无 / None

评估信息 Assessment Information

评估年份 Year Assessed	2020
评定人 Assessor(s)	周开亚 / Kaiya Zhou
审定人 Reviewer(s)	张先锋、王克雄、王丁、祝茜、蒋志刚 / Xianfeng Zhang, Kexiong Wang, Ding Wang, Qian Zhu, Zhigang Jiang
其他贡献人 Other Contributor(s)	李立立、丁晨晨 / Lili Li, Chenchen Ding

理由 Justification: 弗氏海豚分布于全球热带海洋，其在热带东太平洋的数量约 289,500 头。因此，列为无危等级 / Fraser's Dolphin is distributed throughout the tropical oceans and its population in the tropical east Pacific Ocean is about 289,500 individuals. Thus, it is listed as Least Concern

地理分布 Geographical Distribution

国内分布 Domestic Distribution	属于以下省份或地区的海域：台湾、广东、香港 / Oceans of the provinces or regions: Taiwan, Guangdong, Hong Kong
世界分布 World Distribution	北纬20°到南纬30°热带海洋深水区 / Deep waters of tropical oceans between 20°N and 30°S
分布标注 Distribution Note	非特有种 / Non-endemic

国内分布图
Map of Domestic Distribution

种群 Population

种群数量 Population Size	中国海域弗氏海豚的种群数量未知 / The population size of Fraser's Dolphin in Chinese waters is unknown
种群趋势 Population Trend	未知 / Unknown

生境与生态系统 Habitat(s) and Ecosystem(s)

生　　境 Habitat(s)	海洋 / Ocean
生态系统 Ecosystem(s)	海洋生态系统 / Ocean Ecosystem

威胁 Threat(s)

主要威胁 Major Threat(s)	刺网兼捕、船舶撞击、污染 / Catch in Gill Net, Boat Strike, Pollution

保护级别与保护行动 Protection Category and Conservation Action(s)

国家重点保护野生动物等级 (2021) Category of National Key Protected Wild Animals (2021)	二级 / Category II
IUCN 红色名录 (2020-2) IUCN Red List (2020-2)	无危 / LC
CITES 附录 (2019) CITES Appendix (2019)	II
保护行动 Conservation Action(s)	法律保护物种 / Legally protected species

相关文献 Relevant References

Burgin *et al.*, 2018; Jiang *et al.* (蒋志刚等), 2015a; Mittermeier and Wilson, 2014; Culik, 2011; Wang (王丕烈), 2011; Zhou (周开亚), 2008, 2004

弗氏海豚 *Lagenodelphis hosei*

里氏海豚 *Grampus griseus*

无危 LC

| 数据缺乏 DD | 无危 LC | 近危 NT | 易危 VU | 濒危 EN | 极危 CR | 区域灭绝 RE | 野外灭绝 EW | 灭绝 EX |

分类地位 Taxonomic Status

动物界 Animalia	脊索动物门 Chordata	哺乳纲 Mammalia	鲸目 Cetacea	海豚科 Delphinidae

学名 Scientific Name	*Grampus griseus*
命名人 Species Authority	G. Cuvier, 1812
英文名 English Name(s)	Risso's Dolphin
同物异名 Synonym(s)	Monk Dolphin
种下单元评估 Infra-specific Taxa Assessed	无 / None

评估信息 Assessment Information

评估年份 Year Assessed	2020
评定人 Assessor(s)	周开亚 / Kaiya Zhou
审定人 Reviewer(s)	张先锋、王克雄、王丁、祝茜、蒋志刚 / Xianfeng Zhang, Kexiong Wang, Ding Wang, Qian Zhu, Zhigang Jiang
其他贡献人 Other Contributor(s)	李立立、丁晨晨 / Lili Li, Chenchen Ding

理由 Justification: 里氏海豚在热带和温带海洋广泛分布。因此，列为无危等级 / Risso's Dolphins is widely distributed in tropical and temperate ocean. Thus, it is listed as Least Concern

地理分布 Geographical Distribution

国内分布 Domestic Distribution

属于以下省份或地区的海域：辽宁、浙江、福建、台湾、广东、香港、海南 / Oceans of the provinces or regions: Liaoning, Zhejiang, Fujian, Taiwan, Guangdong, Hong Kong, Hainan

世界分布 World Distribution

世界上温带和热带海洋中靠近陆地较深的水域 / Worldwide deeper waters close to land in temperate and tropical oceans

分布标注 Distribution Note

非特有种 / Non-endemic

国内分布图
Map of Domestic Distribution

种群 Population

种群数量 Population Size	中国海域里氏海豚的种群数量未知 / The population size of Risso's Dolphin in Chinese waters is unknown
种群趋势 Population Trend	未知 / Unknown

生境与生态系统 Habitat(s) and Ecosystem(s)

生　　境 Habitat(s)	海洋 / Ocean
生态系统 Ecosystem(s)	海洋生态系统 / Ocean Ecosystem

威胁 Threat(s)

主要威胁 Major Threat(s)	渔业兼捕、海军声呐噪声、污染 / By-catch in Fishery, Naval Sonar Noise, Pollution

保护级别与保护行动 Protection Category and Conservation Action(s)

国家重点保护野生动物等级 (2021) Category of National Key Protected Wild Animals (2021)	二级 / Category II
IUCN 红色名录 (2020-2) IUCN Red List (2020-2)	无危 / LC
CITES 附录 (2019) CITES Appendix (2019)	II
保护行动 Conservation Action(s)	法律保护物种 / Legally protected species

相关文献 Relevant References

Burgin *et al.*, 2018; Jiang *et al.* (蒋志刚等), 2015a; Mittermeier and Wilson, 2014; Culik, 2011; Wang (王丕烈), 2011; Zhou (周开亚), 2008, 2004

里氏海豚 *Grampus griseus*

太平洋斑纹海豚
Lagenorhynchus obliquidens

无危 LC

数据缺乏 DD	无危 LC	近危 NT	易危 VU	濒危 EN	极危 CR	区域灭绝 RE	野外灭绝 EW	灭绝 EX

分类地位 Taxonomic Status

动物界 Animalia	脊索动物门 Chordata	哺乳纲 Mammalia	鲸目 Cetacea	海豚科 Delphinidae

学名 Scientific Name	*Lagenorhynchus obliquidens*
命名人 Species Authority	Gill, 1865
英文名 English Name(s)	Pacific White-sided Dolphin
同物异名 Synonym(s)	*Lagenorhynchus ognevi* (Sleptsov, 1955)
种下单元评估 Infra-specific Taxa Assessed	无 / None

评估信息 Assessment Information

评估年份 Year Assessed	2020
评定人 Assessor(s)	周开亚 / Kaiya Zhou
审定人 Reviewer(s)	张先锋、王克雄、王丁、祝茜、蒋志刚 / Xianfeng Zhang, Kexiong Wang, Ding Wang, Qian Zhu, Zhigang Jiang
其他贡献人 Other Contributor(s)	李立立、丁晨晨 / Lili Li, Chenchen Ding

理由 Justification: 太平洋斑纹海豚数量多，在北太平洋冷温带海域广泛分布。因此，列为无危等级 / Pacific White-sided Dolphin has large populations, and it is widely distributed in the cool-temperate waters of the North Pacific Ocean. Thus, it is listed as Least Concern

地理分布 Geographical Distribution

国内分布 Domestic Distribution
属于以下省份或地区的海域：江苏、福建、广西 / Oceans of these provinces or regions: Jiangsu, Fujian, Guangxi
世界分布 World Distribution
横跨北太平洋从冷到暖的水域 / Across the cool to temperate waters of the North Pacific Ocean
分布标注 Distribution Note
非特有种 / Non-endemic

国内分布图
Map of Domestic Distribution

种群 Population

种群数量 Population Size	中国海域太平洋斑纹海豚的种群数量未知 / The population of Pacific White-sided Dolphin in Chinese waters is unknown
种群趋势 Population Trend	未知 / Unknown

生境与生态系统 Habitat(s) and Ecosystem(s)

生境 Habitat(s)	海洋 / Ocean
生态系统 Ecosystem(s)	海洋生态系统 / Ocean Ecosystem

威胁 Threat(s)

主要威胁 Major Threat(s)	渔业兼捕、污染 / By-catch in Fishery, Pollution

保护级别与保护行动 Protection Category and Conservation Action(s)

国家重点保护野生动物等级 (2021) Category of National Key Protected Wild Animals (2021)	二级 / Category II
IUCN 红色名录 (2020-2) IUCN Red List (2020-2)	无危 / LC
CITES 附录 (2019) CITES Appendix (2019)	II
保护行动 Conservation Action(s)	法律保护物种 / Legally protected species

相关文献 Relevant References

Burgin *et al.*, 2018; Jiang *et al.* (蒋志刚等), 2017; Mittermeier and Wilson, 2014; Culik, 2011; Wang (王丕烈), 2011; Zhou (周开亚), 2008, 2004

太平洋斑纹海豚 *Lagenorhynchus obliquidens*

瓜头鲸
Peponocephala electra
无危 LC

| 数据缺乏 DD | 无危 LC | 近危 NT | 易危 VU | 濒危 EN | 极危 CR | 区域灭绝 RE | 野外灭绝 EW | 灭绝 EX |

分类地位 Taxonomic Status

| 动物界 Animalia | 脊索动物门 Chordata | 哺乳纲 Mammalia | 鲸目 Cetacea | 海豚科 Delphinidae |

学名 Scientific Name	*Peponocephala electra*
命名人 Species Authority	Gray, 1846
英文名 English Name(s)	Melon-headed Whale
同物异名 Synonym(s)	Electra Dolphin; Little Killer Whale; Many-toothed Blackfish; *Electra electra* (Gray, 1846); *Lagenorhynchus electra* (Gray, 1846)
种下单元评估 Infra-specific Taxa Assessed	无 / None

评估信息 Assessment Information

评估年份 Year Assessed	2020
评定人 Assessor(s)	周开亚 / Kaiya Zhou
审定人 Reviewer(s)	张先锋、王克雄、王丁、祝茜、蒋志刚 / Xianfeng Zhang, Kexiong Wang, Ding Wang, Qian Zhu, Zhigang Jiang
其他贡献人 Other Contributor(s)	李立立、丁晨晨 / Lili Li, Chenchen Ding

理由 Justification: 瓜头鲸的全球种群数量最低估计数为 50,000 头。因此，列为无危等级 / Melon-headed Whale has populations with more than 50,000 individuals world-wide. Thus, it is listed as Least Concern

地理分布 Geographical Distribution

国内分布 Domestic Distribution	台湾海域 / Taiwan Water
世界分布 World Distribution	北纬 40° 至南纬 35° 之间的热带和亚热带深海海域 / Deep tropical and subtropical oceanic waters between 40°N to 35°S
分布标注 Distribution Note	非特有种 / Non-endemic

国内分布图
Map of Domestic Distribution

种群 Population

种群数量 Population Size	未知 / Unknown
种群趋势 Population Trend	未知 / Unknown

生境与生态系统 Habitat(s) and Ecosystem(s)

生　　境 Habitat(s)	海洋 / Ocean
生态系统 Ecosystem(s)	海洋生态系统 / Ocean Ecosystem

威胁 Threat(s)

主要威胁 Major Threat(s)	捕杀、渔业兼捕、军事声呐噪声、污染 / Catch, By-catch in Fishery, Military Sonar Noise, Pollution

保护级别与保护行动 Protection Category and Conservation Action(s)

国家重点保护野生动物等级 (2021) Category of National Key Protected Wild Animals (2021)	二级 / Category II
IUCN 红色名录 (2020-2) IUCN Red List (2020-2)	无危 / LC
CITES 附录 (2019) CITES Appendix (2019)	II
保护行动 Conservation Action(s)	法律保护物种 / Legally protected species

相关文献 Relevant References

Burgin *et al.*, 2018; Jiang *et al.* (蒋志刚等), 2015a; Mittermeier and Wilson, 2014; Culik, 2011; Wang (王丕烈), 2011; Zhou (周开亚), 2008, 2004

瓜头鲸 *Peponocephala electra*

金背松鼠 *Callosciurus caniceps*

无危 LC

| 数据缺乏 DD | 无危 LC | 近危 NT | 易危 VU | 濒危 EN | 极危 CR | 区域灭绝 RE | 野外灭绝 EW | 灭绝 EX |

分类地位 Taxonomic Status

动物界 Animalia	脊索动物门 Chordata	哺乳纲 Mammalia	啮齿目 Rodentia	松鼠科 Sciuridae

学名 Scientific Name	*Callosciurus caniceps*
命名人 Species Authority	Gray, 1842
英文名 English Name(s)	Gray-bellied Squirrel
同物异名 Synonym(s)	Red-bellied Tree Squirrel; *Callosciurus altinsularis* (Miller, 1903); *bentincanus* (Miller, 1903); *chrysonotus* (Blyth, 1847); *fallax* (Robinson and Kloss, 1914); *fluminalis* (转下页)
种下单元评估 Infra-specific Taxa Assessed	无 / None

评估信息 Assessment Information

评估年份 Year Assessed	2020
评定人 Assessor(s)	蒋志刚、刘少英 / Zhigang Jiang, Shaoying Liu
审定人 Reviewer(s)	马勇、鲁长虎、鲍毅新 / Yong Ma, Changhu Lu, Yixin Bao
其他贡献人 Other Contributor(s)	李立立、丁晨晨 / Lili Li, Chenchen Ding

理由 Justification: 金背松鼠的种群数量较多。因此，列为无危等级 / Gray-bellied Squirrel has large populations. Thus, it is listed as Least Concern

地理分布 Geographical Distribution

国内分布 Domestic Distribution
云南 / Yunnan
世界分布 World Distribution
中国；南亚 / China; South Asia
分布标注 Distribution Note
非特有种 / Non-endemic

国内分布图
Map of Domestic Distribution

种群 Population

种群数量 Population Size	较多 / Relatively abundant
种群趋势 Population Trend	稳定 / Stable

生境与生态系统 Habitat(s) and Ecosystem(s)

生境 Habitat(s)	种植园、耕地、次生林、森林 / Plantation, Arable Land, Secondary Forest, Forest
生态系统 Ecosystem(s)	森林生态系统、农田生态系统 / Forest Ecosystem, Cropland Ecosystem

威胁 Threat(s)

主要威胁 Major Threat(s)	无 / None

保护级别与保护行动 Protection Category and Conservation Action(s)

国家重点保护野生动物等级 (2021) Category of National Key Protected Wild Animals (2021)	无 / NA
IUCN 红色名录 (2020-2) IUCN Red List (2020-2)	无危 / LC
CITES 附录 (2019) CITES Appendix (2019)	无 / NA
保护行动 Conservation Action(s)	无 / None

相关文献 Relevant References

Burgin *et al*., 2018; Jiang *et al*.（蒋志刚等）, 2017; Wilson *et al*., 2016; Smith *et al*.（史密斯等）, 2009; Pan *et al*.（潘清华等）, 2007; Wang（王应祥）, 2003

（接上页）
(Robinson and Wroughton, 1911); *hastilis* (Thomas, 1923); *helgei* (Gyldenstolpe, 1917); *helvus* (Shamel, 1930); *inexpectatus* (Kloss, 1916); *lancavensis* (Miller, 1903); *lucas* (Miller, 1903); *mapravis* (Thomas and Robinson, 1921); *matthaeus* (Miller, 1903); *moheius* (Thomas and Robinson, 1921); *mohillius* (Thomas and Robinson, 1921); *nakanus* (Thomas and Robinson, 1921); *panjioli* (Thomas and Robinson, 1921); *panjius* (Thomas and Robinson, 1921); *pipidonis* (Thomas and Robinson, 1921); *samuiensis* (Robinson and Kloss, 1914); *sullivanus* (Miller, 1903); *tabaudius* (Thomas, 1922); *tacopius* (Thomas and Robinson, 1921); *telibius* (Thomas and Robinson, 1921); *terutavensis* (Thomas and Wroughton, 1909)

金背松鼠 *Callosciurus caniceps*

赤腹松鼠
Callosciurus erythraeus

无危 LC

| 数据缺乏 DD | 无危 LC | 近危 NT | 易危 VU | 濒危 EN | 极危 CR | 区域灭绝 RE | 野外灭绝 EW | 灭绝 EX |

分类地位 Taxonomic Status

动物界 Animalia	脊索动物门 Chordata	哺乳纲 Mammalia	啮齿目 Rodentia	松鼠科 Sciuridae

学名 Scientific Name	*Callosciurus erythraeus*
命名人 Species Authority	Pallas, 1779
英文名 English Name(s)	Pallas's Squirrel
同物异名 Synonym(s)	*Callosciurus cinnamomeiventris* (Swinhoe, 1862); *cucphuongis* (Dao Van Tien, 1965); *dabshanensis* (Xu and Chen, 1989); *gongshanensis* (Wang, 1981); *griseopectus* (Blyth, (转下页)
种下单元评估 Infra-specific Taxa Assessed	无 / None

评估信息 Assessment Information

评估年份 Year Assessed	2020
评定人 Assessor(s)	蒋志刚、刘少英 / Zhigang Jiang, Shaoying Liu
审定人 Reviewer(s)	马勇、鲁长虎、鲍毅新 / Yong Ma, Changhu Lu, Yixin Bao
其他贡献人 Other Contributor(s)	李立立、丁晨晨 / Lili Li, Chenchen Ding

理由 Justification: 赤腹松鼠的发生区大、种群数量多。因此，列为无危等级 / Pallas's Squirrel has large area of occurrence and populations. Thus, it is listed as Least Concern

地理分布 Geographical Distribution

国内分布 Domestic Distribution

河南、陕西、湖北、四川、贵州、云南、广西、广东、海南、安徽、江苏、上海、浙江、湖南、江西、福建、西藏、台湾、重庆 / Henan, Shaanxi, Hubei, Sichuan, Guizhou, Yunnan, Guangxi, Guangdong, Hainan, Anhui, Jiangsu, Shanghai, Zhejiang, Hunan, Jiangxi, Fujian, Tibet (Xizang), Taiwan, Chongqing

世界分布 World Distribution

中国；南亚、东南亚 / China; South Asia, Southeast Asia

分布标注 Distribution Note

非特有种 / Non-endemic

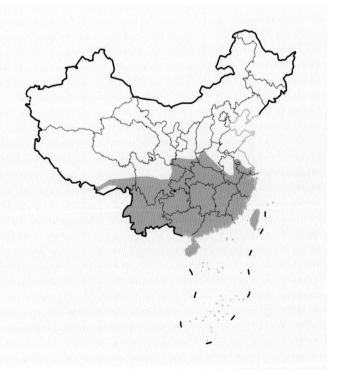

国内分布图
Map of Domestic Distribution

种群 Population

种群数量 Population Size	数量多 / Abundant
种群趋势 Population Trend	稳定 / Stable

生境与生态系统 Habitat (s) and Ecosystem (s)

生 境 Habitat(s)	热带和亚热带湿润低地森林、泰加林、针阔混交林 / Tropical and Subtropical Moist Lowland Forest, Taiga Forest, Coniferous and Broad-leaved Mixed Forest
生态系统 Ecosystem(s)	森林生态系统 / Forest Ecosystem

威胁 Threat (s)

主要威胁 Major Threat(s)	无 / None

保护级别与保护行动 Protection Category and Conservation Action (s)

国家重点保护野生动物等级 (2021) Category of National Key Protected Wild Animals (2021)	无 / NA
IUCN 红色名录 (2020-2) IUCN Red List (2020-2)	无危 / LC
CITES 附录 (2019) CITES Appendix (2019)	无 / NA
保护行动 Conservation Action(s)	无 / None

相关文献 Relevant References

Burgin *et al.*, 2018; Jiang *et al.* (蒋志刚等), 2017; Wilson *et al.*, 2016; Zheng *et al.* (郑智民等), 2012; Kong *et al.* (孔令雪等), 2011; Wen *et al.* (温知新等), 2010; Jiang (蒋志刚), 2009; Smith *et al.* (史密斯等), 2009

(接上页)
1847); *quinlingensis* (Xu and Chen, 1989); *tsingtauensis* (Hilzheimer, 1906); *wuliangshanensis* (Li and Wang, 1981)

赤腹松鼠 *Callosciurus erythraeus*

印支松鼠
Callosciurus inornatus

无危 LC

分类地位 Taxonomic Status

动物界 Animalia	脊索动物门 Chordata	哺乳纲 Mammalia	啮齿目 Rodentia	松鼠科 Sciuridae

学名 Scientific Name	*Callosciurus inornatus*
命名人 Species Authority	Gray, 1867
英文名 English Name(s)	Inornate Squirrel
同物异名 Synonym(s)	*Callosciurus imitator* (Thomas, 1925)
种下单元评估 Infra-specific Taxa Assessed	无 / None

评估信息 Assessment Information

评估年份 Year Assessed	2020
评定人 Assessor(s)	蒋志刚、刘少英 / Zhigang Jiang, Shaoying Liu
审定人 Reviewer(s)	马勇、鲁长虎、鲍毅新 / Yong Ma, Changhu Lu, Yixin Bao
其他贡献人 Other Contributor(s)	李立立、丁晨晨 / Lili Li, Chenchen Ding

理由 Justification: 印支松鼠的种群数量较多。因此，列为无危等级 / Inornate Squirrel has large populations. Thus, it is listed as Least Concern

地理分布 Geographical Distribution

国内分布 Domestic Distribution
云南 / Yunnan

世界分布 World Distribution
中国、老挝、越南 / China, Laos, Viet Nam

分布标注 Distribution Note
非特有种 / Non-endemic

国内分布图
Map of Domestic Distribution

种群 Population

种群数量 Population Size	较多 / Relatively abundant
种群趋势 Population Trend	稳定 / Stable

生境与生态系统 Habitat(s) and Ecosystem(s)

生境 Habitat(s)	灌丛、森林 / Shrubland, Forest
生态系统 Ecosystem(s)	森林生态系统、灌丛生态系统 / Forest Ecosystem, Shrubland Ecosystem

威胁 Threat(s)

主要威胁 Major Threat(s)	无 / None

保护级别与保护行动 Protection Category and Conservation Action(s)

国家重点保护野生动物等级 (2021) Category of National Key Protected Wild Animals (2021)	无 / NA
IUCN 红色名录 (2020-2) IUCN Red List (2020-2)	无危 / LC
CITES 附录 (2019) CITES Appendix (2019)	无 / NA
保护行动 Conservation Action(s)	无 / None

相关文献 Relevant References

Burgin *et al.*, 2018; Jiang *et al.* (蒋志刚等), 2017; Wilson *et al.*, 2016; Zheng *et al.* (郑智民等), 2012; Smith *et al.* (史密斯等), 2009; Wang (王应祥), 2003

印支松鼠 *Callosciurus inornatus*

黄足松鼠
Callosciurus phayrei

无危 LC

| 数据缺乏 DD | 无危 LC | 近危 NT | 易危 VU | 濒危 EN | 极危 CR | 区域灭绝 RE | 野外灭绝 EW | 灭绝 EX |

分类地位 Taxonomic Status

| 动物界 Animalia | 脊索动物门 Chordata | 哺乳纲 Mammalia | 啮齿目 Rodentia | 松鼠科 Sciuridae |

学名 Scientific Name	*Callosciurus phayrei*
命名人 Species Authority	Blyth, 1856
英文名 English Name(s)	Phayre's Squirrel
同物异名 Synonym(s)	*Callosciurus blanfordii* (Blyth, 1862); *heinrichi* (Tate, 1954)
种下单元评估 Infra-specific Taxa Assessed	无 / None

评估信息 Assessment Information

评估年份 Year Assessed	2020
评定人 Assessor(s)	蒋志刚、刘少英 / Zhigang Jiang, Shaoying Liu
审定人 Reviewer(s)	马勇、鲁长虎、鲍毅新 / Yong Ma, Changhu Lu, Yixin Bao
其他贡献人 Other Contributor(s)	李立立、丁晨晨 / Lili Li, Chenchen Ding

理由 Justification: 黄足松鼠的种群数量较多。因此，列为无危等级 / Phayre's Squirrel has large populations. Thus, it is listed as Least Concern

地理分布 Geographical Distribution

国内分布 Domestic Distribution	
云南 / Yunnan	
世界分布 World Distribution	
中国、缅甸 / China, Myanmar	
分布标注 Distribution Note	
非特有种 / Non-endemic	

国内分布图
MAP of Domestic Distribution

种群 Population

种群数量 Population Size	较多 / Relatively abundant
种群趋势 Population Trend	稳定 / Stable

生境与生态系统 Habitat (s) and Ecosystem (s)

生　　境 Habitat(s)	森林 / Forest
生态系统 Ecosystem(s)	森林生态系统 / Forest Ecosystem

威胁 Threat (s)

主要威胁 Major Threat(s)	无 / None

保护级别与保护行动 Protection Category and Conservation Action (s)

国家重点保护野生动物等级 (2021) Category of National Key Protected Wild Animals (2021)	无 / NA
IUCN 红色名录 (2020-2) IUCN Red List (2020-2)	无危 / LC
CITES 附录 (2019) CITES Appendix (2019)	无 / NA
保护行动 Conservation Action(s)	无 / None

相关文献 Relevant References

Burgin *et al*., 2018; Jiang *et al*.（蒋志刚等）, 2017; Zheng *et al*.（郑智民等）, 2012; Smith *et al*.（史密斯等）, 2009; Wang（王应祥）, 2003

黄足松鼠 *Callosciurus phayrei*

蓝腹松鼠
Callosciurus pygerythrus

无危 LC

| 数据缺乏 DD | 无危 LC | 近危 NT | 易危 VU | 濒危 EN | 极危 CR | 区域灭绝 RE | 野外灭绝 EW | 灭绝 EX |

分类地位 Taxonomic Status

动物界 Animalia	脊索动物门 Chordata	哺乳纲 Mammalia	啮齿目 Rodentia	松鼠科 Sciuridae

学 名 Scientific Name	*Callosciurus pygerythrus*
命 名 人 Species Authority	I. Geoffroy Saint Hilaire, 1832
英 文 名 English Name(s)	Irrawaddy Squirrel
同物异名 Synonym(s)	Hoary-bellied Squirrel
种下单元评估 Infra-specific Taxa Assessed	无 / None

评估信息 Assessment Information

评估年份 Year Assessed	2020
评 定 人 Assessor(s)	蒋志刚、刘少英 / Zhigang Jiang, Shaoying Liu
审 定 人 Reviewer(s)	马勇、鲁长虎、鲍毅新 / Yong Ma, Changhu Lu, Yixin Bao
其他贡献人 Other Contributor(s)	李立立、丁晨晨 / Lili Li, Chenchen Ding

理由 Justification: 蓝腹松鼠的种群数量较大。因此，列为无危等级 / Irrawaddy Squirrel has large populations. Thus, it is listed as Least Concern

地理分布 Geographical Distribution

国内分布 Domestic Distribution
云南、西藏 / Yunnan, Tibet (Xizang)
世界分布 World Distribution
中国；南亚 / China; South Asia
分布标注 Distribution Note
非特有种 / Non-endemic

国内分布图
Map of Domestic Distribution

种群 Population

种群数量 Population Size	较多 / Relatively abundant
种群趋势 Population Trend	稳定 / Stable

生境与生态系统 Habitat(s) and Ecosystem(s)

生　　境 Habitat(s)	亚热带湿润低地森林、灌丛、种植园 / Subtropical Moist Lowland Forest, Shrubland, Plantation
生态系统 Ecosystem(s)	森林生态系统、灌丛生态系统、农田生态系统 / Forest Ecosystem, Shrubland Ecosystem, Cropland Ecosystem

威胁 Threat(s)

主要威胁 Major Threat(s)	耕种、伐木、火灾、住宅区及商业发展、狩猎 / Plantation, Logging, Fire, Residential and Commercial Development, Hunting

保护级别与保护行动 Protection Category and Conservation Action(s)

国家重点保护野生动物等级 (2021) Category of National Key Protected Wild Animals (2021)	无 / NA
IUCN 红色名录 (2020-2) IUCN Red List (2020-2)	无危 / LC
CITES 附录 (2019) CITES Appendix (2019)	无 / NA
保护行动 Conservation Action(s)	无 / None

相关文献 Relevant References

Burgin *et al*., 2018; Jiang *et al*. (蒋志刚等), 2017; Wilson *et al*., 2016; Zheng *et al*. (郑智民等), 2012; Smith *et al*. (史密斯等), 2009

蓝腹松鼠 *Callosciurus pygerythrus*

明纹花松鼠
Tamiops macclellandii

无危 LC

| 数据缺乏 DD | 无危 LC | 近危 NT | 易危 VU | 濒危 EN | 极危 CR | 区域灭绝 RE | 野外灭绝 EW | 灭绝 EX |

分类地位 Taxonomic Status

| 动物界 Animalia | 脊索动物门 Chordata | 哺乳纲 Mammalia | 啮齿目 Rodentia | 松鼠科 Sciuridae |

学名 Scientific Name	Tamiops macclellandii
命名人 Species Authority	Horsfield, 1840
英文名 English Name(s)	Himalayan Striped Squirrel
同物异名 Synonym(s)	Western Striped Squirrel, Burmese Striped Squirrel
种下单元评估 Infra-specific Taxa Assessed	无 / None

评估信息 Assessment Information

评估年份 Year Assessed	2020
评定人 Assessor(s)	蒋志刚、刘少英 / Zhigang Jiang, Shaoying Liu
审定人 Reviewer(s)	马勇、鲁长虎、鲍毅新 / Yong Ma, Changhu Lu, Yixin Bao
其他贡献人 Other Contributor(s)	李立立、丁晨晨 / Lili Li, Chenchen Ding

理由 Justification: 明纹花松鼠的种群数量较大。因此，列为无危等级 / Himalayan Striped Squirrel has large populations. Thus, it is listed as Least Concern

地理分布 Geographical Distribution

国内分布 Domestic Distribution

云南、西藏、广西 / Yunnan, Tibet (Xizang), Guangxi

世界分布 World Distribution

中国；南亚、东南亚 / China; South Asia, Southeast Asia

分布标注 Distribution Note

非特有种 / Non-endemic

国内分布图
Map of Domestic Distribution

种群 Population

种群数量 Population Size	较多 / Relatively abundant
种群趋势 Population Trend	稳定 / Stable

生境与生态系统 Habitat(s) and Ecosystem(s)

生 境 Habitat(s)	热带和亚热带湿润山地森林、种植园 / Tropical and Subtropical Moist Mountain Forest, Plantation
生态系统 Ecosystem(s)	森林生态系统、农田生态系统 / Forest Ecosystem, Cropland Ecosystem

威胁 Threat(s)

主要威胁 Major Threat(s)	火灾、耕种、狩猎 / Fire, Plantation, Hunting

保护级别与保护行动 Protection Category and Conservation Action(s)

国家重点保护野生动物等级 (2021) Category of National Key Protected Wild Animals (2021)	无 / NA
IUCN 红色名录 (2020-2) IUCN Red List (2020-2)	无危 / LC
CITES 附录 (2019) CITES Appendix (2019)	无 / NA
保护行动 Conservation Action(s)	无 / None

相关文献 Relevant References

Burgin *et al.*, 2018; Jiang *et al.*（蒋志刚等）, 2017; Wilson *et al.*, 2016; Zheng *et al.*（郑智民等）, 2012; Smith *et al.*（史密斯等）, 2009; Moore and Tate, 1965

明纹花松鼠 *Tamiops macclellandii*

倭花鼠
Tamiops maritimus

无危 LC

| 数据缺乏 DD | 无危 LC | 近危 NT | 易危 VU | 濒危 EN | 极危 CR | 区域灭绝 RE | 野外灭绝 EW | 灭绝 EX |

分类地位 Taxonomic Status

| 动物界 Animalia | 脊索动物门 Chordata | 哺乳纲 Mammalia | 啮齿目 Rodentia | 松鼠科 Sciuridae |

学名 Scientific Name	*Tamiops maritimus*
命名人 Species Authority	Bonhote, 1900
英文名 English Name(s)	Maritime Striped Squirrel
同物异名 Synonym(s)	Eastern Striped Squirrel
种下单元评估 Infra-specific Taxa Assessed	无 / None

评估信息 Assessment Information

评估年份 Year Assessed	2020
评定人 Assessor(s)	蒋志刚、刘少英 / Zhigang Jiang, Shaoying Liu
审定人 Reviewer(s)	马勇、鲁长虎、鲍毅新 / Yong Ma, Changhu Lu, Yixin Bao
其他贡献人 Other Contributor(s)	李立立、丁晨晨 / Lili Li, Chenchen Ding

理由 Justification: 倭花鼠种群数量较大。因此，列为无危等级 / Maritime Striped Squirrel has large populations. Thus, it is listed as Least Concern

地理分布 Geographical Distribution

国内分布 Domestic Distribution

贵州、福建、海南、台湾、浙江、安徽、江西、湖南、广东、广西、云南 / Guizhou, Fujian, Hainan, Taiwan, Zhejiang, Anhui, Jiangxi, Hunan, Guangdong, Guangxi, Yunnan

世界分布 World Distribution

中国、老挝、越南 / China, Laos, Viet Nam

分布标注 Distribution Note

非特有种 / Non-endemic

国内分布图
Map of Domestic Distribution

种群 Population

种群数量 Population Size	较多 / Relatively abundant
种群趋势 Population Trend	稳定 / Stable

生境与生态系统 Habitat (s) and Ecosystem (s)

生　　境 Habitat(s)	次生林、针阔混交林 / Secondary Forest, Coniferous Broad-leaved Mixed Forest
生态系统 Ecosystem(s)	森林生态系统 / Forest Ecosystem

威胁 Threat (s)

主要威胁 Major Threat(s)	无 / None

保护级别与保护行动 Protection Category and Conservation Action (s)

国家重点保护野生动物等级 (2021) Category of National Key Protected Wild Animals (2021)	无 / NA
IUCN 红色名录 (2020-2) IUCN Red List (2020-2)	无危 / LC
CITES 附录 (2019) CITES Appendix (2019)	无 / NA
保护行动 Conservation Action(s)	无 / None

相关文献 Relevant References

Burgin *et al*., 2018; Jiang *et al.* (蒋志刚等), 2017; Wilson *et al*., 2016; Zheng *et al.* (郑智民等), 2012; Smith *et al*., 2009; Thorington and Hoffmann, 2005

倭花鼠 *Tamiops maritimus*

隐纹花松鼠 *Tamiops swinhoei*

无危 LC

| 数据缺乏 DD | 无危 LC | 近危 NT | 易危 VU | 濒危 EN | 极危 CR | 区域灭绝 RE | 野外灭绝 EW | 灭绝 EX |

分类地位 Taxonomic Status

动物界 Animalia	脊索动物门 Chordata	哺乳纲 Mammalia	啮齿目 Rodentia	松鼠科 Sciuridae

学名 Scientific Name	*Tamiops swinhoei*
命名人 Species Authority	Milne-Edwards, 1874
英文名 English Name(s)	Swinhoe's Striped Squirrel
同物异名 Synonym(s)	无 / None
种下单元评估 Infra-specific Taxa Assessed	无 / None

评估信息 Assessment Information

评估年份 Year Assessed	2020
评定人 Assessor(s)	蒋志刚、刘少英 / Zhigang Jiang, Shaoying Liu
审定人 Reviewer(s)	马勇、鲁长虎、鲍毅新 / Yong Ma, Changhu Lu, Yixin Bao
其他贡献人 Other Contributor(s)	李立立、丁晨晨 / Lili Li, Chenchen Ding

理由 Justification: 隐纹花松鼠的种群数量较多。因此，列为无危等级 / Swinhoe's Striped Squirrel has large populations. Thus, it is listed as Least Concern

地理分布 Geographical Distribution

国内分布 Domestic Distribution
云南、西藏、四川、北京、河北、河南、陕西、山西、甘肃、宁夏、湖北 / Yunnan, Tibet (Xizang), Sichuan, Beijing, Hebei, Henan, Shaanxi, Shanxi, Gansu, Ningxia, Hubei
世界分布 World Distribution
中国、老挝、越南 / China, Laos, Viet Nam
分布标注 Distribution Note
非特有种 / Non-endemic

国内分布图
Map of Domestic Distribution

种群 Population

种群数量 Population Size	较多 / Relatively abundant
种群趋势 Population Trend	稳定 / Stable

生境与生态系统 Habitat(s) and Ecosystem(s)

生　　境 Habitat(s)	森林 / Forest
生态系统 Ecosystem(s)	森林生态系统 / Forest Ecosystem

威胁 Threat(s)

主要威胁 Major Threat(s)	无 / None

保护级别与保护行动 Protection Category and Conservation Action(s)

国家重点保护野生动物等级 (2021) Category of National Key Protected Wild Animals (2021)	无 / NA
IUCN 红色名录 (2020-2) IUCN Red List (2020-2)	无危 / LC
CITES 附录 (2019) CITES Appendix (2019)	无 / NA
保护行动 Conservation Action(s)	无 / None

相关文献 Relevant References

Burgin *et al.*, 2018; Jiang *et al.*（蒋志刚等）, 2017; Wilson *et al.*, 2016; Zheng *et al.*（郑智民等）, 2012; Smith *et al.*（史密斯等）, 2009

隐纹花松鼠 *Tamiops swinhoei*

珀氏长吻松鼠
Dremomys pernyi

无危 LC

| 数据缺乏 DD | 无危 LC | 近危 NT | 易危 VU | 濒危 EN | 极危 CR | 区域灭绝 RE | 野外灭绝 EW | 灭绝 EX |

分类地位 Taxonomic Status

| 动物界 Animalia | 脊索动物门 Chordata | 哺乳纲 Mammalia | 啮齿目 Rodentia | 松鼠科 Sciuridae |

学名 Scientific Name	*Dremomys pernyi*
命名人 Species Authority	Milne-Edwards, 1867
英文名 English Name(s)	Perny's Long-nosed Squirrel
同物异名 Synonym(s)	无 / None
种下单元评估 Infra-specific Taxa Assessed	无 / None

评估信息 Assessment Information

评估年份 Year Assessed	2020
评定人 Assessor(s)	蒋志刚、刘少英 / Zhigang Jiang, Shaoying Liu
审定人 Reviewer(s)	马勇、鲁长虎、鲍毅新 / Yong Ma, Changhu Lu, Yixin Bao
其他贡献人 Other Contributor(s)	李立立、丁晨晨 / Lili Li, Chenchen Ding

理由 Justification: 珀氏长吻松鼠的种群数量较大。因此，列为无危等级 / Perny's Long-nosed Squirrel has large populations. Thus, it is listed as Least Concern

地理分布 Geographical Distribution

国内分布 Domestic Distribution

陕西、甘肃、四川、贵州、云南、西藏、安徽、浙江、湖北、湖南、江西、福建、台湾、重庆、广西、广东 / Shaanxi, Gansu, Sichuan, Guizhou, Yunnan, Tibet (Xizang), Anhui, Zhejiang, Hubei, Hunan, Jiangxi, Fujian, Taiwan, Chongqing, Guangxi, Guangdong

世界分布 World Distribution

中国；南亚、东南亚 / China; South Asia, Southeast Asia

分布标注 Distribution Note

非特有种 / Non-endemic

国内分布图
Map of Domestic Distribution

种群 Population

种群数量 Population Size	数量多 / Abundant
种群趋势 Population Trend	稳定 / Stable

生境与生态系统 Habitat (s) and Ecosystem (s)

生　　境 Habitat(s)	森林、泰加林 / Forest, Taiga Forest
生态系统 Ecosystem(s)	森林生态系统 / Forest Ecosystem

威胁 Threat (s)

主要威胁 Major Threat(s)	未知 / Unknown

保护级别与保护行动 Protection Category and Conservation Action (s)

国家重点保护野生动物等级 (2021) Category of National Key Protected Wild Animals (2021)	无 / NA
IUCN 红色名录 (2020-2) IUCN Red List (2020-2)	无危 / LC
CITES 附录 (2019) CITES Appendix (2019)	无 / NA
保护行动 Conservation Action(s)	无 / None

相关文献 Relevant References

Burgin *et al*., 2018; Jiang *et al*.（蒋志刚等）, 2017; Wilson *et al*., 2016; Zheng *et al*.（郑智民等）, 2012; Smith *et al*.（史密斯等）, 2009; Men *et al*.（门兴元等）, 2006, Wang（王应祥）, 2003

珀氏长吻松鼠 *Dremomys pernyi*

红颊长吻松鼠
Dremomys rufigenis

无危 LC

| 数据缺乏 DD | 无危 LC | 近危 NT | 易危 VU | 濒危 EN | 极危 CR | 区域灭绝 RE | 野外灭绝 EW | 灭绝 EX |

分类地位 Taxonomic Status

动物界 Animalia	脊索动物门 Chordata	哺乳纲 Mammalia	啮齿目 Rodentia	松鼠科 Sciuridae

学名 Scientific Name	*Dremomys rufigenis*
命名人 Species Authority	Blanford, 1878
英文名 English Name(s)	Asian Red-cheeked Squirrel
同物异名 Synonym(s)	无 / None
种下单元评估 Infra-specific Taxa Assessed	无 / None

评估信息 Assessment Information

评估年份 Year Assessed	2020
评定人 Assessor(s)	蒋志刚、刘少英 / Zhigang Jiang, Shaoying Liu
审定人 Reviewer(s)	马勇、鲁长虎、鲍毅新 / Yong Ma, Changhu Lu, Yixin Bao
其他贡献人 Other Contributor(s)	李立立、丁晨晨 / Lili Li, Chenchen Ding

理由 Justification: 红颊长吻松鼠的种群数量较大。因此，列为无危等级 / Asian Red-cheeked Squirrel has large populations. Thus, it is listed as Least Concern

地理分布 Geographical Distribution

国内分布 Domestic Distribution
云南、广西、安徽、湖南 / Yunnan, Guangxi, Anhui, Hunan

世界分布 World Distribution
中国、印度、老挝、马来西亚、缅甸、泰国、越南 / China, India, Laos, Malaysia, Myanmar, Thailand, Viet Nam

分布标注 Distribution Note
非特有种 / Non-endemic

国内分布图
Map of Domestic Distribution

种群 Population

种群数量 Population Size	较多 / Relatively abundant
种群趋势 Population Trend	稳定 / Stable

生境与生态系统 Habitat (s) and Ecosystem (s)

生　　境 Habitat(s)	森林 / Forest
生态系统 Ecosystem(s)	森林生态系统 / Forest Ecosystem

威胁 Threat (s)

主要威胁 Major Threat(s)	无 / None

保护级别与保护行动 Protection Category and Conservation Action (s)

国家重点保护野生动物等级 (2021) Category of National Key Protected Wild Animals (2021)	无 / NA
IUCN 红色名录 (2020-2) IUCN Red List (2020-2)	无危 / LC
CITES 附录 (2019) CITES Appendix (2019)	无 / NA
保护行动 Conservation Action(s)	无 / None

相关文献 Relevant References

Burgin *et al.*, 2018; Jiang *et al.* (蒋志刚等), 2017; Wilson *et al.*, 2016; Zheng *et al.* (郑智民等), 2012; Smith *et al.* (史密斯等), 2009; Moore and Tate, 1965

红颊长吻松鼠 *Dremomys rufigenis*

条纹松鼠 Menetes berdmorei

无危 LC

分类地位 Taxonomic Status

动物界 Animalia	脊索动物门 Chordata	哺乳纲 Mammalia	啮齿目 Rodentia	松鼠科 Sciuridae

学名 Scientific Name	*Menetes berdmorei*
命名人 Species Authority	Blyth, 1849
英文名 English Name(s)	Indochinese Ground Squirrel
同物异名 Synonym(s)	Berdmore's Ground Squirrel
种下单元评估 Infra-specific Taxa Assessed	无 / None

评估信息 Assessment Information

评估年份 Year Assessed	2020
评定人 Assessor(s)	蒋志刚、刘少英 / Zhigang Jiang, Shaoying Liu
审定人 Reviewer(s)	马勇、鲁长虎、鲍毅新 / Yong Ma, Changhu Lu, Yixin Bao
其他贡献人 Other Contributor(s)	李立立、丁晨晨 / Lili Li, Chenchen Ding

理由 Justification: 条纹松鼠的种群数量较大。因此，列为无危等级 / Indochinese Ground Squirrel has large populations. Thus, it is listed as Least Concern

地理分布 Geographical Distribution

国内分布 Domestic Distribution
云南 / Yunnan
世界分布 World Distribution
柬埔寨、中国、老挝、缅甸、泰国、越南 / Cambodia, China, Laos, Myanmar, Thailand, Viet Nam
分布标注 Distribution Note
非特有种 / Non-endemic

国内分布图
Map of Domestic Distribution

种群 Population

种群数量 Population Size	较多 / Relatively abundant
种群趋势 Population Trend	稳定 / Stable

生境与生态系统 Habitat(s) and Ecosystem(s)

生境 Habitat(s)	森林、耕地 / Forest, Arable Land
生态系统 Ecosystem(s)	森林生态系统、农田生态系统 / Forest Ecosystem, Cropland Ecosystem

威胁 Threat(s)

主要威胁 Major Threat(s)	耕种 / Plantation

保护级别与保护行动 Protection Category and Conservation Action(s)

国家重点保护野生动物等级 (2021) Category of National Key Protected Wild Animals (2021)	无 / NA
IUCN红色名录 (2020-2) IUCN Red List (2020-2)	无危 / LC
CITES 附录 (2019) CITES Appendix (2019)	无 / NA
保护行动 Conservation Action(s)	无 / None

相关文献 Relevant References

Burgin *et al.*, 2018; Jiang *et al.* (蒋志刚等), 2017; Wilson *et al.*, 2016; Zheng *et al.* (郑智民等), 2012; Smith *et al.* (史密斯等), 2009; Pan *et al.* (潘清华等), 2007; Wang (王应祥), 2003

条纹松鼠 *Menetes berdmorei*

中国生物多样性红色名录
岩松鼠
Sciurotamias davidianus

无危 LC

| 数据缺乏 DD | 无危 LC | 近危 NT | 易危 VU | 濒危 EN | 极危 CR | 区域灭绝 RE | 野外灭绝 EW | 灭绝 EX |

分类地位 Taxonomic Status

| 动物界 Animalia | 脊索动物门 Chordata | 哺乳纲 Mammalia | 啮齿目 Rodentia | 松鼠科 Sciuridae |

学名 Scientific Name	*Sciurotamias davidianus*
命名人 Species Authority	Milne-Edwards, 1867
英文名 English Name(s)	Père David's Rock Squirrel
同物异名 Synonym(s)	Chinese Rock Squirrel; *Sciurotamias collaris* (Heude, 1898)
种下单元评估 Infra-specific Taxa Assessed	无 / None

评估信息 Assessment Information

评估年份 Year Assessed	2020
评定人 Assessor(s)	蒋志刚、刘少英 / Zhigang Jiang, Shaoying Liu
审定人 Reviewer(s)	马勇、鲁长虎、鲍毅新 / Yong Ma, Changhu Lu, Yixin Bao
其他贡献人 Other Contributor(s)	李立立、丁晨晨 / Lili Li, Chenchen Ding

理由 Justification: 岩松鼠的种群数量较多。因此，列为无危等级 / Père David's Rock Squirrel has large populations. Thus, it is listed as Least Concern

地理分布 Geographical Distribution

国内分布 Domestic Distribution

辽宁、河北、天津、北京、河南、安徽、山西、陕西、四川、重庆、宁夏、甘肃、云南、贵州、湖南、湖北 / Liaoning, Hebei, Tianjin, Beijing, Henan, Anhui, Shanxi, Shaanxi, Sichuan, Chongqing, Ningxia, Gansu, Yunnan, Guizhou, Hunan, Hubei

世界分布 World Distribution

中国 / China

分布标注 Distribution Note

特有种 / Endemic

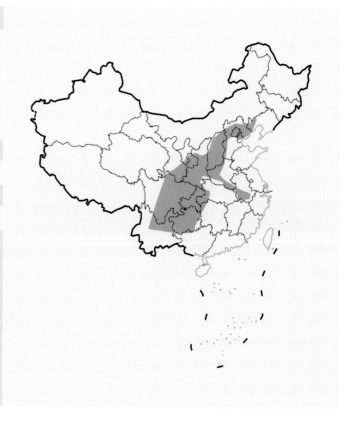

国内分布图
Map of Domestic Distribution

种群 Population

种群数量 Population Size	较多 / Relatively abundant
种群趋势 Population Trend	稳定 / Stable

生境与生态系统 Habitat(s) and Ecosystem(s)

生　　境 Habitat(s)	森林 / Forest
生态系统 Ecosystem(s)	森林生态系统 / Forest Ecosystem

威胁 Threat(s)

主要威胁 Major Threat(s)	无 / None

保护级别与保护行动 Protection Category and Conservation Action(s)

国家重点保护野生动物等级 (2021) Category of National Key Protected Wild Animals (2021)	无 / NA
IUCN 红色名录 (2020-2) IUCN Red List (2020-2)	无危 / LC
CITES 附录 (2019) CITES Appendix (2019)	无 / NA
保护行动 Conservation Action(s)	无 / None

相关文献 Relevant References

Burgin *et al*., 2018; Jiang *et al*.（蒋志刚等），2017; Wilson *et al*., 2016; Zheng *et al*.（郑智民等），2012; Smith *et al*.（史密斯等），2009; Guo（郭建荣），2003; Wang（王应祥），2003

岩松鼠 *Sciurotamias davidianus*

侧纹岩松鼠 Rupestes forresti

无危 LC

| 数据缺乏 DD | 无危 LC | 近危 NT | 易危 VU | 濒危 EN | 极危 CR | 区域灭绝 RE | 野外灭绝 EW | 灭绝 EX |

分类地位 Taxonomic Status

动物界 Animalia	脊索动物门 Chordata	哺乳纲 Mammalia	啮齿目 Rodentia	松鼠科 Sciuridae

学名 Scientific Name	*Rupestes forresti*
命名人 Species Authority	Thomas, 1922
英文名 English Name(s)	Forrest's Rock Squirrel
同物异名 Synonym(s)	无 / None
种下单元评估 Infra-specific Taxa Assessed	无 / None

评估信息 Assessment Information

评估年份 Year Assessed	2020
评定人 Assessor(s)	蒋志刚、刘少英 / Zhigang Jiang, Shaoying Liu
审定人 Reviewer(s)	马勇、鲁长虎、鲍毅新 / Yong Ma, Changhu Lu, Yixin Bao
其他贡献人 Other Contributor(s)	李立立、丁晨晨 / Lili Li, Chenchen Ding

理由 Justification: 侧纹岩松鼠的种群数量较多。因此，列为无危等级 / Forrest's Rock Squirrel has large populations. Thus, it is listed as Least Concern

地理分布 Geographical Distribution

国内分布 Domestic Distribution
云南、四川 / Yunnan, Sichuan
世界分布 World Distribution
中国 / China
分布标注 Distribution Note
特有种 / Endemic

国内分布图
Map of Domestic Distribution

种群 Population

种群数量 Population Size	较多 / Relatively abundant
种群趋势 Population Trend	稳定 / Stable

生境与生态系统 Habitat (s) and Ecosystem (s)

生　　境 Habitat(s)	灌丛 / Shrubland
生态系统 Ecosystem(s)	灌丛生态系统 / Shrubland Ecosystem

威胁 Threat (s)

主要威胁 Major Threat(s)	无 / None

保护级别与保护行动 Protection Category and Conservation Action (s)

国家重点保护野生动物等级 (2021) Category of National Key Protected Wild Animals (2021)	无 / NA
IUCN 红色名录 (2020-2) IUCN Red List (2020-2)	无危 / LC
CITES 附录 (2019) CITES Appendix (2019)	无 / NA
保护行动 Conservation Action(s)	无 / None

相关文献 Relevant References

Burgin *et al.*, 2018; Jiang *et al.* (蒋志刚等), 2017; Wilson *et al.*, 2016; Zheng *et al.* (郑智民等), 2012; Smith *et al.* (史密斯等), 2009; Wang and Xie (汪松和解焱), 2004

侧纹岩松鼠 *Rupestes forresti*

北花松鼠
Tamias sibiricus
无危 LC

| 数据缺乏 DD | 无危 LC | 近危 NT | 易危 VU | 濒危 EN | 极危 CR | 区域灭绝 RE | 野外灭绝 EW | 灭绝 EX |

分类地位 Taxonomic Status

动物界 Animalia	脊索动物门 Chordata	哺乳纲 Mammalia	啮齿目 Rodentia	松鼠科 Sciuridae

学名 Scientific Name	*Tamias sibiricus*
命名人 Species Authority	Laxmann, 1769
英文名 English Name(s)	Siberian Chipmunk
同物异名 Synonym(s)	Common Chipmunk
种下单元评估 Infra-specific Taxa Assessed	无 / None

评估信息 Assessment Information

评估年份 Year Assessed	2020
评定人 Assessor(s)	蒋志刚、刘少英 / Zhigang Jiang, Shaoying Liu
审定人 Reviewer(s)	马勇、鲍毅新 / Yong Ma, Yixin Bao
其他贡献人 Other Contributor(s)	李立立、丁晨晨 / Lili Li, Chenchen Ding

理由 Justification: 北花松鼠分布区大、种群数量多。因此，列为无危等级 / Siberian Chipmunk has an extensive range and large populations. Thus, it is listed as Least Concern

地理分布 Geographical Distribution

国内分布 Domestic Distribution

山西、陕西、内蒙古、河南、新疆、青海、北京、天津、河北、辽宁、吉林、黑龙江、四川、甘肃、宁夏 / Shanxi, Shaanxi, Inner Mongolia (Nei Mongol), Henan, Xinjiang, Qinghai, Beijing, Tianjin, Hebei, Liaoning, Jilin, Heilongjiang, Sichuan, Gansu, Ningxia

世界分布 World Distribution

中国、日本、哈萨克斯坦、朝鲜、韩国、俄罗斯 / China, Japan, Kazakhstan, Korea (the Democratic People's Republic of), Korea (the Republic of), Russia

分布标注 Distribution Note

非特有种 / Non-endemic

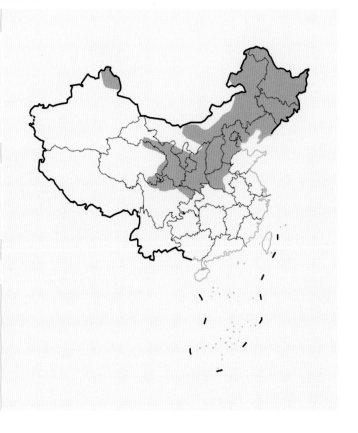

国内分布图
Map of Domestic Distribution

种群 Population

种群数量 Population Size	较多 / Relatively abundant
种群趋势 Population Trend	稳定 / Stable

生境与生态系统 Habitat (s) and Ecosystem (s)

生境 Habitat(s)	针叶林、针阔混交林 / Coniferous Forest, Coniferous and Broad-leaved Mixed Forest
生态系统 Ecosystem(s)	森林生态系统 / Forest Ecosystem

威胁 Threat (s)

主要威胁 Major Threat(s)	无 / None

保护级别与保护行动 Protection Category and Conservation Action (s)

国家重点保护野生动物等级 (2021) Category of National Key Protected Wild Animals (2021)	无 / NA
IUCN 红色名录 (2020-2) IUCN Red List (2020-2)	无危 / LC
CITES 附录 (2019) CITES Appendix (2019)	无 / NA
保护行动 Conservation Action(s)	无 / None

相关文献 Relevant References

Burgin *et al.*, 2018; Jiang *et al.* (蒋志刚等), 2017; Wilson *et al.*, 2016; Smith *et al.* (史密斯等), 2009; Pan *et al.* (潘清华等), 2007; Wang (王应祥), 2003

北花松鼠 *Tamias sibiricus*

阿拉善黄鼠
Spermophilus alashanicus

无危 LC

| 数据缺乏 DD | 无危 LC | 近危 NT | 易危 VU | 濒危 EN | 极危 CR | 区域灭绝 RE | 野外灭绝 EW | 灭绝 EX |

分类地位 Taxonomic Status

动物界 Animalia	脊索动物门 Chordata	哺乳纲 Mammalia	啮齿目 Rodentia	松鼠科 Sciuridae
学名 Scientific Name		*Spermophilus alashanicus*		
命名人 Species Authority		Büchner, 1888		
英文名 English Name(s)		Alashan Ground Squirrel		
同物异名 Synonym(s)		*Spermophilus dilutus* (Formozov, 1929); *obscurus* (Büchner, 1888); *siccus* (G. M. Allen, 1925)		
种下单元评估 Infra-specific Taxa Assessed		无 / None		

评估信息 Assessment Information

评估年份 Year Assessed	2020
评定人 Assessor(s)	蒋志刚、刘少英 / Zhigang Jiang, Shaoying Liu
审定人 Reviewer(s)	马勇、刘伟、鲍毅新 / Yong Ma, Wei Liu, Yixin Bao
其他贡献人 Other Contributor(s)	李立立、丁晨晨 / Lili Li, Chenchen Ding

理由 Justification: 阿拉善黄鼠分布广，种群数量大。因此，列为无危等级 / Alashan Ground Squirrel has a large distribution range and large populations. Thus, it is listed as Least Concern

地理分布 Geographical Distribution

国内分布 Domestic Distribution
宁夏、甘肃、青海、陕西、内蒙古 / Ningxia, Gansu, Qinghai, Shaanxi, Inner Mongolia (Nei Mongol)
世界分布 World Distribution
中国、蒙古 / China, Mongolia
分布标注 Distribution Note
非特有种 / Non-endemic

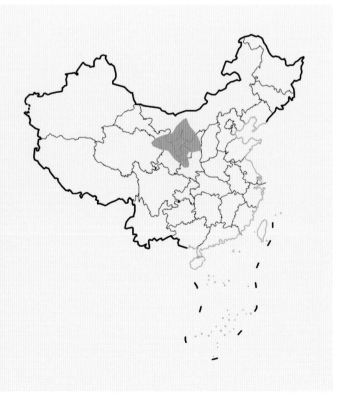

国内分布图
Map of Domestic Distribution

种群 Population

种群数量 Population Size	较多 / Relatively abundant
种群趋势 Population Trend	下降 / Decreasing

生境与生态系统 Habitat(s) and Ecosystem(s)

生境 Habitat(s)	沙漠、干旱草地 / Desert, Arid Grassland
生态系统 Ecosystem(s)	草地生态系统、荒漠生态系统 / Grassland Ecosystem, Desert Ecosystem

威胁 Threat(s)

主要威胁 Major Threat(s)	无 / None

保护级别与保护行动 Protection Category and Conservation Action(s)

国家重点保护野生动物等级 (2021) Category of National Key Protected Wild Animals (2021)	无 / NA
IUCN 红色名录 (2020-2) IUCN Red List (2020-2)	无危 / LC
CITES 附录 (2019) CITES Appendix (2019)	无 / NA
保护行动 Conservation Action(s)	无 / None

相关文献 Relevant References

Burgin *et al.*, 2018; Jiang *et al.*（蒋志刚等）, 2017; Wilson *et al.*, 2016; Li *et al.*（李国军等）, 2013; Smith *et al.*（史密斯等）, 2009; Wang（王应祥）, 2003; Qin（秦长育）, 1985

阿拉善黄鼠 *Spermophilus alashanicus*

短尾黄鼠
Spermophilus brevicauda

无危 LC

| 数据缺乏 DD | 无危 LC | 近危 NT | 易危 VU | 濒危 EN | 极危 CR | 区域灭绝 RE | 野外灭绝 EW | 灭绝 EX |

🦌 分类地位 Taxonomic Status

动物界 Animalia	脊索动物门 Chordata	哺乳纲 Mammalia	啮齿目 Rodentia	松鼠科 Sciuridae

学名 Scientific Name	*Spermophilus brevicauda*
命名人 Species Authority	Brandt, 1843
英文名 English Name(s)	Brandt's Ground Squirrel
同物异名 Synonym(s)	*Spermophilus carruthersi* (Thomas, 1912); *intermedius* (Brandt, 1844); *ilensis* (Belyaev, 1945); *saryarka* (Selevin, 1937); *selevini* (Argyropolu, 1941)
种下单元评估 Infra-specific Taxa Assessed	无 / None

🦌 评估信息 Assessment Information

评估年份 Year Assessed	2020
评定人 Assessor(s)	蒋志刚、刘少英 / Zhigang Jiang, Shaoying Liu
审定人 Reviewer(s)	马勇、刘伟、鲍毅新 / Yong Ma, Wei Liu, Yixin Bao
其他贡献人 Other Contributor(s)	李立立、丁晨晨 / Lili Li, Chenchen Ding

理由 Justification: 短尾黄鼠分布区大，种群数量较多。因此，列为无危等级 / Brandt's Ground Squirrel occupies a large geographic range and has large populations. Thus, it is listed as Least Concern

🦌 地理分布 Geographical Distribution

国内分布 Domestic Distribution
新疆 / Xinjiang
世界分布 World Distribution
中国、哈萨克斯坦 / China, Kazakhstan
分布标注 Distribution Note
非特有种 / Non-endemic

国内分布图
Map of Domestic Distribution

种群 Population

种群数量 Population Size	较多 / Relatively abundant
种群趋势 Population Trend	稳定 / Stable

生境与生态系统 Habitat (s) and Ecosystem (s)

生　　境 Habitat(s)	草地、半荒漠、灌丛 / Grassland, Semi-desert, Shrubland
生态系统 Ecosystem(s)	灌丛生态系统、草地生态系统、荒漠生态系统 / Shrubland Ecosystem, Grassland Ecosystem, Desert Ecosystem

威胁 Threat (s)

主要威胁 Major Threat(s)	无 / None

保护级别与保护行动 Protection Category and Conservation Action (s)

国家重点保护野生动物等级 (2021) Category of National Key Protected Wild Animals (2021)	无 / NA
IUCN 红色名录 (2020-2) IUCN Red List (2020-2)	无危 / LC
CITES 附录 (2019) CITES Appendix (2019)	无 / NA
保护行动 Conservation Action(s)	无 / None

相关文献 Relevant References

Burgin *et al.*, 2018; Jiang *et al.* (蒋志刚等), 2017; Wilson *et al.*, 2016; Zheng *et al.* (郑智民等), 2012; Smith *et al.* (史密斯等), 2009; Wang and Xie (汪松和解焱), 2004

短尾黄鼠 *Spermophilus brevicauda*

达乌尔黄鼠
Spermophilus dauricus

无危 LC

| 数据缺乏 DD | 无危 LC | 近危 NT | 易危 VU | 濒危 EN | 极危 CR | 区域灭绝 RE | 野外灭绝 EW | 灭绝 EX |

分类地位 Taxonomic Status

| 动物界 Animalia | 脊索动物门 Chordata | 哺乳纲 Mammalia | 啮齿目 Rodentia | 松鼠科 Sciuridae |

学名 Scientific Name	*Spermophilus dauricus*
命名人 Species Authority	Brandt, 1843
英文名 English Name(s)	Daurian Ground Squirrel
同物异名 Synonym(s)	*Spermophilus mongolicus* (Milne-Edwards, 1867); *ramosus* (Thomas, 1909); *umbratus* (Thomas, 1908); *yamashinae* (Kuroda, 1939)
种下单元评估 Infra-specific Taxa Assessed	无 / None

评估信息 Assessment Information

评估年份 Year Assessed	2020
评定人 Assessor(s)	蒋志刚、刘少英 / Zhigang Jiang, Shaoying Liu
审定人 Reviewer(s)	马勇、刘伟、鲍毅新 / Yong Ma, Wei Liu, Yixin Bao
其他贡献人 Other Contributor(s)	李立立、丁晨晨 / Lili Li, Chenchen Ding

理由 Justification: 达乌尔黄鼠分布区广，种群数量较多。因此，列为无危等级 / Daurian Ground Squirrel has an extensive range and large populations. Thus, it is listed as Least Concern

地理分布 Geographical Distribution

国内分布 Domestic Distribution	黑龙江、吉林、辽宁、内蒙古、河北、北京、天津、河南、山东、陕西、山西 / Heilongjiang, Jilin, Liaoning, Inner Mongolia (Nei Mongol), Hebei, Beijing, Tianjin, Henan, Shandong, Shaanxi, Shanxi
世界分布 World Distribution	中国、蒙古、俄罗斯 / China, Mongolia, Russia
分布标注 Distribution Note	非特有种 / Non-endemic

国内分布图
Map of Domestic Distribution

种群 Population

种群数量 Population Size	较多 / Relatively abundant
种群趋势 Population Trend	稳定 / Stable

生境与生态系统 Habitat(s) and Ecosystem(s)

生境 Habitat(s)	沙漠 / Desert
生态系统 Ecosystem(s)	荒漠生态系统 / Desert Ecosystem

威胁 Threat(s)

主要威胁 Major Threat(s)	家畜放牧、干旱 / Livestock Ranching, Drought

保护级别与保护行动 Protection Category and Conservation Action(s)

国家重点保护野生动物等级 (2021) Category of National Key Protected Wild Animals (2021)	无 / NA
IUCN 红色名录 (2020-2) IUCN Red List (2020-2)	无危 / LC
CITES 附录 (2019) CITES Appendix (2019)	无 / NA
保护行动 Conservation Action(s)	无 / None

相关文献 Relevant References

Burgin *et al*., 2018; Jiang *et al.* (蒋志刚等), 2017; Wilson *et al*., 2016; Zheng *et al.* (郑智民等), 2012; Smith *et al.* (史密斯等), 2009; Mi *et al.* (米景川等), 2003; Zhang and Wang (张晓华和王广仁), 2002

达乌尔黄鼠 *Spermophilus dauricus*

淡尾黄鼠
Spermophilus pallidicauda
无危 LC

| 数据缺乏 DD | 无危 LC | 近危 NT | 易危 VU | 濒危 EN | 极危 CR | 区域灭绝 RE | 野外灭绝 EW | 灭绝 EX |

分类地位 Taxonomic Status

动物界 Animalia	脊索动物门 Chordata	哺乳纲 Mammalia	啮齿目 Rodentia	松鼠科 Sciuridae

学名 Scientific Name	*Spermophilus pallidicauda*
命名人 Species Authority	Satunin, 1903
英文名 English Name(s)	Pallid Ground Squirrel
同物异名 Synonym(s)	无 / None
种下单元评估 Infra-specific Taxa Assessed	无 / None

评估信息 Assessment Information

评估年份 Year Assessed	2020
评定人 Assessor(s)	蒋志刚、刘少英 / Zhigang Jiang, Shaoying Liu
审定人 Reviewer(s)	马勇、刘伟、鲍毅新 / Yong Ma, Wei Liu, Yixin Bao
其他贡献人 Other Contributor(s)	李立立、丁晨晨 / Lili Li, Chenchen Ding

理由 Justification: 淡尾黄鼠的种群数量较多。因此，列为无危等级 / Pallid Ground Squirrel has large populations. Thus, it is listed as Least Concern

地理分布 Geographical Distribution

国内分布 Domestic Distribution	内蒙古、甘肃 / Inner Mongolia (Nei Mongol), Gansu
世界分布 World Distribution	中国、蒙古 / China, Mongolia
分布标注 Distribution Note	非特有种 / Non-endemic

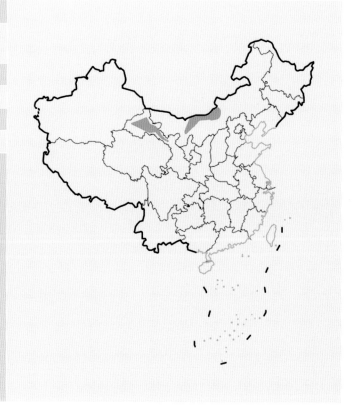

国内分布图
Map of Domestic Distribution

种群 Population

种群数量 Population Size	较多 / Relatively abundant
种群趋势 Population Trend	稳定 / Stable

生境与生态系统 Habitat(s) and Ecosystem(s)

生　　境 Habitat(s)	干旱草地 / Dry Steppe
生态系统 Ecosystem(s)	草地生态系统 / Grassland Ecosystem

威胁 Threat(s)

主要威胁 Major Threat(s)	耕种、干旱 / Plantation, Drought

保护级别与保护行动 Protection Category and Conservation Action(s)

国家重点保护野生动物等级 (2021) Category of National Key Protected Wild Animals (2021)	无 / NA
IUCN 红色名录 (2020-2) IUCN Red List (2020-2)	无危 / LC
CITES 附录 (2019) CITES Appendix (2019)	无 / NA
保护行动 Conservation Action(s)	无 / None

相关文献 Relevant References

Burgin *et al.*, 2018; Jiang *et al.*（蒋志刚等）, 2017; Wilson *et al.*, 2016; Zheng *et al.*（郑智民等）, 2012; Smith *et al.*（史密斯等）, 2009; Wang and Xie（汪松和解焱）, 2004

淡尾黄鼠 *Spermophilus pallidicauda*

长尾黄鼠 *Spermophilus parryii*

中国生物多样性红色名录

无危 LC

| 数据缺乏 DD | 无危 LC | 近危 NT | 易危 VU | 濒危 EN | 极危 CR | 区域灭绝 RE | 野外灭绝 EW | 灭绝 EX |

分类地位 Taxonomic Status

动物界 Animalia	脊索动物门 Chordata	哺乳纲 Mammalia	啮齿目 Rodentia	松鼠科 Sciuridae
学名 Scientific Name		*Spermophilus parryii*		
命名人 Species Authority		Richardson, 1825		
英文名 English Name(s)		Arctic Ground Squirrel		
同物异名 Synonym(s)		*Citellus buxtoni* (Allen, 1903); *C. eversmanni* (Ognev, 1937) subsp. *janensis*; *C. parryi* (Tchernyavsky, 1972) subsp. *tshuktschorum*; *C. stejnegeri* (Allen, 1903);（转下页）		
种下单元评估 Infra-specific Taxa Assessed		无 / None		

评估信息 Assessment Information

评估年份 Year Assessed	2020
评定人 Assessor(s)	蒋志刚、刘少英 / Zhigang Jiang, Shaoying Liu
审定人 Reviewer(s)	马勇、刘伟、鲍毅新 / Yong Ma, Wei Liu, Yixin Bao
其他贡献人 Other Contributor(s)	李立立、丁晨晨 / Lili Li, Chenchen Ding

理由 Justification：长尾黄鼠的种群数量较多。因此，列为无危等级 / Arctic Ground Squirrel has large populations. Thus, it is listed as Least Concern

地理分布 Geographical Distribution

国内分布 Domestic Distribution
黑龙江、新疆 / Heilongjiang, Xinjiang
世界分布 World Distribution
中国、加拿大、俄罗斯、美国 / China, Canada, Russia, United States
分布标注 Distribution Note
非特有种 / Non-endemic

国内分布图
Map of Domestic Distribution

种群 Population

种群数量 Population Size	较多 / Relatively abundant
种群趋势 Population Trend	稳定 / Stable

生境与生态系统 Habitat(s) and Ecosystem(s)

生境 Habitat(s)	沙漠、草地、灌木、草甸、湿地 / Desert, Grassland, Shrubland, Meadow, Wetland
生态系统 Ecosystem(s)	森林生态系统、草地生态系统、荒漠生态系统、湿地生态系统 / Forest Ecosystem, Grassland Ecosystem, Desert Ecosystem, Wetland Ecosystem

威胁 Threat(s)

主要威胁 Major Threat(s)	狩猎、家畜放牧、干旱 / Hunting, Livestock Ranching, Drought

保护级别与保护行动 Protection Category and Conservation Action(s)

国家重点保护野生动物等级 (2021) Category of National Key Protected Wild Animals (2021)	无 / NA
IUCN 红色名录 (2020-2) IUCN Red List (2020-2)	无危 / LC
CITES 附录 (2019) CITES Appendix (2019)	无 / NA
保护行动 Conservation Action(s)	无 / None

相关文献 Relevant References

Burgin *et al.*, 2018; Jiang *et al.*（蒋志刚等）, 2017; Zheng *et al.*（郑智民等）, 2012; Smith *et al.*（史密斯等）, 2009; Wang（王应祥）, 2003; Lin *et al.*（林纪春等）, 1989

（接上页）
C. undulatus (Portenko, 1963) subsp. *coriakorum*; *Spermophilus brunniceps* (Kittlitz, 1858); *S. leucosticus* (Brandt, 1844)

长尾黄鼠 *Spermophilus parryii*

By Lanare Sévi

天山黄鼠
Spermophilus relictus
无危 LC

DD	LC	NT	VU	EN	CR	RE	EW	EX
数据缺乏	无危	近危	易危	濒危	极危	区域灭绝	野外灭绝	灭绝

分类地位 Taxonomic Status

动物界 Animalia	脊索动物门 Chordata	哺乳纲 Mammalia	啮齿目 Rodentia	松鼠科 Sciuridae

学名 Scientific Name	*Spermophilus relictus*
命名人 Species Authority	Kashkarov, 1923
英文名 English Name(s)	Tien Shan Ground Squirrel
同物异名 Synonym(s)	Relict Ground Squirrel; *Citellus relictus* (Kashkarov, 1923); *Urocitellus relictus* (Kashkarov, 1923)
种下单元评估 Infra-specific Taxa Assessed	无 / None

评估信息 Assessment Information

评估年份 Year Assessed	2020
评定人 Assessor(s)	蒋志刚、刘少英 / Zhigang Jiang, Shaoying Liu
审定人 Reviewer(s)	马勇、刘伟、鲍毅新 / Yong Ma, Wei Liu, Yixin Bao
其他贡献人 Other Contributor(s)	李立立、丁晨晨 / Lili Li, Chenchen Ding

理由 Justification: 天山黄鼠有一定的分布区，且种群数量大。因此，列为无危等级 / Tien Shan Ground Squirrel is distributed in certain region, and has large populations. Thus, it is listed as Least Concern

地理分布 Geographical Distribution

国内分布 Domestic Distribution
新疆 / Xinjiang
世界分布 World Distribution
中国；中亚 / China; Central Asia
分布标注 Distribution Note
非特有种 / Non-endemic

国内分布图
Map of Domestic Distribution

种群 Population

种群数量 Population Size	较多 / Relatively abundant
种群趋势 Population Trend	稳定 / Stable

生境与生态系统 Habitat(s) and Ecosystem(s)

生　　境 Habitat(s)	草地 / Steppe
生态系统 Ecosystem(s)	草地生态系统 / Grassland Ecosystem

威胁 Threat(s)

主要威胁 Major Threat(s)	无 / None

保护级别与保护行动 Protection Category and Conservation Action(s)

国家重点保护野生动物等级 (2021) Category of National Key Protected Wild Animals (2021)	无 / NA
IUCN 红色名录 (2020-2) IUCN Red List (2020-2)	无危 / LC
CITES 附录 (2019) CITES Appendix (2019)	无 / NA
保护行动 Conservation Action(s)	无 / None

相关文献 Relevant References

Burgin *et al.*, 2018; Jiang *et al.* (蒋志刚等), 2017; Wilson *et al.*, 2016; Zheng *et al.* (郑智民等), 2012; Smith *et al.* (史密斯等), 2009; Wang (王应祥), 2003

天山黄鼠 *Spermophilus relictus*　　　　　　张永 摄　By Yong Zhang

灰旱獭 *Marmota baibacina*

无危 LC

| 数据缺乏 DD | 无危 LC | 近危 NT | 易危 VU | 濒危 EN | 极危 CR | 区域灭绝 RE | 野外灭绝 EW | 灭绝 EX |

分类地位 Taxonomic Status

动物界 Animalia	脊索动物门 Chordata	哺乳纲 Mammalia	啮齿目 Rodentia	松鼠科 Sciuridae
学 名 Scientific Name		*Marmota baibacina*		
命 名 人 Species Authority		Kastschenko, 1899		
英 文 名 English Name(s)		Altai Marmot		
同物异名 Synonym(s)		Gray Marmot; *Marmota baibacina* (Stroganov and Yudin, 1956) subsp. *kastschenkoi*; *M. baibacina* (Skalon, 1950) subsp. *ognevi*; *M. centralis* (Thomas, 1909)		
种下单元评估 Infra-specific Taxa Assessed		无 / None		

评估信息 Assessment Information

评 估 年 份 Year Assessed	2020
评 定 人 Assessor(s)	蒋志刚、刘少英 / Zhigang Jiang, Shaoying Liu
审 定 人 Reviewer(s)	马勇、刘伟、鲍毅新 / Yong Ma, Wei Liu, Yixin Bao
其他贡献人 Other Contributor(s)	李立立、丁晨晨 / Lili Li, Chenchen Ding

理由 Justification: 灰旱獭分布区大，种群数量大。因此，列为无危等级 / Altai Marmot has a large range and large populations. Thus, it is listed as Least Concern

地理分布 Geographical Distribution

国内分布 Domestic Distribution
新疆 / Xinjiang
世界分布 World Distribution
中国、哈萨克斯坦、吉尔吉斯斯坦、蒙古、俄罗斯 / China, Kazakhstan, Kyrgyzstan, Mongolia, Russia
分布标注 Distribution Note
非特有种 / Non-endemic

国内分布图
Map of Domestic Distribution

🦌 种群 Population

种群数量 Population Size	数量多 / Abundant
种群趋势 Population Trend	稳定 / Stable

🦌 生境与生态系统 Habitat (s) and Ecosystem (s)

生　　境 Habitat(s)	草甸、森林、草地 / Meadow, Forest, Grassland
生态系统 Ecosystem(s)	森林生态系统、草地生态系统 / Forest Ecosystem, Grassland Ecosystem

🦌 威胁 Threat (s)

主要威胁 Major Threat(s)	狩猎、家畜放牧 / Hunting, Livestock Ranching

🦌 保护级别与保护行动 Protection Category and Conservation Action (s)

国家重点保护野生动物等级 (2021) Category of National Key Protected Wild Animals (2021)	无 / NA
IUCN 红色名录 (2020-2) IUCN Red List (2020-2)	无危 / LC
CITES 附录 (2019) CITES Appendix (2019)	无 / NA
保护行动 Conservation Action(s)	无 / None

🦌 相关文献 Relevant References

Burgin *et al.*, 2018; Jiang *et al.*（蒋志刚等）, 2017; Wilson *et al.*, 2016; Zheng *et al.*（郑智民等）, 2012; Smith *et al.*（史密斯等）, 2009; Wang（王应祥）, 2003

灰旱獭 *Marmota baibacina*

长尾旱獭 Marmota caudata

无危 LC

分类地位 Taxonomic Status

动物界 Animalia	脊索动物门 Chordata	哺乳纲 Mammalia	啮齿目 Rodentia	松鼠科 Sciuridae

学 名 Scientific Name	*Marmota caudata*
命 名 人 Species Authority	Geoffroy, 1844
英 文 名 English Name(s)	Golden Marmot
同物异名 Synonym(s)	Long-tailed Marmot
种下单元评估 Infra-specific Taxa Assessed	无 / None

评估信息 Assessment Information

评 估 年 份 Year Assessed	2020
评 定 人 Assessor(s)	蒋志刚、刘少英 / Zhigang Jiang, Shaoying Liu
审 定 人 Reviewer(s)	马勇、刘伟、鲍毅新 / Yong Ma, Wei Liu, Yixin Bao
其他贡献人 Other Contributor(s)	李立立、丁晨晨 / Lili Li, Chenchen Ding

理由 Justification: 长尾旱獭种群数量较多。因此，列为无危等级 / Golden Marmot has large populations. Thus, it is listed as Least Concern

地理分布 Geographical Distribution

国内分布 Domestic Distribution	
新疆 / Xinjiang	
世界分布 World Distribution	
中国；中亚 / China; Central Asia	
分布标注 Distribution Note	
非特有种 / Non-endemic	

国内分布图
Map of Domestic Distribution

种群 Population

种群数量 Population Size	较多 / Relatively abundant
种群趋势 Population Trend	稳定 / Stable

生境与生态系统 Habitat(s) and Ecosystem(s)

生　　境 Habitat(s)	干旱草地 / Dry Steppe
生态系统 Ecosystem(s)	草地生态系统 / Grassland Ecosystem

威胁 Threat(s)

主要威胁 Major Threat(s)	家畜养殖或放牧、耕种、狩猎 / Livestock Farming or Ranching, Plantation, Hunting

保护级别与保护行动 Protection Category and Conservation Action(s)

国家重点保护野生动物等级 (2021) Category of National Key Protected Wild Animals (2021)	无 / NA
IUCN 红色名录 (2020-2) IUCN Red List (2020-2)	无危 / LC
CITES 附录 (2019) CITES Appendix (2019)	III
保护行动 Conservation Action(s)	无 / None

相关文献 Relevant References

Burgin *et al.*, 2018; Jiang *et al.* (蒋志刚等), 2017; Wilson *et al.*, 2016; Zheng *et al.* (郑智民等), 2012; Smith *et al.* (史密斯等), 2009; Wang (王应祥), 2003

长尾旱獭 *Marmota caudata*

喜马拉雅旱獭
Marmota himalayana

无危 LC

| 数据缺乏 DD | 无危 LC | 近危 NT | 易危 VU | 濒危 EN | 极危 CR | 区域灭绝 RE | 野外灭绝 EW | 灭绝 EX |

分类地位 Taxonomic Status

动物界 Animalia	脊索动物门 Chordata	哺乳纲 Mammalia	啮齿目 Rodentia	松鼠科 Sciuridae

学名 Scientific Name	*Marmota himalayana*
命名人 Species Authority	Hodgson, 1841
英文名 English Name(s)	Himalayan Marmot
同物异名 Synonym(s)	无 / None
种下单元评估 Infra-specific Taxa Assessed	无 / None

评估信息 Assessment Information

评估年份 Year Assessed	2020
评定人 Assessor(s)	蒋志刚、刘少英 / Zhigang Jiang, Shaoying Liu
审定人 Reviewer(s)	马勇、刘伟、鲍毅新 / Yong Ma, Wei Liu, Yixin Bao
其他贡献人 Other Contributor(s)	李立立、丁晨晨 / Lili Li, Chenchen Ding

理由 Justification: 喜马拉雅旱獭种群数量较多。因此，列为无危等级 / Himalayan Marmot has large populations. Thus, it is listed as Least Concern

地理分布 Geographical Distribution

国内分布 Domestic Distribution
甘肃、青海、新疆、四川、云南、西藏、内蒙古 / Gansu, Qinghai, Xinjiang, Sichuan, Yunnan, Tibet (Xizang), Inner Mongolia (Nei Mongol)
世界分布 World Distribution
中国；南亚 / China; South Asia
分布标注 Distribution Note
非特有种 / Non-endemic

国内分布图
Map of Domestic Distribution

种群 Population

种群数量 Population Size	较多 / Relatively abundant
种群趋势 Population Trend	稳定 / Stable

生境与生态系统 Habitat(s) and Ecosystem(s)

生　　　境 Habitat(s)	草甸 / Meadow
生态系统 Ecosystem(s)	草地生态系统 / Grassland Ecosystem

威胁 Threat(s)

主要威胁 Major Threat(s)	家畜放牧、狩猎 / Livestock Farming or Ranching, Hunting

保护级别与保护行动 Protection Category and Conservation Action(s)

国家重点保护野生动物等级 (2021) Category of National Key Protected Wild Animals (2021)	无 / NA
IUCN 红色名录 (2020-2) IUCN Red List (2020-2)	无危 / LC
CITES 附录 (2019) CITES Appendix (2019)	III
保护行动 Conservation Action(s)	无 / None

相关文献 Relevant References

Burgin *et al.*, 2018; Jiang *et al.* (蒋志刚等), 2017; Wilson *et al.*, 2016; Zheng *et al.* (郑智民等), 2012; Xu *et al.* (徐金会等), 2009; Wang (王应祥), 2003

喜马拉雅旱獭 *Marmota himalayana*

西伯利亚旱獭
Marmota sibirica

无危 LC

DD	LC	NT	VU	EN	CR	RE	EW	EX
数据缺乏	无危	近危	易危	濒危	极危	区域灭绝	野外灭绝	灭绝

分类地位 Taxonomic Status

动物界 Animalia	脊索动物门 Chordata	哺乳纲 Mammalia	啮齿目 Rodentia	松鼠科 Sciuridae

学名 Scientific Name	*Marmota sibirica*
命名人 Species Authority	Radde, 1862
英文名 English Name(s)	Mongolian Marmot
同物异名 Synonym(s)	Tarbagan Marmot; *Marmota sibirica* (Bannikov and Skalon, 1949) subsp. *caliginosus*; *M. sibirica* (Dybowski, 1922) subsp. *dahurica*
种下单元评估 Infra-specific Taxa Assessed	无 / None

评估信息 Assessment Information

评估年份 Year Assessed	2020
评定人 Assessor(s)	蒋志刚、刘少英 / Zhigang Jiang, Shaoying Liu
审定人 Reviewer(s)	马勇、刘伟、鲍毅新 / Yong Ma, Wei Liu, Yixin Bao
其他贡献人 Other Contributor(s)	李立立、丁晨晨 / Lili Li, Chenchen Ding

理由 Justification: 西伯利亚旱獭分布区广，种群数量较大，为常见种。因此，列为无危等级 / Mongolian Marmot is a common species with a large range and large populations. Thus, it is listed as Least Concern

地理分布 Geographical Distribution

国内分布 Domestic Distribution
内蒙古 / Inner Mongolia (Nei Mongol)
世界分布 World Distribution
中国、蒙古、俄罗斯 / China, Mongolia, Russia
分布标注 Distribution Note
非特有种 / Non-endemic

国内分布图
Map of Domestic Distribution

种群 Population

种群数量 Population Size	数量多 / Abundant
种群趋势 Population Trend	下降 / Decreasing

生境与生态系统 Habitat (s) and Ecosystem (s)

生　　境 Habitat(s)	干旱草地 / Dry Steppe
生态系统 Ecosystem(s)	草地生态系统 / Grassland Ecosystem

威胁 Threat (s)

主要威胁 Major Threat(s)	狩猎、耕种 / Hunting, Plantation

保护级别与保护行动 Protection Category and Conservation Action (s)

国家重点保护野生动物等级 (2021) Category of National Key Protected Wild Animals (2021)	无 / NA
IUCN 红色名录 (2020-2) IUCN Red List (2020-2)	濒危 / EN
CITES 附录 (2019) CITES Appendix (2019)	无 / NA
保护行动 Conservation Action(s)	无 / None

相关文献 Relevant References

Burgin *et al.*, 2018; Jiang *et al.* (蒋志刚等), 2017; Wilson *et al.*, 2016; Zheng *et al.* (郑智民等), 2012; Wang (王应祥), 2003; Ma and Li (马建章和李津友), 1979

西伯利亚旱獭 *Marmota sibirica*

毛耳飞鼠
Belomys pearsonii
无危 LC

| DD 数据缺乏 | LC 无危 | NT 近危 | VU 易危 | EN 濒危 | CR 极危 | RE 区域灭绝 | EW 野外灭绝 | EX 灭绝 |

分类地位 Taxonomic Status

| 动物界 Animalia | 脊索动物门 Chordata | 哺乳纲 Mammalia | 啮齿目 Rodentia | 松鼠科 Sciuridae |

学 名 Scientific Name	*Belomys pearsonii*
命 名 人 Species Authority	Gray, 1842
英 文 名 English Name(s)	Hairy-footed Flying Squirrel
同物异名 Synonym(s)	无 / None
种下单元评估 Infra-specific Taxa Assessed	无 / None

评估信息 Assessment Information

评估年份 Year Assessed	2020
评 定 人 Assessor(s)	蒋志刚、刘少英 / Zhigang Jiang, Shaoying Liu
审 定 人 Reviewer(s)	马勇、鲍毅新 / Yong Ma, Yixin Bao
其他贡献人 Other Contributor(s)	李立立、丁晨晨 / Lili Li, Chenchen Ding

理由 Justification: 毛耳飞鼠的种群数量较多。因此，列为无危等级 / Hairy-footed Flying Squirrel has large populations. Thus, it is listed as Least Concern

地理分布 Geographical Distribution

国内分布 Domestic Distribution
河南、云南、贵州、广西、广东、海南、台湾 / Henan, Yunnan, Guizhou, Guangxi, Guangdong, Hainan, Taiwan
世界分布 World Distribution
中国；南亚、东南亚 / China; South Asia, Southeast Asia
分布标注 Distribution Note
非特有种 / Non-endemic

国内分布图
Map of Domestic Distribution

种群 Population

种群数量 Population Size	较多 / Relatively abundant
种群趋势 Population Trend	稳定 / Stable

生境与生态系统 Habitat(s) and Ecosystem(s)

生境 Habitat(s)	亚热带湿润低地森林、针阔混交林 / Subtropical Moist Lowland Forest, Coniferous and Broad-leaved Mixed Forest
生态系统 Ecosystem(s)	森林生态系统 / Forest Ecosystem

威胁 Threat(s)

主要威胁 Major Threat(s)	耕种、伐木、狩猎、火灾 / Plantation, Logging, Hunting, Fire

保护级别与保护行动 Protection Category and Conservation Action(s)

国家重点保护野生动物等级 (2021) Category of National Key Protected Wild Animals (2021)	无 / NA
IUCN红色名录 (2020-2) IUCN Red List (2020-2)	数据缺乏 / DD
CITES 附录 (2019) CITES Appendix (2019)	无 / NA
保护行动 Conservation Action(s)	无 / None

相关文献 Relevant References

Burgin *et al*., 2018; Jiang *et al*. (蒋志刚等), 2017; Wilson *et al*., 2016; Zheng *et al*. (郑智民等), 2012; Smith *et al*. (史密斯等), 2009; Jin *et al*. (金一等), 2007; Wang (王应祥), 2003

毛耳飞鼠 *Belomys pearsonii*

红白鼯鼠
Petaurista alborufus
无危 LC

分类地位 Taxonomic Status

动物界 Animalia	脊索动物门 Chordata	哺乳纲 Mammalia	啮齿目 Rodentia	松鼠科 Sciuridae

学 名 Scientific Name	*Petaurista alborufus*
命 名 人 Species Authority	Milne-Edwards, 1870
英 文 名 English Name(s)	Red and White Giant Flying Squirrel
同物异名 Synonym(s)	无 / None
种下单元评估 Infra-specific Taxa Assessed	无 / None

评估信息 Assessment Information

评估年份 Year Assessed	2020
评 定 人 Assessor(s)	蒋志刚、刘少英 / Zhigang Jiang, Shaoying Liu
审 定 人 Reviewer(s)	马勇、鲍毅新 / Yong Ma, Yixin Bao
其他贡献人 Other Contributor(s)	李立立、丁晨晨 / Lili Li, Chenchen Ding

理由 Justification: 红白鼯鼠分布区大，种群数量较多。因此，列为无危等级 / Red and White Giant Flying Squirrel has a large range and large populations. Thus, it is listed as Least Concern

地理分布 Geographical Distribution

国内分布 Domestic Distribution

陕西、甘肃、四川、贵州、云南、湖北、湖南、广东、广西、台湾、重庆 / Shaanxi, Gansu, Sichuan, Guizhou, Yunnan, Hubei, Hunan, Guangdong, Guangxi, Taiwan, Chongqing

世界分布 World Distribution

中国 / China

分布标注 Distribution Note

特有种 / Endemic

国内分布图
Map of Domestic Distribution

种群 Population

种群数量 Population Size	较多 / Relatively abundant
种群趋势 Population Trend	稳定 / Stable

生境与生态系统 Habitat (s) and Ecosystem (s)

生　　境 Habitat(s)	森林 / Forest
生态系统 Ecosystem(s)	森林生态系统 / Forest Ecosystem

威胁 Threat (s)

主要威胁 Major Threat(s)	无 / None

保护级别与保护行动 Protection Category and Conservation Action (s)

国家重点保护野生动物等级 (2021) Category of National Key Protected Wild Animals (2021)	无 / NA
IUCN红色名录 (2020-2) IUCN Red List (2020-2)	无危 / LC
CITES 附录 (2019) CITES Appendix (2019)	无 / NA
保护行动 Conservation Action(s)	无 / None

相关文献 Relevant References

Burgin *et al.*, 2018; Jiang *et al.* (蒋志刚等), 2017; Wilson *et al.*, 2016; Zheng *et al.* (郑智民等), 2012; Smith *et al.* (史密斯等), 2009; Wang (王应祥), 2003

红白鼯鼠 *Petaurista alborufus*

灰头小鼯鼠
Petaurista caniceps

无危 LC

分类地位 Taxonomic Status

动物界 Animalia	脊索动物门 Chordata	哺乳纲 Mammalia	啮齿目 Rodentia	松鼠科 Sciuridae

学 名 Scientific Name	*Petaurista caniceps*
命 名 人 Species Authority	Gray, 1842
英 文 名 English Name(s)	Spotted Giant Flying Squirrel
同物异名 Synonym(s)	*Petaurista elegans caniceps* (Gray, 1842)
种下单元评估 Infra-specific Taxa Assessed	无 / None

评估信息 Assessment Information

评 估 年 份 Year Assessed	2020
评 定 人 Assessor(s)	蒋志刚、刘少英 / Zhigang Jiang, Shaoying Liu
审 定 人 Reviewer(s)	李松、马勇 / Song Li, Yong Ma
其他贡献人 Other Contributor(s)	李立立、丁晨晨 / Lili Li, Chenchen Ding

理由 Justification: 灰头小鼯鼠的种群数量较多。因此，列为无危等级 / Spotted Giant Flying Squirrel has large populations. Thus, it is listed as Least Concern

地理分布 Geographical Distribution

国内分布 Domestic Distribution

四川、重庆、陕西、湖北、湖南、广东、广西、贵州、云南、西藏、甘肃 / Sichuan, Chongqing, Shaanxi, Hubei, Hunan, Guangdong, Guangxi, Guizhou, Yunnan, Tibet (Xizang), Gansu

世界分布 World Distribution

中国；南亚、东南亚 / China; South Asia, Southeast Asia

分布标注 Distribution Note

非特有种 / Non-endemic

国内分布图
Map of Domestic Distribution

种群 Population

种群数量 Population Size	较多 / Relatively abundant
种群趋势 Population Trend	稳定 / Stable

生境与生态系统 Habitat(s) and Ecosystem(s)

生境 Habitat(s)	泰加林、温带森林 / Taiga Forest, Temperate Forest
生态系统 Ecosystem(s)	森林生态系统 / Forest Ecosystem

威胁 Threat(s)

主要威胁 Major Threat(s)	无 / None

保护级别与保护行动 Protection Category and Conservation Action(s)

国家重点保护野生动物等级 (2021) Category of National Key Protected Wild Animals (2021)	无 / NA
IUCN 红色名录 (2020-2) IUCN Red List (2020-2)	未列入 / NA
CITES 附录 (2019) CITES Appendix (2019)	I
保护行动 Conservation Action(s)	无 / None

相关文献 Relevant References

Burgin *et al*., 2018; Jiang *et al*.（蒋志刚等），2017; Wilson *et al*., 2016; Li *et al*., 2013; Li *et al*.（李艳红等），2013; Zheng *et al*.（郑智民等），2012; Smith *et al*.（史密斯等），2009

灰头小鼯鼠 *Petaurista caniceps*

霜背大鼯鼠
Petaurista philippensis

无危 LC

| 数据缺乏 DD | 无危 LC | 近危 NT | 易危 VU | 濒危 EN | 极危 CR | 区域灭绝 RE | 野外灭绝 EW | 灭绝 EX |

分类地位 Taxonomic Status

| 动物界 Animalia | 脊索动物门 Chordata | 哺乳纲 Mammalia | 啮齿目 Rodentia | 松鼠科 Sciuridae |

学名 Scientific Name	*Petaurista philippensis*
命名人 Species Authority	Elliot, 1839
英文名 English Name(s)	Large Brown Flying Squirrel
同物异名 Synonym(s)	Indian Giant Flying Squirrel, Common Giant Flying Squirrel
种下单元评估 Infra-specific Taxa Assessed	无 / None

评估信息 Assessment Information

评估年份 Year Assessed	2020
评定人 Assessor(s)	蒋志刚、刘少英 / Zhigang Jiang, Shaoying Liu
审定人 Reviewer(s)	马勇、鲍毅新 / Yong Ma, Yixin Bao
其他贡献人 Other Contributor(s)	李立立、丁晨晨 / Lili Li, Chenchen Ding

理由 Justification: 霜背大鼯鼠的种群数量较多。因此，列为无危等级 / Large Brown Flying Squirrel has large populations. Thus, it is listed as Least Concern

地理分布 Geographical Distribution

国内分布 Domestic Distribution

台湾、海南、云南、四川、贵州、重庆、甘肃、陕西、河南、湖北、湖南、广西、广东 / Taiwan, Hainan, Yunnan, Sichuan, Guizhou, Chongqing, Gansu, Shaanxi, Henan, Hubei, Hunan, Guangxi, Guangdong

世界分布 World Distribution

中国；东南亚 / China; Southeast Asia

分布标注 Distribution Note

非特有种 / Non-endemic

国内分布图
Map of Domestic Distribution

种群 Population

种群数量 Population Size	较多 / Relatively abundant
种群趋势 Population Trend	稳定 / Stable

生境与生态系统 Habitat(s) and Ecosystem(s)

生　　境 Habitat(s)	森林 / Forest
生态系统 Ecosystem(s)	森林生态系统 / Forest Ecosystem

威胁 Threat(s)

主要威胁 Major Threat(s)	无 / None

保护级别与保护行动 Protection Category and Conservation Action(s)

国家重点保护野生动物等级 (2021) Category of National Key Protected Wild Animals (2021)	无 / NA
IUCN 红色名录 (2020-2) IUCN Red List (2020-2)	无危 / LC
CITES 附录 (2019) CITES Appendix (2019)	无 / NA
保护行动 Conservation Action(s)	无 / None

相关文献 Relevant References

Burgin *et al.*, 2018; Jiang *et al.*（蒋志刚等）, 2017; Wilson *et al.*, 2016; Zheng *et al.*（郑智民等）, 2012; Smith *et al.*（史密斯等）, 2009; Wang（王应祥）, 2003

霜背大鼯鼠 *Petaurista philippensis*

白斑小鼯鼠
Petaurista punctatus

无危 LC

| 数据缺乏 DD | 无危 LC | 近危 NT | 易危 VU | 濒危 EN | 极危 CR | 区域灭绝 RE | 野外灭绝 EW | 灭绝 EX |

分类地位 Taxonomic Status

| 动物界 Animalia | 脊索动物门 Chordata | 哺乳纲 Mammalia | 啮齿目 Rodentia | 松鼠科 Sciuridae |

学名 Scientific Name	*Petaurista punctatus*
命名人 Species Authority	Thomas, 1912
英文名 English Name(s)	Lesser Giant Flying Squirrel
同物异名 Synonym(s)	*Petaurista caniceps* (Gray, 1842)
种下单元评估 Infra-specific Taxa Assessed	无 / None

评估信息 Assessment Information

评估年份 Year Assessed	2020
评定人 Assessor(s)	蒋志刚、刘少英 / Zhigang Jiang, Shaoying Liu
审定人 Reviewer(s)	马勇、鲍毅新 / Yong Ma, Yixin Bao
其他贡献人 Other Contributor(s)	李立立、丁晨晨 / Lili Li, Chenchen Ding

理由 Justification: 白斑小鼯鼠在其生境中常见，种群数量较多。因此，列为无危等级 / Lesser Giant Flying Squirrel is a common species in its habitat with large populations. Thus, it is listed as Least Concern

地理分布 Geographical Distribution

国内分布 Domestic Distribution	广西、云南 / Guangxi, Yunnan
世界分布 World Distribution	中国；南亚、东南亚 / China; South Asia, Southeast Asia
分布标注 Distribution Note	非特有种 / Non-endemic

国内分布图
Map of Domestic Distribution

种群 Population

种群数量 Population Size	较多 / Relatively abundant
种群趋势 Population Trend	稳定 / Stable

生境与生态系统 Habitat(s) and Ecosystem(s)

生　　境 Habitat(s)	森林 / Forest
生态系统 Ecosystem(s)	森林生态系统 / Forest Ecosystem

威胁 Threat(s)

主要威胁 Major Threat(s)	无 / None

保护级别与保护行动 Protection Category and Conservation Action(s)

国家重点保护野生动物等级 (2021) Category of National Key Protected Wild Animals (2021)	无 / NA
IUCN红色名录 (2020-2) IUCN Red List (2020-2)	无危 / LC
CITES 附录 (2019) CITES Appendix (2019)	无 / NA
保护行动 Conservation Action(s)	无 / None

相关文献 Relevant References

Burgin *et al.*, 2018; Jiang *et al.* (蒋志刚等), 2017; Li *et al.*, 2013; Smith *et al.* (史密斯等), 2009; Wang (王应祥), 2003

白斑小鼯鼠 *Petaurista punctatus*　　　　　蒋志刚 绘　Drawn by Zhigang Jiang

橙色小鼯鼠
Petaurista sybilla

无危 LC

| 数据缺乏 DD | 无危 LC | 近危 NT | 易危 VU | 濒危 EN | 极危 CR | 区域灭绝 RE | 野外灭绝 EW | 灭绝 EX |

分类地位 Taxonomic Status

动物界 Animalia	脊索动物门 Chordata	哺乳纲 Mammalia	啮齿目 Rodentia	松鼠科 Sciuridae

学名 Scientific Name	*Petaurista sybilla*
命名人 Species Authority	Thomas and Wroughton, 1916
英文名 English Name(s)	Chindwin Giant Flying Squirrel
同物异名 Synonym(s)	无 / None
种下单元评估 Infra-specific Taxa Assessed	无 / None

评估信息 Assessment Information

评估年份 Year Assessed	2020
评定人 Assessor(s)	蒋志刚、刘少英 / Zhigang Jiang, Shaoying Liu
审定人 Reviewer(s)	马勇、鲍毅新 / Yong Ma, Yixin Bao
其他贡献人 Other Contributor(s)	李立立、丁晨晨 / Lili Li, Chenchen Ding

理由 Justification: 橙色小鼯鼠的种群数量较多。因此，列为无危等级 / Chindwin Giant Flying Squirrel has large populations. Thus, it is listed as Least Concern

地理分布 Geographical Distribution

国内分布 Domestic Distribution	云南、贵州、四川、重庆 / Yunnan, Guizhou, Sichuan, Chongqing
世界分布 World Distribution	中国、缅甸 / China, Myanmar
分布标注 Distribution Note	非特有种 / Non-endemic

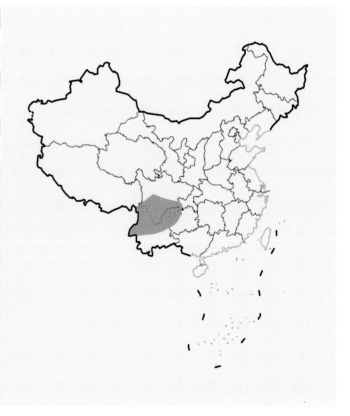

国内分布图
Map of Domestic Distribution

种群 Population

种群数量 Population Size	较多 / Relatively abundant
种群趋势 Population Trend	稳定 / Stable

生境与生态系统 Habitat (s) and Ecosystem (s)

生　　境 Habitat(s)	森林 / Forest
生态系统 Ecosystem(s)	森林生态系统 / Forest Ecosystem

威胁 Threat (s)

主要威胁 Major Threat(s)	无 / None

保护级别与保护行动 Protection Category and Conservation Action (s)

国家重点保护野生动物等级 (2021) Category of National Key Protected Wild Animals (2021)	无 / NA
IUCN 红色名录 (2020-2) IUCN Red List (2020-2)	未列入 / NA
CITES 附录 (2019) CITES Appendix (2019)	无 / NA
保护行动 Conservation Action(s)	无 / None

相关文献 Relevant References

Burgin *et al*., 2018; Jiang *et al*. (蒋志刚等), 2017; Wilson *et al*., 2016; Li *et al*., 2013; Pan *et al*. (潘清华等), 2007; Wang (王应祥), 2003

橙色小鼯鼠 *Petaurista sybilla*

灰鼯鼠 *Petaurista xanthotis*

无危 LC

| DD 数据缺乏 | **LC 无危** | NT 近危 | VU 易危 | EN 濒危 | CR 极危 | RE 区域灭绝 | EW 野外灭绝 | EX 灭绝 |

分类地位 Taxonomic Status

| 动物界 Animalia | 脊索动物门 Chordata | 哺乳纲 Mammalia | 啮齿目 Rodentia | 松鼠科 Sciuridae |

学名 Scientific Name	*Petaurista xanthotis*
命名人 Species Authority	Milne-Edwards, 1872
英文名 English Name(s)	Chinese Giant Flying Squirrel
同物异名 Synonym(s)	*Petaurista buechneri* (Matschie, 1907); *filchnerinae* (Matschie, 1907)
种下单元评估 Infra-specific Taxa Assessed	无 / None

评估信息 Assessment Information

评估年份 Year Assessed	2020
评定人 Assessor(s)	蒋志刚、刘少英 / Zhigang Jiang, Shaoying Liu
审定人 Reviewer(s)	马勇、鲍毅新 / Yong Ma, Yixin Bao
其他贡献人 Other Contributor(s)	李立立、丁晨晨 / Lili Li, Chenchen Ding

理由 Justification: 灰鼯鼠的种群数量较多。因此，列为无危等级 / Chinese Giant Flying Squirrel has large populations. Thus, it is listed as Least Concern

地理分布 Geographical Distribution

国内分布 Domestic Distribution	甘肃、青海、四川、云南、西藏 / Gansu, Qinghai, Sichuan, Yunnan, Tibet (Xizang)
世界分布 World Distribution	中国 / China
分布标注 Distribution Note	特有种 / Endemic

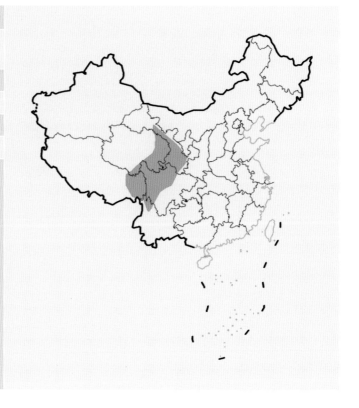

国内分布图 Map of Domestic Distribution

种群 Population

种群数量 Population Size	较多 / Relatively abundant
种群趋势 Population Trend	未知 / Unknown

生境与生态系统 Habitat(s) and Ecosystem(s)

生　　境 Habitat(s)	森林 / Forest
生态系统 Ecosystem(s)	森林生态系统 / Forest Ecosystem

威胁 Threat(s)

主要威胁 Major Threat(s)	无 / None

保护级别与保护行动 Protection Category and Conservation Action(s)

国家重点保护野生动物等级 (2021) Category of National Key Protected Wild Animals (2021)	无 / NA
IUCN 红色名录 (2020-2) IUCN Red List (2020-2)	无危 / LC
CITES 附录 (2019) CITES Appendix (2019)	无 / NA
保护行动 Conservation Action(s)	无 / None

相关文献 Relevant References

Burgin *et al*., 2018; Jiang *et al*.（蒋志刚等），2017; Wilson *et al*., 2016; Liu *et al*.（刘正祥等），2013; Zheng *et al*.（郑智民等），2012; Smith *et al*.（史密斯等），2009

灰鼯鼠 *Petaurista xanthotis*

海南小飞鼠 *Hylopetes phayrei*

无危 LC

| 数据缺乏 DD | 无危 LC | 近危 NT | 易危 VU | 濒危 EN | 极危 CR | 区域灭绝 RE | 野外灭绝 EW | 灭绝 EX |

分类地位 Taxonomic Status

| 动物界 Animalia | 脊索动物门 Chordata | 哺乳纲 Mammalia | 啮齿目 Rodentia | 松鼠科 Sciuridae |

学名 Scientific Name	*Hylopetes phayrei*
命名人 Species Authority	Blyth, 1859
英文名 English Name(s)	Indochinese Flying Squirrel
同物异名 Synonym(s)	无 / None
种下单元评估 Infra-specific Taxa Assessed	无 / None

评估信息 Assessment Information

评估年份 Year Assessed	2020
评定人 Assessor(s)	刘少英、蒋志刚 / Shaoying Liu, Zhigang Jiang
审定人 Reviewer(s)	马勇、鲍毅新 / Yong Ma, Yixin Bao
其他贡献人 Other Contributor(s)	李立立、丁晨晨 / Lili Li, Chenchen Ding

理由 Justification: 海南小飞鼠的种群数量较多。因此，列为无危等级 / Indochinese Flying Squirrel has large populations. Thus, it is listed as Least Concern

地理分布 Geographical Distribution

国内分布 Domestic Distribution	海南、福建、广西、贵州 / Hainan, Fujian, Guangxi, Guizhou
世界分布 World Distribution	中国、缅甸、泰国、越南 / China, Myanmar, Thailand, Viet Nam
分布标注 Distribution Note	非特有种 / Non-endemic

国内分布图
Map of Domestic Distribution

种群 Population

种群数量 Population Size	较多 / Relatively abundant
种群趋势 Population Trend	稳定 / Stable

生境与生态系统 Habitat(s) and Ecosystem(s)

生　　境 Habitat(s)	森林 / Forest
生态系统 Ecosystem(s)	森林生态系统 / Forest Ecosystem

威胁 Threat(s)

主要威胁 Major Threat(s)	未知 / Unknown

保护级别与保护行动 Protection Category and Conservation Action(s)

国家重点保护野生动物等级 (2021) Category of National Key Protected Wild Animals (2021)	无 / NA
IUCN 红色名录 (2020-2) IUCN Red List (2020-2)	无危 / LC
CITES 附录 (2019) CITES Appendix (2019)	无 / NA
保护行动 Conservation Action(s)	无 / None

相关文献 Relevant References

Burgin *et al.*, 2018; Jiang *et al.* (蒋志刚等), 2017; Wilson *et al.*, 2016; Zheng *et al.* (郑智民等), 2012; Smith *et al.* (史密斯等), 2009; Wang and Xie (汪松和解焱), 2004

海南小飞鼠 *Hylopetes phayrei*

黑线仓鼠 *Cricetulus barabensis*

无危 LC

| | 数据缺乏 DD | 无危 LC | 近危 NT | 易危 VU | 濒危 EN | 极危 CR | 区域灭绝 RE | 野外灭绝 EW | 灭绝 EX |

分类地位 Taxonomic Status

动物界 Animalia	脊索动物门 Chordata	哺乳纲 Mammalia	啮齿目 Rodentia	仓鼠科 Cricetidae

学名 Scientific Name	*Cricetulus barabensis*
命名人 Species Authority	Pallas, 1773
英文名 English Name(s)	Striped Dwarf Hamster
同物异名 Synonym(s)	Chinese Striped Hamster; *Cricetulus ferrugineus* (Argyropulo, 1941); *fumatus* (Thomas, 1909); *furunculus* (Pallas, 1779); *griseus* (Milne-Edwards, 1867); *manchuricus* (转下页)
种下单元评估 Infra-specific Taxa Assessed	无 / None

评估信息 Assessment Information

评估年份 Year Assessed	2020
评定人 Assessor(s)	刘少英、蒋志刚 / Shaoying Liu, Zhigang Jiang
审定人 Reviewer(s)	马勇、路纪琪、鲍毅新 / Yong Ma, Jiqi Lu, Yixin Bao
其他贡献人 Other Contributor(s)	李立立、丁晨晨 / Lili Li, Chenchen Ding

理由 Justification: 黑线仓鼠的种群数量多。因此，列为无危等级 / Striped Dwarf Hamster has large populations. Thus, it is listed as Least Concern

地理分布 Geographical Distribution

国内分布 Domestic Distribution

黑龙江、吉林、辽宁、内蒙古、河北、北京、天津、山东、河南、山西、甘肃、宁夏、安徽、江苏 / Heilongjiang, Jilin, Liaoning, Inner Mongolia (Nei Mongol), Hebei, Beijing, Tianjin, Shandong, Henan, Shanxi, Gansu, Ningxia, Anhui, Jiangsu

世界分布 World Distribution

中国；东北亚 / China; Northeast Asia

分布标注 Distribution Note

非特有种 / Non-endemic

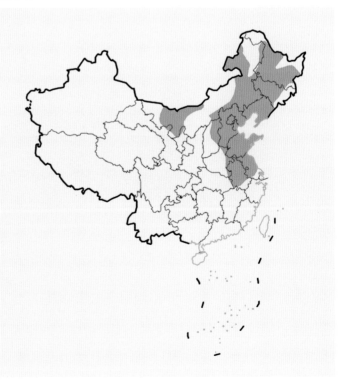

国内分布图
Map of Domestic Distribution

🦌 种群 Population

种群数量 Population Size	数量多 / Abundant
种群趋势 Population Trend	相对稳定 / Relatively stable

🦌 生境与生态系统 Habitat (s) and Ecosystem (s)

生　　境 Habitat(s)	耕地、荒漠 / Arable Land, Desert
生态系统 Ecosystem(s)	农田生态系统、荒漠生态系统 / Cropland Ecosystem, Desert Ecosystem

🦌 威胁 Threat (s)

主要威胁 Major Threat(s)	无 / None

🦌 保护级别与保护行动 Protection Category and Conservation Action (s)

国家重点保护野生动物等级 (2021) Category of National Key Protected Wild Animals (2021)	无 / NA
IUCN红色名录 (2020-2) IUCN Red List (2020-2)	无危 / LC
CITES 附录 (2019) CITES Appendix (2019)	无 / NA
保护行动 Conservation Action(s)	无 / None

🦌 相关文献 Relevant References

Burgin *et al*., 2018; Jiang *et al*.（蒋志刚等）, 2017; Wilson *et al*., 2017; Zheng *et al*.（郑智民等）, 2012; Smith *et al*.（史密斯等）, 2009; Wu *et al*.（武文华等）, 2007; Wang（王应祥）, 2003; Luo（罗泽珣）, 2000

（接上页）
(Mori, 1930); *mongolicus* (Thomas, 1888); *obscurus* (Milne-Edwards, 1867); *pseudogriseus* (Iskhakova, 1974); *pseudogriseus* (Orlov and Iskhakova, 1975); *tuvinicus* (Iskhakova, 1974); *xinganensis* (Wang, 1980)

黑线仓鼠 *Cricetulus barabensis*

长尾仓鼠
Cricetulus longicaudatus

无危 LC

| 数据缺乏 DD | 无危 LC | 近危 NT | 易危 VU | 濒危 EN | 极危 CR | 区域灭绝 RE | 野外灭绝 EW | 灭绝 EX |

分类地位 Taxonomic Status

| 动物界 Animalia | 脊索动物门 Chordata | 哺乳纲 Mammalia | 啮齿目 Rodentia | 仓鼠科 Cricetidae |

学名 Scientific Name	*Cricetulus longicaudatus*
命名人 Species Authority	Milne-Edwards, 1867
英文名 English Name(s)	Long-tailed Dwarf Hamster
同物异名 Synonym(s)	*Cricetulus andersoni* (Thomas, 1908); *chiumalaiensis* (Wang and Cheng, 1973); *dichrootis* (Satunin, 1902); *griseiventris* (Satunin, 1903); *kozhantschikovi* (Vinogradov, (转下页)
种下单元评估 Infra-specific Taxa Assessed	无 / None

评估信息 Assessment Information

评估年份 Year Assessed	2020
评定人 Assessor(s)	刘少英、蒋志刚 / Shaoying Liu, Zhigang Jiang
审定人 Reviewer(s)	马勇、路纪琪、鲍毅新 / Yong Ma, Jiqi Lu, Yixin Bao
其他贡献人 Other Contributor(s)	李立立、丁晨晨 / Lili Li, Chenchen Ding

理由 Justification: 长尾仓鼠的种群大。因此，列为无危等级 / Long-tailed Dwarf Hamster has large populations. Thus, it is listed as Least Concern

地理分布 Geographical Distribution

国内分布 Domestic Distribution

内蒙古、河北、北京、天津、河南、山西、陕西、甘肃、宁夏、四川、青海、西藏、新疆 / Inner Mongolia (Nei Mongol), Hebei, Beijing, Tianjin, Henan, Shanxi, Shaanxi, Gansu, Ningxia, Sichuan, Qinghai, Tibet (Xizang), Xinjiang

世界分布 World Distribution

中国；中亚 / China; Central Asia

分布标注 Distribution Note

非特有种 / Non-endemic

国内分布图
Map of Domestic Distribution

种群 Population

种群数量 Population Size	较多 / Relatively abundant
种群趋势 Population Trend	相对稳定 / Relatively stable

生境与生态系统 Habitat (s) and Ecosystem (s)

生　　境 Habitat(s)	沙漠、灌丛、森林、草地 / Desert, Shrubland, Forest, Grassland
生态系统 Ecosystem(s)	森林生态系统、灌丛生态系统、草地生态系统、荒漠生态系统 / Forest Ecosystem, Shrubland Ecosystem, Grassland Ecosystem, Desert Ecosystem

威胁 Threat (s)

主要威胁 Major Threat(s)	无 / None

保护级别与保护行动 Protection Category and Conservation Action (s)

国家重点保护野生动物等级 (2021) Category of National Key Protected Wild Animals (2021)	无 / NA
IUCN 红色名录 (2020-2) IUCN Red List (2020-2)	无危 / LC
CITES 附录 (2019) CITES Appendix (2019)	无 / NA
保护行动 Conservation Action(s)	无 / None

相关文献 Relevant References

Burgin *et al.*, 2018; Jiang *et al.* (蒋志刚等), 2015a; Wilson *et al.*, 2016; Zheng *et al.* (郑智民等), 2012; Smith *et al.* (史密斯等), 2009; Luo (罗泽珣), 2000; Shi and Lang (史荣耀和郎彩琴), 2000; Shi *et al.* (施银柱等), 1991

(接上页)
1927); *nigrescens* (G. M. Allen, 1925)

长尾仓鼠 *Cricetulus longicaudatus*

灰仓鼠
Cricetulus migratorius
无危 LC

分类地位 Taxonomic Status

动物界 Animalia	脊索动物门 Chordata	哺乳纲 Mammalia	啮齿目 Rodentia	仓鼠科 Cricetidae

学 名 Scientific Name	*Cricetulus migratorius*
命 名 人 Species Authority	Pallas, 1773
英 文 名 English Name(s)	Gray Dwarf Hamster
同物异名 Synonym(s)	Migratory Hamster; *Cricetulus accedula* (Pallas, 1779); *arenarius* (Pallas, 1773); *atticus* (Nehring, 1902); *bellicosus* (Scharleman, 1915); *caesius* (Kashkarov, 1923); (转下页)
种下单元评估 Infra-specific Taxa Assessed	无 / None

评估信息 Assessment Information

评估年份 Year Assessed	2020
评 定 人 Assessor(s)	刘少英、蒋志刚 / Shaoying Liu, Zhigang Jiang
审 定 人 Reviewer(s)	马勇、路纪琪、鲍毅新 / Yong Ma, Jiqi Lu, Yixin Bao
其他贡献人 Other Contributor(s)	李立立、丁晨晨 / Lili Li, Chenchen Ding

理由 Justification: 灰仓鼠的种群大。因此，列为无危等级 / Gray Dwarf Hamster has large populations. Thus, it is listed as Least Concern

地理分布 Geographical Distribution

国内分布 Domestic Distribution

内蒙古、新疆、宁夏、甘肃、青海 / Inner Mongolia (Nei Mongol), Xinjiang, Ningxia, Gansu, Qinghai

世界分布 World Distribution

阿富汗、阿塞拜疆、保加利亚、中国、印度、伊朗、伊拉克、以色列、约旦、哈萨克斯坦、黎巴嫩、摩尔多瓦、蒙古、巴基斯坦、罗马尼亚、俄罗斯、叙利亚、土耳其、乌克兰 / Afghanistan, Azerbaijan, Bulgaria, China, India, Iran, Iraq, Israel, Jordan, Kazakhstan, Lebanon, Moldova, Mongolia, Pakistan, Romania, Russia, Syrian Arab Republic, Turkey, Ukraine

分布标注 Distribution Note

非特有种 / Non-endemic

国内分布图
Map of Domestic Distribution

种群 Population

种群数量 Population Size	数量多 / Abundant
种群趋势 Population Trend	相对稳定 / Relatively stable

生境与生态系统 Habitat (s) and Ecosystem (s)

生　　境 Habitat(s)	沙漠、半荒漠、草地、草甸 / Desert, Semi-desert, Grassland, Meadow
生态系统 Ecosystem(s)	草地生态系统、荒漠生态系统 / Grassland Ecosystem, Desert Ecosystem

威胁 Threat (s)

主要威胁 Major Threat(s)	无 / None

保护级别与保护行动 Protection Category and Conservation Action (s)

国家重点保护野生动物等级 (2021) Category of National Key Protected Wild Animals (2021)	无 / NA
IUCN 红色名录(2020-2) IUCN Red List (2020-2)	无危 / LC
CITES 附录 (2019) CITES Appendix (2019)	无 / NA
保护行动 Conservation Action(s)	无 / None

相关文献 Relevant References

Burgin *et al.*, 2018; Jiang *et al.* (蒋志刚等), 2017; Zheng *et al.* (郑智民等), 2012; Smith *et al.* (史密斯等), 2009; Tursun *et al.* (阿不都热合曼·吐尔逊等), 2008; Wang (王应祥), 2003

(接上页)

cinerascens (Wagner, 1848); *cinereus* (Kashkarov, 1926); *coerulescens* (Severtzov, 1879); *elisarjewi* (Afanasiev, 1953); *falzfeini* (Matschie, 1918); *fulvus* (Blanford, 1875); *griseiventris* (Thomas, 1917); *griseus* (Kashkarov, 1923); *isabellinus* (de Filippi, 1865); *murinus* (Severtzov, 1876); *myosurus* (Argyropulo, 1932); *neglectus* (Ognev, 1916); *ognevi* (Argyropulo, 1932); *ognevi* (Argyropulo, 1941); *pamirensis* (Ognev, 1923); *phaeus* (Pallas, 1779); *pulcher* (Ognev, 1924); *sviridenkoi* (Pidoplitschka, 1928); *tauricus* (Satunin, 1908); *vernula* (Thomas, 1917); *zvierezombi* (Pidoplitschka, 1928)

灰仓鼠 *Cricetulus migratorius*

大仓鼠
Tscherskia triton

无危 LC

| 数据缺乏 DD | 无危 LC | 近危 NT | 易危 VU | 濒危 EN | 极危 CR | 区域灭绝 RE | 野外灭绝 EW | 灭绝 EX |

分类地位 Taxonomic Status

| 动物界 Animalia | 脊索动物门 Chordata | 哺乳纲 Mammalia | 啮齿目 Rodentia | 仓鼠科 Cricetidae |

学名 Scientific Name	*Tscherskia triton*
命名人 Species Authority	De Winton, 1899
英文名 English Name(s)	Greater Long-tailed Hamster
同物异名 Synonym(s)	*Tscherskia albipes* (Ognev, 1914); *arenosus* (Mori, 1939); *bampensis* (Kishida, 1929); *collinus* (G. M. Allen, 1925); *fuscipes* (G. M. Allen, 1925); *incanus* (Thomas,（转下页）
种下单元评估 Infra-specific Taxa Assessed	无 / None

评估信息 Assessment Information

评估年份 Year Assessed	2020
评定人 Assessor(s)	刘少英、蒋志刚 / Shaoying Liu, Zhigang Jiang
审定人 Reviewer(s)	马勇、路纪琪、鲍毅新 / Yong Ma, Jiqi Lu, Yixin Bao
其他贡献人 Other Contributor(s)	李立立、丁晨晨 / Lili Li, Chenchen Ding

理由 Justification: 大仓鼠的种群大。因此，列为无危等级 / Greater Long-tailed Hamster has large populations. Thus, it is listed as Least Concern

地理分布 Geographical Distribution

国内分布 Domestic Distribution

黑龙江、吉林、辽宁、内蒙古、河北、北京、天津、山东、河南、山西、陕西、宁夏、甘肃、江苏、安徽、浙江 / Heilongjiang, Jilin, Liaoning, Inner Mongolia (Nei Mongol), Hebei, Beijing, Tianjin, Shandong, Henan, Shanxi, Shaanxi, Ningxia, Gansu, Jiangsu, Anhui, Zhejiang

世界分布 World Distribution

中国、朝鲜、韩国、俄罗斯 / China, Korea (the Democratic People's Republic of), Korea (the Republic of), Russia

分布标注 Distribution Note

非特有种 / Non-endemic

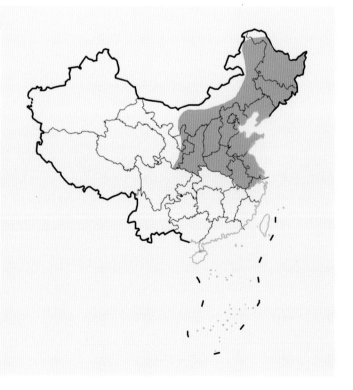

国内分布图
Map of Domestic Distribution

种群 Population

种群数量 Population Size	数量多 / Abundant
种群趋势 Population Trend	相对稳定 / Relatively stable

生境与生态系统 Habitat (s) and Ecosystem (s)

生　　境 Habitat(s)	草原、农田、山地林缘 / Grassland, Gropland, Montane Forest Edge
生态系统 Ecosystem(s)	森林生态系统、草地生态系统、农田生态系统 / Forest Ecosystem, Grassland Ecosystem, Cropland Ecosystem

威胁 Threat (s)

主要威胁 Major Threat(s)	无 / None

保护级别与保护行动 Protection Category and Conservation Action (s)

国家重点保护野生动物等级 (2021) Category of National Key Protected Wild Animals (2021)	无 / NA
IUCN 红色名录(2020-2) IUCN Red List (2020-2)	无危 / LC
CITES 附录 (2019) CITES Appendix (2019)	无 / NA
保护行动 Conservation Action(s)	无 / None

相关文献 Relevant References

Burgin *et al*., 2018; Jiang *et al.* (蒋志刚等), 2017; Wilson *et al*., 2016; Zheng *et al.* (郑智民等), 2012; Smith *et al.* (史密斯等), 2009; Gao *et al.* (高倩等), 2008; Luo (罗泽珣), 2000

(接上页)
1908); *meihsienensis* (Ho, 1935); *nestor* (Thomas, 1907); *ningshaanensis* (Song, 1985); *yamashinai* (Kishida, 1929)

大仓鼠 *Tscherskia triton*

无斑短尾仓鼠
Allocricetulus curtatus
无危 LC

| 数据缺乏 DD | 无危 LC | 近危 NT | 易危 VU | 濒危 EN | 极危 CR | 区域灭绝 RE | 野外灭绝 EW | 灭绝 EX |

分类地位 Taxonomic Status

动物界 Animalia	脊索动物门 Chordata	哺乳纲 Mammalia	啮齿目 Rodentia	仓鼠科 Cricetidae

学　　名 Scientific Name	*Allocricetulus curtatus*
命 名 人 Species Authority	G. M. Allen, 1925
英 文 名 English Name(s)	Mongolian Hamster
同物异名 Synonym(s)	无 / None
种下单元评估 Infra-specific Taxa Assessed	无 / None

评估信息 Assessment Information

评 估 年 份 Year Assessed	2020
评 定 人 Assessor(s)	刘少英、蒋志刚 / Shaoying Liu, Zhigang Jiang
审 定 人 Reviewer(s)	马勇、路纪琪、鲍毅新 / Yong Ma, Jiqi Lu, Yixin Bao
其他贡献人 Other Contributor(s)	李立立、丁晨晨 / Lili Li, Chenchen Ding

理由 Justification: 无斑短尾仓鼠的种群数量较多。因此，列为无危等级 / Mongolian Hamster has large populations. Thus, it is listed as Least Concern

地理分布 Geographical Distribution

国内分布 Domestic Distribution

宁夏、甘肃、内蒙古、新疆 / Ningxia, Gansu, Inner Mongolia (Nei Mongol), Xinjiang

世界分布 World Distribution

中国、蒙古、俄罗斯 / China, Mongolia, Russia

分布标注 Distribution Note

非特有种 / Non-endemic

国内分布图
Map of Domestic Distribution

种群 Population

种群数量 Population Size	较多 / Relatively abundant
种群趋势 Population Trend	未知 / Unknown

生境与生态系统 Habitat(s) and Ecosystem(s)

生境 Habitat(s)	草地、半荒漠 / Grassland, Semi-desert
生态系统 Ecosystem(s)	草地生态系统、荒漠生态系统 / Grassland Ecosystem, Desert Ecosystem

威胁 Threat(s)

主要威胁 Major Threat(s)	无 / None

保护级别与保护行动 Protection Category and Conservation Action(s)

国家重点保护野生动物等级 (2021) Category of National Key Protected Wild Animals (2021)	无 / NA
IUCN 红色名录 (2020-2) IUCN Red List (2020-2)	无危 / LC
CITES 附录 (2019) CITES Appendix (2019)	无 / NA
保护行动 Conservation Action(s)	无 / None

相关文献 Relevant References

Burgin *et al.*, 2018; Jiang *et al.* (蒋志刚等), 2017; Zheng *et al.* (郑智民等), 2012; Wang (王应祥), 2003; Xu and Huang (徐肇华和黄文几), 1982

无斑短尾仓鼠 *Allocricetulus curtatus*

短尾仓鼠 Allocricetulus eversmanni

无危 LC

| 数据缺乏 DD | 无危 LC | 近危 NT | 易危 VU | 濒危 EN | 极危 CR | 区域灭绝 RE | 野外灭绝 EW | 灭绝 EX |

分类地位 Taxonomic Status

动物界 Animalia	脊索动物门 Chordata	哺乳纲 Mammalia	啮齿目 Rodentia	仓鼠科 Cricetidae

学名 Scientific Name	*Allocricetulus eversmanni*
命名人 Species Authority	Brandt, 1859
英文名 English Name(s)	Eversmann's Hamster
同物异名 Synonym(s)	*Allocricetulus beljawi* (Argyropulo, 1933); *belajevi* (Selevin, 1934); *beljaevi* (Kuznetzov, 1944); *microdon* (Ognev, 1925); *pseudocurtatus* (Vorontsov and Kryukova, 1969)
种下单元评估 Infra-specific Taxa Assessed	无 / None

评估信息 Assessment Information

评估年份 Year Assessed	2020
评定人 Assessor(s)	刘少英、蒋志刚 / Shaoying Liu, Zhigang Jiang
审定人 Reviewer(s)	马勇、路纪琪、鲍毅新 / Yong Ma, Jiqi Lu, Yixin Bao
其他贡献人 Other Contributor(s)	李立立、丁晨晨 / Lili Li, Chenchen Ding

理由 Justification: 短尾仓鼠的种群数量较多。因此，列为无危等级 / Eversmann's Hamster has large populations. Thus, it is listed as Least Concern

地理分布 Geographical Distribution

国内分布 Domestic Distribution
新疆 / Xinjiang
世界分布 World Distribution
中国、哈萨克斯坦、俄罗斯 / China, Kazakhstan, Russia
分布标注 Distribution Note
非特有种 / Non-endemic

国内分布图
Map of Domestic Distribution

种群 Population

种群数量 Population Size	较多 / Relatively abundant
种群趋势 Population Trend	未知 / Unknown

生境与生态系统 Habitat(s) and Ecosystem(s)

生　　境 Habitat(s)	草地、半荒漠 / Grassland, Semi-desert
生态系统 Ecosystem(s)	草地生态系统、荒漠生态系统 / Grassland Ecosystem, Desert Ecosystem

威胁 Threat(s)

主要威胁 Major Threat(s)	无 / None

保护级别与保护行动 Protection Category and Conservation Action(s)

国家重点保护野生动物等级 (2021) Category of National Key Protected Wild Animals (2021)	无 / NA
IUCN 红色名录 (2020-2) IUCN Red List (2020-2)	无危 / LC
CITES 附录 (2019) CITES Appendix (2019)	无 / NA
保护行动 Conservation Action(s)	无 / None

相关文献 Relevant References

Burgin *et al.*, 2018; Jiang *et al.*（蒋志刚等）, 2017; Wilson *et al.*, 2016; Zheng *et al.*（郑智民等）, 2012; Smith *et al.*（史密斯等）, 2009; Fu *et al.*（付和平等）, 2005; Wang（王应祥）, 2003; Luo（罗泽珣）, 2000

短尾仓鼠 *Allocricetulus eversmanni*

小毛足鼠
Phodopus roborovskii

无危 LC

| 数据缺乏 DD | 无危 LC | 近危 NT | 易危 VU | 濒危 EN | 极危 CR | 区域灭绝 RE | 野外灭绝 EW | 灭绝 EX |

分类地位 Taxonomic Status

| 动物界 Animalia | 脊索动物门 Chordata | 哺乳纲 Mammalia | 啮齿目 Rodentia | 仓鼠科 Cricetidae |

学名 Scientific Name	*Phodopus roborovskii*
命名人 Species Authority	Satunin, 1903
英文名 English Name(s)	Roborovski's Desert Hamster
同物异名 Synonym(s)	Roborovski Hamster; Desert Hamster; Roborovski Dwarf Hamster; *Phodopus bedfordiae* (Thomas, 1908); *praedilectus* (Mori, 1930); *przewalskii* (Vorontsov and Kriukova, 1969)
种下单元评估 Infra-specific Taxa Assessed	无 / None

评估信息 Assessment Information

评估年份 Year Assessed	2020
评定人 Assessor(s)	刘少英、蒋志刚 / Shaoying Liu, Zhigang Jiang
审定人 Reviewer(s)	马勇、路纪琪、鲍毅新 / Yong Ma, Jiqi Lu, Yixin Bao
其他贡献人 Other Contributor(s)	李立立、丁晨晨 / Lili Li, Chenchen Ding

理由 Justification: 小毛足鼠的种群大。因此，列为无危等级 / Roborovski's Desert Hamster has large populations. Thus, it is listed as Least Concern

地理分布 Geographical Distribution

国内分布 Domestic Distribution	吉林、辽宁、内蒙古、陕西、山西、甘肃、青海、新疆、北京、河北 / Jilin, Liaoning, Inner Mongolia (Nei Mongol), Shaanxi, Shanxi, Gansu, Qinghai, Xinjiang, Beijing, Hebei
世界分布 World Distribution	中国、哈萨克斯坦、蒙古、俄罗斯 / China, Kazakhstan, Mongolia, Russia
分布标注 Distribution Note	非特有种 / Non-endemic

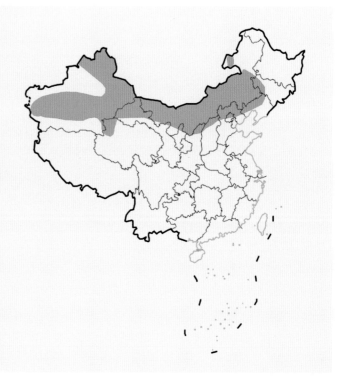

国内分布图
Map of Domestic Distribution

种群 Population

种群数量 Population Size	数量多 / Abundant
种群趋势 Population Trend	稳定 / Stable

生境与生态系统 Habitat(s) and Ecosystem(s)

生　　境 Habitat(s)	草地、半荒漠 / Grassland, Semi-desert
生态系统 Ecosystem(s)	草地生态系统、荒漠生态系统 / Grassland Ecosystem, Desert Ecosystem

威胁 Threat(s)

主要威胁 Major Threat(s)	无 / None

保护级别与保护行动 Protection Category and Conservation Action(s)

国家重点保护野生动物等级 (2021) Category of National Key Protected Wild Animals (2021)	无 / NA
IUCN红色名录 (2020-2) IUCN Red List (2020-2)	无危 / LC
CITES 附录 (2019) CITES Appendix (2019)	无 / NA
保护行动 Conservation Action(s)	无 / None

相关文献 Relevant References

Burgin *et al.*, 2018; Jiang *et al.*（蒋志刚等）, 2017; Wilson *et al.*, 2016; Zheng *et al.*（郑智民等）, 2012; Smith *et al.*（史密斯等）, 2009; Hou *et al.*（侯希贤等）, 2003; Luo（罗泽珣）, 2000

小毛足鼠 *Phodopus roborovskii*

鼹形田鼠
Ellobius talpinus

无危 LC

| 数据缺乏 DD | 无危 LC | 近危 NT | 易危 VU | 濒危 EN | 极危 CR | 区域灭绝 RE | 野外灭绝 EW | 灭绝 EX |

分类地位 Taxonomic Status

| 动物界 Animalia | 脊索动物门 Chordata | 哺乳纲 Mammalia | 啮齿目 Rodentia | 仓鼠科 Cricetidae |

学名 Scientific Name	*Ellobius talpinus*
命名人 Species Authority	Pallas, 1770
英文名 English Name(s)	Northern Mole Vole
同物异名 Synonym(s)	*Ellobius ciscaucasicus* (Sviridenko, 1936); *murinus* (Pallas, 1770); *rufescens* (Eversmann, 1850); *tanaiticus* (Zubko, 1940); *transcaspiae* (Thomas, 1912)
种下单元评估 Infra-specific Taxa Assessed	无 / None

评估信息 Assessment Information

评估年份 Year Assessed	2020
评定人 Assessor(s)	刘少英、蒋志刚 / Shaoying Liu, Zhigang Jiang
审定人 Reviewer(s)	马勇、路纪琪、鲍毅新 / Yong Ma, Jiqi Lu, Yixin Bao
其他贡献人 Other Contributor(s)	李立立、丁晨晨 / Lili Li, Chenchen Ding

理由 Justification: 鼹形田鼠的种群数量较多，因此，列为无危等级 / Northern Mole Vole has large populations. Thus, it is listed as Least Concern

地理分布 Geographical Distribution

国内分布 Domestic Distribution	内蒙古、陕西、甘肃、宁夏、新疆 / Inner Mongolia (Nei Mongol), Shaanxi, Gansu, Ningxia, Xinjiang
世界分布 World Distribution	中国、哈萨克斯坦、俄罗斯、土库曼斯坦、乌克兰、乌兹别克斯坦 / China, Kazakhstan, Russia, Turkmenistan, Ukraine, Uzbekistan
分布标注 Distribution Note	非特有种 / Non-endemic

国内分布图
Map of Domestic Distribution

种群 Population

种群数量 Population Size	较多 / Relatively abundant
种群趋势 Population Trend	下降 / Decreasing

生境与生态系统 Habitat(s) and Ecosystem(s)

生境 Habitat(s)	草地、半荒漠、溪流边、绿洲、耕地 / Grassland, Semi-desert, Near Stream, Oasis, Arable Land
生态系统 Ecosystem(s)	草地生态系统、荒漠生态系统、湖泊河流生态系统、农田生态系统 / Grassland Ecosystem, Desert Ecosystem, Lake and River Ecosystem, Cropland Ecosystem

威胁 Threat(s)

主要威胁 Major Threat(s)	无 / None

保护级别与保护行动 Protection Category and Conservation Action(s)

国家重点保护野生动物等级 (2021) Category of National Key Protected Wild Animals (2021)	无 / NA
IUCN 红色名录 (2020-2) IUCN Red List (2020-2)	无危 / LC
CITES 附录 (2019) CITES Appendix (2019)	无 / NA
保护行动 Conservation Action(s)	无 / None

相关文献 Relevant References

Burgin *et al.*, 2018; Jiang *et al.*(蒋志刚等), 2017; Wilson *et al.*, 2016; Zheng *et al.*(郑智民等), 2012; Wang(王应祥), 2003; Luo(罗泽珣), 2000; Wang *et al.*(王思博等), 2000

鼹形田鼠 *Ellobius talpinus*

By Михайло Колесніков

灰棕背䶄 Myodes centralis

无危 LC

| DD 数据缺乏 | LC 无危 | NT 近危 | VU 易危 | EN 濒危 | CR 极危 | RE 区域灭绝 | EW 野外灭绝 | EX 灭绝 |

分类地位 Taxonomic Status

动物界 Animalia	脊索动物门 Chordata	哺乳纲 Mammalia	啮齿目 Rodentia	仓鼠科 Cricetidae

学名 Scientific Name	*Myodes centralis*
命名人 Species Authority	Miller, 1906
英文名 English Name(s)	Tien Shan Red-backed Vole
同物异名 Synonym(s)	Red-backed Vole; *Clethrionomys centralis* (Miller, 1906); *C. frater* (Thomas, 1908)
种下单元评估 Infra-specific Taxa Assessed	无 / None

评估信息 Assessment Information

评估年份 Year Assessed	2020
评定人 Assessor(s)	刘少英、蒋志刚 / Shaoying Liu, Zhigang Jiang
审定人 Reviewer(s)	马勇、路纪琪、鲍毅新 / Yong Ma, Jiqi Lu, Yixin Bao
其他贡献人 Other Contributor(s)	李立立、丁晨晨 / Lili Li, Chenchen Ding

理由 Justification: 灰棕背䶄的种群大。因此，列为无危等级 / Tien Shan Red-backed Vole has large populations. Thus, it is listed as Least Concern

地理分布 Geographical Distribution

国内分布 Domestic Distribution
新疆 / Xinjiang
世界分布 World Distribution
中国、吉尔吉斯斯坦 / China, Kyrgyzstan
分布标注 Distribution Note
非特有种 / Non-endemic

国内分布图 Map of Domestic Distribution

种群 Population

种群数量 Population Size	数量多 / Abundant
种群趋势 Population Trend	稳定 / Stable

生境与生态系统 Habitat(s) and Ecosystem(s)

生　　境 Habitat(s)	森林 / Forest
生态系统 Ecosystem(s)	森林生态系统 / Forest Ecosystem

威胁 Threat(s)

主要威胁 Major Threat(s)	无 / None

保护级别与保护行动 Protection Category and Conservation Action(s)

国家重点保护野生动物等级 (2021) Category of National Key Protected Wild Animals (2021)	无 / NA
IUCN 红色名录 (2020-2) IUCN Red List (2020-2)	无危 / LC
CITES 附录 (2019) CITES Appendix (2019)	无 / NA
保护行动 Conservation Action(s)	无 / None

相关文献 Relevant References

Burgin *et al*., 2018; Jiang *et al*.（蒋志刚等），2017; Wilson *et al*., 2016; Zheng *et al*.（郑智民等），2012; Smith *et al*.（史密斯等），2009; Wang（王应祥），2003; Luo（罗泽珣），2000

灰棕背䶄 *Myodes centralis*

天山林䶄
Myodes frater

无危 LC

| 数据缺乏 DD | 无危 LC | 近危 NT | 易危 VU | 濒危 EN | 极危 CR | 区域灭绝 RE | 野外灭绝 EW | 灭绝 EX |

分类地位 Taxonomic Status

| 动物界 Animalia | 脊索动物门 Chordata | 哺乳纲 Mammalia | 啮齿目 Rodentia | 仓鼠科 Cricetidae |

学名 Scientific Name	*Myodes frater*
命名人 Species Authority	Thomas, 1908
英文名 English Name(s)	Tian Shan Vole
同物异名 Synonym(s)	无 / None
种下单元评估 Infra-specific Taxa Assessed	无 / None

评估信息 Assessment Information

评估年份 Year Assessed	2020
评定人 Assessor(s)	刘少英、蒋志刚 / Shaoying Liu, Zhigang Jiang
审定人 Reviewer(s)	马勇、路纪琪、鲍毅新 / Yong Ma, Jiqi Lu, Yixin Bao
其他贡献人 Other Contributor(s)	李立立、丁晨晨 / Lili Li, Chenchen Ding

理由 Justification: 天山林䶄的种群大。因此，列为无危等级 / Tian Shan Vole has large populations. Thus, it is listed as Least Concern

地理分布 Geographical Distribution

国内分布 Domestic Distribution
新疆 / Xinjiang

世界分布 World Distribution
中国 / China

分布标注 Distribution Note
特有种 / Endemic

国内分布图
Map of Domestic Distribution

种群 Population

种群数量 Population Size	数量多 / Abundant
种群趋势 Population Trend	稳定 / Stable

生境与生态系统 Habitat (s) and Ecosystem (s)

生　　境 Habitat(s)	森林 / Forest
生态系统 Ecosystem(s)	森林生态系统 / Forest Ecosystem

威胁 Threat (s)

主要威胁 Major Threat(s)	无 / None

保护级别与保护行动 Protection Category and Conservation Action (s)

国家重点保护野生动物等级 (2021) Category of National Key Protected Wild Animals (2021)	无 / NA
IUCN 红色名录 (2020-2) IUCN Red List (2020-2)	无危 / LC
CITES 附录 (2019) CITES Appendix (2019)	无 / NA
保护行动 Conservation Action(s)	无 / None

相关文献 Relevant References

Burgin *et al*., 2018; Jiang *et al*.（蒋志刚等）, 2017; Wilson *et al*., 2016; Zheng *et al*.（郑智民等）, 2012; Smith *et al*.（史密斯等）, 2009; Wang and Xie（汪松和解焱）, 2004; Wang（王应祥）, 2003; Luo（罗泽珣）, 2000

天山林䶄 *Myodes frater*　　蒋志刚 绘　Drawn by Zhigang Jiang

棕背䶄
Myodes rufocanus
无危 LC

| 数据缺乏 DD | 无危 LC | 近危 NT | 易危 VU | 濒危 EN | 极危 CR | 区域灭绝 RE | 野外灭绝 EW | 灭绝 EX |

分类地位 Taxonomic Status

动物界 Animalia	脊索动物门 Chordata	哺乳纲 Mammalia	啮齿目 Rodentia	仓鼠科 Cricetidae
学 名 Scientific Name		*Myodes rufocanus*		
命 名 人 Species Authority		Sundevall, 1846		
英 文 名 English Name(s)		Gray Red-backed Vole		
同物异名 Synonym(s)		Gray-sided Vole; *Myodes akkeshii* (Imaizumi, 1949); *arsenjevi* (Dukelsky, 1928); *bargusinensis* (Turov, 1924); *bedfordiae* (Thomas, 1905); *bromleyi* [Kostenko, (转下页)		
种下单元评估 Infra-specific Taxa Assessed		无 / None		

评估信息 Assessment Information

评估年份 Year Assessed	2020
评 定 人 Assessor(s)	刘少英、蒋志刚 / Shaoying Liu, Zhigang Jiang
审 定 人 Reviewer(s)	马勇、路纪琪、鲍毅新 / Yong Ma, Jiqi Lu, Yixin Bao
其他贡献人 Other Contributor(s)	李立立、丁晨晨 / Lili Li, Chenchen Ding

理由 Justification: 棕背䶄的种群大。因此，列为无危等级 / Gray Red-backed Vole has large populations. Thus, it is listed as Least Concern

地理分布 Geographical Distribution

国内分布 Domestic Distribution	黑龙江、吉林、辽宁、内蒙古、新疆 / Heilongjiang, Jilin, Liaoning, Inner Mongolia (Nei Mongol), Xinjiang
世界分布 World Distribution	亚洲、欧洲 / Asia, Europe
分布标注 Distribution Note	非特有种 / Non-endemic

国内分布图
Map of Domestic Distribution

种群 Population

种群数量 Population Size	数量多 / Abundant
种群趋势 Population Trend	下降 / Decreasing

生境与生态系统 Habitat (s) and Ecosystem (s)

生　　境 Habitat(s)	森林、灌丛 / Forest, Shrubland
生态系统 Ecosystem(s)	森林生态系统 / Forest Ecosystem

威胁 Threat (s)

主要威胁 Major Threat(s)	无 / None

保护级别与保护行动 Protection Category and Conservation Action (s)

国家重点保护野生动物等级 (2021) Category of National Key Protected Wild Animals (2021)	无 / NA
IUCN红色名录 (2020-2) IUCN Red List (2020-2)	无危 / LC
CITES 附录 (2019) CITES Appendix (2019)	无 / NA
保护行动 Conservation Action(s)	无 / None

相关文献 Relevant References

Burgin *et al*., 2018; Jiang *et al*.（蒋志刚等）, 2017; Wilson *et al*., 2016; Zheng *et al*.（郑智民等）, 2012; Smith *et al*.（史密斯等）, 2009; Wilson and Reeder, 2005; Wang（王应祥）, 2003; Luo（罗泽珣）, 2000

(接上页)
(date unknown)]; *changbaishanensis* (Jang, Ma, and Luo, 1993); *irkutensis* (Ognev, 1924); *kamtnschaticus* (Poliakov, 1881); *kolymensis* (Ognev, 1922); *kurilensis* (Tokuda, 1932); *latastei* (J. A. Allen, 1903); *microtinus* (Kuzyakin, 1963); *siberica* (Poliakov, 1881); *sikotanensis* (Tokuda, 1935); *wosnessenskii* (Poliakov, 1881); *yesomontanus* (Kishida, 1931)

棕背䶄 *Myodes rufocanus*　　蒋志刚 绘　Drawn by Zhigang Jiang

红背䶄

Myodes rutilus

无危 LC

| 数据缺乏 DD | 无危 LC | 近危 NT | 易危 VU | 濒危 EN | 极危 CR | 区域灭绝 RE | 野外灭绝 EW | 灭绝 EX |

分类地位 Taxonomic Status

动物界 Animalia	脊索动物门 Chordata	哺乳纲 Mammalia	啮齿目 Rodentia	仓鼠科 Cricetidae

学名 Scientific Name	*Myodes rutilus*
命名人 Species Authority	Pallas, 1779
英文名 English Name(s)	Northern Red-backed Vole
同物异名 Synonym(s)	*Myodes alascensis* (Miller, 1898); *albiventer* (Hall and Gilmore, 1932); *amurensis* (Schrenk, 1859); *aikalensis* (Ognev, 1924); *dawsoni* (Merriam, 1888); (转下页)
种下单元评估 Infra-specific Taxa Assessed	无 / None

评估信息 Assessment Information

评估年份 Year Assessed	2020
评定人 Assessor(s)	刘少英、蒋志刚 / Shaoying Liu, Zhigang Jiang
审定人 Reviewer(s)	马勇、路纪琪、鲍毅新 / Yong Ma, Jiqi Lu, Yixin Bao
其他贡献人 Other Contributor(s)	李立立、丁晨晨 / Lili Li, Chenchen Ding

理由 Justification: 红背䶄发生区较大，种群大。因此，列为无危等级 / Northern Red-backed Vole has relatively large area of occurrence, it also has large populations. Thus, it is listed as Least Concern

地理分布 Geographical Distribution

国内分布 Domestic Distribution
黑龙江、吉林、内蒙古、新疆 / Heilongjiang, Jilin, Inner Mongolia (Nei Mongol), Xinjiang
世界分布 World Distribution
环北极圈 / Circum-Arctic
分布标注 Distribution Note
非特有种 / Non-endemic

国内分布图
Map of Domestic Distribution

种群 Population

种群数量 Population Size	数量多 / Abundant
种群趋势 Population Trend	稳定 / Stable

生境与生态系统 Habitat(s) and Ecosystem(s)

生　　境 Habitat(s)	森林、灌木 / Forest, Shrubland
生态系统 Ecosystem(s)	森林生态系统 / Forest Ecosystem

威胁 Threat(s)

主要威胁 Major Threat(s)	无 / None

保护级别与保护行动 Protection Category and Conservation Action(s)

国家重点保护野生动物等级 (2021) Category of National Key Protected Wild Animals (2021)	无 / NA
IUCN 红色名录 (2020-2) IUCN Red List (2020-2)	无危 / LC
CITES 附录 (2019) CITES Appendix (2019)	无 / NA
保护行动 Conservation Action(s)	无 / None

相关文献 Relevant References

Burgin *et al.*, 2018; Jiang *et al.* (蒋志刚等), 2017; Wilson *et al.*, 2016; Zheng *et al.* (郑智民等), 2012; Smith *et al.* (史密斯等), 2009; Giraudoux *et al.*, 2008; Wilson and Reeder, 2005

(接上页)

dorogostaiskii (Vinogradov, 1933); *finmarchius* (Siivonen, 1967); *glacialis* (Orr, 1945); *hintoni* (Vinogradov, 1933); *hintoni* (Zolotarev, 1936); *insularis* (Heller, 1910); *jacutensis* (Vinogradov, 1927); *jochelsoni* (J. A. Allen, 1903); *laticeps* (Ognev, 1924); *latigriseus* (Argyropulo and Afanasiev, 1939); *lenaensis* (Koljuschev, 1936); *mikado* (Thomas, 1905); *mollessonae* (Kastschenko, 1910); *narymensis* (Argyropulo and Afanasiev, 1939); *orca* (Merriam, 1900); *otus* (Turov, 1924); *parvidens* (Ognev, 1924); *platycephalus* (Manning, 1957); *rjabovi* (Beljaeva, 1953); *rossicus* (Dukelsky, 1928); *russatus* (Rae, 1862); *salairicus* (Egorin, 1936); *tugarinovi* (Vinogradov, 1933); *tundrensis* (Bolshakov and Schwarz, 1965); *uralensis* (Vinogradov, 1933); *uralensis* (Koljusch, 1936); *vinogradovi* (Naumov, 1933); *volgensis* (Kaplanov and Raevsky, 1928); *washburni* (Hanson, 1952); *watsoni* (Orr, 1945)

红背䶄 *Myodes rutilus*

西南绒鼠
Eothenomys custos
无危 LC

| 数据缺乏 DD | 无危 LC | 近危 NT | 易危 VU | 濒危 EN | 极危 CR | 区域灭绝 RE | 野外灭绝 EW | 灭绝 EX |

分类地位 Taxonomic Status

动物界 Animalia	脊索动物门 Chordata	哺乳纲 Mammalia	啮齿目 Rodentia	仓鼠科 Cricetidae
学名 Scientific Name		*Eothenomys custos*		
命名人 Species Authority		Thomas, 1912		
英文名 English Name(s)		Southwest China Red-backed Vole		
同物异名 Synonym(s)		Southwest China Vole; *Eothenomys cangshanensis* (Wang and Li, 2000); *hintoni* (Osgood, 1932); *ninglangensis* (Wang and Li, 2000); *rubelius* (Hinton, 1932); *rubellus* (G. M. Allen, 1924)		
种下单元评估 Infra-specific Taxa Assessed		无 / None		

评估信息 Assessment Information

评估年份 Year Assessed	2020
评定人 Assessor(s)	刘少英、蒋志刚 / Shaoying Liu, Zhigang Jiang
审定人 Reviewer(s)	马勇、路纪琪、鲍毅新 / Yong Ma, Jiqi Lu, Yixin Bao
其他贡献人 Other Contributor(s)	李立立、丁晨晨 / Lili Li, Chenchen Ding

理由 Justification: 西南绒鼠的种群大。因此，列为无危等级 / Southwest China Red-backed Vole has large populations. Thus, it is listed as Least Concern

地理分布 Geographical Distribution

国内分布 Domestic Distribution
四川、云南 / Sichuan, Yunnan
世界分布 World Distribution
中国 / China
分布标注 Distribution Note
特有种 / Endemic

国内分布图
Map of Domestic Distribution

种群 Population

种群数量 Population Size	数量多 / Abundant
种群趋势 Population Trend	未知 / Unknown

生境与生态系统 Habitat(s) and Ecosystem(s)

生境 Habitat(s)	亚热带湿润山地森林、溪流边、灌丛、竹林、草地 / Subtropical Moist Montane Forest, Near Stream, Shrubland, Bamboo Grove, Grassland
生态系统 Ecosystem(s)	森林生态系统、灌丛生态系统、草地生态系统、湖泊河流生态系统 / Forest Ecosystem, Shrubland Ecosystem, Grassland Ecosystem, Lake and River Ecosystem

威胁 Threat(s)

主要威胁 Major Threat(s)	无 / None

保护级别与保护行动 Protection Category and Conservation Action(s)

国家重点保护野生动物等级 (2021) Category of National Key Protected Wild Animals (2021)	无 / NA
IUCN 红色名录 (2020-2) IUCN Red List (2020-2)	无危 / LC
CITES 附录 (2019) CITES Appendix (2019)	无 / NA
保护行动 Conservation Action(s)	无 / None

相关文献 Relevant References

Burgin *et al.*, 2018; Jiang *et al.* (蒋志刚等), 2017; Wilson *et al.*, 2016; Liu *et al.* (刘正祥等), 2013; Liu *et al.*, 2012a; Zheng *et al.* (郑智民等), 2012; Wang (王应祥), 2003; Luo (罗泽珣), 2000; Yang and Yang (杨光荣和杨学时), 1985

西南绒鼠 *Eothenomys custos*

滇绒鼠
Eothenomys eleusis
无危 LC

| 数据缺乏 DD | 无危 LC | 近危 NT | 易危 VU | 濒危 EN | 极危 CR | 区域灭绝 RE | 野外灭绝 EW | 灭绝 EX |

分类地位 Taxonomic Status

动物界 Animalia	脊索动物门 Chordata	哺乳纲 Mammalia	啮齿目 Rodentia	仓鼠科 Cricetidae

学名 Scientific Name	*Eothenomys eleusis*
命名人 Species Authority	Thomas, 1911
英文名 English Name(s)	Small Oriental Vole
同物异名 Synonym(s)	*Eothenomys aurora* (G. M. Allen, 1912)
种下单元评估 Infra-specific Taxa Assessed	无 / None

评估信息 Assessment Information

评估年份 Year Assessed	2020
评定人 Assessor(s)	刘少英、蒋志刚 / Shaoying Liu, Zhigang Jiang
审定人 Reviewer(s)	马勇、路纪琪、鲍毅新 / Yong Ma, Jiqi Lu, Yixin Bao
其他贡献人 Other Contributor(s)	李立立、丁晨晨 / Lili Li, Chenchen Ding

理由 Justification: 滇绒鼠的种群大。因此，列为无危等级 / Small Oriental Vole has large populations. Thus, it is listed as Least Concern

地理分布 Geographical Distribution

国内分布 Domestic Distribution
云南、贵州、重庆、湖北 / Yunnan, Guizhou, Chongqing, Hubei
世界分布 World Distribution
中国 / China
分布标注 Distribution Note
特有种 / Endemic

国内分布图
Map of Domestic Distribution

种群 Population

种群数量 Population Size	数量多 / Abundant
种群趋势 Population Trend	稳定 / Stable

生境与生态系统 Habitat(s) and Ecosystem(s)

生　　境 Habitat(s)	山地森林 / Montane Forest
生态系统 Ecosystem(s)	森林生态系统 / Forest Ecosystem

威胁 Threat(s)

主要威胁 Major Threat(s)	无 / None

保护级别与保护行动 Protection Category and Conservation Action(s)

国家重点保护野生动物等级 (2021) Category of National Key Protected Wild Animals (2021)	无 / NA
IUCN 红色名录 (2020-2) IUCN Red List (2020-2)	无危 / LC
CITES 附录 (2019) CITES Appendix (2019)	无 / NA
保护行动 Conservation Action(s)	无 / None

相关文献 Relevant References

Burgin *et al.*, 2018; Jiang *et al.* (蒋志刚等), 2017; Wilson *et al.*, 2016; Liu *et al.*, 2012a; Pan *et al.* (潘清华等), 2007; Wang (王应祥), 2003; Luo (罗泽珣), 2000; Zhang (张荣祖), 1997

滇绒鼠 *Eothenomys eleusis*　　　　　蒋志刚 绘　Drawn by Zhigang Jiang

黑腹绒鼠
Eothenomys melanogaster
无危 LC

| 数据缺乏 DD | 无危 LC | 近危 NT | 易危 VU | 濒危 EN | 极危 CR | 区域灭绝 RE | 野外灭绝 EW | 灭绝 EX |

分类地位 Taxonomic Status

动物界 Animalia	脊索动物门 Chordata	哺乳纲 Mammalia	啮齿目 Rodentia	仓鼠科 Cricetidae

学 名 Scientific Name	*Eothenomys melanogaster*
命 名 人 Species Authority	Milne-Edwards, 1871
英 文 名 English Name(s)	Père David's Red-backed Vole
同物异名 Synonym(s)	*Eothenomys bonzo* (Cabrera, 1922); *chenduensis* (Wang and Li, 2000); *mucronatus* (G. M. Allen, 1912)
种下单元评估 Infra-specific Taxa Assessed	无 / None

评估信息 Assessment Information

评 估 年 份 Year Assessed	2020
评 定 人 Assessor(s)	刘少英、蒋志刚 / Shaoying Liu, Zhigang Jiang
审 定 人 Reviewer(s)	马勇、路纪琪、鲍毅新 / Yong Ma, Jiqi Lu, Yixin Bao
其他贡献人 Other Contributor(s)	李立立、丁晨晨 / Lili Li, Chenchen Ding

理由 Justification: 黑腹绒鼠的种群大。因此，列为无危等级 / Père David's Red-backed Vole has large populations. Thus, it is listed as Least Concern

地理分布 Geographical Distribution

国内分布 Domestic Distribution
四川、贵州、云南、西藏、陕西、宁夏、甘肃、安徽、浙江、福建、广西、广东、湖北、湖南、江西、台湾、重庆 / Sichuan, Guizhou, Yunnan, Tibet (Xizang), Shaanxi, Ningxia, Gansu, Anhui, Zhejiang, Fujian, Guangxi, Guangdong, Hubei, Hunan, Jiangxi, Taiwan, Chongqing
世界分布 World Distribution
中国；南亚、东南亚 / China; South Asia, Southeast Asia
分布标注 Distribution Note
非特有种 / Non-endemic

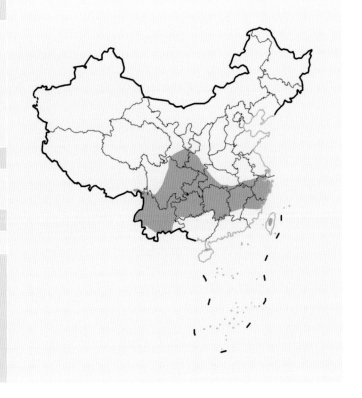

国内分布图
Map of Domestic Distribution

种群 Population

种群数量 Population Size	未知 / Unknown
种群趋势 Population Trend	稳定 / Stable

生境与生态系统 Habitat(s) and Ecosystem(s)

生　　境 Habitat(s)	灌丛 / Shrubland
生态系统 Ecosystem(s)	灌丛生态系统 / Shrubland Ecosystem

威胁 Threat(s)

主要威胁 Major Threat(s)	无 / None

保护级别与保护行动 Protection Category and Conservation Action(s)

国家重点保护野生动物等级 (2021) Category of National Key Protected Wild Animals (2021)	无 / NA
IUCN 红色名录 (2020-2) IUCN Red List (2020-2)	无危 / LC
CITES 附录 (2019) CITES Appendix (2019)	无 / NA
保护行动 Conservation Action(s)	无 / None

相关文献 Relevant References

Burgin *et al*., 2018; Jiang *et al*.（蒋志刚等）, 2017; Wilson *et al*., 2016; Yang *et al*.（杨再学等）, 2013; Liu *et al*., 2012a; Zheng *et al*.（郑智民等）, 2012; Li *et al*.（黎运喜等）, 2011; Luo（罗泽珣）, 2000

黑腹绒鼠 *Eothenomys melanogaster*

大绒鼠
Eothenomys miletus

无危 LC

| 数据缺乏 DD | 无危 LC | 近危 NT | 易危 VU | 濒危 EN | 极危 CR | 区域灭绝 RE | 野外灭绝 EW | 灭绝 EX |

分类地位 Taxonomic Status

动物界 Animalia	脊索动物门 Chordata	哺乳纲 Mammalia	啮齿目 Rodentia	仓鼠科 Cricetidae
学名 Scientific Name		*Eothenomys miletus*		
命名人 Species Authority		Thomas, 1914		
英文名 English Name(s)		Yunnan Red-backed Vole		
同物异名 Synonym(s)		*Eothenomys fidelis* (Hinton, 1923)		
种下单元评估 Infra-specific Taxa Assessed		无 / None		

评估信息 Assessment Information

评估年份 Year Assessed	2020
评定人 Assessor(s)	刘少英、蒋志刚 / Shaoying Liu, Zhigang Jiang
审定人 Reviewer(s)	马勇、路纪琪、鲍毅新 / Yong Ma, Jiqi Lu, Yixin Bao
其他贡献人 Other Contributor(s)	李立立、丁晨晨 / Lili Li, Chenchen Ding

理由 Justification: 大绒鼠的种群大。因此，列为无危等级 / Yunnan Red-backed Vole has large populations. Thus, it is listed as Least Concern

地理分布 Geographical Distribution

国内分布 Domestic Distribution
云南 / Yunnan
世界分布 World Distribution
中国 / China
分布标注 Distribution Note
特有种 / Endemic

国内分布图
Map of Domestic Distribution

种群 Population

种群数量 Population Size	数量多 / Abundant
种群趋势 Population Trend	未知 / Unknown

生境与生态系统 Habitat(s) and Ecosystem(s)

生　　境 Habitat(s)	热带和亚热带湿润山地森林 / Tropical and Subtropical Moist Montane Forest
生态系统 Ecosystem(s)	森林生态系统 / Forest Ecosystem

威胁 Threat(s)

主要威胁 Major Threat(s)	无 / None

保护级别与保护行动 Protection Category and Conservation Action(s)

国家重点保护野生动物等级 (2021) Category of National Key Protected Wild Animals (2021)	无 / NA
IUCN 红色名录 (2020-2) IUCN Red List (2020-2)	无危 / LC
CITES 附录 (2019) CITES Appendix (2019)	无 / NA
保护行动 Conservation Action(s)	无 / None

相关文献 Relevant References

Burgin *et al.*, 2018; Jiang *et al.*（蒋志刚等）, 2017; Wilson *et al.*, 2016; Zheng *et al.*（郑智民等）, 2012; Smith *et al.*（史密斯等）, 2009; Ye *et al.*（叶晓堤等）, 2002; Zhang *et al.*（张云智等）, 2002; Luo（罗泽珣）, 2000

大绒鼠 *Eothenomys miletus*